JPEG2000
Image Compression Fundamentals,
Standards and Practice

THE KLUWER INTERNATIONAL SERIES
IN ENGINEERING AND COMPUTER SCIENCE

JPEG2000
Image Compression Fundamentals, Standards and Practice

DAVID S. TAUBMAN
Senior Lectuer, Electrical Engineering and Telecommunications
The University of New South Wales
Sydney, Australia

MICHAEL W. MARCELLIN
Professor, Electrical and Computer Engineering
The University of Arizona
Tucson, Arizona, USA

Kluwer Academic Publishers
Boston/Dordrecht/London

Distributors for North, Central and South America:
Kluwer Academic Publishers
101 Philip Drive
Assinippi Park
Norwell, Massachusetts 02061 USA
Telephone (781) 871-6600
Fax (781) 681-9045
E-Mail <kluwer@wkap.com>

Distributors for all other countries:
Kluwer Academic Publishers Group
Distribution Centre
Post Office Box 322
3300 AH Dordrecht, THE NETHERLANDS
Telephone 31 78 6392 392
Fax 31 78 6546 474
E-Mail <services@wkap.nl>

 Electronic Services <http://www.wkap.nl>

Library of Congress Cataloging-in-Publication Data

Taubman, David S.
  JPEG2000: image compression fundamentals, standards, and practice / David S.
  Taubman, Michael W. Marcellin.
    p. cm.—(The Kluwer international series in engineering and computer science; SECS 642)
  Includes bibliographical references and index.
  ISBN 0-7923-7519-X (alk. paper)
    1. JPEG (Image coding standard) 2. Image compression. I. Marcellin, Michael W. II.
  Title. III. Series.

  TK6680.5 .T38 2002
  006.6—dc21

2001038589

to Mandy, Samuel and
Joshua,
Therese, Stephanie and
Sarah

Contents

Preface

JPEG2000 is the most recent addition to a family of international standards developed by the Joint Photographic Experts Group (JPEG). The original JPEG image compression standard has found wide acceptance in diverse application areas, including the internet, digital cameras, and printing and scanning peripherals. Image compression plays a central role in modern multi-media communications and compressed images arguably represent the dominant source of internet traffic today. The JPEG2000 standard is intended as the successor to JPEG in many of its application areas. It is motivated primarily by the need for compressed image representations which offer features increasingly demanded by modern applications, while also offering superior compression performance.

This text is written to serve the interests of a wide readership and to facilitate the adoption of the JPEG2000 standard by providing the tools needed to efficiently exploit its capabilities. The book is organized into four parts and is accompanied by a comprehensive software implementation of the standard. The first part provides a thorough grounding in the theoretical underpinnings and fundamental algorithms contributing to the standard. Although the elements of the original JPEG standard are carefully expounded in a large body of existing works, JPEG2000 employs fundamentally different approaches and many recently developed techniques to achieve its goals. This first part of the book provides in-depth coverage of a diverse range of topics, which have not previously been brought together in a single volume. The intent is not only to provide a backdrop to the JPEG2000 standard, but also to serve the needs of students and academics interested in modern image compression techniques.

The second part of the book is devoted to a thorough description of the JPEG2000 standard. This material is intended to serve as a comprehensive reference for implementors of the standard. The authors draw upon their extensive involvement with the development of JPEG2000 to shed light on all technical aspects of JPEG2000 Part 1. Treatment of JPEG2000 Part 2 (extensions) is less comprehensive. Parts I and II of the book are written so as to complement one another. The book offers at least two different perspectives on many of the key concepts, with Part I offering the more theoretical perspective and Part II offering the more practical. As far as possible, Part II of the book strives to provide an accessible description of the standard, which can be comprehended without first absorbing the more theoretical material in Part I.

The third part of the book addresses practical considerations for implementing and efficiently utilizing the standard. The intention is to impart a body of knowledge acquired by the authors through their involvement in developing the standard, including software and hardware implementation strategies and guidelines for selecting the most appropriate parameters for a variety of applications. This part of the book also deals with compliance testing and related matters.

The fourth and final part of the book provides a useful introduction to other image compression standards, namely JPEG and JPEG-LS. The purpose of this material is twofold. In the first place, these much simpler standards provide excellent practical examples of some of the image compression techniques which are treated in Part I of the book, but do not find expression in JPEG2000. Secondly, JPEG and JPEG-LS provide the most important alternatives to JPEG2000 in its two most important fields of application: lossy and lossless compression of continuous tone imagery. Only by describing these standards is the text able to offer meaningful comparisons with JPEG2000. In some cases, particularly those in which scalability and accessibility are not sought-after features, the use of JPEG2000 in preference to JPEG or JPEG-LS may be likened to using a sledge hammer to swat a fly. Part IV of the book should prove a useful guide to application developers wishing to avoid such excesses.

Included with the book is a compact disc, containing documentation, binaries and all source code to the Kakadu software tools. This software provides a complete C++ implementation of JPEG2000 Part 1, demonstrating many of the principles described in the text itself. The software is frequently referenced from the text as an additional resource for understanding complex or subtle aspects of the standard. Conversely, the software makes frequent reference to this text and has been written to mesh with the terminology and notation employed herein. The Kakadu tools have been commercially licensed by a significant number of corporations. Non-commercial licenses are also sold separately by the University of New South Wales and the software may otherwise be obtained only with the purchase of this book. A copy of the non-commercial license granted with this book may be found at the back cover. Provisions are also in place to encourage site-licensing by Universities whose libraries own a copy of the book. For more information in this regard, refer to the compact disc itself and the accompanying license statement.

Acknowledgments

There are many individuals without whom this work would never have come to pass. To our colleagues in the JPEG working group, WG1, we extend our most sincere gratitude. Their cooperative endeavours and determination to see this new standard meet the communication needs of the modern world have shaped JPEG2000. We especially thank the tireless editor of the standard, Martin Boliek, for his instrumental role in initiating the standardization process and his extensive and ongoing contribution in documenting and solidifying the JPEG2000 technology. We thank the WG1 convener, Daniel Lee, and the coeditors, Eric Majani and Charis Christopoulos, for their many labours in keeping the standard on track. Also deserving of special thanks is Thomas Flohr, for his outstanding support of the JPEG2000 Verification Model software.

We would like to thank the many individuals who have encouraged us in this work and especially Michael Gormish, Jim Andrew and Ali Bilgin, whose feedback has significantly improved the quality of the text. Finally, and above all, we acknowledge the encouragement, love and support of our wives and families, whose sacrifice and generosity has enabled us to find the energy to write.

I
FUNDAMENTAL CONCEPTS

Chapter 1

IMAGE COMPRESSION OVERVIEW

1.1 ELEMENTARY CONCEPTS

1.1.1 DIGITAL IMAGES

For our purposes an image is a two dimensional sequence of sample values,

$$x[n_1, n_2], \quad 0 \leq n_1 < N_1, \quad 0 \leq n_2 < N_2,$$

having finite extents, N_1 and N_2, in the vertical and horizontal directions, respectively. The term "pixel," where used here, is to be understood as synonymous with an image sample. The first coordinate, n_1 is understood as the row index, while the second coordinate, n_2, is understood as the column index of the sample or pixel. This is illustrated in Figure 1.1.

The sample value, $x[n_1, n_2]$, represents the intensity (brightness) of the image at location $[n_1, n_2]$. The sample values will usually be B-bit signed or unsigned integers. Thus,

$$x[n_1, n_2] \in \{0, 1, \ldots, 2^B - 1\} \text{ for unsigned imagery}$$

$$x[n_1, n_2] \in \{-2^{B-1}, -2^{B-1} + 1, \ldots, 2^{B-1} - 1\} \text{ for signed imagery}$$

Most commonly encountered digital images have an unsigned $B = 8$ bit representation, although larger bit-depths are frequently encountered in medical, military and scientific applications. In many cases, the B-bit sample values are best interpreted as uniformly quantized representations of real-valued quantities, $x'[n_1, n_2]$, in the range 0 to 1 (unsigned) or $-\frac{1}{2}$ to $\frac{1}{2}$ (signed). Letting $\langle \cdot \rangle$ denote rounding to the nearest integer, the relationship between the real-valued and integer sample values may be written as

$$x[n_1, n_2] = \langle 2^B x'[n_1, n_2] \rangle \tag{1.1}$$

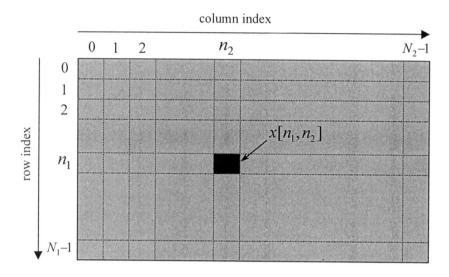

column index

Figure 1.1.   Interpretation of image sample coordinates

Colour images are typically represented with three values per sample location, corresponding to red, green and blue primary colour components. We represent such images with three separate sample sequences, $x_R[n_1, n_2]$, $x_G[n_1, n_2]$ and $x_B[n_1, n_2]$. More generally, we may have an arbitrary collection of image components,

$$x_c[n_1, n_2], \quad c = 1, 2, \ldots, C$$

Images prepared for colour printing often have four colour components corresponding to cyan, magenta, yellow and black dyes; in fact, some colour printers add green and violet for six primary colour components. Hyperspectral satellite images can have hundreds of image components, corresponding to different regions of the spectrum. For the most part we shall restrict our attention to a single image component, with the understanding that it is always possible to apply a compression system separately to each component in turn.

The degree to which an image may be compressed depends upon its content. For this reason, we often refer to particular classes of imagery. Some useful classifications are:

Natural Images representing natural scenes, including photographic images.

Text Images representing scanned or computer generated text, e.g., facsimile images.

Graphics Scanned or computer generated graphics such as line-art and comics.

Compound Images which typically contain a mixture of the above three types of content, e.g., scanned documents.

Unless otherwise stated, we will be primarily concerned with natural images in this text.

1.1.2 LOSSLESS AND LOSSY COMPRESSION

The primary goal of lossless compression is to minimize the number of bits required to represent the original image samples without any loss of information. All B bits of each sample must be reconstructed perfectly during decompression. For image compression, however, some loss of information is usually acceptable for the following three reasons:

- Significant loss can often be tolerated by the human visual system without interfering with perception of the scene content.

- In most cases, digital input to the compression algorithm is itself an imperfect representation of the real-world scene. This is certainly true when the image sample values are quantized versions of underlying real-valued quanties, as expressed in equation (1.1).

- Lossless compression is usually incapable of achieving the high compression requirements of many storage and distribution applications.

Nevertheless, lossless compression is often demanded in medical applications so as to avoid legal disputation over the significance of errors introduced into the imagery. Lossless compression is also often applied in cases where it is difficult to determine how to introduce an acceptable loss which will increase compression. In palettized colour images, for example, a small error in the numeric sample value may have a drastic effect upon the colour representation. The highly structured nature of non-natural imagery such as text and graphics usually renders it more amenable to lossless compression. Finally, lossless compression may be appropriate in applications where the image is to be extensively edited and recompressed so that the accumulation of errors from multiple lossy compression operations may become unacceptable. We note, however, that JPEG2000 allows certain common image editing operations such as cropping and simple geometric manipulations to be performed as often as desired without the accumulation of errors.

LOSSY COMPRESSION AND DISTORTION

By allowing the introduction of small errors, it is natural to expect that we should be able to represent the image approximately using a smaller number of bits than is possible within the constraints of lossless compression. The more distortion we allow, the smaller the compressed representation can be. The primary goal of lossy compression is to minimize the number of bits required to represent an image with an allowable level of distortion. Distortion of course must be assessed in an appropriate manner. Formally, we write $D\left(\mathbf{x}, \hat{\mathbf{x}}\right)$, for the distortion between the original image, $\mathbf{x} \equiv x\left[n_1, n_2\right]$, and the reconstructed image, $\hat{\mathbf{x}} \equiv \hat{x}\left[n_1, n_2\right]$.

The most commonly employed measure of distortion is MSE (Mean Squared Error), defined by

$$\text{MSE} \triangleq \frac{1}{N_1 N_2} \sum_{n_1=0}^{N_1-1} \sum_{n_2=0}^{N_2-1} \left(x\left[n_1, n_2\right] - \hat{x}\left[n_1, n_2\right]\right)^2$$

For image compression, the MSE is most commonly quoted in terms of the equivalent reciprocal measure, PSNR (Peak Signal to Noise Ratio), defined by

$$\text{PSNR} \triangleq 10 \log_{10} \frac{\left(2^B - 1\right)^2}{\text{MSE}} \tag{1.2}$$

The PSNR is expressed in dB (decibels). Good reconstructed images typically have PSNR values of 30 dB or more.

The popularity of MSE as a measure for image distortion derives partly from the ease with which it may be calculated and partly from the tractability of linear optimization problems involving squared error metrics. More appropriate measures of visual distortion are discussed in Sections 4.3.4 and 16.1. At this point, however, it is worth pointing out the importance of non-linearities in the most commonly encountered image representations.

GAMMA CORRECTION

Display devices such as televisions and computer monitors are highly non-linear in that the excitation power delivered to the phosphor is approximately proportional to v^γ, where v is the control voltage applied to the electron gun and γ typically ranges from about 1.8 to 2.8. The image sample values, $x\left[n_1, n_2\right]$, are usually assigned so as to compensate for such a non-linearity.

More specifically, let $x_{\text{lin}}\left[n_1, n_2\right]$ denote the normalized scene radiance at image location $[n_1, n_2]$. The normalization is such that $x_{\text{lin}} = 0$ corresponds to the absence of light and $x_{\text{lin}} = 1$ corresponds to the maximum

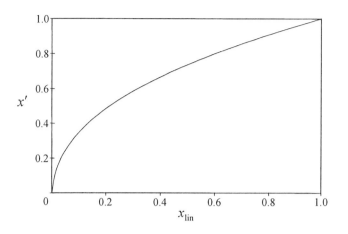

Figure 1.2.   The sRGB gamma function.

intensity level which we expect to encounter in the scene. The so-called "gamma" function, with parameters γ and β, assigns similarly normalized image sample values, $x'[n_1, n_2]$, in the range 0 to 1, according to

$$x'[n_1, n_2] = \begin{cases} gx_{\text{lin}}[n_1, n_2] & \text{if } 0 \leq x_{\text{lin}}[n_{1,n_2}] \leq \varepsilon, \\ (1+\beta)(x_{\text{lin}}[n_1, n_2])^{\frac{1}{\gamma}} - \beta & \text{if } \varepsilon \leq x_{\text{lin}}[n_1, n_2] \leq 1 \end{cases}$$

where the linear breakpoint, ε, and the gradient, g, are defined in terms of γ and β by

$$\varepsilon = \left(\frac{\beta}{(1+\beta)\left(1 - \frac{1}{\gamma}\right)} \right)^{\gamma} \quad \text{and} \quad g = \frac{\beta}{\varepsilon(\gamma - 1)}$$

These definitions ensure that the gamma function has a continuous derivative at the breakpoint, $x_{\text{lin}} = \varepsilon$.

An emerging standard for the representation of colour images is the sRGB (standard RGB) colour space, in which carefully defined linear red, green and blue primaries are each mapped to non-linear RGB sample values through the gamma function described above with parameters $\gamma = 2.4$ and $\beta = 0.055$. The function is plotted in Figure 1.2.

It should be noted that most digital images encountered in practice will be gamma corrected, which affects the interpretation of errors introduced in the image sample values during compression. Ignoring the small linear segment in the gamma function (or assuming that $\beta = 0$), so that $x_{\text{lin}} = (x')^{\gamma}$, we see that a small error, dx', in the gamma corrected value corresponds to a scene radiance error, dx_{lin}, of

$$dx_{\text{lin}} = \gamma \left(x'\right)^{\gamma-1} dx'$$
$$= \gamma \left(x_{\text{lin}}\right)^{1-\frac{1}{\gamma}} dx' \qquad (1.3)$$

Thus, the scene radiance error will be larger in brighter regions of the image. By a most fortunate coincidence (as opposed to design), this behaviour is well matched to a property of the human visual system known as Weber's law. According to Weber's law, the change in scene radiance dx_{lin}, required to effect a just noticeable change in perceived brightness is proportional to x_{lin} itself. For large values of γ, equation (1.3) indicates that $\frac{dx_{\text{lin}}}{x_{\text{lin}}}$ is approximately proportional to dx'. Thus, the gamma corrected values are more perceptually uniform measures of intensity than linear scene radiance, x_{lin}. In this way, the effect of Weber's law is automatically accomodated in simple numerical distortion measures such as MSE, provided they are applied to gamma corrected sample values.

Conversely, MSE turns out to be a much less useful measure of distortion when applied to image samples which have not been gamma corrected. Lossy compression algorithms also yield substantially poorer visual performance when applied to such images. Special care should be taken when working with non-natural image sources; medical X-rays and SAR (Synthetic Aperture Radar) images, for example, are often encountered in linear form.

1.1.3   MEASURES OF COMPRESSION

The purpose of image compression is to represent the image with a string of binary digits or "bits," called the compressed "bit-stream," denoted $\mathbf{c}$. The objective is to keep the length, $||\mathbf{c}||$, as small as possible. In the absence of any compression, we require $N_1 N_2 B$ bits to represent the image sample values, so we define the compression ratio as

$$\text{compression ratio} \triangleq \frac{N_1 N_2 B}{||\mathbf{c}||}$$

Equivalently, we define the compressed bit-rate, expressed in bps (bits per sample), as

$$\text{bit-rate (bps)} \triangleq \frac{||\mathbf{c}||}{N_1 N_2}$$

For lossy compression, bit-rate is arguably a more meaningful performance measure for image compression systems, since the least significant bits of high bit-depth imagery can often be discarded without introducing substantial visual distortion. As a result, the average number of bits

Table 1.1.   Typical compressed bit-rates.

| Lossless | Lossy quality | | |
	High	Moderate	Usable
$B - 3$ bps	1 bps	0.5 bps	0.25 bps

spent in representing each image sample is often the more meaningful measure of compression performance, regardless of the precision with which these samples were originally represented.

If images are to be printed or displayed with a constant physical size regardless of the pixel dimensions, a similar argument to that given above suggests that the size of the bit-stream itself is a more meaningful measure of performance than the bit-rate. In such applications, much of the original image resolution may be lost during display so that the compression algorithm could be applied to a reduced resolution version of the image without incurring substantial distortion. In summary, the bit-rate is a meaningful measure of lossy compression performance only when N_1 and N_2 are proportional to the physical dimensions with which the image is to be printed or displayed.

Table 1.1 provides a rough indication of the compressed bit-rates which may be achieved when compressing natural images, although there can be substantial dependence on the content of the particular image. The assumption here is that lossy reconstructed images are viewed on a computer monitor with a typical resolution of about 90 pixels/inch (22 pixels/mm). Higher compression ratios are usually achievable if the image is to be printed with a much closer dot pitch.

1.2   EXPLOITING REDUNDANCY

Without any compression, the image sample values are represented with $N_1 N_2 B$ bits. Compression is only possible if some of these bits may be understood as redundant. In this section we briefly discuss the nature of this redundancy so as to motivate the operations introduced in Section 1.3 which are common to most image compression systems.

1.2.1   STATISTICAL REDUNDANCY

Consider two B-bit integers, $x_1, x_2 \in [0, 1, \dots, 2^B - 1]$. As an example, these integers might correspond to two adjacent image sample values. Without compression, the two integers are represented using $2B$ bits. Suppose, however, that the decompressor knows *a priori* that the only values which will ever occur are 0 and 1; for example, the image might be known to be bi-level. Clearly, then, it is sufficient to use only one bit in the representation of each of x_1 and x_2 with a compression ratio of $B : 1$. Suppose further that the decompressor knows *a priori* that the two values are always identical. Then, of course, a single bit is sufficient to represent the pair of numbers, with a compression ratio of $2B : 1$.

Of course, the type of prior knowledge described above is uncommon in practice. More commonly, the decompressor might know that some subset of the possible values is more likely than the others. If the decompressor knows that $x_1 \in \{0, 1\}$ with very high probability then we hope to be able to expend little more than 1 bit representing the actual value of x_1. This hope is well founded. As we shall see in Chapter 2, we expect on average to be able to spend as little as $H(X_1)$ bits where X_1 is a random variable, which summarizes the decompressor's a priori knowledge concerning the value of x_1, and $H(X_1)$ is a function of the statistical distribution of X_1, known as its entropy.

Similarly, the decompressor might know that x_1 and x_2 are very likely to be equal, perhaps because the image is smooth so that changes in intensity between adjacent pixels is rare. In this case, we would hope to be able to avoid spending many bits in representing x_2 once the value of x_1 has already been identified in the compressed bit-stream. Again, this hope is well founded. The relevant prior knowledge is summarized by the joint statistics of the random variables, X_1 and X_2, corresponding to the values, x_1 and x_2 and the expected number of bits required to represent x_1 given that x_2 is already known to the decompressor is given by the conditional entropy, $H(X_2|X_1)$.

A thorough development of these concepts, along with practical coding tools which are able to exploit the redundancy is the subject of Chapter 2. For the moment, however, it is sufficient to appreciate that the average number of bits required to represent the image sample values without error depends upon their statistical properties. In the worst case, if all possible combinations of sample values are equally likely, there is no redundancy whatsoever and all $N_1 N_2 B$ bits must be used. At the opposite extreme, when a small number of samples provides sufficient information to predict the remaining samples with high probability, high compression ratios can be achieved.

1.2.2   IRRELEVANCE

The form of redundancy described above allows us to exactly represent the original sample values with a reduced number of bits. In many cases, however, some of the information associated with these sample values may be irrelevant so that it is unnecessary to represent the image exactly. The following examples should help to clarify this point.

Visual irrelevance: If the image sample density exceeds the limits of human visual accuity for any appropriate set of display and viewing conditions, the excess image resolution is irrelevant to the human observer. Visual irrelevance may arise in more complex ways, some of which are addressed in Section 16.1.

Application specific irrelevance: In some applications, particularly in the military and medical arenas, the value of an image may be determined entirely by its usefulness in fulfilling some task; e.g., target recognition or medical diagnosis. Regions of the image which do not contribute to this task may then be taken to be irrelevant.

One way to exploit irrelevance is to transform the original image sample values to a new set of sample values which capture the relevant information using fewer bits. In the simplest case, this transform might involve sub-sampling or discarding the samples corresponding to irrelevant regions of the image. The statistical redundancy of the remaining samples may be further exploited to improve compression.

IRRELEVANCE IN COLOUR IMAGERY

An interesting example of irrelevance occurs in colour imagery, where the human viewer is substantially less sensitive to rapid changes in the hue and saturation properties of the image than to intensity changes. For image compression purposes, this property is usually modeled by mapping the original RGB image samples to a luminance-chrominance space using a linear transform and then sub-sampling the chrominance components.

We illustrate this with the so-called YCbCr transform:

$$
\begin{pmatrix} x_Y \\ x_{C_b} \\ x_{C_r} \end{pmatrix} = \begin{pmatrix} 0.299 & 0.587 & 0.114 \\ -0.169 & -0.331 & 0.5 \\ 0.5 & -0.419 & -0.0813 \end{pmatrix} \cdot \begin{pmatrix} x_R \\ x_G \\ x_B \end{pmatrix}
$$

Note that the first chrominance component, x_{C_b}, is a scaled version of the difference between the original blue channel and the new luminance (intensity) channel, specifically

$$
x_{C_b} = 0.564 \, (x_B - x_Y)
$$

Similarly, x_{C_r} is a red-luminance colour difference,

$$
x_{C_r} = 0.713 \, (x_R - x_Y)
$$

It is common to model the reduced visual sensitivity to rapid colour changes by reducing the resolution of the chrominance channels. Specifically, it is common to work with YCbCr representations in which the chrominance components are sub-sampled by 2 in both the horizontal and vertical directions. In this way, the eliminated samples are deemed irrelevant.

IRRELEVANCE AND DISTORTION

Recall that the goal of a lossy compression system is to minimize distortion for a given bit-rate or, equivalently, to minimize the bit-rate for a given distortion. A suitable distortion measure should reflect the relevance of information in the image. In particular, visually or otherwise irrelevant information should have no impact on the distortion measure whatsoever. That is, $D(\mathbf{x}, \hat{\mathbf{x}}) = 0$ whenever the images $\mathbf{x} \equiv x[n_1, n_2]$ and $\hat{\mathbf{x}} \equiv \hat{x}[x_1, x_2]$ differ only in some irrelevant respect.

Suppose, for example, that $\hat{\mathbf{x}}$ is obtained by interpolating a sub-sampled version of the original image, $\mathbf{x}$. Since $\hat{\mathbf{x}}$ is represented with fewer samples, it should be more easily compressed. Thus, provided the additional spatial resolution associated with $\mathbf{x}$ is visually irrelevant, we should have $D(\mathbf{x}, \hat{\mathbf{x}}) = 0$ and a good lossy compression algorithm will choose to code the sub-sampled representation.

In this way, irrelevance may be "automatically" exploited by a good lossy compression algorithm provided the distortion measure which it is attempting to minimize correctly reflects the relevance of different types of information. This is possible within the framework offered by the JPEG2000 image compression standard; this framework is the subject of Chapter 8.

The role of the distortion measure is to capture the relative significance of different types of distortion. Completely irrelevant information may thus be viewed as an extreme case of the redundancy embodied by the distortion measure. This situation parallels that of statistical redundancy, discussed in Section 1.2.1, where the extreme case corresponds to entirely deterministic relationships among the image samples.

We conclude by pointing out that there should be no need to explicitly sub-sample chrominance components of a YCbCr representation for colour image compression, as suggested in the previous section, provided the distortion measure which is minimized by the lossy compression system correctly models the relative significance of spatial resolution in the luminance and chrominance components. For this very reason, the JPEG2000 standard does not explicitly offer a sub-sampled YCbCr representation for colour image compression. The reader is referred to Section 16.1.2 for a discussion of colour image compression with JPEG2000.

Figure 1.3.   Compression viewed as a global mapping operation.

1.3   ELEMENTS OF A COMPRESSION SYSTEM

Figure 1.3 depicts the compression and subsequent decompression systems as two mappings, M, and $\overline{M^{-1}}$, respectively. For lossless compression, we require $\overline{M^{-1}} = M^{-1}$. For lossy compression, however, M is not invertible, so we use the notation, $\overline{M^{-1}}$, to remind the reader that the decompression system represents an approximate inverse. We can think of the compressor as an enormous lookup table with $2^{N_1 N_2 B}$ entries.

Compression systems may be classified as either "fixed length" or "variable length." In the former case, the compressed bit-stream has a fixed length, $||\mathbf{c}||$, and the reconstructed image distortion, $D(\mathbf{x}, \hat{\mathbf{x}})$, will vary from image to image. In the case of fixed length compression, we can also think of the decompressor as an enormous lookup table with $2^{||\mathbf{c}||}$ entries. Since the image is being compressed, $||\mathbf{c}||$ should be much smaller than $N_1 N_2 B$, so that the decompressor's lookup table is smaller than that used during compression. In fact, an obvious way to construct the compressor, M, is to assign

$$\mathbf{c} = M(\mathbf{x}) = \operatorname*{argmin}_{\mathbf{c}'} D\left(\mathbf{x}, \overline{M^{-1}}(\mathbf{c}')\right) \tag{1.4}$$

Thus, it is sufficient to maintain the smaller lookup table corresponding to $\overline{M^{-1}}$ in both the compressor and the decompressor. The compressor then selects the bit-stream whose reconstructed image will be "closest" to the original in the sense induced by the appropriate distortion measure. This approach has the desirable side effect that $\overline{M^{-1}}$ is a right-inverse of M; i.e.,

$$M\left(\overline{M^{-1}}(\mathbf{c})\right) = \mathbf{c}$$

Such a compression system is said to be "idempotent" because the operations of compression and decompression may be repeated indefinitely without effecting the result produced by their first application.

1.3.1   THE IMPORTANCE OF STRUCTURE

The approach embodied by equation (1.4) is essentially the idea behind Vector Quantization (VQ). The generality of the VQ approach is

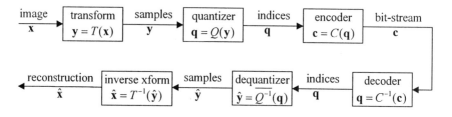

Figure 1.4.   Elements of a structured compression system.

attractive; however, the exponential growth in the size of the lookup table for M^{-1} renders it impractical for all but the smallest of images having only a few samples. For practical image compression, it is necessary to impose additional structure on the form of the maps, M and M^{-1}. Although this can be done in various ways, our objective here is to motivate the structure which is most commonly encountered in image compression systems. This structure is illustrated in Figure 1.4. Some alternative structures are discussed briefly in Section 1.4.

As depicted in Figure 1.4, the first step is to transform the original image samples into a new set of samples, which are more amenable to compression. For this step we write $\mathbf{y} = T(\mathbf{x})$, where $\mathbf{y} \equiv y[k_1, k_2]$ is another finite two dimensional sequence, having $K_1 K_2$ elements. The properties of the operator T will be discussed further shortly, but for the moment we point out that the operator is usually invertible. The decompressor employs the inverse transform, T^{-1}, and no distortion is introduced by this step. The second step is to represent the transform samples approximately using a sequence of quantization indices. For this step we write $\mathbf{q} = Q(\mathbf{y})$, where $\mathbf{q} \equiv q[p_1, p_2]$ denotes the finite two dimensional sequence of quantization indices, having $P_1 P_2$ elements. The set of possible outcomes for each quantization index, $q[p_1, p_2]$, is generally much smaller than that for the transform samples; also, the number of such quantization indices, $P_1 P_2$, is no larger and may be smaller than the number of transform samples, $K_1 K_2$. Thus, the quantization mapping, Q, introduces distortion and the decompressor uses an approximate inverse, Q^{-1}. Finally, the quantization indices are coded to form the final bit-stream. We write $\mathbf{c} = C(\mathbf{q})$. This step is invertible and introduces no distortion so that the decompressor may recover the quantization indices as $\mathbf{q} = C^{-1}(\mathbf{c})$.

1.3.2   CODING

The purpose of coding is to exploit statistical redundancy amongst the quantization indices, $q[p_1, p_2]$, as introduced in Section 1.2.1. The quantization and transform elements are designed in such a way as to ensure that this redundancy is spatially localized. Ideally, the underlying random variables, $Q[p_1, p_2]$, are all statistically independent. In that case, the indices may be coded independently and the only form of statistical redundancy which need be considered is that associated with any non-uniformity in their probability distributions.

As a simple example suppose that there are four possible quantization indices which we label $q = 0, 1, 2, 3$, having probabilities,

$$P(Q = q) = \begin{cases} 1/2 & \text{if } q = 0 \\ 1/4 & \text{if } q = 1 \\ 1/8 & \text{if } q = 2 \\ 1/8 & \text{if } q = 3 \end{cases}$$

An optimal code in this case represents each of the four possible outcomes using the following bit strings:

$$\begin{array}{ccc} q = 0 & & \text{``1''} \\ q = 1 & & \text{``01''} \\ q = 2 & \longrightarrow & \text{``001''} \\ q = 3 & & \text{``000''} \end{array}$$

That is, a prefix of q "0"s precedes the "1" which delineates the codewords corresponding to distinct quantization indices. The average number of bits spent coding each sample is

$$\frac{1}{2} \cdot 1 + \frac{1}{4} \cdot 2 + \frac{1}{8} \cdot 3 + \frac{1}{8} \cdot 3 = 1\frac{3}{4} \text{ bits}$$

which is less than the 2 bits which would be required to distinguish the four possible indices without coding. In practical image compression applications, much larger reductions in bit-rate are often possible.

It is not usually possible to ensure that the quantization indices are statistically independent. However, so long as statistical interactions are confined to the immediate neighbours of any given sample, it is often possible to design efficient coding schemes with manageable complexity. Coding is the subject of Chapter 2.

1.3.3 QUANTIZATION

Quantization is solely responsible for introducing distortion. For lossless compression there should be no quantization. In the simplest case we might map each transform sample, $y[k_1, k_2]$, independently to a corresponding quantization index, $q[k_1, k_2]$. This is known as scalar quantization; it is the simplest and most commonly employed form of quantization. Scalar quantization associates each quantization index with an

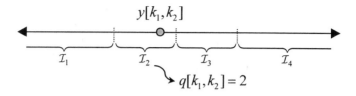

Figure 1.5.   Simple scalar quantizer with four output symbols.

interval on the real line according to

$$q\,[k_1, k_2] = i \text{ if } y\,[k_1, k_2] \in \mathcal{I}_i$$

where the intervals, $\mathcal{I}_i$, are disjoint and cover the real line[1]. As an example, the scalar quantizer of Figure 1.5 maps each transform sample to one of 4 distinct indices.

The approximate inverse quantization operator, $\overline{Q^{-1}}$, maps each index, $q\,[k_1, k_2]$, to some representation level in the corresponding interval, $\hat{y}\,[k_1, k_2]$. In the simplest case, we might select $\hat{y}\,[k_1, k_2]$ as the midpoint of the interval, $\mathcal{I}_{q[k_1, k_2]}$. From this elementary discussion, it should be apparent that $\overline{Q^{-1}}$ is a right-inverse of Q; i.e.,

$$Q\left(\overline{Q^{-1}}\,(\hat{\mathbf{y}})\right) = \hat{\mathbf{y}}$$

Both scalar and more sophisticated quantization schemes are the subject of Chapter 3.

1.3.4   TRANSFORMS

The transform is responsible for massaging the original image samples into a form which enables comparatively simple quantization and coding operations. On the one hand, the transform should capture the essence of statistical dependencies amongst the original image samples so that the transform samples, $y\,[k_1, k_2]$, and hence the quantization indices, $q\,[p_1, p_2]$, exhibit at most only very local dependencies; ideally, they should be statistically independent. On the other hand, the transform should separate irrelevant information from relevant information so that the irrelevant samples can be identified and quantized more heavily or even discarded.

Fortunately, it is possible to construct transforms which at least partially achieve both of these objectives simultaneously. Such transforms

---

[1] Strictly speaking, the $\mathcal{I}_i$ may be any disjoint cover of $\mathbb{R}$; however, there is no practical value in selecting the sets to be anything other than intervals.

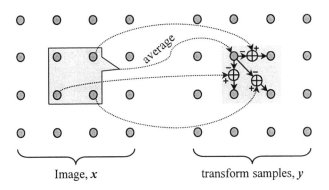

Figure 1.6. Simple image transform example.

are the subject of Chapters 4 and 6. For the moment, however, we motivate the concept with a simple example. Taking the image in 2×2 blocks we may represent each block with the average sample value over the block and the differences between any three of the samples and this average; e.g.,

$$y[m,n] = \begin{cases} \frac{1}{4}\sum_{i,j\in\{0,1\}} x[m+i,n+j] & \text{if } m,n \text{ both even} \\ x[m,n] - y\left[2\left\lfloor\frac{m}{2}\right\rfloor, 2\left\lfloor\frac{n}{2}\right\rfloor\right] & \text{otherwise} \end{cases}$$

This transform is illustrated in Figure 1.6. The inverse transform is obtained by setting

$$\hat{x}[n,m] = \begin{cases} 2\hat{y}[m,n] - \sum_{i,j\in\{0,1\}}\hat{y}[m+i,n+j] & \text{if } m,n \text{ both even} \\ \hat{y}[m,n] + \hat{y}\left[2\left\lfloor\frac{m}{2}\right\rfloor, 2\left\lfloor\frac{n}{2}\right\rfloor\right] & \text{otherwise} \end{cases}$$

The transform samples, $y[m,n]$ with m or n odd, contain high resolution details which may be discarded if the original image resolution exceeds the visual accuity of the intended observer. Thus, this simple transform assists in exposing a potential form of irrelevance in the image. Since most images contain substantial smooth regions, we also expect the detail samples, $y[m,n]$ with m or n odd, to contain values which are close to zero most of the time. As a result, the quantization index corresponding to zero should occur with high probability so that simple codes which operate on each sample independently should be able to exploit at least some of the underlying statistical redundancy.

1.4 ALTERNATIVE STRUCTURES

The image compression algorithms explored in the present text have the structure illustrated in Figure 1.4. Various other structures, however, are of interest in some applications. The most popular of these involve some form of predictive feedback, such as that illustrated in Figure 1.7. In this case, a scalar quantizer is used so that transform samples may be quantized one-by-one following a raster scan from the top row of the sample array through to the bottom row and from left to right within each row. Instead of quantizing sample $y[\mathbf{k}] = y[k_1, k_2]$ directly, we quantize the prediction residual,

$$e[\mathbf{k}] = y[\mathbf{k}] - y_p[\mathbf{k}]$$

where the predictor, $y_p[\mathbf{k}]$, is a function of the reconstructed samples, $\hat{y}[\mathbf{n}]$, at locations $\mathbf{n}$ which have already been visited. That is, either $n_1 < k_1$, or $n_1 = k_1$ and $n_2 < k_2$. These samples have already been reconstructed by the decompressor so that exactly the same predictor, $y_p[\mathbf{k}]$, is computed in both the compressor and the decompressor.

The idea behind predictive feedback is that the prediction residual, $e[\mathbf{k}]$, should generally be close to zero. After quantization, then, the statistical distribution of the quantization indices should be highly skewed toward the index whose quantization interval contains zero. The resulting statistical redundancy is then exploited by appropriate coding algorithms. The simplest predictor is the previous reconstructed sample, $y_p[k_1, k_2] = \hat{y}[k_1, k_2 - 1]$; for historical reasons, this is known as DPCM (Differential Pulse Code Modulation). As another example, we might use the average of the previous samples to the left and above that being predicted; i.e., $y_p[k_1, k_2] = \frac{1}{2}(\hat{y}[k_1, k_2 - 1], \hat{y}[k_1 - 1, k_2])$.

Predictive feedback is of little value unless the transform fails to remove most of the spatial redundancy – recall that a key objective of the transform is to minimize the statistical interaction between samples. For this reason, predictive feedback of the form shown in Figure 1.7 is often used in place of a transform. Alternatively, the feedback loop may include both the quantizer and the transform, as shown in Figure 1.8. In this case, multiple images are to be compressed one after the other and the prediction is formed from previous reconstructed images. The feedback structure of Figure 1.8 is fundamental to popular video compression schemes such as those embodied by the CCITT H.261 and H.263 video telephony standards and the group of ISO/IEC standards developed by the MPEG (Motion Picture Experts Group) working group. Other variants on the predictive feedback concept may be found in lossless compression image algorithms and some of the modes supported by the JPEG image compression standard.

A general characteristic of feedback compression structures is that they rely upon the compressor's ability to precisely replicate some or all of the samples which will be reconstructed by the decompressor. In fact,

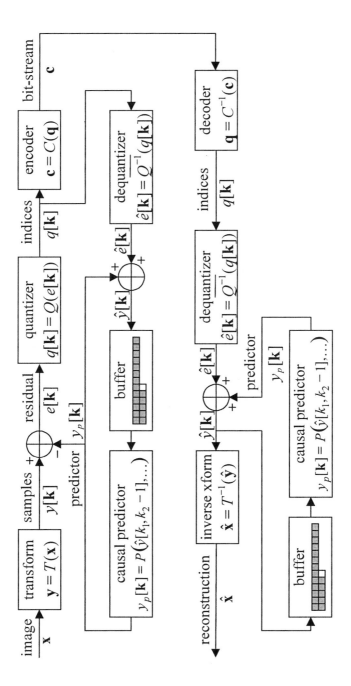

Figure 1.7. Modified compression structure with predictive feedback.

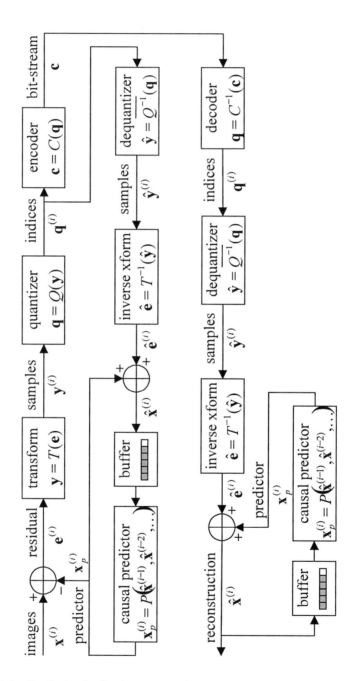

Figure 1.8. Predictive feedback structure for compressing multiple images (video).

Figures 1.7 and 1.8 reveal the fact that the compressor must include an exact copy of certain elements from the decompressor. Therein lies the principle weakness of predictive feedback structures.

A key requirement driving the JPEG2000 standardization process is scalability. A scalable bit-stream is one which may be partially discarded to obtain an efficient representation of the original image or a lower resolution version of it at a different bit-rate. A highly scalable bit-stream may be decompressed in many different ways with different results, depending upon what information has been discarded. It is difficult for the compressor to replicate the state which the decompressor may attain under all possible scalings of the compressed bit-stream. Consequently, the feedforward structure of Figure 1.4 is preferred for scalable compression.

Chapter 2

ENTROPY AND CODING TECHNIQUES

2.1   INFORMATION AND ENTROPY

A binary digit, or "bit," b, takes one of the values $b = 0$ or $b = 1$. A single bit has the ability to convey a certain amount of information – the information corresponding to the outcome of a binary decision, or "event," such as a coin toss. If we have N bits, then we can identify the outcomes of N binary decisions.

Intuitively, the average amount of information associated with a binary decision depends upon prior knowledge which we have concerning the likelihoods of the possible outcomes. For example, there is little informative value to including snow conditions in the weather report during summer – in common parlance, the result is a foregone conclusion. By contrast, the binary events which convey most information on average are those which are equally likely. Similarly, the N-bit sequences which convey most information are those for which each bit has equally likely outcomes, regardless of the outcomes of the other bits in the sequence – loosely speaking, these are "entirely random" sequences of bits.

Source coding is the art of mapping each possible output from a given information source to a sequence of binary digits called "code bits." Ideally, the mapping has the property that the code bits are "entirely random," i.e., statistically independent, taking values of 0 and 1 with equal probability. In this way, the code bits convey the maximum possible amount of information. Then, provided the mapping is invertible, we can identify the number of code bits with the amount of information in the original source output.

The above concepts were formalized in the pioneering work of Claude Shannon [130]. A quantity known as "entropy" is defined in terms of the statistical properties of the information source. The entropy represents a lower bound on the average number of bits required to represent the source output. Moreover, it is possible to approach this lower bound arbitrarily closely. In fact, practical coding algorithms can achieve average bit rates which are extremely close to the entropy in many applications and when they do so the code bits must be entirely random.

2.1.1   MATHEMATICAL PRELIMINARIES

RANDOM VARIABLES AND VECTORS

Let X denote a random variable. Associated with the random variable is a set of possible outcomes, known as the alphabet, $\mathcal{A}_X$. The outcome of the random variable is denoted x, and is one of the elements of $\mathcal{A}_X$. A random variable is said to be discrete if its alphabet is finite or at most countably infinite. That is, we can enumerate the elements of the alphabet,

$$\mathcal{A}_X = \{\alpha_0, \alpha_1, \alpha_2, \ldots\}$$

In this case, the statistical properties of the random variable are described by its probability mass function (PMF)

$$f_X(x) \triangleq P(X = x) \text{ for each } x \in \mathcal{A}_X$$

In words, $f_X(x)$ is the probability of the outcome $X = x$. By contrast, a continuous random variable has uncountably many outcomes, e.g. $\mathcal{A}_X = \mathbb{R}$, the set of all real numbers. In this chapter we will be concerned exclusively with discrete alphabets. As an example, we model binary decisions as random variables whose alphabets have only two entries, usually written $\mathcal{A}_X = \{0, 1\}$. Binary random variables play a special role in coding.

The notion of a random variable is trivially extended to random vectors, $\mathbf{X}$, with alphabet, $\mathcal{A}_\mathbf{X}$ and PMF, $f_\mathbf{X}(\mathbf{x})$, for each vector, $\mathbf{x} \in \mathcal{A}_\mathbf{X}$. An m-dimensional random vector is a collection of m random variables, usually taken as a column vector,

$$\mathbf{X} = \begin{pmatrix} X_0 \\ X_1 \\ \vdots \\ X_{m-1} \end{pmatrix}$$

The PMF, $f_\mathbf{X}(\mathbf{x})$, is sometimes written longhand as

$$f_\mathbf{X}(\mathbf{x}) \equiv f_{X_0, X_1, \ldots, X_{m-1}}(x_0, x_1, \ldots, x_{m-1})$$

It denotes the probability that $X_0 = x_0$, $X_1 = x_1$, ..., and $X_{m-1} = x_{m-1}$ simultaneously. For this reason, it is often called the joint PMF, or joint distribution, for the m random variables.

From the joint distribution of a collection of m random variables, we can obtain the "marginal" distribution of any one of the random variables, X_i, as

$$f_{X_i}(x) = \sum_{\mathbf{x} \ni x_i = x} f_{\mathbf{X}}(\mathbf{x})$$

INDEPENDENCE AND CONDITIONAL PMF'S

We say that two random variables are statistically independent, or simply independent, if their joint distribution is separable; i.e.,

$$f_{X_0, X_1}(x_0, x_1) = f_{X_0}(x_0) f_{X_1}(x_1)$$

That is, the probability that both $X_0 = x_0$ and $X_1 = x_1$ is the product of the two marginal probabilities. As suggested by the introductory comments above, the notion of statistical independence plays an important role in coding.

We define the conditional distribution of X_1, given X_0, by

$$f_{X_1|X_0}(x_1, x_0) \triangleq \frac{f_{X_1, X_0}(x_1, x_0)}{f_{X_0}(x_0)} = \frac{f_{X_1, X_0}(x_1, x_0)}{\sum_x f_{X_1, X_0}(x, x_0)}$$

The function, $f_{X_1|X_0}(\cdot, x_0)$, is interpreted as a modified PMF for X_1, where the modification is to reflect the fact that the outcome $X_0 = x_0$ is already known. If the two random variables are statistically independent, we expect that the outcome of X_0 has no bearing on the distribution of X_1 and indeed we find that

$$f_{X_1|X_0}(x_1, x_0) = f_{X_1}(x_1) \text{ if and only if } X_1, X_0 \text{ are independent}$$

We note that the marginal distribution of X_0 and the conditional distribution of X_1, given X_0, together are equivalent to the joint distribution of X_1 and X_0. More generally, we write $f_{X_n|X_{n-1},...,X_0}(x_n, ..., x_0)$ for the conditional distribution of X_n, given X_0 through X_{n-1}. The joint distribution of all m random variables of an m-dimensional random vector, $\mathbf{X}$, may be recovered from

$$f_{\mathbf{X}}(\mathbf{x}) = f_{X_0}(x_0) f_{X_1|X_0}(x_1, x_0) \cdots f_{X_{m-1}|X_{m-2},...,X_0}(x_{m-1}, ..., x_0) \tag{2.1}$$

and the random variables are said to be mutually independent if

$$f_{\mathbf{X}}(\mathbf{x}) = f_{X_0}(x_0) f_{X_1}(x_1) \cdots f_{X_{m-1}}(x_{m-1})$$

EXPECTATION

The expectation of a random variable, X, is denoted $E[X]$ and defined by

$$E[X] \triangleq \sum_{x \in \mathcal{A}_X} x f_X(x)$$

It represents the statistical average or mean of the random variable X. Here, for the first time, we are concerned with the algebraic properties of random variables. More generally, let $g()$ be any function. We may define $Y = g(X)$ to be the random variable whose outcomes are $y = g(x)$ whenever the outcome of X is x. Consequently, the distribution of Y may be found from

$$f_Y(y) = \sum_{x \ni g(x)=y} f_X(x)$$

It is readily shown that the expectation of the new random variable, Y, satisfies

$$E[Y] = E[g(X)] = \sum_{y \in \mathcal{A}_Y} y f_Y(y) = \sum_{x \in \mathcal{A}_X} g(x) f_X(x) \qquad (2.2)$$

Given two random variables, X_0 and X_1, we may define conditional expectations in the most obvious way as

$$E[X_1 \mid X_0 = x_0] \triangleq \sum_{x \in \mathcal{A}_{X_1}} x f_{X_1|X_0}(x, x_0)$$

and for any function, $g()$, we have

$$E[g(X_1) \mid X_0 = x_0] = \sum_{x \in \mathcal{A}_{X_1}} g(x) f_{X_1|X_0}(x, x_0)$$

RANDOM PROCESSES

We conclude this section by introducing the concept of a discrete random process, denoted $\{X_n\}$. A random process is nothing but a sequence of individual random variables, X_n, $n \in \mathbb{Z}$, all having a common alphabet, $\mathcal{A}_X$. The key distinction from a random vector is that there are infinitely many random variables. The statistics are summarized by the vector PMF's, $f_{\mathbf{X}_{i:j}}()$, for all $i < j \in \mathbb{Z}$, where we use the notation, $\mathbf{X}_{i:j}$, to refer to the $(j-i)$-dimensional random vector formed from the elements, X_k, $i \le k < j$, of the random process. The random process, $\{X_n\}$, is said to be stationary if the vector PMF's satisfy

$$f_{\mathbf{X}_{i:i+m}} = f_{\mathbf{X}_{0:m}} \text{ for all } i, m \in \mathbb{Z}, m > 0$$

That is, all collections of m consecutive random variables from the process have exactly the same joint distribution. Thus, a stationary random process is characterized by the PMF's, $f_{\mathbf{X}_{0:m}}$ for each $m = 1, 2, \ldots$. Alternatively, from equation (2.1) we see that stationary random processes are characterized by the marginal distribution, $f_{X_0} \equiv f_X$, together with the sequence of conditional distributions, $f_{X_m|\mathbf{X}_{0:m}}$, for $m = 1, 2, \ldots$.

In most applications we find that the conditional distributions satisfy

$$f_{X_m|\mathbf{X}_{0:m}} = f_{X_m|X_{m-p:m}} \qquad (2.3)$$

for a sufficiently large value of the parameter, p. That is, the conditional distribution of X_m given X_0 through X_{m-1}, is actually a function of only the p most recent random variables, X_{m-p} through X_{m-1}. We say that X_m is "conditionally independent" of X_0 through X_{m-p-1}. Conditional independence is a phenomenon which we usually expect to encounter in the information sources which we model using random processes. Indeed statistical dependencies among samples taken from natural physical phenomena such as images and audio are generally of a local nature. For stationary processes, conditional independence means that the entire process is described by a finite number of conditional PMF's

$$f_{X_0}, \ f_{X_1|X_0}, \ f_{X_2|\mathbf{X}_{0:2}}, \ \ldots, \ f_{X_p|\mathbf{X}_{0:p}}$$

These are called Markov random processes with parameter p. A Markov-1 random process is entirely described by f_X and $f_{X_1|X_0}$. If $p = 0$, all elements of the random process are statistically independent with identical distribution, f_X. Such a random process is said to be IID (Independent and Identically Distributed). It is also said to be "memoryless."

Stationary random processes with conditional independence properties (i.e. Markov processes) play an extremely important role in coding, precisely because they are described by a finite number of conditional PMF's. By observing the outcomes of the random process over a finite period of time, we can hope to estimate these conditional PMF's and use these estimates to code future outcomes of the random process. In this way, we need not necessarily have any a priori knowledge concerning the statistics in order to effectively code the source output. Adaptive coders are based on this principle.

The technical condition required to enable estimation of the relevant PMF's from a finite number of outcomes is "ergodicity." To be more precise, suppose we observe the outcomes of random variables X_0 through X_{M-1}. For each m-dimensional vector, $\mathbf{y}$, let $K_{\mathbf{y},M}$ denote the number of occurrences of $\mathbf{y}$ as a "sub-string" of the observed sequence,

$\mathbf{x}_{0:M}$; i.e.,

$$K_{\mathbf{y},M} = \|\{i | 0 \le i < M - m \text{ and } \mathbf{x}_{i:i+m} = \mathbf{y}\}\|$$

It is natural to estimate the conditional PMF's according to

$$\hat{f}_{X_m | \mathbf{X}_{0:m}}(x, \mathbf{y}) = \frac{K_{(\mathbf{y},x),M} + \delta}{\sum_{z \in A_X}(K_{(\mathbf{y},z),M} + \delta)} \text{ for each } m \le p < M$$

where δ is a small offset (e.g. $\delta = 1$), included to avoid undefined or ill-conditioned estimates when M is small and $(\mathbf{y}, x)$ denotes the vector formed by appending x to $\mathbf{y}$. If the random process is ergodic, these estimates will converge to the actual conditional PMF's as M increases. Most random processes encountered in practice are ergodic. At least, practical coding algorithms are based on the assumption that the underlying random process is ergodic, so that we can estimate the statistics through observation.

2.1.2   THE CONCEPT OF ENTROPY
ENTROPY OF A RANDOM VARIABLE

The entropy of a random variable, X, is defined as

$$H(X) \triangleq - \sum_{x \in A_X} f_X(x) \log_2 f_X(x)$$

We shall find the following equivalent expression more convenient and intuitive

$$H(X) = E[-\log_2 f_X(X)]$$

To clarify this expression, define the function, $h_X()$, by

$$h_X(x) \triangleq -\log_2 f_X(x)$$

As with any function, then, we may define the random variable, $Y = h_X(X)$, and apply equation (2.2) to see that the entropy of X is the expectation of the new random variable, Y; i.e.,

$$H(X) = E[h_X(X)]$$

As we shall see, the quantity $h_X(x)$ may be interpreted as the amount of information associated with the event $X = x$ and then the entropy may be interpreted as the average (or expected) amount of information conveyed by the outcome of the random variable, X.

Entropy measures information in units of bits. The precise connection between entropy, as defined above, and the average number of bits required to code the outcome of the random variable will be explored in Sections 2.1.3 and 2.1.4. For the moment, however, it is instructive to reflect on the intuitive appeal of the definition, as suggested by the following properties:

1. $h_X(x)$ is a strictly decreasing function of the likelihood, $f_X(x)$. This agrees with the notion that a highly unlikely event carries considerable information when it occurs. For example, the appearance of snow in summer is a newsworthy event in most cities.

2. The entropy is bounded below by 0. Moreover, $H(X) = 0$ occurs if and only if X has a deterministic outcome, say $X = x_0$. That is,

$$f_X(x) = \begin{cases} 1 & \text{if } x = x_0 \\ 0 & \text{if } x \neq x_0 \end{cases}$$

3. For random variables with finite alphabets, the entropy is bounded above by $H(X) \leq \log_2 \|A_X\|$. Moreover, the upper bound occurs if and only if X has a uniform distribution; i.e., $f_X(x) = \frac{1}{\|A_X\|}, \forall x \in A_X$. Of particular interest is the case in which the alphabet consists of B-bit numbers, $A_X = \{0, 1, \ldots, 2^B - 1\}$. Then $H(X) \leq B$ with equality if and only if all 2^B outcomes are equally likely. Put another way, an information source whose outcomes are represented with B bit numbers has an entropy of at most B bits, where this maximum occurs if and only if the B bits are "entirely random."

Example 2.1 *Let X be a binary random variable with $f_X(0) = p$ and $f_X(1) = (1-p)$. Figure 2.1 plots $H(X)$ as a function of the parameter, p. The figure clearly indicates the fact that the entropy is zero when X is deterministic ($p = 0$ or $p = 1$) and is maximized when the two outcomes are equally likely ($p = \frac{1}{2}$).*

JOINT AND CONDITIONAL ENTROPY

The definition of entropy extends naturally to random vectors so that

$$H(\mathbf{X}) \triangleq E[-\log_2 f_X(\mathbf{X})] = E[h_\mathbf{X}(\mathbf{X})]$$

In fact, since the alphabet of the random vector is discrete, we can enumerate its elements, $A_\mathbf{X} = \{\alpha_0, \alpha_1, \ldots\}$, and define a random variable, $K = k$ whenever $\mathbf{X} = \alpha_k$. Since the outcomes of K and $\mathbf{X}$ convey exactly the same information we must insist that $H(\mathbf{X}) = H(K)$. The above definition then follows immediately.

We sometimes use the longhand expression

$$H(\mathbf{X}) \equiv H(X_0, X_1, \ldots, X_{m-1})$$

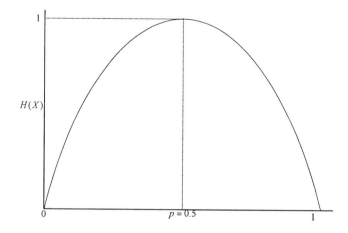

Figure 2.1. Entropy of a binary random variable, X, as a function of $p = f_X(0)$.

where $\mathbf{X}$ is an m-dimensional random vector with elements, X_0 through X_{m-1}. We may also call this the joint entropy of the m random variables since it is a function of their joint PMF.

The conditional entropy of X given Y is defined by

$$H(X \mid Y) \triangleq - \sum_{y \in A_Y} f_Y(y) \sum_{x \in A_X} f_{X|Y}(x,y) \log_2 f_{X|Y}(x,y)$$

$$= - \sum_{y \in A_Y} \sum_{x \in A_X} f_{X,Y}(x,y) \log_2 f_{X|Y}(x,y)$$

$$= E\left[-\log_2 f_{X|Y}(X,Y)\right]$$

$H(X \mid Y)$ may be interpreted as the average additional information we receive from the outcome of X given that the outcome of Y is already known. This interpretation follows directly from the interpretation of $H(X)$ as the average amount of information we receive from the outcome of X.[1] To see this, observe that

$$H(X,Y) = E\left[-\log_2 f_{X,Y}(X,Y)\right]$$

$$= E\left[-\log_2 \left(f_Y(Y) f_{X|Y}(X,Y)\right)\right]$$

$$= E\left[-\log_2 f_Y(Y)\right] + E\left[-\log_2 f_{X|Y}(X,Y)\right]$$

$$= H(Y) + H(X \mid Y) \tag{2.4}$$

---

[1] Although the properties of $H(X)$ suggest an interpretation as a measure of information, the connection with information will be concretely established in Section 2.1.3.

Thus, $H(X \mid Y)$ indeed represents the extra information, $H(X,Y) - H(Y)$. When X and Y are independent random variables we find that

$$H(X \mid Y) = E\left[-\log_2 f_{X|Y}(X,Y)\right]$$
$$= E\left[-\log_2 f_X(X)\right]$$
$$= H(X)$$

which agrees with the fact that the outcome of Y has no bearing on the information conveyed by X.

An important property of the conditional entropy is summarized by the following theorem.

Theorem 2.1 *Let X be a random variable. Let $\mathbf{Y}$ be a random vector with elements Y_0 through Y_{m-1} and let $\mathbf{Y}'$ be a random vector consisting of any subset (possibly empty) of these elements. Then*

$$H(X \mid \mathbf{Y}) \le H(X \mid \mathbf{Y}')$$

with equality if and only if $f_{X|\mathbf{Y}} = f_{X|\mathbf{Y}'}$; i.e., if and only if X is conditionally independent of the elements in $\mathbf{Y}$ which are missing from $\mathbf{Y}'$.

Corollary 2.2 *For random variables, X and Y, $H(X \mid Y) \le H(X)$, with equality if and only if X and Y are independent.*

Corollary 2.3 *From equation (2.4), we also have $H(X,Y) \le H(X) + H(Y)$, with equality if and only if X and Y are independent.*

These results have considerable intuitive appeal. If we know the outcomes of some collection of random variables, Y_0 through Y_{m-1}, which are not statistically independent of X, then this reduces the uncertainty of X and hence the amount of additional information conveyed by the outcome of X. As m increases, the uncertainty in X and hence the conditional entropy, $H(X \mid \mathbf{Y})$, continues to decrease so long as each new random variable Y_m provides some new information about X, which is not already present in the other random variables, Y_0 through Y_{m-1}.

Equation (2.4) is easily generalized to expand the entropy of any random vector, $\mathbf{X}$, as

$$H(\mathbf{X}) = E\left[-\log_2 f_{\mathbf{X}}(\mathbf{X})\right]$$
$$= E\left[-\log_2\left(f_{X_0}(X_0) \cdot \cdots \cdot f_{X_{m-1}|X_{m-2},\ldots,X_0}(X_{m-1},\ldots,X_0)\right)\right]$$
$$= H(X_0) + H(X_1 \mid X_0) + \cdots + H(X_{m-1} \mid X_{m-2},\ldots,X_0)$$
$$\le \sum_{n=0}^{m-1} H(X_n) \tag{2.5}$$

where equality holds if and only if the random variables, X_i, are all independent. As we shall see in Section 2.1.4, this expansion is particularly useful in coding.

ENTROPY RATE

Let $\{X_n\}$ be a discrete stationary random process. Since the random process has infinite extent, the total amount of information conveyed by the outcome of the random process will usually be infinite. In fact, for Markov processes it must be either infinite or zero. Thus, it is more meaningful to introduce the notion of an "information rate" for random processes. A close analogy is the characterization of stationary random processes by their power rather than their energy in the study of linear systems.

We begin by defining the mth order entropy of the random process by

$$H^{(m)}(\{X_n\}) \triangleq \frac{1}{m} H(\mathbf{X}_{0:m}) = \frac{1}{m} H(\mathbf{X}_{i:m+i}), \text{ for all } i \in \mathbb{Z}$$

Thus, the 1st order entropy is simply the entropy of any given random variable, say $X[0]$ (they all have the same distribution), taken individually from the process. The 2nd order entropy is half the joint entropy of any pair of consecutive random variables from the process. From Corollary 2.3,

$$H^{(2)}(\{X_n\}) \leq \frac{1}{2}(H(X_0) + H(X_1))$$
$$= H^{(1)}(\{X_n\})$$

In fact, from Theorem 2.1, we see that

$$m H^{(m)}(\{X_n\}) = H(X_0) + H(X_1 \mid \mathbf{X}_{0:1}) + \cdots + H(X_{m-1} \mid \mathbf{X}_{0:m-1})$$
$$\tag{2.6}$$
$$\geq m H(X_{m-1} \mid \mathbf{X}_{0:m-1})$$

and hence

$$(m+1) H^{(m+1)}(\{X_n\}) = m H^{(m)}(\{X_n\}) + H(X_m \mid \mathbf{X}_{0:m})$$
$$\leq m H^{(m)}(\{X_n\}) + H(X_{m-1} \mid \mathbf{X}_{0:m-1})$$
$$\leq (m+1) H^{(m)}(\{X_n\})$$

So $H^{(m)}(\{X_n\})$ is a monotonically decreasing function of m. Since it is bounded below by 0, the sequence must converge and we define the entropy rate of the random process to be

$$H(\{X_n\}) \triangleq \lim_{m \to \infty} H^{(m)}(\{X_n\}) \tag{2.7}$$

This quantity is interpreted as the average rate at which the random process conveys information as the outcomes $X_n = x_n$ are discovered one by one.

For a Markov-p source process, equation (2.7) may be recast as

$$H(\{X_n\}) = H(X_p \mid \mathbf{X}_{0:p}) \tag{2.8}$$

This very simple form follows from equation (2.6) and the conditional independence property of the random process, according to which

$$H^{(m)}(\{X_n\}) = \frac{1}{m}(H(X_0) + H(X_1 \mid \mathbf{X}_{0:1}) + \cdots + H(X_{p-1} \mid \mathbf{X}_{0:p-1}))$$
$$+ \frac{m-p}{m}H(X_p \mid \mathbf{X}_{0:p})$$
$$\longrightarrow H(X_p \mid \mathbf{X}_{0:p}) \text{ as } m \to \infty$$

2.1.3   SHANNON'S NOISELESS SOURCE CODING THEOREM

In Section 2.1.2 we defined a quantity called entropy, having properties which we would expect of a measure of information. The value of entropy, however, as a tool for understanding and developing practical compression algorithms, arises from a rigorous connection between these definitions and fundamental bounds on coding performance. This connection was first established by Shannon's "noiseless source coding theorem" [130]. The essence of this theorem is that the entropy rate of a random process provides a lower bound on the average number of bits which must be spent in coding each of its outcomes and also that this bound may be approached arbitrarily closely as the complexity of the coding scheme is allowed to grow without bound. Due to the importance of this result, we choose to reproduce Shannon's proof here, for the simple case of a stationary, memoryless random process.

Let $\{X_n\}$ be an IID random process, each element having distribution f_X and entropy $H(X)$. The entropy rate, $H(\{X_n\})$, in this case, is identical to $H(X)$. For ease of expression we shall sometimes refer to the individual source outcomes, $x_0, x_1, \ldots$, as symbols. Then, we assess the information rate of the random process as the average number of bits per symbol required to represent the source output over a period of m consecutive symbols, in the limit as m becomes very large. Specifically, we construct a code which maps outcomes of the random vector, $\mathbf{X}_{0:m}$ to L-bit codewords. This is a "fixed length" code since each block of m symbols is represented by the same number of code-bits, L. Codes of this form are known as (m, L) codes. The ratio, $\frac{L}{m}$ represents the average number of bits spent coding each symbol. The idea is essentially to show that in the limit as m becomes very large, this ratio $\frac{L}{m}$ may be made arbitrarily close to $H(X)$.

There are a total of $\|\mathcal{A}_X\|^m$ possible m-dimensional outcomes, $\mathbf{x}_{0:m}$, and so it is clearly impossible to represent all outcomes perfectly unless $2^L \geq \|\mathcal{A}_X\|^m$; i.e., $\frac{L}{m} \geq \log_2 \|\mathcal{A}_X\|$. But we know that $H(X) \leq \log_2 \|\mathcal{A}_X\|$, attaining this maximum value only when f_X is the uniform distribution,with all outcomes equally likely. Thus, in order to establish a connection between entropy and the bit-rate of a fixed length code, we will need to admit the possibility that the coded representation might not be exact. Let $P_e(m, L)$ denote the probability that our L-bit code does not represent the random vector, $\mathbf{X}_{0:m}$, exactly. The idea behind the noiseless source coding theorem is to show that $P_e(m, L)$ may be made arbitrarily small as m grows, provided the code-rate $\frac{L}{m} > H(X)$.

Theorem 2.4 *Let $\{X_n\}$ be a discrete IID random process having finite entropy, $H(X)$, and consider fixed length (m, L) codes, associating m-element outcome vectors, $\mathbf{x}_{0:m}$, to L-bit codewords. Only one outcome vector may be associated with each codeword, so let $P_e(m, L)$ denote the probability of the event that $\mathbf{X}_{0:m}$ has no associated codeword. Then, by making m sufficiently large, the error probability, $P_e(m, L)$, may be made arbitrarily small, so long as the code-rate satisfies*

$$\frac{L}{m} > H(X)$$

Conversely, the error probability, $P_e(m, L)$, tends to 1 as $m \to \infty$ for codes having

$$\frac{L}{m} < H(X)$$

Proof. Consider the random variable, $h_{\mathbf{X}_{0:m}}(\mathbf{X}_{0:m})$, which we defined to be $-\log_2 f_{\mathbf{X}_{0:m}}(\mathbf{X}_{0:m})$. Since the elements of the random vector, $\mathbf{X}_{0:m}$, are all independent, $f_{\mathbf{X}_{0:m}}$ is separable and we obtain

$$h_{\mathbf{X}_{0:m}}(\mathbf{X}_{0:m}) = -\log_2 \prod_{i=0}^{m-1} f_X(X_i)$$

$$= \sum_{i=0}^{m-1} h_X(X_i)$$

So $h_{\mathbf{X}_{0:m}}(\mathbf{X}_{0:m})$ is a sum of the IID random variables, $h_X(X_i)$. According to the weak law of large numbers, $\frac{1}{m}\sum_{i=0}^{m-1} h_X(X_i)$, converges to $E[h_X(X)] = H(X)$, as $m \to \infty$. Specifically, for any $\delta > 0$, let $\varepsilon(m, \delta)$ denote the probability

$$\varepsilon(m, \delta) = P\left(\left|\frac{1}{m}\sum_{i=0}^{m-1} h_X(X_i) - E[h_X(X)]\right| > \delta\right)$$

$$= P\left(\left|\frac{1}{m} h_{\mathbf{X}_{0:m}}(\mathbf{x}_{0:m}) - H(X)\right| > \delta\right) \tag{2.9}$$

Then the weak law of large numbers states that

$$\lim_{m \to \infty} \varepsilon(m, \delta) = 0$$

Equivalently, let $T(m, \delta)$ be the set of outcomes, $\mathbf{x}_{0:m}$, for which

$$\left| \frac{1}{m} h_{\mathbf{x}_{0:m}}(\mathbf{x}_{0:m}) - H(X) \right| \leq \delta$$

Then $\varepsilon(m, \delta) = P(\mathbf{X}_{0:m} \notin T(m, \delta)) \xrightarrow{m \to \infty} 0$. For small δ and large m so that $\varepsilon(m, \delta)$ is very small, we may think of $T(m, \delta)$ as the set of "typical" outcomes. The idea is to assign codewords only to these typical outcomes, since the probability of anything else becomes vanishingly small as m grows. For each $\mathbf{x}_{0:m} \in T(m, \delta)$, we have

$$H(X) - \delta \leq \frac{1}{m} h_{\mathbf{x}_{0:m}}(\mathbf{x}_{0:m}) \leq H(X) + \delta$$

and hence the probability of each typical outcome is bounded by

$$2^{-m(H(X)+\delta)} \leq f_{\mathbf{x}_{0:m}}(\mathbf{x}_{0:m}) \leq 2^{-m(H(X)-\delta)} \tag{2.10}$$

Letting δ become very small, the typical outcomes all have essentially the same likelihood, so that if we assign codewords only to the typical outcomes, the resulting L-bit codewords will be uniformly distributed, or "entirely random."

Using equation (2.10), we see that

$$P(\mathbf{X}_{0:m} \in T(m, \delta)) = \sum_{\mathbf{x}_{0:m} \in T(m, \delta)} f_{\mathbf{x}_{0:m}}(\mathbf{x}_{0:m})$$

$$\geq 2^{-m(H(X)+\delta)} \, \|T(m, \delta)\|$$

So the number of typical outcomes is bounded above by

$$\|T(m, \delta)\| \leq \frac{1 - \varepsilon(m, \delta)}{2^{-m(H(X)+\delta)}}$$

$$\leq 2^{m(H(X)+\delta)}$$

It follows that so long as we select $L \geq m(H(X) + \delta)$, we can represent all of the typical outcomes with a distinct codeword and then the probability of error, $P_e(m, L)$, must be at most $\varepsilon(m, \delta)$, which tends to 0 as $m \to \infty$. This proves the first statement of the theorem, since $\delta > 0$ is arbitrary.

To prove the converse statement, we use equation (2.10) again to obtain a lower bound for the number of typical outcomes; i.e.,

$$\|T(m, \delta)\| \geq \frac{1 - \varepsilon(m, \delta)}{2^{-m(H(X)-\delta)}}$$

Suppose that $L \leq m(H(X) - 2\delta)$. Let T denote the number of elements from $T(m, \delta)$ which are represented in this code. Then T satisfies

$$\frac{T}{\|T(m, \delta)\|} \leq \frac{2^L}{\|T(m, \delta)\|} \leq \frac{2^{-m\delta}}{1 - \varepsilon(m, \delta)}$$

So the fraction of typical outcomes which can be represented tends to 0 as $m \to \infty$, whenever the code rate is less than $H(X)$. This suggests the validity of the second statement of the theorem. To make the proof rigorous, observe that the total probability associated with the T elements of $T(m, \delta)$ which are represented by the code is at most

$$2^L \cdot 2^{-m(H(X)-\delta)} \leq 2^{-m\delta}$$

and the total probability of all other outcomes is $\varepsilon(m, \delta)$, so

$$P_e(m, L) \geq 1 - \varepsilon(m, \delta) - 2^{-m\delta}$$
$$\xrightarrow{m \to \infty} 1$$

■

Several points are worth noting concerning the noiseless source coding theorem. Firstly, for finite length codes, fixed length coding is incapable of guaranteeing that all source outcomes will be represented exactly. The solution to this dilemma is variable length codes, which are examined next. Despite this obstacle, the noiseless source coding theorem does indeed establish a strong connection between entropy and coding. The entropy clearly partitions the set of code rates into two classes. So long as the code-rate exceeds the entropy, we can make sure that the entire message is coded without error with arbitrarily high confidence; if the code-rate is less than the entropy, long messages will contain errors with probability approaching 1.

As noted in the proof of the theorem, reliable codes whose rate approaches the entropy have the property that their codewords all occur with equal likelihood. That is, the L bit sequences are "entirely random." Recall that we began Section 2.1 with the claim that the representation of source outcomes with entirely random sequences of bits is the goal of source coding. This is perhaps the most important observation arising from the noiseless source coding theorem.

Shannon's original result has been extended over the years to random processes satisfying a variety of technical conditions. For more general random processes than the simple memoryless processes considered above, the key difficulty is to demonstrate convergence of $\varepsilon(m, \delta)$, as defined by equation (2.9). This is known as the entropy-ergodic property. Shannon himself extended the result to Markov processes, while extensions to more general ergodic random processes were developed by McMillan [107] and extended by Breiman [29, 30] and others. The more general result is often known as the Shannon-McMillan-Breiman theorem, or the asymptotic equipartition (AEP) theorem.

2.1.4   ELIAS CODING

As mentioned above, fixed length codes cannot generally guarantee lossless coding. In this section, we consider variable length codes. It is most instructive to describe a particular coding algorithm, whose ability to approach the entropy rate of a stationary markov random process can be demonstrated rather easily. The algorithm is not practical as it stands since its implementation requires infinite precision arithmetic. Nevertheless, it is the basis for a family of highly efficient practical coding techniques, known collectively as arithmetic coding. Indeed one member of this family is at the heart of the JPEG2000 image compression standard (see Section 12.1). Practical arithmetic coding is the subject of Section 2.3. P. Elias is usually credited with conceiving the algorithm shortly after Shannon's original publication on information theory.

MAPPING OUTCOMES TO INTERVALS

Let $\{X_n\}$ be a stationary random process. To begin, we will restrict ourselves to memoryless processes, as in Section 2.1.3. In this case, we hope to be able to code the outcomes of the random process at an average rate of $H(X)$ bits per symbol.

Following the notation developed above, we denote the first n outcomes of the random process by the vector, $\mathbf{x}_{0:n}$. The algorithm is best understood as associating each such length n prefix of the source sequence with a unique interval on the real line,

$$[c_n, c_n + a_n) \subseteq [0, 1)$$

such that the length of this interval is equal to $f_{\mathbf{X}_{0:n}}(\mathbf{x}_{0:n})$. The algorithm is implemented recursively as follows:

Elias Coding Algorithm

Initialize $c_0 = 0$ and $a_0 = 1$.

For each $n = 0, 1, \ldots$

      Update $a_{n+1} \leftarrow a_n f_X(x_n)$

      Update $c_{n+1} \leftarrow c_n + a_n F_X(x_n)$

Here, F_X denotes the cumulative distribution[2],

$$F_X(\alpha_i) \triangleq \sum_{j=0}^{i-1} f_X(\alpha_j) \text{ where } \mathcal{A}_X = \{\alpha_0, \alpha_1, \ldots\}$$

We assume that the encoder and decoder both have access to the underlying distribution function, f_X and hence F_X, or else they both use identical estimates for this function.

---

[2]Note the non-standard definition here, in which the probability of α_i itself is not included in the summation.

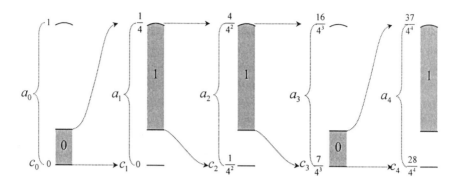

Figure 2.2.   Elias coding for a memoryless binary source.

Example 2.2 *Consider a binary memoryless source with $f_X(0) = \frac{1}{4}$ and $f_X(1) = \frac{3}{4}$ and suppose the source outputs the sequence "01101...". Figure 2.2 indicates the evolution of the intervals $[c_n, c_n + a_n)$.*

The intervals, $[c_n, c_n + a_n)$ have the following easily verified properties:

1. The set of intervals, $[c_n, c_n + a_n)$, corresponding to each distinct vector, $\mathbf{x}_{0:n} \in \|\mathcal{A}_X\|^n$, are disjoint and their union is $[0, 1)$. That is, the set of all possible length n prefixes of the source output induces a partition of the unit interval, $[0, 1)$.

2. The intervals corresponding to successively longer prefixes of the source output sequence are nested; i.e.,

$$[c_{n+1}, c_{n+1} + a_{n+1}) \subseteq [c_n, c_n + a_n)$$

3. The length of the interval associated with $\mathbf{x}_{0:n}$ satisfies

$$a_n = \prod_{i=0}^{n-1} f_X(x_i) = f_{\mathbf{X}_{0:n}}(\mathbf{x}_{0:n})$$

MAPPING INTERVALS TO CODEWORDS

Suppose we apply the recursive algorithm described above for a total of m source output symbols. The key observation behind Elias coding is that the particular outcome, $\mathbf{x}_{0:m}$, may be uniquely identified by any number in the interval $[c_m, c_m + a_m)$, as a consequence of property 1 above. Since the interval has length a_m, it must contain at least one L_m

bit binary fraction of the form

$$0.\underbrace{bbbbb\ldots b}_{L_m}$$

where L_m is any integer satisfying $2^{-L_m} < a_m$. Thus, we conclude that the number of bits in the representation is

$$L_m \approx -\log_2 a_m = h_{\mathbf{X}_{0:m}}\left(\mathbf{x}_{0:m}\right) = \sum_{n=0}^{m-1} h_X\left(x_n\right)$$

In this way, the Elias coding algorithm firmly establishes the connection between $h_{\mathbf{X}}\left(x\right)$ and the amount of information associated with the outcome $X = x$. Each individual outcome, $X_n = x_n$, reduces the interval length by the factor $f_X\left(x_n\right)$, adding exactly $h_X\left(x_n\right) = -\log_2 f_X\left(x_n\right)$ bits to the code length.

This in turn means that the average number of bits required to code m symbols from the source output sequence is $E\left[h_{\mathbf{X}_{0:m}}\left(\mathbf{X}_{0:m}\right)\right] = mH\left(X\right)$.

ELIAS TERMINATION

There is a subtle weakness in the above argument in that the decoder does not know a priori the number of bits, L_m, which are being used to represent the source output, $\mathbf{x}_{0:m}$. Therefore, we ought to provide a mechanism for signalling this length and include the number of bits consumed by this mechanism in the overall bit count. In practical arithmetic coding algorithms, we will usually code a very large number of source outcomes, m, together, so that this cost may often be neglected. Nevertheless, it is worthwhile presenting a particular codeword termination policy suggested by Elias, for which there is no need to explicitly signal the code length, L_m.

Since $c_m \in [0,1)$, it may be represented as a binary fraction of the form

$$0.bbbbb\ldots$$

where the b's denote binary digits, 0 or 1. Now let

$$L_m = \left\lceil \log_2 \frac{1}{a_m} \right\rceil + 1 = \lceil h_{\mathbf{X}_{0:m}}\left(\mathbf{x}_{0:m}\right)\rceil + 1$$

so that

$$2^{-L_m} \leq \frac{1}{2} a_m$$

and let $\hat{c}_m$ be the quantity formed by taking only the first L_m fraction bits of c_m and adding 1 to the least significant bit position; i.e.,

$$\hat{c}_m = 2^{-L_m}\left\lfloor 2^{L_m} c_m + 1 \right\rfloor > c_m$$

Note that

$$\hat{c}_m + 2^{-L_m} \leq c_m + 2 \times 2^{-L_m} \leq c_m + a_m$$

Suppose that the decoder receives any arbitrary string of bits which agrees with $\hat{c}_m$ in its first L_m bit positions. Treating this string of bits as a binary fraction, with value, r, we see that

$$c_m < \hat{c}_m \leq r < \hat{c}_m + 2^{-L_m} \leq c_m + a_m$$

so $r \in [c_m, c_m + a_m)$ uniquely identifies $\mathbf{x}_{0:m}$.

In this way, the outcome, $\mathbf{x}_{0:m}$, is represented exactly using the first L_m bits of an otherwise arbitrary string of bits. Suppose we take the source output in blocks of m symbols at a time, $\mathbf{x}_{mk:m(k+1)}$, and determine the $L_m^{(k)}$-bit representation, $\hat{c}_m^{(k)}$, for each such block. A coded bit-stream may be created by concatenating these representations. The decoder then sees a quantity

$$r^{(0)} = 0.\underbrace{bb\ldots b}_{L_m^{(0)}}\underbrace{bb\ldots b}_{L_m^{(1)}}\ldots$$

It determines the interval, $[c_m, c_m + a_m)$, to which $r^{(0)}$ belongs and hence the first source block, $\mathbf{x}_{0:m}$. Deducing $a_m^{(0)}$ and hence $L_m^{(0)}$, it discards the first $L_m^{(0)}$ bits from the received bit-stream to obtain the quantity

$$r^{(1)} = 0.\underbrace{bb\ldots b}_{L_m^{(1)}}\underbrace{bb\ldots b}_{L_m^{(2)}}\ldots$$

from which the second source block, $\mathbf{x}_{m:2m}$, is decoded, and so forth. In this way, the lengths, $L_m^{(k)}$, need not be transmitted. The average bit-rate is thus

$$\frac{1}{m}E[L_m] = \frac{1}{m}E\left[\left\lceil \log_2 \frac{1}{a_m}\right\rceil + 1\right]$$

$$= \frac{1}{m}E[h_{\mathbf{X}_{0:m}}(\mathbf{X}_{0:m})] + \frac{3}{2m}$$

$$= H(X) + \frac{3}{2m}$$

and in the limit as $m \to \infty$ the bit-rate exceeds the entropy by a negligible margin.

FURTHER OBSERVATIONS ON ELIAS CODING

As it stands, Elias coding is impractical even for moderate values of m since it involves arithmetic operations whose precision is comparable to the number of code bits. Nevertheless, by making suitable approximations, it is relatively straightforward to derive an algorithm which experiences negligible increase in bit-rate and involves only fixed, finite precision arithmetic, for arbitrarily large values of m. In this way, practical algorithms which are able to achieve average bit-rates arbitrarily close to the source entropy do actually exist! These "arithmetic coding" algorithms are discussed further in Section 2.3.

An important property of Elias coding is that it is "incrementally decodable." Given any $r \in [c_m, c_m + a_m)$, we can decode the prefixes $x_{0:n}$ one by one since $r \in [c_n, c_n + a_n)$ for each $n = 1, 2, \ldots, m$. This leads to a recursive algorithm for incrementally decoding the source outputs, $x_0, x_1, \ldots$, which strongly resembles the incremental encoding algorithm already described. We defer further discussion of incremental decoding until Section 2.3.

EXTENSION TO MARKOV RANDOM PROCESSES

The incremental decodability described above permits an easy extension of the Elias coding algorithm to Markov random processes. The modified algorithm becomes

---

Elias Coding Algorithm for Markov-p Sources

Initialize $c_0 = 0$ and $a_0 = 1$.

For each $n = 0, 1, \ldots$

    Let $p' = \min\{p, n\}$

    Update $a_{n+1} \leftarrow a_n f_{X_n | \mathbf{X}_{n-p':n}}\left(x_n, \mathbf{x}_{n-p':n}\right)$

    Update $c_{n+1} \leftarrow c_n + a_n F_{X_n | \mathbf{X}_{n-p':n}}\left(x_n, \mathbf{x}_{n-p':n}\right)$

Here, $F_{X_n | \mathbf{X}_{n-p':n}}$ denotes the cumulative conditional distribution,

$$F_{X_n | \mathbf{X}_{n-p':n}}\left(\alpha_i, \mathbf{x}_{n-p':n}\right) \triangleq \sum_{j=0}^{i-1} f_{X_n | \mathbf{X}_{n-p':n}}\left(\alpha_j, \mathbf{x}_{n:n-p'}\right)$$

Since the random process is assumed to be Markov-p, there are only finitely many conditional distributions, $f_{X_n | \mathbf{X}_{n-p':n}} = f_{X_{p'} | \mathbf{X}_{0:p'}}$, and corresponding cumulative distributions. We assume that the encoder and decoder both have access to these conditional distributions or else they both use identical estimates of the distributions.

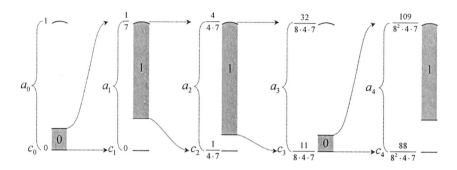

Figure 2.3.   Elias coding for a binary Markov-1 source.

Example 2.3 *Consider a binary Markov-1 source with*

$$
f_{X_1|X_0}(x_1, x_0) = \begin{cases} \frac{1}{4} & if\ x_1 = 0,\ x_0 = 0 \\ \frac{3}{4} & if\ x_1 = 1,\ x_0 = 0 \\ \frac{1}{8} & if\ x_1 = 0,\ x_0 = 1 \\ \frac{7}{8} & if\ x_1 = 1,\ x_0 = 1 \end{cases}
$$

Note that the unconditional (marginal) PMF for this source is $f_X(0) = \frac{1}{7}$, $f_X(1) = \frac{6}{7}$. Figure 2.3 indicates the evolution of the nested sub-intervals $[c_n, c_n + a_n)$ when the source outputs the sequence "01101...".

The algorithm is identical to that described earlier, except that we use conditional distributions to exploit information which is available from previously coded outcomes. To see that any value, $r \in [c_m, c_m + a_m)$, uniquely specifies $\mathbf{x}_{0:m}$, consider the following strategy for incremental decoding. The decoder first reconstructs $x_0 = \mathbf{x}_{0:1}$ based on the interval, $[c_1, c_1 + a_1)$ containing r. This is possible because the partition of $[0, 1)$ into $\|\mathcal{A}_X\|$ sub-intervals corresponding to each possible outcome, x_0, depends only upon f_X, exactly as in the IID case examined earlier. Knowing $\mathbf{x}_{0:1}$, the decoder is able to determine the function, $f_{X_1|X_{0:1}}(\cdot, \mathbf{x}_{0:1})$, and hence the partition of $[c_1, c_1 + a_1)$ which was used to represent x_1. Hence x_1 is decoded from the particular sub-interval, $[c_2, c_2 + a_2)$, containing r, in this partition. By continuing this incremental decoding strategy, the decoder is always able to reconstruct the context, $\mathbf{x}_{0:n}$, which is needed to determine the conditional distribution and hence the sub-interval partition required to recover $\mathbf{x}_{0:n+1}$ for each $n = 0, 1, \ldots, m - 1$.

Exactly as before, a block of m symbols, $\mathbf{x}_{0:m}$, may be represented by an L_m-bit binary fraction where L_m is any integer satisfying

$$L_m \geq \log_2 \frac{1}{a_m}$$

$$= -\log_2 \prod_{n=0}^{m-1} f_{X_n|\mathbf{X}_{0:n}} (x_n, \mathbf{x}_{0:n})$$

$$= -\log_2 f_{\mathbf{X}_{0:m}} (\mathbf{x}_{0:m})$$

$$= h_{\mathbf{X}_{0:m}} (\mathbf{x}_{0:m})$$

Here, we have used the joint PMF expansion formula of equation (2.1). If we employ the Elias termination procedure described above, we conclude that the expected bit-rate is

$$\frac{1}{m} E[L_m] = \frac{1}{m} E[h_{\mathbf{X}_{0:m}} (\mathbf{X}_{0:m})] + \frac{3}{2m}$$

$$= \frac{1}{m} H(\mathbf{X}_{0:m}) + \frac{3}{2m}$$

$$= H^{(m)} (\{X_n\}) + \frac{3}{2m}$$

and so, in the limit as m becomes large, the expected bit-rate approaches the entropy rate, $H(\{X_n\})$, of the random process. Moreover, according to the entropy-ergodic theorem for Markov processes, it is possible to ensure that the actual bit-rate will be arbitrarily close to this expected bit-rate with arbitrarily high probability, by chosing m sufficiently large. This would be of little interest if it were not for the fact that the complexity of practical arithmetic coding algorithms does not grow with m, as we shall see in Section 2.3.

We refer to this modified version of the Elias coding algorithm as conditional coding. Conditional arithmetic coding algorithms are central to the JBIG and JPEG2000 image compression standards.

2.2 VARIABLE LENGTH CODES

In this section, we introduce simple variable length coding techniques which are commonly found as elements of image compression algorithms. Let $\{X_n\}$ be a memoryless random process with alphabet $\mathcal{A}_X$ and distribution, f_X. A variable length code assigns a distinct codeword, c_x to each element, $x \in \mathcal{A}_X$, where c_x is a string of $\|c_x\|$ bits. The sequence of outcomes, x_n, from the random process are represented by concatenated codewords, c_{x_n}. The choice of codewords is clearly constrained by the requirement that the decoder must be able to identify the outcomes,

x_n, from this concatenated sequence of codewords. Codes having this property are said to be uniquely decodable.

Example 2.4 *Consider the quaternary alphabet, $\mathcal{A}_X = \{0,1,2,3\}$, with codewords*

$$c_0 = \text{``0''}$$
$$c_1 = \text{``01''}$$
$$c_2 = \text{``10''}$$
$$c_3 = \text{``11''}$$

Suppose we use this code to represent source outcomes "0, 2, 3, 0, 1". Then the resulting bit-stream is

$$\text{``}\underbrace{0}_{c_0} \ \underbrace{10}_{c_2} \ \underbrace{11}_{c_3} \ \underbrace{0}_{c_0} \ \underbrace{01}_{c_1}\text{''}$$

This same bit-stream may be produced by a different sequence of source outcomes, e.g.

$$\text{``}\underbrace{01}_{c_3} \ \underbrace{0}_{c_0} \ \underbrace{11}_{c_3} \ \underbrace{0}_{c_0} \ \underbrace{01}_{c_1}\text{''}$$

and so it violates the unique decodability requirement.

Amongst all selections of codewords satisfying the unique decodability requirement, we are most interested in those which minimize the average code-rate,

$$R = \sum_{x \in \mathcal{A}_X} \|c_x\| \cdot f_X(x)$$

In view of the fundamental results presented in Section 2.1, we must have

$$R \geq H(X)$$

Example 2.5 *Consider a memoryless source having an alphabet consisting of all non-negative integers, $\mathcal{A}_X = \{0,1,2,\ldots\}$, with distribution*

$$f_X(x) = 2^{-(x+1)}$$

Let c_x be the codeword consisting of $\|c_x\| = x+1$ bits, the initial x of which are "0", with the last bit in each codeword being a "1". Thus, the codewords are

$$c_0 = \text{``1''}$$
$$c_1 = \text{``01''}$$
$$c_2 = \text{``001''}$$

$$\vdots$$

It is easy to see that this code is uniquely decodable, since the codewords are all different and are delimited within the bit-stream by the "1" bit. This is called a "comma" code, since the "1" may be interpreted as a comma separating the codewords corresponding to successive symbols, x_n, in the concatenated bit-stream. In this case, we find that

$$R = \sum_{x \in \mathcal{A}_X} (x+1) \cdot f_X(x)$$

$$= \sum_{x \in \mathcal{A}_X} -\log_2 f_X(x) \cdot f_X(x) = H(X)$$

so this is an optimal code, actually achieving the entropy rate of the source. The comma code is also sometimes called a "unary" code.

To facilitate efficient decoding we are generally interested only in "prefix codes." A prefix code is one in which no codeword is the prefix of any other codeword. The codewords in Example 2.5 clearly satisfy the prefix condition. Any prefix code is also uniquely decodable. This may be seen from the following sequential decoding algorithm:

Sequential Decoding Algorithm for Prefix Codes

For each $l = 0, 1, \ldots$
> For each $\alpha \in \mathcal{A}_X$
>> Compare the first l bits of the received bit-stream with c_α. If a match is found,
>>> The prefix condition guarantees that no other value of l will yield a match and so the first symbol must be $x_0 = \alpha$.
>>> Remove the initial l bits from the bit-stream and apply the algorithm recursively to decode the next symbol.

The following results establish some key properties of variable length codes.

Theorem 2.5 *(McMillan) A necessary condition for unique decodability is that the codeword lengths, $l_x = \|c_x\|$, satisfy*

$$\sum_{x \in \mathcal{A}_X} 2^{-l_x} \leq 1 \tag{2.11}$$

Proof. For a proof, the reader is referred to [106, Thm 10.1]. ■

Theorem 2.6 *(Kraft) Given any set of lengths, l_x, satisfying equation (2.11), there exists a prefix code (it is not unique) having $\|c_x\| = l_x$*

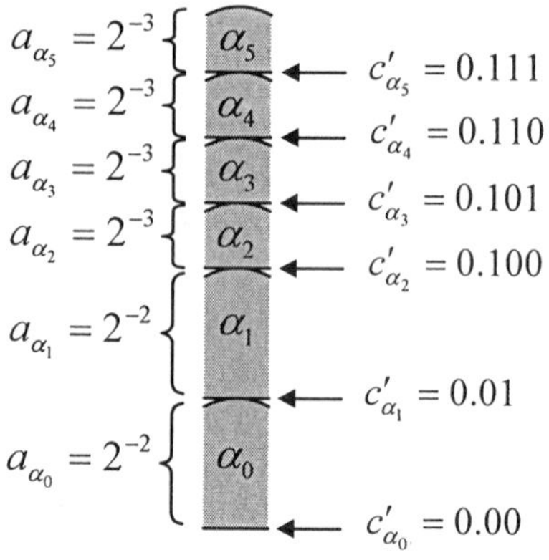

Figure 2.4.   Mapping between prefix codewords and sub-intervals of $[0, 1)$.

for each $x \in \mathcal{A}_X$. So the condition in equation (2.11) is both necessary and sufficient for unique decodability and there is no need to consider anything other than prefix codes.

Proof. Arrange the elements of $\mathcal{A}_X = \{\alpha_0, \alpha_1, \ldots\}$ such that

$$l_{\alpha_0} \leq l_{\alpha_1} \leq l_{\alpha_2} \leq \cdots$$

Then let the codeword, c_{α_i}, be the l_{α_i}-bit integer whose value is

$$c_{\alpha_i} = 2^{l_{\alpha_i}} \sum_{j=0}^{i-1} 2^{-l_{\alpha_j}}$$

To see that these codewords form a prefix code, consider the intervals

$$[c'_x, c'_x + a_x) \subseteq [0, 1)$$
$$c'_x = 2^{-l_x} c_x$$
$$a_x = 2^{-l_x}$$

as illustrated in Figure 2.4. Clearly, the intervals are disjoint. Now consider any sequence of bits having the prefix, c_x, for some $x \in \mathcal{A}_X$, and let $r \in [0, 1)$ be the quantity whose binary fraction representation is formed from these bits; i.e.,

$$r = 0. \underbrace{bb \ldots bb}_{c_x} \ldots$$

Clearly $r \in [c'_x, c'_x + a_x)$. Let $y \neq x \in \mathcal{A}_X$. Since the sub-intervals are disjoint, $c'_y \notin [c'_x, c'_x + a_x)$ and so c_y cannot have c_x as a prefix. ∎

This proof suggests a connection between prefix codes and Elias coding which is by no means coincidental. In fact, if $f_X(x) = 2^{-l_x}$, the variable length code produces exactly the same bit-stream as Elias coding. Of course, this is the special case in which the variable length code achieves the entropy. Thus, variable length coding may be understood as an approximate (and much simpler) form of Elias coding, in which the values assumed by the PMF are approximated by reciprocal powers of 2. This interpretation leads immediately to the next theorem.

Theorem 2.7 *For any distribution, f_X, a prefix code may be found, whose rate satisfies*

$$H(X) \leq R < H(X) + 1 \tag{2.12}$$

Proof. The left hand inequality is a necessary consequence of Shannon's noiseless coding theorem, although a direct proof is not difficult, e.g. [106, Thm 10.3]. For the right hand inequality, simply let $l_x = \lceil -\log_2 f_X(x) \rceil$. Evidently this is a crude approximation of $f_X(x)$ as 2^{-l_x}. These lengths satisfy equation (2.11) and so, by Theorem 2.6, there exists a prefix code with $\|c_x\| = l_x$. The code has rate

$$R = \sum_{x \in \mathcal{A}_X} f_X(x) \lceil -\log_2 f_X(x) \rceil$$
$$< \sum_{x \in \mathcal{A}_X} f_X(x) (1 - \log_2 f_X(x))$$
$$= H(X) + 1$$

∎

2.2.1   HUFFMAN CODING

Given any finite alphabet,

$$\mathcal{A}_X = \{\alpha_0, \alpha_1, \ldots, \alpha_{K-1}\}$$

and associated PMF, f_X, it is reasonable to seek an optimum code, for which the average codeword length is minimized over all uniquely decodable codes. The optimum code is, of course, not unique. In fact, even the codeword lengths, $l_x = \|c_x\|$, need not be unique amongst optimal codes. Huffman [77] developed an algorithm for finding one set of lengths satisfying equation (2.11), which minimize the average code-rate R.

Suppose for convenience that the alphabet is ordered so that

$$f_X(\alpha_0) \leq f_X(\alpha_1) \leq \cdots \leq f_X(\alpha_{K-1})$$

Huffman's algorithm is based on the following key observation.

Lemma 2.8 *Amongst all optimal codes, at least one has $l_{\alpha_0} = l_{\alpha_1} = l_{\max}$, the largest codeword length, with c_{α_0} and c_{α_1} differing only in their last bit.*

Proof. Any optimal code must have $l_{\alpha_0} \geq l_{\alpha_1} \geq \cdots \geq l_{\alpha_{K-1}}$. This intuitive fact is trivially established. Now suppose that we have a prefix code with $l_{\alpha_0} > l_{\alpha_1}$. Then the first l_{α_1} bits of c_{α_0} must differ from all codewords with length l_{α_1} and hence the last bit of c_{α_0} is wasted. We conclude that in an optimal code, $l_{\alpha_0} = l_{\alpha_1}$. The constructive proof of Theorem 2.6 is easily rearranged to show that the prefix code having

$$c_{\alpha_i} = 2^{l_{\alpha_i}} \left(1 - \sum_{j=0}^{i} 2^{-l_{\alpha_j}} \right)$$

is uniquely decodable. This construction yields codewords c_{α_0} and c_{α_1} which differ only in their last bit position. ∎

This observation suggests that we can reduce the optimization problem to that of finding only $K - 1$ codewords: the $l_{\alpha_0} - 1$ bit prefix, $c_{\alpha'_1}$, common to both c_{α_0} and c_{α_1}; and the codewords $c_{\alpha'_i} = c_{\alpha_i}$ for $i = 2$ through $K - 1$. The reduced problem may be stated as follows. Find lengths, $l_{\alpha'_1}$ through $l_{\alpha'_{K-1}}$ satisfying

$$\sum_{i=1}^{K-1} 2^{-l_{\alpha_{i'}}} \leq 1$$

which minimize

$$R = (f_X(\alpha_0) + f_X(\alpha_1)) \left(l_{\alpha'_1} + 1 \right) + \sum_{i=2}^{K-1} f_X(\alpha_i) l_{\alpha_i}$$

$$= f_X(\alpha_0) + f_X(\alpha_1) + \sum_{x' \in \mathcal{A}_{X'}} f_{X'}(x') l_{x'}$$

where X' is a new random variable, having alphabet

$$\mathcal{A}_{X'} = \left\{ \alpha'_1, \alpha_2, \ldots, \alpha_{K-1} \right\}$$

and

$$f_{X'}(x) = \begin{cases} f_X(\alpha_0) + f_X(\alpha_1) & \text{if } x = \alpha'_1 \\ f_X(x) & \text{otherwise} \end{cases} \qquad (2.13)$$

The new problem is thus exactly the same as our original problem, but on a reduced alphabet. This leads naturally to the following algorithm, which recursively reduces the problem of optimal code construction to the trivial case of a binary alphabet.

Huffman Code Construction

Order the elements of the alphabet such that

$$f_X(\alpha_0) \leq f_X(\alpha_1) \leq \cdots \leq f_X(\alpha_{K-1})$$

If $K = 2$,

    Assign $c_{\alpha_0} =$ "0" and $c_{\alpha_1} =$ "1"

Else

    Create a new alphabet, $\mathcal{A}_{X'} = \{\alpha'_1, \alpha_2, \alpha_3, \ldots, \alpha_{K-1}\}$, and probability assignment, $f_{X'}$, satisfying equation (2.13).

    Invoke the code construction algorithm recursively to find an optimal code, $c_{\alpha'_1}, c_{\alpha_2}, \ldots, c_{\alpha_{K-1}}$, for $\mathcal{A}_{X'}$ and $f_{X'}$.

    Extend this code by appending "0" and "1" to $c_{\alpha'_1}$, to obtain c_{α_0} and c_{α_1}, respectively.

Example 2.6 *Suppose $\mathcal{A}_X = \{0, 1, 2, 3\}$ with*

$$f_X(0) = \frac{3}{16}, \quad f_X(1) = \frac{2}{16}, \quad f_X(2) = \frac{2}{16}, \quad f_X(3) = \frac{9}{16}$$

The entropy in this case is

$$H(X) = 1.6697 \ bits/symbol$$

The steps in the recursive algorithm above may be represented in terms of the construction of a binary tree, as in Figure 2.5. Each leaf in the tree corresponds to one of the codewords. The codeword lengths may be read directly from the tree by counting the number of branches between the root and each leaf. A corresponding set of codewords may be obtained by labeling the branches with "0"s and "1"s; and reading the branch labels following the path from the root to each leaf. In this case, the codewords are

$$c_0 = \text{"00"}, \quad c_1 = \text{"010"}, \quad c_2 = \text{"011"}, \quad c_3 = \text{"1"}$$

and the code-rate is

$$R = 2 \cdot \frac{3}{16} + 3 \cdot \frac{2}{16} + 3 \cdot \frac{2}{16} + 1 \cdot \frac{9}{16}$$
$$= 1.6875 \ bits/symbol$$

Notice how closely the code rate approaches the entropy in this case.

LIMITATIONS OF HUFFMAN CODING

Despite the promising performance obtained in Example 2.6, Huffman codes, and hence variable length codes in general, cannot guarantee code-rates which approach the entropy more closely than the bounds indicated in equation (2.12). This performance can be inadequate in some applications. Most notably, when the entropy of the source is much less

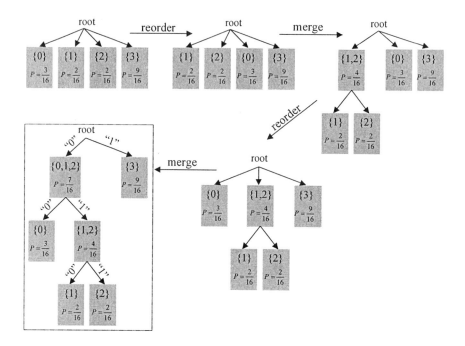

Figure 2.5.   Huffman code construction example.

than 1 bit/symbol, variable length codes are particularly inefficient since at least one bit is consumed by each codeword.

A solution to this problem is to block the source output into m-dimensional vectors, $\mathbf{x}_{km:(k+1)m}$, and assign codewords to each vector. In this way, the inefficiency of up to 1 bit/vector is distributed over all m source symbols in the vector and the Huffman code-rate is bounded by

$$H\left(X\right) \leq R < H\left(X\right) + \frac{1}{m}$$

The problem with this approach is that the size of the alphabet, $\mathcal{A}_{\mathbf{X}_{0:m}}$, grows exponentially with m. The number of codewords which must be maintained in memory grows as $\|\mathcal{A}_X\|^m$.

Up until now we have considered only memoryless random processes. In order to capture the redundancy between successive elements of the random process, the procedure must be modified in one of two ways. One approach is to block the source into m-dimensional vectors, assigning codewords to each vector exactly as described above. In this case the code-rate approaches the mth order entropy of the source, bounded by

$$H^{(m)}\left(\{X_n\}\right) \leq R < H^{(m)}\left(\{X_n\}\right) + \frac{1}{m} \qquad (2.14)$$

The practical limitations described above apply here as well.

A second approach is to construct a separate Huffman code for each of the conditional distributions,

$$f_{X_m|\mathbf{X}_{0:m}}\left(\cdot, \mathbf{x}_{0:m}\right), \quad \mathbf{x}_{0:m} \in \left(\mathcal{A}_X\right)^{m-1}$$

Denote the optimized codewords by $c_{x|\mathbf{x}_{0:m}}$. When we come to code (or decode) symbol x_n, we use the codewords, $c_{\alpha_i|x_{n-m:n}}$, $0 \leq i < K$, where $K = \|\mathcal{A}_X\|$. In this way, there are $\|\mathcal{A}_X\|^{m-1}$ separate codes, each of which has m codewords, so the codeword memory grows as $\|\mathcal{A}_X\|^m$, exactly as in the blocking approach. The code-rate for this conditional Huffman coding strategy, however, is bounded by

$$H\left(X_m \mid X_{0:m}\right) \leq R < H\left(X_m \mid X_{0:m}\right) + 1$$

We note that $H\left(X_m \mid X_{0:m}\right) \leq H^{(m)}\left(\{X_n\}\right)$ with equality only for memoryless processes and, for a Markov-p process, $H\left(X_m \mid X_{0:m}\right) = H\left(\{X_n\}\right)$, provided $m \geq p+1$.

Thus, conditional Huffman coding may approach the entropy rate of the process to within 1 bit per symbol, with a finite (albeit often enormous) codeword memory. Although the conditional entropy, $H\left(X_m \mid X_{0:m}\right)$, approaches the entropy rate of the process much more rapidly than the mth order entropy, $H^{(m)}\left(\{X_n\}\right)$, the blocking approach may still be preferable for sources whose entropy rate is very low, since the upper bound in equation (2.14) tightens as $\frac{1}{m}$. It is worth pointing out that the memory efficiency of conditional Huffman coding may be substantially improved by exploiting the context reduction techniques described in Section 2.4.1.

FAST DECODING ALGORITHMS

From an implementation point of view, Huffman encoding is simply a lookup table; each source symbol is mapped to its codeword with a single table lookup operation. The bit-serial decoding algorithm on Page 45, however, is generally much slower, with a separate operation for each received bit. A faster approach is to use a lookup table with the next

$$L = \max_{x \in \mathcal{A}_X} \{l_x\}$$

bits in the bit stream serving as the table index. The table lookup returns the outcome, $x \in \mathcal{A}_X$, as well as the length, l_x, of the unique codeword, c_x, which forms the l_x-bit prefix of the L-bit index. The initial l_x bits are then removed from the bit-stream in preparation for decoding the next symbol. In this way, a single table lookup suffices to decode each source symbol.

The problem with this method is that the table may need to contain as many as 2^L entries, which can be much larger than the number of codewords. The Huffman algorithm does not constrain the maximum code word length, L, which may end up being as large as $L = \|\mathcal{A}_X\| - 1$. For this reason, "constrained length Huffman codes" have been developed. The Voorhis method [165] is one of the first algorithms developed to find optimal variable length codes, subject to a maximum length constraint.

ADAPTIVE HUFFMAN CODING

Huffman codes are simply variable length codes, optimized for the statistics of a given source. The problem is that the statistics of the source may not be known ahead of time, or they may vary from time to time. Two potential solutions present themselves. The encoder may periodically estimate the statistics of the source, construct an optimal Huffman code and transmit the codewords to the decoder. This approach is used in the JPEG image compression standard (see Chapter 19), where the Huffman codewords are explicitly signalled in the header of each compressed image file.

A second approach is for both the encoder and decoder to periodically estimate the source statistics and construct identical Huffman codes, based on previously encoded source outcomes. This approach, known as adaptive Huffman coding, avoids the overhead of transmitting the codewords to the decoder. On the other hand, only those source outcomes which have already been encoded may be used to estimate source statistics. As a result, the estimates are generally poorer and hence the coding is less efficient than may be obtained if the encoder is free to estimate the source statistics by looking ahead into the source outcomes. Thus there is a trade-off between the cost of explicitly sending Huffman codewords and the reduction in efficiency incurred by adaptively discovering the statistics. The adaptive approach is usually avoided since it also burdens the decoder with the task of periodically implementing the optimal code construction algorithm – not a trivial task.

2.2.2   GOLOMB CODING

As noted above, one of the problems with Huffman coding is that code construction is expensive so that adaptive coding algorithms may require large computational resources as they try to adapt to changing statistics by periodically modifying the code. Golomb coding [71] is an interesting alternative.

Consider a "geometric" source with alphabet $\mathcal{A}_X = \{0, 1, 2, \ldots\} = \mathbb{Z}_+$ and

$$f_X(x) = 2^{-(x+1)}, \quad x \geq 0$$

An optimal prefix code in this case is the "comma" code of Example 2.5, whose codewords, c_x, consist of a string of x consecutive "0"s, terminated by a single "1" (the "comma"). In this isolated case, the comma code achieves the entropy; i.e., $R = H(X)$. More generally, the PMF of a geometric source is given by

$$f_X(x) = (1 - \rho) \rho^x, \text{ with parameter, } 0 < \rho < 1 \qquad (2.15)$$

and the comma code is an optimal variable length code for any geometric source with parameter $\rho \leq \frac{1}{2}$. One way to see this is to apply the Huffman code construction algorithm to such a source.

For geometric sources with parameter $\rho > \frac{1}{2}$, the comma code is no longer so efficient. Suppose, however, that we express each outcome, $x \in \mathbb{Z}_+$, as

$$x = m x_q + x_r$$

where x_q is the quotient and x_r the remainder, upon division of x by the integer, m. That is,

$$x_q = \left\lfloor \frac{x}{m} \right\rfloor$$

$$x_r = x \bmod m$$

Let X_q and X_r denote the random variables whose outcomes are x_q and x_r respectively. Evidently, X_q follows a geometric distribution with parameter ρ^m since

$$f_{X_q}(x_q) = \sum_{i=0}^{m-1} f_X(m x_q + i) = \rho^{m x_q} (1 - \rho) \sum_{i=0}^{m-1} f_X(i)$$

Moreover, it is easily shown that X_q and X_r are independent random variables.

The idea behind Golomb coding is to select the integer divisor, m, such that

$$\rho^m \gtrsim \frac{1}{2}$$

Then the comma code is an efficient code for X_q, while X_r follows an approximately uniform distribution on $\{0, 1, \ldots m - 1\}$. Specifically,

$$f_{X_r}(0) \geq f_{X_r}(1) \geq \cdots \geq f_{X_r}(m - 1) > \frac{1}{2} f_{X_r}(0)$$

Let $k_b = \lfloor \log_2 m \rfloor$ and $k_a = \lceil \log_2 m \rceil$. An optimal variable length code for X_r is the modified binary code, which uses k_b bits to represent outcomes $x_r < 2^{k_a} - m$ and k_a bits to represent the remaining outcomes.[3] It can be shown [68] that the concatenation of the comma code for X_q followed by the modified binary code for X_r yields an optimal variable length code for the geometrically distributed source, subject to suitable choice of the Golomb parameter, m. This is true for any value of the parameter, ρ.

In practice, it is convenient to restrict the Golomb parameter, m, to an exact power of 2, namely

$$m = 2^k$$

so that x_q is trivially formed by discarding the least significant k bits in the binary representation for x and these discarded k bits form the remainder part, x_r. As an example, with parameter $k = 3$, the outcome $x = 21$ would be represented as

$$x = 21 \longrightarrow \text{`` } \underbrace{001}_{x_q=2} \underbrace{101}_{x_r=5} \text{ ''}$$

GOLOMB PARAMETER ESTIMATION

Many sources do exhibit a roughly geometric distribution[4]. With such sources, Golomb coding can achieve close to the optimal variable length coding performance. Since sources are rarely exactly geometric the Golomb parameter, $m = 2^k$, is best optimized experimentally if possible. The scheme is also well suited to adaptive coding, because only a single parameter need be adapted. In this case, simple indicators of the source statistics are formed at the encoder and decoder, based on previously coded outcomes, and these indicators are used to estimate the best value for the parameter, k.

We now describe one suitable adaptation procedure, which is based around estimates of the statistical mean, $E[X]$. Suppose that the source

---

[3] The modified binary code may be obtained as follows. First set $x'_r = 2x_r$ if $x_r < 2^{k_a} - m$ and $x'_r = x_r + 2^{k_a} - m$ otherwise. Next, the fixed length binary code, c'_{x_r}, is formed from the k_a-bit binary representation of x'_r, with the MSB first and LSB last. Finally, observe that the last bit of this code, c'_{x_r}, is guaranteed to be 0 whenever $x_r < 2^{k_a} - m$, allowing us to reduce the codeword length to k_b as claimed.

[4] As an example, consider an IID binary random process for which $f_X(0) = q \approx 1$. We expect the source to produce long runs of 0's, interspersed usually by isolated 1's. Accordingly, it is reasonable to represent the source outcomes via an equivalent sequence of run-lengths, r, indicating the number of consecutive 0's, between each pair of 1's. It is easy to see that the run lengths obey a geometric distribution with parameter, $\rho = 1 - q$, so they are well suited to Golomb coding.

distribution is indeed geometric with unknown parameter, ρ, and observe that

$$E[X] = \sum_{x=0}^{\infty} (1-\rho) x\rho^x = (1-\rho) \sum_{x=1}^{\infty} (x-1) \rho^{x-1}$$

$$= \left\{ (1-\rho) \frac{d}{d\rho} \sum_{x=0}^{\infty} \rho^x \right\} - 1 = \left\{ (1-\rho) \frac{d}{d\rho} \frac{1}{1-\rho} \right\} - 1$$

$$= \frac{\rho}{1-\rho}$$

Suppose further that $(1-\rho) \ll 1$; then

$$\rho^m = (1 - (1-\rho))^m$$
$$\approx 1 - m(1-\rho)$$
$$\approx 1 - \frac{m}{E[X]}$$

Recalling that we want $\rho^m \gtrsim \frac{1}{2}$, this suggests that we should select m so that

$$m = 2^k \gtrsim \frac{1}{2} E[X]$$

An appropriate strategy, then, is to set

$$k = \max \left\{ 0, \left\lceil \log_2 \left(\frac{1}{2} E[X] \right) \right\rceil \right\}$$

Although this policy is derived under the assumption that $\rho \approx 1$, it also yields reasonable Golomb parameters for smaller values of ρ. Further refinements are best derived empirically for the application at hand, since the source is unlikely to be exactly geometric anyway. The following simple algorithm demonstrates the incorporation of the above strategy into an adaptive coding scheme. A very closely related algorithm is employed by the JPEG-LS lossless image compression standard (see Chapter 20) to code prediction residuals.

Adaptive Golomb Coder

Initialize $A = \hat{\mu}_X$ and $N = 1$

> (Here $\hat{\mu}_X$ is an initial estimate for $E[X]$. The ratio, $\frac{A}{N}$, is to be interpreted as an estimate of $E[X]$.)

For each $n = 0, 1, 2, \ldots$

> Set $k = \max \left\{ 0, \left\lceil \log_2 \left(\frac{A}{2N} \right) \right\rceil \right\}$
> Code symbol x_n using the Golomb code with parameter k.

Update Counters

If $N = N_{\max}$ (renormalize counters)
   Set $A \leftarrow \lfloor A/2 \rfloor$ and $N \leftarrow \lfloor N/2 \rfloor$
   Update $A \leftarrow A + x_n$ and $N \leftarrow N + 1$

The decoder updates its own copy of the counters, A and N, following the same procedure as the encoder, so as to deduce the Golomb parameter, k, used to code each source symbol. Larger values of the parameter, $N_{\max}$, yield more stable estimates for $E[X]$, while smaller values enable the algorithm to adapt more rapidly to changing source statistics.

2.3   ARITHMETIC CODING

In Section 2.1.4, we introduced Elias coding. Unlike the variable length codes introduced in Section 2.2, Elias coding incrementally constructs a single codeword for an arbitrarily long sequence of source symbols as they arrive. As we shall see, incremental decoding is possible. In this way, the benefit of very long, highly efficient codes is realized without the delay or the memory required to maintain an enormous collection of codewords. Moreover, the incremental construction is easily adapted to the conditional statistics of Markov sources and it lends itself to adaptive coding in which the relevant conditional probabilities are estimated dynamically from previously coded outcomes of the source process.

The code construction algorithm involves simple arithmetic operations. Unfortunately, these operations involve ever increasing numeric precision, rendering them impractical as is. As a result of this weakness, the Elias coding algorithm remained for quite some time little more than an academic curiosity, before the discovery of finite precision implementations by Rissanen [125] and Pasco [116]. In this section we take the reader through most of the key principles behind modern arithmetic coding algorithms. For a detailed description of the actual arithmetic coding variant employed in the JPEG2000 standard, the reader is referred to Section 12.1.

2.3.1   FINITE PRECISION REALIZATIONS

Recall that the recursive interval sub-division algorithm operates on the lower bound and length of an interval,

$$[c_n, c_n + a_n) \subseteq [0, 1)$$

The interval corresponding to $\mathbf{x}_{0:n}$ is updated to the interval for $\mathbf{x}_{0:n+1}$ by assigning

$$a_{n+1} \leftarrow a_n f_X(x_n) \tag{2.16}$$

$$c_{n+1} \leftarrow c_n + a_n F_X(x_n) \tag{2.17}$$

The key observation required to bound the implementation precision is that we need not use the exact value of a_n produced by these ideal update relationships. Suppose instead that

$$0 < a_{n+1} \lesssim a_n f_X(x_n)$$

Then the sub-intervals corresponding to each potential outcome of X_n remain disjoint so that unique decoding is still guaranteed. There is of course some loss in coding efficiency; in fact, we sacrifice $\log_2 \frac{a_n f_X(x_n)}{a_{n+1}}$ bits in coding the outcome $X_n = x_n$. As we shall see, however, modest arithmetic precision is sufficient to render this loss negligible.

We are now in a position to describe a practical coding algorithm. Let the interval length be represented by an N-bit integer, A_n, together with an exponent, b_n, with

$$a_n = 2^{-b_n}\left(2^{-N}A_n\right)$$

The quantity, $A'_n = 2^{-N}A_n$, is an N-bit binary fraction of the form

$$A'_n = 0.\underbrace{1aa\ldots a}_{N \text{ bits}}$$

and the quantity, b_n, is the number of leading "0"s in the binary fraction representation of a_n; i.e.,

$$a_n = 0.\underbrace{00\ldots0}_{b_n \text{ bits}}\underbrace{1aa\ldots a}_{A_n} \tag{2.18}$$

Next, we represent all probabilities approximately using P-bit integers, p_α, such that

$$f_X(\alpha) \approx p'_\alpha = 2^{-P}p_\alpha, \quad \alpha \in \mathcal{A}_X$$

The interval length is then updated according to equation (2.16) and rounded down, if necessary, to the closest representation of the form in equation (2.18). Together, these operations are embodied by the following algorithm.

Set $T \leftarrow A_n p_{x_n}$ and $b_{n+1} \leftarrow b_n$
(Note that $T' = 2^{-(N+P)}T$ is an $(N+P)$-bit binary fraction with $T' = A'_n p'_{x_n} < 1$)

While $T < 2^{N+P-1}$ (i.e., while $T' < \frac{1}{2}$)
    Increment $b_{n+1} \leftarrow b_{n+1} + 1$
    Shift $T \leftarrow 2T$
Set $A_{n+1} = \lfloor 2^{-P} T \rfloor$

Evidently we have made two approximations which lead to slight losses in coding efficiency: the probabilities are approximated with finite precision representations; and the rounding operation in the last line of the algorithm, which reduces the interval length and hence increases the number of code bits by an amount strictly less than

$$\log_2 \frac{2^{N-1} + 1}{2^{N-1}} \approx 2^{-(N-1)} \log_2 e \text{ bits}$$

We have now only to describe the manipulation of the interval lower bound, c_n. Since the PMF, $f_X(\alpha)$, is approximated by P-bit binary fractions, p'_α, the cumulative distribution is also approximated by P-bit binary fractions,

$$F'_{\alpha_i} = \sum_{j=0}^{i-1} p'_{\alpha_j} = 2^{-P} F_{\alpha_i}$$

From the update equation (2.17) we deduce that c_n is an $(N + P + b_n)$-bit binary fraction of the form

$$c_n = 0.\underbrace{xx \ldots x}_{b_n \text{ bits}} \underbrace{cc \ldots c}_{C_n}$$

Let C_n be the integer formed from the least significant $N + P$ bits of this representation. Then the update operation consists of adding two $(N + P)$-bit integers, $A_n F_{x_n}$ and C_n, and propagating any carry bit into the initial b_n-bit prefix of c_n.

At first glance it appears that the need to resolve a carry will force us to buffer the entire b_n-bit prefix of c_n. Fortunately, however, the carry may effect only the least significant $r_n + 1$ bits of this prefix, where r_n is the number of consecutive least significant 1's. In fact, no future coding operations may have any effect on the more significant bits in the prefix. To see this, observe that

$$c_{n+k} \in [c_n, c_n + a_n) \subset \left[c_n, c_n + 2^{-b_n}\right), \quad \forall k \geq 0$$

so that at most one carry bit may be propagated into the b_n most significant fraction bits of the codeword when augmenting c_n to c_{n+k} for any k.

It follows that the initial $b_n - r_n - 1$ bits of the codeword may be sent to the decoder so we need not allocate storage for them. The binary fraction representation of the evolving codeword then consists of three key segments,

$$0. \underbrace{xxxxxx\ldots x}_{b_n - r_n - 1 \text{ bits}} \underbrace{011\ldots 1}_{r_n + 1 \text{ bits}} \underbrace{cc\ldots c}_{C_n}$$

The encoder need only maintain four state variables, A_n, C_n, r_n and b_n and the complete encoding algorithm is shown below. Note that we drop the sub-scripts in order to better reflect the behaviour of a real coder. Also, note that we need to introduce a special state, identified by $r = -1$, to deal with the possibility that a carry may occur when $r = 0$, causing the 0 bit to flip to a 1 with no subsequent "0" bits. Since future carries can never propagate this far, it is sufficient to flag the unusual condition by setting $r \leftarrow -1$, which has the interpretation that the central segment in the binary fraction representation of c_n is empty and can remain empty until a zero bit is propagated out of C_n.

Finite Precision Arithmetic Coding

Initialize $C = 0$, $A = 2^N$, $r = -1$, $b = 0$

For each $n = 0, 1, \ldots$,
    Set $T \leftarrow Ap_{x_n}$
    Set $C \leftarrow C + AF_{x_n}$
    If $C \geq 2^{N+P}$,

Propagate carry

emit-bit(1)
If $r > 0$,
    execute $r - 1$ times, emit-bit(0)
    Set $r = 0$
else (we can be sure that $r = 0$)
    Set $r = -1$

While $T < 2^{N+P-1}$,

Renormalize once

Increment $b \leftarrow b + 1$
Shift $T \leftarrow 2T$
Shift $C \leftarrow 2C$
If $C \geq 2^{N+P}$ (pushing a "1" bit out of C)
    If $r < 0$,
        emit-bit(1)

else,
    Increment $r \leftarrow r + 1$
else
    If $r \geq 0$
      emit-bit(0)
      execute r times, **emit-bit**(1)
      Set $r = 0$

Set $A_{n+1} = \lfloor 2^{-P} T \rfloor$

After each iteration of the algorithm, the number of bits which have actually been output is given by $b_n - r_n - 1$. If this quantity is of no interest, the state variable, b_n, may be dropped. For simplicity, we will describe the corresponding decoding algorithm only in connection with binary arithmetic coding below.

2.3.2   BINARY ENCODING AND DECODING

Specializing the arithmetic coding procedure to the case of a binary alphabet, $\mathcal{A}_X = \{0, 1\}$, we obtain the following algorithm

---

Binary Arithmetic Encoder

Initialize $C = 0$, $A = 2^N$, $r = -1$, $b = 0$
For each $n = 0, 1, \ldots,$
    Set $T \leftarrow A p_{0,n}$
    If $x_n = 1$
      $C \leftarrow C + T$
      $T \leftarrow 2^P A - T$
    If $C \geq 2^{N+P}$,
      Propagate carry (affects r; outputs bits)
    While $T < 2^{N+P-1}$,
      Renormalize once (affects T, C, b, r; outputs bits)
    Set $A_{n+1} = \lfloor 2^{-P} T \rfloor$

Here, $p_{0,n}$ denotes the P-bit integer which is used to represent the probability that $X_n = 0$. For a stationary memoryless process, $p_{0,n}$ has no dependence on n. For a Markov$-k$ binary random process, $p_{0,n}$ depends upon the previous k outcomes; i.e.,

$$2^{-P} p_{0,n} \approx f_{X_n | \mathbf{X}_{n-k:n}} (0, \mathbf{x}_{n-k:n})$$

In practice, the source random process may not be stationary and we generally have to estimate the probabilities. Consequently, it is convenient to simply write $p'_{0,n} = 2^{-P} p_{0,n}$ for the current estimate of the

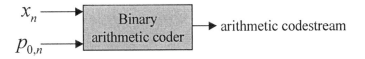

Figure 2.6.   Binary arithmetic coding machine.

probability that $X_n = 0$, given the previously coded source outcomes. The binary arithmetic coder may then be represented by the machine ("black box") illustrated in Figure 2.6. If the statistical properties of the source are known exactly, then we supply the appropriate probabilities, $p'_{0,n}$, with each symbol, x_n, and achieve a code-rate which approaches the entropy rate of the source to within a negligible margin. The operation of the machine, however, is independent of the correctness of these probability estimates.

SUFFICIENCY OF BINARY CODERS

Henceforth, we shall consider only binary arithmetic coders. As it turns out, this does not represent a practical limitation. To see this, suppose that $\mathcal{A}_X$ has 2^K entries for some $K \in \mathbb{Z}$. Then each element of $\mathcal{A}_X$ may be represented by a K bit integer. In this way, the random variable, X, is equivalent to a K-dimensional random vector, $\mathbf{B}$, where

$$\mathbf{B} = \begin{pmatrix} B_0 \text{ (MSB)} \\ B_1 \\ \vdots \\ B_{K-1} \text{ (LSB)} \end{pmatrix}$$

and the B_k are binary random variables representing the binary digits in the K-bit representation of X. Then

$$H(X) = H(\mathbf{B})$$
$$= H(B_0) + H(B_1 \mid B_0) + \cdots + H(B_{K-1} \mid B_0, \ldots, B_{K-2})$$

Now suppose we have a memoryless random process with alphabet $\mathcal{A}_X$ and we wish to code the outcomes at a bit-rate which approaches the entropy-rate of the process. This may be accomplished with the binary arithmetic coding machine of Figure 2.6 by supplying the pairs

$$(b_0, p_{0,0}), (b_1, p_{0,1}), \ldots, (b_{K-1}, p_{0,K-1})$$

when coding each of the successive bits of each source symbol, where the $p_{0,k}$ represent conditional probability estimates

$$p'_{0,k} = 2^{-P} p_{0,k} \approx f_{B_k | \mathbf{B}_{0:k}} (0, \mathbf{b}_{0:k})$$

The total number of conditional probability estimates is

$$1 + 2 + \cdots 2^{K-1} = 2^K - 1$$

which is identical to the total number of unique probabilities describing the original PMF, f_X. An important simplification arising from the use of binary arithmetic coding is that there are often only a few non-trivial conditional probabilities to estimate. As an example, the least significant bits in many numeric quantities often obey an approximately uniform distribution (i.e., they are "entirely" random); then we need only estimate and store $2^{K-U} - 1$ conditional probabilities, where U is the number of uniformly distributed LSBs.

The binary coding approach described above is easily extended to Markov processes and sources with arbitrary alphabets, finite or otherwise; we have only to supply the appropriate conditional probabilities to the binary arithmetic coding machine.

DECODING ALGORITHM

We now describe an incremental decoding algorithm for the binary arithmetic codeword. The decoder maintains an N-bit state variable, A, which represents the current interval width, a_n, exactly as in the encoder, following identical update procedures. The decoder also maintains an $(N + P)$-bit state variable, C; however, the interpretation of this quantity is somewhat different to that in the encoder.

To develop the decoding algorithm, let c denote the value represented by the entire arithmetic codeword, taken as a binary fraction. Then

$$c \in [c_n, c_n + a_n), \quad \forall n$$

Suppose we have correctly decoded x_0 through x_{n-1} and that the decoder has reproduced the evolution of a_n in the encoder. We could keep track of c_n in the decoder and then decode x_n according to

$$x_n = \begin{cases} 0 & \text{if } c < c_n + a_n p'_{0,n} \\ 1 & \text{if } c \geq c_n + a_n p'_{0,n} \end{cases}$$

It is simpler, however, to keep track of $c - c_n$ and then decode x_n according to

$$x_n = \begin{cases} 0 & \text{if } c - c_n < a_n p'_{0,n} \\ 1 & \text{if } c - c_n \geq a_n p'_{0,n} \end{cases} \tag{2.19}$$

To see why this is simpler, note that $c - c_n \in [0, a_n)$ where a_n has the binary fraction representation

$$a_n = 0.\underbrace{00\ldots0}_{b_n \text{ bits}}\underbrace{1aa\ldots a}_{A_n}$$

and $a_n p'_{0,n}$ has the binary fraction representation

$$a_n p'_{0,n} = 0.\underbrace{00\ldots0}_{b_n \text{ bits}}\underbrace{xx\ldots x}_{A_n p_{0,n}}$$

It follows that the b_n-bit prefix of $c - c_n$ is zero and the decision in equation (2.19) may be formed using the next $N + P$ bits of $c - c_n$. This is the quantity managed by the decoder's state variable, C_n. The binary fraction representation of $c - c_n$ has the structure,

$$c - c_n = 0.\underbrace{00\ldots0}_{b_n \text{ bits}}\underbrace{cc\ldots c}_{C_n}bbb\ldots$$

where the suffix, $bbb\ldots$, represents remaining bits in the arithmetic codeword, which have not yet been imported by the decoder. The decoding algorithm follows immediately:

Binary Arithmetic Decoder

Initialize $A = 2^N$, $b = 0$

Import $N + P$ bits from the codeword to initialize C.

For each $n = 0, 1, \ldots$,
    Set $T \leftarrow A p_{0,n}$
    If $C < T$
        Output $x_n = 0$
    else
        Output $x_n = 1$
        $C \leftarrow C - T$
        $T \leftarrow 2^P A - T$
    While $T < 2^{N+P-1}$,

> **Renormalize once**
> Increment $b \leftarrow b + 1$
> Shift $T \leftarrow 2T$
> Shift $C \leftarrow 2C$
> $C \leftarrow C + $ **retrieve-bit**().

Set $A_{n+1} = \lfloor 2^{-P} T \rfloor$

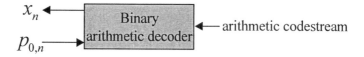

Figure 2.7.   Binary arithmetic decoding machine.

After each iteration of the algorithm, the number of bits which have been imported from the arithmetic codeword is $b_n + N + P$. The decoder is somewhat simpler than the encoder, since it need not deal with the effects of carry propagation. The binary arithmetic decoder may then be represented by the machine illustrated in Figure 2.7.

2.3.3   LENGTH-INDICATED TERMINATION

In Section 2.1.4 we described a particular termination policy which allows the decoder to discover the number of bits occupied by the arithmetic codeword when it is included as part of a larger bit-stream. This Elias termination policy produces a codeword of length

$$L_m^{\text{elias}} = 1 + \left\lceil \log_2 \frac{1}{a_m} \right\rceil$$

It has the advantage that there is no need to explicitly indicate the number of bits in the arithmetic codeword. In some applications, however, the length must be explicitly signalled to fulfill some other objective, e.g. to facilitate manipulation and indexing of a compressed data file.

In this section we discuss a termination strategy which takes advantage of the fact that the length of the codeword, L_m, is explicitly sent to the decoder. We call this length-indicated termination. As with Elias termination, we assume that the decoder knows the number of source outcomes, m, which have been coded.

Since the decoder knows the value of L_m, it can append a known sequence of bits to the L_m bits which it receives in order to construct the quantity, $c \in [c_m, c_m + a_m)$. In particular, we assume that the decoder extends the received string of bits by appending 1's as needed, until all m symbols have been decoded. A simple termination policy for the encoder is to set

$$L_m = b_m + 1$$

outputting the first L_m bits of c_m. The decoder then reconstructs

$$c = 0.\underbrace{bb\ldots b}_{L_m \text{ bits}}111\ldots$$

$$\in \left[c_m, c_m + 2^{-L_m}\right)$$

$$\subseteq \left[c_m, c_m + a_m\right)$$

where the final relationship follows from

$$a_m \in \left[2^{-(b_m+1)}, 2^{-b_m}\right)$$

Now b_m is the smallest integer such that $a_m \geq 2^{-(b_m+1)}$; i.e., $b_m = \left\lceil \log_2 \frac{1}{a_m} \right\rceil - 1$. Consequently, this termination policy yields a code length of

$$L_m = \left\lceil \log_2 \frac{1}{a_m} \right\rceil$$

which is 1 bit less than L_m^{elias}.

The encoder can easily improve on this termination by discarding any trailing 1's from the string of bits sent to the decoder. This policy reduces L_m by 1 bit on average, so if we ignore the often negligible inefficiencies introduced by the finite precision implementation

$$E\left[L_m\right] = E\left[\lceil h_{\mathbf{X}_{0:m}}(\mathbf{X}_{0:m})\rceil\right] - 1$$
$$< H\left(\mathbf{X}_{0:m}\right)$$

This result appears to contradict the noiseless coding theorem. However, we are exploiting the fact that the decoder knows L_m and we are not including the number of bits required to signal its value.

Even more careful termination is possible. The ultimate objective is to compute a minimum length prefix of c_m such that the non-negative error introduced by appending 1's to this prefix is strictly less than a_m. In this way, length-indicated termination can produce codewords approximately $2\frac{1}{2}$ bits shorter than those obtained with Elias termination, bearing in mind that we are not counting the cost of explicitly signalling the value of L_m.

2.3.4   MULTIPLIER-FREE VARIANTS

A significant source of complexity in the binary arithmetic encoding and decoding algorithms described above is the multiplication required to implement

$$T \leftarrow A_n p_{0,n}$$

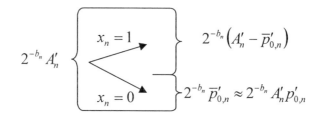

Figure 2.8.   Interval sub-division using the multiplier-free approximation.

In dedicated hardware implementations, fast parallel multiplier circuits consume substantially more silicon real-estate than adders. Modern CPU's do not always incorporate dedicated fast integer multiplication paths and even those CPU's which do offer such features often have higher latencies for multiplication than addition.

As a result, most practical arithmetic coders introduce a further approximation in order to avoid the need for multiplication. The approximation is based on the observation that A_n always lies between 2^{N-1} and 2^N so that

$$A_n p_{0,n} \approx 2^N \alpha p_{0,n} \text{ where } \alpha \approx \frac{3}{4}$$

We defer a discussion of the optimum selection of α for a little while, using the approximate value of $\frac{3}{4}$ for illustrative purposes only. Adopting this approximation, we should be able to fold the factor, $2^N \alpha$, into our probability estimates and simply assign

$$T \leftarrow \bar{p}_{0,n}$$

where

$$\bar{p}_{0,n} = 2^N \alpha p_{0,n} = 2^{N+P} \alpha p'_{0,n}$$

Before proceeding any further we will need to resolve a serious problem with this approximation. If the symbol is $x_n = 1$, the algorithm proceeds to assign

$$T \leftarrow 2^P A_n - T = 2^P \left(A_n - 2^N \alpha p'_{0,n} \right)$$

which can be negative if $p'_{0,n} > \frac{1}{2\alpha} \approx \frac{2}{3}$! The problem may be understood with the aid of Figure 2.8, where we use the use the prime-notation, A'_n, $p'_{0,n}$ $\bar{p}_{0,n}$, to denote the binary fractions represented by A_n, $p_{0,n}$ and $\bar{p}_{0,n}$. The approximation is clearly inappropriate when

$$\bar{p}_{0,n} = \alpha p'_{0,n} \geq A'_n \in \left[\frac{1}{2}, 1 \right)$$

a situation which will arise with any choice of $\alpha > \frac{1}{2}$.

THE MPS-LPS SWITCH

The usual solution to the above dilemma is to flip the roles of the symbols 0 and 1 whenever $p'_{0,n} > \frac{1}{2}$. Specifically, let $s_n \in \{0,1\}$ denote the most probable symbol (MPS) outcome. That is,

$$s_n \triangleq \left\{ \begin{array}{ll} 1 & \text{if } p'_{0,n} \leq \frac{1}{2} \\ 0 & \text{if } p'_{0,n} > \frac{1}{2} \end{array} \right.$$

and let p'_n denote our estimate of the probability that the least probable symbol (LPS) occurs; i.e.,

$$p'_n = \left\{ \begin{array}{ll} p'_{0,n} & \text{if } p'_{0,n} \leq \frac{1}{2} \\ 1 - p'_{0,n} & \text{if } p'_{0,n} > \frac{1}{2} \end{array} \right.$$

Estimating the probability of the zero symbol, $p'_{0,n}$, is equivalent to estimating the identity of the MPS, s_n, and the probability of the LPS, p'_n, so we will work exclusively with these quantities from now on. The binary encoding algorithm now becomes

---

Multiplier-Free Encoder

Initialize $C = 0$, $A = 2^N$, $r = -1$, $b = 0$

For each $n = 0, 1, \ldots$,

    If $x_n = s_n$, (encode an MPS)

        $A \leftarrow A - \bar{p}_n$

        $C \leftarrow C + \bar{p}_n$

    else (encode an LPS)

        $A \leftarrow \bar{p}_n$

    If $C \geq 2^N$,

        Propagate carry (affects r; outputs bits)

    While $A < 2^{N-1}$,

        Renormalize once (affects A, C, b, r; outputs bits)

Notice that there is no longer any need to carry an intermediate $(N + P)$-bit quantity, T, and that all operations are performed directly on the N-bit variable, A. For this to work, we require only that

$$\bar{p}'_n = 2^{-N} \bar{p}_n = \alpha p'_n$$

With these modified conventions, C is now an N-bit quantity and the "renormalize once" routine must be modified in an obvious way. The multiplier-free decoding algorithm becomes

---

Multiplier-Free Decoder

Initialize $A = 2^N$, $b = 0$

Import N bits from the codeword to initialize C.

For each $n = 0, 1, \ldots,$

　　If $C < \bar{p}_n$ (decode an LPS)

　　　　Output $x_n = 1 - s_n$

　　　　$A \leftarrow \bar{p}_n$

　　else (decode an MPS)

　　　　Output $x_n = s_n$

　　　　$A \leftarrow A - \bar{p}_n$

　　　　$C \leftarrow C - \bar{p}_n$

　　While $A < 2^{N-1}$,

　　　　Renormalize once (affects A, C, b; imports bits)

IMPACT ON CODING EFFICIENCY

It is instructive to investigate the impact on coding efficiency of the multiplier-free approximation developed above. To do this, we will need to determine the best value for α. The effect of coding any symbol x_n is to add $\log_2 \frac{a_n}{a_{n+1}}$ bits to the final codeword length. To facilitate analysis, we shall assume here that A'_n is uniformly distributed over the interval $\left[\frac{1}{2}, 1\right)$ and statistically independent of the source process; we shall reconsider this assumption shortly. The expected code rate may then be expressed as

$$R\left(p'_n, \alpha\right) = 2 \int_{\frac{1}{2}}^{1} \left[p'_n \log_2 \left(\frac{x}{\alpha p'_n} \right) + \left(1 - p'_n\right) \log_2 \left(\frac{x}{x - \alpha p'_n} \right) \right] dx$$

which is easily integrated with standard forms. We may find the optimum value of α in the obvious manner by solving $\frac{\partial R}{\partial \alpha} = 0$ for each LPS probability, $p'_n \in \left(0, \frac{1}{2}\right]$. The result is plotted in Figure 2.9.

Evidently, α is a weak function of p'_n. As it turns out, however, the code rate is not highly sensitive to the exact choice of α within the range suggested by Figure 2.9. To illustrate this point, Figure 2.10 shows a plot of the code rate expansion factor,

$$\frac{R\left(p'_n, \alpha\right)}{-p'_n \log_2 p'_n - \left(1 - p'_n\right) \log_2 \left(1 - p'_n\right)}$$

as a function of p'_n for two different fixed choices of α. From the figure, we see that the loss in coding efficiency is quite small, particularly when the

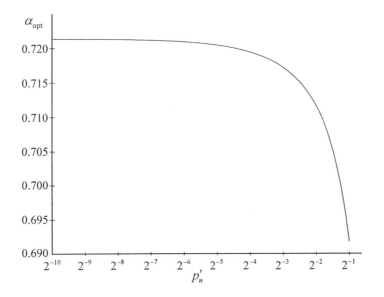

Figure 2.9.   Optimum α as a function of the LPS probability, p'_n, assuming a uniform distribution for A'_n.

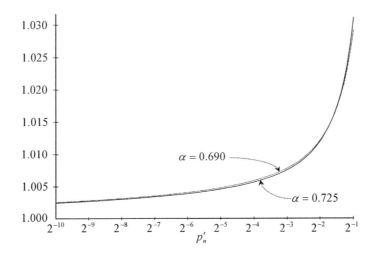

Figure 2.10.   Code rate expansion factor, $R(p'_n, \alpha)/H(p'_n)$, for values of α which bracket α_{opt}.

symbol probabilities are highly skewed, where the highest compression ratios are achieved.

The above analysis is based on the assumption that A'_n is uniformly distributed over $\left[\frac{1}{2}, 1\right)$. This assumption is quite reasonable for mixed context applications in which the LPS probability, p'_n, changes rapidly with n, usually because the arithmetic coder is switching between different contexts, having quite distinct probability models. Most image compression applications involve a mixture of contexts. In single context models, however, where p'_n is at most a slowly varying function of n, A'_n tends to be distributed more toward the lower half of the interval, so that the optimum value for α is close to $\frac{2}{3}$. The reader is referred to [88] for further discussion of this phenomenon. A value of $\alpha = \frac{2}{3}$ is implicitly assumed in many developments of multiplier-free arithmetic coding. Henceforth, we shall adopt the value $\alpha = 0.708$, which is reported in [119] as the experimentally observed optimum value, α_{opt}, for the JBIG application.

CONDITIONAL EXCHANGE

From Figure 2.10, we see that the largest loss in coding efficiency occurs when the LPS probability is close to $\frac{1}{2}$. A mechanism known as conditional exchange was invented to mitigate this loss in the QM coder which is used by the JPEG and JBIG image compression standards [119]. Referring to Figure 2.8, we see that the interval assigned to the MPS is smaller than that assigned to the LPS when

$$\vec{p}'_n > \frac{1}{2} A'_n \in \left[\frac{1}{4}, \frac{1}{2}\right)$$

The conditional exchange mechanism exchanges the roles of the MPS and LPS whenever this happens, so as to ensure that the MPS is always assigned the larger interval. From the above relationship, conditional exchange can affect only those symbols for which $\frac{1}{4} < \vec{p}'_n < \frac{1}{2}$; i.e., those whose distributions are close to uniform. The modified encoding algorithm is

---

Conditional Exchange Encoder

Initialize $C = 0$, $A = 2^N$, $r = -1$, $b = 0$

For each $n = 0, 1, \ldots,$
    $s \leftarrow s_n$
    $A \leftarrow A - \bar{p}_n$
    If $A < \bar{p}_n$
        $s \leftarrow 1 - s$
    If $x_n = s$, (encode an MPS)
        $C \leftarrow C + \bar{p}_n$

else (encode an LPS)

$\qquad A \leftarrow \bar{p}_n$

If $C \geq 2^N$,

$\qquad$ **Propagate carry** (affects r; outputs bits)

While $A < 2^{N-1}$,

$\qquad$ **Renormalize once** (affects A, C, b, r; outputs bits)

The modified decoding algorithm is

Conditional Exchange Decoder

Initialize $A = 2^N$, $b = 0$

Import N bits from the codeword to initialize C.

For each $n = 0, 1, \ldots$,

$\qquad s \leftarrow s_n$

$\qquad A \leftarrow A - \bar{p}_n$

$\qquad$ If $A < \bar{p}_n$

$\qquad\qquad s \leftarrow 1 - s$

$\qquad$ If $C < \bar{p}_n$ (decode an LPS)

$\qquad\qquad$ Output $x_n = 1 - s$

$\qquad\qquad A \leftarrow \bar{p}_n$

$\qquad$ else (decode an MPS)

$\qquad\qquad$ Output $x_n = s$

$\qquad\qquad C \leftarrow C - \bar{p}_n$

$\qquad$ While $A < 2^{N-1}$,

$\qquad\qquad$ **Renormalize once** (affects A, C, b; imports bits)

In the straightforward incarnations illustrated above, the introduction of conditional exchange appears to increase the complexity of the coder by at least one test per symbol. Fortunately, however, the algorithm can be reoganized to test for conditional exchange only once it has been determined that a renormalization is required. To understand this, observe that whenever conditional exchange occurs we must have

$$A - \bar{p}_n < \bar{p}_n = 2^N \alpha p'_n < 2^{N-1} \qquad (2.20)$$

Since A is assigned to one of $A - \bar{p}_n$ or $\bar{p}_n$, conditional exchange must always be accompanied by renormalization.

2.3.5   ADAPTIVE PROBABILITY ESTIMATION

In this section we introduce the reader to some of the considerations involved in estimating the probabilities, $p'_{0,n}$, used to code the source outcome of $X_n \in \{0, 1\}$. We assume here that the source is a stationary Markov-k random process so that we are trying to estimate

$$p'_{0,n} \approx f_{X_n|\mathbf{X}_{n-k:n}}(0, \mathbf{x}_{n-k:n})$$

Now, since the random process is assumed stationary, there are only 2^k distinct conditional probabilities,

$$f_{X_k|\mathbf{X}_{0:k}}(0, \mathbf{x}_{n-k:n}) = f_{X_n|\mathbf{X}_{n-k:n}}(0, \mathbf{x}_{n-k:n})$$

corresponding to the 2^k possible context vectors, $\mathbf{x}_{n-k:n}$. It is helpful to define a context labeling function, $\lambda(\mathbf{x})$, which assigns a unique integer in the range 0 through $2^k - 1$ to each k-dimensional vector, $\mathbf{x}$. An adaptive coder estimates $p'_{0,n}$ based on any or all previous outcomes, x_i, $i < n$, which have occurred within the same context; i.e., $\lambda(\mathbf{x}_{i-k:i}) = \lambda(\mathbf{x}_{n-k:n})$. A natural way to form such estimates is to maintain counts of the number of 0 and 1 symbols which have been observed within each context; i.e.,

$$C_{0,l}[n] = \sum_{i<n} \delta(l, \lambda(\mathbf{x}_{i-k:i})) \cdot (1 - x_i)$$

$$C_{1,l}[n] = \sum_{i<n} \delta(l, \lambda(\mathbf{x}_{i-k:i})) \cdot x_i$$

If the counts are sufficiently large, we should expect any reasonable estimate to satisfy

$$p'_{0,n} \approx \frac{C_{0,\lambda(\mathbf{x}_{n-k:n})}[n]}{C_{0,\lambda(\mathbf{x}_{n-k:n})}[n] + C_{1,\lambda(\mathbf{x}_{n-k:n})}[n]}$$

Although it is conceivable that prior knowledge concerning the interaction of distinct contexts might prove useful in estimating probabilities, we shall assume that probabilities are estimated independently within each context. Associated with each context, then, is a "learning penalty" which arises from the fact that the first few symbols are generally coded using inappropriate probability estimates. As more symbols are coded within any given context, the probability estimates stabilize (assuming an approximately stationary source) and coding becomes more efficient. The learning penalty is a function of the estimation procedure, its initial state and the conditional PMF which is being estimated. Regardless of such variables, however, it is clear that the creation of too many contexts is undesirable. If two distinct coding contexts exhibit identical conditional PMF's, their combined learning penalty may clearly be halved by merging the contexts.

In the ensuing development we shall ignore the existence of multiple contexts and the problem of good context design. This approach simplifies the discussion without sacrificing its applicability to multiple contexts, since the probability estimates for each context are to be adapted independently.

SCALED COUNT ESTIMATORS

Perhaps the most natural estimation strategy is to assign

$$p'_{0,n} = \frac{C_0[n] + 1}{(C_0[n] + 1) + (C_1[n] + 1)}$$

where 1 is added to each of the symbol counts so as to ensure that $p'_{0,n} \in (0,1)$.[5] More generally, we might assign

$$p'_{0,n} = \frac{C_0[n] + \Delta}{(C_0[n] + \Delta) + (C_1[n] + \Delta)}$$

Large values of Δ reflect a conservative policy in which we are reluctant to estimate highly skewed (i.e., non-uniform) distributions until we have observed a large number of outcomes. Conversely, smaller values of Δ reflect a more radical approach. These estimators may be shown to be maximum a posteriori (MAP) estimates for the actual zero-symbol probability, p'_0, subject to certain a priori assumptions on the distribution of the underlying random variable from which p'_0 is drawn [175]. In particular, the selection $\Delta = 1$ yields the MAP estimate if we assume that p'_0 is uniformly distributed on $(0,1)$ a priori, while smaller values of Δ correspond to the assumption that highly skewed probabilities (p'_0 close to 0 or 1) are most likely. This interpretation is useful, since in many applications we have some idea as to whether or not we expect highly skewed distributions.

In practical applications, the statistics are often not stationary, so we prefer to weight the probability estimates toward more recently observed outcomes. This can be done by periodically renormalizing the counts. A simple renormalization strategy is to halve both C_0 and C_1 whenever the count exceeds some limits. There will inevitably be some upper bound, $C_{\max}$, to the counts which can be represented in an implementation so that the need for a renormalization strategy is usually unavoidable. If, however, we wish to track non-stationary statistics, then it is advisable to renormalize as frequently as possible without overly compromising the

---

[5] Without deterministic prior knowledge, we must generally assume that both symbols have some non-zero probability of occuring in every context.

accuracy of the estimates. Intuitively, the minimum of the two counters primarily determines the accuracy of our probability estimate and this may be demonstrated more rigorously [53]. This suggests that we should renormalize whenever either C_0 or C_1 exceeds some lower bound, $C_{\min}$. The following so-called "scaled count" estimation algorithm reflects these considerations.

Scaled Count Probability Estimator

Initialize $C_0 = C_1 = 0$

For $n = 0, 1, \ldots,$

    If $x_n = 1$,

        $C_1 \leftarrow C_1 + 1$

    else

        $C_0 \leftarrow C_0 + 1$

    If $\min\{C_0, C_1\} > C_{\min}$ or $\max\{C_0, C_1\} > C_{\max}$

        $C_0 \leftarrow \lfloor \frac{C_0}{2} \rfloor; C_1 \leftarrow \lfloor \frac{C_1}{2} \rfloor$

    Estimate $p'_{0,n} \leftarrow \frac{C_0 + \Delta}{C_0 + C_1 + 2\Delta}$.

There are, of course, many variations on this basic theme.

FINITE STATE MACHINES FOR PROBABILITY ESTIMATION

The scaled count estimator described above is a finite state machine. The number of states is more apparent if we modify the implementation to count the number of LPS and MPS symbols, $C_L[n]$ and $C_M[n]$, and to keep track of the identity of the MPS; i.e., s_n. Then $C_L[n] \leq C_M[n]$ so the range of these counters is

$$0 \leq C_L[n] \leq C_{\min}$$
$$0 \leq C_M[n] \leq C_{\max}$$

and we have a total of $2(C_{\min} + 1)(C_{\max} + 1)$ states[6]. As an example, we might set $C_{\min} = 15$ and $C_{\max} = 1023$, enabling us to generate reliable estimates over the range of probabilities encountered in many practical applications. The task of computing an estimate of the LPS symbol probability, p'_n, may then be reduced to a table lookup operation, involving a table with 2^{14} entries. The table lookup approach has the added advantage that the mapping from p'_n to $\bar{p}_n = 2^N \alpha p'_n$, for multiplier-free implementations, may be built into the table.

---

[6]The factor of 2 arises from the fact that we must keep track of which symbol is the MPS.

Table 2.1. Probability state transition table for the MQ-coder. LPS probability, p', is estimated using $\alpha = 0.708$.

Σ	Transition			Estimate		Σ	Transition			Estimate	
	Σ_{mps}	Σ_{lps}	X_s	$\bar{p}$ (hex)	$p' = \frac{\bar{p}}{2^{16}\alpha}$		Σ_{mps}	Σ_{lps}	X_s	$\bar{p}$ (hex)	$p' = \frac{\bar{p}}{2^{16}\alpha}$
0	1	1	1	x5601	0.475	24	25	22	0	x1C01	0.155
1	2	6	0	x3401	0.292	25	26	23	0	x1801	0.132
2	3	9	0	x1801	0.132	26	27	24	0	x1601	0.121
3	4	12	0	x0AC1	0.0593	27	28	25	0	x1401	0.110
4	5	29	0	x0521	0.0283	28	29	26	0	x1201	0.0993
5	38	33	0	x0221	0.0117	29	30	27	0	x1101	0.0938
6	7	6	1	x5601	0.475	30	31	28	0	x0AC1	0.0593
7	8	14	0	x5401	0.463	31	32	29	0	x09C1	0.0499
8	9	14	0	x4801	0.397	32	33	30	0	x08A1	0.0476
9	10	14	0	x3801	0.309	33	34	31	0	x0521	0.0283
10	11	17	0	x3001	0.265	34	35	32	0	x0441	0.0235
11	12	18	0	x2401	0.199	35	36	33	0	x02A1	0.0145
12	13	20	0	x1C01	0.155	36	37	34	0	x0221	0.0117
13	29	21	0	x1601	0.121	37	38	35	0	x0141	0.00692
14	15	14	1	x5601	0.475	38	39	36	0	x0111	0.00588
15	16	14	0	x5401	0.463	39	40	37	0	x0085	0.00287
16	17	15	0	x5101	0.447	40	41	38	0	x0049	0.00157
17	18	16	0	x4801	0.397	41	42	39	0	x0025	0.000797
18	19	17	0	x3801	0.309	42	43	40	0	x0015	0.000453
19	20	18	0	x3401	0.292	43	44	41	0	x0009	0.000194
20	21	19	0	x3001	0.265	44	45	42	0	x0005	0.000108
21	22	19	0	x2801	0.221	45	45	43	0	x0001	0.000022
22	23	20	0	x2401	0.199	46	46	46	0	x5601	0.475
23	24	21	0	x2201	0.188						

As it turns out, the complexity associated with probability estimation may be reduced significantly further again by using the renormalization events in the arithmetic coder to probabilistically gate transitions in the state machine. Specifically, the state is updated immediately after any symbol coding operation which involves one or more calls to the "**renormalize once**" routine. The new state identifies the probability estimates to be used for all subsequent symbols, until a subsequent renormalization event induces another transition in the state machine.

We illustrate the process with the MQ coder's state transition table, Table 2.1. The second and third columns in the table indicate the state to which the machine transitions in the event of MPS-induced and LPS-induced renormalizations, respectively, while the fourth column holds a 1 if the symbols associated with the LPS and MPS are to be exchanged upon an LPS-induced renormalization (i.e., if s is to be replaced by $1-s$).

States $\Sigma = 0$ through $\Sigma = 13$ correspond to the "start-up" portion of the transition table, where few symbols have been observed. This part of the table is complicated to analyze. The last state, $\Sigma = 46$ is non-adaptive, since it is impossible to enter this state from any other state or leave it once one has entered; the JPEG2000 application uses this state to code symbols which are known to have an essentially uniform distribution. The remaining states $\Sigma = 14$ through $\Sigma = 45$ represent the non-transient portion of the table. Once entered from one of the start-up states, the state machine can never leave the non-transient portion.

To understand the principles behind renormalization-driven probability estimation, it is instructive to consider the non-transient portion of Table 2.1. The LPS probability is a decreasing function of the state index, Σ. MPS-induced renormalizations tend to drive the machine toward larger state indices, decreasing $\bar{p}$ and hence the relative frequency of MPS-induced renormalizations. Conversely, LPS-induced renormalizations tend to drive the machine toward smaller state indices. In this way, the machine can be expected to converge to an equilibrium state which depends upon the LPS probability, p'.

Now suppose that an MPS-induced renormalization always increments Σ by 1 and let k denote the amount by which an LPS-induced renormalization reduces Σ. We suppose for convenience that all states in the neighbourhood of the equilibrium state have approximately the augend value, $\bar{p}' = \alpha p'$. Then the average downward drift due to LPS-induced renormalization is $p'k$ states per symbol. The average upward drift due to MPS-induced renormalization may be estimated by assuming that A' is uniformly distributed over $[\frac{1}{2}, 1)$. At equilibrium, these effects must balance with

$$p'k = \left(1 - p'\right) \int_{\frac{1}{2}}^{\frac{1}{2}+\bar{p}'} 2dx = \left(1 - p'\right) 2\alpha p'$$

which yields

$$k = 2\alpha \left(1 - p'\right) \approx 1.5 \left(1 - p'\right)$$

This crude analysis suggests that k should be set to 1 when p' is not too small and set to 1 or 2 at the highly skewed end of the table where $p' \ll 1$. This conclusion is supported by the structure of the table, where the transition from $k = 1$ to $k = 2$ occurs at an LPS probability of about $\frac{1}{4}$.

Renormalization-driven probability estimation works well in practice, not so much because the models used to construct the table are accurate, but because the optimum transition step, k, is almost entirely independent of p'. The natural forces driving the state machine to equilibrium will yield an approximately optimal value of $\bar{p}$, regardless of the particular entries which we place in the table, provided they are monotonically decreasing with Σ and cover the range of probabilities which we expect to encounter. Careful modeling is of value only in fine tuning the structure.

2.3.6 OTHER VARIANTS

Multiplier-free arithmetic coder variants may be traced to the "skew coder" [89], some of the history being reproduced in [88]. Multiplier-free operation and renormalization-driven probability estimation are the most distinguishing features of a broad class of arithmetic coding algorithms which includes the Q coder [118], QM coder, [119] MQ coder (see Section 12.1) and Z coder [28]. An alternate method of approximating the multiplication of A by $p_{0,n}$, is embodied in the ELS coder [170].

The Q coder and MQ coder variants incorporate a "bit-stuffing" mechanism which limits the extent of carry propagation in the encoder, simplifying the implementation at the expense of a small loss in coding efficiency. This mechanism is explained in Section 12.1.

It should be noted that we have adopted a number of arbitrary conventions in our discussion of arithmetic coding. The assignment of sub-intervals to specific symbols and the representation of the coding interval in terms of a lower bound and length, rather than an upper bound and length, are arbitrary choices. Various coders and implementations of these coders adopt different conventions.

2.4 IMAGE CODING TOOLS

In the preceding sections we have introduced what we might call "low-level" coding tools. In theory, these are sufficient to fully exploit the statistical redundancy in any data set, including a collection of image sample values. In practice, however, these techniques alone are usually insufficient to exploit the rich structure in images subject to reasonable constraints on implementation complexity. For this reason, we now briefly discuss a selection of image-specific coding tools which are widely used in practice.

2.4.1 CONTEXT ADAPTIVE CODING

Markov random processes are a powerful modeling tool for information sources, including images. We have already seen (equation (2.8)) that the entropy rate of a Markov-p random process is given by the pth order conditional entropy $H(X_p \mid \mathbf{X}_{0:p})$. Moreover, we have seen that practical arithmetic coding algorithms exist which are able to achieve

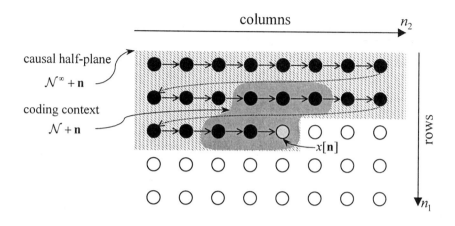

Figure 2.11.   Context coding and associated neighbourhoods.

code rates remarkably close to this bound, provided they are driven by the appropriate conditional probabilities, $f_{X_p|\mathbf{X}_{0:p}}$.

In light of these observations, a natural approach to image compression is to scan the sample values, $x[\mathbf{n}] \equiv x[n_1, n_2]$, into a one dimensional sequence, usually following a lexicogaphical (raster-scan) order, and to code each sample using an appropriate model for the sample's distribution, conditioned on previous samples in the scan. This is illustrated in Figure 2.11.

In two dimensions, the Markov model is parametrized by a causal "neighbourhood," $\mathcal{N}$, rather than a single parameter, p. Specifically, $\mathcal{N}$ is an ordered subset of the causal half-plane,

$$\mathcal{N}^{\infty} \triangleq \{\mathbf{n} \mid n_1 < 0\} \cup \{\mathbf{n} \mid n_1 = 0, \, n_2 < 0\}$$

and the Markov conditional independence property becomes

$$f_{X[\mathbf{n}]|\mathbf{X}_{\mathcal{N}+\mathbf{n}}}(x, \mathbf{x}_{\mathcal{N}+\mathbf{n}}) = f_{X[\mathbf{n}]|\mathbf{X}_{\mathcal{N}^{\infty}+\mathbf{n}}}(x, \mathbf{x}_{\mathcal{N}^{\infty}+\mathbf{n}})$$

The notation used here is a natural extension of that we have been using for one dimensional processes. The vector, $\mathbf{x}_{\mathcal{N}}$, consists of the elements $x[\mathbf{k}]$ for each $\mathbf{k} \in \mathcal{N}$ and the set $\mathcal{N}+\mathbf{n}$ is obtained by adding the displacement vector, $\mathbf{n}$, to each element of $\mathcal{N}$. A typical neighbourhood configuration is illustrated in Figure 2.11.

A straightforward application of these principles to image compression is depicted schematically in Figure 2.12. In the simplest case, the context labeling operation assigns a distinct label,

$$l = \lambda(\mathbf{x}_{\mathcal{N}+\mathbf{n}})$$

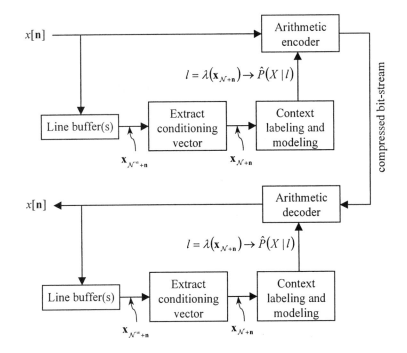

Figure 2.12.   Context-adaptive image compression and decompression.

to each neighbourhood vector, $\mathbf{x}_{\mathcal{N}+\mathbf{n}}$. The arithmetic encoder and decoder are driven by identical estimates, $\hat{P}(X \mid l)$, for the conditional distribution of $X[\mathbf{n}]$, given the context label, $l = \lambda(\mathbf{x}_{\mathcal{N}+\mathbf{n}})$. The estimates may be fixed or adaptive probability estimates. If the underlying random process is indeed Markov-$\mathcal{N}$ and the probability estimates are exact, the compression system will be optimal, achieving the entropy rate of the source to within a negligible margin.

This direct approach is most appropriate for compressing bi-level imagery for which $x[\mathbf{n}] \in \{0, 1\}$, since then the total number of context labels for which probability estimates are required can be quite modest. The JBIG image compression standard operates in this manner, assigning a separate context label to each of the 2^{10} possible neighbourhood vectors arising from the selection of one or the other of the two neighbourhood configurations shown in Figure 2.13. In the JBIG application, the arithmetic coder variant is the QM coder, and estimates of the LPS probability, p'_n, and MPS identity, s_n, are obtained from a renormalization-driven finite state machine, as discussed in Section 2.3.5. When a renormalization event occurs, the state machine updates the state, Σ_l, associated with context label l. Each state may be repre-

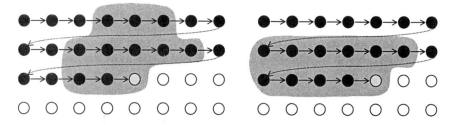

Figure 2.13. Two context generating neighbourhoods used by the JBIG image compression standard.

sented with a single byte so that the entire adaptive probability model is contained within a 1 kbyte memory.

For non-binary alphabets, the direct labeling approach is less attractive. Consider, for example, a direct application to 8-bit imagery. The conditional distribution for each context is characterized by 255 free parameters and there are $256^{\|\mathcal{N}\|}$ distinct contexts. Thus, even for a simple two-element neighbourhood, we must estimate approximately 2^{24} distinct parameters. Quite apart from the storage concerns, images do not contain sufficient samples to reliably estimate this many parameters.

For multi-valued images, therefore, we must employ a context reduction function, $\lambda()$, to reduce the input vector, $\mathbf{x}_{\mathcal{N}+\mathbf{n}}$, to a manageable number of context labels. A common approach is to first apply the predictive techniques of Section 2.4.2, reducing the original image sample values to an equivalent array of prediction residuals, $e[\mathbf{n}]$. The prediction residuals tend to be clustered around zero so that $e[\mathbf{n}]$ is very likely to lie inside a small alphabet, $\overline{\mathcal{A}}_E \subset \mathcal{A}_E$, with $\|\overline{\mathcal{A}}_E\| \ll \|\mathcal{A}_E\|$. An obvious way to reduce the number of context labels without sacrificing much coding efficiency is to assign a unique label,

$$l = \lambda\left(\mathbf{e}_{\mathcal{N}+\mathbf{n}}\right) = \lambda'\left(\mathbf{e}'_{\mathcal{N}+\mathbf{n}}\right),$$

to each $\mathbf{e}'_{\mathcal{N}+\mathbf{n}}$, where the elements of $\mathbf{e}'$ are formed by setting

$$e'_i = \begin{cases} e_i, & \text{if } e_i \in \overline{\mathcal{A}}_E \\ \bar{e}, & \text{otherwise} \end{cases}, \quad i = 1, 2, \ldots \|\mathcal{N}\|$$

Here, $\bar{e}$, is used to collectively represent all symbols in $\mathcal{A}_E \setminus \overline{\mathcal{A}}_E$. The JPEG-LS image compression standard follows this paradigm, as discussed in Chapter 20.

We conclude by noting that the arithmetic coding operation illustrated in Figure 2.12 may be replaced by Huffman coding, Golomb coding, or any of a variety of other coding techniques, adaptive or otherwise, which are able to exploit the statistical redundancy available within each context.

2.4.2 PREDICTIVE CODING

In predictive coding, the image samples, $x[\mathbf{n}]$, are converted into an equivalent array of prediction residuals (or errors), $e[\mathbf{n}]$. Because $x[\mathbf{n}]$ and $e[\mathbf{n}]$ are equivalent sequences, with the same sample rate, they have the same entropy rate, $H(\{X[\mathbf{n}]\}) = H(\{E[\mathbf{n}]\})$. However, $e[\mathbf{n}]$ is generally easier to code efficiently when faced with practical limitations.

The prediction residual sequence is formed by setting

$$e[\mathbf{n}] = x[\mathbf{n}] - \mu_p(\mathbf{x}_{\mathcal{N}+\mathbf{n}})$$

where $\mathcal{N} \subset \mathcal{N}^\infty$ is a causal neighbourhood[7] and $\mu_p()$ is a function of the prediction vector, $\mathbf{x}_{\mathcal{N}+\mathbf{n}}$, whose elements have already been coded. The predictor function, μ_p, is ideally designed to minimize the first order entropy of the residuals, $H^{(1)}(\{E[\mathbf{n}]\}) = H(E)$. In practice, this is usually approximately equivalent to minimizing the variance of the residuals, which may be achieved by setting the predictor equal to the conditional mean; i.e.,

$$\mu_p(\mathbf{x}_{\mathcal{N}+\mathbf{n}}) \approx E[X[\mathbf{n}] \mid X_{\mathcal{N}+\mathbf{n}} = \mathbf{x}_{\mathcal{N}+\mathbf{n}}]$$

This explains our choice of notation, μ_p for the predictor[8]. If a good predictor can be found, most of the residuals will be close to zero.

As a simple example, let

$$\mathcal{N} = \{[0, -1], [-1, 0]\}$$

so that the prediction vector consists of the values of the samples to the left and immediately above $x[\mathbf{n}]$; i.e.,

$$\mathbf{x}_{\mathcal{N}+\mathbf{n}} = (x[n_1, n_2 - 1], x[n_1 - 1, n_2])^t$$

The most obvious predictor function would simply average these two neighbouring samples,

$$\mu_p(\mathbf{x}_{\mathcal{N}+\mathbf{n}}) = \frac{1}{2}(x[n_1, n_2 - 1] + x[n_1 - 1, n_2])$$

---

[7] The reader is referred to Section 2.4.1 for definitions of $\mathcal{N}$, $\mathcal{N}^\infty$ and $\mathbf{x}_{\mathcal{N}+\mathbf{n}}$.
[8] We use the same notation for the predictor in a lossy DPCM feedback loop, as discussed in Section 3.3. Subject to certain assumptions, the optimal predictor for lossy DPCM is well justified as the conditional mean of X_n given its causal neighbours.

and indeed this is one of the predictors supported by the original lossless compression algorithm defined by the JPEG image compression standard[9] [119].

In many images, the $X[\mathbf{n}]$ obey an approximately uniform distribution over the range of sample values, whereas the random variables, $E[\mathbf{n}]$, generally have highly non-uniform distributions, centered about 0. Hence we may conclude that the first order entropy of the prediction residual process, $H(E)$, should be significantly smaller than that of the original image, $H^{(1)}(\{X[\mathbf{n}]\})$. This means that simple arithmetic or Huffman coders, which use no context modeling at all, stand to benefit significantly from a prediction front end. Moreover, if context modeling is also to be used then contexts with a reduced number of states may be formed more easily from the prediction error sequence than from the image samples themselves, as discussed at the end of Section 2.4.1.

2.4.3   RUN-LENGTH CODING

Graphics images, bi-level images and pseudo-images representing class information extracted from a real image (e.g. high activity vs. low activity) often contain large regions of constant sample values. Context-based coding schemes generally have difficulty capturing the statistical behaviour of such regions because memory resources and the need for reliable statistical estimates place practical limits on the size of the context neighbourhoods, $\mathcal{N}$, which may be used. Run-length coding schemes have been adapted to handle this very situation.

The most basic run-length coding scheme may be understood by considering a one dimensional sequence, $\{x_n\}$, of symbols from an alphabet, $\mathcal{A}_X$. Run-length coding replaces $\{x_n\}$ by a sequence of symbol pairs, $\{(a_k, r_k)\}$, representing symbol values, $a_k \in \mathcal{A}_X$, and run-lengths, $r_k \in \mathbb{Z}_+$. The mapping between $\{(a_k, r_k)\}$ and $\{x_n\}$ is obvious; namely, $x_n = a_k$ for all n such that

$$\sum_{j=1}^{k-1} r_j < n \le \sum_{j=1}^{k} r_j$$

where $k = 1, 2, \ldots$ and $n = 1, 2, \ldots$. The value, r_k is normally the longest run of symbols, x_n, $n > \sum_{j=1}^{k-1} r_j$, such that x_n has a constant value, a_n.

The sequence of run-length symbol pairs, $\{(a_k, r_k)\}$, is usually coded using a Huffman code, although arithmetic coding may also be used. In the simplest case, separate codes are constructed for the symbol values,

---

[9] The original lossless algorithm described by the JPEG standard is not to be confused with its more efficient successor, known as JPEG-LS.

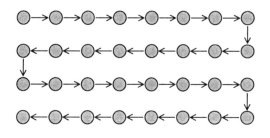

Figure 2.14. A suitable scanning pattern for applying 1D run-length coding to image data.

a_k, and the run-lengths, r_k. When the alphabet is small and the distribution of run-lengths is expected to vary for different symbols, then the pairs, (a_k, r_k), should be coded jointly. Equivalently, we may code a_k and then r_k conditioned on the context established by a_k, since

$$H((A_k, R_k)) = H(A_k) + H(R_k \mid A_k)$$

This is most commonly done for binary images, in which case runs of the "0" symbol are referred to as "black" runs, while runs of the "1" symbol are referred to as "white" runs. Separate conditional distributions are estimated for the black and white runs, from which separate codes are optimized for the black and white run lengths. Note that in the binary case the symbol values, a_k, may not need to be coded provided we can guarantee that we have alternating sequences of black and white runs; i.e., $a_k = 1 - a_{k-1}$. The situation often becomes slightly more complex, as a result of the practical necessity to impose a limit on the maximum run length.

The one dimensional run-length coding schemes discussed above may be applied to images by following an appropriate scanning pattern, such as that illustrated in Figure 2.14. Further improvement in coding efficiency, however, is often achievable by explicitly modifying the run-length coding scheme to exploit statistical dependencies between scan lines. Facsimile codes use such an approach, in which the end of each white run may be specified relative to the beginning of that run on the current line (regular one dimensional run-length coding) or relative to the beginning or end of the nearest white run on the previous scan line. For further information concerning such "two dimensional" run-length coding schemes, the reader is referred to [35].

2.4.4 QUAD-TREE CODING

Quad-tree coding shares many features with run-length coding, but is intended to exploit multi-dimensional dependencies more efficiently.

The idea is most easily explained and most commonly applied with bi-level images, where one symbol, say "0", occurs much more frequently than the other, "1". For simplicity, we assume that the image is square with dimensions $2^T \times 2^T$ for some $T \in \mathbb{Z}_+$.

Let $x[\mathbf{n}] \in \{0,1\}$ denote the image sample values, defined over $0 \leq n_1, n_2 < 2^T$. Define quad-tree node values, $x^{(t)}[\mathbf{n}]$, at each level, t, in a quad-tree as follows. At $t = 0$, the leaf nodes are the image samples themselves; i.e., $x^{(0)}[\mathbf{n}] = x[\mathbf{n}]$. At higher levels in the tree, the node values are defined recursively through the relations

$$x^{(t+1)}[\mathbf{n}] = \max_{0 \leq k_1, k_2 < 2} x^{(t)}[2\mathbf{n} + \mathbf{k}], \quad 0 \leq n_1, n_2 < 2^{T-t}$$

Thus, at the root of the tree we have the single node value

$$x^{(T)}[\mathbf{0}] = \max_{0 \leq k_1, k_2 < 2^T} x[\mathbf{n}]$$

The idea is to emit the quad-tree node values, starting from the root and working down to the leaves of the tree, skipping any nodes whose value may be deduced from a higher level node in the tree. In particular, if $x^{(t+1)}[\mathbf{n}] = 0$, the definition given above implies that all four children, $x^{(t)}[2\mathbf{n} + \mathbf{k}]$, $0 \leq k_1, k_2 < 2$, must also be zero, so these nodes and their descendants will not contribute to the bit-stream. Since the "0" symbol is assumed to occur with high probability, we hope to encounter high level tree nodes whose value is zero, thereby coding a large block of zeros with a single binary digit.

Quad-tree coding may be implemented using the following algorithm, although recursive implementations are also possible and may be more natural. To simplify the description we define $x^{(T+1)}[\mathbf{0}] = 1$.

Quad-Tree Coder

For $t = T, \ldots, 1, 0$
    For each $\mathbf{n}$ over the range $0 \leq n_1, n_2 < 2^{T-t}$
    If $x^{(t+1)}\left[\left\lfloor \frac{n_1}{2} \right\rfloor, \left\lfloor \frac{n_2}{2} \right\rfloor\right] = 1$
        emit-bit$(x^{(t)}[\mathbf{n}])$

While simple and often more efficient than run-length coding, quad-tree coding can be substantially more memory intensive than the latter when T is large. To minimize this difficulty, the image may be divided into smaller blocks to which the quad-tree code is applied independently. Many other variants on the basic algorithm exist, some of which are explored later in this book. Embedded quad-tree codes for non-binary image data are explored in Section 8.3.4.

2.5   FURTHER READING

In Section 2.1.2 we introduced random variables as a tool for refering to the underlying statistical properties which govern the likelihood of different events. More rigorous mathematical treatments of random variables may be found in many texts, including [115]. More comprehensive treatments of the properties of entropy and entropy rate, coding theorems for Markov and more general random processes, and entropy-ergodic theorems may be found in [17], [67], [106] and [74], amongst others. An alternative treatment of arithmetic coding, along with a comprehensive list of references to articles and private communications may be found in [119]. For a useful tutorial article on adaptive probability estimation for arithmetic coders, the reader is referred to [53].

Chapter 3

QUANTIZATION

3.1 RATE-DISTORTION THEORY

Chapter 2 discussed entropy coding algorithms possessing the desirable feature that the data obtained from decompression are identical to the original data. That is, the compression algorithms described in that chapter are lossless. As mentioned in Chapter 1, some applications (such as certain medical imaging systems) require lossless compression, while other applications may tolerate some amount of distortion in the decompressed data in return for a smaller compressed representation. Quantization is the element of lossy compression systems responsible for reducing the precision of data in order to make them more compressible. In most lossy compression systems, it is the only source of distortion.

Chapter 2 introduced the concept of entropy, and established it as the fundamental bound on the performance of lossless compression. In this section, the rate-distortion function is introduced as the fundamental bound on the performance of quantization.

3.1.1 SOURCE CODES

Let $\{X_n\}$ be a discrete random process taking values in $\mathcal{A}_X$. A source code of length m, and size $M = 2^{Rm}$ with alphabet $\mathcal{A}_{\hat{X}}$ is a set of codewords (vectors) $\mathcal{C} = \{\hat{\mathbf{x}}_0, \hat{\mathbf{x}}_1, \ldots, \hat{\mathbf{x}}_{M-1}\}$ with $\hat{\mathbf{x}}_q \in \mathcal{A}_{\hat{X}}^m$ $q = 0, 1, \ldots, M - 1$. Given a data vector $\mathbf{x} = \mathbf{x}_{0:m}$, a source coder (vector quantizer) selects the index of the codeword in $\mathcal{C}$ that minimizes some distortion measure. That is

$$Q(\mathbf{x}) = \underset{q \in \{0,1,\ldots,M-1\}}{\operatorname{argmin}} \rho_m(\mathbf{x}, \hat{\mathbf{x}}_q) \tag{3.1}$$

The corresponding source decoder (dequantizer) is given by

$$\overline{Q^{-1}}(q) = \hat{\mathbf{x}}_q$$

In the discussion above, we have referred to Q as quantization, and Q^{-1} as dequantization. Often in the literature, the end-to-end behavior

$$\hat{\mathbf{x}} = \overline{Q^{-1}}(Q(\mathbf{x}))$$

is referred to as quantization. Even though Q is generally many-to-one, Q^{-1} is one-to-one. Thus, "the quantized version of $\mathbf{x}$" may refer to either q or $\hat{\mathbf{x}}_q$ and the ambiguity in terminology is generally not a problem.

The distortion measure employed in equation (3.1) is often chosen to be of the form

$$\rho_m(\mathbf{x}, \hat{\mathbf{x}}) = \frac{1}{m} \sum_{j=0}^{m-1} \rho(x_j, \hat{x}_j)$$

where ρ is a non-negative measure of distortion between a single sample from each vector. Measures of this form are referred to as *single letter distortion measures*.

The average distortion of a source code $\mathcal{C}$ is given by

$$d(\mathcal{C}) = E\left[\rho_m(\mathbf{X}_{0:m}, \hat{\mathbf{X}}_{0:m})\right] = E\left[\rho_m(\mathbf{X}_{0:m}, \overline{Q^{-1}}(Q(\mathbf{X}_{0:m})))\right] \quad (3.2)$$

Since the index of the codeword selected ($q = Q(\mathbf{x})$) can be represented with $\log_2 M$ bits, and each codeword represents m samples, the rate of the code is given by

$$R = \frac{\log_2 M}{m} \text{ bits/sample}$$

with a resulting compression ratio of

$$\frac{\log_2 \|\mathcal{A}_X\|}{R}$$

3.1.2   MUTUAL INFORMATION AND THE RATE-DISTORTION FUNCTION

From Section 2.1.2, the entropy and conditional entropy of discrete random variables X and Y are given by

$$H(X) = E\left[-\log_2 f_X(X)\right]$$

and

$$H(X|Y) = E\left[-\log_2 f_{X|Y}(X, Y)\right]$$

respectively.

$H(X)$ was established as the average information conveyed (equivalently, the average uncertainty removed) by learning the value of X. Similarly, the conditional entropy is the average uncertainty remaining about X after learning the value of Y. The mutual information between X and Y is then defined by

$$I(X;Y) = H(X) - H(X|Y)$$

and is the difference in uncertainty about X before and after learning Y. In other words, it is the average information conveyed (uncertainty removed) about X by learning Y. Since $H(X|Y) \leq H(X)$, it is clear that $I(X;Y) \geq 0$ with equality if and only if X and Y are independent.

Manipulating the expression for $I(X;Y)$, we have

$$\begin{aligned}
I(X;Y) &= H(X) - H(X|Y) \\
&= -\sum_x f_X(x) \log_2 f_X(x) \\
&\quad + \sum_y f_Y(y) \sum_x f_{X|Y}(x,y) \log_2 f_{X|Y}(x,y) \\
&= \sum_x \sum_y f_{X,Y}(x,y) \log_2 \frac{f_{X,Y}(x,y)}{f_X(x) f_Y(y)} \qquad (3.3) \\
&= \sum_x \sum_y f_{Y|X}(y,x) f_X(x) \log_2 \frac{f_{Y|X}(y,x)}{f_Y(y)} \qquad (3.4)
\end{aligned}$$

From equation (3.3), it is clear that $I(X;Y) = I(Y;X)$. Also, noting that $f_Y(y) = \sum_x f_{Y|X}(y,x) f_X(x)$, we conclude that for a given PMF f_X, the mutual information is a function of the conditional PMF $f_{Y|X}$.

Now, given a single letter distortion measure, and an IID process $\{X_n\}$ with marginal PMF f_X, the rate-distortion function is given by

$$R(D) = \inf_{f_{\hat{X}|X} \in \mathcal{F}_D} I(X;\hat{X})$$

where

$$\mathcal{F}_D = \left\{ f_{\hat{X}|X} : \sum_x \sum_{\hat{x}} f_{\hat{X}|X}(\hat{x},x) f_X(x) \rho(x,\hat{x}) \leq D \right\}$$

The importance of the rate-distortion function is summed up in the two theorems given below. These theorems are offered without rigorous proof. The interested reader is referred to [24, 67, 43]. In these theorems, it is assumed that ρ_m is a single letter distortion measure with $\rho(x, \hat{x}) < \infty$, $\forall (x, \hat{x}) \in \mathcal{A}_X \times \mathcal{A}_{\hat{X}}$. It is also assumed that for every $x \in \mathcal{A}_X$ there is at least one $\hat{x} \in \mathcal{A}_{\hat{X}}$ such that $\rho(x, \hat{x}) = 0$.

Theorem 3.1 *Source Coding Theorem*

For any $\varepsilon > 0$ and any $D \geq 0$, there exists an integer m such that a source code $\mathcal{C}$, of length m, exists having average distortion $d(\mathcal{C}) \leq D$ and rate $R < R(D) + \varepsilon$.

Theorem 3.2 *Converse Source Coding Theorem*

For all $D \geq 0$, there exists no source code $\mathcal{C}$ with average distortion $d(\mathcal{C}) < D$ and rate $R < R(D)$.

Intuition into these results can be obtained by noting that a source coder is a deterministic mapping (i.e., $\hat{\mathbf{X}}_{0:m}$ is uniquely determined as $Q(\mathbf{X}_{0:m})$). Thus, $H(\hat{\mathbf{X}}_{0:m}|\mathbf{X}_{0:m}) = 0$ which yields

$$
\begin{aligned}
I(\mathbf{X}_{0:m}; \hat{\mathbf{X}}_{0:m}) &= I(\hat{\mathbf{X}}_{0:m}; \mathbf{X}_{0:m}) \\
&= H(\hat{\mathbf{X}}_{0:m}) - H(\hat{\mathbf{X}}_{0:m}|\mathbf{X}_{0:m}) \\
&= H(\hat{\mathbf{X}}_{0:m})
\end{aligned}
$$

Thus, $I(\hat{\mathbf{X}}_{0:m}; \mathbf{X}_{0:m}) = H(\hat{\mathbf{X}}_{0:m})$ is the information required to represent $\hat{\mathbf{X}}_{0:m}$ (the quantized version of $\mathbf{X}_{0:m}$) and should be minimized under the constraint that the average distortion should not exceed D. That is,

$$
\begin{aligned}
d(\mathcal{C}) &= E\left[\rho_m(\mathbf{X}_{0:m}, \overline{Q^{-1}}(Q(\mathbf{X}_{0:m})))\right] \\
&= \sum_{\mathbf{x} \in \mathcal{A}_X^m} \sum_{\hat{\mathbf{x}} \in \mathcal{C}} f_{\hat{\mathbf{X}}_{0:m}|\mathbf{X}_{0:m}}(\hat{\mathbf{x}}|\mathbf{x}) f_{\mathbf{X}_{0:m}}(\mathbf{x}) \rho_m(\mathbf{x}, \hat{\mathbf{x}}) \leq D
\end{aligned}
$$

Comparing to the definition of the rate-distortion function, we see that the rate-distortion function is the scalar version of the vector expressions above. It is interesting to note that in the vector version, the deterministic nature of Q implies that $f_{\hat{\mathbf{X}}_{0:m}|\mathbf{X}_{0:m}}(\hat{\mathbf{x}}|\mathbf{x})$ is necessarily either 0 or 1, while the resulting marginal distribution, $f_{\hat{X}|X}(\hat{x}|x)$ is not so constrained.

3.1.3   CONTINUOUS RANDOM VARIABLES

Extension of the theory from discrete random variables to continuous random variables is straightforward. In this case, $\mathcal{A}_X = \mathcal{A}_{\hat{X}} = \mathbb{R}$ and the entropy (as defined previously) is generally infinite. Replacing probability mass functions by probability density functions (PDF) and sums by integrals, the *differential entropy* is defined by

$$h(X) = -E\left[\log_2 f_X(X)\right] = -\int f_X(x)\log_2 f_X(x)dx$$

where f_X is the PDF of X. Differential entropy is a relative measure of uncertainty and, unlike entropy for discrete random variables, can be negative. $h(X)$ can be thought of as being relative to the case when X is distributed uniformly on 0 to 1, for which $h(X) = 0$.

Example 3.1 *Let X be Gaussian with mean 0 and variance σ^2. Then,*

$$f_X(x) = \frac{1}{\sqrt{2\pi\sigma^2}}e^{-x^2/2\sigma^2}$$

and

$$
\begin{aligned}
h(X) &= -\int f_X(x)\log_2 f_X(x)dx \\
&= \int \frac{f_X(x)\left[\frac{x^2}{2\sigma^2} + \ln\sqrt{2\pi\sigma^2}\right]}{\ln 2}dx \qquad (3.5) \\
&= \frac{\left[\frac{1}{2} + \ln\sqrt{2\pi\sigma^2}\right]}{\ln 2} \\
&= \frac{1}{2}\log_2 2\pi e\sigma^2 \qquad (3.6)
\end{aligned}
$$

In fact, the same result is obtained in the example above when the mean is not zero. More generally, the differential entropy of any random variable is unaffected by the addition of a constant to that random variable (change of mean). On the other hand, scaling a random variable by any constant a (change of variance) will add $\log_2|a|$ bits to the differential entropy.

An interesting property of the Gaussian distribution is that it has the highest differential entropy of any continuous distribution of a given variance.

Theorem 3.3 *Let X and Y be random variables each of variance σ^2 and let X be Gaussian. Then $h(X) \geq h(Y)$.*

Proof. Without loss of generality, assume X and Y are zero mean. Then

$$h(Y) - h(X) = h(Y) + \int f_X(x) \log_2 f_X(x) dx$$

$$= h(Y) + \int f_X(x) \left[c_1 x^2 + c_2 \right] dx$$

$$= h(Y) + c_1 \sigma^2 + c_2$$

$$= h(Y) + \int f_Y(y) \left[c_1 y^2 + c_2 \right] dy$$

$$= h(Y) + \int f_Y(y) \log_2 f_X(y) dy$$

$$= \int f_Y(y) \log_2 \frac{f_X(y)}{f_Y(y)} dy$$

where c_1 and c_2 are as in equation (3.5). Applying Jensen's inequality, which states that $E[g(Y)] \leq g(E[Y])$ for convex $\cap$ (concave) functions g,

$$h(Y) - h(X) \leq \log_2 \int f_Y(y) \frac{f_X(y)}{f_Y(y)} dy$$

$$= \log_2 \int f_X(y) dy \leq \log_2 1 = 0$$

The inequality in the second line above results from the fact that the last integral is over the support of f_Y (which may not contain the support of f_X). ∎

Conditional differential entropy is defined as

$$h(X|Y) = -E \left[\log_2 f_{X|Y}(X, Y) \right]$$

$$= - \iint f_{X,Y}(x, y) \log_2 f_{X|Y}(x, y) dx dy$$

and mutual information is defined as

$$I(X; Y) = h(X) - h(X|Y)$$

It should be noted that $I(X; Y)$ is interpreted as in the discrete case. It is an absolute (i.e., not relative) measure of the information that Y provides about X.

For an IID process $\{X_n\}$, and single letter distortion measure, the rate-distortion function is given as

$$R(D) = \inf_{f_{\hat{X}|X} \in \mathcal{F}_D} I(X; \hat{X})$$

where

$$\mathcal{F}_D = \left\{ f_{\hat{X}|X} : \iint f_{\hat{X}|X}(x, \hat{x}) f_X(x) \rho(x, \hat{x}) dx d\hat{x} \leq D \right\}$$

This is the obvious generalization from the discrete case, obtained by substituting PDFs for PMFs, and integrals for sums. Theorems similar to Theorem 3.1 and Theorem 3.2 can be proven for a large class of distortion measures and source distributions, and hence, in the continuous case, the rate-distortion function is again the fundamental limit on performance.

It can be shown [24] that the rate-distortion function is convex $\cup$, continuous and monotonically decreasing on the interval $(0, D_{max})$ where D_{max} is some value of D after which $R(D) = 0$. Hence the rate-distortion function has an inverse which is called the distortion-rate function. The obvious interpretation of the distortion-rate function is that it is the theoretical limit on distortion given a desired encoding rate. The rate-distortion function and the distortion-rate function will be used interchangeably throughout the text as the choice of one or the other is largely a matter of convenience.

A key point in the development of any source coding scheme is the choice of the distortion measure ρ. The most widely used measure is squared error with $\rho(x_i, \hat{x}_i) = (x_i - \hat{x}_i)^2$, and hence, average distortion (equation (3.2)) becomes mean-squared-error (MSE). It should be noted that for continuous random variables coded at a rate of R bits/sample, the compression ratio is infinite, and clearly not meaningful. In this case, the more relevant measure of performance is R for a given distortion D or vice versa.

Theorem 3.4 *Shannon Lower Bound*
For an IID process with variance σ^2, the MSE rate-distortion function is lower bounded by

$$R_L(D) = h(X) - \frac{1}{2} \log_2 2\pi e D \qquad (3.7)$$

Proof. Without loss of generality, assume X is zero mean. We then seek to minimize $I(X; \hat{X})$ subject to the constraint that $E\left[(X - \hat{X})^2\right] \leq D$. Now,

$$\begin{aligned} I(X; \hat{X}) &= h(X) - h(X|\hat{X}) & (3.8) \\ &= h(X) - h(X - \hat{X}|\hat{X}) \\ &\geq h(X) - h(Z) \end{aligned}$$

where $Z = X - \hat{X}$. The second equality is easily established, and follows from the fact that given $\hat{X}$, $X - \hat{X}$ has the same distribution as X (but different mean) and thus the same differential entropy. The inequality follows from $h(Z|\hat{X}) \leq h(Z)$. Now, for a given value of σ_Z^2, $E\left[(X - \hat{X})^2\right] = E\left[Z^2\right] = \sigma_Z^2 + \mu_Z^2$ is smallest when $\mu_Z = 0$. Also, for a given value of σ_Z^2, $h(Z)$ is maximum when Z is Gaussian (Theorem 3.3).

Thus, for $E\left[(X - \hat{X})^2\right] \leq D$, we have

$$I(X; \hat{X}) \geq h(X) - \frac{1}{2} \log_2 2\pi e D$$

■

Corollary 3.5 *For an IID Gaussian process with variance σ^2, the MSE rate-distortion function is given by*

$$
\begin{aligned}
R(D) &= \frac{1}{2} \log_2 2\pi e \sigma^2 - \frac{1}{2} \log_2 2\pi e D \\
&= \frac{1}{2} \log_2 \frac{\sigma^2}{D}
\end{aligned}
\tag{3.9}
$$

Proof. Choose $f_{\hat{X}|X}$ so that X and $\hat{X}$ are jointly Gaussian with

$$f_{X|\hat{X}}(x, \hat{x}) = \frac{1}{\sqrt{2\pi D}} e^{-(x - \hat{x})^2 / 2D}$$

From Example 3.1, we then have

$$
\begin{aligned}
I(X; \hat{X}) &= h(X) - h(X|\hat{X}) \\
&= \frac{1}{2} \log_2 2\pi e \sigma^2 - \frac{1}{2} \log_2 2\pi e D \\
&= \frac{1}{2} \log_2 \frac{\sigma^2}{D}
\end{aligned}
$$

Furthermore,

$$
\begin{aligned}
E\left[\left(X - \hat{X}\right)^2\right] &= \iint (x - \hat{x})^2 f_{X|\hat{X}}(x, \hat{x}) \, dx f_{\hat{X}}(\hat{x}) \, d\hat{x} \\
&= \int D f_{\hat{X}}(\hat{x}) \, d\hat{x} \\
&= D
\end{aligned}
$$

Thus, $I(X; \hat{X})$ achieves the minimum possible value (as given in Theorem 3.4) subject to the constraint that $E\left[(X - \hat{X})^2\right] \leq D$. Hence, $R(D) = \frac{1}{2} \log_2 \frac{\sigma^2}{D}$. It is easily verified, and worth noting, that the correlation coefficient between X and $\hat{X}$ is given by $r = \sqrt{1 - D/\sigma^2}$. Also, $E[\hat{X}] = E[X]$ and $\sigma_{\hat{X}}^2 = \sigma^2 - D$. As should be expected, when $D \longrightarrow 0$, we have $r \longrightarrow 1$ and $\sigma_{\hat{X}}^2 \longrightarrow \sigma^2$. ■

Note that if $D \geq \sigma^2$, $R(D) = 0$. In this case, fixing $\hat{X} = E[X]$ yields $E\left[(X - \hat{X})^2\right] = \sigma^2 \leq D$. Inverting the rate-distortion function, we get the distortion-rate function

$$D(R) = \sigma^2 2^{-2R} \tag{3.10}$$

Table 3.1.   Entropy power by distribution.

Distribution	Entropy Power
Uniform	$\frac{6}{\pi e}\sigma^2 \cong 0.703\sigma^2$
Laplacian	$\frac{e}{\pi}\sigma^2 \cong 0.865\sigma^2$
Gaussian	σ^2

The theoretical bound on signal-to-noise ratio (SNR) is then given by

$$SNR_{D(R)} = 10\log_{10}\frac{\sigma^2}{D(R)}$$
$$= 20R\log_{10}2$$
$$\cong 6.02R \text{ dB}$$

For most distributions, the rate-distortion function cannot be expressed in closed form. In these cases, it must be computed numerically. While most numerical optimization techniques can be used, the iterative technique of [26, 21] is particularly elegant.

The rate-distortion function for any IID distribution can be bounded by

$$R_L(D) \leq R(D) \leq \frac{1}{2}\log_2\frac{\sigma^2}{D} \qquad (3.11)$$

The right hand expression is the Gaussian rate-distortion function, while the expression on the left is the Shannon lower bound as given by equation (3.7) The Shannon lower bound is known to be tight for small D (large R). That is, $D_L(R) \cong D(R)$ when R is large.

Inverting the expressions of equations (3.11) and (3.7), we get

$$D_L(R) = \frac{1}{2\pi e}2^{2h(X)}2^{-2R} \leq D(R) \leq \sigma^2 2^{-2R} \qquad (3.12)$$

The quantity $\frac{1}{2\pi e}2^{2h(X)}$ is known as the entropy power, and has maximum value of σ^2 when X has the Gaussian distribution. Expressions for the entropy power of the uniform, Laplacian (two-sided exponential), and Gaussian (normal) distributions are given in Table 3.1.

3.1.4   CORRELATED PROCESSES

In general, the rate-distortion function is extremely difficult to compute for correlated processes. A notable exception is for stationary Gaussian processes. Assume a mean of zero and denote the autocorrelation function of such a process by

$$R_X[k] = E[X_n X_{n+k}]$$

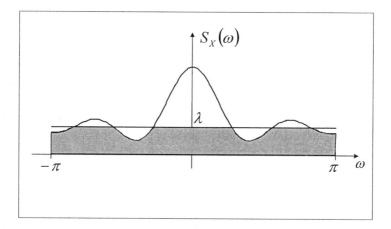

Figure 3.1. Computation of D as water filling. The shaded region has area $2\pi D$.

The power spectral density of the process is the Fourier transform of $R_X[k]$,

$$S_X(\omega) = \sum_{k=-\infty}^{\infty} R_X[k]\, e^{-jk\omega}$$

The rate-distortion function is then given in parametric form [24] by

$$D(\lambda) = \frac{1}{2\pi} \int_{-\pi}^{\pi} \min\{\lambda, S_X(\omega)\}\, d\omega$$

and

$$R(\lambda) = \frac{1}{2\pi} \int_{-\pi}^{\pi} \max\left\{0, \frac{1}{2}\log_2 \frac{S_X(\omega)}{\lambda}\right\} d\omega$$

The rate-distortion function is computed from these expressions by varying $\lambda \in \left(0, \max_{\omega}\{S_X(\omega)\}\right]$ to obtain values of D and the corresponding values of $R(D)$. The computation of D is often described as "water filling." As can be seen in Figure 3.1, D can be pictured as the "area of water" obtained when $S_X(\omega)$ is "filled" to the level of λ. Computation of R can be viewed as a similar water filling but on a level shifted version of $\frac{1}{2}\log_2 S_X(\omega)$.

Note that when $\lambda = \max_{\omega}\{S_X(\omega)\}$, $R(\lambda) = 0$ and $D(\lambda) = D_{\max} = \frac{1}{2\pi}\int_{-\pi}^{\pi} S_X(\omega)\, d\omega = \sigma^2$. Finally, we note that for $\lambda \leq \min_{\omega}\{S_X(\omega)\}$ (small D, large R),

$$D(\lambda) = \lambda \tag{3.13}$$

and

$$R(\lambda) = \frac{1}{2\pi} \int_{-\pi}^{\pi} \frac{1}{2} \log_2 \frac{S_X(\omega)}{\lambda} d\omega$$

$$= \frac{1}{2\pi} \int_{-\pi}^{\pi} \frac{1}{2} \log_2 S_X(\omega) d\omega - \frac{1}{2} \log_2 \lambda \qquad (3.14)$$

Solving equation (3.14) for λ and substituting into equation (3.13) yields

$$D(R) = \gamma_X^2 \sigma^2 2^{-2R} \qquad (3.15)$$

where

$$\gamma_X^2 = \frac{1}{\sigma^2} \exp\left[\frac{1}{2\pi} \int_{-\pi}^{\pi} \ln S_X(\omega) d\omega\right]$$

is known as the spectral flatness measure of $\{X_n\}$. It is easily shown that $\gamma_X^2 \leq 1$ with equality if and only if $S_X(\omega) = \sigma^2 \ \forall \omega$ (i.e., $\{X_n\}$ is IID).

3.2 SCALAR QUANTIZATION

The scalar quantizer (SQ) is the simplest of all lossy compression schemes. It can be described as a function that maps each element in a subset of the real line to a particular value in that subset. Consider partitioning the real line into M disjoint intervals

$$\mathcal{I}_q = [t_q, t_{q+1}), q = 0, 1, \ldots, M - 1$$

with

$$-\infty = t_0 < t_1 < \cdots < t_M = +\infty.$$

Within each interval, a point $\hat{x}_q$ is selected as the output value (or codeword) of $\mathcal{I}_q$. A scalar quantizer is then a mapping from $\mathbb{R}$ to $\{0, 1, \ldots, M - 1\}$. Specifically, for a given x, $Q(x)$ is the index q of the interval $\mathcal{I}_q$ which contains x. The dequantizer is given by

$$\overline{Q^{-1}}(q) = \hat{x}_q$$

Example 3.2 *Let $M = 4$, $t_1 = -1$, $t_2 = 0$, $t_3 = 1$, $\hat{x}_0 = -1.5$, $\hat{x}_1 = -0.5$, $\hat{x}_2 = 0.5$, $\hat{x}_3 = 1.5$. Then, if $x < -1$, the quantized version of x is -1.5 (index $= 0$). Specifically, $Q(x) = 0$ and $\overline{Q^{-1}}(Q(x)) = \overline{Q^{-1}}(0) = -1.5$. Similarly, if $0 \leq x < 1$, the quantized version of x is 0.5 (index $= 2$). This situation is illustrated in Figure 3.2.*

A different depiction of the quantizer from Example 3.2 is found in Figure 3.3a. The more general case is shown in Figure 3.3b. This figure shows that when $x \in \mathcal{I}_q = [t_q, t_{q+1})$, that $\overline{Q^{-1}}(Q(x)) = \overline{Q^{-1}}(q) = \hat{x}_q$. Clearly, the t_q can be thought of as thresholds, or decision boundaries for the $\hat{x}_q$.

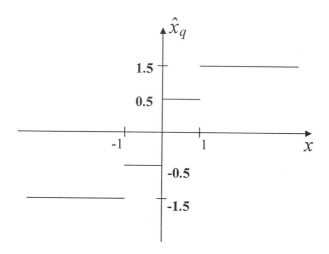

Figure 3.2. Scalar quantizer of Example 3.2.

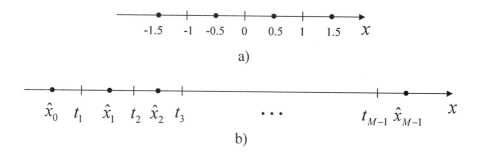

Figure 3.3. Alternate graphical representation of scalar quantization.

3.2.1 THE LLOYD-MAX SCALAR QUANTIZER

In this section, we develop necessary conditions for the optimality of scalar quantizers of a fixed size M. Quantizers satisfying these conditions are known as Lloyd-Max quantizers [97, 105]. Here we assume a stationary process with marginal PDF f_X and the MSE criterion for measuring distortion is adopted. Using the notation developed in previous subsections,

$$d = \text{MSE} = E\left[(X - \hat{X})^2\right] = \sum_{k=0}^{M-1} E\left[(X - \hat{X})^2 | X \in \mathcal{I}_k\right] P(X \in \mathcal{I}_k)$$

$$= \sum_{k=0}^{M-1} \int_{t_k}^{t_{k+1}} (x - \hat{x}_k)^2 f_X(x) dx \qquad (3.16)$$

Setting the partial derivative of d with respect to t_q equal to zero yields $(t_q - \hat{x}_{q-1})^2 f_x(t_q) - (t_q - \hat{x}_q)^2 f_x(t_q) = 0$. Solving for t_q yields

$$t_q = \frac{\hat{x}_{q-1} + \hat{x}_q}{2} \qquad q = 1, 2, \dots, M - 1 \qquad (3.17)$$

Similarly, differentiating with respect to $\hat{x}_q$ yields

$$\hat{x}_q = \frac{\int_{t_q}^{t_{q+1}} x f_X(x) dx}{\int_{t_q}^{t_{q+1}} f_X(x) dx} \qquad q = 0, 1, \dots, M - 1 \qquad (3.18)$$

Equations (3.17) and (3.18) then form necessary conditions for an optimal scalar quantizer.

Equation (3.17) implies that the endpoints of the quantizer decision regions should be halfway between output points. This implies nearest neighbor encoding, which means that the input x is encoded as $\hat{x}_q$, where $\hat{x}_q$ is the codeword closest to x. The denominator of equation (3.18) is the probability that X lies in $\mathcal{I}_q$ and hence, $\hat{x}_q = E[X | X \in \mathcal{I}_q]$. Codewords satisfying this property are called conditional means, or *centroids*.

Easily proven consequences of these properties are that

$$E[(X - \hat{X})] = 0$$

$$\sigma_{\hat{X}}^2 = \sigma_X^2 - E[(X - \hat{X})^2]$$

$$E\left[\left(X - \hat{X}\right)\hat{X}\right] = 0$$

$$E[(X - \hat{X})^2 | X \in \mathcal{I}_j]p_j = E[(X - \hat{X})^2 | X \in \mathcal{I}_q]p_q \quad \forall j, q$$

where $p_q = P(X \in \mathcal{I}_q)$. These properties state respectively, that the quantization error is zero mean (whether or not $E[X] = 0$), quantization reduces the data variance by an amount equal to the MSE. The quantization error is uncorrelated with the quantizer output (but not the input), and that on average, the contribution of all intervals (toward the overall MSE) are equal.

As necessary conditions, equations (3.17) and (3.18) do not guarantee optimality. In fact, examples of suboptimal quantizers satisfying equations (3.17) and (3.18) are easily constructed. Optimality is assured however, for all sources having log-concave PDFs (i.e., $\log f_X$ is a concave, or convex $\cap$, function) [61]. Uniform, Laplacian, and Gaussian distributions all satisfy this property and hence, their corresponding Lloyd-Max quantizers are optimal (as scalar quantizers).

Solving equations (3.17) and (3.18) in closed form is usually quite difficult. One notable exception occurs for symmetric distributions and $M = 2$. Let μ be the point of symmetry, i.e., $f_X(x+\mu) = f_X(-x+\mu) \; \forall x$. Then, $t_1 = \mu$, $\hat{x}_0 = E[X|X < \mu]$, $\hat{x}_1 = E[X|X > \mu] = 2\mu - \hat{x}_0$. Another exception occurs when X is distributed uniformly. Specifically, suppose that X is uniform on $[a, b)$. Then the Lloyd-Max quantizer partitions $[a, b)$ evenly into M intervals

$$\mathcal{I}_q = \left[a + \frac{(b-a)}{M} q, a + \frac{(b-a)}{M} (q+1) \right) \quad q = 0, 1, \ldots, M-1 \quad (3.19)$$

and $\hat{x}_q$ is the center of each such interval. Quantizers of this form (equal length decision regions, with their centers as output points) are known as uniform scalar quantizers.

For many other cases, solution of equations (3.17) and (3.18) must proceed iteratively. The following algorithm can be used to that end.

The Max Algorithm (Lloyd Form II)

1) Choose an initial code $\mathcal{C} = \{\hat{x}_0, \hat{x}_1, \ldots, \hat{x}_{M-1}\}$, set $j = 1, d_0 = \infty$
2) Calculate the t_q according to equation (3.17)
3) Calculate the $\hat{x}_q$ according to equation (3.18)
4) Calculate d_j according to equation (3.16)
5) If $\frac{d_{j-1}-d_j}{d_j} < \varepsilon$ stop; else set $j = j + 1$ and go to step 2

For a given set of $\hat{x}_q$, the t_q computed in step 2 are optimal. Thus, step 2 is guaranteed not to increase distortion. Similarly, for the t_q computed in step 2, the $\hat{x}_q$ computed in step 3 will not increase the distortion. Thus, the sequence of distortions d_j, $j = 1, 2, \ldots$ is monotone non-increasing (and bounded below by 0). Thus, the algorithm is guaranteed to converge. In the case of a log-concave PDF, the resulting code is optimal. Otherwise, the resulting code is at least locally optimal.

This algorithm can be rewritten in the form shown below. This form is easily adapted to estimation of the optimal quantizer in the event that the PDF is unknown, but sample data from $\{X_n\}$ are available.

The Lloyd Algorithm (Form I)

1) Choose an initial code $\mathcal{C} = \{\hat{x}_0, \hat{x}_1, \ldots, \hat{x}_{M-1}\}$, set $j = 1, d_0 = \infty$
2) Let $\mathcal{I}_q = \left\{ x \in \mathbb{R} : (x - \hat{x}_q)^2 < (x - \hat{x}_k)^2 \; \forall \, k \neq q \right\} \; q = 0, 1, \ldots,$
   $M - 1$

3) $\hat{x}_q = E[X|X \in \mathcal{I}_q]$ $q = 0, 1, \ldots, M-1$

4) $d_j = \sum_{q=0}^{M-1} E\left[(X - \hat{x}_q)^2 | X \in \mathcal{I}_q\right] P(X \in \mathcal{I}_q)$

5) If $\frac{d_{j-1} - d_j}{d_j} < \varepsilon$ stop; else set $j = j+1$ and go to step 2

In step 2, we ignore the zero probability event that $(x - \hat{x}_q)^2 = (x - \hat{x}_k)^2$ for $k \neq q$. In this case, x can be assigned to either set with no effect on MSE

Now, in the case where the PDF of $\{X_n\}$ is unknown, but a set of statistically representative "training data" $\mathcal{T}$ is available, the algorithm can be modified by substituting sample averages for expectations and relative frequencies for probabilities. In the limit as the size of the training set goes to infinity, the weak law of large numbers implies that these statistical substitutes will converge to the true underlying ensemble values. The resulting algorithm is as follows.

The Lloyd Algorithm (with Training Data)

1) Given a training set of samples from $\{X_n\}$, say $\mathcal{T} = \{x_0, x_1, \ldots, x_{\|\mathcal{T}\|-1}\}$, choose an initial code $\mathcal{C} = \{\hat{x}_0, \hat{x}_1, \ldots, \hat{x}_{M-1}\}$, set $j = 1$, $d_0 = \infty$

2) Let $\mathcal{B}_q = \{x \in \mathcal{T} : Q(x) = q\} = \{x \in \mathcal{T} : (x - \hat{x}_q)^2 < (x - \hat{x}_k)^2 \ \forall k \neq q\}$ $q = 0, 1, \ldots, M-1$

3) $\hat{x}_q = \frac{1}{\|\mathcal{B}_q\|} \sum_{x \in \mathcal{B}_q} x$ $q = 0, 1, \ldots, M-1$

4) $d_j = \frac{1}{\|\mathcal{T}\|} \sum_{x \in \mathcal{T}} (x - \overline{Q^{-1}(Q(x))})^2 = \frac{1}{\|\mathcal{T}\|} \sum_{q=0}^{M-1} \sum_{x \in \mathcal{B}_q} (x - \hat{x}_q)^2$

5) If $\frac{d_{j-1} - d_j}{d_j} < \varepsilon$ stop; else set $j = j+1$ and go to step 2

As before $d_j \leq d_{j-1}$ and convergence is ensured.

3.2.2   PERFORMANCE OF THE LLOYD-MAX SCALAR QUANTIZER

Table 3.2 gives the values of $\hat{x}_q$ for the zero mean, unit variance Uniform, Laplacian, and Gaussian PDFs at rates $R = 1, 2$, and 3 bits/sample. Table 3.3 gives the SNR values for Lloyd-Max quantization of these sources at rates up to 5 bits/sample. For more extensive tables, see [82].

For large R, it can be shown that the MSE of Lloyd-Max quantization behaves like

$$d(R) \cong \varepsilon^2 \sigma^2 2^{-2R} \tag{3.20}$$

Table 3.2.   Lloyd-Max quantizer reconstruction values.

		Uniform	Laplacian	Gaussian
	1	±0.866	±0.707	±0.798
R	2	±0.433, ±1.299	±0.420, ±1.834	±0.453, ±1.510
	3	±0.217, ±0.650,	±0.233, ±0.833,	±0.245, ±0.756,
		±1.083, ±1.516	±1.673, ±3.087	±1.344, ±2.152

Table 3.3.   MSE performance of Lloyd-Max quantizers (SNR in dB).

		Uniform	Laplacian	Gaussian
	1	6.02	3.01	4.40
	2	12.04	7.54	9.30
R	3	18.06	12.64	14.62
	4	24.08	18.13	20.22
	5	30.10	23.87	26.01

where ε^2 is a function of the particular PDF. For smooth, zero mean, symmetric PDFs

$$\varepsilon^2 \sigma^2 = \frac{2}{3} \left[\int_0^\infty \sqrt[3]{f_X(x)} dx \right]^3$$

The reader is referred to [114] for a proof of this fact. The values of ε^2 for the uniform, Laplacian, and Gaussian sources are given in Table 3.4 as 1, 9/2, and $\sqrt{3}\pi/2 \cong 2.721$, respectively.

It is worth noting that equation (3.20) is somewhat pessimistic at low rates. For example, equation (3.20) predicts 13.72 dB for the Gaussian PDF at $R = 3$, while the actual value from Table 3.3 is 14.62 dB. For $R = 6$, the values predicted by equation (3.20) are all within about 0.2 dB of the exact values. It is also worth noting that equation (3.20) provides the exact correct values for the uniform PDF at all rates.

We close this section by noting that Lloyd-Max scalar quantizers are in fact, special cases of the source codes described in Section 3.1.1. In this case, $m = 1, \mathcal{C} = \{\hat{x}_0, \ldots, \hat{x}_{M-1}\}, \rho(x, \hat{x}) = (x - \hat{x})^2$, and M is usually chosen as a power of 2 so that $\hat{x}_q$ can be signalled using an integer number $(R = \log_2 M)$ of bits.

3.2.3   ENTROPY CODED SCALAR QUANTIZATION

The Lloyd-Max quantizer minimizes the MSE subject to a constraint on the size of the code, M. The presumption is that $R = \log_2 M$ bits will be used to signal the codeword chosen by the quantizer. If $\log_2 M$ is an integer, this is straightforward. If $\log_2 M$ is not an integer, then $R = \lceil \log_2 M \rceil < \log_2 M + 1$ bits may be used to signal one index at a time. Alternatively, we can block L indices $q_0, q_1, \ldots, q_{L-1}$ together to form one "super-index" in $\{0, 1, \ldots, M^L - 1\}$. This super-index can then be signalled using

$$R = \frac{1}{L} \lceil \log_2 M^L \rceil < \log_2 M + \frac{1}{L} \text{ bits/index}$$

Thus, we see that for the Lloyd-Max quantizer $R = \log_2 M$ bits/sample can be approached arbitrarily closely with "fixed length" coding, even when M is not a power of 2.

We now consider the application of variable length codes such as Huffman, or arithmetic (Chapter 2) to the indices produced by a scalar quantizer. Such coding is lossless and has no effect on $Q^{-1}(q) = \overline{Q^{-1}(Q(x))}$. Thus, the MSE is unchanged. However, the rate of the resulting "entropy coded quantizer" can approach the entropy of the indices, which is equal to $H(\hat{X})$. For simplicity, we assume in what follows that efficient entropy coding is used, so that

$$R \cong H(\hat{X}) \leq \log_2 M \tag{3.21}$$

with the inequality achieving equality if and only if all quantizer outputs are equally likely (i.e., $f_{\hat{X}}(\hat{x}_q) = 1/M \quad q = 0, 1, \ldots, M - 1$).

In the case that $H(\hat{X}) < \log_2 M$, the Lloyd-Max quantizer is not the optimal quantizer to be used in the entropy coded scenario. The optimal quantizer minimizes MSE subject to a constraint on entropy. From equation (3.21), we see that we must have $M > 2^R$. M can be chosen arbitrarily largely without concern for performance. In fact, for unbounded random variables (e.g., Laplacian, Gaussian) $M = \infty$ is generally the optimal choice. If a particular choice of M is too large for a given PDF, the optimization techniques discussed below will result in some of the codewords having probability of zero, effectively reducing the value of M.

For a desired rate of R, we now seek to minimize

$$E\left[\left(X - \hat{X}\right)^2\right] = \sum_{q=0}^{M-1} \int_{t_q}^{t_{q+1}} (x - \hat{x}_q)^2 f_X(x) dx \tag{3.22}$$

subject to the constraint that

$$H(\hat{X}) = -\sum_{q=0}^{M-1} p_q \log_2 p_q \leq R \tag{3.23}$$

where

$$p_q = P\left(X \in \mathcal{I}_q\right) = \int_{t_q}^{t_{q+1}} f_X(x)dx$$

Using the technique of Lagrange multipliers, [54] we seek to minimize

$$J(\lambda) = E\left[(X - \hat{X})^2\right] + \lambda H(\hat{X}) \tag{3.24}$$

$$= \left[\sum_{j=0}^{M-1} \int_{t_j}^{t_{j+1}} (x - \hat{x}_j)^2 f_X(x)dx \right.$$

$$\left. -\lambda \sum_{j=0}^{M-1} \int_{t_j}^{t_{j+1}} f_X(x)dx \log_2 \int_{t_j}^{t_{j+1}} f_X(x)dx \right]$$

If there exists a $\lambda \geq 0$ such that the solution to the unconstrained minimization of $J(\lambda)$ yields $H(\hat{X}) = R$, the same solution will satisfy the constrained problem of equations (3.22) and (3.23). Setting $\frac{\partial}{\partial \hat{x}_q} J(\lambda) = 0$ reveals that $\hat{x}_q$ is the conditional mean of $\mathcal{I}_q$ as before (see equation (3.18)). Similarly, differentiating with respect to t_q yields

$$(t_q - \hat{x}_{q-1})^2 - (t_q - \hat{x}_q)^2 - \lambda\left(\log_2 p_{q-1} - \log_2 p_q\right) = 0 \tag{3.25}$$

For a given λ, these equations form the basis for an iterative algorithm which is the generalization of the Max Algorithm of Section 3.2.1. An outer loop can be employed to search for the proper λ such that $H(\hat{X}) = R$. Unfortunately, equation (3.25) can not be solved for t_q in terms of $\hat{x}_q$ and $\hat{x}_{q-1}$ as in equation (3.17), since p_{q-1} and p_q together depend on t_{q-1}, t_q, and t_{q+1}. However, another iterative algorithm can be used to solve for the t_q as a subroutine to the main iteration. The interested reader is referred to [55].

Rather than pursue this algorithm, we examine the generalization of the Lloyd Algorithm (Section 3.2.1) to the entropy constrained case. Here we try to minimize equation (3.24) directly rather than attempting to solve the necessary conditions. This results in the following iterative algorithm [39].

Entropy Coded Scalar Quantizer Design Algorithm

0) Choose an initial $\lambda \geq 0$

1) Choose an initial code $\mathcal{C} = \{\hat{x}_0, \hat{x}_1, \ldots, \hat{x}_{M-1}\}$, with initial probabilities $\mathcal{P} = \{p_0, p_1, \ldots, p_{M-1}\}$, and set $j = 1$, $d_0 = \infty$

2) Let $\mathcal{I}_q = \{x \in \mathbb{R} : (x - \hat{x}_q)^2 - \lambda \log_2 p_q < (x - \hat{x}_k)^2 - \lambda \log_2 p_k \ \forall \ k \neq q\}$ $q = 0, 1, \ldots, M - 1$

3) $\hat{x}_q = E[X | X \in \mathcal{I}_q]$ $q = 0, 1, \ldots, M - 1$

4) $p_q = P(X \in \mathcal{I}_q)$ $q = 0, 1, \ldots, M - 1$

5) $d_j = \sum\limits_{q=0}^{M-1} E\left[(X - \hat{x}_q)^2 | X \in \mathcal{I}_q\right] p_q$

6) If $\frac{d_{j-1} - d_j}{d_j} < \varepsilon$ go to step 7; else set $j = j + 1$ and go to step 2

7) If $H(\hat{X}) = -\sum\limits_{q} p_q \log_2 p_q \notin (R - \tau, R]$, adjust λ and go to step 1

Comments:

1) It is easily shown that $H(\hat{X})$ is non-increasing as a function of λ, which makes the search for the proper λ (such that $H(\hat{X}) = R$) straightforward. One simple approach is to choose two extreme values of λ and then use bisection to iteratively narrow the interval of possible λ.

2) For a given value of λ, the inner loop (steps 2–6) is non-increasing in d_j. Thus, convergence is ensured.

3) Adaptation to training data is straightforward with sample averages replacing expectations, and relative frequencies replacing probabilities in the obvious way, as in the Lloyd Algorithm of Section 3.2.1.

4) At convergence, λ, p_q, and $\hat{x}_q$ are known for all $q = 0, 1, \ldots, M-1$. Equation (3.25) is then easily solved for

$$t_q = \frac{\hat{x}_{q-1} + \hat{x}_q}{2} + \lambda \frac{\log_2 p_{q-1} - \log_2 p_q}{2(\hat{x}_q - \hat{x}_{q-1})}$$

and we see that the entropy constraint effectively introduces a bias to the nearest neighbor thresholds of equation (3.17).

3.2.4 PERFORMANCE OF ENTROPY CODED SCALAR QUANTIZATION

It can be shown that for large $H(\hat{X})$ (small MSE), the optimal scalar quantizer in the entropy coded case is uniform for all "smooth" PDFs [70]. As mentioned previously, the optimum number of levels is generally infinite. In the case of a zero mean, symmetric PDF, it is then convenient to let the quantizer indices range over all integers. The quantizer output values can then be written in the form $q\Delta, q = 0, \pm 1, \pm 2, \ldots$ and the thresholds are the midpoints between the output values. Thus, the quantization intervals are given by

$$\mathcal{I}_q = \left[q\Delta - \frac{\Delta}{2}, q\Delta + \frac{\Delta}{2} \right) \tag{3.26}$$

From equation (3.16),

$$d = \sum_{q=-\infty}^{\infty} \int_{q\Delta - \frac{\Delta}{2}}^{q\Delta + \frac{\Delta}{2}} (x - q\Delta)^2 f_X(x) dx$$

For small Δ, $f_X(x)$ is approximately constant within each interval. That is, $f_X(x) \cong f_X(q\Delta) \ \forall x \in \mathcal{I}_q$. So

$$d \cong \sum_{q=-\infty}^{\infty} f_X(q\Delta) \int_{q\Delta - \frac{\Delta}{2}}^{q\Delta + \frac{\Delta}{2}} (x - q\Delta)^2 \, dx = \sum_{q=-\infty}^{\infty} f_X(q\Delta) \frac{\Delta^3}{12}$$

$$= \frac{\Delta^2}{12} \sum_{q=-\infty}^{\infty} f_X(q\Delta)\Delta \cong \frac{\Delta^2}{12} \int_{-\infty}^{\infty} f_X(x) dx$$

$$= \frac{\Delta^2}{12} \tag{3.27}$$

Similarly,

$$H(\hat{X}) = - \sum_{q=-\infty}^{\infty} p_q \log_2 p_q$$

$$\cong - \sum_{q=-\infty}^{\infty} [f_X(q\Delta)\Delta] \log_2 [f_X(q\Delta)\Delta]$$

$$\cong - \int_{-\infty}^{\infty} f_X(x) \log_2 f_X(x) dx - \int f_X(x) \log_2 \Delta dx$$

$$= h(X) - \log_2 \Delta \tag{3.28}$$

As before, we assume efficient entropy coding so that $R \cong H(\hat{X})$. Then from equation (3.28)

$$\Delta \cong 2^{h(X)-R}$$

Substituting in equation (3.27) yields

$$d(R) \cong \frac{1}{12} 2^{2h(X)} 2^{-2R} \tag{3.29}$$

Comparing equation (3.29) to the Shannon lower bound of equation (3.12), we see that at high rates, entropy coded uniform scalar quantization differs from the distortion-rate function by a factor of only $\pi e/6$, or 1.53 dB.

At low rates, the optimal entropy coded scalar quantizer is no longer uniform. However, a quantizer with uniform intervals (equation (3.26)), but centroid codewords, $\hat{x}_q = E[X|X \in \mathcal{I}_q]$, is very nearly optimal [55]. Often, for zero mean PDFs, a small improvement in the $d(R)$ behavior can be obtained by widening the interval about 0. This interval, $\mathcal{I}_0$, is sometimes called the "zero-bin." Quantizers of this type are usually called "deadzone uniform scalar quantizers." Widening $\mathcal{I}_0$ increases the distortion somewhat, but often decreases $H(\hat{X})$ enough to offset this effect. The intervals of a deadzone quantizer are of the form

$$
\mathcal{I}_q = \begin{cases}
[-(1-\xi)\Delta, (1-\xi)\Delta) & q = 0 \\
[(q-\xi)\Delta, (q+1-\xi)\Delta) & q > 0 \\
[(q-1+\xi)\Delta, (q+\xi)\Delta) & q < 0
\end{cases} \tag{3.30}
$$

where $\xi < 1$ determines the width of $\mathcal{I}_0$.

This quantizer can be implemented as

$$
q = Q(x) = \begin{cases}
\text{sign}(x)\left\lfloor \frac{|x|}{\Delta} + \xi \right\rfloor & \frac{|x|}{\Delta} + \xi > 0 \\
0 & \text{otherwise}
\end{cases} \tag{3.31}
$$

Interesting special cases occur when $\xi = 0$, and $\xi = 1/2$. When $\xi = 1/2$, the previous case of equation (3.26) results, with $\mathcal{I}_0$ having a width of Δ. On the other hand, $\xi = 0$ results in a zero-bin width of 2Δ. This case is particularly important, and is discussed further in Section 3.2.7. Values of $\xi < 0$ are reasonable, and result in further widening of the deadzone.

As before, the optimal reconstruction values are centroids. However, for simplicity, some fixed value within $\mathcal{I}_q$ is often employed. In this case,

$$
\begin{aligned}
\hat{x}_q &= \begin{cases}
0 & q = 0 \\
(q-\xi+\delta)\Delta & q > 0 \\
(q+\xi-\delta)\Delta & q < 0
\end{cases} \\
&= \begin{cases}
0 & q = 0 \\
\text{sign}(q)\,(|q|-\xi+\delta)\Delta & q \neq 0
\end{cases}
\end{aligned}
$$

where $0 \leq \delta < 1$ specifies the placement of $\hat{x}_q$ within $\mathcal{I}_q$. The case with $\delta = 1/2$ yields $\hat{x}_q$ at the center of $\mathcal{I}_q$. The quantizer that results from $\xi = 0$ and $\delta = 1/2$ is depicted in Figure 3.4.

As in the Lloyd-Max case, the high rate asymptotic results are rather pessimistic if applied at low rates. Indeed, as R approaches 0, the MSE of a well designed quantizer should approach $D(0) = \sigma^2$, rather than $\left(2^{2h(X)}\right)/12$ as predicted by equation (3.29).

Figure 3.4.   Uniform scalar quantizer with deadzone.

3.2.5   SUMMARY OF SCALAR QUANTIZER PERFORMANCE

In summary, we see from equations (3.12), (3.20),and (3.29) that for high rates and IID data, the distortion-rate function, as well as the MSE of both Lloyd-Max quantization, and entropy coded uniform quantization, are all of the form

$$d(R) \cong \varepsilon^2 \sigma^2 2^{-2R} \qquad (3.32)$$

More generally,

$$d(R) = g(R)\sigma^2 2^{-2R} \qquad (3.33)$$

where $g(R)$ is a weak function of R. For large R, $g(R) \cong \varepsilon^2$. On the other hand, as R approaches 0, $g(R)$ approaches 1.

The appropriate values of ε^2 for each case are given in Table 3.4. From this table, we see that (as expected) entropy coding is of no benefit for the uniform PDF. This follows from the fact that for this PDF, ε^2 is the same for both Lloyd-Max and entropy coded quantizers. Also as expected, the Gaussian PDF boasts the largest value of ε^2 for the $D(R)$ and entropy-coded cases. This is in support of the previous statement that the distortion-rate function is largest in the Gaussian case. On the other hand, it is interesting to note that the Lloyd-Max MSE is larger for the Laplacian PDF than for the Gaussian PDF. This is indicative of a general trend that Lloyd-Max scalar quantization performance suffers for "heavy tailed" PDFs.

We conclude this subsection with Figure 3.5. This figure shows the distortion-rate function, as well as MSE performance for Lloyd-Max and entropy-coded quantization of IID Gaussian data. As R grows, all three curves become parallel straight lines with slopes of 6.02 dB/bit. The vertical gap between $D(R)$ and entropy coded quantization is $\pi e/6$, or 1.53 dB. The vertical gap between $D(R)$ and Lloyd-Max quantization is $\sqrt{3}\pi/2$, or 4.35 dB. Although not discussed above, the performance for uniform scalar quantization without entropy coding will result in less than 6.02 dB per bit improvement (for all but the uniform PDF). For these cases, the SNR curve will diverge from $D(R)$ as R grows, as shown in Figure 3.5.

Table 3.4. Values of ε^2 for various PDFs.

	$D(R)$	*Lloyd-Max*	*Entropy-Coded*
Uniform	$\frac{6}{\pi e} \cong 0.703$	1	1
Laplacian	$\frac{e}{\pi} \cong 0.865$	$\frac{9}{2}$	$\frac{e^2}{6} \cong 1.232$
Gaussian	1	$\frac{\sqrt{3}\pi}{2} \cong 2.721$	$\frac{\pi e}{6} \cong 1.423$

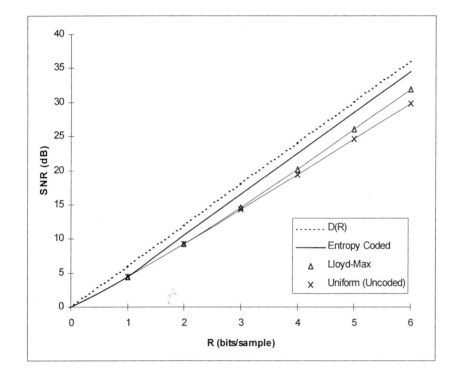

Figure 3.5. MSE performance for scalar quantization of IID Gaussian data.

3.2.6 EMBEDDED SCALAR QUANTIZATION

A very desirable feature of compression systems is the ability to successively refine the reconstructed data as the bit-stream is decoded. In this situation, a (perhaps crude) approximation of the reconstructed data becomes available after decoding only a small subset of the compressed bit-stream. As more of the compressed bit-stream is decoded, the reconstruction can be improved incrementally until full quality reconstruction is obtained upon decoding the entire bit-stream. Compression systems possessing this property are facilitated by embedded quantization.

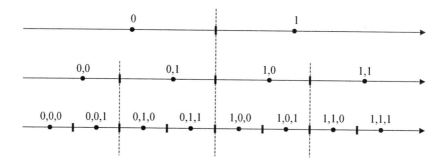

Figure 3.6. Embedded scalar quantizers Q_0, Q_1, and Q_2, of rates $R = 1, 2$, and 3 bits/sample.

In embedded quantization, the intervals of higher rate quantizers are embedded within the intervals of lower rate quantizers. Equivalently, the intervals of lower rate quantizers are partitioned to yield the intervals of higher rate quantizers. Consider a sequence of K embedded scalar quantizers $Q_0, Q_1, Q_2, \ldots, Q_{K-1}$. The intervals of Q_{K-1} are then embedded within the intervals of Q_{K-2}, which in turn are embedded within those of Q_{K-3}, and so on. Equivalently, the intervals of Q_0 are partitioned to get the intervals of Q_1, which in turn are partitioned to get the intervals of Q_2, and so on.

Specifically, each interval of Q_0 ($\mathcal{I}_{q_0}$ $q_0 = 0, 1, \ldots, M_0 - 1$) is partitioned into M_1 intervals $\mathcal{I}_{q_0,q_1}$ $q_1 = 0, 1, \ldots, M_1 - 1$. The total number of intervals of Q_1 is then $M_0 M_1$. Similarly, the intervals of Q_1 are partitioned to obtain the intervals of Q_2 as $\mathcal{I}_{q_0,q_1,q_2}$ $q_2 = 0, 1, \ldots, M_2 - 1$.

In general then, Q_k ($k = 0, 1, \ldots, K - 1$) has $\prod_{j=0}^{k} M_j$ intervals, given by $\mathcal{I}_{q_0,q_1,\ldots,q_k}$.

With this partitioning, it is natural to take the comma separated list $q_0, q_1, \ldots, q_k$ as the "quantizer index" of $\mathcal{I}_{q_0,q_1,\ldots,q_k}$. This situation is illustrated in Figure 3.6 for $K = 3$ and $M_0 = M_1 = M_2 = 2$. It should be clear from this figure, that for the binary case ($M_k = 2$ $k = 0, 1, \ldots, K - 1$), the indices of Q_k (i.e., $q_0, q_1, \ldots, q_k$) can be interpreted as binary representations of the "usual integer" indices.

With the formalism described above, all indices of lower rate quantizations can be obtained by dropping components from (the comma separated) indices of higher rate quantizations. Specifically, the index for Q_k can be obtained by dropping the last component of Q_{k+1}. To see this, let

$$Q_{k+1}(x) = q_0, \ldots, q_k, q_{k+1}$$

This implies that $x \in \mathcal{I}_{q_0,...,q_k,q_{k+1}}$. Now since

$$\mathcal{I}_{q_0,q_1,...,q_k} = \bigcup_{q_{k+1}=0}^{M_{k+1}-1} \mathcal{I}_{q_0,...,q_k,q_{k+1}}$$

we must have $x \in \mathcal{I}_{q_0,q_1,...,q_k}$, which implies that

$$Q_k(x) = q_0, q_1, \ldots, q_k$$

We close this subsection by noting that the embedded quantization framework discussed here suggests a simple implementation trick for scalar quantization. For simplicity, we consider only the case depicted in Figure 3.6. The Q_2 index is easily determined using only three binary threshold comparisons. The first bit q_0 is determined by comparing x to the single threshold of Q_0. Once q_0 is known, q_1 is determined by comparing x to the single threshold within $\mathcal{I}_{q_0}$. Finally q_2 is determined by comparing x to the single threshold within $\mathcal{I}_{q_0,q_1}$.

This process (especially in the non-binary case) can be viewed as a series of dependent quantizations of x. For example, $q_1 = Q_{1,q_0}(x)$ where $Q_{1,q_0}(x)$ represents the quantization of x within $\mathcal{I}_{q_0}$. Similarly, $q_2 = Q_{2,q_0,q_1}(x)$ represents the quantization of x within $\mathcal{I}_{q_0,q_1}$. We will have more to say on this subject in Section 3.4.4. For now, we point out that it is not possible (in general) for more than one of the embedded quantizers to simultaneously satisfy the Lloyd-Max conditions. For example, if the highest rate quantizer is chosen to satisfy the Lloyd-Max conditions, then the thresholds of all lower rate quantizers are fixed. The only design parameter then available for the lower rates are the codewords, which should be chosen as the centroids of their respective intervals.

3.2.7   EMBEDDED DEADZONE QUANTIZATION

A notable example where all embedded quantizers *can* be optimal is the uniform case. Indeed, uniformly subdividing the intervals of a uniform scalar quantizer clearly yields another uniform scalar quantizer. In this way, a family of embedded uniform scalar quantizers may be constructed. These quantizers all satisfy the Lloyd-Max conditions for the uniform distribution.

A particularly elegant (and important) example is the uniform dead-zone quantizer of Section 3.2.4. For the case when $\xi = 0$, we have[1]

$$q = Q(x) = \text{sign}(x) \left\lfloor \frac{|x|}{\Delta} \right\rfloor \tag{3.34}$$

and

$$\hat{x} = \overline{Q^{-1}}(q) = \begin{cases} 0 & q = 0 \\ \text{sign}(q)(|q| + \delta)\Delta & q \neq 0 \end{cases} \tag{3.35}$$

This quantizer has embedded within it, all uniform deadzone quantizers with step sizes $2^p\Delta$ for integer $p \geq 0$.

Assuming that the magnitude of q can be represented with K bits, then q can be written in sign magnitude form as

$$q = Q_{K-1}(x) = s, q_0, q_1, \ldots, q_{K-1} \tag{3.36}$$

Now, let

$$q^{(p)} = s, q_0, q_1, \ldots, q_{K-1-p}$$

be the index obtained by dropping the last p bits of q. Equivalently, $q^{(p)}$ is obtained by right shifting the binary representation of $|q|$ by p bits. It is then easily verified that

$$Q_{K-1-p}(x) = q^{(p)}$$

where Q_{K-1-p} is the uniform deadzone quantizer with step size $2^p\Delta$.

From this discussion, we can deduce that if the p LSBs of $|q|$ are unavailable, we may still dequantize, but at a lower level of quality. In particular, the result will be the same as if quantization were performed using a step size of $2^p\Delta$ (rather than Δ) in the first place. In this situation, the inverse quantization is performed as

$$\hat{x} = \begin{cases} 0 & q^{(p)} = 0 \\ \text{sign}(q^{(p)})(|q^{(p)}| + \delta)2^p\Delta & q^{(p)} \neq 0 \end{cases} \tag{3.37}$$

It is worth noting that when $p = 0$, this yields the full quality dequantization as given by equation (3.35).

We wrap up this section by noting that context dependent binary entropy coding can be used to compress the bits in equation (3.36). In the case of IID data, only the previous bits of the same index (as discussed in Section 2.3) need to be incorporated in the contexts to achieve the performance promised in Section 3.2.4 for each and every partial decoding (with step size $2^p\Delta$ $p \geq 0$). When the data is not IID, compression performance can be improved by including bits from neighboring indices when forming contexts (as described in Section 8.3).

---

[1] The case when $\xi \neq 0$ is discussed in Section 8.3.1.

3.3 DIFFERENTIAL PULSE CODE MODULATION

Throughout our discussion of quantization, we have assumed that $\{X_n\}$ is IID. If the process is not IID (but stationary), scalar quantization performance will be governed by the marginal PDF, and will be identical to the performance that would be achieved for IID data having the same marginal PDF. On the other hand, more sophisticated techniques exist which can exploit the dependencies in non-IID data to improve quantization performance. One such approach is to perform a transform on the data so that the resulting transform coefficients are (at least approximately) IID. Such transforms are discussed in detail in Chapters 4 and 6. Another approach is to employ *context dependent* entropy coding (Section 2.4.1) to scalar quantization indices. A hybrid of these two approaches is discussed in Chapter 8 and forms the basis for JPEG2000. Yet another approach is vector quantization, as discussed in a subsequent section of this chapter.

In this section, we discuss the addition of quantization to the predictive scheme described in Section 2.4.2. The resulting system is known as differential pulse code modulation (DPCM). Figure 3.7 shows the basic block diagram for DPCM. As was the case in Section 2.4.2, a prediction μ_n of the current sample X_n is formed using previously coded values. As before, only previously coded values are used in forming predictions since the decoder must track the procedure performed by the encoder. The prediction error (or prediction residual) is formed as $e_n = x_n - \mu_n$. The algorithm now differs from that discussed previously in that e_n is quantized to get an index q_n (Both entropy and non-entropy coded quantization are possible.) The reconstructed prediction error $\hat{e}_n$ is added back to the prediction at both encoder and decoder to obtain the reconstructed value of the data $\hat{x}_n$.

A fundamental property of DPCM is that the error between x_n and $\hat{x}_n$ is precisely equal to the error introduced into e_n by quantization. This can be seen by examining Figure 3.7 in which

$$e_n = x_n - \mu_n$$

and

$$\hat{x}_n = \hat{e}_n + \mu_n$$

Combining these two equations yields the stated result. Specifically,

$$x_n - \hat{x}_n = e_n - \hat{e}_n$$

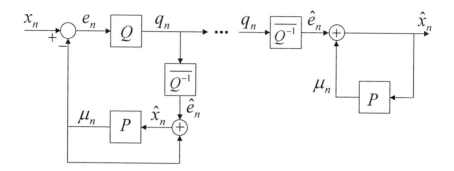

Figure 3.7. Block diagram for differential pulse code modulation.

From this, we have that

$$E\left[(X - \hat{X})^2\right] = E\left[(E - \hat{E})^2\right]$$

From previous sections, we then have $d(R) = \varepsilon_E^2 \sigma_E^2 2^{-2R}$ where ε_E^2 and σ_E^2 are the appropriate values for the PDF of the prediction error. If $\varepsilon_E^2 = \varepsilon_X^2$, we see that the MSE for DPCM is better than that for scalar quantization by a factor (prediction gain) of $G_p = \sigma_X^2/\sigma_E^2$. That is,

$$\text{MSE}_{\text{DPCM}} = \frac{1}{G_p}\text{MSE}_{\text{SQ}}$$

or,

$$\text{SNR}_{\text{DPCM}} = \text{SNR}_{\text{SQ}} + 10\log_{10}G_p$$

It is clear that G_p is maximized by minimizing the prediction error variance. This occurs when μ_n is chosen as the conditional mean of X_n given the neighborhood used in prediction. Calculation of the optimal μ_n is often difficult since the prediction must be based on previously coded (quantized) values of the neighborhood samples. A high rate (small MSE) assumption is often used to overcome this difficulty. In this case, the optimal predictor is formulated assuming that $\hat{x}_n = x_n$. Another common simplification is to restrict the predictor to be a linear combination of the values in the neighborhood. A suitably chosen linear predictor will be optimal in the Gaussian case, but will be suboptimal in general.

Example 3.3 *Suppose $\{X_n\}$ is a Gaussian Markov-1 process with mean 0 and correlation coefficient r. Then*

$$\mu_n = E\left[X_n|\mathbf{x}_{0:n}\right] = E\left[X_n|x_{n-1}\right] = rx_{n-1} \qquad (3.38)$$

Since $e_n = x_n - \mu_n$, E_n *is zero mean and Gaussian. Thus,* $\varepsilon_X^2 = \varepsilon_E^2$ *and*

$$\sigma_E^2 = E\left[E_n^2\right] = E\left[(X_n - rX_{n-1})^2\right] = (1 - r^2)\sigma_X^2$$

In the example above, $G_p = \frac{1}{1-r^2}$. Thus, at high rates, $d_{\text{DPCM}}(R) = \varepsilon^2(1 - r^2)\sigma_X^2 2^{-2R}$. It can be shown, that for this process, the spectral flatness measure is $\gamma_X^2 = (1 - r^2)$. In fact, for any Gaussian Markov process, $G_p = 1/\gamma_X^2$ and thus

$$d_{\text{DPCM}}(R) = \varepsilon^2 \gamma_X^2 \sigma_X^2 2^{-2R} \tag{3.39}$$

Comparing to equations (3.15), (3.10), and (3.32), we see that the high rate performance of DPCM differs from the distortion-rate function by a factor of ε^2, just as in the case of scalar quantization of IID Gaussian data. Specifically, DPCM (using entropy coded uniform quantization) is within 1.53 dB of the distortion-rate function at high rates.

We close this section by noting that in the low rate case, equation (3.39) is overly optimistic. Indeed, as R approaches zero, the MSE of any realizable quantization scheme must be at least σ_X^2 which can be considerably larger than the value of $\varepsilon^2 \gamma_X^2 \sigma_X^2$ as implied by equation (3.39). This should be expected since equation (3.39) was derived assuming that predictions are based on the unquantized data X_n. In practice, we are forced to use $\hat{X}_n$ which can differ from X_n significantly at low rates. This degrades the quality of predictions, and ultimately, the compression performance.

3.4   VECTOR QUANTIZATION

Vector quantization (VQ) is another name for the general case ($m \geq 1$) of the "source codes" discussed in Section 3.1.1. It is a way of quantizing all samples in a vector of data $\mathbf{x}_{0:m}$ jointly rather than individually (as in the previous sections on scalar quantization (SQ)). As a simple example, consider the vector quantizer depicted in Figure 3.8. This figure extends the graphical depiction of 1-D (scalar) quantization to the 2-D case. As before, the heavy "dots" represent quantizer codewords (codevectors). The regions bounded by the dashed lines are generalizations of the decision intervals $\mathcal{I}_q$ from the 1-D case. These "decision regions" are often called Voronoi regions. The 2-D VQ of Figure 3.8 is of size $M = 8$. The rate of this VQ is $R = \frac{1}{m} \log_2 M = 1.5$ bits/sample. This follows from the fact that the index of a codeword can be represented using $\log_2 8 = 3$ bits, and a codeword serves as the reconstruction for $m = 2$ samples.

VQ can outperform SQ, even in the case when $\{X_n\}$ is IID. To see why this is true, it is useful to examine the VQ of Figure 3.9a. This figure

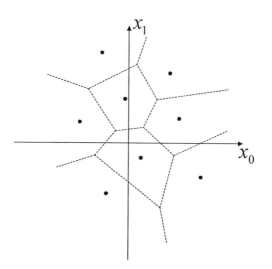

Figure 3.8.   Two-dimensional VQ of rate $R = 1.5$ bits/sample.

depicts a 2-D VQ of size $M = 64$ ($R = 3$ bits/sample). This VQ is particularly interesting in that it yields precisely the same results as the 3 bits/sample uniform SQ of Figure 3.9b. For example, consider a data vector (two consecutive samples) $\mathbf{x} = (x_0, x_1)^t$ that lies in the rectangle bounded by 4 and 6 horizontally, and -4 and -2 vertically. The VQ reconstruction of any such $\mathbf{x}$ is $\hat{\mathbf{x}} = (5, -3)^t$. Note that each such $\mathbf{x}$ has $4 \leq x_0 < 6$ and $-4 \leq x_1 < -2$. Thus, applying the quantizer of Figure 3.9b independently to x_0 and x_1 also yields $\hat{\mathbf{x}} = (5, -3)^t$. Generalizing this idea to higher dimensions, it is easy to see that independent SQ of all samples in a vector is a special case of VQ. In this special case, the (m-dimensional) codewords are constrained to lie on a rectangular grid. Note however, that the grid spacing need not be uniform (e.g., Lloyd-Max scalar quantizer).

The first (and foremost) deficiency of the scalar quantizer is then the fact the boundary of the grid is constrained to be cubic (generalized cube for $m \neq 3$). The second deficiency is that the Voronoi regions are also constrained to be cubic (rectangular for non-uniform scalar quantizers). That the first property is indeed a deficiency can be seen by considering data having a high probability of falling within the circular region shown in Figure 3.10a (e.g., IID Gaussian data). Since the "corner" codewords of Figure 3.9a fall outside this region, they are largely wasted. They can be discarded to yield a VQ of (at least roughly) the same MSE but of lower rate. If the data have an elliptical high probability region as shown

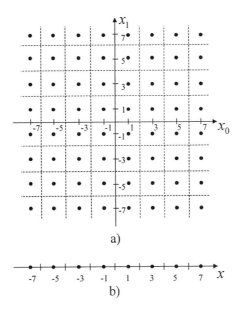

$$a)$$

$$b)$$

Figure 3.9.   Rate 3 quantization: a) 2-D VQ. b) Equivalent SQ.

in Figure 3.10b (e.g., correlated Gaussian data), even more codewords can be discarded to achieve further reduction in rate without significant increase in MSE.

We note here that (joint) entropy coding of SQ indices can be used to similar effect. In particular, such entropy coding can overcome (only) the first deficiency of SQ described above. We will have more to say on this matter in subsequent sections. In what follows however, unless specifically stated otherwise, "SQ" refers to *non-entropy coded* scalar quantization.

3.4.1   ANALYSIS OF VQ

THE TYPICAL REGION

To gain more insight into the first deficiency of SQ, we now examine the high probability (or typical) region of a continuous process. This region is governed by the joint PDF of $\{X_n\}$. In the proof of Theorem 2.4, we saw that for IID discrete processes there is a set of typical vectors (sequences) that are (roughly) equally likely. Furthermore, the probability of getting a sequence outside this set becomes vanishingly small as m gets large. This concept extends naturally to the continuous case.

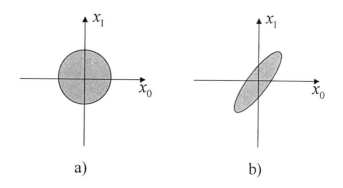

Figure 3.10. High probabilty region: a) Circular region of radius 8. b) Similar, but elliptical.

To this end, let the typical set for a continuous process be given by

$$T(m,\delta) = \left\{ \mathbf{x}_{0:m} : \left| -\frac{1}{m} \log_2 f_{\mathbf{X}_{0:m}}(\mathbf{x}_{0:m}) - h(X) \right| \leq \delta \right\} \qquad (3.40)$$

The volume of this set is

$$\text{Vol}(T(m,\delta)) = \int \cdots \int_{T(m,\delta)} dx_0 dx_1 \ldots dx_{m-1} = \int \cdots \int_{T(m,\delta)} d\mathbf{x}_{0:m}$$

Theorem 3.6 *AEP for Continuous IID Processes*

Let $\{X_n\}$ be a sequence of IID random variables with marginal PDF $f_X(x)$. Then for any $\delta > 0$ and $\varepsilon > 0$, there exists m suitably large so that

1) $P(\mathbf{X}_{0:m} \notin T(m,\delta)) < \varepsilon$

2) $(1-\varepsilon)\, 2^{m(h(X)-\delta)} \leq \text{Vol}(T(m,\delta)) \leq 2^{m(h(X)+\delta)}$

Proof. The first property follows from the weak law of large numbers. Specifically,

$$-\frac{1}{m} \log_2 f_{\mathbf{X}_{0:m}}(\mathbf{X}_{0:m}) = -\frac{1}{m} \sum_{i=0}^{m-1} \log_2 f_X(X_i)$$

$$\longrightarrow E\left[-\log_2 f_X(X)\right] = h(X) \quad \text{as } m \longrightarrow \infty$$

The right hand side of property 2 follows from

$$1 = \int \cdots \int f_{\mathbf{X}_{0:m}}(\mathbf{x}_{0:m})\, d\mathbf{x}_{0:m} \geq \int \cdots \int_{T(m,\delta)} f_{\mathbf{X}_{0:m}}(\mathbf{x}_{0:m})\, d\mathbf{x}_{0:m}$$

$$\geq \int \cdots \int_{T(m,\delta)} 2^{-m(h(X)+\delta)}\, d\mathbf{x}_{0:m} \quad \text{(by equation (3.40))}$$

$$= 2^{-m(h(X)+\delta)} \text{Vol}(T(m,\delta))$$

The left hand side of property 2 follows from

$$1 - \varepsilon \leq P\left(\mathbf{X}_{0:m} \in T\left(m, \delta\right)\right) \quad \text{(from property 1)}$$

$$= \int_{T(m,\delta)} \cdots \int f_{\mathbf{X}_{0:m}}\left(\mathbf{x}_{0:m}\right) d\mathbf{x}_{0:m}$$

$$\leq \int_{T(m,\delta)} \cdots \int 2^{-m(h(X)-\delta)} d\mathbf{x}_{0:m} \quad \text{(by equation (3.40))}$$

$$= 2^{-m(h(X)-\delta)} \text{Vol}(T\left(m, \delta\right))$$

∎

It is worth noting that within the typical region, $\mathbf{X}_{0:m}$ is distributed roughly uniformly. This follows from the definition of $T\left(m, \delta\right)$. That is, for $\mathbf{x}_{0:m} \in T\left(m, \delta\right)$,

$$-\frac{1}{m} \log_2 f_{\mathbf{X}_{0:m}}\left(\mathbf{x}_{0:m}\right) \cong h(X)$$

or

$$f_{\mathbf{X}_{0:m}}\left(\mathbf{x}_{0:m}\right) \cong 2^{-mh(X)}$$

Example 3.4 *Consider the IID Gaussian case for which the typical set satisfies*

$$-\frac{1}{m} \log_2 f_{\mathbf{X}_{0:m}}\left(\mathbf{x}_{0:m}\right) \cong h(X) = \frac{1}{2} \log_2 2\pi e \sigma^2$$

or,

$$-\frac{1}{m} \log_2 \left[\prod_{j=0}^{m-1} \frac{1}{\sqrt{2\pi\sigma^2}} e^{-x_j^2/2\sigma^2} \right] \cong \frac{1}{2} \log_2 2\pi e \sigma^2$$

Some algebra results in

$$\frac{1}{m} \sum_{j=0}^{m-1} x_j^2 = \frac{1}{m} \parallel \mathbf{x}_{0:m} \parallel^2 \cong \sigma^2 \qquad (3.41)$$

Thus, the typical region for the IID Gaussian source consists of all vectors lying near the surface of a sphere having radius $\sqrt{m\sigma^2}$. For large m, the probability of a vector occurring far from this surface is vanishingly small, and the distribution of vectors near this surface is uniform.

The following discussion can be carried out more generally, but for simplicity, we limit ourselves to the IID Gaussian case from the example above.

Since with very high probability, only vectors in the typical region will occur, codewords should be placed only near the sphere surface.

Actually, for large m, essentially all the volume of a sphere lies near its surface, and it is reasonable (and in fact preferable for small to moderate m) to distribute codewords throughout the entire interior of the sphere. This was alluded to in the previous discussion of Figure 3.10a.

The fact that all the volume lies near the surface of the sphere follows from noting that a sphere of radius ρ has volume given by

$$\text{Vol}(\mathcal{S}(\rho)) = \frac{\pi^{m/2}\rho^m}{\Gamma\left(\frac{m}{2}+1\right)}$$

and then considering the fraction of volume lying within $\varepsilon > 0$ of the surface. Specifically

$$\frac{\text{Vol}(\mathcal{S}(\rho)) - \text{Vol}(\mathcal{S}(\rho-\varepsilon))}{\text{Vol}(\mathcal{S}(\rho))} = 1 - \left(\frac{\rho-\varepsilon}{\rho}\right)^m$$

which tends to 1 as m gets large.

VECTOR QUANTIZATION OF IID GAUSSIAN DATA

Consider populating the typical region with codewords having spherical Voronoi regions each of radius

$$\zeta = \sqrt{(m+2)D} \tag{3.42}$$

Since the distribution within the typical region is uniform, the distribution within each (spherical) Voronoi region is also uniform, and we have

$$\frac{1}{m}E\left[\parallel \mathbf{X}_{0:m} - \hat{\mathbf{X}}_{0:m} \parallel^2 | \mathbf{X}_{0:m} \in \mathcal{I}_q\right]$$

$$= \frac{1}{m}\int \cdots \int_{\sum x_i^2 \leq \zeta^2} \frac{(x_0^2 + \cdots + x_{m-1}^2)}{\text{Vol}(\mathcal{I}_q)} d\mathbf{x}_{0:m}$$

$$= \frac{\zeta^2}{m+2} \tag{3.43}$$

The last equality above follows from [72, equation (4.642)], repeated here as

$$\int \cdots \int_{\sum x_i^2 \leq \zeta^2} f\left(\sqrt{x_0^2 + \cdots + x_{m-1}^2}\right) d\mathbf{x}_{0:m}$$

$$= \frac{2\pi^{m/2}}{\Gamma(m/2)} \int_0^\zeta x^{m-1} f(x) dx$$

Substituting equation (3.42) and ignoring the low probability event that $\mathbf{X}_{0:m}$ falls outside the typical region, we have

$$d = \frac{1}{m} \sum_q E\left[\| \mathbf{X}_{0:m} - \hat{\mathbf{X}}_{0:m} \|^2 \,|\mathbf{X}_{0:m} \in \mathcal{I}_q\right] P(\mathbf{X}_{0:m} \in \mathcal{I}_q)$$

$$= D \sum_q P\left(\mathbf{X}_{0:m} \in \mathcal{I}_q\right)$$

$$= D$$

The number of spheres of radius ζ (number of codewords) required to cover the typical region is

$$M \cong \frac{\text{Vol}(\mathcal{T}\,(m,\delta))}{\text{Vol}(\mathcal{S}\,(\zeta))} \cong \frac{2^{mh(X)}}{\left(\frac{\pi^{m/2}\zeta^m}{\Gamma(\frac{m}{2}+1)}\right)}$$

Substituting equations (3.6) and (3.42), and simplifying yields

$$M \cong \frac{(2e)^{m/2}\,\Gamma\left(\frac{m}{2}+1\right)\left(\frac{\sigma^2}{D}\right)^{m/2}}{(m+2)^{m/2}}$$

Substituting Stirling's formula, $\Gamma\,(\alpha+1) \cong \sqrt{2\pi\alpha}\left(\frac{\alpha}{e}\right)^\alpha$ and solving for the required rate yields

$$R = \frac{1}{m}\log_2 M$$

$$\cong \frac{1}{2}\log_2 \frac{\sigma^2}{D} + \frac{1}{2}\log_2 \frac{m}{m+2} + \frac{1}{2m}\log_2 \pi m$$

which converges to

$$R = \frac{1}{2}\log_2 \frac{\sigma^2}{D}$$

as m grows large. Comparing to equation (3.9), we see that a VQ designed in this fashion can achieve the rate-distortion function as m tends to infinity.

In fact, this derivation is not completely rigorous. It ignores the unlikely sequences that fall outside the typical set, and ignores the fact that spheres do not pack. That is, spheres cannot be used to cover the typical region without overlap. Nevertheless, the discussion provides substantial intuition into the superiority of VQ over SQ, and motivates the comparison of spherical vs. rectangular Voronoi regions in the subsection below.

SPHERICAL VORONOI REGIONS

We can now examine the second deficiency stated earlier for SQ. Specifically, the Voronoi regions induced by SQ are restricted to be rectangular. For simplicity, assume cubic Voronoi regions with edges of length a. Considering the equivalent scalar quantizer (with $\Delta = a$), it is then clear that the MSE is $d = a^2/12$. To preserve the same rate as in the previous discussion, we equate the volume of the Voronoi regions so that

$$a^m = \text{Vol}\,(\mathcal{I}_q) = \text{Vol}(\mathcal{S}\,(\zeta))$$
$$= \frac{\pi^{m/2}\,(m+2)^{m/2}\,D^{m/2}}{\Gamma\left(\frac{m}{2}+1\right)}$$

or,

$$a = \frac{\sqrt{(m+2)\,\pi D}}{\Gamma^{1/m}\left(\frac{m}{2}+1\right)}$$

with a resulting MSE of

$$d = \frac{a^2}{12} = \frac{(m+2)\,\pi D}{12\Gamma^{2/m}\left(\frac{m}{2}+1\right)}$$

Applying Stirling's formula and simplifying yields

$$d \cong \frac{e\,(m+2)\,\pi D}{6m\,(m\pi)^{1/m}} \longrightarrow \frac{\pi e}{6}D$$

Thus, as m grows we see that cubic Voronoi regions are inferior to spherical Voronoi regions by a factor of $\pi e/6 = 1.53$ dB.

Although spheres do not pack for any dimension $2 \leq m < \infty$, spherical Voronoi regions can be well approximated for large m. Much effort has been spent investigating lattice structures for codewords with efficient Voronoi regions. In two dimensions, the optimal shape is known to be a regular hexagon. In higher dimensions, the optimal shape is generally unknown, however lattices have been found [42] in dimensions $m = 4, 8, 16$, and 24 that achieve $0.37, 0.65, 0.86$, and 1.03 dB (of the 1.53 dB possible), respectively. Trellis coded quantization (Section 3.5) can achieve substantially all of the 1.53 dB difference.

DISCUSSION

The 1.53 dB difference between VQ (with cubic Voronoi regions) and the distortion-rate function is no coincidence. We saw in Section 3.2.4, that for IID processes, entropy coded uniform SQ also falls 1.53 dB short of the distortion-rate function. As discussed throughout this section, the gap in performance between uncoded uniform SQ and the distortion-rate function is comprised of two portions: a portion associated with the high probability (or typical) region; and a portion associated with the shape of the Voronoi region. In the previous subsection, we showed that this latter portion is 1.53 dB. We can conclude from this (as claimed just before Section 3.4.1) that entropy coding of SQ indices can obtain (only) the gain associated with the typical region. We also conclude that for IID processes, joint (or context dependent) entropy coding is not required.[2]

As a final note, we point out that the optimal (spherical) Voronoi cell shape is a property only of the MSE distortion measure and not the PDF of $\{X_n\}$. The typical region, on the other hand, is a function of the PDF (but is independent of any distortion measure that might be chosen). For example, the typical region for a Gaussian Markov process lies near the surface of an ellipsoid (as hinted by Figure 3.10b), while the typical region for Laplacian data lies near the surface of a "pyramid" [57].

3.4.2   THE GENERALIZED LLOYD ALGORITHM

As in the scalar case $(m = 1)$, iterative techniques can be employed to design VQ codebooks. The direct generalization of the Lloyd Algorithm (with training data) from Section 3.2.1 appeared in [96] and resulted in an explosion of research in the area of vector quantization. The resulting algorithm is given by

---

The Generalized Lloyd Algorithm (with Training Data)

1) Given a set of training vectors, say $\mathcal{T} = \{\mathbf{x}_0, \mathbf{x}_1, \ldots, \mathbf{x}_{\|\mathcal{T}\|-1}\}$, choose an initial code $\mathcal{C} = \{\hat{\mathbf{x}}_0, \hat{\mathbf{x}}_1, \ldots, \hat{\mathbf{x}}_{M-1}\}$, set $j = 1$, $d_0 = \infty$

2) Let $\mathcal{B}_q = \{\mathbf{x} \in \mathcal{T} : Q(\mathbf{x}) = q\} = \{\mathbf{x} \in \mathcal{T} : (\mathbf{x} - \hat{\mathbf{x}}_q)^2 < (\mathbf{x} - \hat{\mathbf{x}}_k)^2 \ \forall k \neq q\}$

3) $\hat{\mathbf{x}}_q = \frac{1}{\|\mathcal{B}_q\|} \sum_{\mathbf{x} \in \mathcal{B}_q} \mathbf{x} \qquad q = 0, 1, \ldots, M - 1$

4) $d_j = \frac{1}{m\|\mathcal{T}\|} \sum_{\mathbf{x} \in \mathcal{T}} \left\| \mathbf{x} - \overline{Q^{-1}}(Q(\mathbf{x})) \right\|^2$

---

[2] For Markov processes, MSE performance within 1.53 dB of the distortion-rate function can still be achieved by entropy coding of uniform SQ indices. However conditional (context dependent) entropy coding is required.

5) If $\frac{d_{j-1}-d_j}{d_j} < \varepsilon$ stop; else set $j = j+1$ and go to step 2

The entropy coded scalar quantizer design algorithm of Section 3.2.3 generalizes in the same way to yield an entropy coded VQ design algorithm [39].

An unfortunate result of these algorithms is the potential for getting caught in local minima. The codebook (and associated MSE performance) can change significantly depending on the choice of the initial codebook in step 1. Strategies for choosing the initial codebook, as well as alternate design algorithms (e.g., simulated annealing) are discussed in [69].

Another problem that plagues most VQ design algorithms is the lack of structure among the resulting codewords. Although fast search techniques may be found for small m, in general, exhaustive search must be employed to implement the nearest neighbor encoding rule. That is, a data vector $\mathbf{x}_{0:m}$ to be encoded must be compared to each codevector $\hat{\mathbf{x}}_q$ $q = 0, 1, \ldots, M = 2^{mR}$ to find the one that minimizes the MSE. Specifically,

$$q = \operatorname*{argmin}_{q \in \{0,1,\ldots,M-1\}} \left[\frac{1}{m} \|\mathbf{x} - \hat{\mathbf{x}}_q\|^2 \right]$$

Thus, the encoding complexity is proportional to the size of codebook $\left(M = 2^{mR} \right)$ and grows exponentially in both m and R.

3.4.3   PERFORMANCE OF VECTOR QUANTIZATION

According to Theorem 3.1, the MSE performance of VQ tends to the distortion-rate function as m tends to infinity. Unfortunately, complexity severely limits practically achievable values for m. For IID Gaussian data at a rate of $R = 2$ bits/sample, $m = 6$ yields performance roughly 1 dB better than Lloyd-Max SQ ($m = 1$). For correlated Gaussian data, improvements can be much more significant. For example, for a correlation coefficient of $r = 0.9$, and a rate of $R = 1$ bit/sample, $m = 6$ provides roughly 6.5 dB improvement over non-entropy coded Lloyd-Max SQ [69]. For comparison, entropy coded DPCM can provide about 7 dB, while uncoded DPCM provides only about 5.5 dB.

The encoder complexity of VQ is significantly higher than that of entropy coded DPCM which in turn is higher than that of uncoded DPCM. On the other hand, the decoder complexity of VQ is negligible. It consists of a simple table look-up operation. VQ also has an advantage in error resiliency. If a bit error occurs in the communication of compressed data, no more than m samples will be affected in the VQ case. Since

DPCM contains a prediction feedback loop, error propagation can be significant. Some form of damping is often included to reduce this effect [82].

3.4.4 TREE-STRUCTURED VQ

Tree-structured VQ (TSVQ) is a technique for reducing the search complexity required in VQ. The basic idea of TSVQ is to search a series of "small" codebooks to "home in" on the choice of a vector in the codebook $\mathcal{C}$. At level 0 (or the root) of the "tree" is a single codebook $\mathcal{C}_0$ of length m and size $M_0 = 2^{mR_0}$. As before, a data vector $\mathbf{x} = \mathbf{x}_{0:m}$ is quantized to yield an index $q_0 = Q_0(\mathbf{x})$.

At level 1 of the tree, there are M_0 codebooks $\mathcal{C}_{1,q_0}$ $q_0 = 0, \ldots, M_0 - 1$. Each of these codebooks is of length m and of size $M_1 = 2^{mR_1}$. As hinted by the notation, if $Q_0(\mathbf{x}) = q_0$, then $\mathbf{x}$ is subjected to $\mathcal{C}_{1,q_0}$ to get a second index $q_1 = Q_{1,q_0}(\mathbf{x})$. At level 2 of the tree, there are M_1 codebooks for each choice of q_0, denoted by $\mathcal{C}_{2,q_0,q_1}$ $q_1 = 0, \ldots, M_1 - 1$, each of length m and of size $M_2 = 2^{mR_2}$. If $Q_{1,q_0}(\mathbf{x}) = q_1$, then $\mathbf{x}$ is subjected to $\mathcal{C}_{2,q_0,q_1}$ to yield $q_2 = Q_{2,q_0,q_1}(\mathbf{x})$, and so on. The union of all codebooks at the final level $(K - 1)$ forms the codebook of the TSVQ.

This situation is depicted in Figure 3.11. In this figure, there are $K = 3$ levels with $M_0 = M_1 = M_2 = 2$ $(R_0 = R_1 = R_2 = 1/m)$. At level 0 (the "root," or "top" level), is one codebook $\mathcal{C}_0$ of size two. The possible index selections are shown (on the two branches leaving the root) as 0 and 1. At level 1, there are two codebooks, $\mathcal{C}_{1,0}$ and $\mathcal{C}_{1,1}$, each of size two, for a total of four codewords. At level 2, there are four codebooks each of size two, for a total of eight codewords. At this level, we have added "leaves" depicting the codewords of the level 2 codebooks to emphasize that these (taken together) form the codebook of the TSVQ. Specifically, $\mathcal{C} = \{\hat{\mathbf{x}}_0, \hat{\mathbf{x}}_1, \ldots, \hat{\mathbf{x}}_7\}$, $M = 8$, and $R = \frac{1}{m} \log_2 8 = 3/m$.

More generally, at level 0, there is one codebook of size $M_0 = 2^{mR_0}$. At level 1, there are M_0 codebooks, each of size $M_1 = 2^{mR_1}$ for a total of $M_0 M_1 = 2^{m(R_0+R_1)}$ codewords. At level 2, there are $M_0 M_1$ codebooks, each of size M_2, for a total of $M_0 M_1 M_2 = 2^{m(R_0+R_1+R_2)}$ codewords. For a K level tree, this continues until level $K - 1$, where there are

$$M = \prod_{k=0}^{K-1} M_k = 2^{m \sum_{k=0}^{K-1} R_k}$$

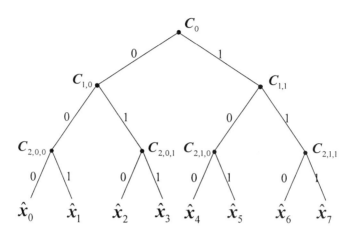

Figure 3.11.   Binary tree structured VQ with $K = 3$ levels.

codewords. The rate of the resulting TSVQ is

$$R = \frac{1}{m} \log_2 M = \sum_{k=0}^{K-1} R_k$$

Since one codebook must be searched at each level, the search complexity of TSVQ is proportional to $\sum_{k=0}^{K-1} M_k$. If the same size codebooks are used at each level then

$$R_k = R/K \qquad (3.44)$$

With this choice, $M_k = 2^{mR_k} = 2^{mR/K}$, so the total search complexity is proportional to

$$\sum_{k=0}^{K-1} M_k = K 2^{mR/K} \qquad (3.45)$$

The smallest complexity is achieved for the binary tree with $M_k = 2$, or $R_k = 1/m$. From equation (3.44), we then require $K = mR$ levels in the tree. Substituting in equation (3.45) results in a complexity proportional to only $2mR$. Thus, we see that the search complexity of TSVQ can grow only linearly in rate and dimension (as opposed to exponentially as in the full search case). The reader should verify however, that in this case, the storage space required for the codewords is roughly doubled.

Many algorithms have been proposed for TSVQ design. One reasonable approach is to use the generalized Lloyd algorithm to design C_0.

The training set $\mathcal{T}$ can then be partitioned into M_0 training sets

$$\mathcal{T}_{q_0} = \{\mathbf{x} \in \mathcal{T} : Q_0(\mathbf{x}) = q_0\} \quad q_0 = 0, 1, \ldots, M_0 - 1$$

These training sets can then be used to design the level 1 codebooks $\mathcal{C}_{1,q_0}$ $q_0 = 0, 1, \ldots, M_0 - 1$. The $\mathcal{T}_{q_0}$ can then be further subdivided based on the level 1 results, for use in training at level 2, and so on.

TSVQ PERFORMANCE

As mentioned in Section 3.2.6, it is generally not possible for more than one of the embedded quantizers to be optimal. When $m = 1$, it is easy to ensure that any single quantizer Q_k satisfies the Lloyd-Max conditions. We discussed this for the highest rate quantizer in Section 3.2.6. This is possible when $m = 1$ because the Voronoi regions are intervals, and highly structured. Unfortunately, in the more general case, it is difficult to ensure optimality for all but the lowest rate quantizer Q_0.

In fact, even if the final TSVQ codebook

$$\mathcal{C} = \bigcup \mathcal{C}_{K-1,q_0,\ldots,q_{K-2}}$$

is optimal, the search strategy imposed by the tree structure can render the overall process suboptimal. Specifically, there is no guarantee that for a given input $\mathbf{x}$, the best $\hat{\mathbf{x}} \in \mathcal{C}$ will be chosen. However, experimental results show that for binary trees ($M_0 = M_1 = \cdots = M_{K-1} = 2$, or $R_0 = R_1 = \cdots = R_{K-1} = 1/m$), TSVQ can perform within about 1.0 dB of "full search" VQ for Gaussian Markov-1 processes. This gap can be decreased considerably by increasing M_k to only 4 [69].

EMBEDDED VQ

We close our discussion of VQ by noting that although TSVQ was described above as a complexity reduction technique, it also provides for embedded quantization. In fact, TSVQ is exactly the generalization (to $m \geq 1$) of the embedded scalar quantizers discussed in Section 3.2.6. Additionally, Figure 3.11 is just another representation of Figure 3.6 when $m = 1$.

As discussed above, $\mathbf{x}$ is first quantized using codebook $\mathcal{C}_0$ to yield an index q_0. This implies that $\mathbf{x}$ is in the Voronoi region $\mathcal{I}_0$ of Q_0. $\mathbf{x}$ is then requantized using codebook $\mathcal{C}_1, q_0$ to yield an index q_1, implying that $\mathbf{x}$ is in the Voronoi region $\mathcal{I}_{q_0,q_1}$ of Q_1. Note that the $\mathcal{I}_{q_0,q_1}$ $q_1 = 0, \ldots, M_1 - 1$ necessarily form a partition of $\mathcal{I}_{q_0}$ so that $\mathcal{I}_{q_0} = \bigcup_{q_1} \mathcal{I}_{q_0,q_1}$. Continuing this process, eventually $\mathbf{x}$ is quantized using codebook $\mathcal{C}_{K-1,q_0,\ldots,q_{K-2}}$ to yield an index q_{K-1}, implying that $\mathbf{x} \in \mathcal{I}_{q_0,q_1,\ldots,q_{K-1}}$ of Q_{K-1}. The partitioning at this final level is such that $\mathcal{I}_{q_0,q_1,\ldots,q_{K-2}} = \bigcup_{q_{K-1}} \mathcal{I}_{q_0,\ldots,q_{K-2},q_{K-1}}$.

From this discussion it should be clear that dropping q_{K-1} from the index representation is equivalent to dropping the last level of the tree. This results in a TSVQ with $K-1$ levels and a rate of $R = \sum_{k=0}^{K-2} R_k$. Continuing to drop one level at a time, we see that embedded in the index from the rate $R = \sum_{k=0}^{K-1} R_k$ TSVQ, is the corresponding index for all TSVQs of rates $R = \sum_{k=0}^{j} R_k$ $j = 0, 1, \ldots, K-1$. It is then clear that dropping index bits in this fashion yields identical results as if TSVQ had been performed at the lower rates in the first place.

3.5   TRELLIS CODED QUANTIZATION

Trellis coded quantization (TCQ) [104] is a special case of trellis coding [56, 169]. TCQ borrows ideas from communication theory to achieve better MSE performance at lower complexity than previous trellis coding systems. Specifically, TCQ employs the trellises and set partitioning ideas from trellis coded modulation [153] to achieve MSE performance very close to that promised by rate-distortion theory.

3.5.1   TRELLIS CODING

A trellis is nothing more than a state transition diagram (that takes time into account) for a finite state machine. Consider the 4 state machine shown in Figure 3.12. In this machine, each of the boxes (labeled t_1 and t_0) are binary storage elements (of a shift register) and the circle containing "+" represents modulo-2 addition, or "exclusive OR" (i.e., $0+0 = 1+1 = 0, 0+1 = 1+0 = 1$). The (binary) input to the machine is labeled u, while z_1 and z_0 are the (binary) outputs. The *state* of the machine is simply the contents of the storage elements. For example, if $t_1 = 1$ and $t_0 = 0$, the state of the machine is written as $t_1 t_0 = 10 = 2$.

Figure 3.13 shows the state transition diagram for the machine of Figure 3.12. In this diagram, each circle represents a state of the machine and the binary number inside each circle is the number or label of that state (i.e., $t_1 t_0$). The arrows represent state transitions, while the labels on each arrow indicate the input necessary to cause that transition, together with the associated output. Each of these labels is of the form $u/z_1 z_0$. For example, if the machine is in state $01 = 1$ and $u = 0$ is input, the next state will be $00 = 0$ and the output will be $z_1 z_0 = 10 = 2$.

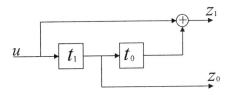

Figure 3.12.   Block diagram of a finite state machine (with four states).

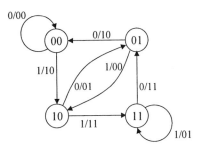

Figure 3.13.   State transistion diagram for the machine of Figure 3.12.

Similarly, if a 1 is input, the next state will be 10=2, and the output will be 00 = 0.

Trellises are used to study sequences of state transitions, or equivalently, sequences of states. A typical trellis is diagrammed in Figure 3.14. Each column of heavy dots (or nodes) represents the four possible states at one point in time. The states are implicitly labeled 0, 1, 2, and 3 from top to bottom. Each branch in the trellis represents a transition from one state to another, at the next point in time (next stage). The reader should verify that the trellis in Figure 3.14 is equivalent to the state machine of Figure 3.12 and the state transition diagram of Figure 3.13.

Specifying a path through the trellis is equivalent to specifying a sequence of states or state transitions. Given an initial state at the left edge of the trellis, such a path can be specified by a sequence of 1's and 0's (the associated binary input sequence, u).

A rich class of trellises having 2 branches entering and leaving each state can be specified by the machine of Figure 3.15. In the communication theory literature, machines of this type are known as rate 1/2 feedback-free convolutional encoders [63]. Trellises derived from the machine in this figure have $N = 2^v$ states. The parameters $h^j = (h^j_v, h^j_{v-1}, \ldots, h^j_0)$, $j = 0, 1$ are called parity check coefficients. The bi-

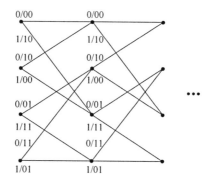

Figure 3.14.   Trellis for the machine of Figure 3.12.

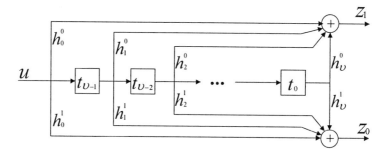

Figure 3.15.   Finite state machine with $N = 2^v$ states.

nary values h_i^j specify whether or not a connection is present in Figure 3.15. The values for $\mathbf{h}^j$ used in this text are from [154], and are given in Table 3.5. It is important to note that the parity check coefficients given in this table are in *octal*. For example, the finite state machine of Figure 3.12 (and hence, the trellis of Figure 3.14) are obtained from $\mathbf{h}^0 = 5 = (1, 0, 1)$ and $\mathbf{h}^1 = 2 = (0, 1, 0)$. Similarly, the eight-state trellis of Figure 3.16 is obtained from $\mathbf{h}^0 = 13 = (1, 0, 1, 1)$ and $\mathbf{h}^1 = 04 = (0, 1, 0, 0)$.

In a fixed rate trellis code, there are 2^R (with $R > 0$ an integer) branches leaving each trellis state and hence, any path through the trellis starting from a given initial state can be specified by a sequence of R bit indices. A method for constructing trellises of this type will be given in Section 3.5.2.

Consider associating a symbol from some output alphabet $\mathcal{A}_{\hat{x}}$ with each branch in such a trellis. Then, a sequence of R bit indices can be

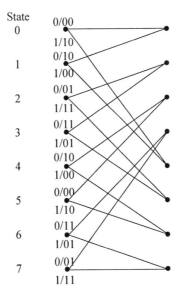

State

0   0/00

1   1/10
    0/10
    1/00

2   0/01
    1/11

3   0/11
    1/01

4   0/10
    1/00

5   0/00
    1/10

6   0/11
    1/01

7   0/01
    1/11

Figure 3.16. Eight-state trellis used in JPEG2000.

Table 3.5. Parity check coefficients for TCQ. Values are given in octal.

	Number of States						
	4	8	16	32	64	128	256
h^0	5	13	23	45	103	235	515
h^1	2	04	04	10	024	126	362

used to specify a sequence (vector) of output symbols associated with a specific trellis path. Hence, populating a trellis (specifying output symbols for each branch) and specifying an initial state yields a set of allowable output sequences, $C = \{\hat{\mathbf{x}}_0, \hat{\mathbf{x}}_1, \ldots, \hat{\mathbf{x}}_{M-1}\}$, where $M = 2^{Rm}$ and m is the number of trellis stages, or equivalently the length of each output sequence. This set of output sequences is known as an R bit per sample trellis code.

For a given data sequence $\mathbf{x}$, a trellis coder outputs the sequence of R bit indices corresponding to the output sequence in C that minimizes $\rho_m(\mathbf{x}, \hat{\mathbf{x}})$. Thinking of the concatenation of the R bit indices as one "super index," $I \in \{0, 1, \ldots, M - 1\}$, we see that trellis coding can be considered a special case of VQ.

The structure of the trellis allows encoding to be done with the Viterbi algorithm [64] which eliminates the need for an exhaustive search over all

2^{Rm} allowable output sequences for the one which minimizes the distortion. The Viterbi algorithm is a clever application of forward dynamic programming [23] that allows encoding to progress from left to right through the trellis with a number of hard decisions being made at each stage. In fact, all but N paths may be discarded at each stage, where N is the number of trellis states (independent of m or R).

For a single letter distortion measure, the Viterbi algorithm is quite simple. Given a data sequence **x**, begin computing the distortion associated with each path emanating from the (given) initial state. Each time two or more paths pass through the same state (at the same time), only the one with the smallest distortion up to that point needs to be searched further. This path is known as a "survivor path."

The number of survivor paths (and hence, the number of paths to be searched at each stage) increases until there is one for each trellis state. The number of survivors remains constant at N from then on. When the end of the data sequence is reached, the sequence in $\mathcal{C}$ that minimizes $\rho_m(\mathbf{x}, \hat{\mathbf{x}})$ is simply the sequence of output symbols associated with the survivor having the lowest final distortion. This process is described more fully (including an example) in Section 3.5.3.

There are many theorems (see [164, 73] for example) guaranteeing the existence of trellis codes with performance converging to the rate-distortion function as the number of trellis states and the length of the sequences to be coded get large. The proofs of these theorems entail random coding arguments, and codes based on these proofs involve populating the trellis in a stochastic manner.

Early trellis coding schemes populated trellises stochastically with output symbols drawn from a continuous distribution (resulting in an uncountable output alphabet $\mathcal{A}_{\hat{X}}$). For long input sequences ($m >> v = \log_2 N$), the computational burden for such a trellis coder is easily shown to be $N2^R$ scalar distortion calculations $\left(\text{e.g., } \rho(x, \hat{x}) = (x - \hat{x})^2\right)$, $N2^R$ additions, and $N\ 2^R$-way compares per data sample. For large N, the computational burden can be quite large. Pearlman and his coauthors showed that good performance is obtainable by stochastically populating a trellis from a finite alphabet of size, say L. This reduces the number of required scalar distortion calculations (per sample) from $N2^R$ to L.

3.5.2   FIXED RATE TCQ

As discussed in the previous subsection, 2^R branches leave each state of a trellis coder. In general, this leads to the computation of distortion information for each such branch at each state, resulting in $N2^R$ scalar distortion calculations, as claimed above. TCQ borrows the set parti-

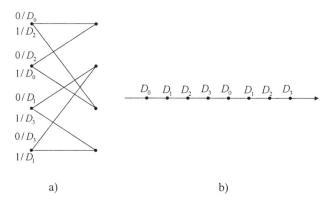

Figure 3.17. Four-state trellis and codebook with subset labeling for $R = 2$ bits/sample.

tioning ideas of [154] to add more structure to the trellis, which results in a reduced computational burden.

To this end, TCQ takes an output alphabet $\mathcal{A}_{\hat{X}}$ (scalar codebook) of size $2^{R+\tilde{R}}$ and partitions it into $2^{\tilde{R}+1}$ subsets. For the sake of simplicity, we take $\tilde{R} = 1$ in what follows. Thus, for an encoding rate of R bits/sample, $\mathcal{A}_{\hat{X}}$ is of size 2^{R+1} which is twice that of a scalar quantizer of the same rate. This codebook is partitioned into four subsets called $D_0, D_1, D_2,$ and D_3, each of size 2^{R-1}. The partitioning is done starting with the left-most codeword and proceeding to the right, labeling consecutive codewords $D_0, D_1, D_2, D_3, D_0, D_1, D_2, D_3, \ldots$, until the right-most codeword is reached. This is illustrated in Figure 3.17b, for the case of $R = 2$.

Subsets obtained in this fashion are then associated with branches of a trellis having only two branches leaving each node. This is illustrated in Figure 3.17a, where the state machine output bits $(s_1 s_0)$ have been used to select which subset is associated with which branch of the trellis in Figure 3.14. Only one stage (or section) of the trellis is shown, as all other stages are identical.

Since each subset contains 2^{R-1} codewords, each branch of the trellis in Figure 3.17 can be thought of as comprising 2^{R-1} parallel branches. With this interpretation, there are 2^R branches leaving each state, and as before, a path through the trellis (sequence of scalar codewords) can be specified by a sequence of R bit indices. Each R bit index can be broken down into two components. The first such component is a single bit u used to specify the next state (as well as the branch followed and its associated subset). The second component consists of $R-1$ bits used to specify a particular scalar codeword within the subset selected.

3.5.3   THE VITERBI ALGORITHM

For a data sequence (vector) $\mathbf{x} = \mathbf{x}_{0:m}$, an N-state trellis of m stages is employed. Such a trellis has $m+1$ columns of states, which we label $\mathcal{S}_{i,l}$ $i = 0, 1, \ldots, m$, $l = 0, 1, \ldots, N-1$. For each state $\mathcal{S}_{i+1,l}$ let $\mathcal{S}_{i,l'}$ and $\mathcal{S}_{i,l''}$ be the two states having branches ending in $\mathcal{S}_{i+1,l}$. Also, let $D^{l',l}$ and $D^{l'',l}$ be the subsets associated with those branches, respectively. Let $c_{l',l}$ and $c_{l'',l}$ be the codewords in $D^{l',l}$ and $D^{l'',l}$ that minimize $\rho(x_i, c) = (x_i - c)^2$, and let $d_{l',l} = (x_i - c_{l',l})^2$ and $d_{l'',l} = (x_i - c_{l'',l})^2$. Finally, let $s_{i+1,l}$ be the "survivor distortion" associated with the survivor path at state $\mathcal{S}_{i+1,l}$.

The ith step $(i = 0, \ldots, m-1)$ in the Viterbi algorithm then consists of setting $s_{i+1,l} = \min\{s_{i,l'} + d_{l',l}, s_{i,l''} + d_{l'',l}\}$, preserving the branch that achieves this minimum, while deleting the other branch from the trellis. If two values compared for minimum survivor distortion are equal, the "tie" can be resolved arbitrarily with no impact on MSE.

When the end of the data is reached $(i = m-1)$, the trellis is traced back from the final state having the lowest survivor distortion, and the corresponding set of TCQ indices are produced. For long data sequences $(m \gg v = \log_2 N)$ the choice of initial state has negligible impact on MSE. Thus, we arbitrarily fix the initial state at 0. This is easily done by setting $s_{0,0} = 0$ and $s_{0,l} = \infty$, $l = 1, 2, \ldots, N-1$.

Example 3.5 *Let $m = 4$, $\mathbf{x} = (-4.1, 2.2, 0.3, -2.5)$, $R = 2$, and let the eight codewords of Figure 3.17b be given by $\mathcal{A}_{\hat{X}} = \{-7, -5, -3, -1, 1, 3, 5, 7\}$. Figure 3.18 shows the results of the encoding steps of the Viterbi algorithm. The initial survivor distortions are shown as $0, \infty, \infty, \infty$ down the left-most set of states.*

The best codewords in D_0, D_1, D_2, D_3 for $x_0 = -4.1$ are $-7, -5, -3,$ and -1, with distortions of $8.41, 0.81, 1.21$, and 9.61, respectively. These distortions are attached to the appropriate branches in the first trellis stage of Figure 3.18. For state $\mathcal{S}_{1,0}$ we compute the survivor distortion as the minimum between $0 + 8.41$ and $\infty + 1.21$. We write the new survivor distortion of 8.41 above $\mathcal{S}_{1,0}$ and delete the losing branch (denoted by $\times$). This process is repeated at $\mathcal{S}_{1,1}$, $\mathcal{S}_{1,2}$, and $\mathcal{S}_{1,3}$ to obtain survivor distortions of $\infty, 1.21$, and ∞, respectively.

For the next sample, $x_1 = 2.2$, the best subset codewords are $1, 3, 5,$ and -1, with distortions of $1.44, 0.64, 7.84$, and 10.24, respectively. These are attached to the appropriate branches and survivor distortions are

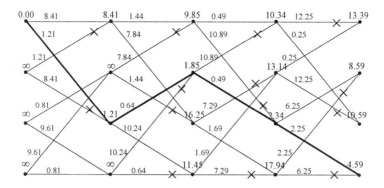

Figure 3.18. Example of the Viterbi algorithm for TCQ with $N = 4$, $m = 4$, and $R = 2$.

computed. For $S_{2,0}$ the winning value is $8.41 + 1.44 = 9.85$. The losing branch $(\infty + 7.84)$ is deleted.

This process is repeated until the end of the data is reached. The smallest survivor distortion at that point is $s_{4,3} = 4.59$. The "best path" through the trellis is "traced back" from that point along the branches not deleted. This is indicated by the heavy lines in the trellis. The codewords associated with this path (from left to right) are $-3, 3, 1, -1$. It is easily verified that the resulting MSE is $\frac{1}{4} \|\mathbf{x} - \hat{\mathbf{x}}\|^2 = \frac{1}{4}(4.59)$, as indicated by the survivor distortion.

Since each subset has two codewords, the codeword chosen within a subset can be specified with $R - 1 = 1$ bit. Denoting the left-most codeword of each subset by 0 (and the other codeword by 1), the set of $R = 2$ bit TCQ indices becomes $10, 01, 11, 10$. This set of indices can be decoded by starting at state 0 and noting that the first bit $(u = 1)$ denotes that the encoder progressed to state 2 and that the subset along the corresponding branch is D_2 (see Figure 3.17). The second bit of 0 denotes the left-most codeword of D_2 which is -3. The next bit of 0 indicates that the encoder progressed from state 2 to state 1 with associated subset D_1. The following 1 bit denotes the right-most codeword of D_1 which is 3. This process can be continued to decode the next two codewords as 1 and -1, respectively.

It should be pointed out that the scalar codewords (chosen from $\mathcal{A}_{\hat{X}}$) are not always the closest to x_i. For example, the first codeword chosen in the example above was -3, while the codeword -5 is actually closer to the data $(x_0 = -4.1)$. However, the entire sequence of codewords (vector codeword $\hat{\mathbf{x}} \in \mathcal{C}$, as allowed by the trellis structure) will always be the one closest to $\mathbf{x}$.

It should also be pointed out that m, N, and R can all be chosen independently. For a given R, increasing m and/or N generally improves MSE performance. The computational requirements and memory usage are both proportional to mN for a data sequence of length m. On a per sample basis however, the computational requirements are independent of m. More specifically, the ith step (corresponding to the ith sample x_i) in the Viterbi algorithm requires essentially four scalar quantizer operations (to find the best codeword in each subset), followed by four add-multiply operations (to compute the distortion associated with each such codeword). Finally, to determine the survivor at each state requires $2N$ adds and N 2-way compares.

The memory requirements can also be made independent of m [103]. Consider a given desired memory depth (or delay) of i_d+1 samples. After completion of the Viterbi processing for sample x_i $i = i_d, i_d+1, i_d+2, \ldots$, the survivor path is traced back i_d steps. The TCQ index for sample x_{i-i_d} (from the "traced back" path) is released for the purpose of bit-stream formation. The path metric (for step i) is then set to infinity for any "inconsistent" survivor paths. An inconsistent survivor path is any path that has not merged with the best path, when traced back i_d steps.

With this procedure, one index is released (with an i_d sample delay) every time one new sample is input to the Viterbi algorithm. It should be clear then that arbitrarily long sequences of data can be processed using memory proportional to only $i_d N$ (rather mN).

Experimental results indicate that negligible loss in performance occurs if i_d is sufficiently large. In particular, whenever $i_d \gtrsim 5 \log_2 N$, the MSE performance of TCQ is substantially preserved. For example, an i_d of 15 to 20 is sufficient for the 8-state trellis of Figure 3.16.

3.5.4   PERFORMANCE OF FIXED RATE TCQ

The Lloyd algorithm is easily adapted to the design of codebooks for TCQ. The resulting algorithm encodes a long sequence of training data over-and-over, replacing each (scalar) codeword in $\mathcal{A}_{\hat{X}}$ by the average of all samples quantized to that codeword at each iteration [142]. Rate $R + 1$ Lloyd-Max scalar quantizer codewords can serve as good initial guesses for the iterative TCQ design procedure.

MSE results for IID uniform and Gaussian data appear in Tables 3.6 and 3.7, respectively. These results were obtained by averaging the 100 MSEs resulting from encoding 100 sequences (each of length $m = 1000$) using codewords designed iteratively as discussed above. Also included for comparison are MSE results for Lloyd-Max scalar quantization, and the distortion-rate function. For both PDFs, the performance of TCQ is

Table 3.6.   TCQ performance for IID uniform data (SNR in dB).

Rate (bits)	Trellis Size (States)							Lloyd-Max Quantizer	Distortion-Rate Function
	4	8	16	32	64	128	256		
1	6.22	6.33	6.39	6.44	6.48	6.55	6.58	6.02	6.79
2	12.62	12.73	12.80	12.85	12.91	12.97	13.00	12.04	13.21
3	18.83	18.94	19.01	19.08	19.13	19.18	19.23	18.06	19.42

Table 3.7.   TCQ performance for IID Gaussian data (SNR in dB).

Rate (bits)	Trellis Size (States)							Lloyd-Max Quantizer	Distortion-Rate Function
	4	8	16	32	64	128	256		
1	5.00	5.19	5.27	5.34	5.43	5.52	5.56	4.40	6.02
2	10.56	10.70	10.78	10.85	10.94	10.99	11.04	9.30	12.04
3	16.19	16.33	16.40	16.47	16.56	16.61	16.64	14.62	18.06

superior to that of scalar quantization for only four states ($N = 4$) and improves steadily as N is increased.

It is interesting to note that for the uniform PDF, TCQ is approaching the distortion-rate function rather quickly as N is increased. In fact, when $N = 256$, TCQ garners all but about 0.2 dB of the gap between SQ and the distortion-rate function. This is also true for larger R, where the optimum TCQ codebook is uniform [104]. As N grows beyond 256, this gap can be further diminished.

Recall from Section 3.4.1, that the 1.53 dB gap for SQ is due solely to the cubic Voronoi cell shape induced by SQ. Evidently, the m dimensional Voronoi regions induced by TCQ are very "sphere-like." Indeed, for small m, these Voronoi regions can be sketched. This is straightforward for the case of $m = 2$. For example, consider all possible pairs of codewords allowed by the trellis structure. Specifically, starting with an initial state of 0 in Figure 3.17, we see that D_0 can be followed by D_0 or by D_2. Similarly, D_2 can be followed by D_1 or by D_3. Thus, the set of all pairs of codewords consistent with these constraints forms the vector

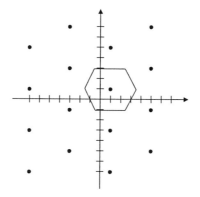

Figure 3.19.   Codewords and Veronoi region for 2-D TCQ ($m = 2$) using Figure 3.17.

codebook with $m = 2$. Specifically,

$$
\begin{aligned}
\mathcal{C} &= (D_0 \times D_0) \cup (D_0 \times D_2) \cup (D_2 \times D_1) \cup (D_2 \times D_3) \\
&= \{(-7, -7), (-7, 1), (1, -7), (1, 1)\} \\
&\cup \{(-7, -3), (-7, 5), (1, -3), (1, 5)\} \\
&\cup \{(-3, -5), (-3, 3), (5, -5), (5, 3)\} \\
&\cup \{(-3, -1), (-3, 7), (5, -1), (5, 7)\}
\end{aligned}
$$

Plotting these codewords in 2-D and sketching the Voronoi region about any of the "interior" codewords yields a hexagon as shown in Figure 3.19.

Unfortunately, the hexagon is not quite regular, as the angles formed at the vertices are 116.57° and 126.86° rather than all 120°, as in the (optimal) hexagonal lattice. This demonstrates that the structure of TCQ results in some suboptimality over unconstrained VQ. However, this same structure allows encoding with very large m which ultimately leads to very high compression efficiency. In fact, the results reported in Table 3.6 for $N = 256$, correspond to a suboptimal (but low complexity) VQ with $m = 1000$. These results are better than theoretically possible for any VQ of dimension less than 69 [104], which would be of astronomical complexity.

3.5.5   ERROR PROPAGATION IN TCQ

Although we have portrayed TCQ as high dimensional VQ in the discussion above, it can also be thought of as time varying scalar quantization. From this point of view, there are four scalar quantizers to be chosen from for each sample. This choice is not arbitrary, but is governed by the structure of the trellis (state machine). From this point of view, we can establish that error propagation is not a serious problem for TCQ.

Recall from previous sections, that if a bit error occurs in the communication of a VQ index, m output samples are affected by the resulting incorrect decoding. Since TCQ employs very large m, this might appear to be of some concern. Fortunately, the number of samples potentially corrupted is considerably less than m for TCQ. Specifically, if a bit error occurs in one of the $(R-1)$ bits used to select codewords from subsets, only one decoded sample will be affected. On the other hand, if a bit error occurs in one of the (single) bits used to select the trellis path, multiple decoded samples can be affected.

This error propagation can be seen by examination of Figure 3.15. This figure shows that one bit in error (at the input u) can affect the selection of the current subset, plus the selection of the next v subsets, as the error propagates through the shift register. It is clear however, that after $v + 1 = \log_2 N + 1$ samples, correct decoding will resume.

Even these "short" sequences of subset errors are less serious than might be expected. For example, consider the $R = 2$ case of Figure 3.17b. Each subset has a positive codeword and a negative codeword. It is easy to see that even when the subset is chosen incorrectly, the sign of the codeword chosen by the decoder will still be correct. For higher rates, the bit assignment (for indices within subsets) can be chosen so that the problem of incorrect subset selection is even less serious.

We close this subsection by noting that the scalar codebook $\mathcal{A}_{\hat{x}}$ can be replaced by a vector codebook to obtain trellis coded vector quantization (TCVQ) [59]. TCVQ can exploit dependencies in non-IID data, as well as achieve non-integer (but rational) encoding rates.

3.5.6 ENTROPY CODED TCQ

As in the case of scalar quantization, TCQ indices can be entropy coded to achieve MSE performance limited only by the Voronoi cell shape. Specifically, 256-state entropy coded TCQ (ECTCQ) can achieve MSE performance within about 0.2 dB of the distortion-rate function at all rates R for any smooth PDF.

The entropy constrained codebook design algorithm of Section 3.2.3 is easily adapted to TCQ. However, as in the case of SQ, uniform thresholds with centroid codewords are very nearly optimal. In scalar quantization, using uniform thresholds with centroid codewords is equivalent to using nearest neighbor encoding with uniform codewords, then substituting centroid codewords at decode time. For TCQ, uniform codewords are

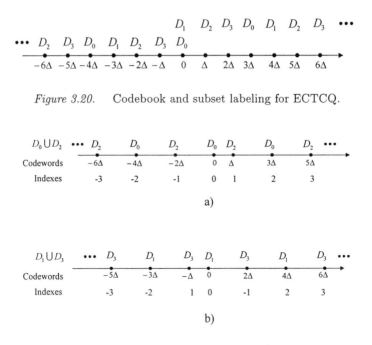

Figure 3.20.   Codebook and subset labeling for ECTCQ.

Figure 3.21.   Union codebooks with ECTCQ indexes.

employed in the Viterbi algorithm while centroid codewords can be substituted at the decoder.

Many variants of ECTCQ have been explored [60, 102, 84, 86, 25]. We describe here only the scheme supported in JPEG2000. The single trellis included in JPEG2000 is shown in Figure 3.16, however, the discussion below applies to any trellis.

The codebook and partition used for ECTCQ is shown in Figure 3.20. Notable features of this figure are that the codebook is uniform with step size Δ, and that the zero codeword appears in two subsets, D_0 and D_1. Encoding proceeds via the Viterbi algorithm as described previously, however, the ECTCQ indices are constructed in a different manner.

All trellises resulting from Figure 3.15 and Table 3.5 (including those of Figures 3.16 and 3.17a) have the property that the subsets associated with the two branches leaving a given state are either D_0 and D_2, or D_1 and D_3. The "union codebooks" $D_0 \cup D_2$ and $D_1 \cup D_3$ are shown in Figure 3.21, along with the ECTCQ index for each codeword. As was the case for entropy coded SQ, the indices are signed.

Note that from the point of view of the decoder, there are two possible codewords for each index. This ambiguity is resolved however by the trellis structure. For example, with an initial state of 0 in the trellis of

Figure 3.17a, the index sequence $(1, 2, -3, -1, 1)$ would be decoded as $(\Delta, 4\Delta, -6\Delta, -2\Delta, -\Delta)$. This can be seen by noting that at state 0, $D_0 \cup D_2$ is the appropriate union codebook. Thus, the index 1 indicates the codeword is Δ, which is in D_2, indicating (by examination of the branch labels) that the next state is 2. At state 2, the appropriate union codebook is $D_1 \cup D_3$. The index 2 then indicates that the codeword is 4Δ, which is in D_1, indicating that the next state is 1, and so on.

At this point, we mention the seemingly strange sign flipping of the ± 1 indices in $D_1 \cup D_3$. In fact, flipping the signs in this fashion for all the indices in $D_1 \cup D_3$ is desirable. For symmetric PDFs, this would cause the indices to have identical distributions in both $D_0 \cup D_2$ and $D_1 \cup D_3$. This in turn allows the trellis state to be omitted from any context model employed for entropy coding of ECTCQ indices. Unfortunately, flipping all the signs destroys any possibility of successive refinement.

A compromise solution is to flip only the sign of ± 1, and omit the trellis state from the context model anyway. Simple analysis and experimental results show that the resulting distribution mismatch results in a negligible loss in entropy coding efficiency.

EMBEDDED ECTCQ

The ECTCQ indices as described above can be entropy coded as integers using Huffman or arithmetic coding [84], as in the case of entropy coded SQ. As mentioned previously, the performance of such a scheme can approach the distortion-rate function quite closely for any IID data with smooth PDF.

As in the case of SQ (Section 3.2.7), the sign magnitude representation of the TCQ indices can be employed to achieve an (approximate) embedding for TCQ. Although inverse TCQ requires exact knowledge of each index to track the state progression through the trellis, a partial inverse is still possible. Examination of Figure 3.21 shows that if the LSB of an ECTCQ index is unknown, the ambiguity is limited to the choice of 4 codewords. For example, if the binary representation (sign-magnitude form) of an index is known to be $+1?$ (where ? denotes the missing LSB), the correct index is guaranteed to be either 2 or 3, and the codeword can be any one of $3\Delta, 4\Delta, 5\Delta$, or 6Δ. One reasonable reconstruction policy in this case would be to chose $\hat{x} = 4\Delta$. Extending this argument to the general case of $p \geq 1$ missing LSBs, a reasonable reconstruction policy is to set the missing LSBs to 0 to obtain an approximate index $\hat{q}$ and then setting $\hat{x} = 2\hat{q}\Delta$. This is in fact the policy recommended for use with JPEG2000 [25].

Employing the notation of Section 3.2.7, we have more generally

$$\hat{x} = \begin{cases} 0 & q^{(p)} = 0 \\ \text{sign}\left(q^{(p)}\right)\left(\left|q^{(p)}\right| + \delta\right)2^{p+1}\Delta & q^{(p)} \neq 0 \end{cases} \qquad (3.46)$$

When there are missing LSBs, this form of ECTCQ provides MSE performance slightly worse than that of embedded ECSQ (Section 3.2.7). However, when all bits are decoded, the full benefit of ECTCQ is realized, and the MSE performance is superior to that of ECSQ.

3.5.7   PREDICTIVE TCQ

TCQ can be inserted into the DPCM structure of Figure 3.7. DPCM decompression is essentially unchanged by this modification. Since inverse TCQ can be performed on a sample-by-sample basis, it is easily substituted for inverse SQ in the DPCM decompression structure.

On the other hand, DPCM compression can be a challenge when TCQ is employed. In fact, optimal encoding is not generally possible. One reasonable approach to compression is to perform a prediction for each of the two paths (emanating from each state) at each step in the Viterbi algorithm [104]. Of course, half of these predictions are immediately discarded by the survivor selection process at each "next" state. The complexity of such an approach is roughly $2N$ times that of DPCM.

Although the encoding of the Predictive TCQ (PTCQ) system described above is suboptimal, the resulting MSE performance is quite good. The MSE gain of PTCQ over DPCM is comparable to the gain of TCQ over SQ. In particular, we saw in Section 3.3 that entropy coded DPCM can achieve MSE performance within 1.53 dB of the distortion-rate function for Gaussian Markov processes. Not surprisingly, entropy coded PTCQ can bridge substantially all of this gap [16].

3.6   FURTHER READING

There are many excellent texts that treat the information theory of quantization. One of the earlier works is the classic book by Gallager [67]. More recent entries include Cover and Thomas [43] and Gray [74]. The special case of rate-distortion theory is covered extensively by Berger in [24]. Although somewhat dated at the time of this writing, the Jayant and Noll text [82] provides fairly comprehensive coverage of speech and image compression, and in particular scalar quantization. Gersho and Gray [69] provide a more recent discussion of compression concepts, and an extensive treatment of vector quantization.

Chapter 4

IMAGE TRANSFORMS

This is the first of two chapters dealing with the extensive subject of image transforms. This first chapter serves as an introduction to block transform methods and the generalization of these methods to subband transforms. As a practical matter, we do not begin with a motivating discussion on the use of transforms for image compression. For a brief review of the role played by linear transforms in image compression applications, the reader is encouraged to re-read Section 1.3. Armed with a knowledge of the structural and mathematical properties of linear image transforms, the reader should be able to appreciate the more comprehensive arguments organized at the end of the chapter, in Section 4.3.

4.1   LINEAR BLOCK TRANSFORMS
4.1.1   INTRODUCTION
ANALYSIS AND SYNTHESIS MATRICES

In this section, we consider finite dimensional linear transforms, which map an n-dimensional input vector, $\mathbf{x}$, into an m-dimensional output vector, $\mathbf{y}$, according to

$$\mathbf{y} = A^*\mathbf{x}, \quad \mathbf{x} = \begin{pmatrix} x_0 \\ x_1 \\ \vdots \\ x_{n-1} \end{pmatrix}, \quad \mathbf{y} = \begin{pmatrix} y_0 \\ y_1 \\ \vdots \\ y_{m-1} \end{pmatrix}$$

Here, A is an $n \times m$ matrix of real or complex coefficients and A^* is its $m \times n$ conjugate transpose. If the transform is real-valued then $A^* = A^t$. An example of a complex-valued finite dimensional transform is the well-known DFT (Discrete Fourier Transform).

We restrict our attention to invertible transforms, writing the inverse as

$$\mathbf{x} = S\mathbf{y}$$

where S is a $n \times m$ left-inverse of A^*; i.e., $SA^* = I$, the $n \times n$ identity matrix. Note that S is not necessarily unique. Henceforth, however, we shall restrict our attention to "non-expansive" transforms, having dimension $m = n$, using m to denote this dimension and reserving the symbol, n, for other purposes. In this case, S is the unique inverse of A^*, which we may write as

$$S^{-1} = A^*, \text{ or } S = A^{-*}$$

Observe that the transform coefficients may be expressed as

$$y_q = \mathbf{a}_q^* \mathbf{x}, \quad q = 0, 1, \ldots, m-1 \tag{4.1}$$

where $\mathbf{a}_q$ is the qth column of the $m \times m$ matrix, A. We refer to $\mathbf{a}_q$ as the qth "analysis vector," since it "analyzes" the original vector $\mathbf{x}$, to determine its qth transform coefficient. Accordingly, we refer to A as the analysis matrix. Also, the inverse transform may be expressed as

$$\mathbf{x} = \sum_{q=0}^{m-1} y_q \mathbf{s}_q \tag{4.2}$$

where $\mathbf{s}_q$ is the qth column of the $m \times m$ matrix, S. We refer to $\mathbf{s}_q$ as the qth "synthesis vector," since $\mathbf{x}$ is "synthesized" from a linear combination of the $\mathbf{s}_q$, with the transform coefficients serving as the weights. Accordingly, the matrix, S, is known as the "synthesis matrix."

Although it is natural to think primarily in terms of the forward transform, A, the inverse transform, S, often provides greater insight. In particular, the transform may be understood as a decomposition of any input vector, $\mathbf{x}$, as a linear combination of the synthesis vectors, $\mathbf{s}_q$, according to equation (4.2). The $\mathbf{s}_q$ may thus be interpreted as "prototype" vectors, with $\mathbf{x}$ a linear combination of these prototypes. As we shall see, transforms which are important for compression are able to represent the source accurately with only a few such prototype vectors. We also refer to the $\mathbf{s}_q$, as "synthesis basis" vectors because they necessarily form a basis for the linear space of all possible input vectors.

In summary, the analysis and synthesis vectors are the columns of the respective analysis and synthesis matrices, A and S. The synthesis vectors form a basis of prototype signals and the transform decomposes its input as a linear combination of these prototypes.

Example 4.1 *The m-point DFT (Discrete Fourier Transform) is de-fined by the equations*[1]

$$y_q = \frac{1}{\sqrt{m}} \sum_{p=0}^{m-1} x_p e^{-j\frac{2\pi}{m}pq}, \quad 0 \le q < m$$

$$x_p = \frac{1}{\sqrt{m}} \sum_{q=0}^{m-1} y_q e^{j\frac{2\pi}{m}pq}, \quad 0 \le p < m$$

In this case, the analysis and synthesis vectors are identical. The elements of $\mathbf{a}_q = \mathbf{s}_q$ are unit spaced samples of a complex sinusoid with frequency $f_q = \frac{q}{m}$; specifically, the pth element is

$$a_{q,p} = s_{q,p} = \frac{1}{\sqrt{m}} e^{j2\pi f_q p}$$

Thus, the DFT decomposes its input as a linear combination of complex exponential waveforms having uniformly spaced frequencies, f_q.

BLOCKING OF 1D AND 2D SIGNALS

In most applications, the length of the signals to be transformed is unbounded, or at least far exceeds the dimension of the transform. Consequently, the source must be processed in blocks and the selection of a finite dimensional transform implies a block transform process. Figure 4.1 illustrates the process for one dimensional signals. We first partition the signal, $x[k]$, into contiguous blocks of m samples each, denoted $\mathbf{x}[n]$. These blocks are then independently transformed into corresponding transform blocks, denoted $\mathbf{y}[n]$, which form an equivalent one dimensional sequence of transform coefficients, $y[k]$.

Two dimensional signals such as images, are usually partitioned into square blocks of $m \times m$ samples each, which we may represent by m^2-dimensional vectors, $\mathbf{x}[n_1, n_2] = \mathbf{x}[n]$, as illustrated in Figure 4.2. The order in which the m^2 samples in each block appear within its vector is unimportant from a conceptual point of view. When necessary, we shall assume that the samples are scanned into the vector in raster-scan order. Often, however, it will be more convenient to employ a two dimensional

---

[1]Here, we use a less common normalization convention, which distributes the scaling factor of $\frac{1}{m}$ between the forward and inverse transforms.

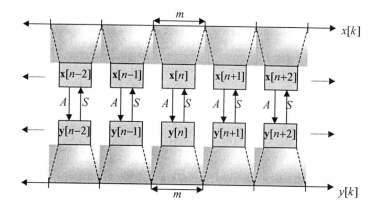

Figure 4.1.   Block transform process for 1D signals.

indexing notation[2] for the elements of each block vector, $\mathbf{x}[\mathbf{n}]$; i.e.,

$$\mathbf{x}[\mathbf{n}] = \left(\begin{array}{c} \left. \begin{array}{c} x_{0,0}[\mathbf{n}] \\ x_{0,1}[\mathbf{n}] \\ \vdots \\ x_{0,m-1}[\mathbf{n}] \end{array} \right\} \text{ first scan line} \\ x_{1,0}[\mathbf{n}] \\ \vdots \\ x_{m-1,m-1}[\mathbf{n}] \end{array} \right)$$

where

$$x_{p_1,p_2}[\mathbf{n}] = x[mn_1 + p_1, mn_2 + p_2], \quad 0 \leq p_1, p_2 < m$$

and p_1 and p_2 denote row and column indices, respectively, of the samples in the block vector. We employ a similar notation to refer to the elements, $y_{q_1,q_2}[\mathbf{n}]$, of each transform vector, $\mathbf{y}[\mathbf{n}] = A^*\mathbf{x}[\mathbf{n}]$. Thus, we have

$$y_{q_1,q_2}[\mathbf{n}] = \mathbf{a}^*_{q_1,q_2}\mathbf{x}[\mathbf{n}]; \text{ and } \mathbf{x}[\mathbf{n}] = \sum_{q_1,q_2} \mathbf{s}_{q_1,q_2} y_{q_1,q_2}[\mathbf{n}]$$

where the analysis and synthesis vectors, $\mathbf{a}_{q_1,q_2}$ and $\mathbf{s}_{q_1,q_2}$, are the $(q_2 + mq_1)^{\text{th}}$ columns of the $m^2 \times m^2$ matrices, A and S, respectively.

---

[2]The most natural notation here is that of tensors, where the vectors are replaced by two dimensional tensors and the analysis and synthesis matrices are replaced by four dimensional tensors. We choose not to burden the reader with tensor notation in this treatment.

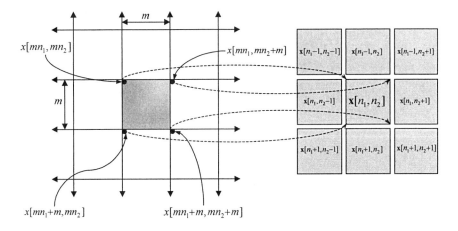

Figure 4.2. Blocking for 2D block transforms.

Amongst the clutter of notation being introduced here, substantial insight may be gained into the behaviour of two dimensional block transforms by observing that they represent each block in the input image as a linear combination of "prototype image blocks," $\mathbf{s}_{q_1,q_2}[\mathbf{n}]$. Not surprisingly, when block transforms are used in image compression, the block structure of the prototype images can sometimes be observed in the reconstructed image. We shall investigate the structure of prototype image blocks further in Section 4.1.3.

SEPARABLE TRANSFORMS

An obvious way to construct a two dimensional transform, A, is by "separable extension" of a one dimensional transform, A'. Each analysis vector, $\mathbf{a}_{q_1,q_2}$, of a separable transform is formed by taking the "tensor product" of the one dimensional analysis vectors, $\mathbf{a}'_{q_1}$ and $\mathbf{a}'_{q_2}$; i.e.,

$$\left(\mathbf{a}_{q_1,q_2}\right)_{p_1,p_2} = \left(\mathbf{a}'_{q_1}\right)_{p_1} \left(\mathbf{a}'_{q_2}\right)_{p_2}, \quad 0 \le q_1, q_2, p_1, p_2 < m$$

Similarly, each synthesis vector, $\mathbf{s}_{q_1,q_2}$, is the tensor product of $\mathbf{s}'_{q_1}$ and $\mathbf{s}'_{q_2}$.

A key practical advantage of separable transforms is that they may be implemented by applying the one dimensional transform first to the rows of the image and then to its columns. To clarify this point, observe

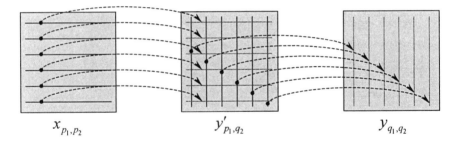

Figure 4.3. Separable transform implementation by one dimensional transformation of the rows and then the columns.

that

$$y_{q_1,q_2} = \mathbf{a}^*_{q_1,q_2}\mathbf{x} = \sum_{p_1,p_2} \left(\mathbf{a}_{q_1,q_2}\right)^*_{p_1,p_2} x_{p_1,p_2}$$

$$= \sum_{p_1} \left(\mathbf{a}'_{q_1}\right)^*_{p_1} \underbrace{\sum_{p_2} \left(\mathbf{a}'_{q_2}\right)^*_{p_2} x_{p_1,p_2}}_{\left(\mathbf{y}'_{q_2}\right)_{p_1}}$$

$$= \mathbf{a}'^*_{q_1}\mathbf{y}'_{q_2}$$

Thus, we first apply the one dimensional transform independently to each row, p_1, of the image, generating an intermediate two dimensional array, y'_{p_1,q_2}, each column of which is a vector, $\mathbf{y}'_{q_2}$, to which we apply the one dimensional transform again. The procedure is illustrated in Figure 4.3.

Of course, the same argument shows that we can apply the transform first to the columns and then to the rows, with identical results. Furthermore, the inverse transform may clearly be implemented in an analogous manner, starting with the rows of the coefficient array, y_{q_1,q_2}, and proceeding to the columns, or vice versa. Separable transforms involve a substantial reduction in complexity. To form each coefficient of a non-separable transform, we require m^2 multiplications and additions. By contrast, only $2m$ multiplications and additions are required to implement a separable transform[3].

---

[3] Further minor simplifications are possible. For example, one of the multiplicands for the column transform may be folded into the row transform. In many cases, symmetry or richer structural properties may be exploited for further simplification.

Example 4.2 *The two dimensional $m \times m$-point DFT is defined by*

$$y_{q_1,q_2} = \frac{1}{m} \sum_{p_1=0}^{m-1} \sum_{p_2=0}^{m-1} x_{p_1,p_2} e^{-j\frac{2\pi}{m}(p_1 q_1 + p_2 q_2)}, \quad 0 \le q_1, q_2 < m$$

$$x_{p_1,p_2} = \frac{1}{m} \sum_{q_1=0}^{m-1} \sum_{q_2=0}^{m-1} y_{q_1,q_2} e^{j\frac{2\pi}{m}(p_1 q_1 + p_2 q_2)}, \quad 0 \le p_1, p_2 < m$$

As in the case of the one dimensional DFT, the analysis and synthesis vectors are identical; i.e., $\mathbf{a}_{q_1,q_2} = \mathbf{s}_{q_1,q_2}$. Their elements are unit spaced samples of a two dimensional complex sinusoid with vertical and horizontal frequencies, $f_{q_1} = \frac{q_1}{m}$ and $f_{q_2} = \frac{q_2}{m}$, respectively. It is easily verified that this transform is the separable extension of the one dimensional DFT in Example 4.1.

VECTOR SPACE PERSPECTIVE

The inner product between two m-dimensional vectors, $\mathbf{v}$ and $\mathbf{w}$, is defined by[4]

$$\langle \mathbf{v}, \mathbf{w} \rangle = \sum_{p=0}^{m-1} v_p w_p^*$$

This is the familiar "dot-product." Thus, the forward transform (or analysis) operation of equation (4.1) may be written as the inner product,

$$y_q = \langle \mathbf{x}, \mathbf{a}_q \rangle$$

Admittedly, this is only notation. Its chief benefit to us is that we shall be able to use the same notation with identical results and interpretations when referring to non-block transforms, including subband and wavelet transforms.

A transform is said to be orthonormal if the analysis vectors are all mutually orthogonal, having unit norm (length); i.e.,

$$\langle \mathbf{a}_i, \mathbf{a}_j \rangle = 0, \quad \forall i \ne j \tag{4.3}$$

$$\langle \mathbf{a}_i, \mathbf{a}_i \rangle = \|\mathbf{a}_i\|^2 = 1, \quad \forall i \tag{4.4}$$

This means that $AA^* = A^*A = I$, so that $S = A$ is a unitary matrix. Equivalently, the analysis and synthesis vectors for orthonormal transforms are identical. The one- and two dimensional DFT's of Examples 4.1 and 4.2 are orthonormal transforms. An orthonormal transform

---

[4]Formally, this is the definition of inner-product on a finite dimensional Hilbert space over the field of complex numbers, $\mathbb{C}$.

performs an orthonormal expansion of the input signal as the sum of its projections onto each of the basis vectors; i.e.,

$$\mathbf{x} = \sum_q y_q \mathbf{s}_q = \sum_q \langle \mathbf{x}, \mathbf{s}_q \rangle \cdot \mathbf{s}_q$$

An important property of orthonormal transforms/expansions is that they are "energy preserving," meaning that

$$\sum_p |x_p|^2 = \|\mathbf{x}\|^2 = \langle \mathbf{x}, \mathbf{x} \rangle = \left\langle \sum_i y_i \mathbf{s}_i, \sum_j y_j \mathbf{s}_j \right\rangle$$

$$= \sum_{i,j} y_i y_j^* \langle \mathbf{s}_i, \mathbf{s}_j \rangle = \sum_q |y_q|^2 = \|\mathbf{y}\|^2$$

In words, the sum of the squares of the input samples (energy of the input), is identical to the sum of the squares of the transform coefficients (energy of the output). This property is often known as Parseval's relation.

To appreciate the significance of this property for compression, consider the compression system of Figure 1.4. Let $\mathbf{e}_y = \mathbf{y} - \hat{\mathbf{y}}$ denote the error introduced into the transform coefficients by quantization. Similarly, let $\mathbf{e}_x = \mathbf{x} - \hat{\mathbf{x}}$ denote the error introduced into the reconstructed image by the entire compression system. By linearity of the transform, $\mathbf{e}_x = S\mathbf{e}_y$ and if the transform is orthonormal, $\|\mathbf{e}_x\|^2 = \|\mathbf{e}_y\|^2$. In words, the error energy in the image domain is identical to the error energy in the transform domain. Minimizing the MSE of the quantized transform coefficients is then identical to minimizing the MSE of the reconstructed image. This property will be exploited in Section 4.3, where we discuss the rate-distortion properties of compression systems involving transforms.

An orthogonal transform is one whose analysis vectors satisfy equation (4.3), but not necessarily equation (4.4). In this case $AA^* = D$, a diagonal matrix, with

$$D_{ii} = \langle \mathbf{a}_i, \mathbf{a}_i \rangle, \text{ and } \mathbf{s}_i = D_{ii}^{-1} \mathbf{a}_i$$

Thus, the synthesis vectors are also orthogonal and the energy preserving property persists in the modified form

$$\sum_p |x_p|^2 = \sum_q D_{qq}^{-1} |y_q|^2$$

More generally, any arbitrary invertible transform has the property that $SA^* = A^*S = I$, which means that the following "biorthogonality" relations must hold

$$\langle \mathbf{s}_i, \mathbf{a}_j \rangle = 0, \quad \forall i \neq j$$
$$\langle \mathbf{s}_i, \mathbf{a}_i \rangle = 1, \quad \forall i$$

The term "biorthogonal transform" may thus be applied to any invertible transform, including orthogonal transforms as a special case.

4.1.2   KARHUNEN-LOEVE TRANSFORM

The KLT (Karhunen-Loève Transform) is an orthonormal transform of substantial theoretical significance. Since information sources are statistical in nature, let $\mathbf{x}$ be the outcome of an underlying random vector, $\mathbf{X}$. Then the transform vector, $\mathbf{y} = K^*\mathbf{x}$, is an outcome of the random vector

$$\mathbf{Y} = K^*\mathbf{X}$$

Let $\mu_{\mathbf{X}} = E[\mathbf{X}]$ denote the mean of the random vector, $\mathbf{X}$, and $C_{\mathbf{X}}$ its covariance matrix; i.e.,

$$C_{\mathbf{X}} = E\left[(\mathbf{X} - \mu_{\mathbf{X}})(\mathbf{X} - \mu_{\mathbf{X}})^*\right] = E[\mathbf{X}\mathbf{X}^*] - \mu_{\mathbf{X}}\mu_{\mathbf{X}}^*$$

Clearly, $C_{\mathbf{X}} = C_{\mathbf{X}}^*$ is a hermitian matrix (symmetric for real-valued random vectors). The element at row i and column j of $C_{\mathbf{X}}$ holds the covariance of the random variables, X_i and X_j; i.e.,

$$\text{cov}(X_i, X_j) = E\left[(X_i - \mu_{X_i})(X_j - \mu_{X_j})^*\right] = E\left[X_i X_j^*\right] - \mu_{X_i}\mu_{X_j}^*$$

Recall that two random variables are said to be "uncorrelated" if their covariance is zero. Accordingly, $C_{\mathbf{X}}$ will be a diagonal matrix if and only if the constituent random variables, $\{X_i\}_{0 \leq i < m}$ are mutually uncorrelated.

Amongst all possible orthonormal transforms, the KLT is the unique[5] transform which decorrelates its input. By this, we mean that the transform vector has a diagonal covariance matrix, $C_{\mathbf{Y}}$. The relationship between $C_{\mathbf{X}}$ and $C_{\mathbf{Y}}$ is

$$\begin{aligned} C_{\mathbf{Y}} &= E\left[(\mathbf{Y} - \mu_{\mathbf{Y}})(\mathbf{Y} - \mu_{\mathbf{Y}})^*\right] \\ &= E\left[(K^*\mathbf{X} - K^*\mu_{\mathbf{X}})(K^*\mathbf{X} - K^*\mu_{\mathbf{X}})^*\right] \\ &= E\left[K^*(\mathbf{X} - \mu_{\mathbf{X}})(\mathbf{X} - \mu_{\mathbf{X}})^* K\right] \\ &= K^* C_{\mathbf{X}} K \end{aligned}$$

---

[5] Actually, any orthonormal decorrelating transform must be identical to the KLT, up to a permutation (re-ordering) and potential sign-flipping of the transform coefficients.

Since C_Y is to be diagonal and the transform is to be orthonormal (i.e., $KK^* = I$), we must have

$$KC_Y = C_X K \text{ or}$$
$$\sigma_{Y_i}^2 k_i = C_X k_i, \quad 0 \le i < m$$

where $\sigma_{Y_i}^2 = \text{cov}(Y_i, Y_i)$ is the variance of the ith transform coefficient, and k_i is the ith column of K; i.e., the ith synthesis/analysis vector of the transform. We conclude that the k_i must be the eigenvectors of the hermitian matrix, C_X, with $\sigma_{Y_i}^2$ the eigenvalue corresponding to k_i. A well-known property of hermitian matrices is that their eigenvectors are mutually orthogonal [166, Corollary 4.4.9]. Thus, the KLT always exists.

A popular tool for finding the KLT matrix is the well-known SVD (Singular Value Decomposition). Specifically, the SVD of C_X is

$$C_X = U\Sigma V^*$$

where U and V are unitary (orthonormal rows/columns) and Σ is a diagonal matrix of singular values. Since C_X is symmetric, its SVD has $U = V$. Setting $K = U$, we obtain

$$C_Y = K^* C_X K = U^* U \Sigma U^* U = \Sigma$$

which is diagonal. Thus, the analysis/synthesis vectors, k_i, of the KLT, are the columns of the unitary matrix, U, from the SVD of C_X and the variance of each transform coefficient, $\sigma_{Y_i}^2$, is the corresponding singular value. Both the KLT and the SVD are defined so as to yield coefficient variances (singular values) in decreasing order,

$$\sigma_{Y_0}^2 \ge \sigma_{Y_1}^2 \ge \cdots \ge \sigma_{Y_{m-1}}^2 \tag{4.5}$$

As a result of this connection, the KLT and SVD are often misunderstood as synonymous.

SIGNIFICANCE OF DECORRELATING TRANSFORMS

Suppose two random variables, X and Y, are statistically independent. Then, by definition, their joint PDF (probability density function) is the separable product of the marginal distributions,

$$f_{X,Y}(x, y) = f_X(x) f_Y(y), \quad \forall x, y$$

and hence

$$E[XY] = \int\int xy f_{X,Y}(x,y) \cdot dx \cdot dy$$

$$= \int x f_X(x) dx \int y f_Y(y) dy$$

$$= \mu_X \mu_Y$$

Thus, $\text{cov}(X,Y) = E[XY] - \mu_X \mu_Y = 0$ and the random variables are uncorrelated. Unfortunately, the converse is not generally true. A common mistake is to refer to two quantities as uncorrelated, as though that were tantamount to statistical independence. The following example demonstrates the shortcomings of such an inference.

Example 4.3 *Let X be a zero-mean random variable, uniformly distributed on the interval, $\left[-\frac{1}{2}, \frac{1}{2}\right]$ and let $Y = |X|$. Then*

$$\text{cov}(X,Y) = E[XY] - \mu_X \mu_Y = E[XY]$$

$$= \int_{-\frac{1}{2}}^{\frac{1}{2}} x|x| dx = 0$$

So X and Y are uncorrelated, but they are certainly not independent; in fact, Y is a deterministic function of X.

There is one important class of distributions for which decorrelation and statistical independence are equivalent, namely Gaussian distributions. An m-dimensional random vector, $\mathbf{X}$, has a Gaussian (normal) distribution if its PDF has the form

$$f_{\mathbf{X}}(\mathbf{x}) = \frac{1}{\sqrt{(2\pi)^m \det(C_{\mathbf{X}})}} e^{-\frac{1}{2}(\mathbf{x}-\mu_{\mathbf{X}})^* C_{\mathbf{X}}^{-1}(\mathbf{x}-\mu_{\mathbf{X}})}$$

When the random variables, $X_0, X_1, \ldots, X_{m-1}$, are uncorrelated, $C_{\mathbf{X}}$ is diagonal and the joint distribution becomes

$$f_{\mathbf{X}}(\mathbf{x}) = \frac{1}{\sqrt{(2\pi)^m \prod_{i=0}^{m-1} \sigma_{X_i}^2}} e^{-\frac{1}{2}\sum_{i=0}^{m-1} \frac{1}{\sigma_{X_i}^2}|x_i - \mu_{X_i}|^2}$$

$$= \prod_{i=0}^{m-1} \frac{1}{\sqrt{2\pi\sigma_{X_i}^2}} e^{-\frac{1}{2\sigma_{X_i}^2}|x_i - \mu_{X_i}|^2}$$

which is a separable product of one dimensional Gaussian distributions. Thus, jointly Gaussian random variables are statistically independent if and only if they are uncorrelated.

Another important property of Gaussian distributions is that the distribution remains Gaussian under any linear transformation. Thus, if $\mathbf{X}$ is Gaussian and $\mathbf{Y} = K^*\mathbf{X}$ is its KLT, then $\mathbf{Y}$ is a Gaussian random vector with diagonal covariance matrix, meaning that the Y_i are statistically independent random variables. In Section 4.3.3 we show that the KLT is in fact the optimal block transform for compression, subject to the somewhat contrived assumption that the source follows a Gaussian distribution.

For non-Gaussian distributions, we cannot hope to decompose the source into statistically independent components by means of a linear transform. Nevertheless, since decorrelation is a necessary if not sufficient condition for statistical independence, the KLT is still an excellent choice.

PRINCIPLE COMPONENTS AND THE KLT

As with any transform, the KLT may be interpreted as decomposing the source as a linear combination of prototypes, $\mathbf{k}_q$. In the case of the KLT, these prototypes are known as the "principle components" of the source. Suppose that we are free to keep only a subset of the coefficients, with indices in $\mathcal{M}$, letting the remaining coefficients default to their mean value (usually zero). We synthesize the source using this reduced set of coefficients as

$$\hat{\mathbf{x}} = \sum_{q \in \mathcal{M}} y_q \mathbf{k}_q + \sum_{q \notin \mathcal{M}} \mu_{Y_q} \mathbf{k}_q$$

and the expected (mean) squared error (MSE) of this approximation is

$$E\left[\left\|\mathbf{X} - \hat{\mathbf{X}}\right\|^2\right] = E\left[\left\|\mathbf{Y} - \hat{\mathbf{Y}}\right\|^2\right] = \sum_{q \notin \mathcal{M}} \sigma_{Y_q}^2$$

Note that we have exploited the energy preserving property of orthonormal expansions. In view of equation (4.5), we should select the first $m' = |\mathcal{M}|$ coefficients,

$$\mathcal{M} = \left\{0, 1, \ldots, m' - 1\right\}$$

so as to minimize the MSE of the approximation. In fact, it can be shown that amongst all linear transforms, the transform for which MSE is minimum, if we keep only the first $m' < m$ coefficients, is the KLT. This is most fortuitous, because the optimum transform does not depend upon the number of coefficients, m', which we choose to keep.

Thus, if we are free to approximate $\mathbf{x}$ as a multiple of only one vector (plus a constant offset), the vector which will minimize the mean square

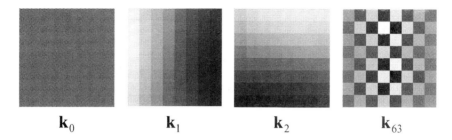

$$\mathbf{k}_0 \qquad\qquad \mathbf{k}_1 \qquad\qquad \mathbf{k}_2 \qquad\qquad \mathbf{k}_{63}$$

Figure 4.4. First three and last principle components for 8×8 blocks taken from the image, "Goldhill."

approximation error is $\mathbf{k}_0$. If we are free to approximate $\mathbf{x}$ as a linear combination of any two vectors (plus a constant offset), the selection which will minimize the mean squared error (MSE) is the vectors, $\mathbf{k}_0$ and $\mathbf{k}_1$, and so on. This explains the name "principle components" for the KLT synthesis vectors, $\mathbf{k}_q$. This property has obvious appeal and has given rise to many compression schemes. Nevertheless, as we shall see in Section 4.3.2, this is not the optimal strategy for exploiting the KLT.

When used as a two dimensional block transform, the synthesis vectors are prototype image blocks. The first prototype image block, $\mathbf{k}_0$, is the first principle component and usually corresponds to a block of constant intensity samples so that $y_0[\mathbf{n}]$ is the "DC" coefficient of image block $\mathbf{x}[\mathbf{n}]$ and $\hat{\mathbf{x}}[\mathbf{n}] = y_0[\mathbf{n}]\mathbf{k}_0$ is a piecewise constant approximation to the original image. In general, the first few principle components usually represent smoothly varying intensity patterns, since images typically contain much more energy at low spatial frequencies than at high frequencies. Figure 4.4 offers evidence for this behaviour. The figure illustrates the intensity patterns associated with the first few and last principle components for blocks of size 8×8, where the relevant covariances are estimated by taking averages over blocks drawn from the 576×720 test image, "Goldhill," shown in Figure 4.5.

4.1.3   DISCRETE COSINE TRANSFORM

The DCT (Discrete Cosine Transform) is a real-valued orthonormal transform, whose analysis/synthesis vectors, $\mathbf{s}_q$, consist of unit spaced samples of cosine functions, having frequencies, $f_q = \frac{q}{2m}$. Specifically,

$$s_{q,p} = c_q \cos\left(2\pi f_q \left(p + \frac{1}{2}\right)\right); \quad f_q = \frac{q}{2m} \qquad (4.6)$$

Figure 4.5. 576×720 test image, "Goldhill."

where the normalization factor is selected to ensure that $\|\mathbf{s}_q\| = 1$; its value is

$$c_q = \begin{cases} \sqrt{\frac{1}{m}} & \text{if } q = 0 \\ \sqrt{\frac{2}{m}} & \text{if } q \neq 0 \end{cases}$$

The orthogonality of these vectors is easily demonstrated.

The DCT and DFT (see Example 4.1) have much in common. Both are orthonormal transforms whose basis vectors are unit sampled sinusoids. In the case of the DCT, the sinusoids are real-valued and their frequencies are spaced at multiples of $\frac{1}{2m}$, while the DFT's sinusoids are complex-valued with frequencies separated by multiples of $\frac{1}{m}$. We will explore the relationship between the DCT and the DFT further shortly.

The two dimensional DCT is the separable extension of the one dimensional DCT, having analysis/synthesis vectors, $\mathbf{s}_{q_1,q_2}$, whose elements are

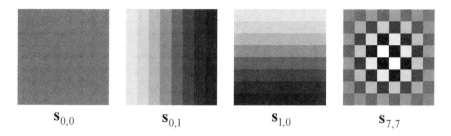

$$\mathbf{s}_{0,0} \qquad\qquad \mathbf{s}_{0,1} \qquad\qquad \mathbf{s}_{1,0} \qquad\qquad \mathbf{s}_{7,7}$$

Figure 4.6.   First few and last basis (prototype) blocks of the 8×8 DCT.

given by

$$\left(\mathbf{s}_{q_1,q_2}\right)_{p_1,p_2} = c_{q_1} c_{q_2} \cos\left(2\pi f_{q_1}\left(p_1 + \frac{1}{2}\right)\right) \cos\left(2\pi f_{q_2}\left(p_2 + \frac{1}{2}\right)\right)$$
(4.7)

These are the prototype vectors of the block-based DCT transform. Figure 4.6 shows the intensity patterns of a few of the lowest frequency prototype blocks for the 8×8 DCT transform, along with the highest frequency prototype block, $\mathbf{s}_{7,7}$. Notice the similarity between these intensity patterns and those of the first few and last principle components from the 8×8 KLT, appearing in Figure 4.4. In fact, a key attribute of the DCT is its similarity to the KLT for natural image sources.

RELATIONSHIP BETWEEN THE DCT AND THE KLT

It can be shown [18] that the DCT approximately diagonalizes the covariance matrix of a first-order Gauss-Markov random process for which

$$\text{cov}\left(X_{p_1,p_2} X_{p_1',p_2'}\right) = \rho^{|p_1 - p_1'| + |p_2 - p_2'|}$$

where ρ is close to 1. In fact, one of the properties of any wide-sense stationary (WSS) random process is that the Fourier coefficients are uncorrelated. That is, as m becomes very large, the DFT, the DCT and other related frequency transforms all diagonalize the source covariance matrix [156]. Thus, these transforms are asymptotically equivalent to the KLT for WSS sources, up to a reordering of the coefficients.

One suitable data independent ordering for the DCT coefficients is the "zig-zag" scan shown in Figure 4.7. This order is based on the observation that the power density spectra of most images tends to decrease rapidly with increasing spatial frequency; it is employed by the JPEG image compression standard and most video compression standards.

Although most images are not well modeled as WSS random processes, the DCT has been found to be a robust approximation to the KLT

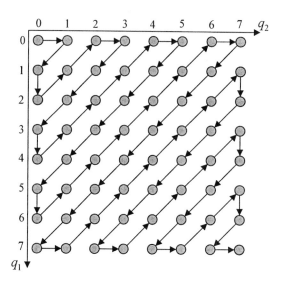

Figure 4.7. Zig-zag scan to visit DCT coefficients in order of roughly decreasing variance.

for natural image sources. From a practical perspective, the DCT has numerous advantages over the KLT: it is a separable transform; highly efficient implementations exist; and there is no need (or opportunity) to adapt the transform to the statistics of the source material.

RELATIONSHIP BETWEEN THE DCT AND THE DFT

It is worth exploring the relationship between the DCT and the DFT further. We restrict our attention to the one dimensional case and note that the input vector, $\mathbf{x}$, consists of a single block from an underlying source sequence. Let $\breve{x}[n]$ be the periodic sequence with period $2m$ defined by

$$\breve{x}[p] = x_p, \quad 0 \le p < m$$
$$\breve{x}[n] = \breve{x}[-1 - n] = \breve{x}[2m - 1 - n], \quad \forall n$$

The first half of each period holds $\mathbf{x}$, while the second half of each period contains a mirror image of $\mathbf{x}$, as illustrated in Figure 4.8. Since $\breve{x}[n]$ is periodic, it may be expanded in a Fourier series as

$$\breve{x}[n] = \frac{1}{\sqrt{2m}} \sum_{q=0}^{2m-1} \breve{y}_q e^{j\frac{2\pi}{2m}nq}$$

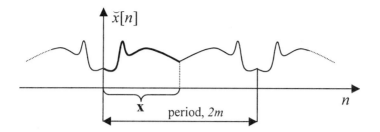

Figure 4.8. Periodic signal whose DFT is a scaled version of the DCT of **x**.

where the Fourier coefficients, $\breve{y}_q$, are given by

$$\breve{y}_q = \frac{1}{\sqrt{2m}} \sum_{n=0}^{2m-1} \breve{x}\,[n]\, e^{-j\frac{2\pi}{2m}nq}$$

which is the $2m$-point DFT of a single period of $\breve{x}\,[n]$. The above expression may be manipulated as follows:

$$\breve{y}_q = \frac{1}{\sqrt{2m}} \left[\sum_{p=0}^{m-1} x_p e^{-j\frac{\pi q}{m}p} + \sum_{p=0}^{m-1} x_p e^{-j\frac{\pi q}{m}(2m-1-p)} \right]$$

$$= \frac{1}{\sqrt{2m}} e^{j\frac{\pi}{2m}q} \sum_{p=0}^{m-1} x_p \left(e^{-j\frac{\pi q}{m}\left(p+\frac{1}{2}\right)} + e^{j\frac{\pi q}{m}\left(p+\frac{1}{2}\right)} \right)$$

$$= \sqrt{\frac{2}{m}} e^{j\frac{\pi}{2m}q} \sum_{p=0}^{m-1} x_p \cos\left(2\pi \frac{q}{2m} \left(p+\frac{1}{2}\right) \right)$$

$$= \left(\sqrt{\frac{2}{m}} c_q^{-1} e^{j\frac{\pi}{2m}q} \right) y_q$$

So the DCT coefficients of **x** are related to the $2m$-point DFT coefficients of the symmetrically extended sequence, $\breve{x}\,[n]$, by a constant scale factor, which is not signal dependent; i.e.,

$$y_q = \left(c_q \sqrt{\frac{m}{2}} e^{-j\frac{\pi}{2m}q} \right) \breve{y}_q, \quad q = 0, 1, \ldots, m-1$$

One of many important consequences of this connection between the DCT and the DFT is that we can exploit efficient FFT (Fast Fourier Transform) algorithms [52] which have been developed for computing the DFT. These algorithms have complexity of order $m \log_2 m$ which represents a significant saving over direct implementation with complexity of order m^2.

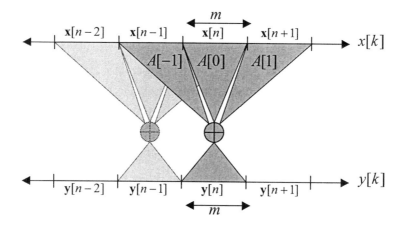

Figure 4.9.   One dimensional convolutional transform.

4.2   SUBBAND TRANSFORMS

The principle limitation of block transforms is that the source signal or image must be processed independently in blocks[6]. Equivalently, the transform decomposes the source as a linear combination of disjoint blocks. Neighbouring source samples may lie within the influence of different blocks. However, we do not usually have any a priori reason to expect these neighbouring samples to be uncorrelated; in fact, for many sources, including natural images, neighbouring samples exhibit the highest correlation. As a result, we can expect the transform coefficients from adjacent blocks to exhibit significant residual correlation, even if the block transform fully decorrelates each block, as in the KLT. Thus, the full decorrelating potential of block transforms is realized only asymptotically as the block size, $m \to \infty$.

4.2.1   VECTOR CONVOLUTION

A natural way to extend block transforms is by adding memory. The idea is illustrated for one dimensional sources in Figure 4.9. Again, the input and output sequences are grouped into m-dimensional vectors, $\mathbf{x}[n]$ and $\mathbf{y}[n]$. In this case, however, each output vector is formed from multiple input vectors, through a sequence of transform matrices, $A[n]$,

---

[6]In accordance with the opening statements of Section 4.1, we consistently restrict our attention to non-expansive transforms, so that redundant transforms involving overlapping blocks are ruled out.

according to

$$\mathbf{y}[n] = \sum_{i \in \mathbb{Z}} A^*[i]\,\mathbf{x}[n-i] \tag{4.8}$$

From one perspective, this is simply the generalization of scalar convolution to m-dimensional vector convolution, suggesting the term "convolutional" transform. From a different perspective, it is the generalization from block transforms to "sliding window" transforms. As we shall see shortly, a frequency domain perspective suggests the term "subband" transform, while a particular implementation suggests the term "maximally decimated filter bank," or simply "filter bank." Throughout the ensuing development, we shall use each of these terms as best befits the context, bearing in mind that they refer to the same fundamental operation.

It turns out that the inverse (synthesis) transform almost always exists and that it has the same form with

$$\mathbf{x}[n] = \sum_{i \in \mathbb{Z}} S[i]\,\mathbf{y}[n-i] \tag{4.9}$$

We point out, however, that unless the analysis matrices are selected carefully, the synthesis system generally has infinite, non-causal support. For some applications, causality is important; i.e., we must have $A[i] = 0$ for $i < 0$, as in the figure. Of particular interest, however, is the case in which both the forward and inverse transforms have finite support. Non-causal transforms with finite support can always be implemented causally by the introduction of sufficient delay, so we will generally consider two-sided sequences, $A[i]$ and $S[i]$.

4.2.2   POLYPHASE TRANSFER MATRICES

The pth elements of each input vector, $\mathbf{x}[n]$, form a sequence, $x_p[n]$, which is known as the pth polyphase component of $x[k]$. Specifically,

$$x_p[n] \triangleq (\mathbf{x}[n])_p = x[nm + p]$$

Similarly, we may refer to the sequences, $y_q[n]$, as the polyphase components of an interleaved sequence of transform coefficients, $y[k]$, given by $y[nm + q] = y_q[n]$.

As for scalar convolution, we may define the Z-transform of the matrix impulse response, $A^*[i]$, as the formal power series

$$H(z) = \sum_{i \in \mathbb{Z}} A^*[i]\,z^{-i}$$

Similarly, we define

$$G(z) = \sum_{i \in \mathbb{Z}} S[i] z^{-i}$$

and then the analysis and synthesis operations of equations (4.8) and (4.9) may be rewritten as

$$\mathbf{y}(z) = H(z) \mathbf{x}(z)$$

and

$$\mathbf{x}(z) = G(z) \mathbf{y}(z)$$

Here, $\mathbf{x}(z)$ and $\mathbf{y}(z)$ denote the vectors formed from the Z-transforms of the respective polyphase components of the input and output sequences.

$H(z)$ and $G(z)$ are known as the "polyphase transfer matrices," or simply the "polyphase matrices," of the analysis and synthesis systems. Evidently, the synthesis system correctly inverts the analysis system if and only if

$$G(z) H(z) = H(z) G(z) = I$$

Equivalently, the analysis and synthesis matrices, $A[i]$ and $S[i]$, must satisfy

$$\sum_{i \in \mathbb{Z}} S[j-i] A^*[i] = \sum_{i \in \mathbb{Z}} A^*[i] S[j-i] = I \cdot \delta[j], \quad \forall j \in \mathbb{Z} \qquad (4.10)$$

The Z-transform representation provides a direct method for deducing the synthesis system from the analysis system or vice-versa. Specifically, we must have

$$G(z) = (H(z))^{-1} = \frac{\text{cofactor}(H(z))}{\det(H(z))}$$

In the simple case when $m = 2$, this relation may be expressed as

$$\begin{pmatrix} G_{00}(z) & G_{01}(z) \\ G_{10}(z) & G_{11}(z) \end{pmatrix} = \frac{\begin{pmatrix} H_{11}(z) & -H_{01}(z) \\ -H_{10}(z) & H_{00}(z) \end{pmatrix}}{H_{00}(z) H_{11}(z) - H_{01}(z) H_{10}(z)}$$

For the analysis system to be practically realizable, the elements of $H(z)$ must be rational polynomials in z, in which case the elements of $G(z)$ will also be rational polynomials in z so long as $\det(H(z))$ is not identically equal to zero. If we select an analysis system at random, the event that $\det(H(z)) \equiv 0$ occurs with zero probability, so it is reasonable to claim that the synthesis system does indeed exist for virtually all choices of the analysis system[7]. If we are interested in analysis and synthesis systems which both have finite support, then the elements of both $H(z)$ and $G(z)$ must be finite polynomials in z. One way to ensure this is by guaranteeing that $\det(H(z))$ is a pure delay, z^{-d} for some $d \in \mathbb{Z}$. As we shall see in Section 6.1, this is also a necessary condition for finite support transforms.

4.2.3   FILTER BANK INTERPRETATION

Equation (4.8) may be rewritten as

$$y[nm+q] = y_q[n] = \sum_i \mathbf{a}_q^*[i]\,\mathbf{x}[n-i]$$

$$= \sum_i \sum_{j=0}^{m-1} (\mathbf{a}_q[i])_j^*\, x[m(n-i)+j]$$

$$= \sum_k h_q[k]\, x[mn-k] = (x \star h_q)[mn] \qquad (4.11)$$

Here, as previously, $\mathbf{a}_q[i]$ denotes the qth column of $A[i]$ and $(\mathbf{a}_q[i])_j^*$ is the complex conjugate of its jth element, which is also $(A^*[i])_{q,j}$, the element at row q and column j of matrix $A^*[i]$. The scalar filter impulse responses, $h_q[k]$, and the analysis matrices are related by

$$(A^*[i])_{q,j} = (\mathbf{a}_q[i])_j^* = h_q[mi-j]; \quad 0 \le j < m \qquad (4.12)$$

Defining the jth polyphase component of the scalar impulse response by

$$h_{q,j}[i] = h_q[mi-j] \qquad (4.13)$$

we see that the Z-transform of this polyphase component is the element at row q and column j of the polyphase analysis matrix, $H(z)$; i.e.,

$$(H(z))_{q,j} = h_{q,j}(z) \qquad (4.14)$$

According to equation (4.11), the forward transform may be implemented by filtering the source sequence, $x[k]$, with a bank of m different

---

[7]For the synthesis system to be practically realizable, the elements of $G(z)$ must be the Z-transforms of realizable filter impulse responses. These may be two-sided IIR filters, for which stable causal realizations are not possible. In image processing applications, however, the relevant signal sequences (rows or columns of the image) generally have finite support. In this case, it is possible to implement anti-causal filtering operations by processing the samples backwards in time. A filter with arbitrary rational Z-transform may then be implemented as the composition of a causal component (poles inside the unit circle) and an anti-causal component (poles outside the unit circle).

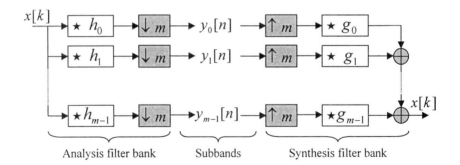

Figure 4.10.   Filter bank realization of a convolutional transform and its inverse.

analysis filters, having impulse responses, $h_q[k]$, and keeping only every m^{th} sample of each filter's output. This analysis filter bank is illustrated in Figure 4.10. Note that the "decimation" or "down-sampling" operator, $\boxed{\downarrow m}$, maps an input sequence, $u[k]$, to an output sequence, $v[k] = u[mk]$. The decimated outputs of the m filters are called subbands, for reasons which shall soon become clear.

We may also rewrite equation (4.9) in terms of a collection of scalar synthesis filters, $g_q[k]$. Specifically,

$$x[nm + p] = (\mathbf{x}[n])_p = \sum_i \sum_{q=0}^{m-1} (\mathbf{s}_q[i])_p \, y_q[n - i]$$

$$= \sum_{q=0}^{m-1} \sum_i y_q[i] \, (\mathbf{s}_q[n - i])_p$$

$$= \sum_{q=0}^{m-1} \sum_i y_q[i] \, g_q[nm + p - mi]$$

$$= \sum_{q=0}^{m-1} (\tilde{y}_q \star g_q)[nm + p] \tag{4.15}$$

where

$$\tilde{y}_q[i] = \begin{cases} y_q\left[\frac{i}{m}\right] & \text{if } m \text{ divides } i \\ 0 & \text{otherwise} \end{cases}$$

and

$$(S[i])_{j,q} = (\mathbf{s}_q[i])_j = g_q[mi + j] \tag{4.16}$$

According to equation (4.15), $x[n]$ is recovered by "up-sampling" the subband sequences, $y_q[n]$, filtering the up-sampled sequences, $\tilde{y}_q[n]$, with a bank of synthesis filters, $g_q[n]$, and adding the results. These operations are also illustrated in Figure 4.10. Note that the "up-sampling" operator, $\boxed{\uparrow m}$, inserts $m-1$ zeros between the elements of $y_q[n]$ to obtain $\tilde{y}_q[n]$.

Defining the jth polyphase component of the synthesis filter, $g_q[k]$, by

$$g_{q,j}[i] = g_q[mi + j] \qquad (4.17)$$

we see that the Z-transform of this polyphase component is the element at row j and column q of the polyphase synthesis matrix, $G(z)$; i.e.,

$$(G(z))_{j,q} = g_{q,j}(z) \qquad (4.18)$$

Note carefully, the difference between the definitions of polyphase components for the analysis and synthesis filter impulse responses in equations (4.13) and (4.17).

FREQUENCY-DOMAIN PERSPECTIVE

The analysis filter bank of Figure 4.10 does not represent an efficient implementation of the vector convolution operation, since the decimation operator discards all but every m^{th} filter output. Similarly, the synthesis filter bank does not represent an efficient implementation of the inverse transform, since $m-1$ out of every m samples processed by the synthesis filters are zero. The filter bank interpretation does, however, allow us to provide a frequency-domain perspective for the most useful class of convolutional transforms. In fact, the term "subband" arises from the fact that the filters, $h_q[n]$, usually correspond to bandpass filters.

Figure 4.11 illustrates stylistic magnitude responses for the analysis filters of an $m = 3$ band filter bank. Note that $\hat{h}_q(\omega)$ denotes the DTFT (Discrete Time Fourier Transform) of the impulse response, $h_q[n]$, given by

$$\hat{h}_q(\omega) = \begin{cases} \sum_n h_q[n] e^{-j\omega n} & \text{if } \omega \in [-\pi, \pi) \\ 0 & \text{otherwise} \end{cases} \qquad (4.19)$$

In general, each frequency band has a nominal bandwidth of $\frac{\pi}{m}$ and, by convention, the bands are uniformly spaced with $q = 0$ corresponding to the DC band and $q = m - 1$ corresponding to the highest frequency band. Although there is nothing in the convolutional transform structure which requires the analysis filters to be bandpass filters, the most useful transforms for image compression and other image processing applications are of this form. We have already seen that the DCT is a good decorrelating transform for images and other signals precisely because it is a frequency transform. As we shall see in Section 4.3, good decorrelating convolutional transforms for these sources all have the frequency

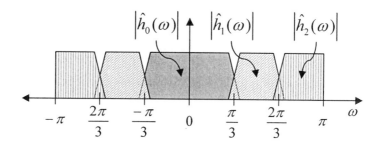

Figure 4.11.   Bandpass filter responses for a $m = 3$ band filter bank.

band structure. In recognition of the importance of this structure, we shall use the term "subband transform" interchangeably with "convolutional transform" and consistently assume the convention that $y_0[n]$ is a DC subband and $y_{m-1}[n]$ is the highest frequency subband.

As illustrated in Figure 4.10, the subband sequences themselves are obtained by decimating the bandpass filtered input sequence. Ideally, the analysis filters are "ideal" bandpass filters with non-overlapping passbands of bandwidth $\frac{\pi}{m}$. In this case, each passband is represented perfectly by its decimated subband sequence, without aliasing, and may be recovered by interpolating the subband samples with "ideal" synthesis filters. This is nothing other than Nyquist's sampling theorem and the ideal bandpass (analysis) and interpolation (synthesis) filters are the modulated sinc functions[8],

$$h_q[n] = g_q[n] = \frac{1}{\sqrt{m}} \begin{cases} \operatorname{sinc}\left(\frac{n}{m}\right) & \text{if } q = 0 \\ \cos\left(\pi \frac{q+\frac{1}{2}}{m} n\right) \operatorname{sinc}\left(\frac{n}{2m}\right) & \text{if } 1 \leq q \leq m - 2 \\ (-1)^n \operatorname{sinc}\left(\frac{n}{m}\right) & \text{if } q = m - 1 \end{cases}$$

$$(4.20a)$$

Of course, it is not possible to implement filters with ideal "brick wall" frequency responses, so aliasing is inevitably incurred during decimation. The effect of the downsampling and upsampling operations of Figure 4.10 is illustrated in Figure 4.12. The figure illustrates the introduction of aliasing and spectral expansion according to the downsampling relationship

$$\hat{y}_q(\omega) = \frac{1}{m} \sum_{k \in \mathbb{Z}} \hat{h}_q\left(\frac{\omega + 2\pi k}{m}\right) \hat{x}\left(\frac{\omega + 2\pi k}{m}\right) \qquad (4.21)$$

---

[8]The normalization factor, $\frac{1}{\sqrt{m}}$, is chosen so that the analysis and synthesis filters will be identical. In this case, each filter will have a passband gain of $\sqrt{m}$.

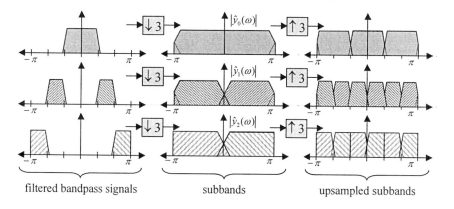

| filtered bandpass signals | subbands | upsampled subbands |

Figure 4.12. Frequency mapping of filtered bandpass signals due to downsampling and upsampling operations.

The figure also illustrates the spectral contraction and replication associated with upsampling, according to

$$\widehat{\widetilde{y}}_q(\omega) = \sum_{k \in \mathbb{Z}} \hat{y}_q(m\omega + 2\pi k) \qquad (4.22)$$

From this perspective, the principle role of the synthesis filters must be to eliminate spectral replicas from the upsampled subband signals, leaving at least an approximate copy of the original bandpass components, which may then be added to reconstruct the original sequence, $x[n]$.

This reasoning suggests that the synthesis filters should be bandpass filters with the same passbands as the corresponding analysis filters. The synthesis filters should also correct for any phase distortion introduced by the corresponding analysis filters. Thus, assuming that both the analysis and synthesis filters are assigned a passband gain of $\sqrt{m}$ to compensate for the factor of $\frac{1}{m}$ in equation (4.21), we should expect to find that

$$\hat{g}_q(\omega) \approx \hat{h}_q^*(\omega) \quad \text{or, equivalently,} \quad g_q[n] \approx h_q^*[-n] \qquad (4.23)$$

Indeed, the ideal analysis and synthesis filters of equation (4.20a) have exactly this form. In order to recover $x[n]$ exactly, however, the synthesis system must also be able to eliminate the aliasing introduced during analysis and so the relationship embodied by equation (4.23) is often only approximately valid. As it turns out, the relationship is exact if and only if the transform is orthonormal. The relationship between analysis and synthesis filters is discussed further below.

4.2.4   VECTOR SPACE INTERPRETATION

As for block transforms, considerable insight may be gained by expressing the forward transform as the inner product of the input with a collection of analysis vectors and the reverse transform as a linear combination of synthesis (or prototype) basis vectors. An appropriate vector space is the Hilbert space of square summable sequences, $\ell^2\,(\mathbb{Z})$. Given any two vectors (sequences), $\mathbf{u} \equiv u\,[k]$ and $\mathbf{v} \equiv v\,[k]$, in $\ell^2\,(\mathbb{Z})$, their inner product is defined by

$$\langle \mathbf{u}, \mathbf{v} \rangle = \sum_{k} u\,[k]\,v^*\,[k]$$

The analysis equation (4.11) may then be massaged into the form of an inner product as follows

$$y_q\,[n] = \sum_{k} h_q\,[k]\,x\,[mn - k] = \sum_{k} a_q^*\,[k]\,x\,[mn + k]$$

$$= \sum_{k} a_q^*\,[k - mn]\,x\,[k] = \left\langle \mathbf{x}, \mathbf{a}_q^{(n)} \right\rangle$$

Here, $\mathbf{a}_q^{(n)}$ signifies a delayed version of the sequence, $\mathbf{a}_q$, by mn samples, and $\mathbf{a}_q \equiv a_q\,[k]$ is given by

$$a_q\,[k] = h_q^*\,[-k]$$

Thus, the analysis vectors are the m-translates of the time-reversed, complex conjugated analysis filter impulse responses.

The synthesis equation (4.15) may be rewritten as

$$x\,[k] = \sum_{q=0}^{m-1} \sum_{n} y_q\,[n]\,g_q\,[k - mn]$$

$$\mathbf{x} = \sum_{q=0}^{m-1} \sum_{n} y_q\,[n]\,\mathbf{s}_q^{(n)} \tag{4.24}$$

where $\mathbf{s}_q^{(n)}$ is a delayed version of the sequence, $\mathbf{s}_q$, by mn samples and $\mathbf{s}_q \equiv s_q\,[k]$ is identical to the qth synthesis filter impulse response, $g_q\,[k]$. Thus, the synthesis vectors are the m-translates of the synthesis filter impulse responses (no time-reversal or complex conjugation).

For completeness, we point out that the analysis vectors, $\mathbf{a}_q$, may be related to the qth columns, $\mathbf{a}_q\,[i]$, of the analysis matrices, $A\,[i]$, through equation (4.12), yielding

$$(A\,[i])_{j,q} = (\mathbf{a}_q\,[i])_j = a_q\,[j - mi]; \quad 0 \leq j < m \tag{4.25}$$

Similarly, the synthesis vectors, s_q, may be related to the qth columns, $s_q[i]$, of the synthesis matrices, $S[i]$, through equation (4.16), yielding

$$(S[i])_{j,q} = (s_q[i])_j = s_q[j + mi] \tag{4.26}$$

For a pure block transform, where $S[i] = 0$ and $A[i] = 0$ for all $i \neq 0$, the analysis and synthesis vectors are exactly the columns of $A[0]$ and $S[0]$, which is consistent with our treatment of block transforms in Section 4.1.

ORTHONORMAL TRANSFORMS

As for pure block transforms, we say that a transform is orthogonal if its synthesis basis vectors are mutually orthogonal and we say that the transform is orthonormal if its synthesis basis vectors also have unit length. Since orthogonal transforms are related to orthonormal transforms through a simple scaling of the subband samples, $y_q[n]$, it is sufficient to consider the orthonormal case.

Orthonormal transforms implement orthonormal expansions of the input signal; i.e.,

$$\mathbf{x} = \sum_{q=0}^{m-1} \sum_{n \in \mathbb{Z}} \left\langle \mathbf{x}, \mathbf{s}_q^{(n)} \right\rangle \mathbf{s}_q^{(n)}$$

It follows that $\left\langle \mathbf{x}, \mathbf{s}_q^{(n)} \right\rangle = y_q^{(n)} = \left\langle \mathbf{x}, \mathbf{a}_q^{(n)} \right\rangle$ for all $\mathbf{x}$, and so the analysis and synthesis vectors are identical. Equivalently, the analysis filters are time-reversed, complex conjugated versions of the synthesis filters and so equation (4.23) is satisfied exactly for orthonormal subband transforms. Orthonormal transforms have the desirable "energy preserving" property, also known as Parseval's relationship, according to which

$$\|\mathbf{x}\|^2 = \langle \mathbf{x}, \mathbf{x} \rangle = \sum_{q=0}^{m-1} \sum_{n \in \mathbb{Z}} |y_q[n]|^2 = \langle \mathbf{y}, \mathbf{y} \rangle = \|\mathbf{y}\|^2$$

Here $\mathbf{y} \equiv y[k]$ is the interleaved sequence, $y[mn + q] = y_q[n]$.

Using equation (4.25), we may express the orthonormality of a convolutional transform in terms of the analysis matrices as

$$\sum_i A^*[i] A[j + i] = I \cdot \delta[j]$$

Also, since the synthesis and analysis vectors must be identical, comparing equations (4.25) and (4.26), we see that the analysis and synthesis matrices of an orthonormal transform are simply related according to

$$A[i] = S[-i] \tag{4.27}$$

Again, these results are consistent with those obtained for block transforms, where there is only one analysis matrix, $A[0]$, and one synthesis matrix, $S[0]$.

Example 4.4 *With the ideal filters of equation (4.20a), the synthesis vectors are given by*

$$s_q^{(n)}[k] = \frac{1}{\sqrt{m}} \begin{cases} \operatorname{sinc}\left(\frac{k-nm}{m}\right) & \text{if } q = 0 \\ \cos\left(\pi\frac{q+\frac{1}{2}}{m}(k-nm)\right) \operatorname{sinc}\left(\frac{k-nm}{2m}\right) & \text{if } 1 \le q \le m-2 \\ (-1)^{k-nm} \operatorname{sinc}\left(\frac{k-nm}{m}\right) & \text{if } q = m-1 \end{cases}$$

To verify that these do indeed form an orthonormal basis, recall Parseval's theorem for the DTFT, according to which

$$\langle \mathbf{u}, \mathbf{v} \rangle = \sum_k u[k]\, v^*[k] = \frac{1}{2\pi} \langle \hat{\mathbf{u}}, \hat{\mathbf{v}} \rangle = \frac{1}{2\pi} \int_{-\pi}^{\pi} \hat{u}(\omega)\, \hat{v}^*(\omega)\, d\omega$$

Now $\hat{s}_q(\omega) = \sqrt{m} I_{\mathcal{R}_q}(\omega)$ where $I_S(\omega)$ denotes the indicator function for set S, equal to 1 for $\omega \in S$ and 0 otherwise, and $\mathcal{R}_q$ denotes the passband region for the qth subband. The passbands are disjoint, with bandwidth $\frac{\pi}{m}$. Thus each passband region, $\mathcal{R}_q$, has area $\frac{2\pi}{m}$ (see Figure 4.11 for an example of the regions when $m = 3$). Hence we may deduce that

$$\left\langle s_{q_1}^{(n_1)}, s_{q_2}^{(n_2)} \right\rangle = \frac{1}{2\pi} \int_{-\pi}^{\pi} m I_{\mathcal{R}_{q_1}}(\omega) I_{\mathcal{R}_{q_2}}(\omega)\, e^{j\omega(n_2-n_1)m} d\omega$$

$$= \begin{cases} 0 & \text{if } q_1 \ne q_2 \\ \sqrt{m} s_q^{(0)}[(n_2 - n_1)m] & \text{if } q_1 = q_2 = q \end{cases}$$

Orthonormality follows from the fact that $s_q^{(0)}[nm] = \frac{1}{\sqrt{m}} \delta[n]$ for all q.

Example 4.5 *Scalar LTI filtering is a degenerate case of a convolutional transform for which $m = 1$. The forward transform is described by a filter, with impulse response $h[k]$. Similarly, the inverse transform is described by a filter, with impulse response $g[k]$. The two filters must satisfy $(h \star g)[k] = \delta[k]$, or equivalently, $\hat{g}(\omega) = \hat{h}^{-1}(\omega)$. The analysis vectors are the sequences, $\mathbf{a}^{(n)} \equiv a^{(n)}[k] = h^*[-k-n]$, and the synthesis vectors are the sequences, $\mathbf{s}^{(n)} \equiv s^{(n)}[k] = g[k-n]$. The transform is orthonormal if and only if*

$$\left\langle \mathbf{a}^{(n_1)}, \mathbf{a}^{(n_2)} \right\rangle = \delta[n_2 - n_1], \quad \forall n_1, n_2 \in \mathbb{Z}$$

which, by Parseval's theorem, is equivalent to the requirement that

$$\delta\left[n_2 - n_1\right] = \frac{1}{2\pi}\int_{-\pi}^{\pi}\hat{h}^*\left(\omega\right)e^{-j\omega n_1}\hat{h}\left(\omega\right)e^{j\omega n_2}d\omega$$

$$= \frac{1}{2\pi}\int_{-\pi}^{\pi}\left|\hat{h}\left(\omega\right)\right|^2 e^{j\omega\left(n_2 - n_1\right)}d\omega$$

We may conclude that scalar filtering is an orthonormal transform if and only if an all-pass filter is used, with $\left|\hat{h}\left(\omega\right)\right| = 1$ for all $\omega \in [-\pi, \pi)$. Then the synthesis and analysis vectors must be identical, meaning that $g\left[k\right] = h^*\left[-k\right]$ is the all-pass filter with the opposite phase response.

LAPPED ORTHOGONAL TRANSFORMS

The two examples of orthonormal subband transforms given above involve filters with infinite impulse responses. Moreover, either one or both of the filters is non-causal and hence not practically realizable in many applications. This is unavoidable in the case of scalar filtering ($m = 1$), except in the trivial case where the transform involves only a single non-zero filter tap (a one dimensional block transform). By contrast, non-trivial orthonormal transforms involving finite support vector filters exist for all $m \geq 2$. In this section we demonstrate this fact by briefly considering a class of orthonormal transforms involving exactly two non-zero analysis matrices, $A\left[0\right]$ and $A\left[1\right]$. For historical reasons, transforms of this form are known as a Lapped Orthogonal Transforms (LOTs).

According to equation (4.25), the analysis vectors, $\mathbf{a}_q \equiv a_q\left[k\right]$, of an LOT are sequences supported on $-m \leq k < m$.[9] We begin by considering perhaps the simplest LOT, in which the analysis vectors are cosines of the form

$$a_q\left[k\right] = c_q\left[k\right] = \begin{cases} \frac{1}{\sqrt{m}}\cos\left(2\pi f_q\left(k - \frac{m-1}{2}\right)\right) & \text{if } -m \leq k < m \\ 0 & \text{otherwise} \end{cases}$$

where the cosine frequencies are

$$f_q = \frac{q + \frac{1}{2}}{2m}; \quad 0 \leq q < m$$

The first few of these cosine functions are plotted in Figure 4.13a for the case $m = 8$. One may easily verify that these cosine vectors are in-

---

[9]Note that the analysis filters are time-reversed versions of the analysis vectors and are hence non-causal in the current development. Most texts, introduce offsets to ensure causality; however, these offsets clutter the notation. Since the system has finite support, there is no need to insist on causality.

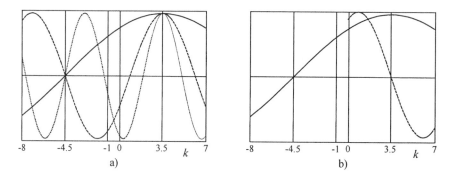

Figure 4.13. First three cosine functions for an LOT with $m = 8$. The discrete time index, k, is treated as though it were continuous to expose the structure of the underlying cosine functions.

deed orthogonal with unit length. For the transform to be orthonormal, however, we require the orthogonality of all the m-translates of these vectors; i.e., all the $a_q^{(n)}$. Since the vectors have length $2m$, it is sufficient to verify that the sequences $c_{q_1}[k]$ and $c_{q_2}[k-m]$ are orthogonal for all q_1, q_2. This follows from the fact that each of the cosines, $c_q[k]$, is symmetric about $k = \frac{m-1}{2}$ within the overlap interval, $0 \le k < m$, while $c_q[k-m]$ exhibits anti-symmetry within the same interval. The overlapping portions of $a_0^{(0)} \equiv c_0[k]$ and $a_1^{(1)} \equiv c_1[k-m]$ are illustrated in Figure 4.13b. It follows that $\left\langle a_{q_1}^{(0)}, a_{q_2}^{(1)} \right\rangle = 0$ and hence the $a_q^{(n)}$ form an orthonormal basis for $\ell^2(\mathbb{Z})$.

A more interesting LOT may be constructed by using the cosine functions described above to modulate a smooth windowing sequence, $w[k]$, supported over $-m \le k < m$. In this case, the analysis vectors are given by

$$a_q[k] = w[k]\, c_q[k]$$

It turns out that the window does not disturb the orthogonality of the transform, provided it satisfies the following symmetry properties:

$$w[k] = w[-1-k]; \quad 0 \le k < m \tag{4.28a}$$

$$w^2[k] + w^2[m-1-k] = 2; \quad 0 \le k < m \tag{4.28b}$$

The first property states the $w[k]$ is an even length symmetric sequence. This property ensures that the condition $\left\langle a_{q_1}^{(0)}, a_{q_2}^{(1)} \right\rangle = 0$ is not destroyed by the presence of the window. To see this, observe that

$$\left\langle \mathbf{a}_{q_1}^{(0)}, \mathbf{a}_{q_2}^{(1)} \right\rangle = \sum_{k=0}^{m-1} c_{q_1}[k]\, c_{q_2}[k-m]\, w[k]\, w[k-m]$$

$$= \sum_{k=0}^{m-1} c_{q_1}[k]\, c_{q_2}[k-m]\, w[k]\, w[m-1-k]$$

$c_{q_1}[k]$ is symmetric, $c_{q_2}[k-m]$ is anti-symmetric, and $w[k][wm-1-k]$ is symmetric on $0 \le k < m$. Thus, their product is anti-symmetric and the above summation yields 0.

The second property, equation (4.28b), ensures that the orthonormality of the vectors, $\mathbf{a}_q$, is not disturbed. To see this, observe that

$$\langle \mathbf{a}_{q_1}, \mathbf{a}_{q_2} \rangle = \sum_{k=0}^{m-1} c_{q_1}[k]c_{q_2}[k]w^2[k] + \sum_{k=0}^{m-1} c_{q_1}[k-m]c_{q_2}[k-m]w^2[k-m]$$

$$= \sum_{k=0}^{\frac{m}{2}-1} c_{q_1}[k]c_{q_2}[k]w^2[k] + \sum_{k=0}^{\frac{m}{2}-1} c_{q_1}[k-m]c_{q_2}[k-m]w^2[k-m]$$

$$+ \sum_{k=0}^{\frac{m}{2}-1} c_{q_1}[m-1-k]c_{q_2}[m-1-k]w^2[m-1-k]$$

$$+ \sum_{k=0}^{\frac{m}{2}-1} c_{q_1}[-1-k]c_{q_2}[-1-k]w^2[-1-k]$$

$$= \sum_{k=0}^{\frac{m}{2}-1} c_{q_1}[k]c_{q_2}[k]w^2[k] + \sum_{k=0}^{\frac{m}{2}-1} c_{q_1}[k-m]c_{q_2}[k-m]w^2[m-1-k]$$

$$+ \sum_{k=0}^{\frac{m}{2}-1} c_{q_1}[k]c_{q_2}[k]w^2[m-1-k] + \sum_{k=0}^{\frac{m}{2}-1} c_{q_1}[k-m]c_{q_2}[k-m]w^2[k]$$

$$= \sum_{k=0}^{\frac{m}{2}-1} \left(c_{q_1}[k]c_{q_2}[k] + c_{q_1}[k-m]c_{q_2}[k-m] \right) \left(w^2[k] + w^2[m-1-k] \right)$$

$$= \langle \mathbf{c}_{q_1}, \mathbf{c}_{q_2} \rangle = \delta(q_1 - q_2)$$

where we have used the fact that $c_{q_1}[k]c_{q_2}[k]$ and $c_{q_1}[k-m]c_{q_2}[k-m]$ are both symmetric functions over $0 \le k < m$.

It is instructive to consider the effect of this windowing operation in the frequency domain. Each analysis sequence, $a_q[k]$, is a cosine modulated version of the window sequence, $w\,[k]$; i.e.,

$$a_q\,[k] = w\,[k]\,\frac{1}{\sqrt{m}}\cos\left(2\pi f_q\left(k - \frac{m-1}{2}\right)\right)$$

Hence its DTFT is given by

$$\hat{a}_q\,(\omega) = \frac{1}{2\sqrt{m}}\left(e^{-j2\pi f_q\frac{m-1}{2}}\,\hat{w}\,(\omega - 2\pi f_q) + e^{j2\pi f_q\frac{m-1}{2}}\,\hat{w}\,(\omega + 2\pi f_q)\right)$$

Noting that the analysis filters satisfy $\hat{h}\,(\omega) = \hat{a}^*\,(\omega)$, we see that the subbands are formed by filtering the input signal with a collection of bandpass filters, whose frequency responses are identical to that of a low-pass prototype, $\hat{w}\,(\omega)$, translated in frequency by $\pm f_q = \pm\frac{2q+1}{4m}$. To obtain good frequency discrimination amongst the subbands, we would like $\hat{w}\,(\omega)$ to be an ideal low-pass function supported on $\omega \in \left[-\frac{2\pi}{4m}, \frac{2\pi}{4m}\right]$. Of course, the window is a finite length sequence, having $2m$ samples, and so we can at best hope to design a low-pass filter, having as little energy outside this band as possible. It is worth noting that the block transform constructed from the DCT is also essentially of this same form, except that its window is a constant over $0 \le k < m$ and zero elsewhere; in the frequency domain, this window is a sinc function, which is well-known for its poor frequency localization.

Example 4.6 *To see that equations (4.28a) and (4.28b) do not prevent us from designing useful low-pass prototypes, consider the case $m = 8$ and the window coefficients identified in Table 4.1 (we have borrowed these from [162]). Figure 4.14 shows the magnitude responses of the first few bandpass filters for this 8-band LOT and compares them with the bandpass filters associated with the 8-band block transform formed by applying the DCT to each block. Evidently, the LOT subbands have superior frequency discrimination to the DCT bands. It is worth noting that even though the lowest frequency cosine in the LOT has a non-zero frequency of $f_0 = \frac{1}{4m}$, the positive and negative frequency contributions, $\hat{w}\left(\omega + \frac{1}{4m}\right)$ and $\hat{w}\left(\omega - \frac{1}{4m}\right)$, merge to produce a response, $\hat{a}_0\,(\omega)$, which peaks at DC.*

Cosine modulated filter banks which implement orthonormal transforms were first discovered by Princen and Bradley[120]. LOT's need not necessarily take the form of a cosine modulated filter bank[100]; however, orthonormal cosine modulated filter banks do possess a number of desirable properties. From their close relationship to the DCT, one might expect that efficient FFT-like algorithms exist for implementing cosine modulated transforms and indeed this is the case[100]. All

Table 4.1. Example window coefficients for the cosine modulated LOT with $m = 8$ bands. Note that $w[-1 - k] = w[k]$ for $0 \le k < 8$.

$w[0]$	$w[1]$	$w[2]$	$w[3]$	$w[4]$	$w[5]$	$w[6]$	$w[7]$
0.1255	0.3347	0.5994	0.8742	1.1117	1.2809	1.3740	1.4086

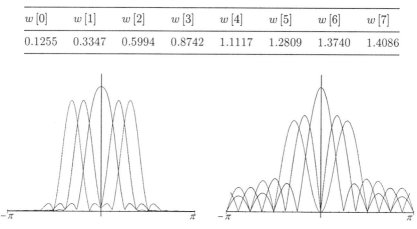

Figure 4.14. Magnitude responses of the first three analysis filters for the 8-band LOT formed by cosine modulating the window sequence of Table 4.1 (left) and for the 8-band DCT (right).

analysis filters have essentially the same passband characteristics, which may be designed through a single low-pass prototype sequence and this greatly simplifies the problem of designing good transforms. It turns out that the ideas presented above may be extended to orthonormal transforms with $2L$ analysis matrices, where L is any integer[99, 87]. This, in turn, allows the design of longer window sequences with improved frequency selectivity.

BIORTHOGONAL TRANSFORMS

Recall that the synthesis system of any convolutional transform inverts the analysis system if and only if equation (4.10) holds. Expressed in vector notation, this becomes

$$\left\langle s_{q_1}^{(n_1)}, a_{q_2}^{(n_2)} \right\rangle = \delta\left[q_1 - q_2\right] \delta\left[n_1 - n_2\right], \quad 0 \le q_1, q_2 < m, \quad n_1, n_2 \in \mathbb{Z} \tag{4.29}$$

That is, each analysis vector, $a_q^{(n)}$, must be orthogonal to every synthesis vector other than $s_q^{(n)}$. This property is known as biorthogonality. We have already seen the biorthogonality relationship in our study of block transforms, in which context it is nothing other than a vector space interpretation of the matrix equation, $A^*S = SA^* = I$.

Even though orthogonal transforms are necessarily biorthogonal, the term "biorthogonal transform" is most commonly employed to indicate that the transform, while invertible, is not orthogonal. As it turns out, biorthogonal transforms play a particularly important role in image compression because all non-trivial orthogonal transforms with $m = 2$ bands (or channels) necessarily involve non-symmetric filter impulse responses or infinite support filters. The importance of two channel transforms and symmetric filters for image compression will become apparent shortly. Since biorthogonal, as opposed to orthogonal, subband transforms are motivated primarily by an interest in symmetric filters, the term "biorthogonal transform" is often understood to refer to a non-orthogonal transform, whose synthesis system correctly inverts the analysis system, where both the analysis and synthesis filter banks employ linear phase filters.

4.2.5 ITERATED SUBBAND TRANSFORMS
SEPARABLE IMAGE TRANSFORMS

Up until now we have considered one dimensional subband transforms, having m subbands, whose sample rates are each $\frac{1}{m}$ times the sample rate of the original input sequence, $x[k]$. Such a transform may be separably extended to a two dimensional transform with m^2 subbands. The separable extension procedure is no different to that described for block transforms in Section 4.1.1 and illustrated in Figure 4.3. Specifically, the one dimensional transform is applied to each row of the image, $x[\mathbf{k}]$, generating a new image, $y'[\mathbf{k}]$, whose rows contain the interleaved subband samples from corresponding rows in $x[\mathbf{k}]$; the one dimensional transform is then applied to each column of $y'[\mathbf{k}]$, generating the transformed image, $y[\mathbf{k}]$, whose columns contain the interleaved subband samples from corresponding columns in $y'[\mathbf{k}]$. As shown previously, it makes no difference whether the one dimensional subband transform is applied first to the rows and then to the columns, or vice-versa. The m^2 subbands are labeled $y_{q_1,q_2}[\mathbf{n}]$ and are related to the interleaved sequence of subband samples, $y[\mathbf{k}]$, according to

$$y_{q_1,q_2}[n_1, n_2] = y[mn_1 + q_1, mn_2 + q_2], \quad 0 \le q_1, q_2 < m$$

In the special case where $m = 2$, the separable two dimensional transform has four subbands; an analysis filter bank is illustrated in Figure 4.15a, while the passband regions are illustrated in Figure 4.15b. In the figure, we use the notation $\star_h$ and $\star_v$ to denote horizontal and vertical convolution along rows and columns of the image, respectively. Similarly, $\downarrow_h$ and $\downarrow_v$ denote horizontal and vertical decimation. In accordance with the convention established for one dimensional transforms,

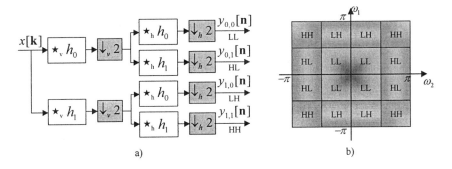

Figure 4.15. Analysis filter bank and passband regions for a 2D separable subband transform with $m = 2$.

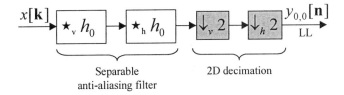

Figure 4.16. LL subband construction, viewed as classic resolution reduction.

$y_{0,0}[\mathbf{n}]$ is the low-pass (or DC) subband and is most commonly identified as the "LL" band. Similarly, $y_{1,1}[\mathbf{n}]$ represents high frequency content in both the horizontal and vertical directions and is identified as the "HH" band. $y_{0,1}[\mathbf{n}]$ represents high frequency content in the horizontal direction and low frequency content in the vertical direction. Accordingly, it is identified as the "HL" band, while $y_{1,0}[\mathbf{n}]$ is the "LH" band.

It is instructive to consider the nature of the image features which we can expect to encounter in each of the four subbands defined above. The LL subband may be understood as arising from the application of a separable low-pass filter to the original image, followed by downsampling in both directions, as illustrated in Figure 4.16. This is the classic paradigm for image resolution reduction, where the low-pass filter is interpreted as an anti-aliasing filter. Accordingly, the LL band is a low resolution version of the original image, which should be largely free from the visually disturbing artifacts of aliasing, so long as the underlying analysis filter, $h_0[k]$, is a good approximation to an ideal half-band filter.

The horizontal high-pass filter used to generate the HL band filters out smooth regions of the image as well as horizontally oriented edges

Figure 4.17.   576×720 cropped, reduced version of the ISO/IEC standard test image, "Bike."

and lines. It responds most strongly to vertical edges and line segments in the image. Similarly, the LH band responds most strongly to horizontal edges and line segments, while the HH band responds primarily to diagonally oriented features. To illustrate this orientation selectivity, we apply a separable subband transform to the image in Figure 4.17; the resulting subband images are shown in Figure 4.18. The underlying one dimensional subband transform employed here is the CDF 9/7 biorthogonal transform developed in Section 6.3.2, having finite support low- and high-pass analysis filters with 9 and 7 taps, respectively. This is one of the two transforms explicitly defined by the JPEG2000 image compression standard.

MULTI-RESOLUTION TRANSFORMS

The one dimensional and separable two dimensional subband transforms described above may be classified as "uniform" transforms, since all subbands have identical sample rates and their passbands all have essentially the same bandwidth. By contrast with uniform subband

Figure 4.18. Separable subbands of the image in Figure 4.17: top left LL; top right HL; bottom left LH; bottom right HH.

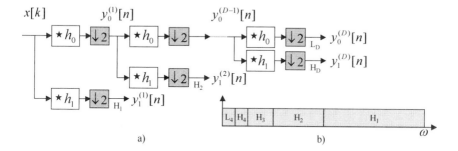

Figure 4.19. One dimensional tree-structured subband transform: (a) analysis filter bank; (b) passband structure for $D = 4$ levels.

transforms, an interesting family of "tree-structured" transforms may be obtained by recursively applying a one or two dimensional subband transform, as appropriate, to its own low-pass subband.

Figure 4.19 illustrates the one dimensional case. The recursive subdivision is continued for D levels, yielding a total of $D+1$ subbands. The

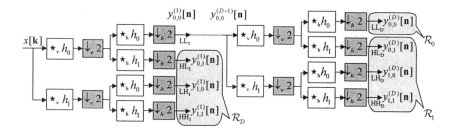

Figure 4.20. Filter bank structure for a 2D tree-structured subband transform with D decomposition levels.

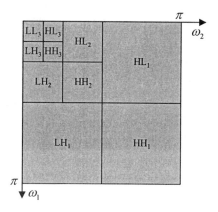

Figure 4.21. Passband structure for a 2D tree-structured subband transform with $D = 3$ levels.

low frequency (DC) subband has sample rate $\frac{1}{2^D}$, while the remaining subbands have rates $\frac{1}{2^d}$ for $0 < d \le D$. The passbands have corresponding bandwidths as suggested by the figure.

A two dimensional tree-structured filter bank is illustrated in Figure 4.20, with the passband structure shown in Figure 4.21. Note that in two dimensions only one of the four subbands, the LL band, is recursively decomposed into further subbands. The recursive sub-division is again continued for D levels, yielding a total of $3D + 1$ subbands, again with non-uniformly spaced passbands. The subbands generated by a $D = 2$ level image transform are illustrated in Figure 4.22, which is obtained using the same source image and subband filters as Figure 4.18.

Recall that the LL band of a uniform subband decomposition is a low resolution version of the original image. It follows that the low-pass subbands, identified as LL_d in Figure 4.20, which are formed during

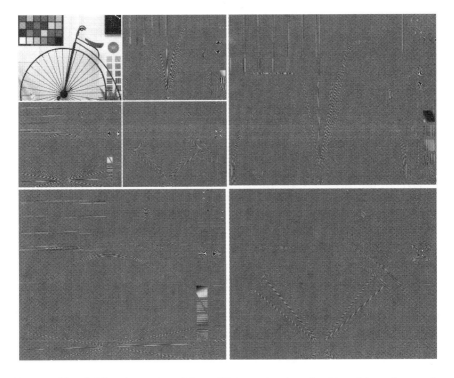

Figure 4.22. Subbands produced by a 2D tree-structured subband transform, when applied to the image in Figure 4.17.

tree-structured analysis, represent a family of successively lower resolution versions of the original image. The sampling density for LL_d is 2^{-d} times that of the original image in each direction, where $d = 1, 2, \ldots, D$. Of course, all but the last of these low resolution images is an intermediate result; only LL_D is actually one of the subbands of the final tree-structured transform. However, each of the images in this multi-resolution family may be recovered by partial application of the synthesis system. LL_{D-1}, for example, may be synthesized from subbands LL_D, LH_D, HL_D and HH_D, while LL_{D-2} may be synthesized from these subbands, together with LH_{D-1}, HL_{D-1} and HH_{D-1}.

This multi-resolution property is particularly interesting for image compression applications, since it provides a mechanism whereby a compressed bit-stream may be partially decompressed to obtain successively higher resolution versions of the original image. To be more specific, let $\mathcal{R}_d$ be the set consisting of subbands LH_{D+1-d}, HL_{D+1-d} and HH_{D+1-d} for $0 < d \leq D$ and let $\mathcal{R}_0$ be the set consisting of only subband LL_D. These groupings are also identified in Figure 4.20. We refer to the $\mathcal{R}_d$

as resolution levels, since $\mathcal{R}_0$ contains the lowest resolution image and each successive resolution level, $\mathcal{R}_d$, contains the additional information required to reconstruct the next member of the multi-resolution family. Suppose now that the elements of each set, $\mathcal{R}_d$, $0 \leq d \leq D$, are compressed independently[10] and their compressed representations are separately identifiably within the compressed image representation. Then, the compressed representation has a property known as "resolution scalability," whereby a compressed representation of any member of the multi-resolution family may be obtained simply by discarding those pieces corresponding to the irrelevant resolution levels, $\mathcal{R}_d$.

Tree-structured subband transforms may be constructed from one dimensional uniform transforms having any number of subbands, m. However, the smallest non-trivial value, $m = 2$, is of principle interest, since it leads to multi-resolution families having the smallest change in resolution from one level to the next. It is common to refer to these as "dyadic" decompositions. For image compression applications, the interest in dyadic decompositions and hence two channel subband transforms is driven primarily by the significance of resolution scalability. Their comparative simplicity is also appealing for practical applications.

4.3   TRANSFORMS FOR COMPRESSION

Many of the fundamental properties of transforms used for image compression have been outlined in the preceding sections. At this point it is appropriate to collect some of the key arguments which have been advanced for the use of transforms in compression.

As discussed in Chapter 3, vector quantization provides a method for achieving the rate-distortion bound of an information source, without the need for additional elements such as image transforms or entropy coding. Unfortunately, the theoretical limit is achieved only asymptotically in the limit as we allow computation and memory to grow without bound, quite apart from the limitations imposed by unknown and/or non-stationary source statistics. In practice, unstructured vector quantization is quite impractical except for very small vectors, where its performance is usually found to be significantly inferior to techniques which impose more structure on the compression system. To remedy this difficulty, a wide variety of structured vector quantization schemes have been developed which permit the use of larger vectors at the expense of generality.

---

[10] Actually, it is sufficient for each $\mathcal{R}_d$ to be compressed in a manner which depends at most on $\mathcal{R}_0$ through $\mathcal{R}_{d-1}$.

The use of a linear transform as part of the compression system, as illustrated in Figure 1.4, represents a particular structuring technique. For the transform to be useful as a structuring technique, it should allow the use of simpler quantization and coding operations without sacrificing compression efficiency. One of our goals in this section is to demonstrate that this is the case. The arguments which follow are inevitably based on assumptions which are at best only approximately valid for real images. Nevertheless, they do help to identify the particular statistical properties of images which linear transforms are able to exploit and the particular characteristics of a transform which lend themselves to this task. We begin in Section 4.3.1 with some informal arguments which are intended to provide an intuitive framework for the more formal results which follow.

4.3.1   INTUITIVE ARGUMENTS
FINITE DIMENSIONAL TRANSFORMS

Consider a random vector, $\mathbf{X}$, of dimension m and let A be the unitary $m \times m$ analysis matrix of an orthonormal transform, which maps outcomes, $\mathbf{x}$, of $\mathbf{X}$ to transform vectors, $\mathbf{y} = A^*\mathbf{x}$, which are outcomes of the random vector, $\mathbf{Y} = A^*\mathbf{X}$. We restrict our attention here to orthonormal transforms because they may be interpreted as rotation operators in the m-dimensional space. We shall also restrict our attention to real-valued transforms for simplicity, although the intuition extends to complex-valued transforms, such as the DFT.

In the simplest case where $m = 2$, all real-valued unitary matrices are of the form

$$A^* = \begin{pmatrix} \cos\theta & -\sin\theta \\ \sin\theta & \cos\theta \end{pmatrix} \qquad (4.30)$$

up to a change of sign in one of the coordinates. We shall ignore the possibility of such sign flips, since they have no impact on the arguments which follow. In general, an m-dimensional real-valued unitary matrix may be factored into a product of $\binom{m}{2}$ one dimensional rotation operators [166, §3.2]. That is,

$$A = \prod_{i=0}^{m-1} \prod_{j=i+1}^{m-1} A^{(i,j)}$$

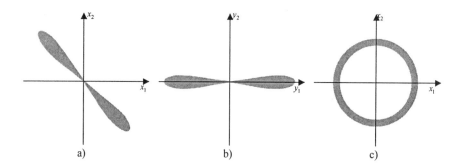

Figure 4.23. Joint distributions: a) a two dimensional random vector, **X**; b) its orthonormal transform, **Y**; and c) a random vector, **X**, which does not benefit from linear transformation.

where $A^{(i,j)}$ implements a rotation in the plane described by coordinates i and j; i.e.,

$$
\left(A^{(i,j)}\right)^* = \begin{pmatrix} I & 0 & 0 & 0 & 0 \\ 0 & \cos\theta^{(i,j)} & 0 & -\sin\theta^{(i,j)} & 0 \\ 0 & 0 & I & 0 & 0 \\ 0 & \sin\theta^{(i,j)} & 0 & \cos\theta^{(i,j)} & 0 \\ 0 & 0 & 0 & 0 & I \end{pmatrix} \begin{matrix} \\ \leftarrow \text{ row } i \\ \\ \leftarrow \text{ row } j \\ \\ \end{matrix}
$$

Rotations in more than 3 dimensions are difficult, if not impossible to visualize. Fortunately, however, substantial intuition may be gained by considering the two dimensional case. Suppose that the distribution, $f_\mathbf{X}$, of the random vector, **X**, is concentrated between the two axes, as shown in Figure 4.23a. In this case, a rotation of $\theta = 45°$, leaves the transformed random vector, **Y**, with a distribution concentrated predominantly about the y_0 axis, as shown in Figure 4.23b. In the extreme case, we can ignore y_1 altogether without introducing substantial distortion and we have only to quantize and code a single quantity, rather than two quantities. Thus, the transform provides us with a simple way to exploit the statistical redundancy in the random vector, **X**. Without the transform, more complex quantization and/or coding techniques would be required to exploit the statistical dependence between X_0 and X_1. More generally, we must quantize and code both quantities. However, we expect to spend very few bits coding the quantized symbols required to represent outcomes of Y_1, so that the bit-rate is approximately half that which we would expect from independent quantization and coding of X_0 and X_1. Of course, a more thorough analysis needs to consider the exact balance of bits spent on the two transform coefficients for a given combined distortion; this is the subject of Section 4.3.2.

A helpful notion for summarizing the effectiveness of an orthonormal transform is that of "energy compaction." The most effective transform will concentrate the maximum amount of the energy from the source vector, $\mathbf{X}$, in a single transform coefficient, say Y_0. That is, $\sigma_{Y_0}^2 \approx \sigma_{X_0}^2 + \sigma_{X_1}^2$. Energy compaction is meaningful only for orthonormal transforms, or transforms which are approximately orthonormal, since these are the energy preserving transforms, with $\sigma_{Y_0}^2 + \sigma_{Y_1}^2 = \sigma_{X_0}^2 + \sigma_{X_1}^2$. An energy compaction ratio may be defined as the ratio $\sigma_{Y_0}^2 / \sigma_{Y_1}^2$.

It is easy to see that linear transforms are not always able to exploit the redundancy in the source. Consider, for example, the distribution indicated in Figure 4.23c. Clearly, there is substantial redundancy, since the distribution is concentrated on a one dimensional manifold (a shell). However, no orthonormal transform is able to achieve any energy compaction. A suitable transform in this case would be a planar to polar coordinate transformation, which is highly non-linear.

With higher dimensions, m, the potential for compression is also higher. We hope to be able to find orthonormal transforms which are able to rotate the source distribution into a low dimensional sub-space, so that the transform vector, $\mathbf{Y}$, is almost entirely described by a few of its elements. In the extreme case, all but one transform coefficient may be ignored and we have reduced the sample rate by a factor of m prior to quantization and coding. Although the concept of energy compaction does not immediately generalize to $m > 2$ dimensions, a related quantity with suitable properties is the ratio of the arithmetic mean of the variances, $\sigma_{Y_i}^2$, to their geometric mean; i.e.,

$$\frac{\text{AM}}{\text{GM}} = \frac{\frac{1}{m} \sum_{i=0}^{m-1} \sigma_{Y_i}^2}{\sqrt[m]{\prod_{i=0}^{m-1} \sigma_{Y_i}^2}} \tag{4.31}$$

The arithmetic mean is dominated by the transform coefficients with the largest variance, while the geometric mean is dominated by the transform coefficients with the smallest variance. As we shall see in Section 4.3.2, this ratio has a useful interpretation as a coding gain.

SUBBAND TRANSFORMS

Orthonormal subband transforms also exhibit the energy preserving property and may be regarded as rotation operators in an infinite dimensional space, $\ell^2(\mathbb{Z})$. Intuition gained in two dimensions extends to these transforms as well. Letting $\sigma_{Y_i}^2$ denote the variance of the subband coefficients, $Y_i[n]$, we may define an energy compaction ratio for $m = 2$ band transforms as $\sigma_{Y_0}^2 / \sigma_{Y_1}^2$, or more generally, an AM/GM ratio as in equation (4.31). Large ratios mean that most of the signal energy is concentrated in a few of the subbands, so that the remaining subbands, may be quantized and coded with very few bits.

Natural images typically have much more power at low frequencies than at high frequencies. One model for the spectral power density of an image, which is supported by some empirical evidence, is

$$S_X(\omega) = \frac{\alpha}{\omega_1^2 + \omega_2^2} \qquad (4.32)$$

so that the power density decreases inversely with the square of radial frequency. As a result, subband image transforms having passbands of the form shown in Figure 4.21 can be expected to yield widely varying subband variances. The higher frequency subbands represent the majority of the subband samples, most of which can be quantized to 0 with negligible distortion. Evidence for this behaviour is also exhibited by the subband images in Figure 4.18, which indicates that the LH, HL and HH subband samples are almost all very close to 0, the exceptions being those samples which lie close to strong edges of the relevant orientation.

IMAGE COMPRESSION EXPERIENCE

The reader may be somewhat sceptical of the extreme nature of the preceding arguments, in which we suggest that most of the transform coefficients might be quantized to 0. As it turns out, however, this is exactly what happens in practical image compression systems. As an example, we note that high quality reconstructed images can usually be obtained at compressed bit-rates of about 0.5 to 1.0 bps (bits per sample). This empirical observation holds for both the JPEG compression standard and its more recent successor, JPEG2000. Since the number of transform coefficients is the same as the number of image samples and we cannot expect to spend less than 2 bits coding a non-zero transform coefficient[11], the majority of the coefficients must be quantized to zero in order to achieve such compression ratios. Experience with the baseline JPEG compression standard, for example, suggests that 70% to 85% of the DCT coefficients can be 0 in a high quality image representation.

4.3.2   CODING GAIN

In this section we attempt to quantify the benefits associated with the use of transforms for image compression.

---

[11] A full bit is generally required to code the sign of the non-zero coefficient (most transform coefficients have zero mean and are largely uncorrelated with one another), while we cannot expect to spend less than one bit coding the magnitude.

CODING GAIN FORMULATION

Coding gain expressions are a traditional means for comparing different compression techniques. The reference scheme is usually simple scalar quantization with independent coding of the source samples; for historical reasons, this is known as PCM. The distortion-rate performance of this simple direct approach can generally be modeled by equation (3.33), which we repeat here as

$$d\left(R\right) = E\left[\left(\hat{X} - X\right)^2\right] = \sigma_X^2 g\left(R\right) e^{-aR} \qquad (4.33)$$

Here X denotes the random variable whose outcome is the source sample being quantized and coded, $\hat{X}$ denotes the output of the dequantizer, $R \geq 0$ is the bit-rate measured in bits per sample, $d\left(R\right)$ is the MSE achieved at rate R, σ_X^2 is the source variance, $a \approx 2\log_e 2$ is a constant, and $g\left(R\right)$ is a weak function of the rate, which we shall take to be a constant, $g\left(R\right) = \varepsilon^2$, for the purposes of this analysis. As it turns out, equation (4.33) is applicable also for more elaborate quantization schemes, such as TCQ (see Section 3.5). On the other hand, we must exclude from our present consideration any quantization and coding schemes whose distortion-rate function depends on the source statistics in a more complex fashion. In particular, the reference PCM scheme should not be able to exploit statistical dependencies between the source samples.

We now consider applying this same PCM scheme to the transform coefficients, instead of the original source samples and adjust the quantization parameters so as to achieve the same overall compressed bit-rate, R, both with and without the transform. The ratio of the resulting distortions (MSE) is identified as the coding gain, G_T, of the transform; i.e.,

$$G_T = \frac{d^{PCM}\left(R\right)}{d^{XFORM}\left(R\right)}$$

The key step in the procedure is the appropriate adjustment of quantization parameters. To understand this step, we begin by defining a transform band as a subset of the transform coefficients which we expect to share the same statistics. The concept of transform bands is natural for both block and subband transforms. For subband transforms, the bands are simply the subbands themselves. For block transforms, the bands contain one coefficient from each source block. Thus, for example, the 8×8 DCT has 64 different bands: a DC band and 63 AC bands. Let

B denote the number of bands[12]. We index the bands by $b = 0$ through $B - 1$. Let $\sigma_{Y_b}^2$ denote the variance in band b and let η_b denote the ratio between the number of coefficients in the band and the total number of samples in the source. We restrict our attention to non-expansive, orthonormal transforms so that

$$\sum_{b=0}^{B-1} \eta_b = 1 \quad \text{and} \quad \sum_{b=0}^{B-1} \eta_b \sigma_{Y_b}^2 = \sigma_X^2$$

Our task is to select the most appropriate operating point on the distortion-rate curve for the PCM quantizer in each band. Equivalently, we must assign rates, R_b, to each band, subject to

$$R = \sum_{b=0}^{B-1} \eta_b R_b$$

Then, since the transform is orthonormal and the quantizer model of equation (4.33) is assumed to hold in each band, the overall MSE will be given by

$$d^{\text{XFORM}}(R) = \sum_{b=0}^{B-1} \eta_b \sigma_{Y_b}^2 \varepsilon^2 e^{-aR_b} \tag{4.34}$$

The most appropriate rate allocation is that which minimizes the overall distortion, subject to the rate constraint. Following the method of "Lagrange multipliers," this constrained minimization problem is equivalent to the solution to the unconstrained minimization problem

$$\underset{R_0, R_1, \ldots, R_{B-1}}{\text{argmin}} \left(\sum_{b=0}^{B-1} \eta_b \sigma_{Y_b}^2 \varepsilon^2 e^{-aR_b} + \lambda \sum_{b=0}^{B-1} \eta_b R_b \right)$$

for some λ. Setting the partial derivatives equal to zero[13] in the usual way, we obtain

$$a\sigma_{Y_b}^2 \varepsilon^2 e^{-aR_b} = \lambda, \quad \forall b$$

Then

$$d^{\text{XFORM}}(\lambda) = \frac{1}{a} \sum_{b=0}^{B-1} \eta_b \lambda = \frac{\lambda}{a}$$

---

[12] We have previously used the symbol m to denote the dimension of a block transform and also the number of bands in a uniform subband transform. In both of these cases, $B = m$. For tree-structured transforms, B is the total number of subbands.

[13] We note that the derivatives exist only for positive rates, R_b, since $d(0) = \sigma_{Y_b}^2$ is the distortion incurred when we code nothing. To extend the result developed here to low bit-rates where some bands might have $R_b = 0$, the Kuhn-Tucker theorem may be employed as in [117].

and

$$R\left(\lambda\right) = \frac{1}{a} \sum_{b=0}^{B-1} \eta_b \log_e \frac{a\sigma_{Y_b}^2 \varepsilon^2}{\lambda}$$

Now, setting the compressed bit-rate for ordinary PCM (i.e., without any transform) equal to $R\left(\lambda\right)$, we obtain

$$d^{\mathrm{PCM}}\left(\lambda\right) = \sigma_X^2 \varepsilon^2 e^{-aR(\lambda)}$$

$$= \sigma_X^2 \varepsilon^2 \prod_{b=0}^{B-1} \left(\frac{\lambda}{a\sigma_{Y_b}^2 \varepsilon^2}\right)^{\eta_b}$$

$$= \frac{\sigma_X^2 \lambda}{a} \prod_{b=0}^{B-1} \left(\frac{1}{\sigma_{Y_b}^2}\right)^{N_b}$$

where the last line follows from the fact that $\sum_{b=0}^{B-1} \eta_b = 1$. Finally, the coding gain becomes

$$G_T = \frac{d^{\mathrm{PCM}}\left(\lambda\right)}{d^{\mathrm{XFORM}}\left(\lambda\right)} = \frac{\sigma_X^2}{\prod_{b=0}^{B-1}\left(\sigma_{Y_b}^2\right)^{\eta_b}} = \frac{\sum_{b=0}^{B-1} \eta_b \sigma_{Y_b}^2}{\prod_{b=0}^{B-1}\left(\sigma_{Y_b}^2\right)^{\eta_b}} \qquad (4.35)$$

The numerator of the expression is a weighted arithmetic mean of the band variances, while the denominator is a weighted geometric mean of the band variances. From the convexity of the log function it is not hard to show that the coding gain is greater than or equal to 1, with equality if and only if all bands have exactly the same variance, $\sigma_{Y_b}^2 = \sigma_X^2$. Thus, the coding gain is a measure of the amount of the diversity amongst the band variances.

CODING GAIN FOR BLOCK TRANSFORMS

Block transforms contribute 1 coefficient to each of the B bands, so that $\eta_b = \frac{1}{B}$. In this case, the coding gain expression simplifies to

$$G_T = \frac{d^{\mathrm{PCM}}\left(\lambda\right)}{d^{\mathrm{XFORM}}\left(\lambda\right)} = \frac{\frac{1}{B} \sum_{b=0}^{B-1} \sigma_{Y_b}^2}{\sqrt[B]{\prod_{b=0}^{B-1} \sigma_{Y_b}^2}} \qquad (4.36)$$

which is the AM/GM ratio of equation (4.31). In this way, the coding gain expression confirms the intuitive arguments developed in Section 4.3.1.

In Section 4.1.2, we developed the KLT as the optimal block transform in the sense of decorrelating random vectors produced by the source. Moreover, based upon its interpretation as a decomposition of the source vectors into their principle components, we speculated that the KLT should be suitable for compression. The following theorem establishes the fact that the KLT is in fact the block transform which maximizes the coding gain expression.

Theorem 4.1 *Out of all orthonormal transforms with block size $m = B$, the KLT maximizes the coding gain expression of equation (4.36).*

Proof. Let K be the analysis (and synthesis) matrix of the KLT and write $\mathbf{y}^{(K)} = K^*\mathbf{x}$ for the vector of KLT transform coefficients. Similarly, let A be the analysis (and synthesis) matrix of any arbitrary $m \times m$ orthonormal transform and write $\mathbf{y}^{(A)} = A^*\mathbf{x}$ for the corresponding transform coefficient vector. Then $\mathbf{y}^{(A)} = (A^*K)\,\mathbf{y}^{(K)}$ and the covariance matrices are related by

$$C_{\mathbf{Y}^{(A)}} = (A^*K)\, C_{\mathbf{Y}^{(K)}}\, (A^*K)^*$$

It follows that

$$\det\left(C_{\mathbf{Y}^{(A)}}\right) = \det\left(A\right)^2 \det\left(K\right)^2 \det\left(C_{\mathbf{Y}^{(K)}}\right) = \prod_{b=0}^{B-1} \sigma_{Y_b^K}^2$$

Finally, we may apply the Hadamard inequality [22, Theorem 3.6.3], which states that the determinant of any symmetric, positive semi-definite matrix is less than or equal to the product of its diagonal elements. This gives us

$$\prod_{b=0}^{B-1} \sigma_{Y_b^A}^2 \geq \prod_{b=0}^{B-1} \sigma_{Y_b^K}^2$$

meaning that the coding gain for transform A is no larger than the coding gain for the KLT. ∎

It is worth restating here the observation that the DCT has similar diagonalizing properties to the KLT, when applied to images and a variety of other sources. This is of great practical value, since the KLT depends upon the source statistics, which can at best only be estimated in practical applications.

CODING GAIN FOR SUBBAND TRANSFORMS

Having established the fact that the KLT is the optimal block transform from the perspective of coding gain, it is natural to enquire as to which subband transforms are likely to yield the largest coding gain. The coding gain expression provides substantial insight in this regard. We shall restrict our attention to image compression. Recall that the power spectrum of a typical image decays rapidly with radial frequency, where a reasonable model is given by equation (4.32). Accordingly, we expect substantial coding gain when the transform divides the source into frequency bands with good frequency selectivity.

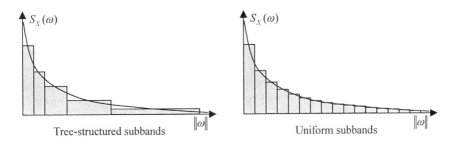

Figure 4.24. One dimensional cross section of image power density spectrum, indicating relative variances of uniform and dyadic tree-structured subbands.

Figure 4.24 illustrates a one dimensional cross section through the image spectrum with two different subband structures, corresponding to a uniform and a dyadic multi-resolution transform. In both cases, substantial coding gain can be expected. In the figure, the uniform subband transform involves many more subbands and will yield a slightly larger coding gain than the dyadic structure. In general, the coding gain increases whenever we divide a subband into smaller subbands. Substantial gain, however, is achieved only when the power spectrum exhibits significant decay over the relevant frequency range. Accordingly, a non-uniform subband structure should always outperform a uniform structure, subject to constraints on the number of subbands or the complexity[14]. The dyadic structure is particularly appealing, both for its efficient exploitation of the decay typically observed in image power spectra, and for its multi-resolution properties which have already been discussed in Section 4.2.5.

LIMITATIONS AND EXTENSIONS

The coding gain formula in equation (4.35) should not be taken too seriously. As a predictive tool for the performance which can be expected in a practical compression system, the expression has a number of weaknesses. Chief among these is the assumption that the quantization and coding techniques do not, themselves, exploit statistical dependencies between the samples. All efficient image compression algorithms employ coding schemes which are able to exploit some of the residual redundancy between the transform coefficients.

---

[14]The complexity of a uniform subband structure is affected not only by the number of subbands, but also by the fact that large, complex filters are required to separate many narrow frequency bands.

Key assumptions in the derivation of the coding gain expression are: 1) that the term, $g(R)$, in equation (4.33), may be taken as a constant, $g(R) \approx \varepsilon^2$; and 2) that both ε^2 and the parameter, a, are identical for all bands and for direct PCM quantization of the original source samples. For well designed quantizers, the parameter $a \approx 2\log_e 2$ for all smooth PDF's. On the other hand, ε^2 generally exhibits a stronger dependence on the source PDF. If we restrict our attention to jointly Gaussian sources, then the transform coefficients also follow a Gaussian distribution and our assumptions are satisfied. In reality, however, untransformed image sample values usually follow a roughly uniform distribution, while the transform coefficients follow a radically different distribution, which is often modeled by a Laplacian or similar generalized Gaussian form.

Finally, we note that the coding gain expression is only strictly applicable to orthonormal transforms. It may be extended to general biorthogonal transforms, subject to the assumption that quantization errors in each subband sample are uncorrelated, with zero mean. Restricting our attention to one dimensional uniform transforms for convenience, the expected energy of the reconstructed signal error may be written using equation (4.24) as

$$E\left[\sum_k \left(\hat{X}[k] - X[k]\right)^2\right] = E\left[\left\|\sum_{q=0}^{m-1}\sum_n \underbrace{\left(\hat{Y}_q[n] - Y_q[n]\right)}_{\delta Y_q[n]} \mathbf{s}_q^{(n)}\right\|^2\right]$$

$$= \sum_{p,q,i,j} E\left[\delta Y_p[i]\, \delta Y_q[j]\right] \left\langle \mathbf{s}_p^{(i)}, \mathbf{s}_q^{(j)}\right\rangle$$

$$= \sum_{q=0}^{m-1}\sum_n E\left[(\delta Y_q[i])^2\right] \cdot \left\|\mathbf{s}_q^{(n)}\right\|^2 \qquad (4.37)$$

Noting that $\left\|\mathbf{s}_q^{(n)}\right\|^2 = G_q$ is independent of n, the reconstructed signal MSE may then be expressed as

$$d^{\text{XFORM}} = \sum_{b=0}^{B-1} \eta_b G_b \sigma_{Y_b}^2 e^{-aR_b} \qquad (4.38)$$

This has the same additive form as equation (4.34), used in the derivation of coding gain. The only change is the introduction of the factors, G_b, which weight the individual subband variances according to the energy expansion properties of the synthesis system. Incorporating these energy gain factors, we readily obtain the following more general expression for coding gain.

$$G_{\mathrm{T}} = \frac{d^{\mathrm{PCM}}(\lambda)}{d^{\mathrm{XFORM}}(\lambda)} = \frac{\sigma_X^2}{\prod_{b=0}^{B-1}\left(G_b\sigma_{Y_b}^2\right)^{\eta_b}}$$

The denominator of the above expression is a weighted geometric mean of the terms, $G_b\sigma_{Y_b}^2$. The numerator, however, is not generally the corresponding weighted arithmetic mean of these same terms. The expression is a ratio of arithmetic to geometric means only for orthogonal transforms. Consequently, for general biorthogonal transforms, the coding gain can actually be less than 1. We also stress the fact that the above expression is based on the assumption of uncorrelated quantization errors. This is a reasonable assumption only at very high bit-rates.

The coding gain expression above is applicable to tree-structured transforms, as well as uniform subband transforms and block transforms. In each case, the synthesis vectors associated with the coefficients in any particular band, b, are all translates of a reference basis vector (sequence), denoted $\mathbf{s}_b \equiv s_b[n]$. The relevant energy gain factor, G_b, is just the squared norm of $\mathbf{s}_b$, i.e., $G_b = \|\mathbf{s}_b\|^2$. For a one dimensional dyadic tree-structured subband transform with D levels, we denote the subbands $\mathbf{L}_D$ and $\mathbf{H}_1$ through $\mathbf{H}_D$, as shown in Figure 4.19. The reference synthesis vector (sequence) may be computed from the low- and high-pass subband synthesis filters, $g_0[n]$ and $g_1[n]$, using the following relations:

$$
\begin{array}{ll}
s_{\mathrm{L}_1}[n] = g_0[n] & s_{\mathrm{H}_1}[n] = g_1[n] \\
s_{\mathrm{L}_d}[n] = \sum_k s_{\mathrm{L}_{d-1}}[k]\,g_0[n-2k] & s_{\mathrm{H}_d}[n] = \sum_k s_{\mathrm{H}_{d-1}}[k]\,g_0[n-2k]
\end{array}
\tag{4.39}
$$

Note that the low-pass synthesis sequences, $s_{\mathrm{L}_d}[n]$, $d < D$, correspond to intermediate low-pass subbands in the tree.

For two dimensional tree-structured subband transforms, such as that shown in Figure 4.20, the subbands are labeled LL_D and HL_d, LH_d, HH_d, for $d = 1, 2, \ldots, D$. The reader may verify that the reference synthesis sequences for one and two dimensional tree-structured transforms are related as follows.

$$
\begin{array}{lll}
s_{\mathrm{LL}_D}[n_1, n_2] = s_{\mathrm{L}_D}[n_1]\,s_{\mathrm{L}_D}[n_2] & \Longrightarrow & G_{\mathrm{LL}_D} = G_{\mathrm{L}_D} \cdot G_{\mathrm{L}_D} \\
s_{\mathrm{HL}_d}[n_1, n_2] = s_{\mathrm{L}_d}[n_1]\,s_{\mathrm{H}_d}[n_2] & \Longrightarrow & G_{\mathrm{HL}_d} = G_{\mathrm{L}_d} \cdot G_{\mathrm{H}_d} \\
s_{\mathrm{LH}_d}[n_1, n_2] = s_{\mathrm{H}_d}[n_1]\,s_{\mathrm{L}_d}[n_2] & \Longrightarrow & G_{\mathrm{LH}_d} = G_{\mathrm{H}_d} \cdot G_{\mathrm{L}_d} \\
s_{\mathrm{HH}_d}[n_1, n_2] = s_{\mathrm{H}_d}[n_1]\,s_{\mathrm{H}_d}[n_2] & \Longrightarrow & G_{\mathrm{HH}_d} = G_{\mathrm{H}_d} \cdot G_{\mathrm{H}_d}
\end{array}
\tag{4.40}
$$

4.3.3   RATE-DISTORTION THEORY

The coding gain arguments advanced above are insufficient to answer some important fundamental questions concerning the use of linear transforms for compression. One such question is whether or not the use of a transform fundamentally limits the achievable compression efficiency. Another such question is under what conditions linear transforms permit the use of simple quantization and coding techniques while approaching the theoretical rate-distortion bound. These questions are the concern of rate-distortion theory.

Unfortunately, results have been derived only under very restrictive assumptions. In particular, the restriction to jointly Gaussian sources, while unrealistic for many applications, dramatically simplifies the formulation of most rate-distortion problems. In this section, our purpose is to indicate without proof some of the conclusions of these information-theoretic studies, along with their incumbent assumptions.

BLOCK TRANSFORMS

Consider an IID vector random process, $\{\mathbf{X}_n\}_{n \in \mathbb{Z}}$. The vectors have dimension m and are statistically independent with PDF, $f_{\mathbf{X}_n} = f_{\mathbf{X}}$. Note that the random variables within each vector are not assumed to be independent. A block transform maps $\{\mathbf{X}_n\}$ to a new random process, $\{\mathbf{Y}_n\}$, also IID, through $\mathbf{Y}_n = A^* \mathbf{X}_n$. In this case, the rate-distortion function for $\{\mathbf{X}_n\}$ is defined by

$$R_{\mathbf{X}}(D) = \inf_{f_{\hat{\mathbf{X}}|\mathbf{X}} \in \mathcal{F}_{\mathbf{X}}(D)} \frac{1}{m} I_f(\mathbf{X}; \hat{\mathbf{X}})$$

where

$$\mathcal{F}_{\mathbf{X}}(D) = \left\{ f_{\hat{\mathbf{X}}|\mathbf{X}} \mid \frac{1}{m} E\left[\left\|\mathbf{X} - \hat{\mathbf{X}}\right\|^2\right] \leq D \right\}$$

where $I_f\left(\mathbf{X}; \hat{\mathbf{X}}\right)$ denotes the mutual information between any random vector, $\mathbf{X}$, from the process, and its distorted reproduction, $\hat{\mathbf{X}}$, for a given conditional PMF, $f_{\hat{\mathbf{X}}|\mathbf{X}}$. Note that the per-element mutual information, $\frac{1}{m} I_f\left(\hat{\mathbf{X}}; \mathbf{X}\right)$, and the per-element distortion, $\frac{1}{m} E\left[\left\|\mathbf{X} - \hat{\mathbf{X}}\right\|^2\right]$, each depend on both $f_{\hat{\mathbf{X}}|\mathbf{X}}$ and the source distribution, $f_{\mathbf{X}}$.

In the particular case of jointly Gaussian distributions, $f_{\mathbf{X}}$, and orthonormal transforms (A a unitary matrix), it is not difficult to show that for each $f_{\hat{\mathbf{X}}|\mathbf{X}}$, there exists an equivalent conditional PMF, $f_{\hat{\mathbf{Y}}|\mathbf{Y}}$, for which $I_f\left(\hat{\mathbf{X}}; \mathbf{X}\right) = I_f\left(\hat{\mathbf{Y}}; \mathbf{Y}\right)$ and $E\left[\left\|\mathbf{X} - \hat{\mathbf{X}}\right\|^2\right] = E\left[\left\|\mathbf{Y} - \hat{\mathbf{Y}}\right\|^2\right]$. It follows that the source vector process, $\{\mathbf{X}_n\}$, and the transform vector process, $\{\mathbf{Y}_n\}$, have identical rate-distortion functions,

$$R_{\mathbf{X}}(D) = R_{\mathbf{Y}}(D)$$

In the specific case where $A = K$ is the KLT, the elements of each transform vector, $\mathbf{Y}_n$, are uncorrelated and hence statistically independent, since the process is assumed Gaussian. Intuitively, then, one would expect that there should be no penalty to considering these elements separately and coding the m IID random processes, $\left\{Y_q^{(n)}\right\}_{n\in\mathbb{Z}}$ independently. Indeed this is the case and we find that $R_{\mathbf{Y}}(D)$ may be expressed as

$$R_{\mathbf{Y}}(D) = \frac{1}{m}\sum_{q=0}^{m-1} R_{Y_q}(D_q)$$

where the distortions, D_q, are chosen so as to minimize the total rate, subject to the constraint

$$\frac{1}{m}\sum_{q=0}^{m-1} D_q \le D$$

The solution to this minimization problem may readily be found using the same Lagrangian techniques used to derive the coding gain expression in Section 4.3.2. For completeness, we summarize the result in parametric form, with parameter λ.

$$D(\lambda) = \frac{1}{m}\sum_{q=0}^{m-1} D_{Y_q}(\lambda) = \frac{1}{m}\sum_{q=0}^{m-1} \min\left\{\sigma_{Y_q}^2, \lambda\right\}$$

$$R(\lambda) = \frac{1}{m}\sum_{q=0}^{m-1} R_q(\lambda) = \frac{1}{m}\sum_{q=0}^{m-1} \max\left\{0, \frac{1}{2}\log_2\frac{\sigma_{Y_q}^2}{\lambda}\right\} \qquad (4.41)$$

In view of Shannon's source coding theorem (see Chapter 3), the significance of this result is that optimal compression of the vector random process, $\{\mathbf{X}_n\}_{n\in\mathbb{Z}}$, is reduced to the simpler problem of optimally compressing the scalar IID processes, $\left\{Y_q^{(n)}\right\}_{n\in\mathbb{Z}}$. Simple scalar quantization and independent coding of the elements, $Y_q^{(n)}$, is well justified under these conditions, although sub-optimal, since the rate-distortion function of an IID random process may be achieved only by exploiting the "Voronoi cell shape gain" or "sphere packing" potential of vector quantization. Fortunately, there is little cause for a pessimistic conclusion, since highly structured vector quantization schemes such as TCQ, are able to approach the rate-distortion function of an IID random process with modest complexity (see Section 3.5).

The assumption of an IID vector process, $\{\mathbf{X}_n\}$, is unrealistic for most sources. Alternatively, we may model the source as a stationary scalar random process, $\{X_k\}$, and consider blocking the random process into vectors of length m, applying the KLT to each block. In this case, the above arguments and definitions refer to the mth order rate-distortion function, $R_X^{(m)}(D)$, of the random process, which approaches the rate-distortion bound, $R_X(D)$, in the limit as $m \rightarrow \infty$.

We have now seen four distinct respects in which the KLT may be described as an optimal block transform and it is worth summarizing these here. As discussed in Section 4.1.2, the KLT is both the optimal decorrelating transform and the transform which yields the minimum MSE when the source must be approximated with only m' transform coefficients per block, for any $m' \leq m$. According to Theorem 4.1, the KLT is also the transform which maximizes the coding gain expression. Note that these three properties are independent of assumptions regarding the source statistics. Finally, in the specific case of Gaussian random processes, the KLT is the transform which maximizes compression performance and asymptotically achieves the rate-distortion bound of the source, if the coefficient sequences, $\left\{Y_q^{(n)}\right\}_{n \in \mathbb{Z}}$, are to be quantized and coded independently as IID processes.

SUBBAND TRANSFORMS

For subband transforms, we shall restrict our attention to stationary Gaussian sources and note that the subband sequences themselves are also stationary Gaussian processes. Without loss of generality, we shall assume that the random processes all have zero mean, so that they are characterized entirely by their auto-correlation sequences, or equivalently, the PDS (Power Density Spectrum), $S_X(\omega)$. For sufficiently high rates, the rate-distortion function is given by[15]

$$R_X(D) = \frac{1}{2\pi} \int_{-\pi}^{\pi} \frac{1}{2} \log_2 \frac{S_X(\omega)}{D} d\omega$$

$$= \frac{1}{2} \log_2 \left(\gamma_X^2 \frac{\sigma_X^2}{D} \right)$$

where

$$\gamma_X^2 \triangleq \frac{\exp \left(\frac{1}{2\pi} \int_{-\pi}^{\pi} \log_e S_X(\omega) \, d\omega \right)}{\frac{1}{2\pi} \int_{-\pi}^{\pi} S_X(\omega) \, d\omega}$$

$$= \frac{\exp \left(\frac{1}{2\pi} \int_{-\pi}^{\pi} \log_e S_X(\omega) \, d\omega \right)}{\sigma_X^2}$$

is known as the "spectral flatness" of the source, $\{X_n\}$. The spectral flatness, γ_X^2, is yet another ratio between arithmetic and geometric means. Note that $\gamma_X^2 \leq 1$ with equality if and only if the power spectrum, $S_X(\omega)$, is flat.

The spectral flatness measure is also useful in evaluating the compression advantage associated with exploiting inter-sample dependencies. In particular, if we apply quantization and coding techniques which are optimized for an IID Gaussian random process (e.g., TCQ), then the best compression we can hope to achieve is given by the first order rate-distortion function and the performance penalty is

$$\Delta R_X(D) = R_X^{(1)}(D) - R_X(D)$$

$$= \frac{1}{2} \log_2 \left(\frac{\sigma_X^2}{D} \right) - \frac{1}{2} \log_2 \left(\gamma_X^2 \frac{\sigma_X^2}{D} \right)$$

$$= \frac{1}{2} \log_2 \frac{1}{\gamma_X^2} \geq 0$$

Now consider an ideal subband transform; i.e., an orthonormal transform whose subbands occupy disjoint frequency bands, as in Example 4.4. Subject to the restriction of a stationary Gaussian source, it can be shown [117] that no penalty (in the rate-distortion sense) is incurred by quantizing and coding the subbands independently. That is,

---

[15] This is just a rearrangement of equation (3.15). The result may be obtained by taking the limit of the expressions in equation (4.41) as $m \to \infty$, noting that the Fourier components of a stationary Gaussian random process are statistically independent. The result is valid for rates sufficiently high that the Lagrangian parameter, λ, of equation (4.41) satisfies $\lambda = D < S_X(\omega)$, for all ω.

provided we optimally balance the contributions of each subband to the overall distortion and rate, the rate-distortion bound of the source may be achieved by quantizing and coding each subband separately. This result is analogous to that for block transforms. In both cases, the use of an orthonormal transform has no impact on the rate-distortion function and the selection of an "ideal" transform ensures that the transform bands may be quantized and coded independently without penalty.

If the individual subbands happen to be IID random processes, then relatively simple techniques such as scalar quantization, or preferably TCQ, may be applied without substantial penalty in compression efficiency. In practice, we cannot expect the subbands to be IID; however, the performance penalty associated with using relatively simple schemes which are adapted to IID sources is given by

$$\Delta R^{\text{subband}}(D) = -\sum_{b=0}^{B-1} \frac{\eta_b}{2} \log_2 \gamma_{Y_b}^2$$

where η_b is the fraction of samples in subband b and $\gamma_{Y_b}^2$ denotes the spectral flatness of subband b. Thus the transform should be selected so as to maximize the average log spectral-flatness of the subbands. This objective is consistent with that deduced from the coding gain formula in Section 4.3.2.

Note that the above result is dependent upon the assumption that the subbands have disjoint passbands. This requires physically unrealizable ideal filters. Fischer [58] showed that relaxing this condition to allow realizable orthonormal subband transforms necessarily implies a rate-distortion penalty, except in the trivial case where the power spectrum of the source is perfectly flat. Further, Wong [172] showed the penalty may be eliminated provided one is prepared to employ a cross-band prediction filter. These and related results are all restricted to stationary Gaussian sources.

DISCUSSION

The theoretical results presented above may be summarized as follows.

- For stationary Gaussian random processes, $\{X_k\}$, the compression performance of linear block transforms (specifically, the KLT), with block size m, is limited only by the difference between $R_X(D)$ and its mth order approximation, $R_X^{(m)}(D)$, assuming that we are prepared to use quantization and coding schemes which are able to approach the rate-distortion bound for the much simpler IID Gaussian random processes.

- Subject to the same assumptions, the compression performance of subband transforms is limited only by the degree to which we are able to approximate ideal bandpass filters and by the lack of spectral flatness within each subband's passband.

- Both block and subband transform methods are able to approach the rate-distortion bound of a stationary Gaussian source, in the limit as $m \to \infty$, and complexity grows without bound.

While limited to the somewhat contrived model of Gaussian random processes, these results serve to reinforce the conclusions drawn from the coding gain arguments in Section 4.3.2.

Interestingly, the asymptotic performance conclusions drawn in this section also apply to DPCM. As discussed at the end of Section 3.3, for a Markov-p Gaussian random process, with optimal linear predictor of order p, the gap between the performance of DPCM and Shannon's lower bound is the factor, ε^2. This is the factor which describes the performance loss of the inner quantizer when applied to IID data. It is the same factor which describes the asymptotic performance loss for block or subband-based compression systems, as discussed above.

In the transform case, we noted that practical techniques such as TCQ may be used to bring ε^2 remarkably close to 1. Unfortunately, the feedback structure of DPCM makes this much more difficult (see Section 3.5.7). More significantly, the DPCM results hold only at high bit-rates. At lower bit-rates, the performance of the predictor degrades substantially. Transform-based approaches do not suffer from this drawback.

4.3.4 PSYCHOVISUAL PROPERTIES

Up until now we have considered the rate-distortion properties of transform coding systems, where distortion is taken as MSE. Unfortunately, MSE is a poor model for the perceptual significance of distortion in images. There remain numerous unanswered questions in regard to modeling of the Human Visual System (HVS). Nevertheless, known properties of the HVS have significant bearing on the selection of suitable transforms for image compression. We identify three such properties below.

CONTRAST SENSITIVITY

Perhaps the best known property of the HVS is its differential sensitivity to spatial frequencies. Figure 4.25 contains a plot of the so-called Contrast Sensitivity Function (CSF), which represents the reciprocal of the detection contrast threshold, $T^{\mathrm{csf}}(f)$, for sinusoidal grating patterns

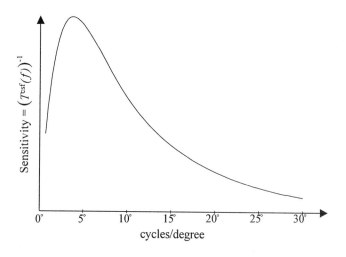

Figure 4.25.   Spatial Contrast Sensitivity Function (CSF).

against a uniform background, as a function of spatial frequency, f. The figure is actually obtained by plotting a specific case of the much more general parametric model proposed in [46]. Not revealed by this simple one dimensional plot, is the fact that the HVS is less sensitive to diagonally oriented features than to horizontally or vertically oriented features.

At high frequencies, sensitivity is limited by the optics of the human eye, by the physical extent of the photo-receptive fields of the "cones," the sensors which determine visual acuity in the HVS, and by primitive neurological processes. We note that the high frequency roll-off of the CSF is steeper for older individuals and at shorter wavelengths (blue portion of the spectrum)[16]. We also note that reported CSF curves occasionally incorporate the high frequency limitations of the display device used for experiments. When using CSF data to optimize an image compression system, the MTF (Modulation Transfer Function) of the intended rendering device should be included, which may differ significantly from that used in the experiments which produced the CSF data.

---

[16]The number of S-cones in the HVS is approximately 1% of the number of M- and L-cones, the latter being sensitive to medium (green) and long (red) wavelengths, respectively. The optical properties of the HVS are adapted to match this difference in sampling densities by attenuating high frequency content in the blue portion of the spectrum more severely than the green and the red.

The behaviour of the CSF at low frequencies may be explained in terms of neurological processes known as "lateral inhibition," which attenuate the sensitivity at any location on the retina in accordance with the response produced at neighbouring locations. Lateral inhibition is responsible for the HVS's remarkable ability to accommodate scenes with internal illumination variations spanning some four orders of magnitude.

The strong dependence of the CSF on spatial frequency may be readily exploited through the use of linear transforms which decompose the image into different spatial frequency bands. As already noted, successful block and subband transforms all possess this structure. A convenient way to incorporate CSF effects into an image compression system is to replace MSE by CSF-Weighted MSE (WMSE) and optimize the parameters of the system to minimize the WMSE for a given bit-rate.

For orthonormal transforms, the MSE distortion measure of equation (4.34), used in the calculation of coding gain, may be replaced by

$$d^{\text{XFORM}} = \sum_{b=0}^{B-1} \eta_b W_b^{\text{csf}} \sigma_{Y_b}^2 e^{-aR_b} \tag{4.42}$$

where the weights, W_b^{csf}, vary as the square of the CSF over the frequency range associated with each band, b. More specifically, we may write

$$W_b^{\text{csf}} = \left(\frac{\alpha}{T_b^{\text{csf}}} \right)^2 \tag{4.43}$$

where T_b^{csf} is some kind of "average" of the detection contrast threshold, $T^{\text{csf}}(f)$, over band b and α is an arbitrary constant. One suitable candidate for the "averaging" operation mentioned here is described in Section 16.1.2.

Using equation (4.42), we find that the coding gain expression becomes

$$G_T^{\text{WMSE}} = \frac{d^{\text{PCM}}(\lambda)}{d^{\text{XFORM}}(\lambda)} = \frac{\sigma_X^2 \sum_{b=0}^{B-1} \eta_b W_b^{\text{csf}}}{\prod_{b=0}^{B-1} \left(W_b^{\text{csf}} \sigma_{Y_b}^2 \right)^{\eta_b}}$$

The factor, $\sum_{b=0}^{B-1} \eta_b W_b^{\text{csf}}$, in the numerator of this expression arises from the assumption that PCM quantization noise is an uncorrelated random process so that the noise appears in every band with equal variance.

The appropriateness of WMSE as a measure of perceived visual distortion depends upon the "flatness" of the CSF within each band and the frequency selectivity of the bands. Block transforms such as the DCT have a uniform frequency band structure with many bands; however, the fact that the blocks do not overlap severely restricts their frequency

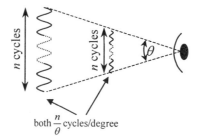

Figure 4.26.   Effect of viewing distance on the relationship between spatial frequency measured in cycles/degree and cycles/pixel.

selectivity, as demonstrated in Example 4.6. Dyadic multi-resolution subband transforms have a logarithmic frequency band structure, as illustrated in Figure 4.21. Frequency selectivity is excellent at low frequencies, but quite poor at high frequencies so that the use of a single CSF weight in the high frequency subbands is far from ideal.

Not all applications lend themselves to exploitation of CSF characteristics. To see this, observe that the horizontal axis in the CSF plot of Figure 4.25 measures spatial frequency in terms of cycles per degree of angle subtended at the observer's eye; Figure 4.26 provides the interpretation for this measure. Changes in viewing distance clearly affect the location of the transform's spatial frequency bands on the CSF curve, so that it is difficult to exploit the CSF characteristic in applications where little can be assumed concerning viewing distance, such as interactive applications where the user is free to "zoom" in and out of the image content.

A particularly convincing demonstration of the CSF characteristic and its dependence on viewing distance is provided by the Campbell and Robson test chart [34], reproduced in Figure 4.27. The frequency of a sinusoidal grating pattern increases from left to right, while the contrast of the pattern increases from the top to the bottom of the test chart. The visual detection threshold at each spatial frequency manifests itself in terms of the height on the chart at which the grating becomes visible. The reader should observe a shift in the location at which the peak sensitivity occurs, as the viewing distance is varied.

VISUAL MASKING

The CSF measures only the detectability of sinusoidal patterns (or artifacts) on a uniform background. Not surprisingly, however, artifacts become less noticeable when superimposed on a non-uniform back-

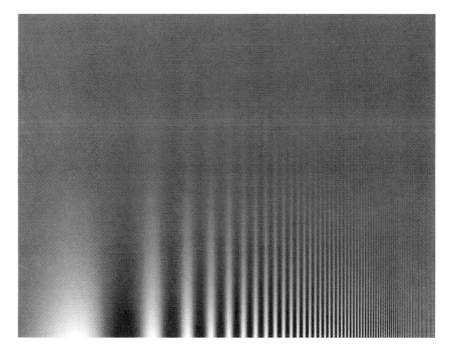

Figure 4.27.   Campbell and Robson CSF test chart.

ground, especially one containing substantial energy at approximately the same spatial frequency and position. This phenomenon, known as "masking," may be exploited with the aid of an image transform whose synthesis vectors have good localization in both space and frequency. To understand this, we begin by reviewing the origin of masking in the HVS.

Early processing of visual stimuli in the part of the brain (visual cortex) known as V1, is commonly modeled in terms of a collection of bandpass filters, whose passbands are distributed logarithmically over the radial frequency spectrum. Typical radial frequency bandwidths for these cortical filters are about 1.4 octaves [49], while typical orientation bandwidths are about 40° [50]. In Watson's "cortex transform," [167] the cortical filter bandwidths are approximated by 1 octave and 45°, leading to the passband structure shown on the left in Figure 4.28. Daly's "Visible Differences Predictor" [46] uses a cortical transform having 6 orientation bands per octave, as illustrated on the right in Figure 4.28.

In Section 1.1.2 we introduced Weber's law, which states that the perceptibility of luminance changes varies inversely with the mean luminance level. A similar process is believed to be at work within each

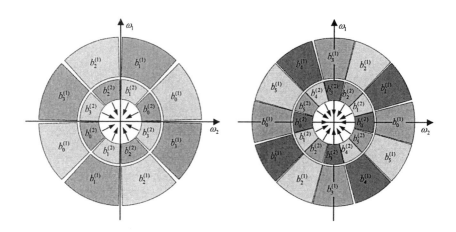

Figure 4.28. Passband structure of the cortical transforms used by Watson [167] (left) and Daly [46] (right). Only the first two resolution levels are shown in each case.

of the cortical bands, whereby the perceptibility of a stimulus whose spatial frequency and orientation falls within some cortical band varies inversely with the local activity within that same band. Like all visual phenomena, masking is a complex process which is not fully understood. This is partly evidenced by the fact that the two sets of cortical transform bands in Figure 4.28 have quite different centre frequencies. Not surprisingly then, signals in one band generally do contribute to the masking of signals in adjacent bands. Nevertheless, we shall restrict our attention to the principle effect of intra-band masking.

Although psychovisual data have been fitted to a variety of empirical models, they differ mainly only in subtle details. A common model for the detection contrast threshold, $T_b(\mathbf{p})$, of an artifact at location $\mathbf{p}$ in cortical band b is [90, 62]

$$T_b(\mathbf{p}) \approx T_b^{\mathrm{csf}} \max\left\{1, \left(\frac{M_b(\mathbf{p})}{T_b^{\mathrm{csf}}}\right)^{\rho}\right\} \qquad (4.44)$$

Here, $M_b(\mathbf{p})$ is a measure of the contrast (amplitude) of the masking signal in cortical band b in the neighbourhood of location $\mathbf{p}$ and T_b^{csf} is the contrast detection threshold for the sinusoidal pattern on a uniform background (no masking). As discussed earlier, the single T_b^{csf} value must be some kind of average of the CSF data over band b. It is related to the CSF energy weighting factors W_b^{csf}, through equation (4.43). The exponent ρ, in equation (4.44), typically varies from 0.6 to 1.0 depending upon the nature of the masking signal and the experience of the observers involved in the psychovisual experiments [139].

It is convenient to define normalized detection and masking contrasts, $t_b(\mathbf{p})$ and $m_p(\mathbf{p})$, as

$$t_b(\mathbf{p}) \triangleq \frac{T_b(\mathbf{p})}{T_b^{\mathrm{csf}}}, \quad \text{and} \quad m_b(\mathbf{p}) \triangleq \frac{M_b(\mathbf{p})}{T_b^{\mathrm{csf}}}$$

Equation (4.44) then assumes the rather simple form

$$t_b(\mathbf{p}) \approx \max\{1, m_b(\mathbf{p})\}^{\rho} \tag{4.45}$$

The normalized detection threshold, $t_b(\mathbf{p})$, is known as the "threshold elevation" factor. It represents the amount by which the detection threshold for an artifact in band b at location $\mathbf{p}$ is scaled (its log is "elevated") relative to the masking-free detection threshold.

Watson [167] constructed an image transform from a bank of cortical filters, followed by appropriate sub-sampling of the cortical subbands, as part of an image compression system capable of exploiting masking. In the proposed compression algorithm, each subband sample, $y_b[\mathbf{n}]$, is subjected to scalar quantization, where the quantizers are designed to keep the quantization error below the relevant detection threshold, $T_b[\mathbf{n}]$. To simplify matters, the masking contrast is taken to be identical to the unquantized amplitude of the same sample whose quantization errors are being masked, i.e., $M_b[\mathbf{n}] = |y_b[\mathbf{n}]|$. With this simplification, the masking effect may be accommodated through appropriate design of the scalar quantizer.

One drawback of Watson's cortical transform is that it cannot be perfectly inverted. Moreover, maximally decimated transforms are not possible without introducing substantial aliasing or significantly distorting the passband characteristics shown in Figure 4.28. Two dimensional tree-structured subband transforms suffer from neither of these ill effects, while possessing a passband structure which bears some useful similarity to that of a cortical transform. This may be seen by comparing Figure 4.28 with Figures 4.15 and 4.21.

Figure 4.29 illustrates the impact of masking within a given subband and between different subbands of a two dimensional tree-structured subband transform. The underlying one dimensional subband transform employed in this example is the CDF 9/7 biorthogonal transform developed in Section 6.3.2. This is one of the two transforms explicitly defined by the JPEG2000 image compression standard. A single quantization error in the middle of subband LH$_2$ passes through the synthesis system to produce the artifact shown without masking as the first image

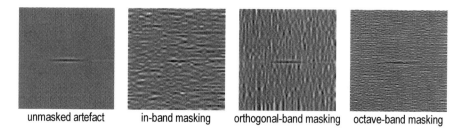

unmasked artefact   in-band masking   orthogonal-band masking   octave-band masking

Figure 4.29. Masking of a single quantization artifact by image content (modeled as Gaussian noise) from the same subband, the subband with opposite orientation, and the subband with the same orientation but in the next resolution level of a two dimensional tree-structured subband transform. All images are offset to allow negative and positive amplitudes to be depicted as deviations from a uniform shade of grey.

in Figure 4.29. The remaining images show this same artifact in the presence of Gaussian noise, which has been added to subbands LH_2 (in-band masking), HL_2 (orthogonal-band masking) and LH_1 (octave-band masking), respectively. The RMS amplitudes of all three synthesized noise images are identical.

Even though Gaussian noise is not a particularly good model for the subband samples produced by typical images, Figure 4.29 does suggest the masking potential of image content from the same subband as the quantization artifact. It also reveals the fact that image content from other subbands has a much weaker masking effect. In Section 16.1.4, we develop these visual masking concepts further in the context of JPEG2000.

VISIBILITY OF BLOCKING ARTIFACTS

Perhaps the greatest disadvantage of block transforms for image compression is that the synthesis (or prototype) vectors are supported within disjoint blocks. After quantization, then, the error image will be a linear combination of disjoint blocks and so will generally exhibit discontinuities at the block boundaries. The HVS is particularly sensitive to the presence of strong gradients which line up in one direction or another, since this is the mechanism by which edges are detected, as the primary source of information for subsequent neurological processing stages. As a result, the alignment of discontinuities at the artificial block boundaries imposed by a block-based transform is a key source of visual disturbance.

To illustrate the phenomenon of blocking artifacts, Figure 4.30 shows decompressed versions of the 256×256 version of the image, "Lenna," at 0.25 bits per sample, where the compression algorithms are JPEG

Figure 4.30. 256×256 image, "Lenna," compressed to 0.25 bps using baseline JPEG (left) and JPEG2000 (right).

and JPEG2000. JPEG is based on the 8×8 DCT, while JPEG2000 is based on a dyadic multi-resolution subband transform. Admittedly, the quantization and coding techniques are somewhat different for the two compression schemes and neither is optimized for the reader's present viewing conditions. Nevertheless, this crude comparison is sufficient to reveal the blocking artifacts which are endemic to block-based transforms.

The visibility of blocking artifacts may be reduced by appropriate post-processing. For example, the decompressed image may be smoothed in a direction perpendicular to the block boundaries. The price paid for such post-processing is both complexity and the introduction of different artifacts, including loss of detail and edge distortion in the neighbourhood of the block boundaries. The problem of blocking artifacts is one of the most significant justifications for the use of subband transforms in place of block transforms.

Chapter 5

RATE CONTROL TECHNIQUES

As discussed in Section 4.3, transform coding involves transforming image samples to get transform coefficients arranged into bands, or subbands. Each band is subjected to quantization and coding. For simplicity, we assume as in Chapter 4 that the quantization and coding of each band is independent of that in other bands, and that quantization and coding do not exploit any dependencies that may exist between coefficients of the same band.

As an example, consider an 8×8 block transform, such as the DCT. All non-overlapping 8×8 blocks of image samples are transformed to get 8×8 blocks of transform coefficients. The $[0,0]$ coefficients from each block comprise band 0, the $[0,1]$ coefficients from each block comprise band 1, and so on. In this way, we have $B = 64$ bands, each with a fraction $\eta_b = 1/64$ of the total coefficients. As another example, consider the $D = 2$ level subband decomposition of Figure 4.22. In this case, we have $B = 3D + 1 = 7$ subbands. The subbands HL_1, LH_1, and HH_1 each have $\eta_b = 1/4$ of the total coefficients, while LL_2, HL_2, LH_2, and HH_2 each have $\eta_b = 1/16$ of the total coefficients.

In this context, rate allocation (or rate control) is the process by which quantization and coding rates, R_b $b = 0, 1, \ldots, B - 1$, are assigned to the various bands. For a given desired overall encoding rate, we may try to minimize MSE, or weighted MSE, while achieving the desired overall encoding rate. On the other hand, for a given desired MSE, we may try to minimize the overall encoding rate while achieving the desired MSE.

5.1   MORE INTUITION
5.1.1   A SIMPLE EXAMPLE

As an intuitive example, consider dividing an image into 2×1 blocks (2-dimensional vectors) $\mathbf{x} = (x_0, x_1)^t$ and transforming each $\mathbf{x}$ to get $\mathbf{y} = (y_0, y_1)^t$ where

$$\mathbf{y} = A^*\mathbf{x} = \frac{1}{\sqrt{2}} \begin{bmatrix} 1 & 1 \\ 1 & -1 \end{bmatrix} \mathbf{x} \tag{5.1}$$

Then all y_0 coefficients comprise band 0 while all y_1 coefficients comprise band 1.

The transform of equation (5.1) can be seen as the special case of equation (4.30) with $\theta = -45°$. The vector $\mathbf{y}$ is easily verified to be a $45°$ rotated (clockwise) version of the vector $\mathbf{x}$. For example, the vector $\mathbf{x} = (1, 1)^t$ has magnitude $\sqrt{2}$ and angle $45°$ while its transformed version $\mathbf{y} = (\sqrt{2}, 0)^t$ has magnitude $\sqrt{2}$ and angle $0°$. Whenever the $\mathbf{x}$ vectors come from a stationary process, equation (5.1) is in fact, the KLT.[1] This is easily verified by noting that in this case, the covariance matrix is given by

$$C_\mathbf{X} = \sigma_X^2 \begin{bmatrix} 1 & r \\ r & 1 \end{bmatrix}$$

where r is the correlation coefficient between x_0 and x_1. The covariance matrix for $\mathbf{Y}$ is then

$$C_\mathbf{Y} = A^*C_\mathbf{X}A = \sigma_X^2 \begin{bmatrix} 1+r & 0 \\ 0 & 1-r \end{bmatrix} \tag{5.2}$$

Now assume that the $\mathbf{x}$ vectors come from a Gaussian Markov process, and consider Figures 3.9 and 3.10b. In Section 3.4, we argued that applying the 2-dimensional VQ of Figure 3.9a to $\mathbf{x} = (x_0, x_1)^t$ is equivalent to applying the SQ of Figure 3.9b independently to x_0 and x_1. In the absence of any entropy coding, the resulting rate is 3 bits/sample. We also argued that for Gaussian vectors $\mathbf{x}$, the codewords of Figure 3.9a that fall outside the elliptical region of Figure 3.10b are largely wasted. These codewords can be discarded to achieve a lower rate without increasing the MSE appreciably. Unfortunately, this leads to an unstructured codebook that is difficult to search for the best codeword $\hat{\mathbf{x}}$.

Consider now applying the transform of equation (5.1) to each point, or vector, in the elliptical high probability region for $\mathbf{x}$. This results in

---

[1] The KLT given by equation (5.1) is entirely independent of the correlation structure of the underlying process. As discussed in Chapter 4, we are not so lucky for vector dimensions greater than 2.

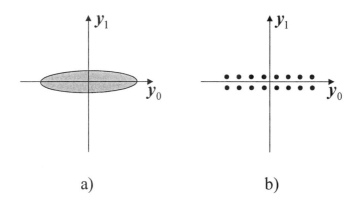

Figure 5.1. a) High probability region for independent (but not identically distributed) Gaussian data. b) R=2 bits/sample VQ.

the high probability region for **y** given in Figure 5.1a. Examination of the rate $R = 2$ bits/coefficient VQ in Figure 5.1b shows that there are no wasted codewords with respect to quantization of **y** vectors. Furthermore, the VQ in this figure is equivalent to independent application of SQ to y_0 and y_1. Specifically, the SQ for y_0 is of rate $R_0 = 3$ bits/coefficient, while the SQ for y_1 is of rate $R_1 = 1$ bit/coefficient. The overall rate is then $R_0 + R_1 = 4$ bits/vector, or $(R_0 + R_1)/2 = 2$ bits/coefficient.

Since each vector of 2 coefficients represents 2 image samples, the rate of the overall transform coding system is also $R = (R_0 + R_1)/2$ bits/sample. Note that the fraction of coefficients in each band is $\eta_0 = \eta_1 = 1/2$, so that R can also be computed as $R = \eta_0 R_0 + \eta_1 R_1$. Finally, when $\hat{\mathbf{y}}$ is inverse transformed to get $\hat{\mathbf{x}}$ via

$$\hat{\mathbf{x}} = \mathbf{A}\hat{\mathbf{y}} = \frac{1}{\sqrt{2}} \begin{bmatrix} 1 & 1 \\ 1 & -1 \end{bmatrix} \hat{\mathbf{y}}$$

the squared error in $\hat{\mathbf{x}}$ is the same as the squared error in $\hat{\mathbf{y}}$. This follows from the fact that the transform of equation (5.1) is orthonormal. Thus, the transform coding system will attain MSE comparable to the $R = 3$ bits/sample SQ of Figure 3.9, while using only $R = 2$ bits/sample.

We conclude this section by stating once again that linear transforms cannot be used to our advantage in the case of uncorrelated data (e.g., Figures 3.10a and 4.23c). Even though there may be striking dependencies of a more general (non-linear) nature (e.g., Figure 4.23c), no benefit can be derived via linear transforms.

5.1.2   AD HOC TECHNIQUES

Many ad hoc techniques have been developed for rate allocation. Principle among these are "zonal coding" and "threshold coding" [81]. In zonal coding, all bands that do not lie within a certain "zone" are discarded (quantized to zero). For example, in the 8×8 DCT case, we might discard all coefficients $y[i, j]$ of each block with $i + j > 3$. In this way, we retain only the coefficients in a triangular zone lying in the upper left corner of each transformed block.

The zone is typically chosen in an ad hoc way, and rates for coefficients within the zone are often chosen to be constant. In this way, the encoding rate is $R = \eta_z R_z$ bits/sample where η_z is the fraction of coefficients in the zone, and R_z is the rate assigned to coefficients within the zone. Assuming $R_z = 4$ in the DCT example of the previous paragraph yields $R = (10/64)4 = 0.625$ bits/sample.

In threshold coding, all coefficients below a certain magnitude threshold are quantized to zero. The coefficients retained are quantized, then encoded via their locations and quantization indices. The SPIHT algorithm (see Chapter 7) was originally described as a very sophisticated form of threshold coding. We will describe SPIHT from a very different point of view in Chapter 7.

5.2   OPTIMAL RATE ALLOCATION

From Section 4.3, we can infer a general expression for the MSE vs. rate of a transform coding system. This expression depends on the MSE and rates assigned to the individual bands as

$$d(R) = \sum_{b=0}^{B-1} \eta_b G_b W_b d_b(R_b) \tag{5.3}$$

where the encoding rate for the image is

$$R = \sum_{b=0}^{B-1} \eta_b R_b \tag{5.4}$$

As before, η_b is the fraction of coefficients in band b, and G_b is the synthesis (inverse transform) gain associated with band b. When orthonormal transforms are employed, $G_b = 1.0 \; \forall b$. The W_b can differ from 1.0 if minimization of "weighted MSE" is desired. For example, these weights might be chosen to match $d(R)$ to the human visual system as discussed in Section 4.3.4.

To find the optimal rates for each band, we may use the method of Lagrange multipliers and set

$$\frac{\partial}{\partial R_b} \left(\sum_{l=0}^{B-1} \eta_l G_l W_l d_l(R_l) + \lambda \sum_{l=0}^{B-1} \eta_l R_l \right) = 0$$

This yields

$$-\lambda = G_b W_b d_b'(R_b) \quad b = 0, 1, \ldots, B-1 \tag{5.5}$$

where $d_b'(R_b)$ is the derivative of the band b distortion, evaluated at R_b bits/coefficient. This quantity represents the slope of d_b at R_b and is non-positive for well designed quantization and coding.

From equation (5.5), the band rates $R_b \ b = 0, 1, \ldots, B-1$ should be chosen so that the weighted distortion-rate slopes, $G_b W_b d_b'(R_b)$, of all bands are equal to the (as yet unknown) constant $-\lambda$.

From equation (3.33), we have

$$d_b(R_b) = g_b(R_b) \sigma_b^2 2^{-2R_b}$$

If all R_b are large enough so that $g_b(R_b) = \varepsilon_b^2$, we have

$$d_b(R_b) = \varepsilon_b^2 \sigma_b^2 2^{-2R_b} \tag{5.6}$$

and

$$\begin{aligned} d_b'(R_b) &= -\varepsilon_b^2 \sigma_b^2 (2 \ln 2) 2^{-2R_b} \\ &= -(2 \ln 2) \, d_b(R_b) \end{aligned} \tag{5.7}$$

Comparing equations (5.5) and (5.7) we see that the weighted distortions of all bands should be equal. That is,

$$G_b W_b d_b(R_b) = \frac{\lambda}{2 \ln 2} \quad b = 0, 1, \ldots, B-1 \tag{5.8}$$

Solving this expression for R_b yields

$$R_b = \frac{1}{2} \log_2 \frac{G_b W_b \varepsilon_b^2 \sigma_b^2 2 \ln 2}{\lambda} \tag{5.9}$$

Substitution of this quantity in equation (5.4) yields

$$2R = \log_2 \prod_{l=0}^{B-1} \left(\frac{G_l W_l \varepsilon_l^2 \sigma_l^2 2 \ln 2}{\lambda} \right)^{\eta_l}$$

Exponentiating both sides (base 2) and noting that $\sum_{l=0}^{B-1} \eta_l = 1$ yields

$$\lambda = 2 \ln 2 \prod_{l=0}^{B-1} \left(G_l W_l \varepsilon_l^2 \sigma_l^2 \right)^{\eta_l} 2^{-2R} \tag{5.10}$$

Substitution in equation (5.9) yields

$$R_b = R + \frac{1}{2}\log_2 \frac{G_b W_b \varepsilon_b^2 \sigma_b^2}{\prod\limits_{l=0}^{B-1}\left(G_l W_l \varepsilon_l^2 \sigma_l^2\right)^{\eta_l}} \tag{5.11}$$

From this expression, we see that any band having weighted variance (numerator of the log expression) equal to the geometric mean of all weighted variances (denominator of the log expression) should receive the nominal R bits/coefficient. Bands with larger or smaller weighted variances should receive more or less than R bits/coefficient, respectively. This is all done in such a way that the weighted average of all band rates is R, as desired.

Substitution of equation (5.11) into equation (5.6) yields

$$d_b(R_b) = \frac{\prod\limits_{l=0}^{B-1}\left(G_l W_l \varepsilon_l^2 \sigma_l^2\right)^{\eta_l}}{G_b W_b} 2^{-2R} \tag{5.12}$$

and we see that as before, the weighted band distortions $G_b W_b d_b(R_b)$ are all equal. In the case of an orthonormal transform and no HVS weighting (i.e., $G_b = W_b = 1$) the band distortions themselves are all equal.

The distortion in the image after inverse quantization and inverse transformation is given by equation (5.3). Substitution of equation (5.12) yields

$$d^{\text{XFORM}}(R) = \prod\limits_{b=0}^{B-1}\left(G_b W_b \varepsilon_b^2 \sigma_b^2\right)^{\eta_b} 2^{-2R} \tag{5.13}$$

We note here that when $W_b \neq 1$ $b = 0, 1, \ldots, B-1$, this expression represents a weighted MSE, and should not be compared directly to the more usual unweighted MSE. Proper treatment of this case is discussed in Section 4.3.4. When $W_b = 1 \; \forall b$, we can compare to the case of direct quantization and coding of the image samples. In this case,

$$d^{\text{PCM}}(R) = \varepsilon_X^2 \sigma_X^2 2^{-2R}$$

and transform coding is superior whenever

$$\prod\limits_{b=0}^{B-1}\left(G_b \varepsilon_b^2 \sigma_b^2\right) < \varepsilon_X^2 \sigma_X^2$$

For efficient quantization of IID Gaussian data, $\varepsilon_b^2 = \varepsilon_X^2 \cong 1$, and the results here are consistent with the discussion in Section 4.3.2.

As an example, consider the two band case from Section 5.1, and let $W_0 = W_1 = 1$. For this case we have $\eta_0 = \eta_1 = 1/2$ and $G_0 = G_1 = 1$. From equation (5.2), we also have $\sigma_0^2 = \sigma_X^2 (1 + r)$ and $\sigma_1^2 = \sigma_X^2 (1 - r)$. Equation (5.11) then yields

$$R_0 = R + \frac{1}{2} \log_2 \sqrt{\frac{1+r}{1-r}} \tag{5.14a}$$

$$R_1 = R - \frac{1}{2} \log_2 \sqrt{\frac{1+r}{1-r}} \tag{5.14b}$$

From this, we can conclude that our intuitive rate allocation of $R_0 = 3$ and $R_1 = 1$ is optimal for $R = 2$ only when $r = 15/17 \cong 0.88$. As r is increased or decreased, the portion of the rate allocated to R_0 increases or decreases accordingly. Note that when $r = 255/257 \cong 0.99$, the limiting case of $R_0 = 4$ and $R_1 = 0$ is reached. In this case, band 1 is discarded entirely, in favor of allocating all available rate to band 0.

From equation (5.12) and equation (5.13),

$$d_0(R_0) = d_1(R_1) = d^{\text{XFORM}}(R) = \sigma_X^2 \left(\sqrt{1 - r^2} \right) 2^{-2R}$$

The ratio of $d^{\text{PCM}}(R)$ to $d^{\text{XFORM}}(R)$ then gives the transform coding gain of equation (4.35). Even for this simple two dimensional example, we get the significant gain

$$G_{TC} = \frac{1}{\sqrt{1 - r^2}} \tag{5.15}$$

which is approximately 3.27 dB for the value $r = 15/17$ discussed above. For the limiting value of $r = 255/257$, we get a transform coding gain of 9.05 dB.

We conclude this section by noting that $r = 255/257$ is only limiting in a very narrow sense. In our example, we chose an encoding rate of $R = 2$ bits/sample. If $r > 255/257$, this value of R is not large enough to satisfy our "high rate" assumption. Specifically, equation (5.14) results in $R_1 < 0$. For larger choices of R, equations (5.14) and (5.15) remain valid for larger values of r.

5.3   QUANTIZATION ISSUES

It should be clear at this point that rate allocation is equivalent to quantizer selection. In the case of uncoded SQ, VQ, or TCQ, choosing a rate R_b is equivalent to choosing a set of quantizer codewords or code-vectors (e.g., Table 3.2). In the case of entropy coded systems designed iteratively (Section 3.2.3), this is also the case.

For uniform ECSQ (Section 3.2.4) or ECTCQ (Section 3.5.6), rate allocation is equivalent to step size selection. For R_b sufficiently large, as assumed throughout this section, we then have

$$d_b(R_b) = c_b \Delta_b^2$$

where c_b is a constant. For example, $c_b = 1/12$ for ECSQ (see equation (3.27)). Comparing with equation (5.6), we see that

$$\Delta_b = \sqrt{\frac{\varepsilon_b^2 \sigma_b^2}{c_b} 2^{-R_b}} \tag{5.16}$$

Finally, since all $G_b W_b d_b(R_b)$ are equal (equation (5.12)), we have

$$G_b W_b d_b(R_b) = G_0 W_0 d_0(R_0) \quad \forall b$$

or

$$G_b W_b c_b \Delta_b^2 = G_0 W_0 c_0 \Delta_0^2$$

Thus,

$$\Delta_b = \sqrt{\frac{G_0 W_0}{G_b W_b}} \Delta_0 \tag{5.17}$$

If an orthonormal transform is employed and we normalize the HVS weights so that $W_0 = 1$, we have

$$\Delta_b = \frac{\Delta_0}{\sqrt{W_b}}$$

This is in fact, the methodology behind the selection of the example "Q-table" (table of step sizes) provided in the original JPEG standard. As a concluding remark, we note that when HVS weights are not employed, all step sizes are equal.

5.3.1   DISTORTION MODELS

We began our discussion of optimal rate allocation with equation (5.3). This equation assumes that the MSE for quantizing coefficients in band b, at a rate of R_b bits/coefficient, is given by $d_b(R_b)$. The constants η_b, G_b, and W_b determine how these "band distortions" combine to yield the overall image distortion $d(R)$. In equation (5.6), we assumed a high-rate approximation of $d_b(R_b)$. This expression requires knowledge (or estimation) of only two parameters, ε_b^2 and σ_b^2. These values can be estimated based on a large collection of typical imagery, or on an image-by-image basis. For a given band of a given image, σ_b^2 can be estimated in the usual way as

$$\hat{\sigma}_b^2 = \frac{1}{N_b - 1} \sum_y (y - \hat{\mu}_b)^2$$

where

$$\hat{\mu}_b = \frac{1}{N_b} \sum_y y$$

The summations above are over all y in band b, and N_b is the number of coefficients in band b.

A common method for estimating ε_b^2 involves modeling the subband data as coming from a generalized Gaussian distribution with PDF

$$f(y) = \frac{\alpha}{2\sigma\Gamma(1/\alpha)} \exp\left\{-\left(\sqrt{\frac{\Gamma(3/\alpha)}{\Gamma(1/\alpha)}} \frac{|y - \mu_Y|}{\alpha}\right)^\alpha\right\} \tag{5.18}$$

Given band b for a particular image, the parameter α_b is estimated and used in turn, to determine an appropriate choice for ε_b^2. For example, with $\alpha_b = 2$, equation (5.18) yields the Gaussian distribution. Similarly, $\alpha_b = 1$ yields the Laplacian distribution. For these choices, appropriate values of ε_b^2 can be found in Table 3.4. For other values of α_b, or for quantization schemes not represented in Table 3.4, suitable values for ε_b^2 can be derived or measured experimentally.

Several strategies for estimating α_b have been proposed. In the case of the DCT, the DC (or $[0,0]$) coefficients are often modeled as Gaussian while all other coefficients are modeled as Laplacian [123]. The maximum likelihood estimate of α_b is discussed in [51], while more ad hoc approaches are discussed in [98], [85], and [86]. In [85], it is shown that there is little difference between any of these approaches.

One approach derives from the fact that for the PDF of equation (5.18),

$$\frac{E\left[(Y - \mu_Y)^4\right]}{(\sigma^2)^2} = \frac{\Gamma(5/\alpha)\Gamma(1/\alpha)}{(\Gamma(3/\alpha))^2} \tag{5.19}$$

Thus, $\hat{\alpha}_b$ can be found from estimates of $E\left[(Y_b - \mu_{Y_b})^4\right]$ and $\sigma_b^2 = E\left[(Y_b - \mu_{Y_b})^2\right]$ using equation (5.19).

5.4   REFINEMENT OF THE THEORY

There are a number of limitations to the rate allocation algorithm discussed in the previous sections. Throughout the discussion, we assumed $R_b \geq 0 \; \forall b$. In fact, we assumed that all R_b are large so that $g_b(R_b) = \varepsilon_b^2$. Additionally, we assumed that quantization and coding can be performed at any rate R_b. In some situations, we may not be able to control these rates precisely. In other situations, we may wish to restrict R_b to integers (e.g., uncoded SQ) or rational numbers with small denominators (e.g., uncoded VQ).

5.4.1   NON-NEGATIVE RATES

We first address the situation when R is not large enough to yield $R_b \geq 0 \; \forall b$. This situation occurs in our two band example when $r = 255/257$ and $R < 2$. The problem arises when λ is so large that equation (5.9) calls for $R_b < 0$. From equation (5.6), this is equivalent to $d_b(R_b)$ exceeding $\varepsilon_b^2 \sigma_b^2$. The solution to this problem[2] is to set

$$R_b(\lambda) = \max \left\{ 0, \frac{1}{2} \log_2 \frac{G_b W_b \varepsilon_b^2 \sigma_b^2 2 \ln 2}{\lambda} \right\} \qquad (5.20)$$

This in turn yields

$$d_b(R_b(\lambda)) = d_b(\lambda) = \min \left\{ \varepsilon_b^2 \sigma_b^2, \frac{\lambda}{G_b W_b 2 \ln 2} \right\}$$

These equations satisfy the Kuhn-Tucker optimality conditions. As before, the value of λ must be adjusted so that the resulting R_b achieve the desired rate. That is,

$$\sum_{b=0}^{B-1} \eta_b R_b(\lambda) = R$$

Unfortunately, for small R, there is generally not a closed form expression for computing the proper λ. From equation (5.20), the R_b are functions of λ, so finding the proper λ is equivalent to finding the root of the non-linear equation

$$f(\lambda) = \sum_{b=0}^{B-1} \eta_b R_b(\lambda) - R = 0$$

Any numerical algorithm for finding roots will suffice. However, the problem is facilitated by the fact that the R_b (and thus $f(\lambda)$) are non-increasing and convex $\cup$ functions of λ (see equation (5.20)). This allows a simple bisection algorithm to be used.

---

[2] In fact, equation (5.6) may not be a reasonable distortion model when R_b is small. We will address this issue in a subsequent section.

5.4.2   DISCRETE RATES

We now treat the case where the rates are constrained to be integers. This constraint occurs when uncoded SQ is employed. Actually, the technique described below [134] is more general, and works for any finite set of allowable rates. For example, rates corresponding to "half integers" $\{0, 1/2, 1, 3/2, \ldots, R_{\max}\}$ arise from the use of 2-dimensional uncoded VQ.

To this end, let $\mathcal{R}_b$ be the set of allowable rates for band b. We proceed as before, and seek to minimize equation (5.3) subject to the constraint of equation (5.4). Additionally, we have the new constraint that $R_b \in \mathcal{R}_b$ $b = 0, 1, \ldots, B - 1$. We form the Lagrangian cost function

$$\sum_{l=0}^{B-1} \eta_l G_l W_l d_l(R_l) + \lambda \sum_{l=0}^{B-1} \eta_l R_l$$

$$= \sum_{l=0}^{B-1} \eta_l \left(G_l W_l d_l(R_l) + \lambda R_l \right) \qquad (5.21)$$

Rather than differentiating as before, we seek to minimize equation (5.21) directly.

Noting that each term in the sum depends on only a single rate, we can minimize the sum by minimizing each term individually. Thus, for a given λ, we perform the B individual minimizations

$$G_b W_b d_b(R_b) + \lambda R_b \quad b = 0, 1, \ldots, B - 1 \qquad (5.22)$$

Since each $\mathcal{R}_b$ is finite in size, we can create B tables of rate-distortion pairs. Specifically, the table for band b contains rate-distortion pairs $(R_b, d_b(R_b))$ for each $R_b \in \mathcal{R}_b$. Each minimization of equation (5.22) can then be performed by testing every pair from the appropriate table.

As in the previous section, λ must be adjusted until the desired rate is achieved. Due to the discrete nature of $\mathcal{R}_b$, it will not generally be possible to achieve the target rate exactly. For example, if integer rates are used in our two-band example, then R must be of the form $\sum_b \eta_b R_b = (R_0 + R_1)/2$ and only "half-integer" rates can be achieved.[3]

---

[3] Actually, "time-sharing" can be used to approach the desired rate more closely. For example, half the coefficients of band 0 can be coded at 3.0 bits/coefficient, while the other half are coded at 2.0 bits/coefficient. The average rate for the band is then $R_0 = 2.5$ bits/coefficient. Applying a similar argument to band 1 and computing $R = (R_0 + R_1)/2$, we see that "quarter-integer" rates can be achieved. The technique is easily generalized to yield finely spaced rates, but non-uniform quality is obtained within each band.

As a concluding remark, we note that the method described in this subsection circumvents all of the limitations discussed at the beginning of Section 5.4. Since only non-negative rates are included in the tables of rate-distortion pairs, there is no need to explicitly ensure $R_b \geq 0$. Also, no explicit expressions are required for the band distortions. The values of $d_b(R_b)$ contained in the tables can be individually calculated, or experimentally measured for each $R_b \in \mathcal{R}_b$. This avoids any reliance on high rate approximations.

5.4.3   BETTER MODELING FOR THE CONTINUOUS RATE CASE

The rate allocation method described above is very general and is particularly appropriate for uncoded SQ and/or VQ, where the achievable rates are inherently discrete. In the case of entropy coded uniform SQ (or TCQ), the step size can be used to adjust distortion and rate in a continuous fashion. In principle, any $R_b \geq 0$, or any distortion $0 \leq d_b(R_b) \leq \sigma_b^2$ can be achieved.

The discrete rate allocation strategy of the previous section can be effectively employed by choosing many finely spaced rates for inclusion in $\mathcal{R}_b$. However, under this strategy the tables of rate-distortion pairs may grow quite large. The tables may be eliminated if expressions for $d_b(R_b)$ are available for $b = 0, 1, \ldots, B-1$. In this case, we can solve the B (continuous) minimization problems of equation (5.22) independently.

Alternatively, we can revisit the continuous rate allocation equations of previous sections. For example, equation (5.11) yields continuous rates, but does not guarantee non-negativity. Equation (5.20) represents an improvement to this situation, as it yields continuous rates and also ensures non-negativity. Unfortunately, both of these schemes employ the simple distortion model of equation (5.6) which is not valid when R_b is small.

On the other hand, equation (5.5) does not depend on a high rate assumption. This expression assumes only that $d_b(R_b)$ is differentiable and convex $\bigcup$. Noting that d_b' is a function of R_b, let h_b be the inverse function. That is,

$$R_b = h_b(d_b') \tag{5.23}$$

From equation (5.5), we then have the rate allocation equations

$$R_b = h_b\left(\frac{-\lambda}{G_b W_b}\right) \quad b = 0, 1, \ldots, B-1 \tag{5.24}$$

As before, λ must be adjusted until the rates satisfy

$$\sum_{b=0}^{B-1} \eta_b R_b = R \tag{5.25}$$

It is worth noting, that for well behaved quantization and coding such as ECSQ or ECTCQ, $d_b(R_b)$ is non-negative, strictly decreasing, convex $\bigcup$, and differentiable. Equivalently, $d_b'(R_b)$ is strictly negative, strictly increasing, and converges to zero as R_b gets large. For example, Figure 5.2 shows a plot of $d_b(R_b)$ vs. R_b for uniform ECSQ[4] of IID Gaussian data. This plot was obtained by quantizing 1,000,000 samples of unit variance pseudo-random Gaussian data and computing the resulting MSE and entropy.[5] This procedure was carried out for many step sizes to obtain very fine spacing (≤ 0.01 bits/coefficient) between samples in the plot. Figure 5.3 shows $d_b'(R_b)$ vs. R_b for IID Gaussian data. This plot was obtained from finite difference calculations of the data from Figure 5.2.

From the discussion of the previous paragraph, we conclude that h_b is non-negative and non-increasing as a function of λ (equation (5.24)). Thus, for every $\lambda > 0$, only non-negative rates are obtained. In fact, for any finite λ, only strictly positive rates are obtained. Thus, with more accurate modeling of the band distortions (than that of equation (5.6)), we will always find that $R_b > 0 \quad b = 0, 1, \ldots, B - 1$, and bands should never be discarded. This is in contrast to zonal coding (Section 5.1) and to the policy obtained via high rate approximation theory (Section 5.2). That being said, R_b becomes negligibly small quite quickly as λ is increased beyond a certain point. Also, for a very small band rate R_b, it is still entirely possible for all coefficients of the band to be quantized to zero in any given image.

For example, if ECSQ is employed, the step size may be chosen so large that only the 0 codeword is chosen. Some portion of the available bits (albeit a small portion) must still be spent to code these zeros. More refined techniques may be useful in this situation. Such techniques include signalling, via a single "flag" bit, that all codewords are zero. This is discussed in Chapters 7 and 8. Even in this case however, bands are not discarded. All bands are assigned finite step sizes for quantization and then coded appropriately.

---

[4]No deadzone was employed. Specifically, quantization was performed using equation (3.31) with $\xi = 0$. Centroids (equation (3.18)) were used for inverse quantization.
[5]As in Chapter 3, we assume efficient entropy coding so that the resulting encoding rate is closely approximated by the entropy.

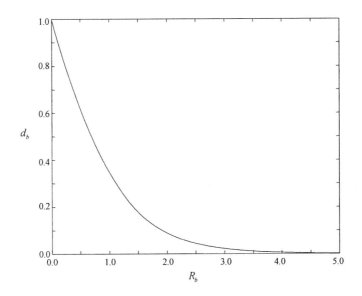

Figure 5.2.   MSE vs. rate for ECSQ of unit variance IID Gaussian data.

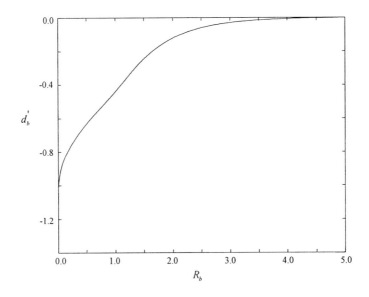

Figure 5.3.   Distortion-rate slope for ECSQ of unit variance IID Gaussian data (i.e., the derivative of Figure 5.2).

5.4.4   ANALYSIS OF DISTORTION MODELS

Considerable insight can be gained by re-plotting Figures 5.2 and 5.3 in the form of Figures 5.4 and 5.5. Figure 5.4 portrays $SNR_b \triangleq 10 \log_{10}\left(\sigma_b^2/d_b(R_b)\right)$ vs. R_b for ECSQ of IID Gaussian data. After a brief transient, the SNR converges to the high rate behavior of $6.02R_b - 1.53$ dB, as expected. Figure 5.5 is essentially a plot of the inverse function h_b (equation (5.23)). Since d_b' is always negative, we have plotted R_b vs. $\log_{10}\left(-d_b'\right)$.

From the high rate approximation of equation (5.7),

$$R_b = -\frac{1}{2} \log_2 \frac{-d_b'}{\varepsilon_b^2 \sigma_b^2 2 \ln 2}$$

$$= -\frac{1}{2 \log_{10} 2} \log_{10}\left(-d_b'\right) + \frac{1}{2} \log_2 \left(\varepsilon_b^2 \sigma_b^2 2 \ln 2\right) \qquad (5.26)$$

Thus, we note that at high rates (i.e., large R_b, small $-d_b'$, small λ), R_b should be linear in $\log_{10}\left(-d_b'\right)$. We note further that the slope is independent of the PDF and is given by

$$-\frac{1}{2 \log_{10} 2} \cong -1.66 \qquad (5.27)$$

The constant term depends on both PDF and quantizer type, through ε_b^2. Varying this term shifts the straight line up or down. As an example, for ECSQ with $\sigma_b^2 = 1$, we have $\varepsilon_b^2 = \pi e/6$ (Table 3.4, IID Gaussian), and the constant term is

$$\frac{1}{2} \log_2 \left(\varepsilon_b^2 \sigma_b^2 2 \ln 2\right) \cong 0.49 \qquad (5.28)$$

The high rate linear approximation given by equations (5.26), (5.27), and (5.28) is included in Figure 5.5, and agrees closely with the measured data. The point at which the line crosses the horizontal axis ($R_b = 0$) occurs at

$$\log_{10}\left(-d_b'\right) = \log_{10}\left(\varepsilon_b^2 \sigma_b^2 2 \ln 2\right) \cong 0.295$$

corresponding to the maximal λ in equation (5.20) beyond which $R_b = 0$. However, as previously noted, this critical slope only appears in the high rate approximation, and never occurs for practical quantization schemes. As λ is increased (equivalently $\log_{10}\left(-d_b'\right)$ is increased), the actual rate becomes nearly zero more quickly than the straight line approximation. However the actual rate never reaches zero, but converges to zero asymptotically.

When ECTCQ is used, plots of SNR_b vs. R_b and R_b vs. $\log_{10}\left(-d_b'\right)$ look very much like those of Figures 5.4 and 5.5. Of course, the gap

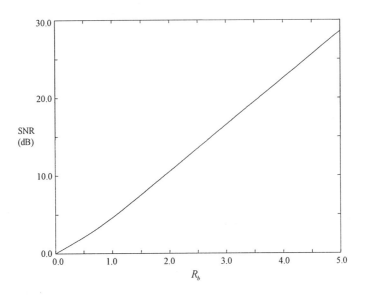

Figure 5.4.   SNR vs. rate for ECSQ of IID Gaussian data.

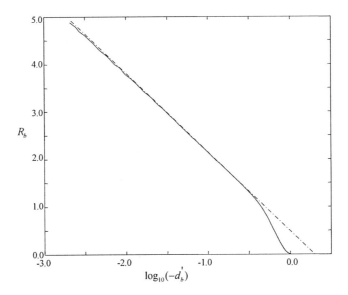

Figure 5.5.   Rate vs. distortion-rate slope for ECSQ of unit variance IID Gaussian data. The dashed line represents the high rate approximation. The solid line represents the actual measured performance.

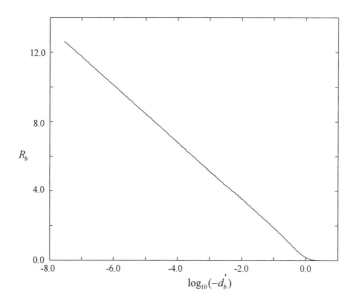

Figure 5.6. Rate vs. distortion-rate slope for ECTCQ of unit variance IID Laplacian data.

between ECTCQ performance and the distortion-rate function is considerably smaller than the 1.53 dB of Figure 5.4. Additionally, the low rate behavior in Figure 5.5 is not quite as precipitous for ECTCQ. Heavier tailed PDFs can also temper the R_b vs. $\log_{10}(-d_b')$ behavior at low rates.

For example, Figure 5.6 shows R_b vs. $\log_{10}(-d_b')$ for 8-state ECTCQ of IID Laplacian data. Note that in both Figures 5.5 and 5.6, the high rate linear segment transitions into a low rate linear segment, followed by an asymptotic convergence to zero. As mentioned in the previous paragraph, the slope of the low rate linear segment is steeper for Gaussian data than for Laplacian data. This is indicative of the heavier tails of the Laplacian PDF. That is, the Gaussian PDF falls off extremely rapidly (exponentially with x^2), while the Laplacian PDF falls off more slowly (exponentially with $|x|$). More generally, smaller α_b in equation (5.18), results in a less precipitous drop in the low rate region of R_b vs. $\log_{10}(-d_b')$.

STEP SIZE VS. DISTORTION-RATE SLOPE

Once λ has been adjusted to satisfy equation (5.25), the distortion-rate slopes (one for each band) are given by $-d_b' = \lambda/(G_b W_b)$, $b = 0, 1, \ldots, B-1$ (equation (5.5)). The required quantization step sizes can

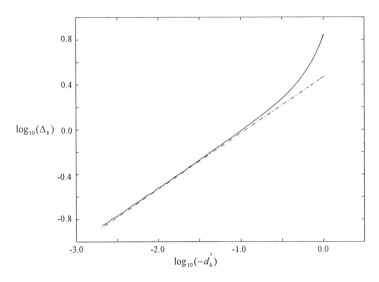

Figure 5.7.   Step size vs. distortion-rate slope for ECSQ of unit variance IID Gaussian data. The dashed line represents the high rate approximation. The solid line represents the actual measured performance.

then be determined as functions of $-d'_b$. Figure 5.7 shows this relationship for ECSQ of unit variance IID Gaussian data. Plots for ECTCQ and/or other PDFs are quite similar.

Substituting equation (5.7) into equation (5.16), we see that the high rate behavior of Δ_b is governed by

$$\Delta_b = \sqrt{\frac{-d'_b}{c_b 2 \ln 2}}$$

or

$$\log_{10} \Delta_b = \frac{1}{2} \log_{10}\left(-d'_b\right) - \frac{1}{2} \log_{10} c_b 2 \ln 2$$

For ECSQ, $c_b = 1/12$, and we have a straight line with slope $1/2$ and intercept $(\log_{10}(-d') = 0)$ of $\log_{10} \Delta_b \cong +0.47$. This high rate linear approximation is included in Figure 5.7 and is in close agreement with the experimentally measured data. At low rates however (i.e., large Δ_b, large λ, large $-d'_b$), the high rate approximation differs significantly from actual performance.

We conclude this section by noting that accurate expressions modeling the behavior of R_b and $\log_{10} \Delta_b$ vs. $\log_{10}(-d'_b)$ were obtained in [86] for ECTCQ. These expressions are based on linear segments as discussed above, but employ hyperbolic sections to model the "curved" convergence behavior at low rates. Similar models are easily developed for ECSQ.

5.4.5   REMAINING LIMITATIONS

In some cases, we may not be able to precisely control the rate achieved in any given band. This situation arises frequently in entropy coded systems. The difference between desired and achieved rate is typically due to errors in the rate-distortion models employed in the allocation process.

One solution to this problem is, of course, better models. Unfortunately, little is currently known in this regard. Another solution is to iteratively encode the image, adjusting the target rate, until the desired rate is achieved. Unfortunately, this is a very high complexity solution. Nevertheless, for many compression systems (e.g., baseline sequential JPEG), there is no other choice when precise rate control is desired.[6]

Embedded quantization and coding provide an elegant solution to this problem. As mentioned in Chapter 3, embedded compression systems result in an incrementally decodable bit-stream. For such systems, precise rate control can be accomplished by simply truncating the bit-stream at the point corresponding to the desired rate. Embedded compression is discussed extensively in Chapters 7 and 8.

5.5   ADAPTIVE RATE ALLOCATION

In the previous sections, rate allocation strategies were all spatially invariant. That is, no effort was made to adapt the quantization and coding to take into account local variations in scene content. For example, in the 8×8 DCT case, all $[0,0]$ coefficients were grouped together into a single band for the purpose of rate allocation, quantization, and coding. More generally for each choice of $i, j \in \{0, 1, \ldots, 7\}$, all $[i, j]$ coefficients were grouped together to form a single band. We then had $B = 64$ bands, each with a fraction $\eta_b = 1/64$ of all coefficients in the image.

We now note that each $[i, j]$ coefficient comes from a different 8×8 image block and thus from a different spatial region. In our previous discussions, we have assumed that the coefficients within a band are IID, and thus stationary (or at least that the quantization and coding do not exploit any dependency or non-stationarity). Since real images

---

[6]The typical policy employed in baseline JPEG is to not attempt any sort of precise rate control. The step size Δ_0 in equation (5.17) is simply set to achieve a desired level of quality (distortion). The rate that results from this process is then not known in advance and can vary widely from image to image.

are non-stationary, it is reasonable to attempt to allocate the available rate in a spatially varying fashion.

To this end, consider classifying image blocks based on local "image activity." One measure of such activity is block variance or "AC energy" given by

$$E_{AC} = \sum_{i=0}^{7}\sum_{j=0}^{7} y^2\,[i,j] - y^2\,[0,0]$$

Consider further, computing E_{AC} for each block and dividing the blocks into classes based on these energies. The simplest case is when each class is chosen to have the same number of blocks [36]. We refer to this case as "percentile" or "uniform" classification.

For a given class and given $i, j \in 0, 1, \ldots, 7$, the $[i, j]$ coefficients from each block within the class are collected together to form a band for the purpose of rate allocation, quantization, and coding. If there are N_c classes, there are then $B' = 64N_c$ bands, each with a fraction $1/(64N_c)$ of the total coefficients. Any of the rate allocation schemes discussed in this chapter can then be applied to these bands, to yield a spatially varying (or spatially adaptive) rate allocation algorithm. The weights W_b, $b = 0, 1, \ldots, B' - 1$ can be chosen to exploit HVS properties and/or spatial masking phenomena as discussed in Section 4.3.4.

These ideas are easily extended to the case of subband transforms [85]. Consider partitioning the image into $N \times N$ non-overlapping blocks. After performing a D-level dyadic subband transform of the entire image, the original $N \times N$ partition induces a $N/2 \times N/2$ coefficient block partition of the HL_1, LH_1, and HH_1 subbands, a $N/4 \times N/4$ partition of the HL_2, LH_2, and HH_2 subbands, and so on. The blocks of each subband can then be classified into N_c classes, based on the variance of coefficients within blocks. All coefficients of all blocks from a given class and subband then comprise one of $B' = (3D + 1)N_c$ "bands" for the purpose of rate allocation, quantization, and coding.

5.5.1   CLASSIFICATION GAIN

For simplicity, we assume the simplified distortion model of equation (5.6) with $W_b = G_b = \varepsilon_b^2 = 1$. As before, we assume that each "original" band is divided into N_c classes resulting in $B' = BN_c$ "bands" for the purpose of rate allocation, quantization and coding. We index these bands by the pair (b, n) $b = 0, 1, \ldots, B - 1$, $n = 0, 1, \ldots, N_c - 1$. We also relax the uniform (or percentile) assumption on class population, and let $p_{b,n}$ be the fraction of coefficients in original band b which belong to band (b, n). Finally, letting $\sigma_{b,n}^2$ be the variance of band (b, n), we have

$$\sigma_b^2 = \sum_{n=0}^{N_c-1} p_{b,n} \sigma_{b,n}^2 \qquad (5.29)$$

As before, we seek to minimize

$$d(R) = \sum_{b=0}^{B-1} \sum_{n=0}^{N_c-1} p_{b,n} \eta_b \sigma_{b,n}^2 2^{-2R_{b,n}}$$

subject to the constraint that

$$\sum_{b=0}^{B-1} \sum_{n=0}^{N_c-1} p_{b,n} \eta_b R_{b,n} = R$$

Proceeding as in previous sections, we employ the Lagrange multiplier method and assume high rates to obtain

$$d(R) = \prod_{b=0}^{B-1} \prod_{n=0}^{N_c-1} \left(\sigma_{b,n}^2 \right)^{p_{b,n} \eta_b} 2^{-2R} \qquad (5.30)$$

which is the obvious generalization of equation (5.13), but with $W_b = G_b = \varepsilon_b^2 = 1$. Taking the ratio of these two expressions results in the classification gain

$$G_C = \frac{\displaystyle\prod_{b=0}^{B-1} \left(\sigma_b^2 \right)^{\eta_b}}{\displaystyle\prod_{b=0}^{B-1} \prod_{n=0}^{N_c-1} \left(\sigma_{b,n}^2 \right)^{p_{b,n} \eta_b}}$$

$$= \prod_{b=0}^{B-1} \left(\frac{\sigma_b^2}{\displaystyle\prod_{n=0}^{N_c-1} \left(\sigma_{b,n}^2 \right)^{p_{b,n}}} \right)^{\eta_b}$$

Substituting equation (5.29), we see that

$$G_C = \prod_{b=0}^{B-1} \left(G_{C,b} \right)^{\eta_b}$$

where

$$G_{C,b} = \frac{\displaystyle\sum_{n=0}^{N_c-1} p_{b,n} \sigma_{b,n}^2}{\displaystyle\prod_{n=0}^{N_c-1} \left(\sigma_{b,n}^2 \right)^{p_{b,n}}}$$

is the classification gain of band b. Since $G_{C,b}$ is a ratio of arithmetic to geometric means, it exceeds 1 to the extent that the $\sigma_{b,n}^2$ differ from σ_b^2. These differences are the mechanism whereby image non-stationarity is exploited. Specifically, when the data is non-stationary, so that local variances differ markedly, the classification gain will be large.

The discussion above omits the overhead associated with the coding of class information. Such information may be required for the decoder to determine class membership and/or encoding rates employed for all coefficients of each "original" band. If R_C bits/coefficient are required for this task, only $R - R_C$ bits are available for coding coefficients. Taking this into account in equation (5.30), the classification gain should be reduced by a factor of 2^{2R_C}.

Several strategies for coding class information are discussed in [85] along with an optimal method for selecting the class population fractions $p_{b,n}$. Experimental results indicate that gains exceeding 1.25 dB are achievable when classification is employed with subband transforms.

In closing we note that JPEG2000 also performs quantization and coding on blocks of subband coefficients. The size of the blocks is not constrained to shrink pyramidally with transform level as described above. Embedded bit-streams are formed for each block, and encoding rates $R_{b,n}$ (or equivalently, truncation points) are chosen for each block as described above. Many such truncation points per block can be chosen to serve a wide variety of purposes. A complete discussion of these ideas, and much more, appears in Chapter 8.

Chapter 6

FILTER BANKS AND WAVELETS

6.1 CLASSIC FILTER BANK RESULTS

In this section we review some of the classic results in the development of filter bank theory and subband transforms. The reader is encouraged not to gloss over this material, since it provides a foundation for intuition concerning subband transforms. We take up from the introduction to subband transforms in Section 4.2. As a departure from the development there, however, we shall assume for simplicity that the analysis and synthesis filters (respectively, the analysis and synthesis vectors) have only real-valued coefficients.

6.1.1 A BRIEF HISTORY

In the development of Section 4.2, we imposed the requirement that the synthesis system must perfectly invert the analysis system of a subband transform. Equivalently, the analysis and synthesis vectors (sequences) must satisfy the biorthogonality relationship of equation (4.29). Moreover, our development of subband transforms began with a natural generalization of block transforms to convolutional, or sliding window transforms, showing that these are equivalent to filter bank systems. While convenient for the present treatment, this is not the historical sequence of events.

Subband transforms were first proposed in the mid-1970's for the coding of speech signals, by Croisier, Esteban and Galand [44, 45], using a special type of analysis-synthesis system, known as a QMF (Quadrature Mirror Filter) filter bank. QMF filter banks do not generally satisfy the so-called "Perfect Reconstruction" (PR) property, whereby the synthesis system perfectly inverts the analysis system. The term "filter bank" here refers not just to an arbitrary collection of filters, but rather a "maximally decimated" system of filters and decimators, of the form shown in Figure 4.10, where the combined rate of the subband samples is identical to the sample rate of the input sequence. The distinctive characteristic of QMF filter banks is that the end-to-end transfer function, through the analysis-synthesis system, is that of an LTI (Linear Time Invariant) operator. That is, the aliasing necessarily introduced during analysis (see Figure 4.12) is perfectly cancelled during synthesis. The introduction of alias-free maximally decimated filter banks, led to the investigation of perfect reconstruction systems, in which the end-to-end transfer function is at most a pure delay. The key results on PR subband systems were developed in the mid-1980's by Smith and Barnwell [137], Mintzer [108], Vetterli [161] and Vaidyanathan [157]. Some of the classic results of these studies are reproduced below.

6.1.2   QMF FILTER BANKS

We are interested in the end-to-end transfer function through the analysis-synthesis system of Figure 4.10. Let $x_q[k]$ be the sequence obtained by filtering the input sequence, $x[n]$, with the qth analysis filter, having impulse response $h_q[k]$. Also, let $x_q'[k]$ be the sequence obtained after decimating $x_q[k]$ by the factor M and then upsampling the resulting subband samples, $y_q[n]$, by the same factor, M. Thus, $x_q'[k]$ is identical to $x_q[k]$ at each $k = mn$ and zero everywhere else. In the Z-transform domain, this may be expressed through the aliasing relationship,

$$x_q'(z) = \frac{1}{m} \sum_{p=0}^{m-1} x_q(W_m^p z)$$

where $W_m^p = e^{-j\frac{2\pi}{m}p}$, $0 \le p < m$ are the mth roots of unity. This result may be verified by observing that

$$\sum_{p=0}^{m-1} x_q(W_m^p z) = \sum_{k\in\mathbb{Z}} x_q[k] z^{-k} \sum_{p=0}^{m-1} W_m^{-pk}$$

and

$$\sum_{p=0}^{m-1} W_m^{-pk} = \sum_{p=0}^{m-1} W_m^{pk} = \begin{cases} m & \text{if } k \text{ is divisible by } m \\ 0 & \text{otherwise} \end{cases}$$

Now let $x'[k]$ denote the sequence produced by the synthesis system. Ideally, this is equal to $x[k]$. In the Z-transform domain it is given by

$$x'(z) = \sum_{q=0}^{m-1} x'_q(z) g_q(z)$$

$$= \frac{1}{m} \sum_{q=0}^{m-1} \sum_{p=0}^{m-1} x(W_m^p z) h_q(W_m^p z) g_q(z)$$

In order to avoid aliasing, all terms in $x(W_m^p z)$ with $p \neq 0$ must be cancelled. That is, the filters must satisfy

$$\sum_{q=0}^{m-1} h_q(W_m^p z) g_q(z) = 0, \quad 1 \leq p < m \tag{6.1}$$

We then have

$$x'(z) = x(z) \frac{1}{m} \sum_{q=0}^{m-1} h_q(z) g_q(z)$$

In the simple case of $m = 2$ subbands, equation (6.1) becomes

$$h_0(-z) g_0(z) + h_1(-z) g_1(z) = 0 \tag{6.2}$$

and

$$x'(z) = \frac{1}{2} x(z) \left(h_0(z) g_0(z) + h_1(z) g_1(z) \right)$$

CLASSICAL QMF DEFINITION

Some confusion exists in the literature concerning the use of the term QMF in describing a filter bank. The original definition proposed by Croisier et al. [44] refers to a two channel system, having analysis and synthesis filters which are related to a single low-pass prototype, $f(z)$, according to

$$h_0(z) = g_0(z) = f(z)$$
$$h_1(z) = -g_1(z) = f(-z)$$

One motivation for this selection is that all filters have the same region of support as the prototype, so that a causal prototype will ensure that all filters are causal. It is trivial to verify that this choice of filters satisfies the alias cancellation condition of equation (6.2). If we also wish to obtain perfect reconstruction, the end-to-end transfer function must be a pure delay, z^{-l}; i.e.,

$$\frac{1}{2} \left(h_0(z) g_0(z) + h_1(z) g_1(z) \right) = \frac{1}{2} \left(f^2(z) - f^2(-z) \right) = z^{-l}$$

Note that $f^2(z) - f^2(-z)$ contains only odd powers of z, so the delay, l, cannot be zero (must be odd). As it turns out [162], this property cannot be satisfied exactly by FIR filters, except in the relatively uninteresting case where $f(z)$ is the 2-tap filter, known as the Haar filter,

$$f(z) = \frac{1}{\sqrt{2}}\left(1 + z^{-1}\right)$$

This case is uninteresting because then the transform becomes a block transform. Nevertheless, FIR filter banks with good frequency discrimination have been designed, which very nearly satisfy the perfect reconstruction requirement. The family of linear phase, even length filters designed by Johnson [83] have been particularly popular in the literature.

ALTERNATIVE QMF DEFINITION

Although the original definition of a QMF filter bank is not compatible with perfect reconstruction using finite support filters, the term is sometimes used in a more generic sense to refer to any two channel alias-free analysis-synthesis system, in which case perfect reconstruction certainly is possible. One popular definition (see [135] for example) is

$$h_0(z) = g_0\left(z^{-1}\right) = f(z)$$
$$h_1(z) = g_1\left(z^{-1}\right) = zf\left(-z^{-1}\right)$$

Equivalently, the impulse responses satisfy

$$h_0[k] = g_0[-k] = f[k]$$
$$h_1[k] = g_1[-k] = (-1)^{k+1} f[-(k+1)] \tag{6.3}$$

Again, we see by substitution that the alias condition of equation (6.2) is satisfied. Moreover, zero delay perfect reconstruction is possible, for which we require

$$2 = h_0(z) g_0(z) + h_1(z) g_1(z)$$
$$= h_0(z) h_0\left(z^{-1}\right) + h_0(-z) h_0\left(-z^{-1}\right) \tag{6.4}$$

Thus, we have only to find a low-pass prototype, $h_0(z)$, whose Fourier transform satisfies the power complementary property,

$$\left|\hat{h}_0(\omega)\right|^2 + \left|\hat{h}_0(\omega \pm \pi)\right|^2 = 2 \tag{6.5}$$

This is the starting point of Smith and Barnwell's popular procedure[1] for designing perfect reconstruction filter banks [136].

---

[1] Outlines of the procedure may be found in many texts, e.g. [162, §3.2.3].

In the time-domain, equation (6.4) becomes

$$2\delta[n] = \sum_k h_0[k]\, h_0[k-n]\,(1+(-1)^n)$$

so the sequence, $h_0[k]$, and all of its 2-translates, $h_0[k-2n]$, are mutually orthonormal. The other filters satisfy the same property, being related through equation (6.3). Moreover, since perfect reconstruction holds, the biorthogonality relations of equation (4.29) must be valid. This in turn means that

$$0 = \sum_k g_0[k]\, h_1[2n-k] = \sum_k h_0[k]\, h_1[k-2n]$$

so that the 2-translates of $h_0[k]$ are orthogonal to all the 2-translates of $h_1[k]$. We conclude that this alternative definition for a QMF filter bank yields an orthonormal subband transform whenever the low-pass prototype filter, h_0, is selected to ensure perfect reconstruction. As we shall see, all 2-band orthonormal subband transforms involving FIR filters are essentially[2] of this form. For transforms with IIR filters, the conditions of equation (6.3) are not the only way to achieve orthonormality[3], but they have convenient properties.

6.1.3   TWO CHANNEL FIR TRANSFORMS

The special case of two channel subband transforms is important for a variety of reasons. These transforms play a central role in the construction of dyadic multi-resolution transforms and hence in implementing resolution-scalable image compression schemes, as pointed out in Section 4.2.5. Moreover, iterative application of a two channel filter bank along one or both of the low- and high-pass subbands provides a conceptually simple way to create a wide variety of different subband structures. For these reasons two channel filter banks and their tree-structured derivatives have always been a primary focus of interest in the literature. Convolutional transforms whose analysis and synthesis systems involve finite support are also of primary interest for a variety of practical reasons. A number of striking statements may be made concerning two channel FIR filter banks exhibiting perfect reconstruction,

---

[2] Orthonormality is preserved if the analysis filter impulse responses are delayed by any number of samples, d, and the synthesis filters are advanced by the same d samples. The present definition corresponds to what we call the "delay-normalized" case.

[3] As an example, note that orthonormality is preserved if the analysis filters are both convolved by any all-pass filter impulse response, $a[n]$, and the synthesis filters are convolved by its inverse, $a[-n]$.

some of which we reproduce here. The statements are either less elegant or simply not true for filter banks involving $m > 2$ bands, or IIR filters.

Since the results which follow concern perfect reconstruction filter banks, it is worth making a few comments concerning end-to-end delay at this point. Perfect reconstruction analysis-synthesis systems are usually permitted to have some non-zero end-to-end delay, since otherwise one or both filters must be non-causal (except in the trivial case of the 2 tap Haar filters). As noted earlier, however, we do not insist on causality, firstly because it is not particularly relevant to image processing applications, and secondly because the addition of sufficient delay to render a finite support filter causal is primarily a matter of implementation. No generality is sacrificed by insisting on zero end-to-end delay, since end-to-end delay can always be absorbed in the synthesis filters. Thus, to simplify the statement and interpretation of results and also for consistency with our development of convolutional transforms, we adopt the convention that the end-to-end delay in a perfect reconstruction system is zero, unless stated otherwise.

ANALYSIS-SYNTHESIS RELATIONSHIPS

Theorem 6.1 *The polyphase analysis and synthesis transfer matrices, $H(z)$ and $G(z)$, of any two channel FIR filter bank having perfect reconstruction must satisfy*

$$\det(H(z)) = \alpha z^{-d}$$
$$\det(G(z)) = \alpha^{-1} z^{d}$$

for some arbitrary delay, $d \in \mathbb{Z}$ and non-zero $\alpha \in \mathbb{R}$. Note that this result is not restricted to $m = 2$ bands.

Proof. Since $G(z) H(z) = I$ for perfect reconstruction, the product of $\det(H(z))$ and $\det(G(z))$ must be equal to 1. But both of these are finite polynomials in z, since both analysis and synthesis systems are FIR. Consequently, they must be monomials of the form given in the statement of the theorem. ∎

Theorem 6.2 *The analysis and synthesis filters of any two channel FIR filter bank with perfect reconstruction are related according to*

$$\begin{aligned} g_0(z) &= -\alpha^{-1} z^{2d-1} h_1(-z) \\ g_1(z) &= \alpha^{-1} z^{2d-1} h_0(-z) \end{aligned} \tag{6.6}$$

where d and α are the delay and scale factor of Theorem 6.1.

Proof. From the definition of analysis polyphase components in equation (4.13) and the association of the elements of $H(z)$ with these polyphase components in equation (4.14), we have

$$h_q(z) = h_{q,0}(z^2) + zh_{q,1}(z^2)$$
$$= H_{q,0}(z^2) + zH_{q,1}(z^2) \tag{6.7}$$

Also, from the definition of the synthesis polyphase components in equation (4.17) and the association of the elements of $G(z)$ with these polyphase components in equation (4.18), we have

$$g_q(z) = g_{q,0}(z^2) + z^{-1}g_{q,1}(z^2)$$
$$= G_{0,q}(z^2) + z^{-1}G_{1,q}(z^2) \tag{6.8}$$

Now, using Theorem 6.1, we have

$$\begin{pmatrix} G_{0,0}(z) & G_{0,1}(z) \\ G_{1,0}(z) & G_{1,1}(z) \end{pmatrix} = \frac{\text{cofactor}(H(z))}{\alpha z^{-d}} = \alpha^{-1}z^d \begin{pmatrix} H_{1,1}(z) & -H_{0,1}(z) \\ -H_{1,0}(z) & H_{0,0}(z) \end{pmatrix}$$

The result then follows by substitution. ∎

TRANSLATED IMPULSE RESPONSES

Before pressing on, it is worth providing a framework to assist in interpreting and remembering the above results. We begin by observing that neither the delay, d, nor the scale factor, α, in the statement of Theorem 6.1 are fundamental properties of the subband transform. The scale factor, α, may trivially be absorbed into the filters. The delay, d, may be eliminated by advancing the analysis filter impulse responses and delaying the synthesis filter impulse responses by exactly d samples. Furthermore, it is easy to see that $\det(H(z))$ and $\det(G(z))$ are unaffected if we delay the low- and high-pass subbands by δ and $-\delta$ samples, respectively, for any $\delta \in \mathbb{Z}$. Thus, without sacrificing any generality, we may restrict our attention to what we shall call "delay-normalized" filter banks (respectively, "delay-normalized" subband transforms), defined as follows.

Definition: We say that an FIR filter bank is "delay-normalized" if the polyphase analysis matrix satisfies $\det(H(z)) = \alpha \in \mathbb{R}$ and the region of support of the low-pass analysis filter's impulse response, $h_0[k]$, is centred about 0 (for odd length filters) or $-\frac{1}{2}$ (for even length filters). Note that this definition is not commonplace in the literature.

The statement and interpretation of Theorem 6.2 may be further simplified by defining translated impulse responses,

$$h_q^t[k] = h_q[k-q]$$
$$g_q^t[k] = g_q[k+q] \tag{6.9}$$

For a delay-normalized transform, equation (6.6) may be restated in terms of these translated impulse responses as

$$g_0^t [k] = \alpha^{-1} (-1)^k h_1^t [k]$$
$$g_1^t [k] = \alpha^{-1} (-1)^k h_0^t [k]$$

(6.10)

Thus, the translated low- and high-pass synthesis filters are obtained by scaling and modulating the translated low- and high-pass analysis filters, and vice-versa. This is an easily remembered form of the statement of Theorem 6.2.

The scale factor, α, may also be found in a simple manner by observing that the polyphase components of the low- and high-pass filters may be expressed as

$$h_{q,0} \left(z^2 \right) = \frac{1}{2} \left(h_q \left(z \right) + h_q \left(-z \right) \right)$$

$$h_{q,1} \left(z^2 \right) = \frac{1}{2} z^{-1} \left(h_q \left(z \right) - h_q \left(-z \right) \right)$$

from which we find, after simple algebraic manipulations, that

$$\alpha = \det \left(H \left(z \right) \right) = \frac{1}{2} z^{-1} \left(h_1 \left(z \right) h_0 \left(-z \right) - h_0 \left(z \right) h_1 \left(-z \right) \right)$$

$$= \frac{1}{2} \left(h_1^t \left(z \right) h_0^t \left(-z \right) + h_0^t \left(z \right) h_1^t \left(-z \right) \right)$$

(6.11)

Setting $z = 1$ and expressing this relationship in the Fourier domain, we obtain

$$\alpha = \frac{1}{2} \left(\hat{h}_0^t \left(0 \right) \hat{h}_1^t \left(\pi \right) + \hat{h}_1^t \left(0 \right) \hat{h}_0^t \left(\pi \right) \right)$$

(6.12)

Substituting equation (6.10) into this expression, we obtain an equivalent relationship in terms of the synthesis filters,

$$\alpha^{-1} = \frac{1}{2} \left(\hat{g}_0^t \left(0 \right) \hat{g}_1^t \left(\pi \right) + \hat{g}_1^t \left(0 \right) \hat{g}_0^t \left(\pi \right) \right)$$

(6.13)

as well as the useful identity,

$$\hat{h}_q^t \left(0 \right) \hat{g}_q^t \left(0 \right) + \hat{h}_q^t \left(\pi \right) \hat{g}_q^t \left(\pi \right) = 2, \quad q = 0, 1$$

(6.14)

Since good low-pass filters will almost invariably have a Nyquist gain close to zero and, likewise, good high-pass filters will almost invariably have a DC gain close to zero, the second term in the above equations (6.12) and (6.13) can usually be neglected, so that α is determined by the product of the DC gain of the low-pass filter and the Nyquist gain of the high-pass filter. In Section 6.3.1, we shall see that this is in fact a necessary condition for the subband transform to arise in connection with a continuous wavelet transform (see Theorem 6.6).

We may specify the regions of support of the finite length translated impulse responses in terms of parameters, L_0^+, L_0^-, L_1^+ and L_1^-, where

$$h_0^t[k], g_1^t[k] = 0 \text{ for } k \notin \left[-L_0^-, L_0^+\right]$$
$$h_1^t[k], g_0^t[k] = 0 \text{ for } k \notin \left[-L_1^-, L_1^+\right]$$

From the definition of a delay-normalized transform, we must have

$$L_0^- = L_0^+ \text{ or } L_0^- = L_0^+ + 1$$

A particularly important class of transforms for image compression involves odd length linear phase filters. In this case, we shall see that the translated impulse responses of a delay-normalized filter bank must all be symmetric about 0; i.e., $L_q^+ = L_q^-$, $q = 0, 1$.

At least in some respects, the translated impulse responses have a more natural interpretation than their non-translated counterparts. To see this, let $y[k]$ be the interleaved sequence of subband samples defined by

$$y[mk + q] = y_q[k]$$

Then the analysis operation may be expressed as the periodically time-varying convolution operation

$$y[k] = \sum_{i \in \mathbb{Z}} h_{k \bmod m}^t[i] \, x[k - i]$$

and synthesis may be expressed as a dual time-varying convolution

$$x[k] = \sum_{i \in \mathbb{Z}} y[i] \, g_{i \bmod m}^t[k - i]$$

Equivalently, suppose we define prototype analysis and synthesis sequences, $\mathbf{a}_q^t \equiv a_q^t[n]$ and $\mathbf{s}_q^t \equiv s_q^t[n]$ and their translates, $\mathbf{a}_{(k)}^t$ and $\mathbf{s}_{(k)}^t$, by

$$a_q^t[n] = h_q^t[-n] \qquad \qquad s_q^t[n] = g_q^t[n]$$
$$a_{(k)}^t[n] = a_{k \bmod m}^t[n - k] \quad \text{and} \quad s_{(k)}^t[n] = s_{k \bmod m}^t[n - k]$$

Then we have

$$y[k] = \left\langle \mathbf{x}, \mathbf{a}_{(k)}^t \right\rangle \quad \text{and} \quad \mathbf{x} = \sum_k y[k] \, \mathbf{s}_{(k)}^t$$

So $\mathbf{a}_{(k)}^t$ and $\mathbf{s}_{(k)}^t$ are the interleaved analysis and synthesis vectors for the subband transform. These equations represent an easily remembered implementation procedure for the analysis and synthesis operations, which

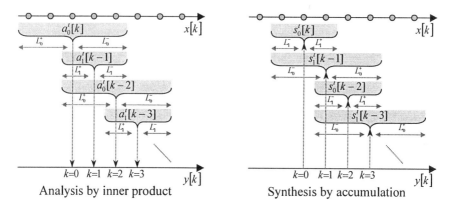

Figure 6.1.   Implementation of subband analysis and synthesis operations.

is illustrated in Figure 6.1. In words, to generate each sample, $y[k]$, in the interleaved sequence of subband samples, we translate the appropriate prototype analysis sequence (vector), $\mathbf{a}_q^t$, k samples to the "right" and take its inner product (dot product) with the input sequence. To synthesize the input sequence, we scan through the interleaved sequence of subband samples, $y[k]$, multiplying the sample value by the relevant prototype synthesis sequence (vector), $\mathbf{s}_q^t$, translated k samples to the "right," and accumulate the result into an output buffer.

ORTHONORMAL FIR TRANSFORMS

In any orthonormal transform, the analysis and synthesis vectors must be identical. Thus,

$$h_q^t[k] = a_q^t[-k] = s_q^t[-k] = g_q^t[k] \qquad (6.15)$$

Together, equations (6.10) and (6.15) reveal the fact that all four filters of a normalized orthonormal filter bank are related to a single prototype according to

$$(-1)^k g_1^t[k] = (-1)^k h_1^t[-k] = g_0^t[-k] = h_0^t[k] \qquad (6.16)$$

This relationship is nothing other than the alternate QMF definition of equation (6.3), in connection with which we found that a necessary and sufficient condition for perfect reconstruction is that $\hat{h}_0(\omega)$ must satisfy the power complementary property of equation (6.5).

Theorem 6.3 *If a two channel FIR filter bank implements an orthonormal subband transform, then all of its filters must have the same* *even length. Moreover, if it is delay-normalized, the regions of support for the translated impulse responses are given by*

$$L_0^- = L_0^+ + 1; \quad L_1^+ = L_0^-; \quad L_1^- = L_0^+ \tag{6.17}$$

Proof. Consider the low-pass synthesis vectors, $\mathbf{s}_{(2k)}^t$, which are the 2-translates of the low-pass prototype, $\mathbf{s}_0^t$. If this sequence has odd length, $2L + 1$, then $\mathbf{s}_{(0)}^t$ and $\mathbf{s}_{(2L)}^t$ have a single non-zero sample of overlap so that $\langle \mathbf{s}_{(0)}^t, \mathbf{s}_{(2L)}^t \rangle \neq 0$, contradicting the requirement that the synthesis vectors be orthonormal. The fact that all filters must have the same even length with the indicated regions of support follows from equation (6.16). ∎

LINEAR PHASE FIR TRANSFORMS

A linear phase filter is one whose impulse response is either symmetric or anti-symmetric. Specifically, one of the following holds:

1. The impulse response is symmetric. Let d denote the centre of symmetry. For odd length filters, $d \in \mathbb{Z}$, while for even length filters, d is an odd multiple of $\frac{1}{2}$. In this case, the impulse response and its Fourier transform (DTFT) satisfy

$$h[d+n] = h[d-n]; \quad \hat{h}(\omega) = \pm \left| \hat{h}(\omega) \right| e^{-j\omega d}$$

and so the phase response is linear in ω.

2. The impulse response is anti-symmetric. Again, letting d denote the centre of symmetry (an odd multiple of $\frac{1}{2}$ for even length filters), we have

$$h[d+n] = -h[d-n]; \quad \hat{h}(\omega) = \pm \left| \hat{h}(\omega) \right| j \operatorname{sgn}(\omega) e^{-j\omega d}$$

and so the phase response is linear in ω, except for a discontinuity at $\omega = 0$.

The following theorem provides a useful characterization of the lengths and regions of support for the filters in a linear phase filter bank.

Theorem 6.4 *Apart from certain pathological cases, perfect reconstruction, linear phase, two channel FIR filter banks involve filters whose lengths are either all odd or all even. Assume a delay-normalized filter bank. Then, in the odd length case, all translated impulse responses are centred about $k = 0$; i.e.,*

$$L_q^+ = L_q^-, \, q = 0, 1$$

In the even length case, the translated low-pass analysis impulse response is centred about $-\frac{1}{2}$ and the translated high-pass analysis impulse response is centred about $+\frac{1}{2}$; i.e.,

$$L_0^- = L_0^+ + 1, \quad and \quad L_1^- = L_1^+ - 1$$

Furthermore, $L_0^+ - L_1^+$ must be odd. In the case of odd length filters, this means that the filter lengths must differ by an odd multiple of 2. For even length filters, the filter lengths must either be identical or differ by an even multiple of 2.

Proof. We assume a delay-normalized filter bank throughout. Combining equations (6.10) and (6.11) we obtain

$$\alpha = \frac{\alpha}{2} \left(g_0^t(-z) h_0^t(-z) + g_0^t(z) h_0^t(z) \right)$$

Let $t(z) = h_0^t(z) g_0^t(z)$. Since $h_0^t(z)$ and $g_0^t(z)$ are both low-pass and linear phase, the impulse responses must be symmetric (anti-symmetry implies a high-pass filter). Hence, $t(z)$ also has linear phase and the finite support impulse response, $t[k]$, is symmetric about its centre of symmetry. Moreover, the above equation implies that $t[2n] = \delta[n]$ and so the even sub-sequence, $t[2n]$, is symmetric about $n = 0$. There are two possibilities: either $t[k]$ is symmetric about $k = 0$; or the odd sub-sequence, $t[2n+1] = \delta[n-p]$ for some $p \in \mathbb{Z}$. The latter possibility leads to the pathological cases in which $t[n]$ has only two non-zero terms; the resulting transforms are of little interest (e.g., filters with lengths 1 and 2). The statement of the theorem excludes these. Thus we conclude that $t(z) = t(z^{-1})$.

If h_0^t is an odd length filter, then $h_0^t(z) = h_0^t(z^{-1})$ from the definition of delay-normalization. Thus, to satisfy $t(z) = t(z^{-1})$, we must have $g_0^t(z) = g_0^t(z^{-1})$, meaning that $g_0^t[k]$ has odd length and $L_1^- = L_1^+$.

If h_0 is an even length filter, then $h_0^t(z) = zh_0^t(z^{-1})$, again from the definition of delay-normalization. Then $g_0^t(z) = z^{-1}g_0^t(z^{-1})$ and so $g_0^t[k]$ has even length and is symmetric about $\frac{1}{2}$; i.e., $L_1^- = L_1^+ - 1$.

To complete the proof we have only to show that $L_0^+ - L_1^+$ is odd. To see this, note that

$$t\left[L_0^+ + L_1^+\right] = \sum_{k=-L_0^-}^{L_0^+} h_0^t[k] g_0^t\left[L_0^+ + L_1^+ - k\right] = h_0^t\left[L_0^+\right] g_0^t\left[L_1^+\right] \neq 0$$

But $t[2n] = 0$ for $n \neq 0$, so $L_0^+ + L_1^+$ and hence $L_0^+ - L_1^+$ must be odd. ■

Linear phase is important for a variety of reasons. Firstly, and perhaps least significantly, the symmetry allows us to reduce the total number of multiplication operations by a factor of two in many implementations (consider the implementation identified in Figure 6.1, for example). A more significant benefit of linear phase filters is that there exist convenient boundary extension formulations, which permit non-expansive transformation of finite length sequences, without significantly complicating the implementation. Boundary handling is the subject of Section 6.5.

Linear phase filters are particularly important for resolution-scalable image compression applications. As discussed in Section 4.2.5, these applications involve a tree-structured transform, constructed by iterative application of a two channel filter bank. Subsets of the bit-stream represent successively lower resolution versions of the original image, obtained by low-pass filtering and sub-sampling, where the low-pass filter is h_0. It is well-known that non-linear phase filtering introduces visually disturbing edge distortion in images [95, §4.1]. In fact, for visual intelligibility, phase alignment of the sinusoidal components in the Fourier representation of an image is far more important than the magnitudes of these components. This explains the interest in linear phase two channel subband transforms for image compression.

Unfortunately, useful two channel linear phase subband transforms with finite support cannot also be orthonormal. We summarize this negative result as follows:

Theorem 6.5 *There are no two channel orthonormal subband transforms having FIR linear phase filters with more than 2 non-zero coefficients in any filter.*

Proof. Suppose that a two channel orthonormal subband transform has finite length linear phase filters. According to Theorem 6.3, all filters must have even length $2L$, so the coefficients of $g_0^t[n]$ must have the form

$$b_0, b_1, \ldots, b_{L-1}, \pm b_{L-1}, \ldots, \pm b_1, \pm b_0$$

where the $\pm$ sign stands for $+$ if the filter is symmetric and $-$ if it is anti-symmetric. For convenience of expression, we define the sequence b_k also in the range $L \leq k < 2L$, by $b_k = \pm b_{2L-1-k}$. Thus, $b_k = g_0^t[k+d]$ for some $d \in \mathbb{Z}$. Note that $b_0 = \pm b_{2L-1}$ is necessarily non-zero; otherwise, the filter length would not be $2L$. We will show that all other coefficients, b_k, $0 < k < 2L - 1$, must be zero; the statement of the theorem then follows from the fact that all filters are related by equation (6.16). By orthogonality, we must have

$$0 = \left\langle s_0^{(0)}, s_0^{(L-l)} \right\rangle, \quad \text{for } l = 1, 2, \ldots L - 1$$

$$= \sum_{i=0}^{2l-1} b_i b_{i+2(L-l)} = \pm \sum_{i=0}^{2l-1} b_i b_{2l-1-i}$$

The fact that $b_1 = 0$ follows by setting $l = 1$ in the above formula, yielding $0 = 2b_1 b_0$ with $b_0 \neq 0$. Proceeding by induction, suppose that the odd terms $b_{2j-1} = 0$ for $1 \leq j < l$, where $l < L$. Then the next odd term, b_{2l-1}, must also be zero, since

$$0 = \sum_{i=0}^{2l-1} b_i b_{2l+1-i} = b_0 b_{2l-1}$$

It follows that $b_{2j-1} = 0$ for all j in the range $1 \leq j < L$. But then $b_{2j} = b_{2L-2j-1} = 0$ for all j in the same range. Thus, the only non-zero filter coefficients are $b_0 = \pm b_{2L-1}$.

∎

We note that this negative result holds only for the two channel case. Nevertheless, the importance of the result is evidenced by the fact that the term "biorthogonal subband transform" is commonly interpreted as a perfect reconstruction filter bank with linear phase filters. As already mentioned, all transforms with inverses possess the biorthogonality property. However, apart from orthogonal transforms, the most interesting practical transforms have been those involving linear phase filters.

6.1.4   POLYPHASE FACTORIZATIONS
ORTHONORMAL FIR TRANSFORMS

Recall (see equation (4.27)) that the analysis and synthesis matrices of an orthonormal convolutional transform satisfy

$$A[i] = S[-i]$$

where this is nothing other than a restatement of the fact that the analysis and synthesis vectors are identical. Now recall that $H(z)$ is the Z-transform of the matrix sequence, $A^*[i]$ and $G(z)$ is the Z-transform of the matrix sequence, $S[i]$. It follows that the polyphase transfer matrices of an orthonormal transform must satisfy

$$H^\dagger(z) = G(z^{-1}) \tag{6.18}$$

where $H^\dagger(z)$ is formed by taking the transpose of $H(z)$ and complex conjugating all the coefficients of the resulting power series in z. Since we are restricting our attention to real-valued coefficients, $H^\dagger(z)$ is simply the transpose, $H^t(z)$.

Combining equation (6.18) with the requirement that

$$G(z)H(z) = H(z)G(z) = I$$

for perfect reconstruction, we see that $H(z)$ must satisfy

$$H^\dagger(z^{-1})H(z) = H(z)H^\dagger(z^{-1}) = I$$

Any matrix of this form is said to be "paraunitary." Of course, $G(z)$ must also be paraunitary. If the convolutional transform is only a block transform (i.e., $A[i] = S[i] = 0$ for $i \neq 0$) then $H(z)$ and $G(z)$ are scalar matrices. In this case $H(z) = H$, $H^\dagger(z) = H^*$ (or simply H^t for real-valued coefficients) and the paraunitary condition reduces to the unitary condition, $H^*H = I$.

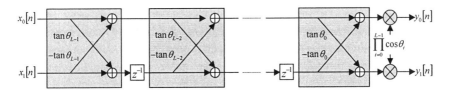

Figure 6.2. Lattice realization of a 2 channel orthonormal subband transform.

It is well-known that $m \times m$ unitary matrices may be factored into a cascade of $\binom{m}{2}$ single-parameter rotation matrices[4]. A similar factorization exists for paraunitary matrices. In the simple case with $m = 2$ subbands, the factorization is normally given for causal filters of length $2L$ samples in the form

$$H(z) = U_0 \prod_{i=1}^{L-1} \left(\begin{array}{cc} 1 & 0 \\ 0 & z^{-1} \end{array} \right) U_i \qquad (6.19)$$

where the U_i are 2×2 unitary matrices. Obviously, each of the factors is paraunitary and so any expansion of this form must be paraunitary. It is less obvious that all paraunitary matrices corresponding to causal FIR filters have such an expansion [157]. Notice that in this factorization, $\det(H(z)) = z^{-(L-1)}$. To obtain a delay-normalized transform from equation (6.19), all analysis filter impulse responses should be advanced by $L - 1$ samples.

Each of the L unitary 2×2 matrices, U_i, in equation (6.19), is characterized by a single rotation parameter, θ_i, with

$$U_i = \left(\begin{array}{cc} \cos\theta_i & -\sin\theta_i \\ \sin\theta_i & \cos\theta_i \end{array} \right) = \cos\theta_i \left(\begin{array}{cc} 1 & -\tan\theta_i \\ \tan\theta_i & 1 \end{array} \right)$$

leading to the lattice implementation depicted in Figure 6.2. The synthesis filter bank has essentially the same structure, except that the operations are performed in reverse order with the angles, θ_i, negated. Notice that only $L+1$ multiplications are required to generate each subband sample. By comparison, a direct implementation of the analysis filter bank of Figure 4.10 involves two $2L$-tap convolutions; i.e., $4L - 2$ multiplications per sample. Of course half of the samples produced by the filters are discarded by the decimators, so a slightly more sophisticated version of the direct implementation would require only $2L - 1$

---

[4] Actually a variety of factorizations exist of which the most widely known are the factorization in terms of Givens rotations and the Householder factorization. Each of these approaches may be extended to paraunitary matrices.

multiplications per sample, still almost twice as large as that required by the lattice implementation. The origin of this computational saving lies in the fact that the filters are far from arbitrary.

The paraunitary factorization provides an obvious design tool, since it exposes all the degrees of freedom in the design (the rotation angles, θ_i). The coefficients of $H(z)$ are multi-variate trigonometric polynomials of degree L and so least square optimization objectives (e.g., minimization of the stop band energy) lead to systems of polynomial equations which may be solved using a variety of standard techniques. Noteworthy among these is a systematic reduction method for systems of polynomial equations due to Buchberger [32].

The factorization of equation (6.19) extends naturally to arbitrary $m \times m$ FIR paraunitary matrices [157]; an alternative factorization [160], however, has been found to be more convenient for optimization. We note that expansions of this form with large m and L can lead to highly complex optimization problems, so that less general formulations are often preferred. For uniform transforms with large m, the modulated cosine structure discussed in Section 4.2.4 leads to a comparatively simply design problem based on a single low-pass prototype.

LINEAR PHASE FIR TRANSFORMS

We have already mentioned the importance of two channel transforms with linear phase filters for image compression and indeed these are central to the JPEG2000 image compression standard. Since there are no useful two channel orthonormal subband transforms with linear phase filters, it is natural to seek alternative factorizations. A large class of useful transforms may be generated by factorizations with similar properties to the paraunitary factorization given above. Unlike the paraunitary case, however, these factorizations do not generate all possible transforms subject to an arbitrary length constraint.

Symmetric filters with odd length have particularly convenient properties when the boundaries of finite length sequences must be taken into account, as discussed in Section 6.5. They are also central to the JPEG2000 image compression standard. For this reason, it is worth reproducing here a factorization due to Vetterli and LeGall [163], which generates symmetric analysis filters of lengths $2L+3$ and $2L+1$.[5] Adapting the original form of the factorization given in [163] to our current

---

[5] Note that odd length symmetric filters must necessarily have lengths which differ by an odd multiple of 2. See, for example, Theorem 6.4.

notation and delay-normalization convention, we have

$$H(z) = \prod_{i=1}^{L} \left(\begin{array}{cc} \dfrac{z + b_i + z^{-1}}{1 + z} & a_i\left(1 + z^{-1}\right) \\ & a_i \end{array} \right)$$

$$= \prod_{i=1}^{L} \left(\begin{array}{cc} \dfrac{z + b_i + z^{-1}}{1 + z} & 1 + z^{-1} \\ & 1 \end{array} \right) \left(\begin{array}{cc} 1 & 0 \\ 0 & a_i \end{array} \right) \qquad (6.20)$$

where the $2L$ parameters, a_i and b_i, are arbitrary apart from the requirement that $a_i \neq 0$ and $b_i \neq 2$. As in the paraunitary case, the factorization represents an efficient lattice implementation of the filter bank, requiring $2(L+1)$ multiplications to generate each pair of subband samples (each of the matrices in equation (6.20) contributes a lattice stage with one multiplication). Thus, implementation complexity is essentially identical to that for orthonormal transforms.

OTHER FACTORIZATIONS AND DESIGNS

We shall encounter a different type of polyphase factorization in Section 6.4, which is particularly interesting for applications in compression. Also, as we shall see in Section 6.2, the wavelet transform provides us with valuable insight for the design of good subband transforms for image compression. Indeed some of the most successful transforms for image compression have been discovered in this way, including the Cohen-Daubechies-Feauveau (CDF) 9/7 biorthogonal transform [41], which is specified by the JPEG2000 image compression standard and has been employed by most proposed image compression algorithms of recent times.

6.2   WAVELET TRANSFORMS

In this section we provide a brief introduction to wavelet transform theory. Our principle objective is to expose the intimate connection between subband transforms and wavelet transforms. So close is this connection that the terms wavelet transform and subband transform are often used interchangeably. More particularly, the Discrete Wavelet Transform (DWT) is generally understood as a dyadic tree-structured subband transform with the multi-resolution structure identified in Figure 4.19 (in the one dimensional case) or Figures 4.20 and 4.21 (in the two dimensional case). As noted in Section 4.2.5, such structures are of great importance for image compression. Compression schemes based upon these tree-structured subband transforms are usually known as wavelet-based schemes. The wavelet transform perspective provides substantial intuition concerning the interpretation of multi-resolution subband transforms, as well as guidance concerning the selection of good subband filters for tree-structured transforms.

As in the case of block and subband transforms, our development focuses primarily on one dimensional signals for simplicity, indicating the generalization to multiple dimensions at the appropriate points.

6.2.1   WAVELETS AND MULTI-RESOLUTION ANALYSIS

Up to this point we have considered discrete signals, $x[k]$. By contrast, the wavelet transform is concerned with functions, $x(t)$. In fact, the importance of the wavelet transform derives from an interest in the regularity (e.g., continuity, differentiability, etc.) of the waveforms which are represented by discrete sequences of samples. We confine our attention to the Hilbert space of square-integrable (finite energy) functions on the real-line, $\mathcal{L}^2(\mathbb{R})$, writing $\mathbf{x} \equiv x(t)$, when we wish to stress the vector-space properties of the function, $x(t)$. Inner products on $\mathcal{L}^2(\mathbb{R})$ are defined by

$$\langle \mathbf{x}, \mathbf{y} \rangle \triangleq \int_{-\infty}^{\infty} x(t)\, y^*(t)\, dt$$

The Fourier transform is well-defined on $\mathcal{L}^2(\mathbb{R})$, through the familiar relations

$$\hat{x}(\omega) = \int_{-\infty}^{\infty} x(t)\, e^{-j\omega t} dt \tag{6.21a}$$

$$x(t) = \frac{1}{2\pi} \int_{-\infty}^{\infty} \hat{x}(\omega)\, e^{j\omega t} d\omega \tag{6.21b}$$

Moreover, we have the generalized Parseval relation, which allows computation of inner products in the Fourier domain according to

$$\langle \mathbf{x}, \mathbf{y} \rangle = \frac{1}{2\pi} \int_{-\infty}^{\infty} \hat{x}(\omega)\, \hat{y}^*(\omega)\, d\omega \tag{6.22}$$

COUNTABLE BASES OF WAVELETS

A wavelet basis for $\mathcal{L}^2(\mathbb{R})$ is a family of functions, $\psi_n^{(m)}(t)$, all derived by translation and dilation (expansion) of a single "mother wavelet," $\psi(t)$, according to

$$\psi_n^{(m)}(t) = \sqrt{2^{-m}}\, \psi\left(2^{-m}t - n\right) \tag{6.23}$$

such that the $\psi_n^{(m)}$ are linearly independent and span $\mathcal{L}^2(\mathbb{R})$. That is, any signal, $\mathbf{x} \in \mathcal{L}^2(\mathbb{R})$, can be written as a linear combination of the form

$$\mathbf{x} = \sum_{m=-\infty}^{\infty} \sum_{n=-\infty}^{n=\infty} y_1^{(m)} [n] \, \psi_n^{(m)} \tag{6.24}$$

where $y_1^{(m)} [n]$ is a sequence of real numbers. Note that increasing the value of m corresponds to increasing the scale (i.e., expansion or dilation) of the wavelet functions, $\psi_n^{(m)} (t)$. The factor, $\sqrt{2^{-m}}$, in equation (6.24) ensures that the wavelet basis signals all have identical norm (energy); i.e., $\|\psi_n^{(m)}\| = \|\psi\|$, $\forall m, n$. This is important if $\{\psi_n^{(m)}\}_{m,n \in \mathbb{Z}}$ is to form an orthonormal basis for $\mathcal{L}^2 (\mathbb{R})$, although wavelet bases need not necessarily be orthonormal.

The fact that only countably many basis functions, $\psi_n^{(m)}$, are required to span the space of all square integrable functions, may seem surprising at first. To alleviate concerns and provide some intuition, we give an example using familiar signal processing concepts.

Example 6.1 *Consider the sub-space, $\mathcal{V}^{(m)} \subset \mathcal{L}^2 (\mathbb{R})$, of finite energy bandlimited signals having Fourier transform supported on $|\omega| < 2^{-m}\pi$. Let $\mathbf{x}^{(m)}$ denote the projection of any $\mathbf{x} \in \mathcal{L}^2 (\mathbb{R})$ onto $\mathcal{V}^{(m)}$. That is, $\mathbf{x} = \mathbf{x}^{(m)} + \mathbf{e}$ where $\mathbf{x}^{(m)} \in \mathcal{V}^{(m)}$ and $\langle \mathbf{x}^{(m)}, \mathbf{e} \rangle = 0$. From the generalized Parseval relation of equation (6.22), we conclude that $x^{(m)} (t)$ is obtained by low-pass filtering $x(t)$ with an ideal low-pass filter with cut-off frequency at $\omega = 2^{-m}\pi$; then $e(t)$ occupies the high frequency remainder and is necessarily orthogonal to $x^{(m)} (t)$.*

Applying Shannon's well-known sampling theorem, we have

$$\mathbf{x}^{(m)} = \sum_{n=-\infty}^{\infty} y_0^{(m)} [n] \frac{\text{sinc} \, (2^{-m}t - n)}{\sqrt{2^m}} = \sum_{n=-\infty}^{\infty} y_0^{(m)} [n] \, \varphi_n^{(m)} \tag{6.25a}$$

$$y_0^{(m)} [n] = \left(x(t) \star \frac{\text{sinc} \, (2^{-m}t)}{\sqrt{2^m}} \right)\Bigg|_{t=2^{-m}n} = \langle \mathbf{x}, \varphi_n^{(m)} \rangle \tag{6.25b}$$

where $\varphi_n^{(m)} (t) = \varphi^{(m)} (t - 2^m n)$ and

$$\varphi^{(m)} (t) = \sqrt{2^{-m}} \, \text{sinc} \, (2^{-m}t)$$

or, equivalently,

$$\hat{\varphi}^{(m)} (\omega) = \begin{cases} \sqrt{2^m} & \text{if } |\omega| < 2^{-m}\pi \\ 0 & \text{if } |\omega| \geq 2^{-m}\pi \end{cases}$$

Recognizing equations (6.25a) and (6.25b) as an orthonormal expansion, we see that $\{\varphi_n^{(m)}\}_{n \in \mathbb{Z}}$ is an orthonormal basis for $\mathcal{V}^{(m)}$ and $\mathbf{x}^{(m)}$ may be represented as a countable linear combination of the basis vectors.

Since **x** *has finite energy,*

$$\|\mathbf{x}\|^2 = 2 \int_0^\infty |\hat{x}(\omega)|^2 \, d\omega$$

the energy of the bandlimited approximation error,

$$\|\mathbf{x}^{(m)} - \mathbf{x}\|^2 = 2 \int_{2^{-m}\pi}^\infty |\hat{x}(\omega)|^2 \, d\omega$$

must converge to 0 as $m \to -\infty$. *Equivalently, the approximation sub-spaces,* $\mathcal{V}^{(m)}$, *converge to* $\mathcal{L}^2(\mathbb{R})$ *as* $m \to -\infty$, *which we may write as*

$$\bigcup_{m \to -\infty} \mathcal{V}^{(m)} = \mathcal{L}^2(\mathbb{R})$$

This suggests that any $\mathbf{x} \in \mathcal{L}^2(\mathbb{R})$ *can be represented to arbitrary accuracy as a countable linear combination of the basis signals* $\{\varphi_n^{(m)}\}_{n \in \mathbb{Z}}$ *for some m. To carry the reasoning further, define* $\mathcal{W}^{(m)}$ *to be the orthogonal complement of* $\mathcal{V}^{(m)}$ *in the higher resolution sub-space,* $\mathcal{V}^{(m-1)}$. *Then* $\mathcal{W}^{(m)}$ *is the set of finite energy signals whose Fourier transform is supported on* $2^{-m}\pi \le |\omega| < 2^{-m+1}\pi$. *Again, applying Shannon's sampling theorem, we find that an orthonormal basis for* $\mathcal{W}^{(m)}$ *is the modulated* sinc *functions,* $\psi_n^{(m)}(t) = \psi^{(m)}(t - 2^m n)$, *with*

$$\psi^{(m)}(t) = \sqrt{2^{-m}} \cos\left(\frac{3}{2}\pi 2^{-m} t\right) \text{sinc}\left(\frac{1}{2} 2^{-m} t\right), \quad \text{or equivalently,}$$

$$\hat{\psi}^{(m)}(\omega) = \begin{cases} \sqrt{2^m} & \text{if } 2^{-m}\pi \le |\omega| < 2^{-m+1}\pi \\ 0 & \text{otherwise} \end{cases}$$

It follows that for any $p \ge m$, $\mathbf{x}^{(m)}$ *can be expressed as*

$$\mathbf{x}^{(m)} = \mathbf{x}^{(p)} + \sum_{k=m+1}^p \sum_{n=-\infty}^\infty y_1^{(m)}[n] \psi_n^{(k)}, \quad \text{where} \quad y_1^{(m)}[n] = \left\langle \mathbf{x}, \psi_n^{(m)} \right\rangle$$

This construction is illustrated in Figure 6.3.

Finally, since the Fourier transform of a finite energy signal cannot grow without bound as $\omega \to 0$, *we must have*

$$\|\mathbf{x}^{(p)}\|^2 = 2 \int_0^{2^{-p}\pi} |\hat{x}(\omega)|^2 \, d\omega < 2^{-p} \left(2\pi B^2\right)$$

for some suitable bound B on $|\hat{x}(\omega)|$. *It follows that* $\lim_{p \to \infty} \|\mathbf{x}^{(p)}\| = 0$, *which may be written as*

$$\bigcap_{m \to \infty} \mathcal{V}^{(m)} = \{\mathbf{0}\}$$

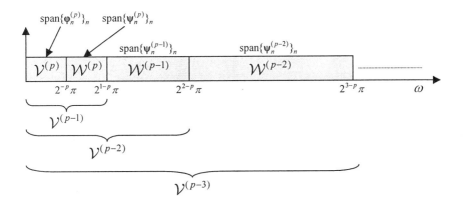

Figure 6.3. Frequency bands corresponding to the subsets, $\mathcal{V}^{(m)}$, and their orthogonal complements, $\mathcal{W}^{(m)}$ in $\mathcal{V}^{(m-1)}$, for the multi-resolution hierarchy generated by the sinc basis functions.

Combining this with the fact that $\mathbf{x}^{(-m)}$ *converges to* $\mathbf{x}$, *we see that any* $\mathbf{x} \in \mathcal{L}^2(\mathbb{R})$ *can be expressed as a linear combination of the orthonormal basis signals,* $\psi_n^{(m)}$, *as in equation (6.24). Moreover, the* $\psi_n^{(m)}$ *are all related to the mother wavelet,* $\psi = \psi_0^{(0)}$, *according to equation (6.23).*

The above example is instructive from a number of perspectives. Firstly, it illustrates the close connection between the wavelet basis and a multi-resolution hierarchy of nested sub-spaces of $\mathcal{L}^2(\mathbb{R})$. The key properties of $\mathcal{L}^2(\mathbb{R})$ which we required to obtain the basis of scaled and dilated versions of a mother wavelet were:

1. Each $\mathbf{x} \in \mathcal{L}^2(\mathbb{R})$ is essentially bandlimited (or "resolution-limited") so that the approximations, $\mathbf{x}^{(m)}$, converge to $\mathbf{x}$ as $m \to -\infty$ and the sub-spaces, $\mathcal{V}^{(m)}$, cover $\mathcal{L}^2(\mathbb{R})$.

2. No non-zero $\mathbf{x} \in \mathcal{L}^2(\mathbb{R})$ can have zero bandwidth (or zero "resolution"), so that the approximations, $\mathbf{x}^{(m)}$, must tend to $\mathbf{0}$ as $m \to \infty$ and the complement spaces, $\mathcal{W}^{(m)}$, must cover $\mathcal{L}^2(\mathbb{R})$.

We point out that convergence of the infinite summations in equation (6.24) is in the $\mathcal{L}^2$ sense. Strictly speaking then, this equation states only that the error between $\mathbf{x}$ and its representation as a linear combination of the wavelet basis signals has zero energy; there is no guarantee of point-wise convergence. The Fourier transform relations of equations (6.21a) and (6.21b) are valid in exactly the same sense, as are most other expressions in the sequel.

MULTI-RESOLUTION ANALYSIS

Motivated by Example 6.1, a "multi-resolution" analysis on $\mathcal{L}^2(\mathbb{R})$ is defined as a set of sub-spaces (note that resolution increases with decreasing m),

$$\cdots \subset \mathcal{V}^{(2)} \subset \mathcal{V}^{(1)} \subset \mathcal{V}^{(0)} \subset \mathcal{V}^{(-1)} \subset \mathcal{V}^{(-2)} \subset \cdots$$

satisfying the following additional properties.

(MR-1) $\bigcup_{m \in \mathbb{Z}} \mathcal{V}^{(m)} = \mathcal{L}^2(\mathbb{R})$. This property, known as "upward completeness" means that every $\mathbf{x} \in \mathcal{L}^2(\mathbb{R})$ is the limit (in the $\mathcal{L}^2$ sense) of its projections, $\mathbf{x}^{(m)}$, onto successively higher resolution spaces, $\mathcal{V}^{(m)}$, $m = -1, -2, \ldots$. Since convergence is only in the $\mathcal{L}^2$ sense, it is formally more correct to say that the union of the resolution spaces is "dense" in $\mathcal{L}^2(\mathbb{R})$, or that $\mathcal{L}^2(\mathbb{R})$ is its closure.

(MR-2) $\bigcap_{m \in \mathbb{Z}} \mathcal{V}^{(m)} = \{0\}$. This property, known as "downward completeness," means that every $\mathbf{x} \in \mathcal{L}^2(\mathbb{R})$ has non-zero resolution so that its projections, $\mathbf{x}^{(m)}$, converge to $\mathbf{0}$, as m tends to ∞, where convergence is again in the $\mathcal{L}^2$ sense.

(MR-3) $x(t) \in \mathcal{V}^{(0)}$ if and only if $x(2^{-m}t) \in \mathcal{V}^{(m)}$. Thus, dilating a signal from resolution space $\mathcal{V}^{(0)}$ by the factor 2^m yields a signal in the lower resolution space $\mathcal{V}^{(m)}$.

(MR-4) $x(t) \in \mathcal{V}^{(0)}$ if and only if $x(t-n) \in \mathcal{V}^{(0)}$ for all $n \in \mathbb{Z}$. Combining this with the above property, we see that translating a signal in resolution space $\mathcal{V}^{(m)}$ by any integer multiple of 2^m does not alter its resolution.

(MR-5) There exists an orthonormal basis, $\{\varphi_n\}_{n \in \mathbb{Z}}$, for $\mathcal{V}^{(0)}$ such that $\varphi_n(t) = \varphi(t-n)$. The function, $\varphi(t)$ is called the "scaling function."

A family of sub-spaces satisfying the above properties is also known as a "multi-scale" analysis. Projections onto larger resolution (or bandwidth) sub-spaces expose features at a smaller scale. The parameter, m, is best understood as a "scale parameter," since it decreases as the scale of signal features decreases (resolution increases). This convention is also convenient in unifying the notation adopted in Section 4.2.5 for tree-structured subband decompositions with that being developed here for multi-resolution analysis and the wavelet transform.

We shall see shortly how this multi-resolution/multi-scale framework leads to wavelets. First, however, properties (MR-4) and (MR-5) deserve some additional comments.

Translation Invariance (MR-4). From a signal processing perspective, the most natural definition of resolution is in terms of bandwidth. Accordingly, the bandlimited spaces of Example 6.1 constitute the most natural multi-resolution analysis. As it turns out, this particular multi-resolution analysis is unique in that **all translates** (whether by integers or otherwise) of a bandlimited signal have the same bandwidth. That is,

$$x\left(t\right) \in \mathcal{V}^{(0)} \Longleftrightarrow x\left(t - \tau\right) \in \mathcal{V}^{(0)}, \quad \forall \tau \in \mathbb{R} \tag{6.26}$$

Intuitively, we expect that the resolution of a signal should be unaffected by translation. By contrast, property (MR-4) states only that the integer translates of each element of $\mathcal{V}^{(0)}$ also belong to $\mathcal{V}^{(0)}$.

This weakening of the translation invariance property is unfortunate, but necessary in order to generalize the multi-resolution concept from strictly bandlimited signals to approximately bandlimited signals for which the basis functions can have more compact support. The sinc basis functions of Example 6.1 decay only slowly as $\frac{1}{t}$. As a result, physically realizable approximations converge only slowly and the wavelet coefficients, $y_1^{(m)}[n]$, exhibit poor localization with respect to features in the underlying signal, $x\left(t\right)$. By contrast, we are chiefly interested in scaling functions with exponential decay and especially finite support. Nevertheless, to the extent that the scaling function approximates the ideal low-pass characteristic of the sinc function, the corresponding multi-resolution analysis will approximately satisfy the ideal of complete translation invariance, as expressed in equation (6.26).

Orthonormal Basis (MR-5). The orthonormality of the basis functions, $\{\varphi_n\}_{n \in \mathbb{Z}}$, is not central to the concept of a multi-resolution analysis. The important property is that the basis functions for $\mathcal{V}^{(0)}$ are integer translates of a single scaling function. In fact, there is a simple orthogonalization procedure which may be used to create an orthonormal basis satisfying this shift property from a non-orthogonal one satisfying the same shift property [98]. The generalization to non-orthogonal bases is discussed in Section 6.2.3. For the moment, however, the simplicity afforded by an orthonormal basis is convenient.

Properties (MR-5) and (MR-3) together indicate that each resolution space, $\mathcal{V}^{(m)}$, has an orthonormal basis, $\{\varphi_n^{(m)}\}_{n \in \mathbb{Z}}$, where

$$\varphi_n^{(m)}\left(t\right) = \varphi^{(m)}\left(t - 2^m n\right) = \sqrt{2^{-m}}\varphi\left(2^{-m}t - n\right)$$

Also, using the Poisson formula, the orthonormality of the functions, $\{\varphi(t-n)\}_{n\in\mathbb{Z}}$, may be expressed in the Fourier domain as

$$\sum_{k\in\mathbb{Z}} |\hat{\varphi}(\omega+2\pi k)|^2 = 1 \tag{6.27}$$

WAVELETS FROM SCALING FUNCTIONS

The introduction of multi-resolution analysis as a tool for interpreting and constructing wavelet bases is due to Mallat [98]. Starting with the scaling function, $\varphi(t)$, which characterizes a multi-resolution analysis, it is possible to construct an orthonormal wavelet basis (it is not unique). Following the procedure of Example 6.1, let $\mathcal{W}^{(m)}$ denote the orthogonal complement of $\mathcal{V}^{(m)}$ in $\mathcal{V}^{(m-1)}$; i.e.,

$$\mathcal{W}^{(m)} \perp \mathcal{V}^{(m)} \quad \text{and} \quad \mathcal{W}^{(m)} \oplus \mathcal{V}^{(m)} = \mathcal{V}^{(m-1)}$$

Our objective is to find an orthonormal basis, $\{\psi_n^{(m)}\}_{n\in\mathbb{Z}}$, for each $\mathcal{W}^{(m)}$ where the basis functions, $\psi_n^{(m)}$ are all translated and dilated versions of a single mother wavelet, ψ. Properties (MR-1) and (MR-2) of the multi-resolution analysis will then ensure that $\{\psi_n^{(m)}\}_{n,m\in\mathbb{Z}}$ is an orthonormal basis for $\mathcal{L}^2(\mathbb{R})$.

Since $\mathcal{V}^{(0)} \subset \mathcal{V}^{(-1)}$, the scaling function, $\varphi(t)$, may be expressed as a linear combination of the functions, $\varphi_n^{(-1)}(t) = \sqrt{2}\varphi(2t-n)$, which span $\mathcal{V}^{(-1)}$. Specifically, we may write

$$\varphi(t) = \sqrt{2} \sum_{n=-\infty}^{\infty} g_0[n]\, \varphi(2t-n) \tag{6.28}$$

for some sequence of weights, $g_0[n]$. The reason for this choice of notation will become apparent shortly. Now, since $\{\varphi_n^{(0)}\}_{n\in\mathbb{Z}}$ and $\{\varphi_n^{(-1)}\}_{n\in\mathbb{Z}}$ are orthonormal bases for $\mathcal{V}^{(0)}$ and $\mathcal{V}^{(-1)}$, respectively, we have

$$\delta[n] = \left\langle \varphi_0^{(0)}, \varphi_n^{(0)} \right\rangle$$

$$= 2 \int_{-\infty}^{\infty} \left(\sum_i g_0[i]\, \varphi(2t-i) \sum_j g_0[j]\, \varphi(2(t-n)-j) \right) dt$$

$$= 2 \sum_{i,j} g_0[i]\, g_0[j-2n] \int_{-\infty}^{\infty} \varphi(2t-i)\, \varphi(2t-j)\, dt$$

$$= \sum_{i,j} g_0[i]\, g_0[j-2n] \left\langle \varphi_i^{(-1)}, \varphi_j^{(-1)} \right\rangle = \sum_i g_0[i]\, g_0[i-2n]$$

That is, treating the coefficient sequence, $g_0[n]$, as a vector in $\ell^2(\mathbb{Z})$, we see that it has unit norm and is orthogonal to all of its 2-translates. As discussed in Section 6.1.2, this is exactly the condition required of the low-pass synthesis filter which defines a two channel orthonormal subband transform. Thus, following equation (6.3) we may set

$$g_1[n] = (-1)^{n+1} g_0[-(n-1)]$$

and the 2-translates of $g_0[n]$, together with the 2-translates of $g_1[n]$, form an orthonormal basis for $\ell^2(\mathbb{Z})$. Now set

$$\psi(t) = \sqrt{2} \sum_{n=-\infty}^{\infty} g_1[n]\, \varphi(2t-n) \qquad (6.29)$$

and observe that $\psi(t)$ satisfies the following properties.

1. $\psi(t)$ and its integer translates are orthonormal, since

$$\langle \psi_0, \psi_n \rangle = 2 \int_{-\infty}^{\infty} \left(\sum_{i,j} g_1[i]\, \varphi(2t-i) \cdot g_1[j]\, \varphi(2(t-n)-j) \right) dt$$

$$= \sum_{i,j} g_1[i]\, g_1[j-2n] \left\langle \varphi_i^{(-1)}, \varphi_j^{(-1)} \right\rangle$$

$$= \sum_i g_1[i]\, g_1[i-2n] = \delta[n]$$

2. $\psi(t)$ and its integer translates are all orthogonal to $\mathcal{V}^{(0)}$, since

$$\langle \psi_n, \varphi_p \rangle = 2 \int_{-\infty}^{\infty} \left(\sum_{i,j} g_1[i]\, \varphi(2(t-n)-i) \cdot g_0[j]\, \varphi(2(t-p)-j) \right) dt$$

$$= \sum_{i,j} g_1[i-2n]\, g_0[j-2p] \left\langle \varphi_i^{(-1)}, \varphi_j^{(-1)} \right\rangle$$

$$= \sum_i g_1[i-2n]\, g_0[i-2p] = 0$$

Combining this with the fact that $\psi(t)$ and its translates are linear combinations of the basis functions, $\varphi_n^{(-1)}$, for $\mathcal{V}^{(1)}$, we see that they all belong to the orthogonal complement of $\mathcal{V}^{(0)}$ in $\mathcal{V}^{(-1)}$; i.e., $\mathcal{W}^{(0)}$.

3. The mutually orthonormal functions, $\{\psi_n\}_{n\in\mathbb{Z}}$ and $\{\varphi_n\}_{n\in\mathbb{Z}}$, together span $\mathcal{V}^{(-1)}$. To see this, note that any $\mathbf{x} \in \mathcal{V}^{(-1)}$ can be written as

$$\mathbf{x} = \sum_{n\in\mathbb{Z}} y_0^{(-1)}[n]\,\varphi_n^{(-1)}$$

$$= \sum_{n\in\mathbb{Z}} \left(\sum_{i\in\mathbb{Z}} y_0^{(0)}[i]\,g_0\,[n - 2i] + \sum_{j\in\mathbb{Z}} y_1^{(0)}[j]\,g_1\,[n - 2j] \right) \varphi_n^{(-1)}$$

$$= \sum_{i\in\mathbb{Z}} y_0^{(0)}[i]\,\varphi_i^{(0)} + \sum_{j\in\mathbb{Z}} y_1^{(0)}[j]\,\psi_j^{(0)}$$

where the second line follows from the fact that the sequence, $y_0^{(-1)}[n]$ has finite energy and so belongs to $\ell^2\,(\mathbb{Z})$, meaning that it can be written as a linear combination of the 2-translates of g_0 and the 2-translates of g_1. It follows that the $\{\psi_n\}_{n\in\mathbb{Z}}$ must span $\mathcal{W}^{(0)}$.

We conclude that $\{\psi_n\}_{n\in\mathbb{Z}}$ is an orthonormal basis for $\mathcal{W}^{(0)}$. Moreover, it is easy to see that the above properties hold also for the dilated versions of $\psi\,(t)$, so that $\{\psi_n^{(m)}\}_{n\in\mathbb{Z}}$ is an orthonormal basis for the orthogonal complement, $\mathcal{W}^{(m)}$, of $\mathcal{V}^{(m)}$ in $\mathcal{V}^{(m-1)}$. In this way, the $\{\psi_n^{(m)}\}_{n,m\in\mathbb{Z}}$ form an orthonormal wavelet basis for $\mathcal{L}^2\,(\mathbb{R})$.

6.2.2   DISCRETE WAVELET TRANSFORM

We have seen that every multi-resolution analysis gives rise to a wavelet basis. Moreover, subband transforms are at the heart of the construction of the wavelet basis. In particular, the key steps in the construction are: 1) write $\varphi = \varphi_0^{(0)}$ as a linear combination of the $\{\varphi_n^{(-1)}\}_{n\in\mathbb{Z}}$, with coefficients, $g_0\,[n]$; 2) identify $g_0\,[n]$ as one of the synthesis filters of a two channel subband transform; and 3) use the other synthesis filter, $g_1\,[n]$, to express $\psi = \psi_0^{(0)}$ as a linear combination of the $\{\varphi_n^{(-1)}\}_{n\in\mathbb{Z}}$. In fact, the underlying filter bank provides a vehicle for implementing the wavelet transform.

Suppose that the input signal, $x\,(t)$, is characterized at some resolution, say $\mathcal{V}^{(0)}$, by the discrete sequence, $y_0^{(0)}[n]$, such that

$$x^{(0)}\,(t) = \sum_{n\in\mathbb{Z}} y_0^{(0)}[n]\,\varphi\,(t - n) \tag{6.30}$$

This sequence may be decomposed into low- and high-pass subband sequences, $y_0^{(1)}[n]$ and $y_1^{(1)}[n]$, using the analysis system of the two channel subband transform and then reconstructed using the synthesis system

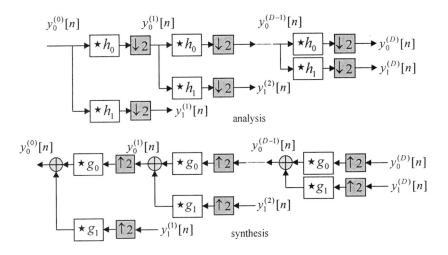

Figure 6.4.   DWT analysis and synthesis with D levels.

of the same transform, yielding

$$x^{(0)}(t) = \sum_{n\in\mathbb{Z}} \left(\sum_{i\in\mathbb{Z}} y_0^{(1)}[i] \, g_0[n-2i] + \sum_{j\in\mathbb{Z}} y_1^{(1)}[j] \, g_1[n-2j] \right) \varphi(t-n)$$

$$= \sum_{i\in\mathbb{Z}} y_0^{(1)}[i] \left(\sum_{n\in\mathbb{Z}} g_0[n] \, \varphi((t-2i)-n) \right)$$

$$+ \sum_{j\in\mathbb{Z}} y_1^{(1)}[j] \left(\sum_{n\in\mathbb{Z}} g_1[n] \, \varphi((t-2j)-n) \right)$$

$$= \underbrace{\sum_{i\in\mathbb{Z}} y_0^{(1)}[i] \, \varphi_n^{(1)}(t)}_{\mathbf{x}^{(1)}\in\mathcal{V}^{(1)}} + \underbrace{\sum_{j\in\mathbb{Z}} y_1^{(1)}[j] \, \psi_n^{(1)}(t)}_{\mathbf{w}^{(1)}\in\mathcal{W}^{(1)}}$$

Thus, the subband transform decomposes $\mathbf{x}^{(0)}$ as a sum of two pieces, $\mathbf{x}^{(1)} \in \mathcal{V}^{(1)}$, and $\mathbf{w}^{(1)} = \mathbf{x}^{(0)} - \mathbf{x}^{(1)} \in \mathcal{W}^{(1)}$.

Now suppose we apply the subband analysis operation recursively to the low-pass subband sequences, $y_0^{(1)}[n]$, $y_0^{(2)}[n]$ and so forth. The result is a tree-structured transform with D levels, as illustrated in Figure 6.4

and $\mathbf{x}^{(0)}$ is decomposed as

$$\mathbf{x}^{(0)} = \sum_{n \in \mathbb{Z}} y_0^{(D)}[n]\, \varphi_n^{(D)} + \sum_{m=1}^{D} \sum_{n \in \mathbb{Z}} y_1^{(m)}[n]\, \psi_n^{(m)}$$

$$= \mathbf{x}^{(D)} + \sum_{m=1}^{D} \sum_{n \in \mathbb{Z}} y_1^{(m)}[n]\, \psi_n^{(m)}$$

$$\xrightarrow[D \to \infty]{} \sum_{m=1}^{D} \sum_{n \in \mathbb{Z}} y_1^{(m)}[n]\, \psi_n^{(m)}$$

In summary, the high-pass subband samples in the tree-structured transform are in fact wavelet transform coefficients and the low-pass subband samples are of vanishing importance as the number of levels in the tree becomes large. For this reason, any tree-structured subband transform of the form illustrated in Figure 6.4 is known as a Discrete Wavelet Transform (DWT).

Before moving on, we make several observations concerning the implementation of the continuous wavelet transform using a tree-structured subband transform; i.e., a DWT. Firstly, we have assumed that the input sequence, $y_0^{(0)}[n]$, supplied to the discrete transform, is related to the underlying continuous input waveform through the basis of translated scaling functions, $\{\varphi_n^{(0)}\}_{n \in \mathbb{Z}}$. There is not usually any reason to believe this. However, physical signals inevitably possess some degree of smoothness[6] so that any smooth interpolation of their samples represents a good approximation, so long as the sampling rate is sufficiently high. Secondly, the DWT is usually applied to signals which have finite power, rather than finite energy. Even when the signals have finite support, as in image compression applications, the DWT is usually applied to appropriately extended, periodic versions of the signals which do not have finite energy (see Section 6.5). As a result, the low-resolution approximations, $\mathbf{x}^{(D)}$, do not generally converge to 0 as $D \to \infty$. Consequently, practical DWT implementations must include the low resolution subband samples, $y_0^{(D)}[n]$, in the reconstruction.

---

[6] Signals with finite energy (respectively power) are essentially bandlimited, meaning that they cannot have non-negligible power at arbitrarily high frequencies. More significantly, physical imaging (more generally, data acquisition) processes necessarily involve signal integration over finite windows in space or time, which is equivalent to low-pass filtering.

6.2.3   GENERALIZATIONS

BIORTHOGONAL WAVELETS

We have seen that every multi-resolution analysis leads to an ortho-normal wavelet basis and that the wavelet transform is associated with a tree-structured orthonormal subband transform. On the other hand, Theorem 6.5 demonstrates that there are no useful two channel ortho-normal subband transforms whose synthesis filters have linear phase and finite support. Linear phase and finite support are important properties[7] for image compression applications, as discussed in Section 6.1.3. This leads us to consider multi-resolution analyses with non-orthogonal bases and hence non-orthogonal wavelet transforms. We begin by modifying property (MR-5) describing a multi-resolution analysis to read

(MR-5)' There exists a basis[8], $\{\varphi_n\}_{n \in \mathbb{Z}}$, (not necessarily orthonormal) for $\mathcal{V}^{(0)}$ such that $\varphi_n(t) = \varphi(t - n)$. The function, $\varphi(t)$ is called the "scaling function."

The scaling function still satisfies equation (6.28), which is known as the "two-scale" equation, and we may identify the coefficient sequence, $g_0[n]$, as the low-pass synthesis impulse response of some two channel subband transform. It is not hard to show that such an association is always possible, although the subband transform is far from unique. Let $g_1[n]$ be the high-pass synthesis impulse response of the transform. Then the 2-translates of g_0 and g_1 together form a basis for $\ell^2(\mathbb{Z})$. Defining the mother wavelet $\psi(t)$ in terms of equation (6.29), we find that any $\mathbf{x} \in \mathcal{V}^{(-1)}$ can be expressed as

$$\mathbf{x} = \sum_{n \in \mathbb{Z}} y_0^{(-1)}[n] \, \varphi_n^{(-1)}$$

$$= \sum_{n \in \mathbb{Z}} \left(\sum_{i \in \mathbb{Z}} y_0^{(0)}[i] \, g_0[n - 2i] + \sum_{j \in \mathbb{Z}} y_1^{(0)}[j] \, g_1[n - 2j] \right) \varphi_n^{(-1)}$$

$$= \sum_{i \in \mathbb{Z}} y_0^{(0)}[i] \, \varphi_i^{(0)} + \sum_{j \in \mathbb{Z}} y_1^{(0)}[j] \, \psi_j^{(0)}$$

exactly as in the orthonormal case. The same property applies at every scale, so that each $\mathbf{x} \in \mathcal{V}^{(m-1)}$ can be written as a sum, $\mathbf{x}^{(m)} + \mathbf{w}^{(m)}$,

---

[7] Actually, if a filter is to have linear phase (symmetric impulse response) then it must also have finite support for practical reasons.

[8] Technically, we require the $\{\varphi_n\}_{n \in \mathbb{Z}}$ to constitute a Riesz basis, meaning that the basis functions are linearly independent and each $\mathbf{x} \in \mathcal{V}^{(0)}$ may be expressed as a linear combination, $\mathbf{x} = \sum_{n \in \mathbb{Z}} y_0^{(0)}[n] \, \varphi_n^{(m)}$, with $\mathbf{y}_0^{(0)} \in \ell^2(\mathbb{Z})$ and $A\|\mathbf{y}_0^{(0)}\| \le \|\mathbf{x}\| \le B\|\mathbf{y}_0^{(0)}\|$, where $A \le 1 \le B$ are two positive constants, independent of $\mathbf{x}$.

where $\mathbf{x}^{(m)}$ is a linear combination of the basis signals, $\{\varphi_n^{(m)}\}_{n\in\mathbb{Z}}$, for $\mathcal{V}^{(m)}$, while $\mathbf{w}^{(m)}$ is a linear combination of the $\{\psi_n^{(m)}\}_{n\in\mathbb{Z}}$. Now define $\mathcal{W}^{(m)}$ to be the linear span of the $\{\psi_n^{(m)}\}_{n\in\mathbb{Z}}$. It follows that

$$\mathcal{V}^{(m-1)} = \mathcal{V}^{(m)} \oplus \mathcal{W}^{(m)}$$

so that $\mathcal{W}^{(m)}$ is a complement of $\mathcal{V}^{(m)}$ in $\mathcal{V}^{(m-1)}$. Then, by upward and downward completeness of the resolution spaces, $\mathcal{V}^{(m)}$, the $\{\psi_n^{(m)}\}_{n,m\in\mathbb{Z}}$ together span $\mathcal{L}^2(\mathbb{R})$. Linear independence of the $\{\psi_n^{(m)}\}_{n\in\mathbb{Z}}$ for each m follows from linear independence of the 2-translates of g_1 in $\ell^2(\mathbb{Z})$. Linear independence of the entire set of wavelet functions, $\{\psi_n^{(m)}\}_{n,m\in\mathbb{Z}}$, may be deduced by observing that

$$\left(\mathcal{W}^{(m+1)} \cap \mathcal{W}^{(m)}\right) \subset \left(\mathcal{V}^{(m)} \cap \mathcal{W}^{(m)}\right)$$

and that $\mathcal{V}^{(m)} \cap \mathcal{W}^{(m)} = \{\mathbf{0}\}$ since the 2-translates of g_0 and the 2-translates of g_1 are linearly independent in $\ell^2(\mathbb{Z})$.

In summary, the "two-scale" equation inherent to any multi-resolution analysis may be used to identify the low-pass synthesis filter of a two channel subband transform, whose high-pass synthesis filter may then be used to generate a wavelet whose translates and dilates form a basis for $\mathcal{L}^2(\mathbb{R})$. For each m, the $\{\psi_n^{(m)}\}_{n\in\mathbb{Z}}$ form a basis for the complement space, $\mathcal{W}^{(m)}$. In general, the complement spaces are neither orthogonal to one another nor to the $\mathcal{V}^{(m)}$, nor are the wavelet basis functions orthonormal.

The construction described above leads to an immediate implementation of the wavelet transform in terms of a tree-structured subband transform, or DWT, exactly as in Section 6.2.2. In fact, we deliberately avoided any dependence on orthonormality in the development of Section 6.2.2. In the orthonormal case, the analysis filters, $h_0[n]$ and $h_1[n]$ (see Figure 6.4), must be time-reversed copies of the synthesis filters, $g_0[n]$ and $g_1[n]$, and the analysis system necessarily implements the inner products

$$y_0^{(m)}[n] = \left\langle \mathbf{x}^{(0)}, \varphi_n^{(m)} \right\rangle, \qquad y_1^{(m)}[n] = \left\langle \mathbf{x}^{(0)}, \psi_n^{(m)} \right\rangle, \qquad \forall n \in \mathbb{Z}, \forall m > 0$$

where $\mathbf{x}^{(0)}$ is defined in terms of the input sequence, $y_0^{(0)}[n]$, by equation (6.30). Also, $\mathbf{x}^{(m)}$ is clearly the orthogonal projection of $\mathbf{x}^{(0)}$ onto $\mathcal{V}^{(m)}$.

In the non-orthogonal case, the analysis and synthesis filters are generally different and the analysis system implements a different set of inner products

$$y_0^{(m)}[n] = \left\langle \mathbf{x}^{(0)}, \tilde{\varphi}_n^{(m)} \right\rangle, \qquad y_1^{(m)}[n] = \left\langle \mathbf{x}^{(0)}, \tilde{\psi}_n^{(m)} \right\rangle, \qquad \forall n \in \mathbb{Z}, \forall m > 0$$

Under appropriate conditions (see Section 6.3.1), the $\{\tilde{\psi}_n^{(m)}\}_{n,m\in\mathbb{Z}}$ constitute a dual wavelet basis for $\mathcal{L}^2(\mathbb{R})$. It is easily checked that these bases satisfy the biorthogonality condition,

$$\left\langle \psi_n^{(m)}, \tilde{\psi}_{\tilde{n}}^{(\tilde{m})} \right\rangle = \delta[n - \tilde{n}]\,\delta[m - \tilde{m}], \qquad \forall n, m, \tilde{n}, \tilde{m} \in \mathbb{Z}$$

which is the analog in $\ell^2(\mathbb{R})$ of the biorthogonality conditions in $\ell^2(\mathbb{Z})$ which must be satisfied by the underlying subband transform (see equation (4.29)). For this reason, we refer to $\{\tilde{\psi}_n^{(m)}\}_{n,m\in\mathbb{Z}}$ and $\{\psi_n^{(m)}\}_{n,m\in\mathbb{Z}}$ as biorthogonal wavelet bases.

Subject to the conditions mentioned above, $\{\tilde{\varphi}_n^{(m)}\}_{n\in\mathbb{Z}}$ and $\{\tilde{\psi}_n^{(m)}\}_{n\in\mathbb{Z}}$ form bases for sub-spaces, $\tilde{\mathcal{V}}^{(m)}$ and $\tilde{\mathcal{W}}^{(m)}$, respectively, such that the $\tilde{\mathcal{V}}^{(m)}$ constitute a dual multi-resolution analysis and $\tilde{\mathcal{W}}^{(m)}$ is the non-orthogonal complement of $\tilde{\mathcal{V}}^{(m)}$ in $\tilde{\mathcal{V}}^{(m-1)}$. Moreover, biorthogonality implies that $\tilde{\mathcal{W}}^{(m)} \perp \mathcal{V}^{(m)}$ and $\mathcal{W}^{(m)} \perp \tilde{\mathcal{V}}^{(m)}$. Note that the succession of lower resolution approximations, $\mathbf{x}^{(m)} \in \mathcal{V}^{(m)}$, no longer represent orthogonal projections of $\mathbf{x}^{(0)}$ onto $\mathcal{V}^{(m)}$. When necessary, we refer to $\tilde{\varphi}(t)$ and $\tilde{\psi}(t)$ as the *analysis* scaling and wavelet functions and to $\varphi(t)$ and $\psi(t)$ as the *synthesis* scaling and wavelet functions.

MULTI-DIMENSIONAL TRANSFORMS

For image compression applications, we are most interested in tree-structured filter banks of the form illustrated in Figure 4.20; the associated passband structure is indicated in Figure 4.21. In this case, the appropriate Hilbert space is $\mathcal{L}^2(\mathbb{R}^2)$, the complete inner product space of all square-integrable two dimensional functions, $\mathbf{x} \equiv x(s_1, s_2) \equiv x(\mathbf{s})$. The multi-resolution analysis consists of a family of nested resolution spaces, $\mathcal{V}^{(m)}$, exhibiting upward and downward completeness on $\mathcal{L}^2(\mathbb{R}^2)$, where $x(s_1, s_2) \in \mathcal{V}^{(0)}$ if and only if its 2^m-fold dilation, $x(2^{-m}s_1, 2^{-m}s_2) \in \mathcal{V}^{(m)}$. The other properties generalize in the obvious way. The scaling function then satisfies a two-scale equation of the form

$$\varphi(s_1, s_2) = 2 \sum_{n_1, n_2 \in \mathbb{Z}} g_0[n_1, n_2]\,\varphi(2s_1 - n_1, 2s_2 - n_2)$$

Since we are generally unwilling to consider implementations involving non-separable subband transforms, we restrict our attention to multi-resolution analyses in which the two-scale equation involves separable coefficient sequences; i.e.,

$$\varphi\left(s_1, s_2\right) = 2 \sum_{n_1, n_2 \in \mathbb{Z}} g_0\left[n_1\right] g_0\left[n_2\right] \varphi\left(2s_1 - n_1, 2s_2 - n_2\right)$$

Identifying $g_0\left[n\right]$ as the low-pass synthesis impulse response of a two channel subband transform, we may use the other synthesis filter, $g_1\left[n\right]$, to construct a wavelet basis. In this case, there are three distinct mother wavelets,

$$_{0,1}\psi\left(s_1, s_2\right) = 2 \sum_{n_1, n_2 \in \mathbb{Z}} g_0\left[n_1\right] g_1\left[n_2\right] \varphi\left(2s_1 - n_1, 2s_2 - n_2\right)$$

$$_{1,0}\psi\left(s_1, s_2\right) = 2 \sum_{n_1, n_2 \in \mathbb{Z}} g_1\left[n_1\right] g_0\left[n_2\right] \varphi\left(2s_1 - n_1, 2s_2 - n_2\right)$$

$$_{1,1}\psi\left(s_1, s_2\right) = 2 \sum_{n_1, n_2 \in \mathbb{Z}} g_1\left[n_1\right] g_1\left[n_2\right] \varphi\left(2s_1 - n_1, 2s_2 - n_2\right)$$

whose translates and dilates together span $\mathcal{L}^2\left(\mathbb{R}^2\right)$. Accordingly, there are three complement spaces, $\mathcal{W}_{0,1}^{(m)}$, $\mathcal{W}_{1,0}^{(m)}$ and $\mathcal{W}_{1,1}^{(m)}$, such that

$$\mathcal{V}^{(m-1)} = \mathcal{V}^{(m)} \oplus \mathcal{W}_{0,1}^{(m)} \oplus \mathcal{W}_{1,0}^{(m)} \oplus \mathcal{W}_{1,1}^{(m)}$$

All other concepts generalize naturally from the one dimensional case and the DWT implementation is simply the two dimensional multi-resolution subband transform of Figure 4.20.

6.3   CONSTRUCTION OF WAVELETS
6.3.1   WAVELETS FROM SUBBAND TRANSFORMS

We have seen that every multi-resolution analysis leads to a wavelet transform (the association is not unique) and that the wavelet transform is associated with a tree-structured subband transform which we call a DWT. It is natural to enquire as to the validity of the converse statements. It is generally acknowledged that every useful wavelet transform arises in connection with a multi-resolution analysis. The key question is whether or not every subband transform can arise from some multi-resolution analysis. The answer is decidedly no.

PRODUCT FORM EXPANSIONS

Suppose we are given the low-pass synthesis filter of a two channel subband transform, having impulse response $g_0\left[n\right]$. If this transform does arise in connection with a multi-resolution analysis, then the scaling function of the multi-resolution analysis must satisfy the two-scale equation (6.28). In the Fourier domain this becomes

$$\hat{\varphi}(\omega) = \sqrt{2} \sum_{n \in \mathbb{Z}} g_0[n] \frac{1}{2} e^{-jn\frac{\omega}{2}} \hat{\varphi}\left(\frac{\omega}{2}\right)$$

$$= \sqrt{2^{-1}} \hat{\varphi}\left(\frac{\omega}{2}\right) \hat{g}_0\left(\frac{\omega}{2}\right), \quad \forall \omega \in \mathbb{R} \tag{6.31}$$

where

$$\hat{g}_0(\omega) = \sum_{n \in \mathbb{Z}} g_0[n] e^{-jn\omega}$$

Note that $\hat{g}_0(\omega)$ is essentially the DTFT of the sequence, $g_0[n]$, except that it is defined for all $\omega \in \mathbb{R}$ as a 2π-periodic function.

Equation (6.31) indicates that $\hat{\varphi}(\omega)$ arises as an infinite product of the form,

$$\hat{\varphi}(\omega) = \beta \prod_{i=1}^{\infty} \hat{m}_0\left(\frac{\omega}{2^i}\right) \tag{6.32}$$

where $\hat{m}_0(\omega)$ is the normalized 2π-periodic function defined by

$$\hat{m}_0(\omega) \triangleq \frac{1}{\sqrt{2}} \hat{g}_0(\omega) \tag{6.33}$$

The infinite product in equation (6.32) can converge at $\omega = 0$ only if $\hat{m}_0(0) \in (-1, 1]$. There are various ways to show that in fact we must have $\hat{m}_0(0) = 1$. One way to see this is as follows. Suppose that $\hat{m}_0(\omega)$ is continuous about $\omega = 0$ (all practical subband synthesis filters will have this property) with $|\hat{m}_0(0)| < 1$. Then there exists $\varepsilon, \delta > 0$ such that $|\hat{m}_0(\omega)| < 1 - \delta$ whenever $|\omega| < \varepsilon$. For any ω, all but a finite number of terms in the infinite product of equation (6.32) have $|2^{-i}\omega| < \varepsilon$ and so the infinite product would converge to 0 everywhere.

The value of β in equation (6.32) determines the norm of the scaling function and hence all of its dilates and translates. In the orthonormal case, we have no option but to select $\beta = 1$.[9] In the general biorthogonal case, we find it convenient to adopt the same value, $\beta = 1$. In the Fourier

---

[9] For an orthonormal transform, equation (6.5) yields $|\hat{m}_0(\omega)|^2 + |\hat{m}_0(\omega + \pi)|^2 = 1$ and so $\hat{m}_0(\pi) = \hat{g}_0(\pi) = 0$. It follows from equation (6.32) that $\hat{\varphi}(2\pi k) = 0$ for all $k \neq 0$. Combining this with the Poisson formula of equation (6.27), we see that $\hat{\varphi}(0) = 1$, so we must set $\beta = 1$.

domain, equation (6.29) becomes

$$\hat{\psi}(\omega) = \frac{1}{\sqrt{2}} \hat{\varphi}\left(\frac{\omega}{2}\right) \hat{g}_1\left(\frac{\omega}{2}\right)$$

$$= e^{-j\frac{\omega}{2}} \hat{m}_1\left(\frac{\omega}{2}\right) \hat{\varphi}\left(\frac{\omega}{2}\right)$$

where we find it convenient to define $\hat{m}_1(\omega)$ as the normalized DTFT of the translated impulse response, $g_1^t[n] = g_1[n+1]$; i.e.,

$$\hat{m}_1(\omega) \triangleq \frac{1}{\sqrt{2}} \hat{g}_1^t(\omega) = \frac{1}{\sqrt{2}} e^{j\omega} \hat{g}_1(\omega)$$

As noted in Section 6.1.3, many statements concerning subband transforms are simpler when expressed in terms of the translated impulse responses, defined through equation (6.9), and this turns out to be true in the present context also.

In summary, given the synthesis filters of a two channel subband transform, and hence $\hat{m}_0(\omega)$ and $\hat{m}_1(\omega)$, the associated scaling function and wavelet, if they exist, may be explicitly constructed using

$$\hat{\varphi}(\omega) = \prod_{i=1}^{\infty} \hat{m}_0\left(\frac{\omega}{2^i}\right) \tag{6.34}$$

$$\hat{\psi}(\omega) = e^{-j\frac{\omega}{2}} \hat{m}_1\left(\frac{\omega}{2}\right) \prod_{i=2}^{\infty} \hat{m}_0\left(\frac{\omega}{2^i}\right) \tag{6.35}$$

We will obviously need to consider the conditions under which the infinite product of equation (6.34) converges. In fact, convergence is not enough to ensure that the limit, $\hat{\varphi}(\omega)$, is the Fourier transform of a valid scaling function, $\varphi(t)$, or that the translates and dilates of this scaling function form bases for a family of resolution spaces, $\mathcal{V}^{(m)}$, satisfying the upward and downward completeness requirements of a multi-resolution analysis. These matters are at the heart of the distinction between wavelet transforms and general multi-resolution tree-structured subband transforms. First, however, it is instructive to consider the following example.

Example 6.2 *(Haar Wavelet) Here we consider one of the simplest linear transforms, in which each pair of samples, $x_0[n] = x[2n]$ and $x_1[n] = x[2n+1]$, is represented by its sum and difference. Specifically, choosing the normalization which leads to an orthonormal transform, we have*

$$\begin{pmatrix} y_0[n] \\ y_1[n] \end{pmatrix} = \frac{1}{\sqrt{2}} \begin{pmatrix} x_0[n] + x_1[n] \\ x_1[n] - x_0[n] \end{pmatrix}$$

Evidently this is an orthonormal block transform with block dimension $m = 2$. It is also a two channel subband transform, where the translated analysis and synthesis impulse responses are given by

$$\begin{pmatrix} h_0^t(z) \\ h_1^t(z) \end{pmatrix} = \begin{pmatrix} g_0^t(z^{-1}) \\ g_1^t(z^{-1}) \end{pmatrix} = \frac{1}{\sqrt{2}} \begin{pmatrix} 1+z \\ 1-z^{-1} \end{pmatrix}$$

Applying equation (6.34) we see that this transform arises in connection with a multi-resolution analysis whose scaling function has Fourier transform,

$$\hat{\varphi}(\omega) = \lim_{d \to \infty} \prod_{i=1}^{d} \frac{1 + e^{-j2^{-i}\omega}}{2} = \lim_{d \to \infty} \prod_{i=1}^{d} e^{-j2^{-i-1}\omega} \cos 2^{-i-1}\omega$$

$$= e^{-j\frac{\omega}{2}} \lim_{d \to \infty} \prod_{i=1}^{d} \frac{\sin 2^{-i}\omega}{2 \sin 2^{-i-1}\omega} = e^{-j\frac{\omega}{2}} \lim_{d \to \infty} \frac{\sin \frac{\omega}{2}}{2^d \sin 2^{-d}\frac{\omega}{2}}$$

$$= e^{-j\frac{\omega}{2}} \frac{\sin \frac{\omega}{2}}{\frac{\omega}{2}} = e^{-j\frac{\omega}{2}} \operatorname{sinc} \frac{\omega}{2\pi} \tag{6.36}$$

In the second line of the above equation, we have used the trigonometric identity, $\sin 2\theta / \sin \theta = 2 \cos \theta$. We conclude that $\varphi(t)$ is a unit pulse, centred at $t = \frac{1}{2}$; i.e.,

$$\varphi(t) = \begin{cases} 1 & \text{if } t \in (0, 1) \\ 0 & \text{if } t \notin (0, 1) \end{cases}$$

From equation (6.35), we deduce that

$$\hat{\psi}(\omega) = \left[e^{-j\frac{\omega}{2}} \frac{1 - e^{j\frac{\omega}{2}}}{2} \right] \cdot \left[\hat{\varphi}\left(\frac{\omega}{2}\right) \right]$$

which is equivalent to the convolution integral,

$$\psi(t) = \int \left[\frac{1}{2} \left(\delta \left(\tau - \frac{1}{2} \right) - \delta(\tau) \right) \right] \cdot [2\varphi(2(t - \tau))] \, d\tau$$

$$= \begin{cases} -1 & \text{if } t \in \left(0, \frac{1}{2}\right) \\ 1 & \text{if } t \in \left(\frac{1}{2}, 1\right) \\ 0 & \text{if } t \notin (0, 1) \end{cases}$$

Note that resolution space, $\mathcal{V}^{(m)}$, consists of functions which are piecewise constant on intervals of length 2^m, so that the projections, $\mathbf{x}^{(m)}$, represent piecewise constant approximations of $\mathbf{x} \in \mathcal{L}^2(\mathbb{R})$. The mother wavelet, $\psi(t)$, is known as the Haar wavelet. Accordingly, the corresponding subband transform and its recursive extension to the DWT are commonly identified as discrete Haar transforms. The discontinuous orthonormal functions, $\varphi(t)$ and $\psi(t)$, are illustrated in Figure 6.5.

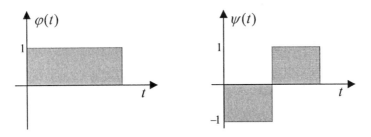

Figure 6.5.   Haar scaling function and wavelet.

CONVERGENCE AND DUALS

Recall that our first concern is with the convergence of the infinite product expression in equation (6.32). To simplify matters, we restrict our attention to FIR subband transforms. Since $m_0[n] = \frac{1}{\sqrt{2}} g_0[n]$ has finite support, its Z-transform, $m_0(z) = \sum_n m_0[n] z^{-n}$, is a finite two-sided polynomial (formally, this is known as a Laurent polynomial). Then $\hat{m}_0(\omega) = m_0(z)|_{z=e^{j\omega}}$ is a finite trigonometric polynomial. It follows that $\hat{m}_0(\omega)$ and hence $|\hat{m}_0(\omega)|$ have bounded derivatives; i.e.,

$$\frac{\partial}{\partial \omega} |\hat{m}_0(\omega)| \leq c, \quad \forall \omega \in \mathbb{R}$$

Combining this with the fact that $\hat{m}_0(0) = 1$, we obtain

$$|\hat{m}_0(\omega)| \leq 1 + c\,|\omega| \leq e^{c|\omega|}, \quad \text{for some } c \geq 0 \qquad (6.37)$$

It follows that the infinite product in equation (6.34) converges uniformly on any compact set to an infinitely differentiable (i.e., C^∞) function bounded by

$$|\hat{\varphi}(\omega)| \leq \exp\left(\sum_{i=1}^\infty \frac{c\,|\omega|}{2^i}\right) = e^{c|\omega|}$$

Of course, this bound is far too loose to guarantee square-integrability of $\hat{\varphi}(\omega)$. Thus, $\hat{\varphi}(\omega)$ is by no means guaranteed to be the Fourier transform of a valid scaling function. Nevertheless, $\hat{\varphi}(\omega)$ and $\hat{\psi}(\omega)$ are at least well-defined by equations (6.34) and (6.35).

At this juncture, it is also appropriate to point out that the dual scaling function, $\tilde{\varphi}(t)$, and the dual wavelet, $\tilde{\psi}(t)$, if they exist, may be constructed through similar infinite product expressions,

$$\widehat{\widetilde{\varphi}}(\omega) = \prod_{i=1}^{\infty} \widehat{\widetilde{m}}_0\left(\frac{\omega}{2^i}\right) \tag{6.38a}$$

$$\widehat{\widetilde{\psi}}(\omega) = e^{-j\frac{\omega}{2}} \widehat{\widetilde{m}}_1\left(\frac{\omega}{2}\right) \prod_{i=2}^{\infty} \widehat{\widetilde{m}}_0\left(\frac{\omega}{2^i}\right) \tag{6.38b}$$

where $\widehat{\widetilde{m}}_0(\omega)$ and $\widehat{\widetilde{m}}_1(\omega)$ are normalized DTFT's of the time-reversed, translated analysis filters, $h_0^t[-n]$ and $h_1^t[-n]$. These duals arise when we exchange the subband transform's analysis and synthesis vectors. As discussed in Section 6.2.3, the analysis of an input signal, $x(t)$, into its multi-resolution coefficients, $y_0^{(m)}[n]$, or its wavelet coefficients, $y_1^{(m)}[n]$, corresponds to taking its inner product with the corresponding translates and dilates of the dual scaling function and wavelet respectively. Of course, in the orthonormal case, we must have $\widetilde{\varphi}(t) = \varphi(t)$, $\widehat{\widetilde{m}}_0(\omega) = \widehat{m}_0(\omega)$, etc. In the biorthogonal case, however, the existence of the duals becomes an additional concern.

Since we are restricting our attention to FIR transforms, we may invoke the analysis-synthesis relationships derived in Section 6.1.3 to express $\widehat{\widetilde{m}}_0(\omega)$ and $\widehat{\widetilde{m}}_1(\omega)$ in terms of $\widehat{m}_1(\omega)$ and $\widehat{m}_0(\omega)$. Specifically, the translated impulse responses are related through equation (6.10), which we reproduce here as

$$h_q^t[k] = \alpha(-1)^k g_{1-q}^t[k], \quad q = 0, 1$$

In the Fourier domain, this becomes

$$\hat{h}_q^t(\omega) = \alpha \hat{g}_{1-q}^t(\omega + \pi), \quad q = 0, 1$$

Theorem 6.6 below shows that a necessary condition for the existence of a suitable scaling function is $\hat{g}_0(\pi) = 0$. Noting that $\hat{g}_0(0) = \sqrt{2}$, equation (6.14) then yields $\hat{h}_0(0) = \sqrt{2}$. Together with equation (6.13), we find easily that

$$\hat{g}_1^t(\pi) = \sqrt{2}\alpha^{-1}, \quad \text{and} \quad \hat{h}_1^t(\pi) = \sqrt{2}\alpha$$

In the orthonormal case, the analysis and synthesis filters must be time-reversed versions of each other, so we are compelled to select $\alpha = \pm 1$. We are free to select any α in the general case; however, a convenient selection is $\alpha = 1$. Putting everything together, we have

$$\begin{aligned} \hat{m}_q(\omega) &= \frac{\hat{g}_q^t(\omega)}{\sqrt{2}} \\ \widehat{\widetilde{m}}_q(\omega) &= \frac{\hat{h}_q^t(-\omega)}{\sqrt{2}} = \hat{m}_{1-q}(\pi - \omega) \end{aligned}, \quad q = 0, 1 \tag{6.39}$$

and

$$\widehat{\tilde{m}}_q (q\pi) = \hat{m}_q (q\pi) = 1, \quad q = 0, 1$$
$$\widehat{\tilde{m}}_{1-q} (q\pi) = \hat{m}_{1-q} (q\pi) = 0, \quad q = 0, 1$$

REGULARITY

We have established the convergence of the infinite product expressions for $\hat{\varphi}(\omega)$ and its dual, $\widehat{\tilde{\varphi}}(\omega)$, in terms of the low-pass synthesis and analysis filters of any FIR subband transform. The important questions now are: 1) whether these well-defined functions are valid Fourier transforms of functions $\varphi(t)$ and $\tilde{\varphi}(t)$ in $\mathcal{L}^2(\mathbb{R})$; and 2) whether $\varphi(t)$ and $\tilde{\varphi}(t)$ are valid scaling functions, whose translates and dilates form bases for sub-spaces $\mathcal{V}^{(m)}$ and $\tilde{\mathcal{V}}^{(m)}$, satisfying the upward and downward completeness conditions of a multi-resolution analysis. A variety of detailed mathematical studies may be found in the literature concerning the conditions under which these questions may be answered in the affirmative. For the mathematically inclined, good entrées to this body of literature may be found in [47, 48, 41].

While of significant theoretical interest, pursuit of the weakest conditions under which wavelets arise from subband transforms has little impact on practical applications, since these conditions admit discontinuous, highly irregular scaling functions and wavelets. For our purposes, it is more fruitful to review the conditions required to produce wavelets and scaling functions with a certain degree of regularity (e.g., continuity, differentiability, etc.). Popular designs for both orthonormal and biorthogonal wavelets are all based on these conditions.

Before plunging into the formal statements of various theorems concerning regularity, it is worth providing a brief summary of the main results. Firstly, the regularity of the wavelet functions is identical to that of their corresponding scaling functions[10], so we may restrict our attention to the latter. The regularity theorems all emphasize the importance of the number of "zeros at π" exhibited by the trigonometric polynomials, $\hat{m}_0(\omega)$ and $\widehat{\tilde{m}}_0(\omega)$. As noted earlier, $\hat{m}_0(\omega)$ may be written as $m_0(z)|_{z=e^{j\omega}}$, where the Z-transform, $m_0(z)$, is a polynomial (generally two-sided) in z. As such, it may be factored as

$$m_0(z) = \left(\frac{1+z}{2}\right)^N r(z)$$

---

[10] The reader is reminded that we are restricting our attention to FIR subband transforms. Consequently, the wavelet functions defined by equations (6.35) and (6.38b) are nothing but finite linear combinations of the corresponding dilated and translated scaling functions.

where N is the number of zeros at $z = -1$ and $r(z)$ represents the remaining factors. Equivalently,

$$\hat{m}_0(\omega) = \left(\frac{1 + e^{-j\omega}}{2}\right)^N \hat{r}(\omega) \qquad (6.40)$$

and we say that $\hat{m}_0(\omega)$ has N zeros at π. Note that $\hat{r}(0) = 1$ since $\hat{m}_0(0) = 1$.

As we shall see, if $\varphi(t)$ is to be a valid scaling function, $\hat{m}_0(\omega)$ must have at least one zero at π. In the case of the Haar wavelet (Example 6.2), we have $N = 1$ and $r(z) = 1$, so it is reasonable to think of this as the simplest possible wavelet transform. When $r(z)$ is more complex, there is no guarantee that a single zero at π is sufficient to obtain a valid scaling function. However, by adding more zeros at π one may ensure the existence of the scaling function and progressively increase its regularity. As discussed in Section 6.3.3, we are interested in scaling functions which are at least continuous, for which at least two zeros at π are required – recall that the Haar wavelet is discontinuous.

REGULARITY THEOREMS

To demonstrate the necessity of a zero at π, we adopt an approach developed by Rioul [124]. This approach has substantial intuitive appeal, being connected with a simple recursive algorithm for constructing the scaling function directly in the signal domain. In what follows, we regard a valid scaling function as one which has at least some degree of regularity (e.g., at most a finite number of discontinuities).

Theorem 6.6 *If* $\hat{\varphi}(\omega) = \prod_{i=1}^{\infty} \hat{m}_0\left(\frac{\omega}{2^i}\right)$ *is the Fourier transform of a valid scaling function,* $\varphi(t) \in \mathcal{L}^2(\mathbb{R})$, *then* $\hat{m}_0(\pi) = 0$.

Proof. To simplify matters, we shall restrict the proof to the case of finite support filters. Write ${}^d\hat{\varphi}(\omega)$ for the dth approximation of $\hat{\varphi}(\omega)$; specifically,

$$
{}^d\hat{\varphi}(\omega) = I_{\left[-2^d\pi, 2^d\pi\right]}(\omega) \cdot \prod_{i=1}^{d} \hat{m}_0\left(\frac{\omega}{2^i}\right)
$$

where $I_X(\omega)$ denotes the indicator function for the set, X. Thus ${}^d\hat{\varphi}(\omega)$ is obtained by expanding the first d terms in the infinite product expression for $\hat{\varphi}(\omega)$ and bandlimiting the resulting $2^{d+1}\pi$-periodic function to the interval, $\left[-2^d\pi, 2^d\pi\right]$. We introduce this bandwidth restriction to ensure that ${}^d\hat{\varphi}(\omega) \in \mathcal{L}^2(\mathbb{R})$; as a result, it must be a valid Fourier transform of some function, ${}^d\varphi(t) \in \mathcal{L}^2(\mathbb{R})$. By assumption, $\hat{\varphi}(\omega) \in \mathcal{L}^2(\mathbb{R})$, so it is effectively bandlimited. Thus, multiplication by $I_{\left[-2^d\pi, 2^d\pi\right]}(\omega)$ does not affect the convergence of ${}^d\hat{\varphi}(\omega)$ to $\hat{\varphi}(\omega)$ and hence ${}^d\varphi(t)$ to $\varphi(t)$.

Now let

$$
{}^d\varphi[k] = {}^d\varphi(t)\Big|_{t=2^{-d}k}
$$

be the Nyquist sampling of $^d\varphi(t)$. Then $^d\hat{\varphi}(\omega)$ is effectively the DTFT of $^d\varphi[k]$ (except for a scaling of the frequency axis) and it is easy to verify (by direct application of the fact that multiplication in the DTFT domain is equivalent to convolution of the discrete sequences) that the sequences, $^d\varphi[k]$, are obtained by the recursive formula

$$^1\varphi[k] = 2m_0[k]$$

$$^d\varphi[k] = \sum_j \left(^{d-1}\varphi[j]\right)(2m_0[k-2j]), \quad d = 2, 3, \dots \tag{6.41}$$

This recursive formula is interesting in its own right, since it is equivalent to obtaining $^d\varphi[k]$ from the synthesis system of a d-level DWT, with the low-pass subband sequence set to $y_0^{(d)}[k] = \delta[k]$ and all high-pass subband sequences set to zero.

By change of variable in equation (6.41), the even and odd sub-sequences of $^d\varphi[k]$ may be expressed as

$$^d\varphi[2k] = \sum_j (2m_0[2j])\left(^{d-1}\varphi[k-j]\right) \tag{6.42}$$

$$^d\varphi[2k+1] = \sum_j (2m_0[2j+1])\left(^{d-1}\varphi[k-j]\right) \tag{6.43}$$

If $\varphi(t)$ exists then $^d\varphi[2k]$ and $^d\varphi[2k+1]$ both converge to samplings of $\varphi(t)$, each having rate 2^{d-1}. Suppose, however, that $\hat{m}_0(\pi) = \varepsilon \neq 0$, meaning that

$$\sum_j m_0[2j] - \sum_j m_0[2j+1] = \varepsilon \neq 0 \tag{6.44}$$

For FIR filters, the summations in equations (6.42) and (6.43) involve only finitely many j. They are thus effectively forming weighted sums of samples from $\varphi(t)$, taken over a narrow region, whose size decreases exponentially with d. It follows that $\varphi(t)$ cannot be continuous. In fact, it cannot be continuous about any point t_0, at which $\varphi(t_0) \neq 0$.

To clarify this, let t_0 be an arbitrary time instant and let $\delta_t > 0$. It is possible to find k and d such that both the left and right hand sides of equations (6.42) and (6.43) involve samples of $\varphi(t)$ which are drawn only from the interval $(t - \delta_t, t + \delta_t)$. If $\varphi(t)$ is continuous at t_0 then we can select any $\delta_\varphi > 0$ and find δ_t sufficiently small such that all of these samples will differ from $\varphi(t_0)$ by no more than δ_φ. Writing $S = \sum_j |m_0[j]|$, the right hand side of equation (6.42) then differs by at most $S\delta_\varphi$ from $\varphi(t_0)\sum_j m_0[2j]$ and the right hand side of equation (6.43) differs by at most $S\delta_\varphi$ from $\varphi(t_0)\sum_j m_0[2j+1]$. This means that

$$|\varphi(t_0)| \cdot \left|\sum_j m_0[2j] - \sum_j m_0[2j+1]\right| \leq 2(S+1)\delta_\varphi$$

Imposing the assumption of equation (6.44) and letting δ_t and δ_φ go to zero, this equation can hold only with $\varphi(t_0) = 0$. That is, unless $\hat{m}_0(\pi) = \varepsilon = 0$, $\varphi(t)$ cannot be continuous about any point which lies within its region of support. ∎

One way to guarantee that $\hat{\varphi}(\omega), \widehat{\tilde{\varphi}}(\omega) \in \mathcal{L}^2(\mathbb{R})$ is to ensure that $|\hat{\varphi}(\omega)|$ and $\left|\widehat{\tilde{\varphi}}(\omega)\right|$ both decay faster than $|\omega|^{-\frac{1}{2}}$ as $|\omega| \to \infty$. Then the scaling functions, $\varphi(t)$ and $\tilde{\varphi}(t)$, will at least exist. In fact, this condition is also sufficient to guarantee that the scaling functions lead to dual multi-resolution analyses and biorthogonal wavelet bases for $\mathcal{L}^2(\mathbb{R})$ (see [41] for a proof).

If $|\hat{\varphi}(\omega)|$ decays faster than $|\omega|^{-(1+\xi)}$ with $\xi \geq 0$, then $\varphi(t)$ not only exists, but is continuous and ξ times differentiable; i.e., $\varphi(t) \in C^\xi$. The parameter, ξ, may then be interpreted as a measure of the regularity of the scaling function. Since we are restricting our attention here to FIR subband transforms, the wavelet is a finite linear combination of dilated and translated scaling functions; it therefore has the same regularity. The following theorem, taken from [48], establishes a useful family of sufficient conditions for such decay, in terms of $\hat{m}_0(\omega)$ and $\widehat{\tilde{m}}_0(\omega)$.

Theorem 6.7 *Suppose $\hat{m}_0(\omega)$ has N zeros at π, with the factorization given in equation (6.40); i.e.,*

$$\hat{m}_0(\omega) = \left(\frac{1 + e^{-j\omega}}{2}\right)^N \hat{r}(\omega)$$

Suppose also that $\hat{r}(\omega)$ is bounded by

$$B = \sup_{\omega \in [0, 2\pi]} |\hat{r}(\omega)| < 2^\gamma$$

Then

$$|\hat{\varphi}(\omega)| \leq C(1 + |\omega|)^{-(N-\gamma)-\varepsilon}$$

where $\varepsilon = \gamma - \log_2 B > 0$.

Proof. We begin by decomposing $|\hat{\varphi}(\omega)|$ as

$$|\hat{\varphi}(\omega)| = |\hat{u}(\omega)|^N \cdot |\hat{v}(\omega)|$$

where the first term is (see equation (6.36))

$$|\hat{u}(\omega)|^N = \left|\prod_{i=1}^\infty \frac{1 + e^{-j2^{-i}\omega}}{2}\right|^N$$

$$= \left|\operatorname{sinc}\left(\frac{\omega}{2\pi}\right)\right|^N \leq C'(1 + |\omega|)^{-N}$$

and the second term is

$$|\hat{v}(\omega)| = \prod_{i=1}^\infty \left|\hat{r}\left(\frac{\omega}{2^i}\right)\right|$$

Now $|\hat{r}(\omega)|$ satisfies the same type of bound as $|\hat{m}_0(\omega)|$ in equation (6.37), since both are trigonometric polynomials with a DC gain of 1. Thus, for some constant, c, we have

$$|\hat{r}(\omega)| \leq 1 + c|\omega| \leq e^{c|\omega|}$$

so that

$$|\hat{v}(\omega)| \leq \exp\left(c\sum_{i=1}^{\infty}\frac{|\omega|}{2^i}\right) = e^{c|\omega|}$$

Clearly, $|\hat{v}(\omega)|$ is bounded for $|\omega| \leq 1$ and we need only consider $|\omega| > 1$. For each $|\omega| > 1$, let J_ω be the non-negative integer for which $2^{J_\omega-1} \leq |\omega| < 2^{J_\omega}$ and write

$$|\hat{v}(\omega)| = \prod_{i=1}^{J_\omega}\left|\hat{r}\left(\frac{\omega}{2^i}\right)\right| \cdot \prod_{i=1}^{\infty}\left|\hat{r}\left(\frac{2^{-J_\omega}\omega}{2^i}\right)\right|$$

$$< \left(\prod_{i=1}^{J_\omega}\left|\hat{r}\left(\frac{\omega}{2^i}\right)\right|\right) \cdot e^c \leq e^c B^{J_\omega} = e^c 2^{J_\omega \log_2 B}$$

$$\leq e^c (2|\omega|)^{\log_2 B} = e^c B |\omega|^{\log_2 B} \leq C'' (1+|\omega|)^{\log_2 B}$$

where $C'' = e^c B$. Combining the two bounds, we obtain

$$|\hat{\varphi}(\omega)| \leq C(1+|\omega|)^{\log_2 B - N} = C(1+|\omega|)^{-(N-\gamma)-\varepsilon}$$

∎

The above theorem suggests a strategy for increasing the regularity of scaling functions and hence wavelets by adding as many zeros at π as possible. In this way, we increase the decay of $|\hat{\varphi}(\omega)|$ and hence the regularity of $\varphi(t)$ and $\psi(t)$. Of course, one cannot simply assume that filters with more zeros necessarily lead to more regular wavelets, because the residual factor, $\hat{r}(\omega)$ is generally non-trivial.

Since the above result represents a sufficient, rather than a necessary condition, one might suspect that it could be possible to construct highly regular wavelets and scaling functions without placing any more than the minimum of one zero at π. The following theorem, however, establishes the fact that there is no way to construct ξ-times differentiable functions without assigning at least $\xi+1$ zeros at π to the trigonometric polynomials, $\hat{m}_0(\omega)$ and $\tilde{m}_0(\omega)$. The reader is referred to [41] for a proof.

Theorem 6.8 *If $\{\psi_n^{(m)}\}_{n,m\in\mathbb{Z}}$ and $\{\tilde{\psi}_n^{(m)}\}_{n,m\in\mathbb{Z}}$ constitute biorthogonal bases for $\mathcal{L}^2(\mathbb{R})$ with continuous mother wavelets, $\psi(t) \in \mathcal{C}^\xi$ and $\tilde{\psi}(t) \in \mathcal{C}^{\tilde{\xi}}$, then the trigonometric polynomials, $\hat{m}_0(\omega)$ and $\widehat{\tilde{m}}_0$ must have at least $\xi+1$ and $\tilde{\xi}+1$ zeros, respectively, at π.*

6.3.2   DESIGN PROCEDURES
ORTHONORMAL WAVELETS

Consider again the regularity condition of Theorem 6.7. As noted above, this theorem suggests that one should design $\hat{m}_0(\omega)$ to have as many zeros at π as possible. Equivalently, $m_0(z)$ should have as many zeros at $z = -1$ as possible. In the orthonormal case, there is only one scaling function and one wavelet and we must select the sequence, $m_0[n] = \frac{1}{\sqrt{2}} g_0[n]$, to satisfy (see equation (6.5))

$$|\hat{m}_0(\omega)|^2 + |\hat{m}_0(\omega + \pi)|^2 = 1$$

or, in the Z-transform domain (see equation (6.4)),

$$m_0(z) m_0(z^{-1}) + m_0(-z) m_0(-z^{-1}) = 1$$

Restricting our attention to FIR filters with N zeros at $\omega = \pi$, we may write

$$\hat{m}_0(\omega) = \left(\frac{1 + e^{-j\omega}}{2}\right)^N \hat{r}(\omega) = e^{-jN\frac{\omega}{2}} \left(\cos\frac{\omega}{2}\right)^N \hat{r}(\omega)$$

Then, noting that $|\hat{r}(\omega)|^2 = P(\cos\omega)$ for some polynomial, $P(x)$, we have

$$1 = |\hat{m}_0(\omega)|^2 + |\hat{m}_0(\omega + \pi)|^2$$
$$= \left(\cos^2\frac{\omega}{2}\right)^N P(\cos\omega) + \left(\cos^2\frac{\omega + \pi}{2}\right)^N P(\cos(\omega + \pi))$$

Our objective is to find the polynomial, $P(x)$, after which $\hat{r}(\omega)$ may be obtained by spectral factorization. Specifically,

$$|\hat{r}(\omega)|^2 = P\left(\frac{e^{j\omega} + e^{-j\omega}}{2}\right) \implies r(z) r(z^{-1}) = P\left(\frac{z + z^{-1}}{2}\right)$$

so that we have only to find the zeros of $P\left(\frac{1}{2}(z + z^{-1})\right)$, and distribute them between $r(z)$ and $r(z^{-1})$. Note that the particular choice of spectral factorization (distribution of the zeros) affects only the phase response of $\hat{m}_0(\omega)$; it has no impact on the regularity of the scaling function or wavelet.

To find the polynomial, $P(x)$, observe that $\sin^2\frac{\omega}{2} = \frac{1}{2}(1 - \cos\omega)$. So $P(\cos\omega) = Q\left(\sin^2\frac{\omega}{2}\right)$, where $Q(x)$ is another polynomial, from which we may recover $P(x) \equiv Q\left(\frac{1}{2}(1 - x)\right)$. To satisfy the orthonormality condition we require

$$\left(\cos^2\frac{\omega}{2}\right)^N Q\left(\sin^2\frac{\omega}{2}\right) + \left(\cos^2\frac{\omega + \pi}{2}\right)^N Q\left(\sin^2\frac{\omega + \pi}{2}\right) = 1$$

or, letting u be a placeholder for $\cos^2\frac{\omega}{2}$,

$$u^N Q(1 - u) + (1 - u)^N Q(u) = 1 \tag{6.45}$$

Table 6.1. Daubechies family of orthonormal subband/wavelet filters, with the minimum phase spectral factorization. Note that we have selected the delay-normalized convention for which the support of $g_0[n]$ is given by Theorem 6.3.

$g_0[n] = h_0[-n]$	$N = 2$ zeros	$N = 3$ zeros	$N = 4$ zeros
$g_0[-3]$			0.230378
$g_0[-2]$		0.332670	0.714847
$g_0[-1]$	0.482963	0.806891	0.630881
$g_0[0]$	0.836516	0.459877	-0.027984
$g_0[1]$	0.224144	-0.135011	-0.187035
$g_0[2]$	-0.129410	-0.085440	0.030841
$g_0[3]$		0.035220	0.032883
$g_0[4]$			-0.010597

It follows that all orthonormal subband transforms whose low-pass filter has at least N zeros at π are obtained by spectral factorization, based on a polynomial $Q(x)$, satisfying equation (6.45).

Using Bezout's theorem, Daubechies [48] showed that all solutions to this problem are of the form

$$Q(x) = \sum_{n=0}^{N-1} \binom{N+n-1}{n} x^n + x^N R\left(\frac{1}{2} - x\right)$$

where $R(x)$ is any polynomial containing only odd powers of x. To minimize the length of the filters, Daubechies selected $R(x) \equiv 0$, obtaining a unique family (up to distribution of the spectral factors) of orthonormal subband/wavelet transforms, parametrized by N, the number of zeros at π. The Daubechies family of orthonormal subband/wavelet transforms has enjoyed tremendous popularity in a wide variety of applications, where the minimal phase factorization is most commonly encountered. The first few members of the family are given in Table 6.1 using this factorization. For greater accuracy, however, the reader can and should reproduce the construction directly from the procedure described.

LINEAR PHASE WAVELETS

Recall that orthonormality is incompatible with linear phase filters (see Theorem 6.5) and linear phase plays a particularly important role in image compression. Following techniques closely related to those involved in the construction of orthonormal wavelets above, Cohen, Daubechies and Feauveau [41] have developed families of biorthogonal transforms involving linear phase filters. Some of their designs are amongst the most successful and widely deployed transforms for image compression. We briefly outline their approach here, together with some important examples.

In the biorthogonal case, the sequences $m_0[n] = \frac{1}{\sqrt{2}} g_0[n]$ and $\tilde{m}_0[n] = \frac{1}{\sqrt{2}} h_0[-n]$ must be designed jointly. The choice of $m_0[n]$ and $\tilde{m}_0[n]$ is constrained by the biorthogonality condition (see equation (4.29)), which may be expressed in the Z-transform domain as

$$m_0(z)\tilde{m}_0(z^{-1}) + m_0(-z)\tilde{m}_0(-z^{-1}) = 1$$

or in the Fourier domain as

$$\hat{m}_0(\omega)\widehat{\tilde{m}}_0^*(\omega) + \hat{m}_0(\omega+\pi)\widehat{\tilde{m}}_0^*(\omega+\pi) = 1$$

To simplify matters, we restrict our attention to delay-normalized transforms having odd length FIR filters with linear phase (consult [41] for the parallel development with even length filters). Then $m_0[n]$ and $\tilde{m}_0[n]$ are both symmetric about $n = 0$ and their Fourier transforms must be polynomials in $\cos\omega$. Noting that

$$\left(\frac{e^{j\frac{\omega}{2}} + e^{-j\frac{\omega}{2}}}{2}\right)^N = \left(\cos\frac{\omega}{2}\right)^N = \left(\frac{1}{2}(1+\cos\omega)\right)^{\frac{N}{2}}$$

is a polynomial in $\cos\omega$ if and only if N is even, we see that $\hat{m}_0(\omega)$ and $\widehat{\tilde{m}}_0(\omega)$ must each have an even number of zeros at π, say $N = 2L$ and $\tilde{N} = 2\tilde{L}$, respectively. We may then write

$$\hat{m}_0(\omega) = \left(\cos\frac{\omega}{2}\right)^{2L} p_0(\cos\omega), \quad \text{and} \quad \widehat{\tilde{m}}_0(\omega) = \left(\cos\frac{\omega}{2}\right)^{2\tilde{L}} \tilde{p}_0(\cos\omega)$$

and the biorthogonality condition becomes

$$\left(\cos^2\frac{\omega}{2}\right)^M P(\cos\omega) + \left(\cos^2\frac{\omega+\pi}{2}\right)^M P(\cos(\omega+\pi)) = 1$$

where $M = L + \tilde{L}$ and $P(x) \equiv p_0(x)\tilde{p}_0(x)$.

We encountered exactly the same constraint in the orthonormal case described above. The solution is obtained by setting $P(x) \equiv Q(\frac{1}{2}(1-x))$ where

$$Q(x) = \sum_{n=0}^{M-1}\binom{M+n-1}{n}x^n + x^M R\left(\frac{1}{2} - x\right)$$

and $R(x)$ is any odd polynomial. As in the orthonormal case, for minimum length filters we select $R(x) \equiv 0$, which leads to a unique family of biorthogonal linear phase subband/wavelet transforms, whose members are distinguished only by the choice of parameters, L and $\tilde{L}$, and the way in which the factors of $P(x)$ are distributed between $p_0(x)$ and $\tilde{p}_0(x)$.

Interestingly, the regularity of the wavelet and its dual generally differ substantially. In the extreme case we may set $p_0(x) \equiv 1$ and $\tilde{p}_0(x) \equiv P(x)$, so that all $2L$ roots of $m_0(z)$ are at $z = -1$. This leads to a family of transforms, parametrized by L and $\tilde{L}$, in which the synthesis scaling function, $\varphi(t)$ is the B-spline of order $N = 2L$. To see this, note that

$$\hat{\varphi}(\omega) = \prod_{i=1}^{\infty} \left(\cos \frac{\omega}{2^{i+1}} \right)^N = \left(\frac{\sin \frac{\omega}{2}}{\frac{\omega}{2}} \right)^N = \left(\operatorname{sinc} \frac{\omega}{2\pi} \right)^N$$

Thus, $\varphi(t)$ is the N-fold convolution of a unit pulse with itself, which is, by definition, the B-spline of order N.

One particularly attractive feature of this family is that the sequences, $m_0[n]$ and $\tilde{m}_0[n]$, both consist entirely of dyadic fractions (rationals whose denominators are powers of 2). This means that the filtering operations of the DWT may be effected using low precision integer multiplication and bit shifting operations.

Example 6.3 *(Spline 5/3 transform) The first member of the spline family described above is obtained by setting $L = \tilde{L} = 1$ so that $\hat{m}_0(\omega)$ and $\widehat{\tilde{m}}_0(\omega)$ each have $N = \tilde{N} = 2$ zeros at π. The analysis and synthesis prototype sequences are given by*

$$\begin{pmatrix} m_0(z) \\ \tilde{m}_0(z) \end{pmatrix} = \begin{pmatrix} \frac{1}{4}z^{-1} + \frac{1}{2} + \frac{1}{4}z \\ -\frac{1}{8}z^{-2} + \frac{1}{4}z^{-1} + \frac{3}{4} + \frac{1}{4}z - \frac{1}{8}z^2 \end{pmatrix} \tag{6.46}$$

We refer to this as the "spline 5/3" or simply the "5/3" transform, since the low- and high-pass analysis filters have 5 and 3 taps, respectively. Its remarkable simplicity recommends it for practical image compression applications and it is one of the two transforms which must be implemented by every compliant JPEG2000 decompressor. The synthesis and analysis scaling functions, $\varphi(t)$ and $\tilde{\varphi}(t)$, and their wavelets, $\psi(t)$ and $\tilde{\psi}(t)$, are plotted in Figure 6.6.

In order to increase the regularity of the dual wavelet and scaling functions in spline-based designs, the support of $\tilde{m}_0[n]$ must be made much larger than that of $m_0[n]$. This is because more zeros at π must be added to compensate for the effect of assigning all factors of $P(x)$ to $\tilde{p}_0(x)$. Cohen, Daubechies and Feauveau [41] describe an alternate design strategy in which the analysis and synthesis filters have "least dissimilar" lengths. Recall from Theorem 6.4 that the filter lengths must differ by an odd multiple of 2. In this design strategy, both $\hat{m}_0(\omega)$

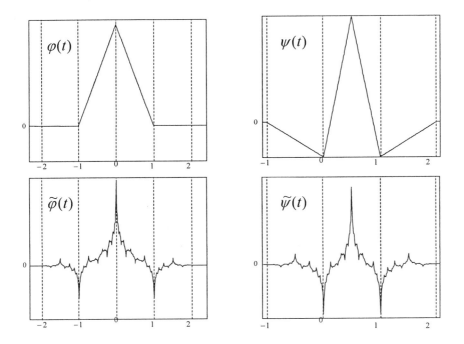

Figure 6.6. Synthesis and analysis scaling and wavelet functions for the spline 5/3 transform.

and $\widehat{m}_0(\omega)$ are assigned the same number of zeros at π and the factors of $P(x)$ are divided as equally as possible between $p_0(x)$ and $\tilde{p}_0(x)$.

Example 6.4 *(CDF 9/7 transform) The first member of the above family has $L = \tilde{L} = 2$ so that there are $N = \tilde{N} = 4$ zeros at π. In this case, $P(x)$ has only 3 factors and the subband/wavelet transform is unique up to exchanging the roles of the analysis and synthesis functions. In particular, we have*

$$m_0(z) = \left(\frac{1+z}{2}\right)^2 \left(\frac{1+z^{-1}}{2}\right)^2 \left(\frac{R_1 - \frac{z+z^{-1}}{2}}{R_1 - 1}\right)$$

$$\tilde{m}_0(z) = \left(\frac{1+z}{2}\right)^2 \left(\frac{1+z^{-1}}{2}\right)^2 \left(\frac{R_2^+ - \frac{z+z^{-1}}{2}}{R_2^+ - 1}\right) \left(\frac{R_2^- - \frac{z+z^{-1}}{2}}{R_2^- - 1}\right)$$

where the roots are

$$R_1 = \frac{4}{3} + \left(\frac{S}{15} - \frac{7}{3S} \right)$$

$$R_2^{\pm} = \frac{4}{3} - \frac{1}{2} \left(\frac{S}{15} - \frac{7}{3S} \right) \pm j \frac{\sqrt{3}}{2} \left(\frac{S}{15} + \frac{7}{3S} \right)$$

$$\text{with } S = \sqrt[3]{\left(350 + 105\sqrt{15} \right)}$$

Evaluating these irrational coefficients numerically yields

$$
\begin{array}{ll}
m_0(z) = & \tilde{m}_0(z) = \\
\quad 0.557543526229 & \quad 0.602949018236 \\
+\ 0.295635881557 \left(z^1 + z^{-1} \right) & +\ 0.266864118443 \left(z^1 + z^{-1} \right) \\
-\ 0.028771763114 \left(z^2 + z^{-2} \right) & -\ 0.078223266529 \left(z^2 + z^{-2} \right) \\
-\ 0.045635881557 \left(z^3 + z^{-3} \right) & -\ 0.016864118443 \left(z^3 + z^{-3} \right) \\
 & +\ 0.026748757411 \left(z^4 + z^{-4} \right)
\end{array}
\tag{6.47}
$$

We refer to this as the "CDF 9/7" transform, since the low- and high-pass analysis filters have 9 and 7 taps respectively. This transform has been found to yield optimal or near optimal performance in image compression applications and has enjoyed widespread popularity in the image compression community. It is one of the two transforms which must be implemented by every compliant JPEG2000 decompressor, the other being that of Example 6.3.

The synthesis and analysis scaling functions, $\varphi(t)$ and $\tilde{\varphi}(t)$, and their wavelets, $\psi(t)$ and $\tilde{\psi}(t)$, are plotted in Figure 6.7. Note that the analysis and synthesis functions here resemble each other much more closely than in the spline 5/3 case; this is a clear indication of the fact that the wavelet and multi-resolution bases are approximately orthonormal.

6.3.3   COMPRESSION CONSIDERATIONS

We have seen that multi-resolution analysis is intimately connected with the wavelet transform and that both are realized in practice through the DWT. Moreover, the DWT is nothing other than a tree-structured multi-resolution filter bank constructed by iterative application of a one dimensional, two channel subband transform. We have also seen that not all subband transforms may be associated with wavelets in this way. The obvious question is whether the narrower conditions associated with the wavelet transform are important for image compression.

To answer this question, consider the synthesis system of a one dimensional tree-structured filter bank with D levels, as shown in Figure 6.8.

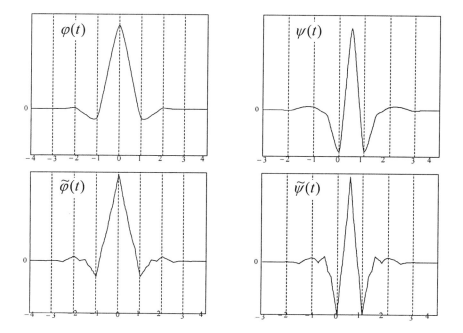

Figure 6.7. Synthesis and analysis scaling and wavelet functions for the CDF 9/7 transform.

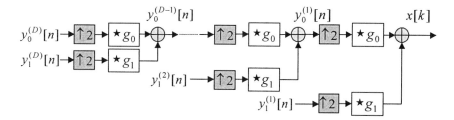

Figure 6.8. Synthesis system for a one dimensional tree-structured subband transform with D levels.

Suppose, for convenience that the synthesis filters are given by

$$g_q[n] = 2m_q[n], \quad q = 0, 1$$

Thus, the synthesis filters are scaled up by $\sqrt{2}$ relative to the normalization represented by equation (6.33), meaning that the analysis filters must be scaled down by the same factor, yielding

$$h_q[n] = \tilde{m}_q[-n], \quad q = 0, 1$$

Table 6.2.   Relationship between scaling and wavelet functions and the synthesis basis sequences of a tree-structured subband transform.

Subband sequence:	$y_0^{(D)}[n]$	$y_1^{(d)}[n],\quad 1 \le d \le D$
Basis vectors:	$^D\varphi_n \equiv {}^D\varphi\left[k - 2^D n\right]$	$^d\psi_n \equiv {}^d\psi\left[k - 2^d n\right]$
Function $\in \mathcal{L}^2(\mathbb{R})$:	$^D\varphi(t) \longrightarrow \varphi(t)$	$^d\psi(t) \longrightarrow \psi(t)$
Sampling:	$^D\varphi[k] = {}^D\varphi(t)\big|_{t=2^{-D}k}$	$^d\psi[k] = {}^d\psi(t)\big|_{t=2^{-d}k}$

Now the synthesis vectors corresponding to the low-pass subband sequence, $y_0^{(D)}[n]$, may be written $^D\varphi_n \equiv {}^D\varphi\left[k - 2^D n\right]$, where $^D\varphi[k]$ is obtained by the recursive up-sampling and filtering algorithm of equation (6.41). As demonstrated in the proof of Theorem 6.6, $^D\varphi[k]$ is a sampling at rate 2^D of the function, $^D\varphi(t)$, which converges to $\varphi(t)$ as $D \to \infty$. Similarly, each high-pass subband sequence, $y_0^{(d)}[n]$, has synthesis vectors, $^d\psi_n \equiv {}^d\psi\left[k - 2^d n\right]$, which are obtained through the recursion

$$^1\psi[k] = 2m_1[k-1]$$
$$^d\psi[k] = \sum_j {}^{d-1}\psi[j]\left(2m_0[k-2j]\right), \quad d = 2, 3, \ldots$$

and represent rate 2^d samplings of a function, $^d\psi(t)$, which converges[11] to $\psi(t)$ as $d \to \infty$. These relationships are summarized in Table 6.2.

We conclude that the synthesis basis vectors of a tree-structured subband transform converge to samplings of functions in $\mathcal{L}^2(\mathbb{R})$ if and only if the scaling and wavelet functions exist. Moreover, the regularity of the synthesis basis vectors[12] is that of the underlying scaling and wavelet functions. For images, the synthesis basis vectors associated with the various subbands of the dyadic multi-resolution subband transform shown in Figure 4.20 are the appropriate translates of the separable products,

$$^D\varphi[k_1, k_2] = {}^D\varphi[k_1] \cdot {}^D\varphi[k_2]$$
$$_{0,1}^d\psi[k_1, k_2] = {}^d\varphi[k_1] \cdot {}^d\psi[k_2]$$
$$_{1,0}^d\psi[k_1, k_2] = {}^d\psi[k_1] \cdot {}^d\varphi[k_2]$$
$$_{1,1}^d\psi[k_1, k_2] = {}^d\psi[k_1] \cdot {}^d\psi[k_2]$$

---

[11] If the scaling function exhibits some regularity, then $^d\varphi(t)$ and $^d\psi(t)$ are typically excellent approximations to $\varphi(t)$ and $\psi(t)$, even for quite small values of d, such as 3 or 4.

[12] We use the term "regularity" losely here to refer to the regularity of the functions obtained by interpolating the discrete synthesis basis sequences.

and the same regularity considerations apply.

Suppose now that some subband sample, say $y_{1,0}^{(d)}[\mathbf{n}]$, is in error by an amount δ, due to quantization during compression. In the image domain, this individual sample error manifests itself as the error vector (image),

$$\delta \cdot {}_{1,0}^{d}\psi_{\mathbf{n}}[\mathbf{k}] = \delta \left({}^{d}\psi(s_1) \cdot {}^{d}\varphi(s_2) \right) \Big|_{\mathbf{s}=2^{-d}\mathbf{k}-\mathbf{n}}$$

$$\approx \delta \cdot \psi \left(2^{-d}k_1 - n_1 \right) \cdot \varphi \left(2^{-d}k_2 - n_2 \right)$$

Thus, if the compression artifacts in the reconstructed image are to have a smooth regular appearance, the underlying subband transform must satisfy the constraints associated with the existence of a sufficiently regular scaling function. By contrast, highly disturbing discontinuous image artifacts arise if the associated scaling function (and hence the wavelet) is discontinuous or does not exist. Differentiability and even greater regularity of the scaling and wavelet functions is also desirable.

The above arguments suggest that good multi-resolution subband transforms for image compression should lead to synthesis wavelets with substantial regularity. When the analysis and synthesis wavelets are different, which is necessarily the case for FIR linear phase transforms, the regularity of the synthesis wavelet is generally more important than that of the analysis wavelet.

6.4 LIFTING AND REVERSIBILITY

In Section 6.1.4, we introduced some useful factorizations for the polyphase analysis or synthesis matrix of an FIR subband transform and showed that these factorizations are convenient both for design and implementation of the transform. In this section we introduce the so-called "lifting factorization," which is particularly important for image compression applications. We confine our attention entirely to one dimensional two channel transforms, since common multi-resolution transforms, including the DWT, are all obtained by iterative application of this basic building block.

6.4.1 LIFTING STRUCTURE

Let $x[k]$ denote the input sequence and $y_0[n]$ and $y_1[n]$ the subband sequences of a two channel subband transform. In the simplest possible transform, the subband sequences are simply the even and odd subsequences of $x[k]$. Call these $y_0^{\{0\}}[n] = x[2n]$ and $y_1^{\{0\}}[n] = x[2n+1]$. This is sometimes called the "lazy wavelet" transform, although it cannot formally arise in connection with a valid wavelet transform. It turns out

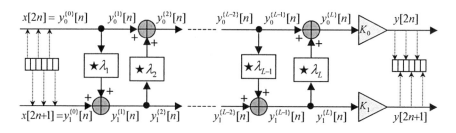

Figure 6.9.   Lifting implementation of subband analysis.

(see Section 6.4.3) that any subband transform may be implemented by successively updating these trivial subband sequences, $y_0^{\{0\}}[n]$ and $y_1^{\{0\}}[n]$, in a sequence of so-called "lifting steps."

The lifting procedure is illustrated in Figure 6.9. In each lifting step, one sub-sequences is updated by the addition of a filtered version of the other sub-sequence. We use indices, $l = 1, 2, \ldots, L$, to identify the lifting steps. For odd indices, l, the odd sub-sequence is updated and we have

$$y_0^{\{l\}}[n] = y_0^{\{l-1\}}[n]$$
$$y_1^{\{l\}}[n] = y_1^{\{l-1\}}[n] + \sum_i \lambda_l[i]\, y_0^{\{l-1\}}[n-i]$$

For even l, the even sub-sequence is updated and we have

$$y_0^{\{l\}}[n] = y_0^{\{l-1\}}[n] + \sum_i \lambda_l[i]\, y_1^{\{l-1\}}[n-i]$$

$$y_1^{\{l\}}[n] = y_1^{\{l-1\}}[n]$$

It is convenient to express these relationships more compactly as

$$y_{1-p(l)}^{\{l\}}[n] = y_{1-p(l)}^{\{l-1\}}[n]$$
$$y_{p(l)}^{\{l\}}[n] = y_{p(l)}^{\{l-1\}}[n] + \sum_i \lambda_l[i]\, y_{1-p(l)}^{\{l-1\}}[n-i] \qquad (6.48)$$

where $p(l)$ denotes the parity of l (0 if l is even, 1 if l is odd).

Upon completion of all L lifting steps, the sub-sequences $y_0^{\{L\}}[n]$ and $y_1^{\{L\}}[n]$ need only be scaled by some factors, K_0 and K_1, to recover the low- and high-pass subbands, $y_0[n]$ and $y_1[n]$. Just as $y_0^{\{0\}}[n]$ and $y_1^{\{0\}}[n]$ are the even and odd sub-sequences of $x[k]$, so $y_0^{\{L\}}[n]$ and $y_1^{\{L\}}[n]$ are essentially the even and odd sub-sequences of the interleaved subband sequence, $y[k]$. We saw in Section 6.1.3 how subband

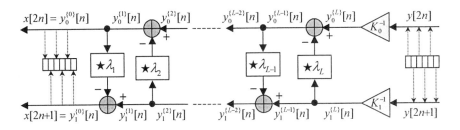

Figure 6.10. Lifting implementation of subband synthesis.

analysis and synthesis may both be understood as time-varying convolution operations mapping $x[k]$ to the interleaved sequence, $y[k]$, and vice-versa (see Figure 6.1). From both perspectives, it is natural to associate even indexed samples with the low-pass subband and odd indexed samples with the high-pass subband. This association provides a convenient mnemonic which will be particularly useful when we consider the boundaries of finite length sequences in Section 6.5.

The lifting structure of Figure 6.9 is trivial to invert. In particular, after inverting the subband gains, K_0 and K_1, we have only to apply the lifting steps in reverse order, flipping the sign of the filter impulse responses, as shown in Figure 6.10. Each successive inverse lifting step, $l = L, \ldots, 1$, implements

$$y_{1-p(l)}^{\{l-1\}}[n] = y_{1-p(l)}^{\{l\}}[n]$$
$$y_{p(l)}^{\{l-1\}}[n] = y_{p(l)}^{\{l\}}[n] - \sum_i \lambda_l[i] \, y_{1-p(l)}^{\{l\}}[n-i]$$
$$= y_{p(l)}^{\{l\}}[n] - \sum_i \lambda_l[i] \, y_{1-p(l)}^{\{l-1\}}[n-i]$$

which trivially inverts the corresponding forward step. Evidently, invertibility of the structure is unaffected if we replace the convolution operators of Figures 6.9 and 6.10 with arbitrary operators, whether linear or non-linear, fixed or time-varying. As we shall see in Section 6.4.2, this flexibility is precisely what is required to construct a single image compression system, capable of achieving both efficient lossy compression and efficient lossless compression.

Example 6.5 *Consider the spline 5/3 biorthogonal transform of Example 6.3. Adopting the convention that the low- and high-pass analysis filters should be normalized to have unit gain at $\omega = 0$ and $\omega = \pi$,* *respectively, we have*

$$
\begin{pmatrix} h_0^t(z) \\ h_1^t(z) \end{pmatrix} = \begin{pmatrix} \tilde{m}_0(z^{-1}) \\ \tilde{m}_1(z^{-1}) \end{pmatrix} = \begin{pmatrix} \tilde{m}_0(z^{-1}) \\ m_0(-z^{-1}) \end{pmatrix}
$$
$$
= \begin{pmatrix} -\frac{1}{8}z^{-2} + \frac{1}{4}z^{-1} + \frac{3}{4} + \frac{1}{4}z - \frac{1}{8}z^2 \\ -\frac{1}{4}z^{-1} + \frac{1}{2} - \frac{1}{4}z \end{pmatrix}
$$

and

$$
\begin{pmatrix} g_0^t(z) \\ g_1^t(z) \end{pmatrix} = 2 \begin{pmatrix} m_0(z) \\ m_1(z) \end{pmatrix} = 2 \begin{pmatrix} m_0(z) \\ \tilde{m}_0(-z) \end{pmatrix}
$$
$$
= \begin{pmatrix} \frac{1}{2}z^{-1} + 1 + \frac{1}{2}z \\ -\frac{1}{4}z^{-2} - \frac{1}{2}z^{-1} + \frac{3}{2} - \frac{1}{2}z - \frac{1}{4}z^2 \end{pmatrix} \tag{6.49}
$$

The transform may be implemented using $L = 2$ lifting steps, where $\lambda_1(z) = -\frac{1}{2}(1 + z)$ and $\lambda_2(z) = \frac{1}{4}(1 + z^{-1})$. To see this, we may expand the lifting steps as follows

$$
y_1^{\{2\}}[n] = y_1^{\{1\}}[n] = y_1^{\{0\}}[n] - \frac{1}{2}\left(y_0^{\{0\}}[n] + y_0^{\{0\}}[n+1]\right)
$$
$$
= x[2n+1] - \frac{1}{2}(x[2n] + x[2n+2])
$$
$$
= 2\left(x \star h_1^t\right)[2n+1] = 2y[2n+1]
$$

and

$$
y_0^{\{2\}}[n] = y_0^{\{1\}}[n] + \frac{1}{4}\left(y_1^{\{1\}}[n] + y_1^{\{1\}}[n-1]\right)
$$
$$
= x[2n] + \frac{1}{4}\begin{pmatrix} x[2n+1] - \frac{1}{2}(x[2n] + x[2n+2]) \\ + x[2n-1] - \frac{1}{2}(x[2n-2] + x[2n]) \end{pmatrix}
$$
$$
= \frac{3x[2n]}{4} + \frac{x[2n+1] + x[2n-1]}{4} - \frac{x[2n+2] + x[2n-2]}{8}
$$
$$
= \left(x \star h_0^t\right)[2n] = y[2n]
$$

Thus, the gain factors are $K_0 = 1$ and $K_1 = \frac{1}{2}$. The complete implementation is illustrated in Figure 6.11.

The first lifting step in the above example has an obvious interpretation in terms of prediction. Specifically, the output of filter λ_1, is $\frac{1}{2}(x[2n] + x[2n+2])$, which may be interpreted as a reasonable predictor for $x[2n+1]$. The first lifting step, then, converts the odd subsequence into a prediction residual, which is essentially the high-pass subband sequence. While this behaviour is shared by the lifting implementations of a number of useful subband transforms, the following

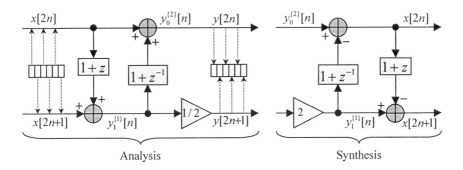

Figure 6.11. Lifting implementation of the spline 5/3 transform.

important example shows that such simple interpretations are not always possible.

Example 6.6 *Consider the CDF 9/7 biorthogonal transform of Example 6.4. Adopting the convention that the low- and high-pass analysis filters should be normalized to have unit gain at $\omega = 0$ and $\omega = \pi$, respectively, we have*

$$\left(\begin{array}{c} h_0^t(z) \\ h_1^t(z) \end{array} \right) = \left(\begin{array}{c} \tilde{m}_0\left(z^{-1}\right) \\ \tilde{m}_1\left(z^{-1}\right) \end{array} \right) = \left(\begin{array}{c} \tilde{m}_0\left(z^{-1}\right) \\ m_0\left(-z^{-1}\right) \end{array} \right)$$

where $m_0(z)$ and $\tilde{m}_0(z)$ are given by equation (6.47). It can be shown, either by direct expansion or by the factorization method of Section 6.4.3, that four lifting steps are required to implement this transform, where

$$\lambda_1(z) = -1.586134342\,(1+z)$$
$$\lambda_2(z) = -0.052980118\,\left(1+z^{-1}\right)$$
$$\lambda_3(z) = 0.882911075\,(1+z)$$
$$\lambda_4(z) = 0.443506852\,\left(1+z^{-1}\right)$$

$$K_0 = \frac{1}{K}, \quad K_1 = \frac{K}{2}, \quad \text{where } K = 1.230174105$$

Notice that only 6 multiplications are required to produce each pair of subband samples. If we are prepared to accept an unusual normalization for the subband samples then the final gains may be omitted and only 4 multiplications are required to produce each pair of subband samples. These may be compared with the 16 multiplications required for a direct implementation as in Figure 6.1, which reduces to 9 multiplications if we are careful to exploit symmetry and 7 if we are prepared to accept an unusual normalization for the subband samples, with the central filter taps scaled to 1.

LIFTING WHAT?

The lifting structure described above was first proposed by Sweldens [145], who coined the term "lifting" in consideration of its usefulness for designing biorthogonal wavelet bases. Specifically, suppose one is given dual scaling functions, $\varphi(t)$ and $\tilde{\varphi}(t)$, and wavelets, $\psi(t)$ and $\tilde{\psi}(t)$, constituting biorthogonal wavelet bases, all four functions having compact support. Associated with this biorthogonal system, there must be an FIR subband transform whose analysis and synthesis filters are appropriately modulated, time-reversed and translated versions of the two finite sequences, $m_0[n]$ and $\tilde{m}_0[n]$, according to equation (6.39). From Theorem 6.6, $m_0(z)$ and $\tilde{m}_0(z)$ must each have at least one zero at $z = -1$.

Now suppose we start with a lifting factorization of this initial transform and add a single lifting step, $\lambda_l(z)$, with l odd. Recall that this affects only the odd output sub-sequence produced by the transform, i.e., the high-pass subband samples. The new lifting structure thus implements an FIR subband transform with a different high-pass analysis filter, but the same low-pass analysis filter. Equivalently, we have modified only $m_0(z) = \tilde{m}_1(-z)$. By suitable choice of $\lambda_l(z)$, we can ensure that the new $m_0(z)$ has at least one extra zero at $z = -1$. Recall that the number of zeros at $z = -1$ is intimately connected with the regularity of the scaling function, $\varphi(t)$ and its wavelet, $\psi(t)$, as established by Theorems 6.7 and 6.8.

In the same way, an even-indexed lifting step can always be designed to "lift" the number of zeros assumed by $\tilde{m}_0(z)$ at $z = -1$ and hence the regularity of the dual functions, $\tilde{\varphi}(t)$ and $\tilde{\psi}(t)$. In this way, the lifting scheme provides a framework for alternately "lifting" the regularity of the synthesis and analysis wavelets and their scaling functions, each building upon the other. Although the number of zeros assumed by $m_0(z)$ and $\tilde{m}_0(z)$ at $z = -1$ does determine a lower bound for the regularity of the respective wavelets, it is not the sole determining factor. Furthermore, the strategy described here is only one particular example of the use of lifting for design.

6.4.2   REVERSIBLE TRANSFORMS

Linear transforms are inherently ill-suited to lossless compression. Many transforms involve irrational coefficients so that the transform samples cannot be precisely represented with any finite number of bits. Two quite different examples of this are the block DCT of Section 4.1.3 and the CDF 9/7 subband/wavelet transform of Example 6.4. In Section 6.3.2, we described a family of biorthogonal subband/wavelet transforms based upon the B-splines. This family has the desirable property that the transform coefficients are dyadic fractions, which means that the subband samples may be represented with finite precision. Even in the simple case of the 5/3 transform, however, the precision of the subband samples increases by approximately 3 bits for each level in a one dimensional DWT and by 6 bits for each level in a two dimensional DWT.

It is instructive to consider the Haar transform of Example 6.2, which is the simplest linear transform with any merit for compression. After suitable normalization, we may write the low- and high-pass subband samples as

$$\begin{pmatrix} y_0[n] \\ y_1[n] \end{pmatrix} = \begin{pmatrix} \frac{1}{2}(x_0[n] + x_1[n]) \\ x_1[n] - x_0[n] \end{pmatrix} \tag{6.50}$$

where $x_0[n] = x[2n]$ and $x_1[n] = x[2n+1]$ are the polyphase components of the input sequence. This is the simplest non-trivial block transform, with block size 2. It is also a degenerate subband transform[13] and it implements the most primitive Haar wavelet transform. Even in this case, the precision of the subband samples increases by 1 bit for each level in a one dimensional DWT and by 2 bits for each level in a two dimensional DWT. This is a substantial obstacle to efficient lossless compression.

The fundamental problem is that linear transforms introduce redundancy into the least significant bits of the transform sample values. This redundancy cannot be fully eliminated by subsequent coding, no matter how elaborate, unless the transform bands are coded jointly. As stated in Section 4.3, however, one of the key benefits which transforms bring to compression is the simplification which results from quantizing and coding each transform band (or even each sample) independently.

In the specific case of the Haar transform, the problem of expanding numeric precision can be solved in the form of the S (Sequential) transform, by the introduction of a slight non-linear perturbation to equation (6.50). Specifically,

$$y_1[n] = x_1[n] - x_0[n]$$
$$y_0[n] = x_0[n] + \left\lfloor \frac{y_1[n]}{2} \right\rfloor \approx \frac{1}{2}(x_0[n] + x_1[n])$$

This construction may be interpreted in terms of the lifting structure of Figure 6.12, whence the inverse transform is easily seen to be

---

[13] Degenerate, because the basis sequences associated with successive subband samples do not overlap – i.e., it is only a block transform.

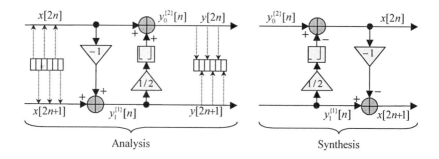

Figure 6.12.   Lifting steps for the S transform.

$$x_0\left[n\right] = y_0\left[n\right] - \left\lfloor \frac{y_1\left[n\right]}{2} \right\rfloor$$

$$x_1\left[n\right] = y_1\left[n\right] + x_0\left[n\right]$$

Suppose that the input sequence, $x\left[k\right]$, consists of B-bit integers. Then $y_0\left[n\right]$ is also a sequence of B-bit integers. The high-pass subband samples, $y_1\left[n\right]$, require a $(B+1)$-bit representation, but the values are mostly close to zero and can usually be coded with a small number of bits. The S transform has been proposed for compression of medical images [75], where truly lossless performance is considered important to minimize the risk of liability in malpractice lawsuits.

The S transform was generalized independently and in related ways by the S+P transform of Said and Pearlman [127] and the TS transform of Zandi et al. [174]. These transforms stimulated some interest in the possibility of a unified framework for both lossy and lossless compression of images; however, they represented only isolated examples of reversible transforms. Calderbank et al. [33] later demonstrated that reversible transforms could be synthesized from the lifting implementation of any two channel FIR subband transform. In particular, if we eliminate the subband gain factors, K_0 and K_1, and replace equation (6.48) with the non-linear approximation,

$$y_{p(l)}^{\{l\}}\left[n\right] = y_{p(l)}^{\{l-1\}}\left[n\right] + \left\lfloor \frac{1}{2} + \sum_i \lambda_l\left[i\right] y_{1-p(l)}^{\{l-1\}}\left[n - i\right] \right\rfloor \qquad (6.51)$$

then integer-valued input samples are always mapped to integer-valued subband samples. The inherent invertibility of the lifting structure of Figure 6.9 is clearly preserved under such non-linear modifications. The

TS transform and some parametrizations of the S+P transform may be understood as special cases of this general construction[14].

The term "reversible" has sometimes been used in the literature to refer to the perfect reconstruction property of filter banks which implement subband transforms. In more recent usage, however, a reversible transform is understood as one which satisfies two conditions: 1) the transform samples must have bounded, finite precision representations, depending upon the input sample bit-depth; and 2) it must be possible to recover the input samples exactly from the subband samples, using finite precision arithmetic with bounded complexity. These are clearly necessary conditions for lossless compression. The first condition is satisfied by the modifications to the lifting structure described above. To satisfy the second condition, any irrational terms, $\lambda_l[k]$, in equation (6.51) must be replaced with rational approximations. In fact, the coefficients in a reversible lifting transform are usually expected to be dyadic fractions for convenience of implementation.

The fact that the transform maps integers to integers does not prevent significant expansion in the dynamic range of the subband samples. For this, it is usually sufficient to insist that the DC gain of the equivalent (linearized) low-pass analysis filter be close to 1. Consider, for example, the lifting implementation of the CDF 9/7 subband/wavelet transform in Example 6.6. To create a reversible version of the transform, we may approximate the four lifting step coefficients with suitable dyadic fractions and eliminate the subband gain factors, K_0 and K_1. The DC gain of the equivalent low-pass analysis filter is then $K \approx 1.23$. Thus, assuming that the original source signal has most of its energy at low frequencies (almost invariably true for images), the dynamic range of the subband samples can be expected to grow by approximately 0.3 bits for each level in a one dimensional DWT, or 0.6 bits for each level in a two dimensional DWT. This difficulty does not arise with the spline 5/3 subband/wavelet transform of Example 6.5, where K_0 is already equal to 1.

6.4.3   FACTORIZATION METHODS

In this section we show that any two channel subband transform involving FIR filters may be implemented using the lifting structure of Figure 6.9, subject to the delay-normalization convention defined on Page 237. Moreover, we show how the lifting steps may be found from

---

[14]The recommended choice of parameters for the S+P transform actually corresponds to a subband system with FIR analysis filters and recursive IIR synthesis filters, which are not compatible with the lifting structure as it is presented here.

the filters of the subband transform. First observe that each lifting step may be written in matrix form as

$$
\begin{pmatrix} y_0^{\{l\}}(z) \\ y_1^{\{l\}}(z) \end{pmatrix} = \Lambda_l(z) \cdot \begin{pmatrix} y_0^{\{l-1\}}(z) \\ y_1^{\{l-1\}}(z) \end{pmatrix}
$$

where

$$
\Lambda_l(z) = \begin{cases} \begin{pmatrix} 1 & \lambda_l(z) \\ 0 & 1 \end{pmatrix} & \text{if } l \text{ is even} \\[2ex] \begin{pmatrix} 1 & 0 \\ \lambda_l(z) & 1 \end{pmatrix} & \text{if } l \text{ is odd} \end{cases}
$$

It follows that the lifting system may be expressed in matrix form as

$$
\begin{pmatrix} y_0(z) \\ y_1(z) \end{pmatrix} = K \cdot \Lambda_L(z) \cdots \Lambda_2(z) \cdot \Lambda_1(z) \cdot \begin{pmatrix} x_0(z) \\ x_1(z) \end{pmatrix}
$$

where K is the diagonal matrix of subband gain factors,

$$
K = \begin{pmatrix} K_0 & 0 \\ 0 & K_1 \end{pmatrix}
$$

Evidently, the lifting steps represent a factorization of the polyphase analysis matrix as

$$
H(z) = K \cdot \Lambda_L(z) \cdots \Lambda_2(z) \cdot \Lambda_1(z) \tag{6.52}
$$

Theorem 6.9 *Suppose that $H(z)$ is the polyphase analysis matrix of a delay-normalized two channel subband transform with FIR filters. Then $H(z)$ may be factored into lifting steps as in equation (6.52).*

Proof. The proof is constructive, although the constructed factorization is not unique. Let $G(z) = (H(z))^{-1}$ be the polyphase synthesis matrix. Since the transform is in delay-normalized form, we have $\det(G(z)) = \alpha^{-1}$. We seek lifting steps which satisfy

$$
\Lambda_L(z) \cdots \Lambda_2(z) \cdot \Lambda_1(z) \cdot G(z) = K^{-1}
$$

Without loss of generality, we may assume that the number of lifting steps, L, is even – if necessary, the last lifting step may be empty (i.e., $\lambda_L(z) = 0$ so that $\Lambda_L(z) = I$). Then, the above expression may be rewritten as

$$
\Lambda_{L-1}(z) \cdots \Lambda_2(z) \cdot \Lambda_1(z) \cdot G(z) = \begin{pmatrix} 1 & -\lambda_L(z) \\ 0 & 1 \end{pmatrix} K^{-1} \tag{6.53}
$$

Let $G^{\{l\}}(z)$ denote the partial products, defined recursively by $G^{\{0\}}(z) = G(z)$ and

$$
G^{\{l\}}(z) = \Lambda_l(z) \cdot G^{\{l-1\}}(z), \quad l \geq 1
$$

The idea is to select the lifting steps, $\Lambda_l(z)$, so as to successively reduce the degree of the polynomials in the first column of $G^{\{l\}}(z)$.[15] Specifically, write

$$G^{\{l\}}(z) = \begin{pmatrix} G_{00}^{\{l\}}(z) & G_{01}^{\{l\}}(z) \\ G_{10}^{\{l\}}(z) & G_{11}^{\{l\}}(z) \end{pmatrix}$$

The entries in the first column of $G^{\{l\}}(z)$ (this is all that concerns us) are generally two-sided (Laurent) polynomials of the form

$$G_{q0}^{\{l\}}(z) = \sum_{i=A_{q0}^{\{l\}}}^{B_{q0}^{\{l\}}} \beta_{q0}^{\{l\}}[i] \, z^{-i}$$

with degree $\deg(G_{q0}^{\{l\}}(z)) = B_{q0}^{\{l\}} - A_{q0}^{\{l\}}$. Note that $\deg(0) = -1$. For convenience, we also define the degree of the whole matrix, $G^{\{l\}}(z)$, to be the sum of the degrees of the polynomials in its first column; i.e., $\deg(G^{\{l\}}(z)) = \deg(G_{00}^{\{l\}}(z)) + \deg(G_{10}^{\{l\}}(z))$.

Suppose l is even; then

$$\begin{pmatrix} G_{00}^{\{l\}}(z) \\ G_{10}^{\{l\}}(z) \end{pmatrix} = \begin{pmatrix} 1 & \lambda_l(z) \\ 0 & 1 \end{pmatrix} \begin{pmatrix} G_{00}^{\{l-1\}}(z) \\ G_{10}^{\{l-1\}}(z) \end{pmatrix} = \begin{pmatrix} G_{00}^{\{l-1\}}(z) + \lambda_l(z) G_{10}^{\{l-1\}}(z) \\ G_{10}^{\{l-1\}}(z) \end{pmatrix}$$

So long as $\deg(G_{00}^{\{l-1\}}(z)) \geq \deg(G_{10}^{\{l-1\}}(z))$, we can always choose $\lambda_l(z)$ so as to ensure that $\deg(G^{\{l\}}(z)) < \deg(G^{\{l-1\}}(z))$. Otherwise, let $\lambda_l(z) = 0$ and there is no change in the degree. Similarly, for l odd we have

$$\begin{pmatrix} G_{00}^{\{l\}}(z) \\ G_{10}^{\{l\}}(z) \end{pmatrix} = \begin{pmatrix} 1 & 0 \\ \lambda_l(z) & 1 \end{pmatrix} \begin{pmatrix} G_{00}^{\{l-1\}}(z) \\ G_{10}^{\{l-1\}}(z) \end{pmatrix} = \begin{pmatrix} G_{00}^{\{l-1\}}(z) \\ G_{10}^{\{l-1\}}(z) + \lambda_l(z) G_{00}^{\{l-1\}}(z) \end{pmatrix}$$

and so long as $\deg(G_{10}^{\{l-1\}}(z)) \geq \deg(G_{00}^{\{l-1\}}(z))$, we can choose $\lambda_l(z)$ to ensure that $\deg(G^{\{l\}}(z)) < \deg(G^{\{l-1\}}(z))$. Thus, subject to suitable choice of the $\lambda_l(z)$, $\deg(G^{\{l\}}(z))$ is non-decreasing in l and decreases at least on every other lifting step, until eventually either $G_{00}^{\{l\}}(z) = 0$ or $G_{10}^{\{l\}}(z) = 0$. In the latter case, we may add extra lifting steps to move the zero from $G_{10}^{\{l\}}(z)$ to $G_{00}^{\{l\}}(z)$. In this way, we arrive at some point, L', with

$$G^{\{L'\}}(z) = \begin{pmatrix} 0 & G_{01}^{\{L'\}}(z) \\ G_{10}^{\{L'\}}(z) & G_{11}^{\{L'\}}(z) \end{pmatrix}$$

But $\det(G(z)) = \alpha^{-1}$ and $\det(\Lambda_l(z)) = 1$ for all l, so we must have

$$G_{10}^{\{L'\}}(z) \cdot G_{01}^{\{L'\}}(z) = -\alpha^{-1}$$

This means that $\deg(G_{10}^{\{L'\}}(z)) = \deg(G_{01}^{\{L'\}}(z)) = 0$ and $G^{\{L'\}}(z)$ has the form

$$G^{\{L'\}}(z) = \begin{pmatrix} 0 & K_1^{-1} z^P \\ -K_0^{-1} z^{-P} & G_{11}^{\{L'\}}(z) \end{pmatrix}$$

---

[15] The choice of the first column, instead of the second, is entirely arbitrary.

for some $p \in \mathbb{Z}$ and some factors, $K_0, K_1 \in \mathbb{R}$, satisfying

$$K_0 K_1 = \alpha = \det(H(z))$$

Setting $L = L' + 3$, we add two further lifting steps to obtain

$$G^{\{L-1\}}(z) = \begin{pmatrix} 1 & 0 \\ z^{-p} & 1 \end{pmatrix} \begin{pmatrix} 1 & -z^p \\ 0 & 1 \end{pmatrix} \begin{pmatrix} 0 & K_1^{-1} z^p \\ -K_0^{-1} z^{-p} & G_{11}^{\{L'\}}(z) \end{pmatrix}$$

$$= \begin{pmatrix} K_0^{-1} & G_{01}^{\{L-1\}}(z) \\ 0 & K_1^{-1} \end{pmatrix}$$

Equation (6.53) is then satisfied by setting $\lambda_L(z) = -K_1 G_{01}^{\{L-1\}}(z)$. ∎

The idea behind the above proof is very simple: select the lifting steps so as to successively reduce the degree of either $G_{00}^{\{l\}}(z)$ or $G_{10}^{\{l\}}(z)$, as appropriate. The proof is complicated only by the final steps, in which we show that differential delay in the subband sequences can always be absorbed into lifting steps. In many cases, these additional steps are not required and the number of lifting steps is substantially smaller than one might suspect from the proof. The following simple example is instructive.

Example 6.7 *Let us start with the spline 5/3 filters of Example 6.5. The translated synthesis impulse responses are given by equation (6.49), which may be rewritten as*

$$\begin{pmatrix} g_0(z) \\ g_1(z) \end{pmatrix} = \begin{pmatrix} g_0^t(z) \\ z^{-1} g_1^t(z) \end{pmatrix} \begin{pmatrix} \frac{1}{2} z^{-1} + 1 + \frac{1}{2} z \\ -\frac{1}{4} z^{-3} - \frac{1}{2} z^{-2} + \frac{3}{2} z^{-1} - \frac{1}{2} - \frac{1}{4} z \end{pmatrix}$$

whence we find that the polyphase synthesis matrix is (see equation (6.8))

$$G(z) = \begin{pmatrix} 1 & -\frac{1}{2}(1 + z^{-1}) \\ \frac{1}{2}(1 + z) & \frac{3}{2} - \frac{1}{4}(z^{-1} + z) \end{pmatrix}$$

Since $\deg(G_{00}(z)) \leq \deg(G_{10}(z))$, *we can reduce* $\deg(G_{10}(z))$ *in the first lifting step. In fact, because the inequality is strict, we can reduce the degree by 2. Specifically, setting* $\lambda_1(z) = -\frac{1}{2}(1 + z)$, *we obtain*

$$G^{\{1\}}(z) = \begin{pmatrix} 1 & 0 \\ -\frac{1}{2}(1 + z) & 1 \end{pmatrix} \begin{pmatrix} 1 & -\frac{1}{2}(1 + z^{-1}) \\ \frac{1}{2}(1 + z) & \frac{3}{2} - \frac{1}{4}(z^{-1} + z) \end{pmatrix}$$

$$= \begin{pmatrix} 1 & -\frac{1}{2}(1 + z^{-1}) \\ 0 & 2 \end{pmatrix}$$

and we are finished, with $K_0 = 1$, $K_1 = \frac{1}{2}$ *and* $\lambda_2(z) = -K_1 G_{01}^{\{1\}}(z) = \frac{1}{4}(1 + z^{-1})$ *satisfying equation (6.53). These are the same lifting steps with which we began in Example 6.5.*

In the above example, we saw that it is possible sometimes to reduce the degree of one of the polynomials by more than 1 in any given lifting step. On the other hand, we could have chosen $\lambda_1(z) = -\frac{1}{2}z$ for the first lifting step; by so doing we would have reduced the degree by only 1 and the total number of lifting steps would have been larger. To minimize the complexity of the implementation it is usually desirable to select each $\lambda_l(z)$ so as to minimize $\deg(G^{\{l\}}(z))$.

6.4.4   ODD LENGTH SYMMETRIC FILTERS

In the case of odd length filters, $\deg(G_{00}(z)) - \deg(G_{10}(z))$ must be an odd integer. It is then easy to verify that $\deg(G(z))$ can be reduced by an even integer (perhaps 0) in the first lifting step, leaving $\deg(G_{00}^{\{1\}}(z)) - \deg(G_{10}^{\{1\}}(z))$ also an odd integer. Proceeding in this way, every non-empty lifting step reduces the degree by a multiple of 2 and the lifting factorization involves relatively few steps.

In the specific case of transforms having odd length filters with linear phase, the translated analysis and synthesis impulse responses are symmetric about the origin (see Theorem 6.4). Now observe that $G_{00}(z)$ and $G_{10}(z)$ are Z-transforms of the even and odd sub-sequences of $g_0[n] = g_0[-n]$. Consequently, they must both be symmetric polynomials of even and odd degree, respectively. Specifically,

$$G_{00}(z) = G_{00}(z^{-1}), \quad \text{and} \quad G_{10}(z) = zG_{10}(z^{-1}) \qquad (6.54)$$

The reader may verify this in the specific case of Example 6.7.

It follows that an efficient lifting step, which reduces $\deg(G_{10}(z))$ by the even integer $\deg(G_{10}(z)) - (\deg(G_{00}(z)) - 1)$, must involve a symmetric polynomial, $\lambda_1(z)$, of odd degree. Moreover, the remainder polynomial, $G_{10}^{\{1\}}(z) = G_{10}(z) - \lambda_1(z)G_{00}(z)$, must have exactly the same symmetry as $G_{10}(z)$. Using equation (6.54), one may easily verify that $\lambda_1(z) = z\lambda_1(z^{-1})$, so that $\lambda_1[n]$ is an even length sequence, symmetric about $n = -\frac{1}{2}$.

Since the symmetry properties have not changed, an efficient choice for the second lifting step is the one which reduces $\deg(G_{00}^{\{1\}}(z))$ by the even integer $\deg(G_{00}^{\{1\}}(z)) - (\deg(G_{10}^{\{1\}}(z)) - 1)$. Again, this involves a symmetric polynomial, $\lambda_2(z)$, of odd degree and leaves $G_{00}^{\{2\}}(z)$ with exactly the same symmetry as $G_{00}(z)$, from which one may verify that $\lambda_2(z) = z^{-1}\lambda_2(z^{-1})$. Proceeding in this way, we find that the sequence of efficient lifting steps always involves symmetric lifting filters, $\lambda_l[n]$, of even length. The centre of symmetry is $-\frac{1}{2}$ for l odd and $\frac{1}{2}$ for l even;

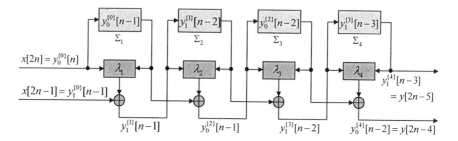

Figure 6.13. Analysis lifting state machine for the special case of 2-tap symmetric lifting filters. Final subband gain factors not shown.

that is,

$$\lambda_l[n] = \begin{cases} \lambda_l[1-n] & \text{if } l \text{ even} \\ \lambda_l[-1-n] & \text{if } l \text{ odd} \end{cases} \quad (6.55)$$

It is also easily shown that any lifting factorization whose lifting filters have even length with the symmetries of equation (6.55) corresponds to an FIR subband transform with odd-length symmetric filters. Thus, we have a compact description of all odd length linear phase subband transforms in the lifting domain. The reader is invited to verify that this description includes that proposed by Vetterli and LeGall [163] (see equation (6.20)), which essentially restricts the lifting steps to length 2.

The specific case of 2-tap symmetric lifting filters, $\lambda_l[n]$, is of particular interest for JPEG2000, since the two wavelet kernels defined for Part 1 of the standard both have this form. All two channel FIR subband transforms having odd length, symmetric filters with least dissimilar lengths (filter lengths differ by 2) may be factored into lifting steps of this form. With this restriction, the analysis lifting network of Figure 6.9 reduces to the simple state machine shown in Figure 6.13. We emphasize the fact that this structure is applicable only to the case of odd length symmetric subband filters whose lengths differ by exactly 2.

The state machine involves L state variables, Σ_l, one for each lifting step (four lifting steps are shown). These state variables are identified by the lightly shaded boxes in the figure. Each state variable holds one of the two inputs required by its lifting step filter, where the other input is the output from the previous lifting step. The lifting step filters (darkly shaded boxes in the figure) add their two inputs and multiply the result by λ_l, the value of the two identical non-zero taps in the lifting filter, $\lambda_l[n]$. For reversible transforms, the result is also rounded to an integer, while for irreversible transforms, the results may need to be scaled by factors K_0 and K_1. Neither of these details are shown in the figure.

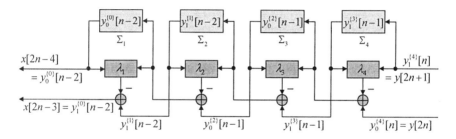

Figure 6.14. Synthesis lifting state machine for 2-tap symmetric lifting filters. Initial subband gain factors not shown.

The analysis state machine of Figure 6.13 may be implemented by the following simple algorithm. In this algorithm, A and Σ_{new} are two temporary variables, which may be interpreted respectively as the "augend" for the next lifting step and the new value which will be assumed by the state variable for that lifting step. At this point, we deliberately avoid specifying initial and terminal boundary conditions, since these are the subject of the ensuing Section 6.5.

Lifting Analysis Algorithm (for 2-tap symmetric lifting filters)

For $n = \ldots, 0, 1, 2, \ldots,$
$$\Sigma_{\text{new}} \leftarrow x\,[2n]$$
$$A \leftarrow x\,[2n-1]$$
For $l = 1, 2, \ldots, L,$
$$A \leftarrow A + \lambda_l\,(\Sigma_{\text{new}} + \Sigma_l)$$
$$A \leftarrow \Sigma_l \leftarrow \Sigma_{\text{new}} \leftarrow A$$
(rotates contents of the triplet, $A, \Sigma_l, \Sigma_{\text{new}}$, to the left)
$$y\,[2n - L - 1] \leftarrow A$$
$$y\,[2n - L] \leftarrow \Sigma_{\text{new}}$$

Not surprisingly, the synthesis network of Figure 6.10 reduces to a similar lifting state machine and algorithm. The synthesis lifting state machine is depicted in Figure 6.14.

6.5   BOUNDARY HANDLING

In our development of subband and wavelet transforms, starting in Chapter 4 and continuing throughout the present chapter, we have consistently ignored the fact that real signals have boundaries. In some one dimensional applications (e.g., sampled time waveforms), the source produces a practically unbounded number of samples and we might be able to ignore boundary effects. By contrast, two dimensional sources such as images usually have modest dimensions, bounded by physical constraints[16]. In this section, we address practical methods for adapting subband/wavelet transform techniques to finite length source sequences. Since our two dimensional transforms have all been constructed by the separable application of one dimensional building blocks, it is sufficient to restrict the ensuing discussion to one dimension.

6.5.1   SIGNAL EXTENSIONS

In general, a one dimensional source produces samples, $x[k]$, over a finite range of indices,

$$E \leq k < F \tag{6.56}$$

It is often convenient to set $E = 0$, with F the length of the sequence. Subband transforms, however, are cyclo-stationary operators so that the location of the source samples has an impact on the subband sample values which are produced. There are important applications in which the ability to control the location of the source sequence relative to the periodicity of the transform is most beneficial. One such application is efficient editing of JPEG2000 compressed images. For this reason, we shall maintain arbitrary lower and upper bounds for the source sequence, as in equation (6.56).

We may consider the finite sequence, $x[k]$, as being embedded in an infinite extended sequence, $\breve{x}[k]$, where

$$x[k] = \breve{x}[k], \quad E \leq k < F \tag{6.57}$$

and we are free to choose the remaining samples of $\breve{x}[k]$. Let $\breve{y}[k]$ be the interleaved sequence of subband samples generated by applying an m band subband transform to the extended sequence, $\breve{x}[k]$. This interleaved sequence is defined by

$$\breve{y}[mn + q] = \breve{y}_q[n], \quad 0 \leq q < m, \forall n \in \mathbb{Z}$$

The relationship between $\breve{x}[k]$ and $\breve{y}[k]$ may be simply expressed in terms of the translated impulse responses defined in equation (6.9). We repeat the time-varying convolution expressions here for convenience.

$$\breve{y}[k] = \sum_{i \in \mathbb{Z}} h_{k \bmod m}^t[i]\, \breve{x}[k - i]$$

$$\breve{x}[k] = \sum_{i \in \mathbb{Z}} \breve{y}[i]\, g_{i \bmod m}^t[k - i]$$

---

[16]One exception to this is satelite imagery, which is often acquired by continuously scanning the earth's surface with a linear sensor.

From these, one deduces easily that so long as $\breve{x}[k]$ is periodic with period divisible by m, $\breve{y}[k]$ will exhibit the same property. This suggests a periodic extension procedure in which the $F - E$ samples of $x[k]$ are first padded out to a multiple of m and then periodically extended as necessary to form $\breve{x}[k]$ and thence $\breve{y}[k]$. Only one period of subband samples must be preserved, so that the total number of subband samples exceeds $F - E$ by at most $m - 1$.

Periodic extension has a number of serious drawbacks. A practical concern is that some of the initial source samples must be buffered in memory until the end of the sequence is encountered, where they are required to perform the extension. In the case of images, the amount of memory required for this buffering is non-trivial, involving some number of rows or columns of the image.

Another serious objection to periodic extension is that the synthesis basis vectors span the boundary between the end of one period and the start of the next. As a result, quantization artifacts introduced at the end of the reconstructed source sequence are highly correlated with those introduced at the beginning of the sequence, which can be quite disturbing. A related difficulty is that the periodically extended sequence, $\breve{x}[k]$, typically contains high frequency transients at the boundaries between successive periods, which manifest themselves in unusually large high frequency subband samples, degrading compression efficiency. These difficulties may all be overcome by increasing the period, at the expense of additional redundant subband samples.

6.5.2   SYMMETRIC EXTENSION.

Suppose we restrict our attention to delay-normalized FIR subband transforms having odd length, linear phase filters. For convenience, we will further restrict our consideration to the important case of $m = 2$ subbands. A discussion of the importance of such transforms for image compression appears toward the end of Section 6.1.3. From Theorem 6.4, we know that the translated impulse responses are all symmetric about the origin; i.e.,

$$h_q^t[k] = h_q^t[-k], \quad \text{and} \quad g_q^t[k] = g_q^t[-k], \quad q = 0, 1$$

Now let $\breve{x}[k]$ be the symmetrically extended sequence defined by equation (6.57), together with

$$\breve{x}[E - k] = \breve{x}[E + k], \quad \forall k \in \mathbb{Z} \tag{6.58a}$$

$$\breve{x}[F - 1 - k] = \breve{x}[F - 1 + k], \quad \forall k \in \mathbb{Z} \tag{6.58b}$$

For $F - E \geq 2$, these constraints uniquely identify $\breve{x}[k]$ as a periodic sequence with period $2(F - E - 1)$, symmetric about each of the original end points, $k = E$ and $k = F - 1$. The following simple algorithm may be used to construct this unique sequence.

Symmetric Extension Algorithm

Assign $\breve{x}[k] \leftarrow x[k]$ for $E \le k < F$

For each $i = 1, 2, \dots$

    Assign $\breve{x}[E - i] \leftarrow \breve{x}[E + i]$.

    (Note that $\breve{x}[E + i]$ is certain to have been assigned earlier)

    Assign $\breve{x}[F - 1 + i] \leftarrow \breve{x}[F - 1 - i]$.

    (Note that $\breve{x}[F - 1 - i]$ is certain to have been assigned earlier)

Now observe that the interleaved sequence of subband samples, $\breve{y}[k]$, exhibits exactly the same symmetries as $\breve{x}[k]$. Specifically,

$$
\begin{aligned}
\breve{y}[E - k] &= \sum_{i \in \mathbb{Z}} h^t_{(E-k) \bmod 2}[i]\, \breve{x}[(E - k) - i] \\
&= \sum_{i \in \mathbb{Z}} h^t_{(E-k) \bmod 2}[-i]\, \breve{x}[E - (k - i)] \\
&= \sum_{i \in \mathbb{Z}} h^t_{(E-k) \bmod 2}[i]\, \breve{x}[(E + k) - i] \\
&= \sum_{i \in \mathbb{Z}} h^t_{(E+k) \bmod 2}[i]\, \breve{x}[(E + k) - i] \\
&= \breve{y}[E + k]
\end{aligned}
$$

and, by similar reasoning, $\breve{y}[F - 1 - k] = \breve{y}[F - 1 + k]$. Thus, it is sufficient to keep the $F - E$ subband samples,

$$
y[k] = \breve{y}[k], \quad E \le k < F \tag{6.59}
$$

from which $\breve{y}[k]$ may be symmetrically extended and inverse transformed to recover the original source samples, $x[k]$. Although $\breve{x}[k]$ and $\breve{y}[k]$ are formally infinite length sequences, the subband filters have only finite extent, so it is sufficient in practice to extend $x[k]$ and $y[k]$ by only a few samples at each end.

Evidently, the symmetric extension policy introduces no redundant subband samples. In fact, it avoids all of the problems cited above in connection with periodic extension. The approach may be extended to subband transforms having even length linear phase filters [138]. However, the symmetry conditions for even length filters are not nearly so elegant as in the odd length case presented above. Moreover, they exhibit a complex dependence on the parity of E.[17] For this reason, we shall not consider the even length case further.

As mentioned several times throughout this chapter, it is appealing to formally associate the even (respectively odd) indexed samples of $x[k]$ with the low-pass (respectively high-pass) subband samples. This connection is strongly reinforced by the symmetric extension procedure described above. In particular, suppose $F - E$ is odd so that the number of low- and high-pass subband samples must differ. According to equation (6.59), the interleaved sequence of subband samples, $y[k]$, has exactly the same region of support as the original source sequence, $x[k]$. Consequently the low-pass (respectively high-pass) subband receives more samples precisely when there are more even (respectively odd) indexed source samples. This association is a direct consequence of the delay-normalization convention, as defined on Page 237. We have consistently adopted this appealing convention for results which depend upon the relative displacement of the source and subband sequences.

6.5.3 BOUNDARIES AND LIFTING

The symmetric extension procedure described above is appropriate only for subband transforms involving linear phase filters. While nonredundant boundary handling methods may be developed for more general subband transforms, they lack the elegant simplicity of symmetric extension. Within the context of a lifting implementation, however, a rich family of boundary handling policies may be realized with ease for any two channel FIR subband transform.

ARBITRARY EXTENSIONS

Recall that every two channel FIR subband transform may be realized through the lifting structure of Figure 6.9 and that the structure is trivially inverted, as in Figure 6.10, even if the lifting step filters, $\lambda_l(z)$, are replaced by non-stationary operators. Let $y^{\{l\}}[k]$ be formed by interleaving the even and odd sub-sequences, $y_0^{\{l\}}[n]$ and $y_1^{\{l\}}[n]$, produced by each lifting step. Since the input sequence, $x[k]$, is supported on $E \leq k < F$, the most reasonable policy is to define the $y^{\{l\}}[k]$ as

---

[17] To deduce the relevant relationships, observe firstly that in the even length case, $h_0^t[k] = h_0^t[-1-k]$ and $h_1^t[k] = -h_1^t[1-k]$ (see Theorem 6.4). Next, construct $\tilde{x}[k]$ to satisfy $\tilde{x}[E+k] = \tilde{x}[E-1-k]$ and $\tilde{x}[F+k] = \tilde{x}[F-1-k]$. The subband symmetries may then be deduced by following essentially the same procedure as that used in the odd length case. Note, however, that the low- and high-pass subband samples need to be treated separately. Also, one encounters a complication at the left boundary when E is odd. The problem can be resolved by changing the delays of $h_0^t[k]$ and $h_1^t[k]$, but then complications arise at the right boundary for odd values of F.

sequences supported over the same interval. Equivalently, $y_0^{\{l\}}[n]$ is supported over $E \leq 2n < F$ and $y_1^{\{l\}}[n]$ is supported over $E \leq 2n+1 < F$. This policy has the agreeable implication that the final interleaved sequence of subband samples, $y[k]$, will have the same region of support as the input sequence.

To operate on these finite sequences, the lifting procedure can be modified in an obvious manner near the boundaries. In particular, each filter, $\lambda_l[k]$, may be applied to an extended version, $\breve{y}_{1-p(l)}^{\{l-1\}}[k]$, of the finite sequence, $y_{1-p(l)}^{\{l-1\}}[k]$, where any arbitrary extension may be employed without sacrificing the invertibility of the system, so long as the same extension procedure is applied during synthesis. The extension procedure may equivalently be defined on the interleaved sequences, $\breve{y}^{\{l\}}[k]$.

SYMMETRIC EXTENSION

In the particular case of a delay-normalized FIR subband transform with linear phase odd length filters, the symmetric extension procedure of Section 6.5.2 may be applied directly in the lifting domain, in the manner described above. To see this, recall that the subband transform may be factored into lifting steps satisfying the symmetry conditions of equation (6.55); in fact, these are the efficient factorizations in the sense that they minimize the total number of lifting steps. Adopting such a lifting factorization, it is easy to see that each individual lifting step represents a "single stage subband transform" which takes the extended interleaved sequence $\breve{y}^{\{l-1\}}[k]$ as its input and produces the interleaved sequence, $\breve{y}^{\{l\}}[k]$. Moreover, this single stage is itself a delay-normalized transform with linear phase odd length filters.

The arguments advanced in Section 6.58 indicate that the interleaved sequences, $\breve{y}^{\{l\}}[k]$, must all satisfy the symmetry conditions of equations (6.58a) and (6.58b). The symmetry conditions for the individual sub-sequences, $\breve{y}_0^{\{l\}}[n]$ and $\breve{y}_1^{\{l\}}[n]$, are trivially derived from those for $\breve{y}^{\{l\}}[k]$, but the expressions are less elegant. In a practical implementation, the symmetric extension procedure need not be applied explicitly. Instead, the convolution operation of equation (6.48) can simply be modified at the boundaries of the finite sequence, $y_{1-p(l)}^{\{l-1\}}[n]$.

In the special case of 2-tap symmetric lifting filters, symmetric extension in the lifting domain requires the synthesis of at most one sample beyond the boundary. Within each lifting step then, the extension operation may be implemented simply by replicating the boundary samples of the sub-sequence which is being filtered. The state-based lifting analysis algorithm on Page 295 may then be completed as shown below. In this algorithm, the temporary variables, A and Σ_{new}, as well as the state variables, Σ_l, are each associated with an "existence flag," which indicates whether or not the corresponding sample position lies within the region of support, $E \leq n < F$. The **exists()** operator identifies whether or not this flag is set and the **assign()** operator marks the relevant variable as non-existent whenever an attempt is made to assign a sample value which lies outside the region of support.

Complete Lifting Analysis Algorithm

Mark all state variables, Σ_l, as non-existent.

For $n = \left\lfloor \frac{E}{2} \right\rfloor, \ldots, \left\lceil \frac{F+L}{2} \right\rceil$,

     assign($x[2n]$ to Σ_{new}) (marks Σ_{new} as non-existent if $2n \notin [E, F)$)

     assign($x[2n-1]$ to A) (marks A as non-existent if $2n - 1 \notin [E, F)$)

     For $l = 1, 2, \ldots, L$,

         if **exists**(A) and **exists**(Σ_{new}) and **exists**(Σ_l)

             $A \leftarrow A + \lambda_l (\Sigma_{\text{new}} + \Sigma_l)$

         else if **exists**(A) and **exists**(Σ_{new})

             $A \leftarrow A + 2\lambda_l \Sigma_{\text{new}}$ (implements symmetric extension at left)

         else if **exists**(A)

             $A \leftarrow A + 2\lambda_l \Sigma_l$ (implements symmetric extension at right)

         $A \leftarrow \Sigma_l \leftarrow \Sigma_{\text{new}} \leftarrow A$

             (rotates values and existence states of $A, \Sigma_l, \Sigma_{\text{new}}$ to the left)

     if **exists**(A)

         $y[2n - L - 1] \leftarrow A$

     if **exists**(Σ_{new})

         $y[2n - L] \leftarrow \Sigma_{\text{new}}$

6.6 FURTHER READING

An early tutorial on subband transforms is that by Vaidyanathan [158], which still makes good introductory reading. Likewise, the classic paper by Mallat [98] provides an excellent introduction to the wavelet transform. For more in depth treatment of these subjects, the book by Vetterli and Kovačević [162] and that by Strang and Nguyen [143] are to be recommended, while the book by Vaidyanathan [159] is a comprehensive reference on subband transforms and filter banks in general. The material on lifting and reversible transforms in Section 6.4 may not be found in those references due to its more recent advent.

We have deliberately limited the scope of the material presented in this chapter in accordance with the types of transforms which have been found most effective and/or most practical for image compression. In particular, there is no discussion whatsoever of non-separable subband and wavelet transforms, or transforms involving recursive, IIR filters. Moreover, our treatment exhibits a decided bias toward two channel subband transforms having odd length linear phase filters.

Chapter 7

ZERO-TREE CODING

Embedded zero-tree coding of wavelet coefficients (EZW) was introduced by Shapiro [132]. At that time, it produced state-of-the-art compression performance at a relatively modest level of complexity. The bit-stream produced by EZW is also embedded. Every prefix of the compressed bit-stream is itself a compressed bit-stream, but at a lower rate (quality). As we will see, EZW achieves its embedding via binary bit-plane coding of deadzone scalar quantizer indices.

Embedding via bit-plane coding had been previously studied, and even included as part of the original JPEG standard [119]. However, most previous "mainstream" bit-plane coding systems followed a fixed scan pattern. For example, [121] describes a raster scan of bit-planes of image sample data, while the progressive mode of JPEG can employ a "zig zag" scan of DCT coefficient bit-planes [119]. In contrast, the bit-plane coding of EZW allows data dependent departures from a raster scan of transform coefficients. These departures are in the form of "zero-trees," which allow for the coding of large numbers of zeros using very few compressed bits.

The state-of-the-art MSE performance of EZW, together with its modest complexity and embedded bit-stream captured the interest of the compression research community, and a tremendous research effort was ignited by its publication. In [126], Shapiro's zero-trees were generalized, and set partitioning techniques were introduced to effectively code these generalized trees of zeros. The resulting technique is known as set partitioning in hierarchical trees (SPIHT — pronounced "spite"). The SPIHT algorithm produces results superior to those of EZW at even lower levels of complexity. Example software for SPIHT was made available by the authors, and SPIHT quickly became the yardstick by which all image compression methods are measured.

7.1   GENEALOGY OF SUBBAND COEFFICIENTS

The EZW algorithm was proposed for use with hierarchical subband transforms as developed in Chapters 4 and 6. As discussed in those chapters, the basic premise of all transform coding systems is that most of the image energy is compacted into only a few transform coefficients. For dyadic subband (or wavelet) transforms, the highest energy coefficients reside in the lowest frequency subbands. As can be seen in Figure 4.22, subband LL_2 consists of coefficients having a large dynamic range (in fact, LL_2 looks a lot like the original image). As we look at higher frequency subbands, we see that few coefficients differ significantly from zero.[1]

In addition to the energy compaction property of transform coding, zero-tree coding is founded on the premise that if a coefficient is small in magnitude, then coefficients corresponding to the same spatial location (in corresponding higher frequency subbands) will also tend to be small. For example, examining subband HL_2 of Figure 4.22, we see a large region of near-zero coefficients within the left portion of the bicycle wheel. Examining subband HL_1 shows a corresponding region of near-zero coefficients. Examining the LH and HH subbands reveals similar spatially commensurate regions.

In an attempt to exploit the dependencies embodied in these "replicated" regions of zeros, the EZW algorithm employs a parent-child relationship among the coefficients from subbands of the same orientation. For example, each coefficient in subband HL_d is considered to be the parent of 4 children in subband HL_{d-1}. Parents and children are defined similarly for the LH and HH orientations. These relationships hold for each $d = 2, \ldots, D$ where D is the number of levels in the transform. The coefficients at the highest frequency bands ($d = 1$) have no children. The coefficients in the lowest frequency subband, LL_D each have only 3 children, comprised of the single coefficient at the same spatial location in HL_D, LH_D, and HH_D.

The "descendants" of a coefficient consist of all its children, grandchildren, great-grandchildren, etc. In general, a coefficient in subband HL_d, LH_d, or HH_d has 4 children, 16 grandchildren, 64 great-grandchildren,

---

[1]Note that for all subbands other than LL_2 in Figure 4.22, zero is represented by a medium shade of gray, while darker and lighter shades represent negative and positive coefficients, respectively.

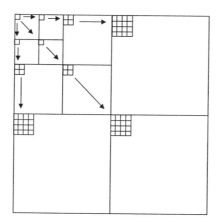

Figure 7.1. EZW descendant structure for one sample in the LL₃ subband.

etc. Thus, each such coefficient has a total number of descendants given by

$$4 + 4^2 + \ldots + 4^{d-1} = \frac{4^d - 4}{3}$$

Similarly, each coefficient in LL_D has 3 children, 12 grandchildren, 48 great-grandchildren, etc., for a total of $4^D - 1$ descendants. All descendants of the $[0,0]$ coefficient of LL_D are shown in Figure 7.1, for $D = 3$.

7.2   SIGNIFICANCE OF SUBBAND COEFFICIENTS

In the EZW algorithm, the sample mean of all image samples is subtracted from each sample prior to performing the wavelet transform. This results in all subbands having a mean of zero. Since the high-pass filter taps sum to zero, all subbands except LL_D are zero-mean with or without the subtraction. Once the level shifting has been performed, the premise of zero-tree coding is that descendants of small coefficients tend to also be small. This idea is made more precise in [132] by comparing coefficients to a series of thresholds, T_k $k = 0, 1, \ldots, K - 1$.

Assuming an orthonormal transform, the zero-tree premise is that coefficients that are less than T_k (in magnitude) tend to have descendants that are less than T_k (in magnitude). In [132] only thresholds of the form $T_k = T_0 2^{-k}$ are considered, where T_0 is chosen to satisfy

$$\frac{|y|_{\max}}{2} < T_0 \leq |y|_{\max} \tag{7.1}$$

Here, $|y|_{\max}$ is taken as the maximum magnitude over all coefficients in all subbands of the transformed image.

A particular coefficient $y = y\,[i,j]$ is said to be significant with respect to T_k if $|y| \geq T_k$, or equivalently, if

$$\frac{|y|}{T_{K-1}} \geq \frac{T_k}{T_{K-1}}$$

Noting that $T_k/T_{K-1} = 2^{K-k-1}$ is an integer, we then have that y is significant with respect to T_k if and only if

$$|q| \triangleq \left\lfloor \frac{|y|}{T_{K-1}} \right\rfloor \geq 2^{K-k-1} \tag{7.2}$$

Comparing equation (7.2) to equation (3.34), we see that $|q|$ is the magnitude of the index resulting from deadzone scalar quantization of y using a step size of $T_{K-1} = 2^{-(K-1)}T_0$.

Furthermore, from the left hand side of equation (7.1), we see that

$$\left\lfloor \frac{|y|}{T_{K-1}} \right\rfloor = \left\lfloor \frac{|y|}{2^{-(K-1)}T_0} \right\rfloor < 2^K$$

So that K bits are sufficient to represent $|q|$. As in equation (3.36), we let the binary representation of q be given by

$$q = s, q_0, q_1, \ldots, q_{K-1} \tag{7.3}$$

From equation (7.2) we conclude that y is significant with respect to T_k if and only if the magnitude of its quantization index is significant with respect to 2^{K-k-1}, or equivalently, if and only if at least one of $q_0, q_1, \ldots, q_k$ is 1.

In what follows, we will sometimes refer to q (or equivalently y) becoming significant in bit k'. By this, we mean that $q_{k'} = 1$ and $q_k = 0$ $\forall k < k'$. We will sometimes call this first 1 bit $(q_{k'})$ the "significance bit" of q. All bits following the significance bit (i.e., q_k $k > k'$) are called "refinement bits."

7.3   EZW

Having established the equivalence of threshold comparisons to deadzone scalar quantization, we now describe the EZW algorithm as a form of entropy coding applied to bit-planes of scalar quantizer indices. Although this description is quite different from that in [132] it leads to a more unified and cohesive discussion of related bit-plane coding techniques.

Recall from Chapter 3, that for entropy coded scalar quantization at high rates, uniform quantization is optimal, and incurs (for step size Δ) a MSE of $\Delta^2/12$. Recall also from Chapter 5, that (at least for high rates and perfectly decorrelated Gaussian data) each coefficient of an orthonormal transform should contribute equally to overall image MSE. Thus, for such transforms, the same step size should be employed for every coefficient of every subband.[2]

For simplicity, we assume an original image of size $N \times N$ where $N = 2^n$ for some integer $n \geq D$. We assume that this image is transformed with D levels of a dyadic tree-structured subband transform, normalized to be (at least approximately) orthonormal.[3] This transform yields $3D+1$ subbands of transform coefficients. Each of the coefficients is then replaced by the index (in sign-magnitude form) resulting from uniform deadzone quantization with step size $T_{K-1} = T_0 2^{-(K-1)}$, where T_0 is selected to satisfy equation (7.1).

The resulting $N \times N$ array of (integer) indices is then "sliced" into $K + 1$ binary arrays (or bit-planes). The first such bit-plane consists of the sign bit of each index, denoted by $s[i,j]$ $i,j = 0,1,\ldots,N-1$. The next bit-plane consists of the MSBs of each magnitude, denoted by $q_0[i,j]$ $i,j = 0,1,\ldots,N-1$. The next bit-plane contains $q_1[i,j]$, and so on, until the final bit-plane contains $q_{K-1}[i,j]$ $i,j = 0,1,\ldots,N-1$. The EZW algorithm then codes all bits of $q_0[i,j]$ followed by all bits of $q_1[i,j]$, and so on. The sign bit of a particular index is coded together with the significance bit of that index. That is, $s[i,j]$ is coded together with $q_{k'}[i,j]$, where $q_{k'}[i,j] = 1$ and $q_k[i,j] = 0$ $\forall k < k'$.

The primary differences between zero-tree coding and previous bit-plane coding schemes, are the order in which the bits within a plane are coded, and the clever extension of "zero run" coding (see Section 2.4.3) to zero-trees of wavelet coefficients.

Each of the K magnitude bit-planes is coded in two passes. The first pass codes all refinement bits, while the second pass codes everything else (all bits not coded in the first pass). More specifically, the first pass (which we call the "refinement pass") codes a bit for each coefficient that is significant (due to its significance bit having been coded in a previous bit-plane). The second pass (which we call the "significance pass") codes a bit for each coefficient that is not yet significant. The name *refinement*

---

[2] We note that many of the assumptions used to arrive at this conclusion do not necessarily hold in real imagery. However, the result is simple, elegant, and works well in practice. More sophisticated techniques are addressed in Chapter 5.

[3] If the transform is not orthonormal, then the step sizes (thresholds) should be adjusted via the synthesis weights G_b as discussed in Section 5.3. Adjustments for HVS weights, W_b, can also be performed if desired.

pass derives from the fact that all bits coded in this pass are *refinement bits* as defined above. The name *significance pass* is derived from the fact that a 1 coded in this pass indicates a coefficient becoming newly significant. When the first bit-plane (i.e., $q_0 = q_0 [\cdot, \cdot]$) is coded, there are no significant coefficients yet. Thus the refinement pass for q_0 is skipped (it can also be thought of as being present, but empty).

In [132], the refinement pass was called the "subordinate pass," while the significance pass was called the "dominant pass." This terminology stemmed from the fact that the significance (dominant) pass is the first to result in coded data, and determines which coefficients are visited in the refinement (subordinate) pass of subsequent bit-planes.

7.3.1   THE SIGNIFICANCE PASS

In the significance pass, each insignificant coefficient is visited in raster order (left-to-right, top-to-bottom), first within LL_D, then within HL_D, then LH_D, then HH_D, then HL_{D-1}, and so on, up to and including HH_1. Coding is accomplished via a 4-ary alphabet:

1. POS = Significant Positive. This symbol is equivalent to a 1 followed immediately by a corresponding "positive" sign bit.

2. NEG = Significant Negative. This symbol is equivalent to a 1 followed immediately by a corresponding "negative" sign bit.

3. ZTR = Zero Tree Root. This symbol indicates that the current bit of a particular coefficient is 0, and that the corresponding bit in each of its descendants is also 0.

4. IZ = Isolated Zero. This symbol indicates that the bit is 0, but at least one descendant has a corresponding 1 bit.

As mentioned previously, the three highest frequency subbands HL_1, LH_1, and HH_1 have no children. Thus, when coding bits from these subbands, the ZTR and IZ symbols are replaced by the single symbol Z.

As the scan of insignificant coefficients progresses through subbands, any bit known already to be zero (by virtue of inclusion in a zero-tree from a previous subband) is not coded again. Also, for the purpose of determining if a bit is a zero-tree root, only insignificant descendants are examined. Equivalently, (refinement) bits of significant descendants are treated as if 0 for the purpose of ZTR formation.

7.3.2   THE REFINEMENT PASS

In the refinement pass, a refinement bit is coded for each significant coefficient. A coefficient is significant if it has been coded POS or NEG

63	-34	49	10	7	13	-12	7
-31	23	14	-13	3	4	6	-1
15	14	3	-12	5	-7	3	9
-9	-7	-14	8	4	-2	3	2
-5	9	-1	47	4	6	-2	2
3	0	-3	2	3	-2	0	4
2	-3	6	-4	3	6	3	6
5	11	5	6	0	3	-4	4

Figure 7.2. Three level dyadic subband decomposition of an 8×8 image.

in a previous bit-plane. Its current refinement bit is simply its corresponding bit in the current bit-plane. The order in which significant coefficients are visited during the refinement pass is: 1st by magnitude (as best as can be determined by previously coded bits), 2nd by raster within subbands in order LL_D, HL_D, LH_D, HH_D, $HL_{D-1}, \ldots, HH_1$.

More specifically, when coding bit-plane $q_k = q_k [\cdot, \cdot]$, k magnitude bits have been previously coded for each coefficient. Equivalently $p = K - k$ magnitude bits remain uncoded. Interpreting the k coded magnitude bits as an unsigned integer $\left| q^{(p)} [i, j] \right|$, all coefficients having $\left| q^{(p)} [i, j] \right| = 2^k - 1$ are visited in raster order by subband, followed by all coefficients having $\left| q^{(p)} [i, j] \right| = 2^k - 2$ in raster order by subband, and so on, until all coefficients having $\left| q^{(p)} [i, j] \right| = 1$ are visited. Note that coefficients with $\left| q^{(p)} [i, j] \right| = 0$ are not visited as they are still insignificant, and thus will be coded during the significance pass. With this description, the refinement pass can thus by thought of as $2^k - 1$ subpasses, or as a single pass through a sorted list of significant coefficients.

Example 7.1 *Consider the subband coefficients shown in Figure 7.2. We choose a quantizer step size of $\Delta = T_{K-1} = 1$, so that the quantizer indices are also as shown in this figure. This choice yields $K = 6$ magnitude bit-planes. The first three of these bit-planes (q_0, q_1, q_2) are shown along with the sign plane, s, in Figure 7.3.*

Coding begins with a significance pass through q_0. The single bit in the LL_3 subband of q_0 has a value of 1. Its corresponding sign is +, so the first coded symbol is POS. Similarly, the single bit in HL_3 of the q_0 bit-plane is coded as NEG. The single bit in LH_3 is coded as IZ, since it is zero but has a non-zero descendant in LH_1 of the q_0 bit-plane. The bit in HH_3 is coded as ZTR indicating that it and all its descendants are 0.

s

+	-	+	+	+	+	-	+
-	+	+	-	+	+	+	-
+	+	+	-	+	-	+	+
-	-	-	+	+	-	+	+
-	+	-	+	+	+	-	+
+	+	-	+	+	-	+	+
+	-	+	-	+	+	+	+
+	+	+	+	+	+	-	+

q_0

1	1	1	0	0	0	0	0
0	0	0	0	0	0	0	0
0	0	0	0	0	0	0	0
0	0	0	0	0	0	0	0
0	0	0	1	0	0	0	0
0	0	0	0	0	0	0	0
0	0	0	0	0	0	0	0
0	0	0	0	0	0	0	0

q_1

1	0	1	0	0	0	0	0
1	1	0	0	0	0	0	0
0	0	0	0	0	0	0	0
0	0	0	0	0	0	0	0
0	0	0	0	0	0	0	0
0	0	0	0	0	0	0	0
0	0	0	0	0	0	0	0
0	0	0	0	0	0	0	0

q_2

1	0	0	1	0	1	1	0
1	0	1	1	0	0	0	0
1	1	0	1	0	0	0	1
1	0	1	1	0	0	0	0
0	1	0	1	0	0	0	0
0	0	0	0	0	0	0	0
0	0	0	0	0	0	0	0
0	1	0	0	0	0	0	0

Figure 7.3. Sign plane and first three magnitude bitplanes for the quantized coefficients of Figure 7.2.

Note that no further coding of HH_2 and HH_1 will be done for bit-plane q_0.

Proceeding to HL_2, the first bit is coded POS, while the remaining three bits are coded with three ZTR symbols. Similarly, LH_2 is coded as ZTR, IZ, ZTR, ZTR. Again, HH_2 has already been coded as "all zero," and is skipped. Now in HL_1, twelve of sixteen bits have already been coded as zero, by virtue of zero-trees emanating from HL_2. The four remaining bits are coded Z, Z, Z, Z. Similarly, there are only four uncoded bits in LH_1 which are coded Z, POS, Z, Z. Thus, the code symbols for the significance pass of q_0 are given by

$$POS,NEG,IZ,ZTR,$$
$$POS,ZTR,ZTR,ZTR,ZTR,IZ,ZTR,ZTR,$$
$$Z,Z,Z,Z,Z,POS,Z,Z$$

The string of code symbols above is broken into three lines for clarity of the text. The first line contains all symbols from LL_3, HL_3, LH_3, and HH_3. The second line contains all symbols from HL_2, LH_2, and HH_2. Finally, the third line contains all symbols from HL_1, LH_1, and HH_1.

By virtue of the q_0 significance pass, there are now four coefficients known to be significant. Specifically, $q[0,0], q[0,1], q[0,2]$, and $q[4,3]$ are significant. The q_1 bit for each of these coefficients is coded in the q_1 refinement pass. Since only one bit-plane (of six) has been coded, $\left|q^{(p)}\right| = \left|q^{(K-1)}\right| = \left|q^{(5)}\right| = 1$ for each of these four coefficients, and no sorting is required. The code symbols for this pass are thus

$$q_1[0,0], q_1[0,1], q_1[0,2], q_1[4,3] = 1,0,1,0 \qquad (7.4)$$

The significance pass for q_1 then codes all remaining bits in the q_1 bit-plane. The code symbols resulting from this pass are given by

NEG,POS,

ZTR,ZTR,ZTR,ZTR,ZTR,ZTR,ZTR,ZTR,ZTR,ZTR,ZTR,

Z,Z,Z,Z

As before, the code symbols are broken into three lines according to transform level. Recall that when testing for the presence of zero-trees, significant coefficients are treated as if 0 (whether they are or not). Thus, LH_2 would still be coded using four ZTR symbols even if the refinement bit at $q_1[4,3]$ were 1.

The q_1 significance pass has added two more significant coefficients ($q[1,0]$ and $q[1,1]$). One bit must be coded for each of the (now six) significant coefficients during the q_2 refinement pass. First, all coefficients with $\left|q^{(4)}\right| = 3$ are visited, followed by those with $\left|q^{(4)}\right| = 2$, and finally by those with $\left|q^{(4)}\right| = 1$. The symbols coded in the refinement pass are thus

$$q_2[0,0], q_2[0,2], q_2[0,1], q_2[4,3], q_2[1,0], q_2[1,1]$$
$$= 1,0,0,1,1,0 \qquad (7.5)$$

Coding continues with the q_2 significance pass, the q_3 refinement pass, and so on.

7.3.3   ARITHMETIC CODING OF EZW SYMBOLS

Context dependent arithmetic coding is used to losslessly compress the sequences of symbols resulting from the procedures discussed above. Shapiro employed the arithmetic coder from [171]. This arithmetic coder codes M-ary symbols directly (without binarization) and employs scaled count probability model adaptation (Section 2.3.5). For a given context, counts are maintained for each symbol. As each symbol is coded, its corresponding counter is incremented. When the sum of the counters for the context reaches a given maximum count value, each counter is incremented and divided by 2.

EZW employs five contexts, each with a maximum count value of 255. All counts in all contexts are initialized to 1. A single context is used in the refinement pass, while four contexts are used in the significance pass. Which of the four context labels is to be used for a particular coefficient is determined by the significance states of two coefficients. These two coefficients are the neighbor immediately previous in the nominal scan order, and the parent coefficient.

Encoding can halt at any point. Reasonable stopping criteria include achieved MSE, or achieved rate. Stopping via achieved rate is particularly simple. For a desired *encoding* rate of R bits/sample, the algorithm is halted when $N^2 R$ arithmetic coded bits have been produced. Similarly, *decoding* can be halted when $N^2 R'$ bits have been consumed from the compressed bit-stream for any $R' \leq R$. Also, the compressed bit-stream can be "truncated" by discarding all but the first $N^2 R'$ bits, to yield a compressed bit-stream at rate $R' \leq R$. Clearly, this truncated bit-stream is identical to that as if encoding were halted at R' bits/sample in the first place. In this way, the prefix of every EZW bit-stream is an EZW bit-stream itself.

With such partial decoding, the number of bits available for inverse quantization can vary coefficient-by-coefficient. For a coefficient with k bits decoded ($p = K - k$ bits missing), dequantization is performed using equation (3.37). For example, consider decoding the bit-stream of Example 7.1 to the end of the q_1 refinement pass (equation (7.4)). The four significant coefficients have each been decoded to a depth of 2 bits to get

$$q^{(4)}[0,0], q^{(4)}[0,1], q^{(4)}[0,2], q^{(4)}[4,3], = 3, -2, 3, 2$$

respectively. These coefficients are dequantized using a step size of $2^4 = 16$ to get[4]

$$\hat{y}[0,0], \hat{y}[0,1], \hat{y}[0,2], \hat{y}[4,3],$$
$$= 56, -40, 56, 40$$

respectively. All insignificant coefficients have all their decoded bits equal to 0 and thus, dequantize to 0. In this case, the MSE of the dequantized coefficients is given by

$$MSE = \frac{1}{N^2} \sum_i \sum_j (y[i,j] - \hat{y}[i,j])^2$$
$$\cong 68.6 \tag{7.6}$$

As another example, consider decoding through the first half of the q_2 refinement pass. The first three significant coefficients (recall the order of equation (7.5)) have $k = 3$ magnitude bits decoded, while the remaining three have only $k = 2$ magnitude bits decoded. Thus the first three coefficients are dequantized using a step size of $2^{6-3} = 8$ to get 60, 52, and -36, respectively. The remaining three coefficients are decoded

---

[4]For simplicity, we let $\delta = 1/2$ in equation (3.37).

using a step size of $2^{6-2} = 16$ to get 40, -24, and 24, respectively. As before, all other coefficients dequantize to zero. We see that decoding an extra pass and half has updated the estimates of five coefficients ($\hat{y}[4, 3]$ remained unchanged). Also, the MSE of equation (7.6) has been lowered to a value of 44.4.

In both cases, the inverse transform follows inverse quantization to obtain the corresponding decompressed image $\hat{x}[\cdot, \cdot]$. Due to our assumption of orthonormality, the MSEs of the decompressed images will be the same as those computed for the transform coefficients above. Specifically,

$$MSE = \frac{1}{N^2} \sum_i \sum_j (x[i, j] - \hat{x}[i, j])^2$$

$$= \frac{1}{N^2} \sum_i \sum_j (y[i, j] - \hat{y}[i, j])^2$$

7.4   SPIHT

As mentioned at the beginning of this chapter, the SPIHT algorithm was motivated by, and has several features in common with, the EZW algorithm. In this section, we employ many of the assumptions from the previous section. Specifically, we describe the SPIHT algorithm in terms of bit-plane coding of signed indices arising from deadzone SQ of subband transform coefficients. As before, we assume an orthonormal transform, and a single step size $T_{K-1} = T_0 2^{-(K-1)}$, with T_0 satisfying equation (7.1). We also assume D levels of a dyadic orthonormal transform are applied to an $N \times N$ image. For simplicity we assume $N = 2^n$ for some integer $n \geq D + 1$.

Although the SPIHT algorithm has many similarities to the EZW algorithm, there are also several significant differences. In particular, the order of the significance and refinement passes is reversed, the parent-child relationship in the LL_D subband is altered, there are two types of zero-trees, the coding order is driven more by the significance of previously coded trees (i.e., SPIHT is less raster-based), more inter-bit-plane memory is exploited (via sorted lists of trees), and all SPIHT output symbols are binary.

7.4.1   THE GENEALOGY OF SPIHT

As mentioned above, the parent-child relationships for SPIHT are the same as those for EZW except for the LL_D subband. Specifically, each coefficient in HL_d, LH_d, or HH_d $d = 2, 3, \ldots, D$ has 4 children, while each coefficient in HL_1, LH_1, or HH_1 has 0 children, as was the case for EZW.

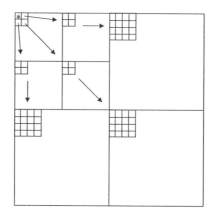

Figure 7.4.   Parent-child relationships in SPIHT.

However in SPIHT, one fourth of the coefficients in LL_D have no children, while the remaining coefficients each have four children. Specifically, coefficients in LL_D with even vertical and horizontal coordinates have no children, while all other coefficients in LL_D have four children. Figure 7.4 depicts this situation for a group of four coefficients in the LL_2 subband. The coefficient marked with an asterisk has no children.

7.4.2   ZERO-TREES IN SPIHT

In EZW, a zero-tree is defined by a "root" coefficient and its descendants *all* having a value of 0 within a bit-plane. Such zero-trees are used to signal the event that all insignificant coefficients (within the footprint of the zero-tree) remain insignificant. SPIHT can be thought of as employing two types of zero-trees. The first type consists of a single root coefficient having all descendants 0 within a given bit-plane. This differs from the EZW zero-tree in that the root itself need not be zero. In fact, although the zero-tree is specified by the coordinates of the root, the root itself is not included in the tree. The second type of zero-tree is similar but also excludes the four children of the root.

These two types of zero-trees are used to signal the continued existence (bit-plane-by-bit-plane) of insignificant sets of coefficients of Type A and Type B, respectively. Specifically, if all descendants of a given root coefficient are insignificant, they comprise an insignificant set of Type A. Type B insignificant sets are similar but do not contain the children of the root. That is, Type B insignificant sets contain only grandchildren, great grandchildren, etc.[5]

7.4.3 LISTS IN SPIHT

SPIHT is best explained using three ordered lists:

1. List of significant coefficients.

2. List of insignificant coefficients.

3. List of insignificant sets of coefficients.

As the name suggests, the list of significant coefficients (LSC) contains the coordinates of all coefficients that are significant. The list of insignificant sets of coefficients (LIS) contains the coordinates of the roots of insignificant sets of coefficients (of Type A, or Type B). Finally, the list of insignificant coefficients (LIC) contains a list of the coordinates of all coefficients that are insignificant, but do not reside within one of the two types of insignificant sets.

7.4.4 THE CODING PASSES

As mentioned previously, the order of significance and refinement passes are reversed from that of EZW. Thus, each bit-plane is coded by a significance pass (called a sorting pass in [126]), followed by a refinement pass. As in EZW, the refinement pass codes a refinement bit for each coefficient that was significant at the end of the *previous* bit-plane. In particular, coefficients that became significant via the significance pass of the *current* bit-plane are not refined until the next bit-plane.

Prior to the significance pass for bit-plane q_0, initialization begins by adding each coefficient in LL_D to the LIC. The coordinates of each coefficient in LL_D with descendants are then added to the LIS as roots of insignificant sets of Type A. In this way, every coefficient in every subband is initialized to the insignificant state.[6] The LSC is initially empty, but becomes populated as significant coefficients are identified.

The q_0 bit of each coefficient in the LIC is coded first. If any bit is coded as 1, its corresponding sign bit is coded immediately thereafter,

---

[5]It is worth noting that a Type B set can be thought of as the union of four Type A sets. This can be seen by noting that the root of a Type B set has four children each of which has only insignificant descendants.

[6]Note that a Type A set indicates that all descendants of its root are insignificant. It says nothing about the root itself. Thus, the root of a Type A set must belong to either the LIC or LSC. Similarly, the root of a Type B set (and the four children of the root) must each belong (individually) to either the LIC or LSC.

and the coefficient is moved to the LSC. Next, each set in the LIS is examined, in order of appearance in the list.

For a given set in the LIS, the coding proceeds as follows. If each coefficient in the set remains insignificant (all bits of q_0 in the appropriate tree are zero), a single 0 is coded and processing proceeds to the next set in the LIS. Otherwise, a 1 is coded. If this occurs, one of the following procedures is executed: 1) If the set is of Type A, it is changed to Type B and sent to the bottom of the LIS.[7] The q_0 bit of each child is coded (with any required sign bit). The child is then sent to the end of the LIC or LSC, as appropriate; 2) If the set is of Type B, it is deleted from the LIS, and each child is added to the end of the LIS as a set of Type A.

Processing continues in this fashion until the end of the LIS is reached. This specifically includes new entries added to the end of the LIS during the current pass. In this way, a magnitude bit eventually gets coded (in the current pass) for every coefficient of q_0. At the completion of the significance pass, the following refinement pass is empty (or skipped) as there were no significant coefficients at the beginning of the q_0 bit-plane.

The significance pass for q_1 proceeds in the same fashion as the q_0 significance pass, using the lists as they were at the end of the q_0 bit-plane. This results in coding of the q_1 bits for all insignificant coefficients (all coefficients indicated in the LIC and LIS). The q_1 refinement pass is then executed using the LSC as it was at the end of the q_0 bit-plane. This procedure is continued bit-plane-by-bit-plane until a target encoding rate is achieved, or all magnitude bit-planes, q_k $k = 0, \ldots, K - 1$ have been processed.

7.4.5   THE SPIHT ALGORITHM

The SPIHT algorithm can be described more precisely using the following notation. For a given coefficient at location $[i, j]$ let $\mathcal{C}[i, j]$ be the set of its children, let $\mathcal{D}[i, j]$ be the set of its descendants, and let $\mathcal{G}[i, j]$ be the set of its grandchildren, great grandchildren, etc. (i.e., $\mathcal{G}[i, j] = \mathcal{D}[i, j] - \mathcal{C}[i, j]$). Let $S_k(\cdot)$ be a mapping from any set of coefficients to $\{0, 1\}$. Specifically, if $\mathcal{B}$ is a set of coefficients, $S_k(\mathcal{B}) = 0$ if every coefficient in $\mathcal{B}$ has $q_k = 0$. Otherwise, at least one coefficient in $\mathcal{B}$ has $q_k = 1$, and $S_k(\mathcal{B}) = 1$. The SPIHT algorithm is then defined as follows:

The SPIHT Algorithm

---

[7]If when changed to Type B, the set is empty (i.e., the root has no grandchildren), then the entry is deleted from the list.

0) Initialization

- Set $k = 0$, LSC $= \phi$, LIC $= \{$all coordinates $[i, j]$ of coefficients in LL$_D\}$, LIS $= \{$all coordinates of coefficients from LIC that have children$\}$. Set all entries of the LIS to Type A.

1) Significance Pass

- For each $[i, j] \in$ LIC, do:
  - Output $q_k[i, j]$. If $q_k[i, j] = 1$, output $s[i, j]$ and move $[i, j]$ to the end of the LSC.
- For each $[i, j]$ in the LIS do:
  - If the set is of Type A, output $S_k(\mathcal{D}[i, j])$. If $S_k(\mathcal{D}[i, j]) = 1$, then
    * For each $[l, m] \in \mathcal{C}[i, j]$, output $q_k[l, m]$. If $q_k[l, m] = 0$, add $[l, m]$ to the LIC. Else, output $s[l, m]$ and add $[l, m]$ to the LSC.
    * If $\mathcal{G}[i, j] \neq \phi$, move $[i, j]$ to the end of the LIS as a set of Type B. Else delete $[i, j]$ from the LIS.
  - If the set is of Type B, output $S_k(\mathcal{G}[i, j])$. If $S_k(\mathcal{G}[i, j]) = 1$, then add each $[l, m] \in \mathcal{C}[i, j]$ to the end of the LIS (as sets of Type A) and delete $[i, j]$ from the LIS.

2) Refinement Pass

- For each $[i, j] \in$ LSC, output $q_k[i, j]$. For this step, the LSC as it was before the most recent significance pass should be used. That is, coefficients that were added to the LSC in the most recent significance pass should not be refined.

3) Set $k = k + 1$ and go to Step 1).

As mentioned previously, Step 1) proceeds until all sets in the LIS (including the new ones being added) have been processed.

Example 7.2 *The sign plane and first three magnitude bit-planes for a $D = 2$ level dyadic orthonormal transform are shown in Figure 7.5. For this case, the lists are initialized to*

LSC	LIC	LIS
	$[0, 0]$	$[0, 1]$ A
	$[0, 1]$	$[1, 0]$ A
	$[1, 0]$	$[1, 1]$ A
	$[1, 1]$	

s

+	-	+	+	+	-	+	+
-	+	+	+	+	+	+	+
+	+	+	+	+	+	+	+
+	+	+	-	+	+	-	+
-	+	+	+	+	+	+	+
+	-	+	+	+	+	+	+
+	+	+	+	+	+	-	+
+	+	+	+	+	+	+	-

q_0

1	1	0	0	0	0	0	0
0	1	0	0	0	0	0	0
0	0	0	0	0	0	0	0
1	0	0	0	0	0	0	0
0	0	0	0	0	0	0	0
0	0	0	0	0	0	0	0
0	0	0	0	0	0	0	0
0	0	0	0	0	0	0	0

q_1

1	0	0	0	0	0	0	0
1	0	0	1	0	0	0	0
1	0	0	0	0	0	0	0
0	0	0	0	0	0	1	0
0	0	0	0	0	0	0	0
0	0	0	0	0	0	0	0
0	0	0	0	0	0	0	0
0	0	0	0	0	0	0	0

q_2

1	1	1	0	1	0	0	0
1	1	0	0	0	1	0	0
0	0	0	0	0	0	0	1
1	0	0	1	0	0	1	0
0	0	0	0	0	0	0	0
0	0	0	0	0	0	0	0
0	0	0	0	0	0	0	0
0	0	0	0	0	0	0	0

Figure 7.5.   An example sign plane along with three magnitude bitplanes.

The q_0 significance pass begins by coding a bit (with signs, as appropriate) for each coefficient in the LIC. Thus, $q_0[0,0] = 1$ is output, followed immediately by $s[0,0] = +$. Similarly, $q_0[0,1] = 1$ is output, followed by $s[0,1] = -$. Next $q_0[1,0] = 0$ is output, followed by $q_0[1,1] = 1$, and its sign, $s[1,1] = +$. As a result of this coding, $[0,0], [0,1],$ and $[1,1]$ are all moved to the LSC.

Processing then proceeds to the LIS. The descendants of $[0,1]$ comprise all coefficients in HL_2 and HL_1. A single 0 is coded to convey the fact that all these coefficients have $q_0 = 0$ i.e., $S_0(\mathcal{D}[0,1]) = 0$. On the other hand $[1,0]$ has a descendant with $q_0 = 1$. Thus, a 1 is output, followed by a bit (with signs, as appropriate) for each coefficient in $\mathcal{C}[1,0] = \{[2,0],[2,1],[3,0],[3,1]\}$. The resulting output values are $0,0,1,+,0,$ respectively. $[2,0],[2,1],$ and $[3,1]$ are each added to the end of the LIC, while $[3,0]$ is added to the end of the LSC. $[1,0]$ is moved to the end of the LIS as an entry of Type B.

All descendants of $[1,1]$ have $q_0 = 0$ i.e., $S_0(\mathcal{D}([1,1])) = 0$, so a single 0 is output. The last remaining entry in the LIS at this point is $[1,0]B$. A single 0 is output to denote that all grandchildren, great grandchildren, etc., of $[1,0]$ have $q_0 = 0$, i.e., $S_0(\mathcal{G}[1,0]) = 0$. The coded values up to this point are

$$1, +, 1, -, 0, 1, +, 0, 1, 0, 0, 1, +, 0, 0, 0$$

and all values of $q_0[\cdot,\cdot]$ can be deduced from this string of symbols.[8]

As mentioned previously, the q_0 refinement pass is empty (as all values of q_0 have already been coded). Thus, processing proceeds to the q_1 significance pass. The initial lists for this pass are the final lists from

---

[8] We note that $\{+, -\}$ can be coded as $\{0, 1\}$ in a practical implementation. We leave them as $\{+, -\}$ here for clarity of the text.

the coding of the q_0 bit-plane. Specifically,

$$
\begin{array}{lll}
LSC & LIC & LIS \\
[0,0] & [1,0] & [0,1]\,A \\
[0,1] & [2,0] & [1,1]\,A \\
[1,1] & [2,1] & [1,0]\,B \\
[3,0] & [3,1] &
\end{array}
\tag{7.7}
$$

The coefficients from the LIC are coded as $1,-,1,+,0,0$ resulting in a move of $[1,0]$ and $[2,0]$ to the LSC. Processing then proceeds to the LIS, where $S_1\left(\mathcal{D}\left([0,1]\right)\right) = 1$ is output. This is followed by $0,0,0,1,+$ representing the four children of $[0,1]$. The first three children $\{[0,2],[0,3], [1,2]\}$ are added to the LIC, while the fourth child $[1,3]$ is added to the LSC. $[0,1]$ is then moved to the end of the LIS as an entry of Type B. The next two output bits are $0,0$ to denote that $S_1\left(\mathcal{D}\left[1,1\right]\right) = S_1\left(\mathcal{G}\left[1,0\right]\right) = 0$.

At this point the "last" entry at the bottom of the LIS is $[0,1]B$. A 1 is output, to indicate that $S_1\left(\mathcal{G}\left[0,1\right]\right) = 1$. $[0,1]$ is removed from the LIS, while its children $\{[0,2],[0,3],[1,2],[1,3]\}$ are added to the LIS as entries of Type A. The first three of these new entries result in outputs of $0,0,0$. The last entry of $[1,3]$ is coded as 1 followed by its children and their signs as $0,0,1,-,0$. Three of these children, $\{[2,6],[2,7],[3,7]\}$ are added to the LIC while $[3,6]$ is added to the LSC. Since $[1,3]$ has no grandchildren, i.e., $\mathcal{G}\left[1,3\right] = \phi$, it is not changed to Type B, but is deleted from the LIS. The resulting outputs of the q_1 significance pass are thus

$$1,-,1,+,0,0,1,0,0,0,1,+,0,0,1,0,0,0,1,0,0,1,-,0$$

The q_1 refinement pass outputs $q_1\left[i,j\right]\ \forall\left[i,j\right]$ in the LSC as shown above in equation (7.7) (i.e., at the beginning of the q_1 significance pass) This results in outputs given by

$$1,0,0,0$$

At this point, all bits in q_1 have been coded and the lists are given by

$$
\begin{array}{lll}
LSC & LIC & LIS \\
[0,0] & [2,1] & [1,1]\,A \\
[0,1] & [3,1] & [1,0]\,B \\
[1,1] & [0,2] & [0,2]\,A \\
[3,0] & [0,3] & [0,3]\,A \\
[1,0] & [1,2] & [1,2]\,A \\
[2,0] & [2,6] & \\
[1,3] & [2,7] & \\
[3,6] & [3,7] &
\end{array}
$$

The reader is invited to verify that the q_2 significance pass results in the following outputs

$$0, 0, 1, +, 0, 0, 0, 1, +, 0, 1, 0, 0, 0, 1, -, 0, 1, 1, +, 0, 0, 1, +, 0, 0, 0$$

7.4.6   ARITHMETIC CODING OF SPIHT SYMBOLS

Arithmetic coding can be used to losslessly compress the output symbols from SPIHT. Unlike EZW, the refinement bits are not arithmetically coded. That is, the data resulting from refinement passes are deposited into the final bit-stream "raw." Thus, in SPIHT, only symbols from significance passes are arithmetically coded.

In an EZW significance pass, the sign information is lumped with significance information in the form of the POS and NEG symbols. In contrast, SPIHT significance passes contain the symbols 0 (coefficient or set of coefficients remains insignificant), 1 (coefficient or set of coefficients become significant), "+" (if it was a coefficient that became significant, its sign is positive), and "−" (if it was a coefficient that became significant, its sign is negative). Even in arithmetically coded SPIHT, the signs are not lumped with significance information. They are not even arithmetically coded, but deposited in the bit-stream raw. Thus, only 0 (remains insignificant) and 1 (newly significant) are arithmetically coded in SPIHT.

To this end, coordinates in the LIC and LIS are arithmetically coded in 2×2 blocks. In this way, local (spatial) dependencies can be exploited. Recall that only insignificant coefficients (or sets) are coded in the significance pass. Thus, if any coordinate in a 2×2 block does not belong to the LIC or LIS, it is not coded. The number of bits to be coded in a 2×2 block is then $b \in \{0, 1, 2, 3, \text{ or } 4\}$. Of course no coding is done when $b = 0$. Different coding contexts are employed depending on which bits are to be coded. Using the appropriate context, a single index in $\{0, 1, \ldots, M - 1\}$ is coded (using M-ary arithmetic coding), to indicate which of the $M = 2^b$ possible bit patterns is present in the block.

As in the case of EZW, SPIHT encoding can be halted when any desired encoding rate (file size) has been achieved. Similarly, decoding can be halted after any desired amount of compressed data has been read (consumed). Similarly, a bit-stream can be truncated to yield a bit-stream of lower rate. The bit-stream will be identical to the bit-stream obtained if coding were performed to the lower rate in the first place.

7.5 PERFORMANCE OF ZERO-TREE COMPRESSION

Figure 7.6 shows the MSE performance of the SPIHT algorithm for a 512×512 portion of the "Goldhill" image (see Figure 4.5). The 512×512 portion used is grayscale with a bit depth of $B = 8$ bits/sample, and is often employed for performance comparisons in the compression literature. Such results are typically reported by plotting peak-signal-to-noise ratio (PSNR) vs. encoding rate R. As defined in Chapter 1, for an $N_1 \times N_2$ grayscale image of bit depth B,

$$\text{PSNR} = 10 \log_{10} \frac{\left(2^B - 1\right)^2}{\text{MSE}}$$

where

$$\text{MSE} = \frac{1}{N_1 N_2} \sum_{n_1=0}^{N_1-1} \sum_{n_2=0}^{N_2-1} \left(x\left[n_1, n_2\right] - \hat{x}\left[n_1, n_2\right]\right)^2$$

The results of Figure 7.6 are for $B = 8$, so that

$$\text{PSNR} = 10 \log_{10} \frac{(255)^2}{\text{MSE}} \cong 48.13 - 10 \log_{10} \text{MSE} \quad \text{dB}$$

Each plot in Figure 7.6 represents PSNR vs. Encoding Rate, R, in bits/sample (the equivalent compression ratio for a given R is $8/R$). As expected from the theoretical results presented in Chapters 3 and 5, each plot becomes roughly linear as R gets large. Furthermore, the slope of these linear segments quickly approaches 6 dB/bit.

The top and bottom plots in Figure 7.6 represent the performance of the SPIHT algorithm with and without arithmetic coding. We refer to these variants as SPIHT-AC and SPIHT-NC, respectively. Although not shown, the results of EZW are somewhat lower than those for SPIHT-NC. The performance gap between EZW and SPIHT-NC tends to be similar to the gap between SPIHT-NC and SPIHT-AC. As a representative example, at $R = 1$ bit/sample, SPIHT-AC outperforms SPIHT-NC by about 0.5 dB, while SPIHT-NC outperforms EZW by about 0.5 dB.

A thorough comparison of various embedded compression systems is deferred until Chapter 8 (Table 8.5). In this chapter, it suffices to say that the complexity of SPIHT-AC is roughly the same as that of EZW. However, the complexity of SPIHT-NC is considerably lower, since arithmetic coding is not employed. The small performance loss of SPIHT-

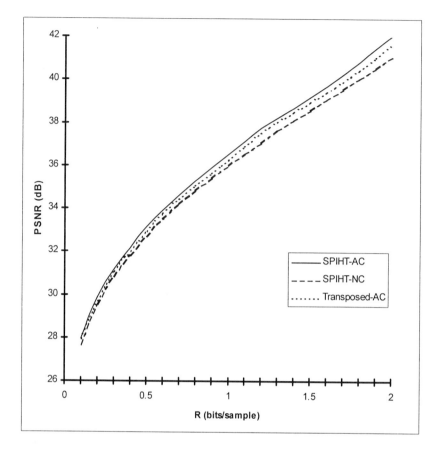

Figure 7.6. MSE performance of SPIHT for the 512 × 512 "Goldhill" image.

NC with respect to SPIHT-AC is typical for a broad class of natural imagery.[9]

On the other hand, significant differences can be observed between SPIHT-AC and SPIHT-NC for "non-natural" imagery. For example, Figure 7.7 represents the MSE performance of the two SPIHT variants on the "Chart" image of Figure 8.24. This image contains continuous tone (grayscale) regions, as well as computer generated graphics, and text. For this image, differences are around 2.5 dB rising to as much as 6.0 dB at the lowest rates.

---

[9] This is in stark contrast to EZW, for which arithmetic coding is required to achieve good performance.

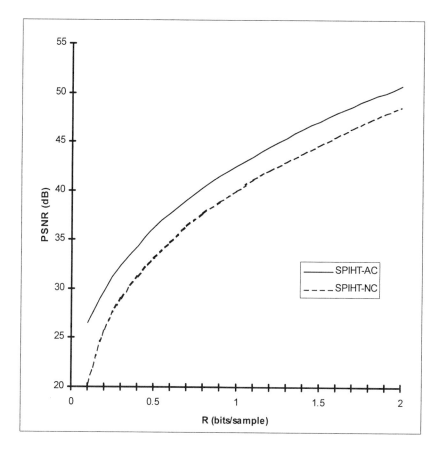

Figure 7.7. MSE performance of SPIHT for the "Chart" image.

7.6 QUANTIFYING THE PARENT-CHILD CODING GAIN

Much has been said in the compression literature of the parent-child dependency present in subband (wavelet) transforms. These dependencies are generally credited for the excellent MSE performance of zero-tree based compression algorithms such as EZW and SPIHT.

Recall the basic premise of zero-tree coding: children of insignificant coefficients tend to be insignificant. Indeed, examination of subband coefficients (e.g., Figure 4.22) supports this premise. More careful analysis however reveals that this is more a property of the energy compaction property of good transforms, and less a property of specific image data dependencies.

In [132], it is acknowledged that wavelet coefficients are largely uncorrelated across bands. On the other hand, an experimentally measured (estimated) correlation coefficient between parent and child *magnitudes* is then quoted at approximately $r = 0.35$. This non-zero correlation is then used as motivation for zero-tree coding.

We point out here, that if this amount of correlation were present in a Gaussian Markov-1 process, the maximum benefit obtainable by exploiting the resulting dependency would be given by the reciprocal of the spectral flatness measure (or prediction gain G_P) given in Sections 3.1 and 3.3 by

$$G_P = 1/\gamma_X = 10 \log_{10} \frac{1}{1 - r^2} \cong 0.57 \ \ dB \qquad (7.8)$$

Of course, a Gaussian Markov-1 assumption for the parent-child magnitude relationship is dubious at best. Even if the magnitudes did satisfy this assumption, the rate-distortion theory of γ_X was not developed in the context of magnitude only, and thus is not applicable. However, this small predicted gain provides motivation to explore this issue further.

It is a simple matter to devise an experiment to remove the ability of zero-tree coders to exploit the parent-child dependence. The third (middle) curve of Figure 7.6 is the result of such an experiment.

Each of the three plots in Figure 7.6 were obtained by transforming the 512×512 Goldhill image with the CDF 9/7 biorthogonal transform of Section 6.3.2. In each case a dyadic pyramid of 6 levels was employed. As discussed previously, the top and bottom plots were obtained by applying SPIHT (with and without arithmetic coding) to the resulting subband coefficients.

The middle plot was also obtained by applying SPIHT (with arithmetic coding) to these same subband coefficients. However, subbands HL_i, LH_i, and HH_i $i \in \{1, 3, 5\}$, were all transposed prior to coding. By transposing all subbands (in every other level) of the transform, the parent-child dependencies in the tree structures are essentially destroyed. Thus, the resulting compression performance must be largely independent of parent-child relationships.

As can be seen in Figure 7.6, there is a definite gain in compression performance due to parent-child dependencies. In what follows, we refer to this gain as the parent-child coding gain, G_{PC}. This gain is somewhat rate dependent (more gain for higher rates), but in general, is rather small. For natural images, G_{PC} is typically less than 0.6 dB with an average value of roughly 0.25 dB. These values are "within the ballpark" of what is implied by equation (7.8).

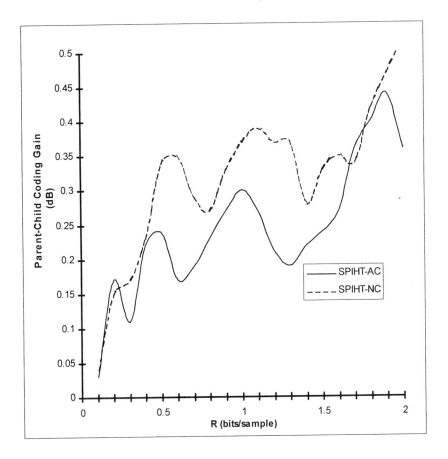

Figure 7.8.   Parent-child coding gain (G_{PC}) for the Goldhill image.

The parent-child coding gain is plotted as a function of R (for the Goldhill image) in Figure 7.8. Both SPIHT-AC and SPIHT-NC are included there. All conditions used to generate this figure are the same as described above. In fact, the SPIHT-AC plot of Figure 7.8 is just the difference between the top two plots of Figure 7.6. As pointed out above, G_{PC} is greater than zero, and in fact is non-trivial. However, it is significantly smaller than what is widely believed at the time of this writing.

As a concluding remark, we note that the SPIHT algorithm can be used with reversible transforms as well. In [127], the S+P transform (see Section 6.4.2) was introduced for this purpose. The resulting compression algorithm is capable of progressive lossy to lossless compression and decompression. If the transposition experiment is performed in this framework, the lossless file sizes are largely unchanged. For the 512×512 Goldhill image, the lossless encoding rates are 4.78 bits/sample and 4.83 bits/sample for "normal" and "transposed" coding, respectively. Similar comparisons for the "Barbara" and "Lenna" images (see Figure 8.23) are 4.71 vs. 4.75, and 4.19 vs. 4.23 bits/sample, respectively. In each case, the difference is roughly 1%.

Chapter 8

HIGHLY SCALABLE COMPRESSION WITH EMBEDDED BLOCK CODING

8.1 EMBEDDING AND SCALABILITY
8.1.1 THE DISPERSION PRINCIPLE

Following the notation of Chapter 1, we write $\mathbf{c}$ for the compressed bit-stream representing the source, $\mathbf{x}$. An "elementary embedded bit-stream" (or simply "embedded bit-stream"), $\mathbf{c}$, has the property that every L-bit prefix, $\mathbf{c}_{0:L}$, is itself an efficient compressed representation of the source. The distortion, $D^{(L)}$, should be comparable to that which might be expected from any practical compression scheme, embedded or otherwise, producing a bit-stream with a similar length, L. A natural and in fact inevitable tool for the construction of embedded bit-streams is embedded quantization, as discussed in Sections 3.2.6 and 3.4.4.

The attraction of embedded bit-streams is that the desired level of compression may be determined after the source has been compressed. Rate control may be achieved simply by truncating the bit-stream to the desired length. Embedded bit-streams have obvious appeal for remote browsing applications, in which compressed images are interactively retrieved over a low bandwidth channel. In this case, the image quality may be successively refined as the received prefix of the bit-stream grows.

An important consequence of embedding is that the information content from any given spatial region within the source must generally be "dispersed" throughout the compressed representation. We demonstrate this "dispersion principle" by considering the following counter-example. Suppose that the compressor processes the source samples locally, say from the top to the bottom of the image, appending its output to the end of the bit-stream as it goes. Assuming that the underlying random process is stationary, an L-bit prefix can be expected to contain a

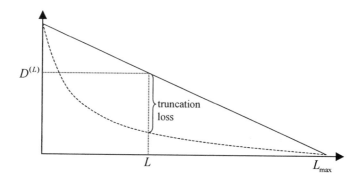

Figure 8.1. Expected distortion due to truncation of a non-embedded bit-stream (solid line), compared with the distortion-rate characteristic of the same compression algorithm (dashed curve).

complete compressed representation of the initial approximately $L/L_{\max}$ fraction of the source[1], where $L_{\max}$ is the length of the complete bit-stream. It follows that $D^{(L)}$ is approximately linearly dependent upon L. By contrast, the distortion-rate function for a stationary random process is known to be convex $\cup$ [67, §9.8] and the same characteristic is generally exhibited by practical compression systems. Thus, as seen from Figure 8.1, no L-bit prefix of the bit-stream can be expected to be an efficient compressed representation of the source, except when $L \approx L_{\max}$.

The algorithms described in Chapter 7, based on zero-tree coding of subband samples, provide excellent examples of embedded image compression. Shapiro's original EZW algorithm [131] and the significant enhancements introduced by Said and Pearlman's SPIHT algorithm [126], have received tremendous attention by the image and even the video compression community. These algorithms provide clear evidence of the dispersion principle mentioned above. Rather than processing the image locally, the compressor passes through the subband samples multiple times, halving the step size of the underlying deadzone quantizer on each iteration. Moreover, the order in which samples are visited within each pass is unpredictable a priori. For these algorithms, the dispersion principle means that the compressor (and also the decompressor) must have access to the entire image, usually through a random access buffer which

---

[1] The reader may already have experienced the behaviour of the baseline JPEG compression algorithm when the bit-stream is truncated to some fraction of its original length – only a corresponding fraction of the image is typically decompressed.

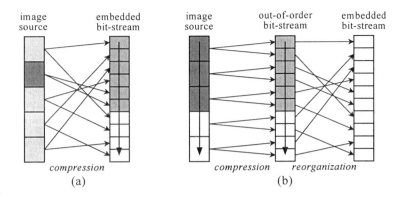

Figure 8.2. Two approaches to generating an embedded bit-stream: (a) in-order generation of the bit-stream with random access to the source; and (b) out-of-order generation of the bit-stream with local access to the source. Shading indicates the processed portion of each data type at some point during the compression process, with darker shading identifying a greater degree of completion.

contains all of the subband samples. For large images, the cost of such a buffer may be prohibitive.

8.1.2   SCALABILITY AND ORDERING

The high buffering cost of embedded compression is unavoidable so long as we insist on generating the embedded bit-stream in order. An alternate approach, however, is to process the image or subband samples locally[2] while producing the embedded bit-stream out of order. The buffer required to hold the out-of-order bit-stream prior to a final reorganization step can be significantly smaller than the image itself, assuming that compression is achieved. These two different approaches to constructing embedded bit-streams are illustrated in Figure 8.2.

For some applications, the bit-stream is best left in its initial out-of-order form so that the decompressor can also be implemented with substantially local processing. In fact, manipulation of the order of the compressed bit-stream is best viewed as a secondary task which is not part of the compression process itself, but may be delayed until the most appropriate order is known. We may think of the out-of-order bit-stream as a "generalized embedded" representation.

---

[2]Note that subband and wavelet transforms are essentially filtering operations, which may be performed incrementally using well established techniques for memory efficient two dimensional filtering. Thus, local processing within subbands is essentially equivalent to local processing of the source image. Neither the image, nor any quantity of comparable size, need be buffered.

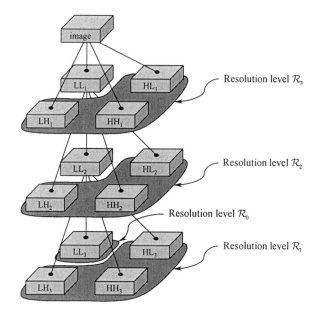

Figure 8.3.   Resolution levels within a dyadic tree-structured subband decomposition with depth $D = 3$.

The important difference between generalized embeddings and an elementary embedded bit-stream is the need for additional structural information to identify the location of the elements corresponding to successively higher quality (lower distortion) versions of the original source. The bit-streams produced by EZW and SPIHT contain no such structural information. In fact, the dependencies hardwired into their coding schemes impose tight constraints on the order in which the bit-stream must be presented and decompressed.

RESOLUTION AND DISTORTION SCALABILITY

Generalized embeddings are clearly more powerful than elementary embedded bit-streams. Their application extends beyond the conservation of implementation memory. To see this, we shall restrict our attention to compression systems which are based upon multi-resolution transforms such as the wavelet transform. In particular, we assume a dyadic tree-structured filter bank of the form shown in Figure 4.20, with D levels of decomposition and $3D + 1$ subbands.

The tree structure and organization of the subbands into resolution levels are revealed in Figure 8.3. As discussed in Section 4.2.5, the lowest resolution level, $\mathcal{R}_0$, consists only of the lowest frequency subband,

LL_D; if the low-pass subband analysis filter is a good approximation to the ideal half-band filter then this subband will contain a good low resolution rendition of the original image. The next lowest resolution level, $\mathcal{R}_1$, contains the additional subbands required to synthesize LL_{D-1}. In general, levels $\mathcal{R}_0$ through $\mathcal{R}_r$ together contain the subbands required to synthesize the reduced resolution image, LL_{D-r}, where the original image is interpreted as LL_0.

We say that a compressed bit-stream is "resolution scalable" if it contains identifiable elements which represent the subbands in each resolution level, $\mathcal{R}_r$, without any dependence on the higher resolution levels, $\mathcal{R}_{r+i}$, $i > 0$. Here and in the sequel, we use the term "identifiable elements" to mean that the bit-stream contains sufficient auxiliary information for an application to readily extract the relevant elements. This auxiliary information may involve marker codes, pointers stored in a lookup table, or some less explicit mechanism. A resolution scalable bit-stream has the property that the compressed representation of LL_{D-r} may be obtained simply by discarding the elements corresponding to $\mathcal{R}_{r+1}$ through $\mathcal{R}_D$. Moreover, this is exactly the same compressed representation which one would have obtained if LL_{D-r} were the image that had been compressed in the first place.

We say that a compressed bit-stream is "distortion scalable" if it can be reorganized into an embedded bit-stream, as defined above. Specifically, it must contain identifiable elements, which successively augment the quality (reduce the distortion) of the compressed representation. We shall call these elements quality layers and denote them $\mathcal{Q}_0$ through $\mathcal{Q}_{\Lambda-1}$. Distortion scalability is inevitably connected with embedded quantization. It is commonly called "SNR scalability" and sometimes "quality scalability" or "rate scalability." A bit-stream which is both distortion and resolution scalable must contain identifiable elements which successively reduce the distortion of the subbands in each resolution level. This is illustrated stylistically in Figure 8.4, where the relevant elements lie at the intersection of each quality layer with each resolution level.

PROGRESSIONS

We use the term "progression" to identify the ordering of elements within a scalable bit-stream. A "quality progressive" bit-stream (also called "SNR progressive") is one in which the dominant ordering is based on quality layers, as illustrated in Figure 8.4a. Similarly, a "resolution progressive" bit-stream is one in which the dominant ordering is based on resolution, as illustrated in Figure 8.4b.

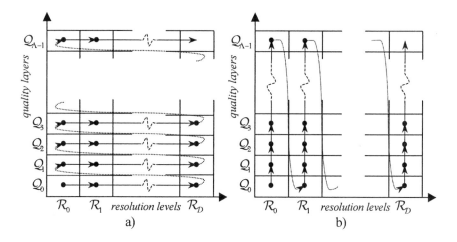

Figure 8.4.   Elements of a distortion and resolution scalable bit-stream, with a) quality progressive and b) resolution progressive orderings.

To enable local processing, a scalable bit-stream may also support spatially oriented progressions. This in turn suggests further structuring of the bit-stream into identifiable elements which correspond to spatial regions (e.g., lines, stripes or blocks) within each quality layer of each resolution level. The most appropriate progression is usually dependent upon the target application.

In summary, the generalized embeddings suggested above are usually known as "scalable" bit-streams. Multiple dimensions of scalability may be supported simultaneously, despite the fact that a linear bit-stream can have only one ordering known as its progression. The JPEG2000 image compression standard supports four dimensions of scalability and defines five native progression schemes, with the ability to inter-mix these progressions. Unlike elementary embeddings, scalable bit-streams contain the necessary information to identify their various elements so as to support reorganization into different progressions. The compressor may utilize whichever progression minimizes its implementation complexity, while external agents may reorder the bit-stream as necessary to suit particular decompression applications.

CODING VERSUS ORDERING

The wavelet transform is an important tool in the construction of resolution scalable bit-streams, while embedded quantization is central to the construction of distortion scalable representations. It is important to observe, however, that dependencies introduced whilst coding the embedded quantization indices (e.g., bit-planes) can destroy one or more degrees of scalability. For example, the zero-tree coding structure employed by the EZW and SPIHT algorithms introduces downward dependencies between resolution levels which interfere with resolution scalability.

The arguments advanced above suggest that one should endeavour to separate the process of efficiently coding quantization indices from the process of ordering the compressed bit-stream. The extent to which this is possible is limited by dependencies introduced during coding. Sources of such dependencies include the use of conditional coding contexts, indivisible codes (e.g., vector, run-length or quad-tree codes) and adaptive probability models. Coding efficiency demands that some or all of these be present at least to some extent.

A number of compression schemes have been proposed based on bit-plane coding of subband samples, which exhibit both resolution and distortion scalability in a single bit-stream. Examples include Taubman and Zakhor's 3D subband video coder (also applied to images) [150], Zandi et al.'s "CREW" (Compression with Reversible Embedded Wavelets) algorithm [174], the TCQ based wavelet compression algorithm of Sementilli et al. [128] and the low complexity embedded Golomb coding scheme developed by Ordentlich et al. [113]. In each of these algorithms the ordering of information within the final bit-stream is still heavily influenced by dependencies introduced by the bit-plane coding process. Of course, support for arbitrary reordering at the level of individual quantization indices is both unwarranted and impractical, since the signalling overhead required to identify the relevant information within the bit-stream would be quite unwieldly.

A natural compromise is to partition the subband samples into small blocks and to code each block independently. The various dependencies described above may exist within a block but not between different blocks. The size of the blocks determines the degree to which one is prepared to sacrifice coding efficiency in exchange for flexibility in the ordering of information within the final compressed image representation.

8.1.3 THE EBCOT PARADIGM

INDEPENDENT EMBEDDED BLOCKS

The coding and ordering techniques adopted by JPEG2000 are based on the concept of Embedded Block Coding with Optimal Truncation (EBCOT), which is the subject of this chapter. Each subband is partitioned into relatively small blocks (e.g., 64×64 or 32×32 samples)

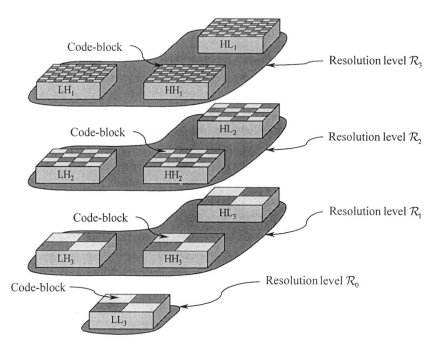

Figure 8.5. Division of subbands into code-blocks, having the same dimensions in every subband. For illustrative purposes, all subbands are depicted with the size and the code-blocks appear to have different sizes.

which we call "code-blocks." This is illustrated in Figure 8.5. Each code-block, $\mathcal{B}_i$, is coded independently, producing an elementary embedded bit-stream, $\mathbf{c}_i$. Although any prefix of length, L_i, should represent an efficient compression of the block's samples at the corresponding rate, embedded coding algorithms inevitably possess a collection of natural truncation points, $L_i^{(z)}$, at which the distortion, $D_i^{(z)}$, can be expected to lie closest to the convex rate-distortion function for the relevant source. This is discussed further in Section 8.3.3. In any event it is convenient to restrict our attention to a finite number of allowable truncation points, $Z_i + 1$, for code-block $\mathcal{B}_i$, having lengths, $L_i^{(z)}$, with

$$0 = L_i^{(0)} \leq L_i^{(1)} \leq \cdots \leq L_i^{(Z_i)}$$

In the present development we are not concerned with the details of the embedded block coding algorithm; these are the subject of Section 8.3.

We assume that the overall reconstructed image distortion can be represented as a sum of distortion contributions from each of the code-blocks and let $D_i^{(z)}$ denote the distortion contributed by block $\mathcal{B}_i$, if its elementary embedded bit-stream is truncated to length $L_i^{(z)}$. Calculation or estimation of $D_i^{(z)}$ depends upon the subband to which block $\mathcal{B}_i$ belongs; when necessary, we identify this subband with the label b_i. For most of the ensuing discussion, however, we may simply consider the image as being composed of a collection of blocks, $\mathcal{B}_i$, without regard for the subbands to which their samples belong.

Since the code-blocks are compressed independently, we are free to use any desired policy for truncating their embedded bit-streams. If the overall length of the final compressed bit-stream is constrained by $L_{\max}$, we are free to select any set of truncation points, $\{z_i\}$, such that

$$\sum_i L_i^{(z_i)} \leq L_{\max}$$

Of course, the most attractive choice is that which minimizes the overall distortion,

$$D = \sum_i D_i^{(z_i)}$$

The selection of truncation points may be deferred until after all of the code-blocks have been compressed at which point the available truncation lengths, $L_i^{(z)}$, and the associated distortions, $D_i^{(z)}$, should all be known. For this reason, we refer to the optimal truncation strategy as one of post-compression rate-distortion optimization (PCRD-opt). A PCRD-opt algorithm is described in Section 8.2.

A chief disadvantage of independent block coding would appear to be that it is unable to exploit redundancy between different blocks within a subband or between different subbands. In fact, an important premise of the zero-tree schemes described in Chapter 7 is that substantial redundancy exists between "parent" and "child" samples within the subband hierarchy. Somewhat surprisingly, these disadvantages are more than compensated by the fact that the contributions of each code-block to the final bit-stream may be independently optimized by the PCRD-opt algorithm.

Exploiting parent-child relationships within the subband hierarchy can indeed lead to small improvements in the coding efficiency within a block. However, the conditional coding techniques required to exploit this redundancy impose constraints on the allowable truncation points for parent and child code-blocks. As it happens, these dependencies so constrain the PCRD-opt algorithm that overall performance is usually degraded, even though individual code-blocks are coded more efficiently. Fortuitously, then, the simpler approach without inter-subband dependencies generally yields superior compression efficiency.

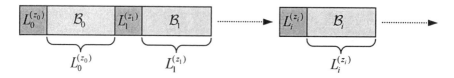

Figure 8.6. A simple pack-stream formed by concatenating the optimally truncated code-block bit-streams.

QUALITY LAYERS AND STRUCTURE

Since each code-block has its own embedded bit-stream, $\mathbf{c}_i$, it is convenient to use a separate term to refer to the overall compressed bit-stream, $\mathbf{c}$. We shall call this a "pack-stream," because it is inevitably constructed by packing contributions from the various code-block bit-streams together in some fashion. The simplest pack-stream organization consistent with the EBCOT paradigm is illustrated in Figure 8.6. In this case, the optimally truncated block bit-streams are simply concatenated. Length tags are inserted to identify the contribution from each code-block.

This simple pack-stream is resolution scalable, since each resolution level consists of a well-defined collection of code-blocks, explicitly identified by the length tags. The pack-stream also possesses a degree of spatial scalability. So long as the subband synthesis filters have finite support, each code-block influences only a finite region in the reconstructed image. Thus, given a spatial region of interest, the relevant code-blocks may be identified and extracted from the pack-stream.

Interestingly, the simple pack-stream of Figure 8.6 is not distortion scalable, even though its individual code-blocks have embedded representations. The problem is that the pack-stream offers no information to assist in the construction of a smaller pack-stream whose code-block contributions, $L_i^{(z_i)}$, minimize the associated distortion.

Figure 8.7 illustrates the quality layer abstraction introduced by the EBCOT algorithm [149] to resolve the above difficulty. The first quality layer, $\mathcal{Q}_0$, contains optimized code-block contributions, having lengths $L_i^{(z_i^0)}$, which minimize the distortion, $D^0 = \sum_i D_i^{(z_i^0)}$, subject to a length constraint, $\sum_i L_i^{(z_i^0)} \leq L_{\max}^0$. Subsequent layers, $\mathcal{Q}_l$, contain additional contributions from each code-block, having lengths $L_i^{(z_i^l)} - L_i^{(z_i^{l-1})}$, which minimize the distortion,

$$D^l = \sum_i D_i^{(z_i^l)}$$

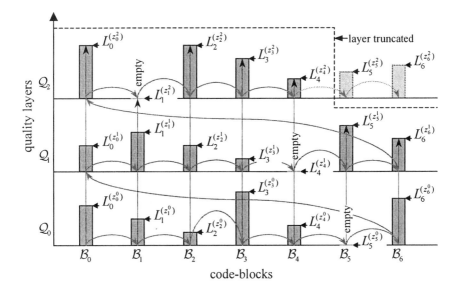

Figure 8.7. Code block contributions to quality layers, indicating a quality progressive pack-stream ordering and mid-layer truncation.

subject to a length constraint,

$$\sum_i L_i^{\left(z_i^l\right)} \leq L_{\max}^l$$

Although each quality layer notionally contains a contribution from every code-block, we emphasize the fact that some or even all of these contributions may be empty. A distortion scalable pack-stream may be constructed by including sufficient information to identify the contribution made by each code-block to each quality layer. Moreover, quality progressive organizations are clearly supported by ordering the information in the manner suggested by the arrows in Figure 8.7.

If a quality progressive pack-stream is truncated at an arbitrary point then the decoder can expect to receive some number of complete quality layers, followed by some fraction of the blocks from the next layer, as suggested by the figure. In this case, the received prefix will not be strictly optimal in the PCRD-opt sense. However, this sub-optimality may be rendered negligible by employing a large number of layers. On the other hand, more layers implies a larger overhead to identify the contributions made by each code-block to each layer. When a large number of layers is involved, some care must be invested in efficiently coding the auxiliary information which identifies the various code-block contributions. These matters are discussed further in Section 8.4.

EBCOT ADVANTAGES

At this point, it is worth summarizing some of the benefits offered by the EBCOT paradigm, which contributed to its adoption in JPEG2000.

Flexible organization: EBCOT pack-streams possess resolution scalability, distortion scalability (so long as multiple quality layers are used) and a degree of spatial scalability. When multiple image components are compressed (e.g., colour components), these components form a fourth dimension of scalability. Progressions along all four dimensions are supported by the JPEG2000 standard.

Custom quality interpretations: Since each quality layer may contain arbitrary contributions from each of the code-blocks, the notion of quality is easily adapted to application specific measures of significance. By contrast with EZW, SPIHT and other embedded compression algorithms, the EBCOT paradigm allows code-blocks to be marginalized or entirely suppressed in lower quality layers when the corresponding spatial regions or frequency bands are known to be less significant for some application.

Local processing: Independent coding allows local processing of the samples in each code-block, which is especially advantageous for hardware implementations. Independent coding also introduces the possibility of highly parallel implementations, where multiple code-blocks are encoded or decoded simultaneously. For large images, spatially oriented progressions of the pack-stream may be used in conjunction with incremental processing of the subband/wavelet transform to facilitate "streaming." In this case, it is sufficient to buffer only a local window into the pack-stream, the image and its subbands. In this way, the implementation memory can be much smaller than the image which is being compressed or decompressed. This same property allows for efficient rotation and flipping of the image during decompression.

Efficient compression: As noted above, the PCRD-opt algorithm can more than compensate for the small efficiency losses arising from the imposition of independent block coding. The algorithm is also able to accommodate spatially varying and/or image dependent measures of distortion. One interesting example arises in visual perception, where local activity can mask the visibility of certain types of compression artifacts. A masking-sensitive distortion measure and promising experimental results are provided in [149].

Error resilience: Errors encountered in any code-block's bit-stream will clearly have no influence on the other blocks. This, together with the natural prioritization of information induced by embedded block coding and quality layers, allows for the construction of powerful differential protection strategies for error prone environments.

8.2   OPTIMAL TRUNCATION
8.2.1   THE PCRD-OPT ALGORITHM

In this section we describe an algorithm which may be used to optimize the set of code-block truncation points, $\{z_i\}$, so as to minimize the overall distortion, D, subject to an overall length constraint, $L_{\max}$. The same algorithm may be used to minimize the overall length subject to a distortion constraint if desired. We refer to this optimization strategy as post-compression rate-distortion optimization (PCRD-opt). The algorithm is implemented by the compressor, which is expected to compute or estimate length and distortion contributions, $L_i^{(z)}$ and $D_i^{(z)}$, for each truncation point, $z = 0, 1, \ldots, Z_i$. This information will not normally be explicitly included in the pack-stream. As a result, the algorithm is not easily reapplied to a previously constructed pack-stream. This need not be a substantial limitation, since the pack-stream may contain many quality layers. Layers $\mathcal{Q}_0$ through $\mathcal{Q}_l$ together embody a PCRD-optimized collection of code-block contributions with length constraint $L_{\max}^l$, for each $l = 0, 1, \ldots, \Lambda - 1$.

To simplify the discussion, we shall temporarily ignore the overhead required to identify the sizes of the code-block contributions themselves. The length constraint is then simply

$$L = \sum_i L_i^{(z_i)} \leq L_{\max} \tag{8.1}$$

We also assume an additive distortion measure so that the overall distortion can be expressed as

$$D = \sum_i D_i^{(z_i)} \tag{8.2}$$

This assumption is revisited in Section 8.3.5, where we also discuss the computation of the $D_i^{(z_i)}$.

Let $\{z_{i,\lambda}\}$ be any set of truncation points which minimizes

$$D(\lambda) + \lambda L(\lambda) = \sum_i \left(D_i^{(z_{i,\lambda})} + \lambda L_i^{(z_{i,\lambda})} \right) \tag{8.3}$$

for some $\lambda > 0$. It is easy to see that these truncation points are optimal in the sense that the distortion, D, cannot be further reduced without also increasing the length, L, or vice-versa. Thus, if we can find a value of λ such that the $\{z_{i,\lambda}\}$ which minimize equation (8.3) yield $L(\lambda) = L_{\max}$, this set of truncation points must be a solution to our optimization problem. Since the set of available truncation points is discrete, we shall not generally be able to find such a λ. Nevertheless, since the code-blocks are relatively small and there are typically many truncation points, it is sufficient in practice to find the smallest value of λ such that $L(\lambda) \leq L_{\max}$. Note that the solution to this modified optimization problem is not strictly guaranteed to minimize D subject to (8.1).

For any given λ, minimization of equation (8.3) reduces to an independent minimization task for each code-block. The following simple algorithm finds the smallest $z_{i,\lambda}$ which minimizes $D_i^{(z_{i,\lambda})} + \lambda L_i^{(z_{i,\lambda})}$. The algorithm can be seen as equivalent to testing $D_i^{(z_i)} + \lambda L_i^{(z_i)}$ for each z_i, as suggested by the discussion in Section 5.4. The indirect approach adopted here, however, will prove helpful for subsequent refinements.

Optimization of $z_{i,\lambda}$ for given λ

Initialize $z^{\mathrm{opt}} = 0$

For $t = 1, 2, \ldots, Z_i$

    Set $\Delta L = L_i^{(t)} - L_i^{(z^{\mathrm{opt}})}$ and $\Delta D = D_i^{(z^{\mathrm{opt}})} - D_i^{(t)}$

    <small>(ΔL is the amount by which the length is increased and ΔD is the amount by which distortion is decreased if we replace the current value of z^{opt} by t. We expect both to be non-negative, although it can happen that distortion increases; that is, ΔD might occasionally be negative.)</small>

    If $\Delta D / \Delta L > \lambda$, replace $z^{\mathrm{opt}} \leftarrow t$

    <small>(This step guarantees that $D_i^{(z^{\mathrm{opt}})} + \lambda L_i^{(z^{\mathrm{opt}})} \leq D_i^{(z)} + \lambda L_i^{(z)}$ for all $z \leq t$.)</small>

Set $z_{i,\lambda} = z^{\mathrm{opt}}$

Since the number of code-blocks may be very large and the algorithm must be executed for many different values of λ, it makes sense to precompute as many quantities as possible. To this end, it is helpful to first characterize the set of feasible truncation points, $\mathcal{H}_i$, defined by

$$\mathcal{H}_i \triangleq \{z \mid z = z_{i,\lambda} \text{ for some } \lambda > 0\}$$

This is the set of truncation points for code-block $\mathcal{B}_i$ which may be produced by the above optimization algorithm for at least one $\lambda > 0$.

It is worth pointing out that given any particular λ, there can be multiple z_i which minimize $D_i^{(z_i)} + \lambda L_i^{(z_i)}$. Our convention is to select only the smallest such z_i for inclusion in $\mathcal{H}_i$. Although multiple minimizing solutions occur rarely in practice, our need to adopt a consistent convention regarding such cases is responsible for a number of subtleties in the ensuing development.

FEASIBLE TRUNCATION POINTS

We begin by defining a quantity, $\lambda_i(z)$, which will turn out to have the interpretation of a "distortion-length slope." Specifically, we define

$$\lambda_i(0) \triangleq \infty$$

$$\lambda_i(z) \triangleq \min_{z' < z} \frac{D_i^{(z')} - D_i^{(z)}}{L_i^{(z)} - L_i^{(z')}}, \quad \text{for } z > 0 \tag{8.4}$$

Lemma 8.1 *Suppose $z = z_{i,\lambda}$ is the optimal index determined by the algorithm described above, for some $\lambda > 0$. Then $\lambda_i(z) > \lambda$.*

Proof. Since $\lambda_i(0) = \infty$ we may restrict our attention to $z > 0$. Denote by $k \geq 0$ the index stored in z^{opt} immediately before it is replaced by z in step z of the above algorithm. It follows that

$$D_i^{(k)} + \lambda L_i^{(k)} \leq D_i^{(t)} + \lambda L_i^{(t)}, \quad \forall t < z \quad \text{and} \quad \frac{D_i^{(k)} - D_i^{(z)}}{L_i^{(z)} - L_i^{(k)}} > \lambda \tag{8.5}$$

Also, let $0 \leq z' < z$ be any truncation point which achieves the minimum in equation (8.4) so that

$$\lambda_i(z) = \frac{D_i^{(z')} - D_i^{(z)}}{L_i^{(z)} - L_i^{(z')}}$$

The first inequality in equation (8.5) applies to $t = z'$ (this produces the first line in the equation below). Then, assuming (for the purpose of contradiction) that $\lambda_i(z) \leq \lambda$, we obtain

$$0 \leq \left(D_i^{(z')} + \lambda L_i^{(z')} \right) - \left(D_i^{(k)} + \lambda L_i^{(k)} \right)$$

$$= \left(D_i^{(z')} - D_i^{(z)} \right) - \left(D_i^{(k)} - D_i^{(z)} \right) + \lambda \left(L_i^{(z')} - L_i^{(z)} \right) - \lambda \left(L_i^{(k)} - L_i^{(z)} \right)$$

$$< \lambda \left(L_i^{(z)} - L_i^{(z')} \right) - \lambda \left(L_i^{(z)} - L_i^{(k)} \right) + \lambda \left(L_i^{(z')} - L_i^{(z)} \right) - \lambda \left(L_i^{(k)} - L_i^{(z)} \right) = 0$$

This contradiction proves the assertion that $\lambda_i(z) > \lambda$. ∎

Theorem 8.2 *Necessary and sufficient conditions for $z \in \mathcal{H}_i$ are that*

$$\lambda_i(z) > 0 \text{ and } \lambda_i(z) > \max_{t > z} \frac{D_i^{(z)} - D_i^{(t)}}{L_i^{(t)} - L_i^{(z)}} \tag{8.6}$$

Proof. (Sufficiency) Suppose equation (8.6) holds. Select $\lambda > 0$ such that $\lambda_i(z) > \lambda \geq \max_{t>z} \frac{D_i^{(z)} - D_i^{(t)}}{L_i^{(t)} - L_i^{(z)}}$. In step $t = z$ of the above algorithm we have $z^{\mathrm{opt}} < z$ and

$$\Delta D / \Delta L = \frac{D_i^{(z^{\mathrm{opt}})} - D_i^{(z)}}{L_i^{(z)} - L_i^{(z^{\mathrm{opt}})}} \geq \min_{z' < z} \frac{D_i^{(z')} - D_i^{(z)}}{L_i^{(z)} - L_i^{(z')}} = \lambda_i(z) > \lambda$$

The inequality here follows immediately from the fact that the minimization over all $z' < z$ includes the case $z' = z^{\mathrm{opt}}$. We conclude that the test $\Delta D / \Delta L > \lambda$ must succeed, causing the assignment $z^{\mathrm{opt}} \leftarrow z$. On the other hand, the same test must fail in every subsequent step $t > z$, at which

$$\Delta D / \Delta L = \frac{D_i^{(z^{\mathrm{opt}})} - D_i^{(t)}}{L_i^{(t)} - L_i^{(z^{\mathrm{opt}})}} = \frac{D_i^{(z)} - D_i^{(t)}}{L_i^{(t)} - L_i^{(z)}} \leq \lambda$$

(Necessity) Obviously $z = 0$ satisfies the conditions in equation (8.6), since $\lambda_i(0) = \infty$. Therefore, we need only consider the non-zero members of $\mathcal{H}_i$. That is, $z > 0$ and $z = z_{i,\lambda}$ for some $\lambda > 0$. From Lemma 8.1 we have $\lambda_i(z) > \lambda$. Noting that the value of z^{opt} in the above algorithm may not be changed in any step $t > z$, we conclude that

$$\lambda_i(z) > \lambda \geq \max_{t>z} \frac{D_i^{(z)} - D_i^{(t)}}{L_i^{(t)} - L_i^{(z)}}$$

This validates the conditions of equation (8.6). ∎

Corollary 8.3 *Let* $0 = h_i^0 < h_i^1 < h_i^2 < \cdots$ *be an enumeration of the truncation points, z, in $\mathcal{H}_i$. Then*

$$\lambda_i(h_i^0) > \lambda_i(h_i^1) > \lambda_i(h_i^2) > \cdots > \lambda_i\left(h_i^{\|\mathcal{H}_i\|-1}\right) > 0 \qquad (8.7)$$

and

$$\lambda_i(h_i^n) = \frac{D_i^{(h_i^{n-1})} - D_i^{(h_i^n)}}{L_i^{(h_i^n)} - L_i^{(h_i^{n-1})}}, \qquad \forall n \geq 1 \qquad (8.8)$$

Proof. From equations (8.6) and (8.4) we have, for each $h_i^n > 0$,

$$\lambda_i(h_i^{n-1}) > \max_{t>h_i^{n-1}} \frac{D_i^{(h_i^{n-1})} - D_i^{(t)}}{L_i^{(t)} - L_i^{(h_i^{n-1})}} \geq \frac{D_i^{(h_i^{n-1})} - D_i^{(h_i^n)}}{L_i^{(h_i^n)} - L_i^{(h_i^{n-1})}} \geq \lambda_i(h_i^n)$$

which validates equation (8.7). For equation (8.8) we must show that the inequality on the right hand side of the above equation is actually an equality. The proof is by contradiction. Suppose the inequality is strict and set $\lambda = \lambda_i(h_i^n)$ so that

$$\lambda_i(h_i^{n-1}) > \frac{D_i^{(h_i^{n-1})} - D_i^{(h_i^n)}}{L_i^{(h_i^n)} - L_i^{(h_i^{n-1})}} > \lambda = \lambda_i(h_i^n) > \lambda_i(h_i^{n+1}) > \cdots$$

By Lemma 8.1 we must have $\lambda_i(z_{i,\lambda}) > \lambda$, from which we conclude that $z_{i,\lambda} \in \{h_i^0, h_i^1, \ldots, h_i^{n-1}\}$. Also, because $\lambda < \lambda_i(h_i^{n-1})$ the test $\Delta D / \Delta L > \lambda$ must succeed

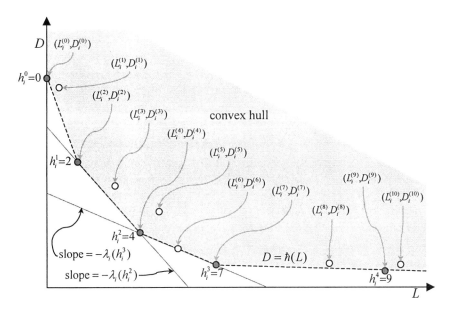

Figure 8.8.   Convex hull formed by the feasible truncation points for code-block $\mathcal{B}_i$.

in step $t = h_i^{n-1}$ of the optimization algorithm. Thus $z_{i,\lambda} = h_i^{n-1}$, meaning that the test $\Delta D/\Delta L > \lambda$ must fail in every step $t > h_i^{n-1}$. In step $t = h_i^n$, however, we find that $\frac{\Delta D}{\Delta L} = \frac{D_i^{\left(h_i^{n-1}\right)} - D_i^{\left(h_i^n\right)}}{L_i^{\left(h_i^n\right)} - L_i^{\left(h_i^{n-1}\right)}} > \lambda$, which contradicts the hypothesis. ■

CONVEX HULL INTERPRETATION

We may now assign the following interpretation to the set of feasible truncation points, $\mathcal{H}_i$. The length-distortion coordinates, $\left(L_i^{(h)}, D_i^{(h)}\right)$, taken over all $h \in \mathcal{H}_i$, define the vertices of a piecewise linear, convex $\cup$ function, $\hbar_i(L)$, as illustrated in Figure 8.8. The gradient of this function is given by $\frac{d}{dL}\hbar_i(L) = -\lambda_i(h_i^n)$ for $L_i^{\left(h_i^{n-1}\right)} < L \leq L_i^{\left(h_i^n\right)}$. Given any truncation point, z, the point $\left(L_i^{(z)}, D_i^{(z)}\right)$ must lie above or on the curve, $\hbar_i(L)$.[3] We refer to $\hbar_i(L)$ as the "convex hull" of the distortion-length characteristic for code-block $\mathcal{B}_i$. We loosely say that

---

[3] To see this, consider any tangent to $\hbar_i(L)$, say the tangent with slope $\lambda_i(h)$ at $\left(L_i^{(h)}, D_i^{(h)}\right)$. If $p < h$, the definition of $\lambda_i()$ yields $D_i^{(p)} - D_i^{(h)} \geq \lambda_i(h)\left(L_i^{(h)} - L_i^{(p)}\right)$. If $p > h$, equation (8.6) yields $D_i^{(p)} - D_i^{(h)} > \lambda_i(h)\left(L_i^{(h)} - L_i^{(p)}\right)$. Thus, each trunctation point lies above or on every tangent to $\hbar_i(L)$, meaning that it lies above or on the curve $\hbar_i(L)$ itself.

the feasible truncation points are those which lie on the convex hull and we loosely refer to $\lambda_i(h)$ as the distortion-length "slope" at $h \in \mathcal{H}_i$.

It is worth remarking on two phenomena exhibited by Figure 8.8. Truncation points $z = 5$, 8 and 10 in the figure violate our intuitive expectation that distortion should decrease as the size of the compressed bit-stream increases. In fact, the effective quantization step size associated with any given sample in the code-block is a non-increasing function of z. It does not seem reasonable that smaller quantization step sizes should produce more distortion. It is true that the expected (ensemble average) distortion should decrease as the step size decreases, subject to appropriate dequantizer design. However, when averaging over a small set of samples, the opposite behaviour is occasionally observed in practice.

Truncation point $z = 6$ in the figure actually lies on the convex hull, but does not belong to the set $\mathcal{H}_i$. When multiple truncation points have the same slope, $\lambda_i(z)$, Theorem 8.2 tells us that only the largest of these points, z, may belong to $\mathcal{H}_i$. Oddly enough, this is quite consistent with our convention that each $z \in \mathcal{H}_i$ must be the smallest truncation point which minimizes $D_i^{(z)} + \lambda L_i^{(z)}$ for some λ.

A SIMPLE OPTIMIZATION ALGORITHM

Since an optimal truncation point, z_i, which minimizes $D_i^{(z_i)} + \lambda L_i^{(z_i)}$, is guaranteed to belong to $\mathcal{H}_i$, there is no harm in restricting the optimization algorithm to just this set. Using Corollary 8.3, the algorithm simplifies to a search for the largest $h \in \mathcal{H}_i$ such that $\lambda_i(h) > \lambda$, as follows.

Optimization of $z_{i,\lambda}$ **over** $\mathcal{H}_i = \left\{ h_i^0 < h_i^1 < \cdots < h_i^{\|\mathcal{H}_i\|-1} \right\}$

Initialize $j = 0$

While $j + 1 < \|\mathcal{H}_i\|$ and $\lambda_i\left(h_i^{j+1}\right) > \lambda$

    Increment j

Set $z_{i,\lambda} = h_i^j$

Note that the distortion-length "slopes," $\lambda_i\left(h_i^k\right)$, are independent of λ. It is also now obvious that $z_{i,\lambda}$ must be a non-increasing function of λ. Thus, as λ decreases the overall length, $L(\lambda)$, must either increase or stay the same. The global optimization problem may then be described as

$$\lambda^{\text{opt}} = \min\left\{\lambda \mid L\left(\lambda\right) \leq L_{\max}\right\}$$

$$= \min\left\{\lambda \mid \sum_i L_i^{\left(z_{i,\lambda}\right)} \leq L_{\max}\right\}$$

$$= \min\left\{\lambda \mid \left(\sum_i L_i^{\left(z\right)}\Big|_{z=\max\{h\in\mathcal{H}_i|\lambda_i(h)>\lambda\}}\right) \leq L_{\max}\right\} \quad (8.9)$$

The fact that $L\left(\lambda\right)$ is monotonic in λ means that the search for λ^{opt} is particularly simple. The well-known bisection method, for example, may be used to successively halve a working interval, $\left(\lambda^{\min}, \lambda^{\max}\right)$, in which λ^{opt} is known to reside.

8.2.2   IMPLEMENTATION SUGGESTIONS

The characterization provided by Theorem 8.2 may be applied directly to identify the elements of the set $\mathcal{H}_i$ and the distortion-length slopes, $\lambda_i\left(h_i^n\right)$. The direct algorithm is as follows.

Direct Computation of Convex Hull and Slopes

Set $\lambda_i\left(0\right) \leftarrow \infty$
For $z = 1, 2, \ldots, Z_i$
    Initialize $\lambda_i\left(z\right) \leftarrow \infty$
    For $t = z - 1$ down to 0
        Set $\Delta D \leftarrow D_i^{(t)} - D_i^{(z)}$ and $\Delta L \leftarrow L_i^{(z)} - L_i^{(t)}$
        If $\Delta D < \lambda_i\left(z\right)\Delta L$
            Set $\lambda_i\left(z\right) \leftarrow \frac{\Delta D}{\Delta L}$ (note that this could be $-\infty$)
Initialize $\mathcal{H}_i = \emptyset$ (the empty set)
For $z = 0, 1, \ldots, Z_i$
    For $t = z + 1, \ldots, Z_i$
        Set $\Delta D \leftarrow D_i^{(z)} - D_i^{(t)}$ and $\Delta L \leftarrow L_i^{(t)} - L_i^{(z)}$
        If $\Delta D \geq \lambda_i\left(z\right)\Delta L$
            $z \notin \mathcal{H}_i$, so break out of inner loop and go to next z
    If $\lambda_i\left(z\right) > 0$,
        Update $\mathcal{H}_i \leftarrow \mathcal{H}_i \cup \{z\}$

This algorithm involves the computation of up to $\frac{1}{2}Z_i\left(Z_i + 1\right)$ quotients, $\frac{\Delta D}{\Delta L}$, and $Z_i\left(Z_i + 1\right)$ products, $\lambda_i\left(z\right)\Delta L$. A significantly more efficient algorithm may be realized on the basis of Corollary 8.3. Let $\mathcal{H}_i^{(z)}$ denote the convex hull set corresponding to the first $z < Z_i$ truncation points. The more efficient algorithm incrementally constructs

$\mathcal{H}_i^{(z+1)}$ from $\mathcal{H}_i^{(z)}$. From Theorem 8.2 one easily deduces that $\mathcal{H}_i^{(z+1)}$ cannot possibly contain any truncation point, $t \leq z$, which does not belong to $\mathcal{H}_i^{(z)}$. The new set might contain the extra point, $z+1$, and it may exclude one or more of the elements from $\mathcal{H}_i^{(z)}$. Theorem 8.2 tells us that the points $h \in \mathcal{H}_i^{(z)}$ which must be excluded from $\mathcal{H}_i^{(z+1)}$ are those for which

$$\lambda_i(h) \leq \frac{D_i^{(h)} - D_i^{(z+1)}}{L_i^{(z+1)} - L_i^{(h)}} \tag{8.10}$$

Let $h^{(z)} = \left\| \mathcal{H}_i^{(z)} \right\| - 1$ be the last element of $\mathcal{H}_i^{(z)}$, and let $\Delta D = D_i^{(h^{(z)})} - D_i^{(z+1)}$ and $\Delta L = L_i^{(z+1)} - L_i^{(h^{(z)})}$ denote the changes in distortion and length between $h^{(z)}$ and $z+1$. If $\Delta D \leq 0$, then $\lambda_i(z+1) \leq 0$ and $z+1$ cannot belong to $\mathcal{H}_i^{(z+1)}$. Moreover, in this case no $h \in \mathcal{H}_i^{(z)}$ will satisfy the exclusion condition of equation (8.10); we conclude that $\mathcal{H}_i^{(z+1)} = \mathcal{H}_i^{(z)}$. Otherwise $\Delta D > 0$, meaning that truncation point $z+1$ yields a smaller distortion than any other point seen thus far. Then $\lambda_i(z+1) > 0$ and $z+1$ belongs to $\mathcal{H}_i^{(z+1)}$. We need not explicitly test the exclusion condition of equation (8.10) for each $h \in \mathcal{H}_i^{(z)}$. From equation (8.7), one easily deduces that $h \in \mathcal{H}_i^{(z)}$ can be excluded from $\mathcal{H}_i^{(z+1)}$ only if all elements $h' > h$ of $\mathcal{H}_i^{(z)}$ are also excluded. Moreover, the new slope value, $\lambda(z+1)$, may be calculated in a single step, using equation (8.8). The incremental construction approach is embodied by the following algorithm.

Incremental Computation of Convex Hull and Slopes

Set $\lambda_i(0) \leftarrow \infty$, $\mathcal{H}_i \leftarrow \{0\}$ and $h^{\text{last}} \leftarrow 0$
For $z = 1, 2, \ldots, Z_i$
    Set $\Delta D \leftarrow D_i^{(h^{\text{last}})} - D_i^{(z)}$ and $\Delta L \leftarrow L_i^{(z)} - L_i^{(h^{\text{last}})}$
    If $\Delta D > 0$
        While $\Delta D \geq \lambda_i(h^{\text{last}}) \Delta L$
            Set $\mathcal{H}_i \leftarrow \mathcal{H}_i \setminus \{h^{\text{last}}\}$ (exclude last element of current hull set)
            Set $h^{\text{last}} \leftarrow \max \mathcal{H}_i$ (get last element in new hull set)
            Set $\Delta D \leftarrow D_i^{(h^{\text{last}})} - D_i^{(z)}$ and $\Delta L \leftarrow L_i^{(z)} - L_i^{(h^{\text{last}})}$
        Set $h^{\text{last}} \leftarrow z + 1$
        Set $\mathcal{H}_i \leftarrow \mathcal{H}_i \cup \{h^{\text{last}}\}$
        Set $\lambda(h^{\text{last}}) \leftarrow \frac{\Delta D}{\Delta L}$

The incremental algorithm computes at most Z_i quotients, $\frac{\Delta D}{\Delta L}$. If the convex hull set consists of all Z_i truncation points, the exclusion condition must be tested only once for each point, requiring Z_i multiplications, $\lambda_i \left(h^{\text{last}} \right) \Delta L$. More generally, one additional multiplication is required each time a truncation point is excluded from the evolving convex hull set, $\mathcal{H}_i$. We can exclude at most Z_i points in total and so the algorithm requires no more than $2Z_i$ multiplications. The complexity of the algorithm is thus linear in the number of truncation points, rather than quadratic, as was the case for the direct implementation.

It is worth noting that the incremental construction algorithm given above can be derived more directly from our original optimization procedure, given on Page 340. The approach adopted in this text, however, provides a more comprehensive characterization of $\mathcal{H}_i$, including its interpretation as the set of bounding points for the distortion-length convex hull.

EFFICIENT REPRESENTATIONS

The convex hull and slope computation procedure described above is generally executed immediately after each block has been coded, at which point the relevant distortion and length quantities should be available. We need store only the lengths $\{L_i^{(h)}\}_{h \in \mathcal{H}_i}$, the slopes $\{\lambda_i(h)\}_{h \in \mathcal{H}_i}$, and the embedded bit-stream, $\mathbf{c}_i$, until all code-blocks have been compressed. At that point we may invoke the algorithm embodied by equation (8.9).

To minimize the amount of information which must be stored, it is desirable to employ a logarithmic representation for the slope values, $\lambda_i(h)$. This is justified by Figure 5.5 which indicates that on average, the code length varies roughly linearly with the log of the distortion-length slope. The PCRD-opt algorithm involves only the comparison of slopes, which may equivalently be conducted in the log domain. Low precision fixed point representations of $\log_2 \lambda_i(h)$ are quite sufficient to achieve all the benefits of the PCRD-opt algorithm. Moreover, conversion to such a representation is trivially implemented in hardware[4].

Logarithmic representation of $\lambda_i(h)$ has the additional advantage that weighting factors for the distortion in any given code-block or subband may be applied by simply adding appropriate offsets to the log-slope values. Examples of such weighting factors are the subband energy gain

---

[4]The integer part of $\log_2 \lambda_i(h)$ is essentially the location of the most significant bit in a fixed point representation of $\lambda_i(h)$, while the first 5 fraction bits of $\log_2 \lambda_i(h)$ may be accurately estimated using a very small lookup table, indexed by the 5 or 6 bits below this most significant bit position.

factors, G_b, introduced at the end of Section 4.3.2 and the contrast sensitivity weights, W_b^{csf}, introduced in Section 4.3.4.

ACCOUNTING FOR SIGNALLING OVERHEAD

Up to this point, we have deliberately ignored the overhead required to signal code-block contributions within the pack-stream. While somewhat sub-optimal, a reasonable way to deal with this overhead is to simply adjust the length constraint, $L_{\max}$, without recomputing the distortion-length slope values. As each λ is considered in the search for λ^{opt}, we may first determine the optimal truncation points, $\{z_{i,\lambda}\}$, in the absence of any signalling overhead; we then check whether the sum of the corresponding code-block lengths, $L_i^{(z_i,\lambda)}$, and the associated signalling overhead exceeds $L_{\max}$. If so, we try a larger value of λ; otherwise, we try a smaller value of λ. Notice that this approach works even when the signalling overhead exhibits complex dependencies on the ensemble of code-block contributions, which is often the case[5].

Before closing this section it is worth noting that for memory constrained applications the PCRD-opt algorithm may actually be executed incrementally, before all code-block bit-streams have been generated. In this case we must determine the best slope threshold, λ^{opt}, to apply to a collection of code-blocks, before the distortion and length information for all code-blocks is available. If the image statistics are stable and the number of code-block bit-streams which we can afford to buffer is moderately large, this locally determined slope threshold can be close to the globally optimal value and little will be lost by incrementally truncating the code-blocks.

8.3   EMBEDDED BLOCK CODING

The purpose of this section is to make concrete the notion of independent embedded block coding. To this end, we describe two broad classes of embedded block coding techniques. The first is based on context-adaptive coding, as introduced in Section 2.4.1, while the second is based on an embedded extension to the quad-tree coding principles introduced in Section 2.4.4. Both approaches have merits and both were investigated during the development of the JPEG2000 image compression standard; the context-adaptive approach was ultimately selected.

---

[5] It is possible for complex dependencies introduced by coding the signalling overhead to disturb the monotonicity of $L(\lambda)$ in λ. This possibility does not appear to have a detrimental effect in practical applications of the algorithm.

8.3.1   BIT-PLANE CODING

Since each code-block is to be represented by an efficient embedded bit-stream, prefixes of the bit-stream must correspond to successively finer quantizations of the block's sample values. In fact, the underlying quantizers are inevitably embedded [151, §4B]. In the ensuing discussion we restrict our attention to the practically appealing case of embedded deadzone quantization, as introduced in Section 3.2.7.

Recall that a deadzone quantizer with step size Δ produces quantization indices

$$q = \text{sign}(y) \begin{cases} \left\lfloor \frac{|y|}{\Delta} + \tau \right\rfloor & \frac{|y|}{\Delta} + \tau > 0 \\ 0 & \text{otherwise} \end{cases} \qquad (8.11)$$

where y denotes a subband sample from the code-block and τ is a parameter controlling the width of the central deadzone. When $\tau = \frac{1}{2}$, the quantizer is uniform, while $\tau = 0$ corresponds to the case in which the deadzone width is 2Δ. The possibility of negative values for τ has already been discussed in connection with equation (3.31). The quantization intervals, denoted $\mathcal{I}_q^{(0)}$, are illustrated in Figure 8.9.

Let $\chi = \text{sign}(y)$ and $v = |q|$ denote the sign and magnitude of q, respectively[6]. Also, let

$$v^{(p)} = \left\lfloor \frac{v}{2^p} \right\rfloor$$

denote the value formed by dropping p LSBs (Least Significant Bits) from $v = v^{(0)}$. Employing the identity,

$$\left\lfloor \frac{\lfloor a \rfloor}{b} \right\rfloor = \left\lfloor \frac{a}{b} \right\rfloor, \quad \forall a \in \mathbb{R} \text{ and } b \in \mathbb{N} \text{ (i.e., } b \text{ a positive integer)}$$

we see that χ and $v^{(p)}$ are the sign and magnitude of the index, $q^{(p)}$, obtained using the coarser quantizer

$$q^{(p)} = \text{sign}(y) \begin{cases} \left\lfloor \frac{|y|}{2^p \Delta} + \frac{\tau}{2^p} \right\rfloor & \frac{|y|}{2^p \Delta} + \frac{\tau}{2^p} > 0 \\ 0 & \text{otherwise} \end{cases}$$

Figure 8.9 illustrates the corresponding quantization intervals, $\mathcal{I}_{q^{(p)}}^{(p)}$.

This family of deadzone quantizers has three notable characteristics: 1) the step sizes are given by $\Delta^{(p)} = 2^p \Delta$; 2) the deadzone width parameters, $\tau^{(p)} = 2^{-p}\tau$, rapidly converge to 0 as p increases; and 3)

---

[6] Strictly speaking, when $q = 0$ the sign of q is indeterminate; this will be reflected in the fact that it is not coded. It is convenient here to associate χ with the sign of the original subband sample, y, which is the same as that of q whenever $q \neq 0$.

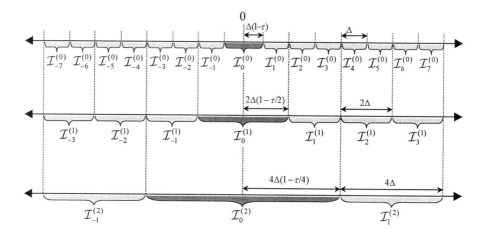

Figure 8.9.   Family of embedded deadzone quantizers.

each quantization interval, $\mathcal{I}_{q^{(p)}}^{(p)}$, is embedded within a coarser interval, $\mathcal{I}_{q^{(p+1)}}^{(p+1)}$. In view of property (2), it makes sense to restrict our attention to the case $\tau = 0$. In this case, all quantizers have the same structure, with a deadzone twice as wide as the other intervals. It is worth noting, however, that the coding techniques described in this chapter are applicable to the more general case in which $\tau \neq 0$. In fact the JPEG2000 standard supports such general deadzone quantizers.

Let $y[\mathbf{j}] \equiv y[j_1, j_2]$ denote the sequence of subband samples belonging to the relevant code-block, having height J_1 and width J_2 so that $0 \leq j_1 < J_1$ and $0 \leq j_2 < J_2$. Similarly, let $\chi[\mathbf{j}]$ and $v^{(p)}[\mathbf{j}]$ denote the sign and magnitude of the embedded quantization indices. Suppose that K is a sufficient number of bits to represent any of the quantization index magnitudes, meaning that $v^{(K)}[\mathbf{j}] = 0$ for all $\mathbf{j}$. Finally, let $v^p[\mathbf{j}] \in \{0, 1\}$ be the LSB of $v^{(p)}[\mathbf{j}]$, which is also bit p of $v[\mathbf{j}]$. We say that bits $v^p[\mathbf{j}]$ from all samples in the code-block constitute "magnitude bit-plane" p.[7] There are at most K non-trivial magnitude bit-planes.

An embedded bit-stream may be formed in the manner suggested by Figure 8.10. The most significant magnitude bit-plane, $v^{K-1}[\mathbf{j}]$, is coded first, together with the sign, $\chi[\mathbf{j}]$, of any sample for which $v^{K-1}[\mathbf{j}] \neq 0$. If the bit-stream is truncated at this point, the decoder can reconstruct

---

[7]To avoid confusion, we point out that bit-planes were numbered in the opposite order for the purpose of describing zero-tree coding principles in Chapter 7.

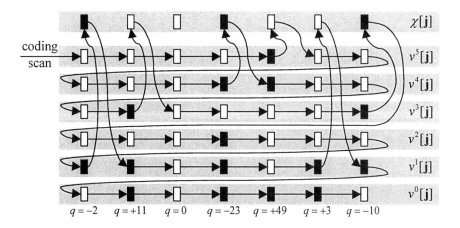

Figure 8.10. Bit-plane coding procedure. Non-zero magnitude bits and negative signs are identified by black boxes.

the coarsest quantization indices, $q^{(K-1)}[\mathbf{j}]$. The next most significant magnitude bit-plane, $v^{K-2}[\mathbf{j}]$, is then coded together with the sign of any sample for which $v^{K-2}[\mathbf{j}] = 1$ and $v^{(K-1)}[\mathbf{j}] = 0$. The process continues in this way for each magnitude bit-plane, p, including the sign of those samples for which $v^p[\mathbf{j}]$ is the most significant non-zero bit. We refer to this process as bit-plane coding and we use the term "bit-plane" loosely to refer to both the magnitude and the associated sign information. If the bit-stream is truncated at the end of bit-plane p, the decoder can reconstruct quantization indices, $q^{(p)}[\mathbf{j}]$.

A variety of techniques may be employed to code the magnitude and sign bits. An efficient bit-plane coder, however, should exploit the substantial redundancy which generally exists between successive bit-planes. A straightforward application of the entropy expansion formula of equation (2.5) yields

$$
\begin{aligned}
H\left(\mathbf{Q}^{(p)}\right) &= H\left(\mathbf{Q}^{(K-1)}, \mathbf{Q}^{(K-2)}, \ldots, \mathbf{Q}^{(p)}\right) = H\left(\mathbf{Q}^{(K-1)}\right) \\
&+ H\left(\mathbf{Q}^{(K-2)} \mid \mathbf{Q}^{(K-1)}\right) + \cdots + H\left(\mathbf{Q}^{(p)} \mid \mathbf{Q}^{(K-1)}, \ldots, \mathbf{Q}^{(p+1)}\right) \\
&= H\left(\mathbf{Q}^{(K-1)}\right) + H\left(\mathbf{Q}^{(K-2)} \mid \mathbf{Q}^{(K-1)}\right) + \cdots + H\left(\mathbf{Q}^{(p)} \mid \mathbf{Q}^{(p+1)}\right)
\end{aligned}
$$

where $\mathbf{Q}^{(p)}$ is the random vector whose outcome represents the quantization indices, $q^{(p)}[\mathbf{j}]$, for all samples in the code-block. Thus, so long as we exploit all available information from the previous bit-planes when coding the magnitude and sign information for a current bit-plane, the total number of bits associated with the most significant $K - p$ bit-planes will be identical to the number of bits required to code the quantization indices $q^{(p)}[\mathbf{j}]$ directly.

In view of the preceding arguments, a natural set of truncation points for the embedded bit-stream is the set of bit-plane end-points. If the truncation lengths are measured in bits and the coding is efficient, then this policy would yield $Z = K$ non-zero truncation points with $L^{(0)} = 0$ and $L^{(z)} \gtrsim H\left(\mathbf{Q}^{(K-z)}\right)$, $z = 1, 2, \ldots, K$. In practice, we shall find it convenient to measure length in bytes rather than bits.

Early bit-plane coders [150, 174] processed the samples following a deterministic scan (line by line) within each bit-plane. In this case, the bit-plane end-points are the only natural truncation points for the embedded bit-stream. To achieve a finer embedding with many more useful truncation points, a data dependent processing order is called for, in which we first code those bits which yield the largest reduction in distortion relative to the increase in length [93, 113, 133, 149].

In the ensuing sub-sections we first discuss conditional coding techniques, whereby each bit is efficiently coded by exploiting information which has already been coded in the same or previous bit-planes. We then turn our attention to the question of the order in which the information bits should be coded. In particular, Section 8.3.2 describes the bit-plane coding primitives adopted by JPEG2000, while Section 8.3.3 describes JPEG2000's data dependent scan. Section 8.3.4 describes an alternative quad-tree coding approach which utilizes the list processing techniques introduced by SPIHT to achieve an appropriate data dependent processing order.

8.3.2   CONDITIONAL CODING OF BIT-PLANES

In this section we describe the bit-plane coding primitives defined by the JPEG2000 image compression standard. At any given sample location, $\mathbf{j}$, in any given bit-plane, p, we must code the value of $v^p[\mathbf{j}]$ and possibly also the sign, $\chi[\mathbf{j}]$, as suggested by Figure 8.10. These are binary events and we employ an adaptive binary arithmetic coder. The specific arithmetic coding variant employed by JPEG2000 is the MQ coder whose details are deferred until Section 12.1. For our present discussion it is sufficient to understand the arithmetic coder as a "machine" such as that illustrated in Figure 2.6, which efficiently represents a sequence of binary outcomes subject to the provision of good probability estimates. The adaptive probability models evolve within a number of distinct contexts which depend upon information which has already been coded. Any of the arithmetic coding procedures and probability estimation techniques discussed in Section 2.3 would be appropriate and should yield comparable compression efficiency.

Ideally, the coding of each binary decision should be conditioned upon the sign and magnitude information which has already been coded for both the sample at hand and all other samples in the code-block. In practice, however, the number of distinct coding contexts must be restricted for practical reasons. Perhaps even more importantly, each additional coding context entails a learning penalty, as binary decisions are coded inefficiently until sufficient information is available to form good probability estimates. This problem is discussed in Section 2.3.5 and is a particular concern in the present application since each code-block is relatively small and is coded independently.

Image subband samples tend to exhibit distributions which are heavily skewed toward small amplitudes. As a result, when $v^{(p+1)}[\mathbf{j}] = 0$, meaning that $y[\mathbf{j}] \in \mathcal{I}_0^{(p+1)}$, we can expect that $y[\mathbf{j}]$ is also very likely to be found in the smaller deadzone, $\mathcal{I}_0^{(p)}$. Equivalently, the conditional PMF, $f_{V^p|V^{(p+1)}}(v^p, 0)$, is heavily skewed toward the outcome $v^p = 0$. For this reason, an important element in the construction of efficient coding contexts is the so-called "significance" of a sample, defined by

$$\sigma^{(p)}[\mathbf{j}] = \begin{cases} 1 & \text{if } v^{(p)}[\mathbf{j}] > 0 \\ 0 & \text{if } v^{(p)}[\mathbf{j}] = 0 \end{cases}$$

To help decouple our description of the coding operations from the order in which they are applied, we introduce the notion of a binary "significance state," $\sigma[\mathbf{j}]$. At any point in the coding process, $\sigma[\mathbf{j}]$ assumes the value of $\sigma^{(p)}[\mathbf{j}]$ where p is the most recent (least significant) bit for which information concerning sample $y[\mathbf{j}]$ has been coded up to that point. Equivalently, we initialize the significance state of all samples in the code-block to 0 at the beginning of the coding process and then toggle the state to $\sigma[\mathbf{j}] = 1$ immediately after coding the first non-zero magnitude bit for sample $y[\mathbf{j}]$.

Given the importance of the condition $\sigma[\mathbf{j}] = 0$, we identify three different types of primitive coding operations as follows. If $\sigma[\mathbf{j}] = 0$ we refer to the task of coding $v^p[\mathbf{j}]$ as "significance coding," since $v^p[\mathbf{j}] = 1$ if and only if the significance state transitions to $\sigma[\mathbf{j}] = 1$ in this coding step. In the event that the sample does become significant, we must invoke a "sign coding" primitive to identify $\chi[\mathbf{j}]$. For samples which are already significant, the value of $v^p[\mathbf{j}]$ serves to refine the decoder's knowledge of the non-zero sample magnitude. Accordingly, we invoke a "magnitude refinement coding" primitive.

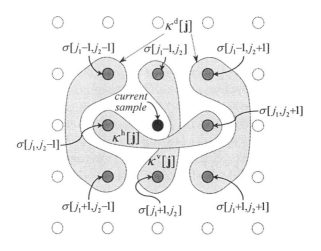

Figure 8.11.   Formation of significance coding contexts..

SIGNIFICANCE CODING (NORMAL MODE)

The significance coding primitive involves a normal mode and a run mode. We describe the normal mode first. In this mode, one of 9 different contexts is used to code the significance (i.e., the value of $v^p[\mathbf{j}]$) of a sample which is currently insignificant (i.e., $v^{(p+1)}[\mathbf{j}] = 0$). Context selection is based upon the significance of the sample's 8 immediate neighbours, as shown in Figure 8.11. The context label, $\kappa^{\text{sig}}[\mathbf{j}]$, is formed from three intermediate quantities,

$$\kappa^{\text{h}}[\mathbf{j}] = \sigma[j_1, j_2 - 1] + \sigma[j_1, j_2 + 1]$$
$$\kappa^{\text{v}}[\mathbf{j}] = \sigma[j_1 - 1, j_2] + \sigma[j_1 + 1, j_2]$$
$$\kappa^{\text{d}}[\mathbf{j}] = \sum_{k_1 = \pm 1} \sum_{k_2 = \pm 1} \sigma[j_1 + k_1, j_2 + k_2]$$

Samples which lie beyond the boundaries of the relevant code-block are regarded as insignificant for the purpose of constructing these three quantities.

Table 8.1 shows how $\kappa^{\text{sig}}[\mathbf{j}]$ is derived from the intermediate quantities, $\kappa^{\text{h}}[\mathbf{j}]$, $\kappa^{\text{v}}[\mathbf{j}]$ and $\kappa^{\text{d}}[\mathbf{j}]$. The principles underlying this context design are as follows. If the code-block belongs to an LH (vertically high-pass) subband, the significant samples are most likely to arise from horizontally oriented features in the image, as suggested by Figure 4.18. Accordingly, significant horizontal neighbours are considered most indicative of the current sample's significance. After the horizontal neighbours, vertical neighbours are considered the next most important indicators of

Table 8.1. Assignment of context labels for significance coding.

$\kappa^{\text{sig}}[\mathbf{j}]$	LL and LH blocks			HL blocks			HH blocks	
	$\kappa^h[\mathbf{j}]$	$\kappa^v[\mathbf{j}]$	$\kappa^d[\mathbf{j}]$	$\kappa^h[\mathbf{j}]$	$\kappa^v[\mathbf{j}]$	$\kappa^d[\mathbf{j}]$	$\kappa^d[\mathbf{j}]$	$\kappa^h[\mathbf{j}]+\kappa^v[\mathbf{j}]$
8	2	x[a]	x	x	2	x	≥ 3	x
7	1	≥ 1	x	≥ 1	1	x	2	≥ 1
6	1	0	≥ 1	0	1	≥ 1	2	0
5	1	0	0	0	1	0	1	≥ 2
4	0	2	x	2	0	x	1	1
3	0	1	x	1	0	x	1	0
2	0	0	≥ 2	0	0	≥ 2	0	≥ 2
1	0	0	1	0	0	1	0	1
0	0	0	0	0	0	0	0	0

[a] "x" means "don't care."

significance. The more distant diagonal neighbours are of interest only if at most one of the four most immediate neighbours is already significant. Context formation for the HL (horizontally high-pass) subband is analogous to that for the LH subband, with the roles of horizontal and vertical neighbours interchanged. If the code-block belongs to an HH (horizontally and vertically high-pass) subband, we expect to encounter diagonally oriented regions of significance, as suggested by Figure 4.18. For these code-blocks, diagonal neighbours are the most important indicators of significance in the current sample.

Although the above arguments go some way to explaining the context design rules given in Table 8.1, we hasten to point out that these rules are actually the product of substantial empirical studies. They are also motivated in part by the need for simple, regular constructions which are amenable to efficient implementation strategies in both hardware and software.

It is worth pointing out that the amount of information represented by the significance of any of the eight neighbours in Figure 8.11 depends upon the order in which the bits are coded. If each bit-plane, p, is coded by visiting the samples in raster scan order, then the four neighbours in the current sample's causal past will have their significance determined by the deadzone interval, $\mathcal{I}_0^{(p)}$, while the significance state of the remaining four neighbours will be determined in relation to the larger deadzone, $\mathcal{I}_0^{(p+1)}$. That is, for each $\mathbf{k} \in \mathcal{N} = \{(0,-1),(-1,-1),$ $(-1,0),(-1,1)\}$, $\sigma[\mathbf{j}+\mathbf{k}] = \sigma^{(p)}[\mathbf{j}+\mathbf{k}]$, while for each for $\mathbf{k} \in -\mathcal{N}$,

$\sigma [\mathbf{j} + \mathbf{k}] = \sigma^{(p+1)} [\mathbf{j} + \mathbf{k}]$. With the data dependent scanning order described in Section 8.3.3, no such simple interpretation is possible[8].

SIGNIFICANCE CODING (RUN MODE)

At moderate to high compression ratios, most of the subband samples are insignificant in all of the bit-planes which are actually included in the final pack-stream. To see why this must be so, observe that whenever a sample becomes significant we must code the significance event (usually with respect to a conditional PMF skewed heavily toward insignificance) and also the sign. The combined cost of these two binary events is unlikely to be less than 2 bits and is usually considerably more. Thus, at a compressed bit-rate of 0.5 bits/sample (according to Table 1.1, this is sufficient to achieve moderate image quality) we may conservatively estimate that more than $\frac{3}{4}$ of the samples are insignificant in all of the available bit-planes. Empirical observations suggest that this estimate is indeed very conservative. Also, even those samples which are eventually coded as significant, may be insignificant for many of the initial bit-planes.

Since code-block samples are expected to be predominantly insignificant, a run mode is introduced to dispatch multiple insignificant samples with a single binary symbol. The run mode serves primarily to reduce complexity, although very small improvements in compression performance are also typical. The run mode is entered if and only if the following three conditions hold simultaneously.

1. Four consecutive samples (following the scan shown in Figure 8.15) must currently be insignificant. That is, $\sigma [\mathbf{j}_r] = 0$ for $0 \le r < 4$, where $\mathbf{j}_0 = \mathbf{j}$ and $\mathbf{j}_r$ is the rth position beyond $\mathbf{j}$ in the scan.

2. All four samples must currently have insignificant neighbourhoods. That is, $\kappa^{\text{sig}} [\mathbf{j}_r] = 0$ for $0 \le r < 4$.

3. The group of four samples must be aligned on a four sample boundary within the scan. As we shall see in Section 8.3.3, the scanning pattern itself works column by column on stripes of four rows at a time. This means that the samples must constitute a single stripe column, where the stripe is required to have the full height of four sample rows.

---

[8]Considering only the horizontal neighbours, for example, it can happen that only the left neighbour has already been coded in bit-plane p, but it can also happen that both of the horizontal neighbours have already been coded in bit-plane p (this is possible in a cleanup coding pass). It can even happen that neither of the horizontal neighbours has been coded in bit-plane p (this is possible in a significance propagation pass).

In run mode, a binary "run interruption" symbol is coded to indicate whether or not all four samples remain insignificant in the current bit-plane, p. Insignificance is identified by the symbol 0, while a value of 1 means that at least one of the four samples becomes significant. The run interruption symbol is coded within its own context, denoted $\kappa^{\text{run}} = 9$.

If one or more of the four samples becomes significant during the current bit-plane, p, the insignificant run length, r, must also be coded, followed by the sign of the first significant sample, $\chi[j_r]$. Experience shows that the run length is nearly uniformly distributed, which is also to be expected if samples transition to significance with very low probability. For this reason the 2-bit run-length, r, is coded one bit at a time, starting with the most significant bit, using a non-adaptive uniform probability model. The MQ arithmetic coder employed by JPEG2000 has a special non-adaptive state in its probability estimation state machine (state 46 of Table 2.1), which is used for exactly this purpose. After run interruption, the significance of the remaining samples is coded in normal mode, until the conditions for run mode are encountered again.

It is worth pointing out that the restriction to appropriately aligned groups of samples (condition (3) above) exists only to facilitate efficient implementation. Studies leading to the introduction of the run mode in [147] revealed that taking the samples in groups of four at a time roughly minimizes the number of symbols which must be arithmetically coded. The interested reader is also invited to consider parallels between JPEG2000 and the run mode in JPEG-LS (see Section 20.3).

SIGN CODING

The sign coding primitive is invoked at most once for any sample, $y[j]$, immediately after the significance coding operation in which the sample first becomes significant. Most algorithms proposed for coding subband sample values, whether embedded or otherwise, treat the sign as an independent, uniformly distributed random variable, devoting 1 bit to coding its outcome. It turns out, however, that the signs of neighbouring sample values exhibit significant statistical redundancy. Some arguments to suggest that this should be the case are presented in [149]. The early bit-plane coding algorithm described in [150] employed conditional sign coding, with 81 contexts derived from the signs of significant neighbours[9]. Wu [173] advocates similar context modeling techniques for the coding of sign bits.

---

[9]Details of the sign coding procedure did not appear in the paper itself. However, the use of conditional sign coding is evident from the source code which was publically released in mid-1994 and is still available from "http://www-video.eecs.berkeley.edu/download/scalable/".

In the interest of minimizing both complexity and the learning penalty, JPEG2000's sign coding primitive employs a meagre 5 contexts. Context design is based upon the relevant sample's immediate four neighbours, each of which may be in one of three states: significant and positive; significant and negative; or insignificant. There are thus 81 unique neighbourhood configurations. The following observations allow us to dramatically reduce this number. Firstly, it is reasonable to expect horizontal and vertical symmetry in the joint distribution of the signs. Further, if the left and right neighbours of a sample are both significant but with opposite signs then they offer no more indication as to the sign of the current sample than a pair of insignificant neighbours. Similarly, disagreeing vertical neighbours might as well be insignificant. Together, these arguments suggest that most if not all of the useful neighbourhood information should be embodied by the quantities

$$\chi^{h}\left[\mathbf{j}\right] = \chi\left[j_1, j_2 - 1\right]\sigma\left[j_1, j_2 - 1\right] + \chi\left[j_1, j_2 + 1\right]\sigma\left[j_1, j_2 + 1\right]$$
$$\chi^{v}\left[\mathbf{j}\right] = \chi\left[j_1 - 1, j_2\right]\sigma\left[j_1 - 1, j_2\right] + \chi\left[j_1 + 1, j_2\right]\sigma\left[j_1 + 1, j_2\right]$$

These quantities represent the "net sign bias" of the horizontal and vertical neighbours, respectively. Both quantities lie in the range $-2 \leq \chi^{h}\left[\mathbf{j}\right], \chi^{v}\left[\mathbf{j}\right] \leq 2$, resulting in 25 neighbourhood configurations.

It is also reasonable to assume that the conditional distribution of $\chi\left[\mathbf{j}\right]$ given any particular neighbourhood, should be identical to the distribution of $-\chi\left[\mathbf{j}\right]$ given a neighbourhood in which the signs of all neighbours are flipped. In this way we may identify $\left\lceil\frac{25}{2}\right\rceil = 13$ unique neighbourhood configurations, which are reduced to 5 by truncating the horizontal and vertical sign biases to the range -1 through 1. Specifically, define the truncated bias terms by

$$\bar{\chi}^{h}\left[\mathbf{j}\right] = \mathrm{sign}\left(\chi^{h}\left[\mathbf{j}\right]\right)\min\left\{1, \left|\chi^{h}\left[\mathbf{j}\right]\right|\right\}$$
$$\bar{\chi}^{v}\left[\mathbf{j}\right] = \mathrm{sign}\left(\chi^{v}\left[\mathbf{j}\right]\right)\min\left\{1, \left|\chi^{v}\left[\mathbf{j}\right]\right|\right\}$$

The context label, κ^{sign}, and sign-flipping factor, χ^{flip}, are then given by Table 8.2. The single binary symbol which is coded with respect to context κ^{sign} takes the value 0 if $\chi\left[\mathbf{j}\right] \cdot \chi^{\mathrm{flip}} = 1$ and 1 if $\chi\left[\mathbf{j}\right] \cdot \chi^{\mathrm{flip}} = -1$.

MAGNITUDE REFINEMENT CODING

The magnitude refinement primitive is used to code the next magnitude bit, $v^{p}\left[\mathbf{j}\right]$, of a sample which is already significant; i.e., $\sigma^{(p+1)}\left[\mathbf{j}\right] = 1$. This information refines the coarser quantization index, $q^{(p+1)}\left[\mathbf{j}\right]$, to the next finer index, $q^{(p)}\left[\mathbf{j}\right]$. As already noted, subband samples tend to exhibit symmetric distributions, $f_Y\left(y\right)$, which are heavily skewed

Table 8.2. Assignment of context labels and flipping factor for sign coding.

$\bar{\chi}^{\mathrm{h}}\,[\mathbf{j}]$	$\bar{\chi}^{\mathrm{v}}\,[\mathbf{j}]$	κ^{sign}	χ^{flip}
1	1	14	1
1	0	13	1
1	−1	12	1
0	1	11	1
0	0	10	1
0	−1	11	−1
−1	1	12	−1
−1	0	13	−1
−1	−1	14	−1

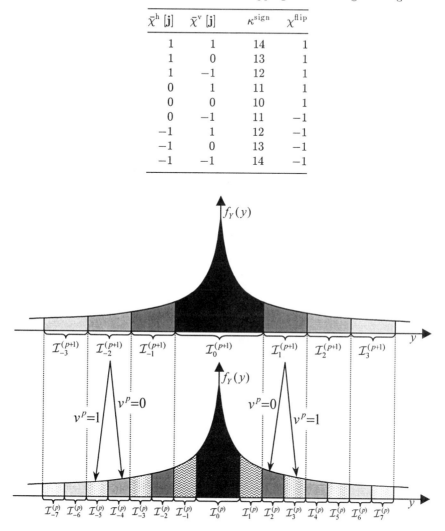

Figure 8.12. Typical subband sample PDF, f_Y, and its impact on the relative probabilities of quantization intervals $\mathcal{I}_{q^{(p+1)}}^{(p+1)}$ and $\mathcal{I}_{q^{(p)}}^{(p)}$.

toward $y = 0$. As suggested by Figure 8.12, the conditional PMF $f_{V^p|Q^{(p+1)}}(v^p \mid q^{(p+1)})$ typically exhibits the following characteristics: 1) it is independent of the sign of $q^{(p+1)}$; 2) $f_{V^p|Q^{(p+1)}}(0 \mid q^{(p+1)}) > \frac{1}{2}$ for all $q^{(p+1)}$; and 3) $f_{V^p|Q^{(p+1)}}(0 \mid q^{(p+1)}) \approx \frac{1}{2}$ for large $|q^{(p+1)}|$. As a result, it is desirable to condition the coding of $v^p\,[\mathbf{j}]$ upon the value of

Table 8.3.   Assignment of context labels for magnitude refinement coding.

$\overleftarrow{\sigma}\,[\mathbf{j}]$	$\kappa^{\mathrm{sig}}\,[\mathbf{j}]$	κ^{mag}
0	0	15
0	> 0	16
1	$\mathrm{x^a}$	17

[a] "x" means "don't care."

$v^{(p+1)}\,[\mathbf{j}]$ when $v^{(p+1)}\,[\mathbf{j}]$ is small. We also find that it can be useful to exploit redundancy between adjacent sample magnitudes when $v^{(p+1)}\,[\mathbf{j}]$ is small.

The above observations serve to justify the assignment of a magnitude refinement coding context, κ^{mag}, according to the rules expressed in Table 8.3. Here, $\overleftarrow{\sigma}\,[\mathbf{j}]$ denotes the value of the significance state variable, $\sigma\,[\mathbf{j}]$, "delayed" by one bit-plane. Both $\sigma\,[\mathbf{j}]$ and $\overleftarrow{\sigma}\,[\mathbf{j}]$ are initialized to 0; when the sample first becomes significant, $\sigma\,[\mathbf{j}]$ is toggled to 1, but $\overleftarrow{\sigma}\,[\mathbf{j}]$ remains zero until after the first magnitude refinement bit has been coded. For subsequent refinement bits, $\overleftarrow{\sigma}\,[\mathbf{j}] = 1$. In this way, $\overleftarrow{\sigma}\,[\mathbf{j}]$ provides a crude indicator of the coarser index magnitude, $v^{(p+1)}\,[\mathbf{j}]$, when we come to code $v^p\,[\mathbf{j}]$. In particular, $\overleftarrow{\sigma}\,[\mathbf{j}] = 1$ if and only if $v^{(p+1)}\,[\mathbf{j}] \geq 2$ (we already know that $v^{(p+1)}\,[\mathbf{j}] > 0$; otherwise, we would be using the significance coding primitive to code $v^p\,[\mathbf{j}]$).

SUMMARY REMARKS

The bit-plane coding primitives described above involve arithmetic coding with a total of 18 different contexts, identified by the unique labels of κ^{sig}, κ^{run}, κ^{sign} and κ^{mag}. Of course, the exact value of these labels is immaterial. What is important is that there are sufficiently few contexts for the states of all 18 probability models to be maintained in high speed registers within a dedicated hardware implementation. The bit-plane coding primitives which we have described are exactly those of the JPEG2000 standard. However, it should be noted that JPEG2000 permits several mode variations which may be employed to increase parallelism or reduce complexity at high bit-rates. These various modes are described carefully in Section 12.4.

8.3.3   DYNAMIC SCAN

FRACTIONAL BIT-PLANES

As mentioned previously, early bit-plane coders processed each bit-plane following a deterministic (e.g., line by line) scan through the sub-

band samples. It is instructive to consider the rate-distortion characteristic described by the resulting embedded bit-stream under truncation.
To this end, it is helpful to introduce temporary notation $L_{bp}^{(p)}$ and $D_{bp}^{(p)}$ to
denote the length and distortion which result when the least significant
p magnitude bit-planes are discarded from the embedded bit-stream.
Suppose now that we truncate the bit-stream at an arbitrary point, L,
between $L_{bp}^{(p+1)}$ and $L_{bp}^{(p)}$. The truncated bit-stream represents a fraction,
ρ, of the samples down to bit-plane p while the remaining samples are
represented only down to bit-plane $p + 1$. Since the scan has no dependence on the sample values themselves and the statistics are assumed
stationary over the code-block, the expected length and distortion are
given by

$$\mu_L = \rho L_{bp}^{(p)} + (1 - \rho) L_{bp}^{(p+1)}$$
$$\mu_D = \rho D_{bp}^{(p)} + (1 - \rho) D_{bp}^{(p+1)}$$

Although the actual length and distortion will vary about these expected
values, the law of large numbers ensures that $L \approx \mu_L$ and $D \approx \mu_D$ for
sufficiently large code-blocks. Consequently the rate-distortion function
of a deterministically scanned bit-plane coder under truncation should
follow the piecewise-linear characteristic illustrated in Figure 8.13. This
is just the convex interpolation of the length and distortion values associated with the bit-plane end-points.

Following the arguments at the end of Section 8.3.1, an efficient bit-
plane coder has the property that $L_{bp}^{(p)}$ and $D_{bp}^{(p)}$ are the length and distortion which could be expected from an efficient non-embedded coding of
the same scalar quantization indices, $q^{(p)}$ [j]. Consequently, the bit-plane
end-point coordinates actually lie on the rate-distortion curve swept out
by an entropy coded deadzone scalar quantizer (ECDZQ) as the step size
is modulated. This is represented by the convex curve in Figure 8.13.
Although the bit-plane end-points are optimal with respect to ECDZQ,
truncation to intermediate lengths is necessarily sub-optimal since the
ECDZQ rate-distortion curve is convex. This is the same phenomenon
which accounts for the much more devastating loss of performance encountered when truncating a non-embedded bit-stream, as illustrated in
Figure 8.1.

In order to improve the rate-distortion performance of bit-plane coding under truncation, we should first code those bits which are likely
to result in the largest reduction in distortion relative to the increase
in code length. In particular, we employ the notion of "fractional bit-
plane" coding passes. JPEG2000 employs three coding passes for each
bit-plane, p, denoted $\mathcal{P}^{(p,0)}$, $\mathcal{P}^{(p,1)}$ and $\mathcal{P}^{(p,2)}$. The sample locations

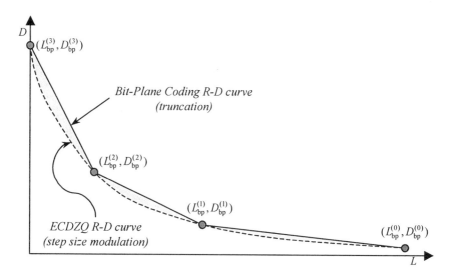

Figure 8.13. Rate-distortion characteristic of deterministically scanned bit-plane coding under truncation.

$\mathbf{j} \in \mathcal{P}^{(p,0)}$ are expected to yield the largest reduction in distortion relative to the increase in code length, as quantization indices $q^{(p+1)}[\mathbf{j}]$ are refined to the finer indices, $q^{(p)}[\mathbf{j}]$. At the other extreme, $\mathcal{P}^{(p,2)}$ contains those sample locations whose distortion-length "slope" is expected to be smallest.

Each sample location in the code-block appears in exactly one of the three coding passes, $\mathcal{P}^{(p,0)}$ through $\mathcal{P}^{(p,2)}$, which together represent all information for bit-plane p. Accordingly, the distortion at the end of coding pass $\mathcal{P}^{(p,2)}$ should be exactly $D_{\mathrm{bp}}^{(p)}$. Also, since the combined information embodied by the three coding passes is identical to that which is conveyed by coding bit-plane p line by line (or in any other order, for that matter), we expect the code length at the end of pass $\mathcal{P}^{(p,2)}$ to be very close to $L_{\mathrm{bp}}^{(p)}$. This last point is contingent on the assumption that the coding techniques are able to exploit most of the available statistical redundancy, approaching the entropy of the quantization indices which are coded, regardless of the order in which they are coded. In practice, the order in which samples are coded has some effect upon the interpretation of the limited conditional coding contexts described in Section 8.3.2, as well as the evolution of the adaptive probability models used by the arithmetic coder. Nevertheless, our experience indicates that the introduction of coding passes has very little impact at all on the efficiency with which the information in each bit-plane is coded.

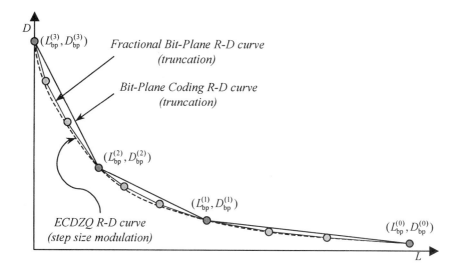

Figure 8.14. Fractional bit-plane rate-distortion characteristic under truncation.

The samples belonging to each coding pass are visited in a deterministic order, so that the rate-distortion characteristic of fractional bit-plane coding under truncation is obtained by convex interpolation of the length and distortion values at the coding pass end-points. This is illustrated in Figure 8.14. The key observation to make here is that it is possible to get much closer to the convex rate-distortion characteristic of step size modulated ECDZQ by separating the information in each bit-plane into multiple coding passes with decreasing distortion-length slopes. The embedding is substantially improved, even though coding efficiency at the bit-plane end-points is substantially unaffected.

THE JPEG2000 CODING PASSES

We turn our attention now to the specific implementation of fractional bit-planes in JPEG2000 and the deterministic scan followed within each coding pass. For each of the three passes, the coder scans through the samples of the code-block following the stripe-oriented scan shown in Figure 8.15. Each stripe contains four sample rows, with the possible exception of the last stripe in the block. Within each stripe, the samples are visited column by column, from left to right.

Membership of each of the three coding passes is determined dynamically, based upon the significance state of each sample's eight immediate neighbours, as depicted in Figure 8.15. These are the same neighbours

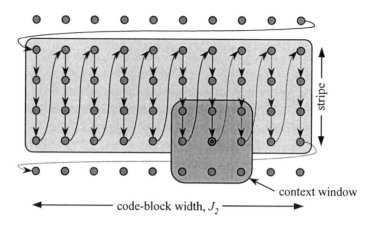

Figure 8.15. Stripe-oriented scanning pattern followed within each coding pass.

which are used to determine the conditional coding contexts described in Section 8.3.2.

Significance Propagation Pass: This is the first coding pass, $\mathcal{P}_i^{(p,0)}$, in each bit-plane p.[10] Sample location $\mathbf{j}$ belongs to this pass if it is insignificant, but has a significant neighbourhood; that is, at least one of its eight neighbours must be significant. Using the definition of $\kappa^{\mathrm{sig}}[\mathbf{j}]$ in Table 8.1, membership in $\mathcal{P}_i^{(p,0)}$ may be expressed by the conditions $\sigma[\mathbf{j}] = 0$ and $\kappa^{\mathrm{sig}}[\mathbf{j}] > 0$. These conditions are designed to include those samples which are most likely to become significant in bit-plane p. Moreover, for a broad class of probability models, the samples in this coding pass are likely to yield the largest decrease in distortion relative to the increase in code length [113].

Each sample in the pass is coded using the significance coding primitive described in Section 8.3.2. Note that the conditions for run mode will never be satisfied in this coding pass. The sign coding primitive is invoked immediately after any significance coding step in which the sample becomes significant; i.e., $v^p[\mathbf{j}] = 1$. It is worth noting that samples which become significant in this pass may give rise to waves of significance determination events which propagate along connected image features such as edges. This is because membership of the coding pass is assessed incrementally. Once a sample becomes significant, the four neighbours which have not yet been visited in the

---

[10] Actually, the definitions of the first two coding passes leave them empty in the code-block's most significant bit-plane, $p = K_i - 1$, which therefore has only one coding pass, $\mathcal{P}^{(K_i-1,2)}$.

scan then also have significant neighbourhoods, and will be included in $\mathcal{P}_i^{(p,0)}$ unless they are already significant. We name this the "significance propagation pass" to remind the reader that its members are assessed dynamically.

Magnitude Refinement Pass: This is the second coding pass, $\mathcal{P}_i^{(p,1)}$, for bit-plane p of code-block $\mathcal{B}_i$. In this coding pass, the magnitude refinement primitive is used to code magnitude bit $v^p[\mathbf{j}]$ of any sample which was already significant in the previous bit-plane; i.e., $\sigma^{(p+1)}[\mathbf{j}] = 1$. Equivalently, $\mathcal{P}_i^{(p,1)}$ includes any sample whose significance state is $\sigma[\mathbf{j}] = 1$, which was not already included in $\mathcal{P}_i^{(p,0)}$.

Cleanup Pass: This final coding pass, $\mathcal{P}_i^{(p,2)}$, includes all samples for which information has not already been coded in bit-plane p. From the definitions of $\mathcal{P}_i^{(p,0)}$ and $\mathcal{P}_i^{(p,1)}$, we see that samples coded in this pass must be insignificant. The significance coding primitive is used to code $v^p[\mathbf{j}]$ for all samples belonging to this pass. We note that the conditions for run mode may occur only in this coding pass. As explained in Section 8.3.2, run mode is entered if an entire stripe column contains insignificant samples with entirely insignificant neighbours. The significance of all of these samples is coded in $\mathcal{P}_i^{(p,2)}$, using the run mode to identify the first if any of the samples which becomes significant in bit-plane p. Coding of any remaining samples in the stripe column proceeds in normal mode. As always, the sign coding primitive is invoked for any sample which becomes significant, immediately after its significance is coded.

Let K_i be a sufficient number of magnitude bit-planes to represent the samples in $\mathcal{B}_i$. As explained in Section 8.3.1, K_i is any integer satisfying

$$2^{K_i} > \max_{\mathbf{j} \in \mathcal{B}_i} v_i[\mathbf{j}]$$

The selected value of K_i is signalled separately, as part of the summary information for each code-block, $\mathcal{B}_i$; this is discussed further in Section 8.4. Then the first potentially non-empty coding pass is $\mathcal{P}_i^{(K_i-1,2)}$, while the last is $\mathcal{P}_i^{(0,2)}$. The sequence of coding passes which constitutes the embedded block bit-stream is indicated in Figure 8.16.

A natural set of truncation points for the embedded bit-stream is formed by the coding pass end-points. Thus, the number of non-zero truncation points for code-block $\mathcal{B}_i$ is

$$Z_i = 3K_i - 2 \tag{8.12}$$

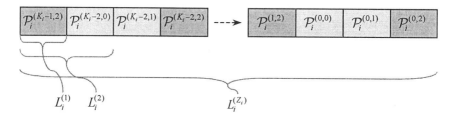

Figure 8.16.   Sequence of fractional bit-plane coding passes constituting the embedded bit-stream for code-block $\mathcal{B}_i$.

For each truncation point, $z \in \{1, 2, \ldots, Z_i\}$, the length, $L_i^{(z)}$, identifies the smallest prefix of the embedded bit-stream which is sufficient to correctly decode all symbols up to the end of coding pass $\mathcal{P}_i^{(p,k)}$, where p, k and z are related through

$$z = 3\left(K_i - p\right) + k - 4, \quad 0 \leq p < K_i, \quad 0 \leq k < 3$$

These lengths are identified in Figure 8.16. The first truncation point, $z = 0$, always corresponds to discarding the entire bit-stream so that $L_i^{(0)} = 0$.

One way to assess the suitability of a particular set of definitions for the fractional bit-plane coding passes, is to measure the frequency with which each of the truncation points, z, belongs to the set $\mathcal{H}_i$, introduced in Section 8.2.1. $\mathcal{H}_i$ is the set of useful truncation points; it also defines the convex hull of the length and distortion values corresponding to the coding pass end-points. If the coding passes successfully sort the code-block samples into subsets with decreasing distortion-length slopes, then we expect most of their end-points to contribute to the convex hull, as suggested by Figure 8.14. If the coding pass definitions are poor, however, we expect $\mathcal{H}_i$ to consist primarily of the bit-plane end-points, meaning that the set of useful truncation points would be no different from that produced by a deterministically scanned bit-plane coder.

Figure 8.17 provides experimentally observed frequencies with which each of the three types of coding pass, $\mathcal{P}_i^{(p,0)}$ through $\mathcal{P}_i^{(p,2)}$, yields a truncation point in $\mathcal{H}_i$. The results are obtained by applying the JPEG2000 algorithm to the three large natural images depicted in Figure 8.21, using code-blocks of 64×64 samples each, with the CDF 9/7 wavelet transform kernel in a 5 level DWT. Average convex hull occupancy over all bit-planes and all three images is quoted as a function of the average compressed bit-rate. Experiments are run for various quantizer step sizes, so as to cover the most interesting range of overall bit-rates, mea-

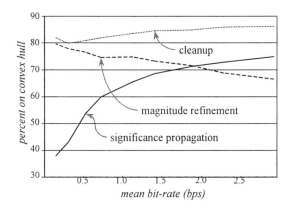

Figure 8.17. Convex hull occupancy rates for each of the three types of fractional bit-plane coding passes defined by JPEG2000.

sured in bits/sample (bps). The bit-plane end-points coincide with the end-points of the cleanup pass, $\mathcal{P}_i^{(p,2)}$, and we note that these do lie on the convex hull most frequently. The other two coding passes, however, also contribute to the convex hull more often than not.

The results in Figure 8.17 also provide justification for the fact that the magnitude refinement pass is best performed before the cleanup pass in JPEG2000. The distortion-length slope in $\mathcal{P}_i^{(p,1)}$ must be steeper than that in $\mathcal{P}_i^{(p,2)}$ whenever the former contributes to $\mathcal{H}_i$ (see Theorem 8.2). Figure 8.17 indicates that this happens much more often than not, so we may conclude that magnitude refinement coding usually exhibits a steeper distortion-length slope than cleanup coding. If the two passes were performed in the opposite order, the slope for the third pass would usually be steeper than that for the second. In this case, the second pass would rarely contribute to $\mathcal{H}_i$ and the embedding would be weakened. This argument relies on the assumption that the distortion-length slopes associated with magnitude refinement and cleanup coding operations are not affected by their order, which is largely the case[11].

---

[11] The magnitude refinement coding context, κ^{mag}, does have some dependence upon the significance of the sample's neighbours and hence the order in which the refinement and cleanup passes are performed. It turns out, however, that this effect is usually quite small. In any event, the effect tends to strengthen the present argument, since delaying the magnitude refinement pass ensures that more information is available for coding, so that its distortion-rate slope can be even steeper.

OTHER VARIATIONS

The idea of sequencing bit-plane coding steps in accordance with their anticipated distortion-length slope was conceived independently by Li and Lei [93] and Ordentlich et al. [113]. It should be noted, however, that the lists maintained by the SPIHT [126] algorithm serve a similar purpose, representing one of its most significant innovations over Shapiro's EZW algorithm [131]. In a separate work, Li et al. [92] proposed a reordering of EZW's coding steps in accordance with their anticipated distortion-length slopes.

The work of Li and Lei [93] is particularly interesting, if less practical, in that their coding passes are not confined to bit-plane boundaries. The distortion-length slope is explicitly estimated for each neighbourhood context configuration, based on distortion models and probability estimates available from the adaptive arithmetic coder. Each coding pass is then composed of those samples which are expected to yield similar slopes. In this way, an arbitrarily fine embedding may be supported and it can happen that some particularly favourable samples are much more finely quantized than others at any given point in the embedding.

Ordentlich et al. [113] explored fractional bit-plane coding passes very similar to those defined above for JPEG2000, in the context of a simple bit-plane coding scheme involving Golomb encoded run lengths. They defined coding passes whose membership is based on information available from previous bit-planes only. The ideas in [113] were combined with conditional arithmetic coding of the bit-planes by Sheng et al. [133].

The above works were based on the coding of subbands as a whole. Independent code-blocks and incremental assessment of membership in the "significance propagation" pass were introduced by Taubman in [148, 149]. That work also investigated more general fractional bit-plane assignment rules. A four pass model incorporating a novel backward scanning pass was found to yield superior embedding to the three pass approach. Not only are there more truncation points, but these truncation points also contribute to the convex hull (these are the useful truncation points) with greater frequency than that observed in Figure 8.17. It is in this form that the EBCOT paradigm was first adopted by the JPEG2000 committee. In most cases, however, the four pass model was found to offer little improvement over the simpler three pass approach described above. A careful study of this and other refinements leading to the final form of the JPEG2000 algorithm may be found in [101].

8.3.4 QUAD-TREE CODING APPROACHES

In this section we briefly describe an alternative class of embedded block coding schemes, based on an embedded extension of the quad-tree coding technique introduced in Section 2.4.4. The principles described here are common to a number of proposed image compression strategies, including [20, 80, 141, 94].

BIT-PLANE CODING WITH QUAD-TREES

The quad-tree provides a simple structure for efficiently representing two dimensional binary sequences. As we shall see, the idea is readily adapted to the coding of significance information, $\sigma^{(p)}[\mathbf{j}]$, in each bit-plane, p. Significance coding is the most important of the bit-plane coding operations described in Section 8.3.2. Sign and magnitude refinement information are then easily incorporated into the quad-tree coding process in a manner which naturally gives rise to an embedded bit-stream.

Recall that a subband sample, $y[\mathbf{j}]$, is said to be insignificant in bit-plane p ($\sigma^{(p)}[\mathbf{j}] = 0$) if the corresponding quantization index magnitude, $v[\mathbf{j}]$, is strictly less than 2^p; otherwise the sample is significant ($\sigma^{(p)}[\mathbf{j}] = 1$). For simplicity, we assume that the code-block dimensions are identical powers of 2; i.e., $J_1 = J_2 = 2^T$ for some T. We introduce the notation, $\sigma^{(p,t)}[\mathbf{j}]$, to identify significance at each level $t = 0, 1, \ldots, T$ of an induced quad-tree. Specifically, $\sigma^{(p,t)}[\mathbf{j}]$ is defined recursively through the relations

$$\sigma^{(p,0)}[\mathbf{j}] = \sigma^{(p)}[\mathbf{j}], \quad 0 \le j_1, j_2 < 2^T \quad \text{and}$$
$$\sigma^{(p,t+1)}[\mathbf{j}] = \max_{k_1,k_2 \in \{0,1\}} \sigma^{(p,t)}[2j_1 + k_1, 2j_2 + k_2], \quad 0 \le j_1, j_2 < 2^{T-t}$$

The interpretation of $\sigma^{(p,t)}[\mathbf{j}] = 1$ is that at least one of the samples in the $\mathbf{j}^{\text{th}}$ "quad" of size $2^t \times 2^t$ is significant in bit-plane p.

Equivalently, $\sigma^{(p,t)}[\mathbf{j}] = 1$ if and only if $v^{(p,t)}[\mathbf{j}] > 0$, where

$$v^{(p,t)}[\mathbf{j}] = \bigvee_{0 \le k_1, k_2 < 2^t} v^{(p)}[2^t j_1 + k_1, 2^t j_2 + k_2] = \left\lfloor \frac{v^{(0,t)}[\mathbf{j}]}{2^p} \right\rfloor$$

and $a \bigvee b$ denotes the integer formed by taking the bit-wise logical "OR" of the binary representations of a and b. This suggests an efficient software implementation strategy, in which the $v^{(0,t)}[\mathbf{j}]$ are computed recursively from the K-bit index magnitudes, $v[\mathbf{j}]$, using a total of $2^{2T} - 1$ logical OR operations. The significance, $\sigma^{(p,t)}[\mathbf{j}]$, of any node in the tree in any given bit-plane, p, may be deduced as needed by testing the

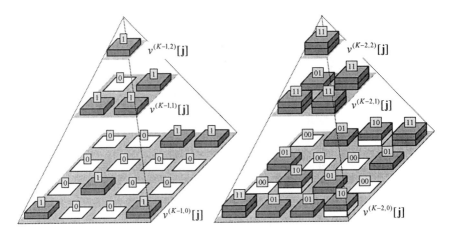

Figure 8.18. Embedded quad-tree structure, shown for the two most significant magnitude bit-planes, $p = K - 1$ (left) and $p = K - 2$ (right).

most significant $K - p$ bits of $v^{(0,t)}[\mathbf{j}]$. Figure 8.18 provides a graphical illustration of this quad-tree construction.

The binary significance map for the most significant bit-plane, $p = K - 1$, may be efficiently represented by direct application of the quad-tree coding algorithm described in Section 2.4.4. The same algorithm may be used in each subsequent bit-plane, $p = K - 2, \ldots, 0$, provided we are careful to avoid coding redundant information. In particular, we know that $\sigma^{(p,t)}[\mathbf{j}] \geq \sigma^{(p+1,t)}[\mathbf{j}]$, so the quad-tree coding step for $\sigma^{(p,t)}[\mathbf{j}]$ should be skipped whenever $\sigma^{(p+1,t)}[\mathbf{j}] = 1$. Introducing sign and magnitude refinement information at the appropriate points, one obtains the bit-plane coding algorithm described below. To simplify the description, we define $\sigma^{(p,T+1)}[\mathbf{0}] = 1$ for all p.

Simple Quad-Tree Bit-Plane Coder

For $p = K - 1, \ldots, 1, 0$

    Significance Coding Pass

    For $t = T, \ldots, 1, 0$

        For each $\mathbf{j}$ over $0 \leq j_1, j_2 < 2^{T-t}$

            If $v^{(p,t)}[\mathbf{j}] = 1$

                Emit a "1" (significant for first time in bit-plane p)

                If $t = 0$, emit a sign bit to identify $\chi[\mathbf{j}]$

            else if $v^{(p,t)}[\mathbf{j}] = 0$ and $\sigma^{(p,t+1)}\left[\left\lfloor \frac{j_1}{2} \right\rfloor, \left\lfloor \frac{j_2}{2} \right\rfloor\right] = 1$

                Emit a "0" (insignificant with significant parent)

    Magnitude Refinement Pass

> For each $\mathbf{j}$ over $0 \leq j_1, j_2 < 2^T$
>   If $v^{(p,0)}[\mathbf{j}] > 1$
>   Emit the LSB of $v^{(p,0)}[\mathbf{j}]$

This simple algorithm exhibits two coding passes per bit-plane, where the coding pass end-points are natural places to consider truncating the embedded bit-stream. We will shortly consider alternate ways of sequencing the quad-tree coding steps, so as to produce more finely embedded bit-streams. Before doing so, however, we observe that the quad-tree code contains some obvious redundancies. In particular, whenever $\sigma^{(p,t+1)}[\mathbf{j}] = 1$ we know that $\sigma^{(p,t)}[2\mathbf{j}+\mathbf{k}] = 1$ for at least one $\mathbf{k} \in \{0,1\}^2$; this is how we defined $\sigma^{(p,t+1)}[\mathbf{j}]$. A convenient way to exploit this redundancy, along with other statistical biases in the significance map, is to modify the algorithm so that the significance of the four nodes which share the same parent are coded together.

Modified Quad-Tree Bit-Plane Coder

For $p = K - 1, \ldots, 1, 0$

   Significance Coding Pass

   If $v^{(p,T)}[\mathbf{0}] \leq 1$
     Emit the value of $v^{(p,T)}[\mathbf{0}]$
   For $t = T, \ldots, 1$
     For each $\mathbf{j}$ over $0 \leq j_1, j_2 < 2^{T-t}$
       If $v^{(p,t)}[\mathbf{j}] = 1$
         code-together$(v^{(p,t-1)}[2\mathbf{j}+\mathbf{k}]$ over $0 \leq k_1, k_2 < 2)$
       else if $v^{(p,t)}[\mathbf{j}] \neq 0$
         For each $\mathbf{k}$ over $0 \leq k_1, k_2 < 2$
           If $v^{(p,t-1)}[2\mathbf{j}+\mathbf{k}] \leq 1$
             Emit the value of $v^{(p,t-1)}[2\mathbf{j}+\mathbf{k}]$
   For each $\mathbf{j}$ over $0 \leq j_1, j_2 < 2^T$
     If $v^{(p,0)}[\mathbf{j}] = 1$
       Emit a sign bit to identify $\chi[\mathbf{j}]$

   Magnitude Refinement Pass

   For each $\mathbf{j}$ over $0 \leq j_1, j_2 < 2^T$
     If $v^{(p,0)}[\mathbf{j}] > 1$
       Emit the LSB of $v^{(p,0)}[\mathbf{j}]$

Joint coding of the four nodes is easiest and also most effective when their parent first becomes significant. This is the condition under which the **code-together** routine is invoked by the above algorithm. The four values, $v^{(p,t-1)}[2\mathbf{j}+\mathbf{k}]$, $0 \leq k_1, k_2 < 2$, supplied to **code-together** are each either 0 or 1. Moreover, at least one of them must be non-zero. Thus, 15 combinations are possible and a variable length code (VLC) may be employed to exploit the redundancy. A reasonable VLC design strategy is to assign the shortest four codewords (3 bits each) to the singleton events; i.e., the events in which only one of the four nodes becomes significant. The remaining 11 combinations are considered less likely and assigned longer codewords.

ORDERING FOR EMBEDDING

In order to improve the embedding, we may split the significance coding pass in two, considering first those samples which are most likely to become significant in the current bit-plane. These can be expected to yield a steeper distortion-length slope. Natural candidates for the first coding pass are individual samples whose parent was significant in the previous bit-plane. The algorithm described above may be modified in a straightforward manner to process these samples in a first pass. In this way, we obtain three coding passes per bit-plane with similar interpretations to the JPEG2000 coding passes described in Section 8.3.3.

Rather than explicitly passing through all nodes in the tree in each coding pass, an attractive alternative is to maintain lists which identify those nodes for which information must be coded. The first list identifies the individual samples which are insignificant themselves, but have a significant parent. Accordingly, this list is called the LIC (List of Insignificant Coefficients). A second list identifies individual samples which were found to be significant in a previous bit-plane. This is the magnitude refinement list, or LSC (List of Significant Coefficients). Finally, the LIS (List of Insignificant Sets) identifies the non-leaf nodes whose significance must be coded during the current bit-plane, p. Each such node represents a non-trivial block (or set) of insignificant samples. Specifically, this list consists of the identifiers $(t, \mathbf{j})$, such that $t > 0$, $\sigma^{(p+1,t)}[\mathbf{j}] = 0$ and $\sigma^{(p+1,t+1)}\left[\left\lfloor \frac{j_1}{2} \right\rfloor, \left\lfloor \frac{j_2}{2} \right\rfloor\right] = 1$. With these definitions, one implementation of the embedded quad-tree coder would proceed as follows.

List-Based Quad-Tree Bit-Plane Coder

Initialize LIC, LSC and TEMP to empty

Initialize LIS with the single element $(T, \mathbf{0})$

For $p = K - 1, \ldots, 1, 0$

    Primary Significance Coding Pass

    For each $\mathbf{j} \in$ LIC

        Emit the value of $v^{(p,0)}[\mathbf{j}]$ (must be 1 or 0)

If $v^{(p,0)}[\mathbf{j}] = 1$
    Emit a sign bit to identify $\chi[\mathbf{j}]$
    Move $\mathbf{j}$ from LIC to TEMP (will move to LSC)

Secondary Significance Coding Pass

For each $(t, \mathbf{j}) \in$ LIS
    Emit the value of $v^{(p,t)}[\mathbf{j}]$ (must be 1 or 0)
    If $v^{(p,t)}[\mathbf{j}] = 1$
        Remove $(t, \mathbf{j})$ from LIS
        code-together$(v^{(p,t-1)}[2\mathbf{j} + \mathbf{k}]$ over $0 \leq k_1, k_2 < 2)$
        For each $\mathbf{k}$ over $0 \leq k_1, k_2 < 2$
            process-significance$(t - 1, 2\mathbf{j} + \mathbf{k}, p)$

Magnitude Refinement Pass

For each $\mathbf{j} \in$ LSC
    Emit the LSB of $v^{(p,0)}[\mathbf{j}]$

Prepare for Next Bit-Plane

For each $\mathbf{j} \in$ TEMP
    Move $\mathbf{j}$ from TEMP to LSC

The routine, **process-significance**, manages the more complex task of determining how to proceed with new nodes whose parent has just become significant. The significance of the node passed to this routine has already been coded by the **code-together** routine, but further coding steps may be required in the current bit-plane. The most natural implementation is recursive, as follows.

---

process-significance$(t, \mathbf{j}, p)$

If $v^{(p,t)}[\mathbf{j}] = 0$ (insignificant node goes to LIS or LIC)
    If $t > 0$,
        Add $(t, \mathbf{j})$ to front of LIS
        (prevents further processing in current bit-plane)
    else add $\mathbf{j}$ to LIC
else (new significant node)
    If $t > 0$
        code-together$(v^{(p,t-1)}[2\mathbf{j} + \mathbf{k}]$, over $0 \leq k_1, k_2 < 2)$
        For each $\mathbf{k}$ over $0 \leq k_1, k_2 < 2$
            process-significance$(t - 1, 2\mathbf{j} + \mathbf{k}, p)$
    else (new significant sample)
        Emit a sign bit to identify $\chi[\mathbf{j}]$
        Add $\mathbf{j}$ to TEMP (will move to LSC)

Although the list processing approach may be viewed simply as an implementation strategy for quad-tree based bit-plane coding, it brings a number of important benefits. Chief among these is the fact that the amount of processing required is directly related to the number of bits which are generated (encoded or decoded). This is because the lists provide a means of skipping over all nodes in the tree for which nothing will be coded. Ignoring the small improvements in efficiency which may be introduced by the **code-together** routine, the algorithm emits at least one bit for each node which is processed.

The list processing approach is also readily adapted to support different coding orders. In [94, SBHP], for example, the LIS is sorted so as to ensure that nodes at lower levels, t, in the tree are processed first. This improves the embedding within the secondary significance pass itself, since these are the nodes which are more likely to yield steeper distortion-length slopes. This generalizes the idea of processing the insignificant nodes at level $t = 0$ (these reside in the LIC) in a primary significance pass.

The SPECK algorithm [80] introduces an interesting modification to the quad-tree structure described here, where samples in each upper left hand sub-block of size $2^{T'} \times 2^{T'}$ reside within a tree of depth T', for each $T' = 0, 1, \ldots, T$. This modification minimizes the number of bits required to identify significant samples near the upper left hand corner of the block, which improves coding efficiency when the entire wavelet transform pyramid is packed into a single block in the manner suggested by Figure 4.22. When applied to small code-blocks taken from distinct subbands, this modified quad-tree structure imposes very little burden on overall coding efficiency. This simple and versatile coding approach was investigated during the JPEG2000 standardization activities.

WEAKNESSES OF QUAD-TREE CODING

For all of its advantages, quad-tree coding approaches do achieve less compression than the conditional coding approach described in Section 8.3.2. One reason for this is the use of very simple coding techniques; magnitude refinement and sign bits, for example, are simply moved directly to the bit-stream. Another important distinction is that quad-tree coders can exploit redundancy only within quads with predetermined boundaries, while conditional coding uses a moving window to exploit redundancy between neighbouring samples, unimpeded by artificial boundaries.

Quad boundaries also interfere with the identification of good candidates for the first coding pass. As a result, the secondary significance coding pass contains a mixture of coding steps which are likely to exhibit large distortion-length slopes and those which are not. This has an adverse impact on the embedding and also explains why the secondary significance pass takes precedence over the magnitude refinement pass in the quad-tree coding algorithm described above. From an implementation perspective, structured coding in quads is the primary appeal of the quad-tree approach. From the perspective of compression efficiency and adaptability to diverse sources, however, the quad structure is its most fundamental deficiency.

Various quad-tree coding schemes tested by the JPEG2000 committee yielded average compression losses of about 7% to 10% (about 0.5 to 1.0 dB in reconstructed MSE) relative to the conditional coding strategy which was finally adopted. These averages were obtained with natural photographic images such as those illustrated in Figures 8.21 and 8.22. Much larger losses were observed with non-natural sources such as those illustrated in Figure 8.24. On the other hand, careful software implementations of list based embedded quad-tree coding exhibited execution speeds several times higher than that of the JPEG2000 block coder. We conduct a comparison between the JPEG2000 algorithm and perhaps its closest quad-tree based contender [94, SBHP] in Section 8.5.2.

8.3.5   DISTORTION COMPUTATION

As mentioned in Section 8.3.3, the candidate truncation points for the embedded bit-stream representing each code-block, $\mathcal{B}_i$, correspond to the end-points of each of its coding passes. During compression we must assess the number of bytes, $L_i^{(z)}$, required to represent all coded symbols up to each truncation point, z, as well as the distortion, $D_i^{(z)}$, incurred by truncating the block bit-stream at that point. Actually, distortion estimation is not strictly necessary to generate a meaningful pack-stream, but it is important to the PCRD-opt algorithm described in Section 8.2; this algorithm is used to obtain all of the JPEG2000 results reported in this text. The task of computing $L_i^{(z)}$ is discussed thoroughly in Section 12.3. Our purpose here is to point out that the distortion estimation task need not add substantial complexity to the implementation of the compression system.

We begin by observing that the convex hull analysis and slope calculation algorithm described in Section 8.2.2 depends on distortion changes, $D_i^{(z_1)} - D_i^{(z_2)}$. Thus, it is sufficient and indeed preferable to determine the amount, $\Delta D_i^{(p,k)}$, by which each coding pass, $\mathcal{P}_i^{(p,k)}$, reduces distortion[12]. As we shall see, this computation can be performed with the aid of two small lookup tables which do not depend upon the coding pass, bit-plane or subband involved.

In this brief discussion we restrict our attention to total squared error (essentially MSE) as the measure of distortion. As shown at the end of Section 4.3.2, we may model the reconstructed image distortion as a linear combination of the squared error distortions from each subband sample, so long as the transform is orthogonal or the quantization errors are uncorrelated. In practice, neither of these requirements is satisfied exactly: we use linear phase (symmetric) subband transforms, which cannot be entirely orthogonal (see Theorem 6.5); and quantization errors usually exhibit significant correlation at low bit-rates. Nevertheless, we proceed on the assumption that at least one of the conditions is approximately valid.

Following equation (4.37), we model the distortion contributions from code-block $\mathcal{B}_i$ as

$$D_i^{(z)} = G_{b_i} \sum_{\mathbf{j}} \left(\hat{y}_i^{\left(p_i^{(z)}[\mathbf{j}] \right)}[\mathbf{j}] - y_i[\mathbf{j}] \right)^2 \qquad (8.13)$$

Here b_i identifies the subband to which code-block $\mathcal{B}_i$ belongs and G_{b_i} is the squared norm (energy) of the synthesis basis vectors for this subband. G_{b_i} is known as the energy gain factor for subband b_i. Its value is readily calculated using equations (4.39) and (4.40). If the objective is to minimize more perceptually relevant measures of distortion than MSE, additional weighting factors may be included such as the contrast sensitivity weights, $W_{b_i}^{\text{csf}}$, from Section 4.3.4.

The rather involved but convenient notation, $p_i^{(z)}[\mathbf{j}]$, refers to the index of the bit-plane to which sample $y_i[\mathbf{j}]$ has been coded prior to truncation point z. The dequantizer reconstructs $\hat{y}_i^{(p)}[\mathbf{j}]$ from the quantization index, $q_i^{(p)}[\mathbf{j}]$. Noting that all samples in coding pass $\mathcal{P}^{(p,k)}$ are refined from bit-plane $p+1$ to bit-plane p, we have

$$\Delta D_i^{(p,k)} = G_{b_i} \sum_{\mathbf{j} \in \mathcal{P}_i^{(p,k)}} \left[\left(\hat{y}_i^{(p+1)}[\mathbf{j}] - y_i[\mathbf{j}] \right)^2 - \left(\hat{y}_i^{(p)}[\mathbf{j}] - y_i[\mathbf{j}] \right)^2 \right] \quad (8.14)$$

Although the decompressor is generally free to choose reconstruction levels as it sees fit, the compressor must make some assumption concern-

---

[12] Note that although distortion will usually be reduced, it can sometimes be increased, in which case $\Delta D_i^{(p,k)}$ will be a negative quantity. This does happen from time to time in typical images. The corresponding truncation points will, of course, never be included in $\mathcal{H}_i$.

ing the dequantization policy in order to estimate distortion. JPEG2000 does not mandate any particular reconstruction policy, except in the case of lossless compression. However, a mid-point reconstruction rule is recommended and we proceed on this assumption. Substituting $\delta = \frac{1}{2}$ into equation (3.37) yields the dequantization formula

$$\hat{y}_i^{(p)}[j] = \text{sign}\,(y_i\,[j]) \begin{cases} 2^p \Delta_i \left(v_i^{(p)}[j] + \frac{1}{2} \right) & \text{if } v_i^{(p)}[j] > 0 \\ 0 & \text{if } v_i^{(p)}[j] = 0 \end{cases}$$

Hence, the distortion attributable to sample, $y_i\,[j]$, when coded down to bit-plane p, is given by

$$G_{b_i} \cdot \left(\hat{y}^{(p)}[j] - y\,[j] \right)^2$$

$$= G_{b_i} \cdot \begin{cases} \left(|y_i\,[j]| - 2^p \Delta_i \left(v_i^{(p)}[j] + \frac{1}{2} \right) \right)^2 & \text{if } v_i^{(p)}[j] > 0 \\ \left(|y_i\,[j]| \right)^2 = \left(|y_i\,[j]| - 2^p \Delta_i v_i^{(p)}[j] \right)^2 & \text{if } v_i^{(p)}[j] = 0 \end{cases} \qquad (8.15)$$

Let $\tilde{v}_i^{(p)}[j]$ denote the fraction part of the binary fraction representing $\frac{|y_i\,[j]|}{2^p \Delta_i}$. Specifically,

$$\tilde{v}_i^{(p)}[j] = \frac{|y_i\,[j]|}{2^p \Delta_i} - v_i^{(p)}[j] \qquad (8.16)$$

$$= \frac{|y_i\,[j]|}{2^p \Delta_i} - \left\lfloor \frac{|y_i\,[j]|}{2^p \Delta_i} \right\rfloor \in [0, 1)$$

Substituting (8.16) and (8.15) into (8.14) yields

$$\Delta D_i^{(p,k)} = G_{b_i} \times \Delta_i^2 \times 2^{2p}$$

$$\times \left\{ \begin{array}{l} \displaystyle\sum_{j \in \mathcal{P}_i^{(p,k)},\, v_i^{(p)}[j]=1} \left[\left(2\tilde{v}_i^{(p+1)}[j] \right)^2 - \left(\tilde{v}_i^{(p)}[j] - \frac{1}{2} \right)^2 \right] \\ + \displaystyle\sum_{j \in \mathcal{P}_i^{(p,k)},\, v_i^{(p)}[j]>1} \left[\left(2\tilde{v}_i^{(p+1)}[j] - 1 \right)^2 - \left(\tilde{v}_i^{(p)}[j] - \frac{1}{2} \right)^2 \right] \end{array} \right\}$$

Here we have used the fact that $v_i^{(p)}[j] = 1$ implies $v_i^{(p+1)}[j] = 0$ and $v_i^{(p)}[j] > 1$ implies $v_i^{(p+1)}[j] > 0$.

Now observe that $\tilde{v}_i^{(p)}[j]$ is actually a function of $\tilde{v}_i^{(p+1)}[j]$. This is most easily seen from the fact that $\tilde{v}_i^{(p)}[j]$ is the fraction part of $\frac{|y_i\,[j]|}{2^p \Delta_i}$ and so $2\tilde{v}_i^{(p+1)}[j]$ is formed from the least significant integer bit and the fraction part of the same quantity. Then

$$\tilde{v}_i^{(p)}[\mathbf{j}] = 2\tilde{v}_i^{(p+1)}[\mathbf{j}] - \left\lfloor 2\tilde{v}_i^{(p+1)}[\mathbf{j}] \right\rfloor$$

Also, when a sample first becomes significant in bit-plane p, we have $v_i^{(p)}[\mathbf{j}] = 1$ and $2\tilde{v}_i^{(p+1)}[\mathbf{j}] \geq 1$. The distortion contribution from coding pass $\mathcal{P}_i^{(p,k)}$ may then be written as

$$\Delta D_i^{(p,k)} = G_{b_i} \times \Delta_i^2 \times 2^{2p}$$

$$\times \left\{ \sum_{\mathbf{j} \in \mathcal{P}_i^{(p,k)}, v_i^{(p)}[\mathbf{j}]=1} T_s\left(\tilde{v}_i^{(p+1)}[\mathbf{j}]\right) + \sum_{\mathbf{j} \in \mathcal{P}_i^{(p,k)}, v_i^{(p)}[\mathbf{j}]>1} T_m\left(\tilde{v}_i^{(p+1)}[\mathbf{j}]\right) \right\}$$

where

$$T_s(\tilde{v}) = (2\tilde{v})^2 - \left(2\tilde{v} - \tfrac{3}{2}\right)^2 \quad \text{and}$$
$$T_m(\tilde{v}) = (2\tilde{v} - 1)^2 - \left(2\tilde{v} - \left\lfloor 2\tilde{v}_i^{(p+1)}[\mathbf{j}] \right\rfloor - \tfrac{1}{2}\right)^2$$

The operators, $T_s(\tilde{v})$ and $T_m(\tilde{v})$, may be implemented as small lookup tables, indexed by the first few fraction bits in the binary representation of $\tilde{v} \in [0,1)$. The first table, T_s, is invoked whenever a sample becomes significant during the significance propagation or cleanup pass, while the second table, T_m, is employed by the magnitude refinement pass. The relevant fraction bits can generally be made available to the bit-plane coder with little or no overhead. For example, the explicit quantization operation which assigns

$$v_i[\mathbf{j}] = \left\lfloor \frac{|y_i[\mathbf{j}]|}{\Delta} \right\rfloor$$

may be replaced by

$$\bar{v}_i[\mathbf{j}] = \left\lfloor \frac{|y_i[\mathbf{j}]|}{2^{-3}\Delta} \right\rfloor$$

Then $v_i^{(p)}[\mathbf{j}]$ is obtained by discarding the least significant $p+3$ bits from $\bar{v}_i[\mathbf{j}]$ and the value $\tilde{v}_i^{(p+1)}[\mathbf{j}]$, which must be passed to T_s or T_m to compute distortion in bit-plane p, is well approximated by bit positions p through $p+3$ of $\bar{v}_i[\mathbf{j}]$.

The table lookup approach is easily adapted to accommodate dequantizer behaviour which differs from the simple mid-point reconstruction policy described here. This is important when a reversible transform (see Section 6.4.2) is used to produce a scalable lossless representation of the image. In this case, the subband samples are integers, the step

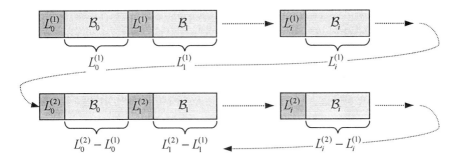

Figure 8.19. Deterministic interleaving of block coding passes as one means of creating a globally embedded representation from embedded block bit-streams.

size is $\Delta_i = 1$, and samples for which all bit-planes are available are reconstructed without error. In the final bit-plane then, the distortion lookup tables should be modified to return the values

$$T_s(\tilde{v}) = (2\tilde{v})^2 \quad \text{and} \quad T_m(\tilde{v}) = (2\tilde{v} - 1)^2$$

8.4 ABSTRACT QUALITY LAYERS
8.4.1 FROM BIT-PLANES TO LAYERS

It is important to appreciate the distinction between the local embedding associated with each code-block's bit-stream, c_i, and the global embedding of the final pack-stream, c. As pointed out in Section 8.1.3, the pack-stream might not be embedded. Simply concatenating an optimally truncated set of code-block bit-streams, as in Figure 8.6, results in a pack-stream whose distortion does not degrade gracefully as it is truncated; entire blocks will be dropped one by one.

One way to construct a globally embedded representation of the image is to interleave the coding passes from each code-block in some predictable fashion. Specifically, the pack-stream might be constructed by including the first pass ($L_i^{(1)}$ bytes) from every code-block, $\mathcal{B}_i$, followed by the second coding pass ($L_i^{(2)} - L_i^{(1)}$ bytes) from each block, and so forth. This is illustrated in Figure 8.19. As the pack-stream is progressively truncated, the distortion associated with each code-block should degrade gracefully. On the other hand, deterministic pack-stream construction rules of this form cannot generally be optimal at any bit-rate. Nevertheless, deterministic interleaving of coding passes or bit-planes is a natural approach, which has been followed by almost all embedded image compression algorithms.

At the opposite extreme, one might consider pack-streams in which the code-block contributions are interleaved in a truly optimal fashion. In this scenario, each contribution from code-block $\mathcal{B}_i$ should advance that block's representation to the next point in the convex hull set, $\mathcal{H}_i$, and the contributing blocks should be globally sequenced according to their distortion-length slopes. Specifically, let $0 = h_i^0 < h_i^1 < \cdots$ denote the elements of $\mathcal{H}_i$ and $\lambda_i \left(h_i^k \right)$ the corresponding distortion-length slopes, as defined in Section 8.2.1. At any given point in the construction of the pack-stream, we may identify the cumulative contribution from each code-block, $\mathcal{B}_i$, by an index, $k_i \geq 0$, into the elements of $\mathcal{H}_i$, such that $h_i^{k_i}$ is the relevant truncation point for the block's embedded bit-stream. The next contribution may come from any code-block, $\mathcal{B}_i$, whose distortion-length slope satisfies $\lambda_i \left(h_i^{k_i+1} \right) \geq \lambda_j \left(h_j^{k_j+1} \right)$, $\forall j$. A pack-stream constructed in this fashion has the property that each prefix corresponds to a rate-distortion optimal representation of the original image.

This optimal sequencing strategy suffers from one very serious disadvantage: each code-block contribution must be accompanied by auxiliary information identifying the particular code-block from which the contribution is drawn. The per-block cost of this signalling information grows logarithmically with the number of code-blocks which must be distinguished. This means that larger images cannot be compressed as efficiently as smaller images! Interestingly, as the image dimensions increase we are more likely to find a large number of code-blocks with similar distortion-length slopes so that the benefit of strictly sequencing the code-block contributions according to their slope values becomes negligible. In order to strike a compromise between the benefits and the costs of optimal sequencing, it makes sense to collect the code-block contributions into "bins" with similar slope values, sequencing the contributions deterministically within each bin so as to avoid the overhead of explicitly identifying the order. This is precisely the thinking which gave rise to the concept of quality layers, which is fundamental to JPEG2000.

As defined previously, quality layers $\mathcal{Q}_0$ through $\mathcal{Q}_l$ together contain the first $L_i^{\left(z_i^l \right)}$ bytes from code-block $\mathcal{B}_i$, where the truncation points, z_i^l, are selected to minimize the overall reconstructed image distortion,

$$D^l = \sum_i D_i^{\left(z_i^l \right)}$$

subject to a length constraint,

$$\sum_i L_i^{\left(z_i^l \right)} \leq L_{\max}^l$$

The PCRD-opt algorithm described in Section 8.2 may be used to select these truncation points, in which case each quality layer, Q_l, has an associated slope threshold, $\lambda_{\min}^l$, and the truncation points are given by

$$z_i^l = \max\left\{h \in \mathcal{H}_i \mid \lambda_i(h) > \lambda_{\min}^l\right\}$$

In this way, the contributions contained in layer Q_l, have distortion-rate slopes in the range $\lambda_{\min}^{l-1} \geq \lambda_i(z_i^l) > \lambda_{\min}^l$ and the quality layers are exactly the "slope bins" mentioned above.

If the pack-stream is truncated to one of the lengths $L_{\max}^l$, the resulting representation will be optimal in the PCRD-opt sense. Truncation to any intermediate length, $L_{\max}^{l-1} < L < L_{\max}^l$, will not generally yield an optimal representation. However, the degree of sub-optimality may be controlled through the size of the slope bins, or equivalently through the ratio $L_{\max}^l/L_{\max}^{l-1}$. The per-block overhead associated with quality layer signalling depends on the number of layers, but not the size of the image. The number of layers may then be selected to match the intended application. The following three scenarios are of interest and guide the experimental results reported in Section 8.5.

Single Layer If distortion scalability is unimportant, only a single layer is required and the pack-stream organization effectively reduces to that depicted in Figure 8.6.

Targeted Layers In some applications there may be a small known set of bit-rates at which the pack-stream is to be made available for decompression. The quality layers may be targeted to these bit-rates.

Generic If distortion scalability is important and nothing can be known a priori concerning the length of the representation which will be available for decompression, the pack-stream may be constructed from a large number of quality layers, whose slope thresholds are judiciously spaced so as to optimize the trade-off between sub-optimal sequencing and signalling overhead. Experience shows that the range of bit-rates from 0.05 through 2.0 bits per sample may be effectively covered using approximately 50 quality layers when working with large images and code-blocks of size 64×64. This number of layers should be reduced somewhat when working with small images or with smaller code-blocks.

It is worth noting that the contribution of each code-block to each quality layer is explicitly signalled within the pack-stream. As a result, there is no requirement that the contributions be sequenced in accordance with rate-distortion criteria. In particular, there is nothing preventing the compressor from reproducing the deterministic interleaving policy of Figure 8.19 by including one coding pass from each code-block in each quality layer. As mentioned, the coding passes or bit-planes provide a natural (though sub-optimal) ordering of the data into layers corresponding to progressively higher overall image quality. The quality layers employed by JPEG2000 introduce an extra level of abstraction, which serves to separate the roles of coding and ordering of information.

8.4.2   MANAGING OVERHEAD

In this section we discuss more specifically the overhead information which must be included in the pack-stream to identify code-block contributions to each quality layer. We also describe mechanisms which may be used to compress this information; these are the mechanisms employed by the JPEG2000 image compression standard, which were adopted directly from the EBCOT algorithm described in [149].

CODE-BLOCK TAGS

Every code-block notionally contributes to every quality layer, although the contribution may be empty. The block signalling information may be understood in terms of "tags." The information conveyed by the tag for code-block $\mathcal{B}_i$ in quality layer $\mathcal{Q}_l$ may be summarized as follows:

Inclusion The tag first identifies whether or not any information is included from the code-block at all; i.e., whether or not $z_i^l > z_i^{l-1}$. If not, the tag is complete. Recall that z_i^l denotes the truncation point for code-block $\mathcal{B}_i$ in quality layer $\mathcal{Q}_l$.

Coding passes Depending upon the details of the block coding algorithm, it may or may not be necessary to identify both the number of coding passes and the number of code bytes contributed to the layer, since one may sometimes be deduced from the other. In JPEG2000, the number of new coding passes, $\Delta z_i^l = z_i^l - z_i^{l-1}$, is identified explicitly.

Code bytes The number of new code bytes, $\Delta L_i^l = L_i^{\left(z_i^l\right)} - L_i^{\left(z_i^{l-1}\right)}$, is also identified explicitly. Some implementations of the block coding algorithm could deduce this quantity during decoding. However, explicitly supplying the length of the contribution allows pack-stream parsers to efficiently eliminate unwanted elements.

Missing MSBs As noted in Section 8.3.3, we must also signal the number of magnitude bit-planes, K_i, used to represent the samples in code-block $\mathcal{B}_i$. The first coding pass, having truncation point $z = 1$,

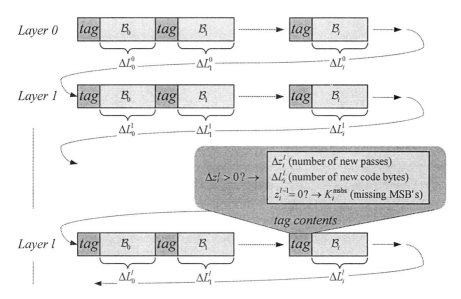

Figure 8.20. Code-block tag information.

codes the most significant of these K_i magnitude bits. We may define a quantity, $K_{b,\max}$, which denotes the maximum number of bits
required to represent quantization index magnitudes for any of the
subband samples in subband b. This quantity depends upon the
quantization step size, Δ_b, for the subband. The value of K_i may
then be signalled through $K_i^{\text{msbs}} \triangleq K_{b_i}^{\max} - K_i$, where b_i denotes the
subband containing code-block $\mathcal{B}_i$. In the first quality layer to which
code-block $\mathcal{B}_i$ makes a non-empty contribution, i.e., when $z_i^{l-1} = 0$
and $z_i^l \neq 0$, the value of K_i^{msbs} is included in the block's tag.

Figure 8.20 illustrates the anatomy of a code-block tag. The figure
suggests that tags might be distributed throughout the pack-stream so
that each tag is immediately followed by the code bytes which are being
contributed. Indeed such an organization should maximize the reconstructed image quality associated with an arbitrarily truncated pack-
stream. The JPEG2000 standard employs a slightly different organization, in which collections of related code-block contributions are assembled into so-called packets and all of the tags associated with a packet
appear together in the packet's header. This packet-based organization
has a number of practical advantages which are discussed further in
Section 12.5.

TAG TREES

Since a pack-stream may involve many quality layers and there may be many code-blocks within each quality layer, it is beneficial to devote some effort to efficiently coding the tag information described above. JPEG2000 employs an embedded structure known as a "tag tree" to exploit residual redundancy between code-blocks. It is convenient to describe tag tree coding first in general terms and later indicate how it is used to code tag information.

Let $w[\mathbf{n}] \equiv w[n_1, n_2]$ denote a two dimensional array of non-negative integers whose values are to be coded progressively using the tag tree algorithm. In our application, there is one number for each code-block and the array includes all code-blocks whose tags are to be coded jointly. In JPEG2000, this may be the set of all code-blocks in the image which belong to the same subband; however, the coding may also be restricted to smaller regions, known as "precincts." Coding proceeds progressively by comparing $w[\mathbf{n}]$ against a sequence of thresholds, $\overline{w} = 1, 2, 3, \ldots$. The tag tree coding process may be understood in terms of a procedure, $W^{\mathrm{enc}}(\overline{w}, \mathbf{n}, \mathcal{W})$, whose objective is to efficiently code whether or not $w[\mathbf{n}] \geq \overline{w}$, given the current state, $\mathcal{W}$, of the tag tree. The procedure produces only those code bits which are required to represent the binary event $w[\mathbf{n}] \geq \overline{w}$, for the particular entry indexed by $\mathbf{n}$. In order to exploit redundancy within a variable length coding framework, some such decisions will involve the generation of multiple code bits, while others (hopefully most) will involve the production of no code bits at all. This is possible because the state, $\mathcal{W}$, reflects the information signalled by previous invocations of the tag tree coding procedure.

Before describing the algorithm itself, it is worth providing a context to motivate the tag tree concept. As we shall see, tag trees will be used to code the quantity K_i^{msbs} (amongst other things), which is interpreted as the number of missing most significant magnitude bit-planes for code-block $\mathcal{B}_i$. In this case, we will define $w[\mathbf{n}_i] = K_i^{\mathrm{msbs}}$, where $\mathbf{n}_i$ is the location in the tag tree array corresponding to code-block $\mathcal{B}_i$. It is reasonable to expect substantial redundancy amongst neighbouring elements in the $w[\mathbf{n}]$ array. However, the process of exploiting this redundancy is somewhat complicated by the fact that we do not want to code all of the values, $w[\mathbf{n}]$, at once. Recall that K_i^{msbs} is to be coded only in the quality layer to which code-block $\mathcal{B}_i$ contributes for the first time, which may be different for each code-block. For this reason, we need a procedure which generates code bits for only one code-block, but is able to exploit redundancy through the state, $\mathcal{W}$, produced by previous coding operations, regardless of the order in which the coding operations are performed.

Let T be the smallest integer such that the indices, $\mathbf{n}$, of all coding quantities, $w[\mathbf{n}]$, are entirely contained within the range $0 \leq n_1, n_2 < 2^T$. Since the tag tree coding procedure is only invoked to code those quantities, $w[\mathbf{n}]$, which are of interest, there is no harm in defining the tree over the entire $2^T \times 2^T$ array, setting $w[\mathbf{n}] = \infty$ at any location in this array for which original data is not available. At each tree level, t, we construct an array of node values, $w^{(t)}[\mathbf{n}]$, as follows. The leaf nodes in the tree are at level $t = 0$, where we define $w^{(0)}[\mathbf{n}] = w[\mathbf{n}]$. At subsequent levels in the tree, the node value is defined as the minimum of its descendants' node values; i.e.,

$$w^{(t)}[\mathbf{n}] = \min_{0 \leq k_1, k_2 < 2} w^{(t-1)}[2\mathbf{n} + \mathbf{k}] \qquad (8.17)$$

$$= \min_{0 \leq k_1, k_2 < 2^t} w[2^t \mathbf{n} + \mathbf{k}], \quad 0 \leq n_1, n_2 < 2^{T-t}, \quad 0 \leq t \leq T$$

Thus, the root of the tree is at level $t = T$, where $w^{(T)}[\mathbf{0}] = \min_{\mathbf{n}} w[\mathbf{n}]$.

The state of the tag tree, $\mathcal{W}$, consists of the node values, $w^{(t)}[\mathbf{n}]$, and a set of corresponding thresholds, $\overline{w}^{(t)}[\mathbf{n}]$. The interpretation of these thresholds is that sufficient information has already been coded to identify whether or not $w^{(t)}[\mathbf{n}] \geq \overline{w}$ for all $\overline{w}$ in the range $0 \leq \overline{w} \leq \overline{w}^{(t)}[\mathbf{n}]$. In particular, if $w^{(t)}[\mathbf{n}] < \overline{w}^{(t)}[\mathbf{n}]$, sufficient information has been coded to precisely identify the value of $w^{(t)}[\mathbf{n}]$. There is thus no need to record values of $\overline{w}^{(t)}[\mathbf{n}]$ which are larger than $w^{(t)}[\mathbf{n}] + 1$ and it is convenient to adhere to this limit. The tag tree is initialized by setting $\overline{w}^{(t)}[\mathbf{n}] = 0, \forall t, \mathbf{n}$. Since the node values are non-negative integers, this is indeed the "zero information" state. The tag tree encoding procedure, $W^{\mathrm{enc}}(\overline{w}, \mathbf{n}, \mathcal{W})$, is given by the following algorithm.

Tag Tree Encoding Procedure, $W^{\mathrm{enc}}(\overline{w}, \mathbf{n}, \mathcal{W})$

Initialize $\overline{w}^{\min} = 0$

> (At each successive level t, this quantity will have the interpretation of a lower bound on the relevant node threshold $\overline{w}^{(t)}[\mathbf{n}^{(t)}]$ at that level, based solely on the information coded at higher levels. The definition of $\mathbf{n}^{(t)}$ appears below. We will always have $w^{(t)}[\mathbf{n}] \geq \overline{w}^{\min}$.)

For $t = T, \ldots, 1, 0$

    Set $n_1^{(t)} = \lfloor \frac{n_1}{2^t} \rfloor$, $n_2^{(t)} = \lfloor \frac{n_2}{2^t} \rfloor$

    ($\mathbf{n}^{(t)} \equiv \left[n_1^{(t)}, n_2^{(t)} \right]$ is the location of the relevant level t ancestor)

    If $\overline{w}^{(t)}[\mathbf{n}^{(t)}] < \overline{w}^{\min}$

        Update $\overline{w}^{(t)}[\mathbf{n}^{(t)}] \leftarrow \overline{w}^{\min}$

    While $w^{(t)}[\mathbf{n}^{(t)}] \geq \overline{w}^{(t)}[\mathbf{n}^{(t)}]$ and $\overline{w}^{(t)}[\mathbf{n}^{(t)}] < \overline{w}$

Increment $\overline{w}^{(t)}\left[\mathbf{n}^{(t)}\right]$

If $w^{(t)}\left[\mathbf{n}^{(t)}\right] \geq \overline{w}^{(t)}\left[\mathbf{n}^{(t)}\right]$

   emit-bit(0)

else

   emit-bit(1)

Update $\overline{w}^{\min} \leftarrow \min\left\{w^{(t)}\left[\mathbf{n}^{(t)}\right], \overline{w}^{(t)}\left[\mathbf{n}^{(t)}\right]\right\}$

The corresponding decoding procedure, $W^{\mathrm{dec}}\left(\overline{w}, \mathbf{n}, \mathcal{W}\right)$, may be implemented using the following algorithm. Here, the tag tree state, $\mathcal{W}$, has the same interpretation as it does during encoding except that the value stored as $w^{(t)}\left[\mathbf{n}\right]$ is the minimum of the original value at the encoder and the current coding threshold, $\overline{w}^{(t)}\left[\mathbf{n}\right]$. For decoding, the state is initialized by setting $w^{(t)}\left[\mathbf{n}\right] = \overline{w}^{(t)}\left[\mathbf{n}\right] = 0$, $\forall t, \mathbf{n}$.

Tag Tree Decoding Procedure, $W^{\mathrm{dec}}\left(\overline{w}, \mathbf{n}, \mathcal{W}\right)$

Initialize $\overline{w}^{\min} = 0$

For $t = T, \ldots, 1, 0$

 Set $n_1^{(t)} = \left\lfloor \frac{n_1}{2^t} \right\rfloor$, $n_2^{(t)} = \left\lfloor \frac{n_2}{2^t} \right\rfloor$

 If $\overline{w}^{(t)}\left[\mathbf{n}^{(t)}\right] < \overline{w}^{\min}$

  Update $\overline{w}^{(t)}\left[\mathbf{n}^{(t)}\right] \leftarrow \overline{w}^{\min}$

  Update $w^{(t)}\left[\mathbf{n}^{(t)}\right] \leftarrow \overline{w}^{\min}$

 While $w^{(t)}\left[\mathbf{n}^{(t)}\right] = \overline{w}^{(t)}\left[\mathbf{n}^{(t)}\right]$ and $\overline{w}^{(t)}\left[\mathbf{n}^{(t)}\right] < \overline{w}$

  Increment $\overline{w}^{(t)}\left[\mathbf{n}^{(t)}\right]$

  If **retrieve-bit**$() = 0$

   Increment $w^{(t)}\left[\mathbf{n}^{(t)}\right]$

 Update $\overline{w}^{\min} \leftarrow \min\left\{w^{(t)}\left[\mathbf{n}^{(t)}\right], \overline{w}^{(t)}\left[\mathbf{n}^{(t)}\right]\right\}$

Example 8.1 *To see how the tag tree coder can exploit redundancy, consider a 4×4 array of identical values, $w\left[\mathbf{n}\right] = 2$. Suppose now that we wish to use the tag tree coder to encode the values of $w\left[0, 0\right]$ and $w\left[1, 1\right]$, respectively. In this particular example, we will code $w\left[0, 0\right]$ by invoking $W^{\mathrm{enc}}\left(\overline{w}, \left[0, 0\right], \mathcal{W}\right)$ repeatedly for each $\overline{w} \in \{1, 2, \ldots, w\left[0, 0\right]+1\}$ and then do the same to code $w\left[1, 1\right]$.*

Invoking $W^{\mathrm{enc}}\left(1, \left[0, 0\right], \mathcal{W}\right)$ yields the output

$$(0), (), ()$$

Here, the code bits produced at each of the three levels in the tree are delimited by parentheses and empty parentheses mean that no code-bits were emitted at that level. The procedure updates the state variables

$\overline{w}^{(2)}[\mathbf{0}]$, $\overline{w}^{(1)}[\mathbf{0}]$, *and* $\overline{w}^{(0)}[\mathbf{0}]$ *to* 1. *Invoking* $W^{enc}(2,[0,0],\mathcal{W})$ *again yields the output*

$$(0),(),()$$

Finally, invoking $W^{enc}(3,[0,0],\mathcal{W})$ *yields the output,*

$$(1),(1),(1)$$

at which point $\overline{w}^{(2)}[\mathbf{0}] = \overline{w}^{(1)}[\mathbf{0}] = \overline{w}^{(0)}[\mathbf{0}] = 3$.

Proceeding to the next node of interest, we invoke $W^{enc}(1,[1,1],\mathcal{W})$. *This yields no output whatsoever, but updates state variable* $\overline{w}^{(0)}[1,1]$ *from* 0 *to* 2. $W^{enc}(2,[1,1],\mathcal{W})$ *produces no output and no change in state. Finally, invoking* $W^{enc}(3,[1,1],\mathcal{W})$ *produces*

$$(),(),(1)$$

At this point the state variables, $\overline{w}^{(2)}[\mathbf{0}]$, $\overline{w}^{(1)}[\mathbf{0}]$, $\overline{w}^{(0)}[\mathbf{0}]$ *and* $\overline{w}^{(0)}[1,1]$ *are all equal to* 3 *and the remaining* 17 *state variables are all still equal to* 0.

In all, 6 *bits have been used to code the two quantities. This is a little less than the* 8 *bits which would be required to code the values separately using a comma code (see Example 2.5). Note that the tag tree code reduces to a comma code when* $T = 0$. *The reader should verify that if we proceed to code all* 16 *of the* $w[\mathbf{n}]$ *values, a total of only* 23 *bits will be produced. This total number of code bits does not depend on the order in which the samples are coded. It is also independent of the order in which the individual invocations of the tag tree coding procedure are issued. Thus, one might invoke* $W^{enc}(1,[1,1],\mathcal{W})$, *followed by* $W^{enc}(1,[0,0],\mathcal{W})$, *followed by* $W^{enc}(2,[1,1],\mathcal{W})$, *followed by* $W^{enc}(2,[0,0],\mathcal{W})$, *etc. The order of invocation affects the number of bits used to code any particular outcome, but not the total number of bits required to code all outcomes. Note, however, that the decoder must use exactly the same order as the encoder.*

TAG INFORMATION CODING

We are now in a position to describe the mechanisms employed by JPEG2000 to code each of the four types of information which might be found in the tag for code-block $\mathcal{B}_i$ in quality layer $\mathcal{Q}_l$.

Inclusion If $\mathcal{B}_i$ has not yet contributed to any quality layer, i.e., $z_i^{l-1} = 0$, we use an inclusion tag tree to efficiently code whether or not $z_i^l > 0$. The quantities coded through the inclusion tag tree are the indices of the quality layer to which the code-block first makes a contribution. Specifically,

$$w\,[\mathbf{n}_i] = \min\left\{l \mid z_i^l > 0\right\}$$

where $\mathbf{n}_i$ is the index of the element in the tag tree array which corresponds to code-block $\mathcal{B}_i$. We invoke the tag tree procedure $W^{\mathrm{enc}}\,(l+1, \mathbf{n}_i, \mathcal{W})$, which generates sufficient bits to indicate whether or not $w\,[\mathbf{n}_i] \geq l+1$; i.e., whether or not $z_i^l = 0$. In this way, we are able to exploit expected redundancy amongst the indices of the layers in which neighbouring code-blocks first contribute to the pack-stream.

If $\mathcal{B}_i$ has already contributed to at least one quality layer, i.e., $z_i^{l-1} > 0$, we have less reason to expect significant redundancy between code-blocks. Accordingly, we simply emit a "1" if $z_i^l > z_i^{l-1}$ and a "0" otherwise.

Missing MSBs If $z_i^{l-1} = 0$ and $z_i^l > 0$, we use another tag tree to signal the number of magnitude bit-planes, K_i. The quantities coded through this tag tree are $w\,[\mathbf{n}_i] = K_i^{\mathrm{msbs}}$, the number of missing MSBs for the code-block. The information is coded by invoking the tag tree procedure, $W^{\mathrm{enc}}\,(k, \mathbf{n}_i, \mathcal{W})$, for each successive k from 1 through $K_i^{\mathrm{msbs}}+1$. The decoder invokes $W^{\mathrm{dec}}\,(k, \mathbf{n}_i, \mathcal{W})$ with $k = 1, 2, \ldots$, until $w^{(t)} < \overline{w}^{(t)}$; of course, this condition will occur when $k = K_i^{\mathrm{msbs}}+1$.

It is worth noting that this particular use of tag trees effectively extends the embedded coding of magnitude bit-planes from the code-block to the entire subband. The important difference between low level block coding and the higher level tag coding techniques is that the order in which information is embedded in the pack-stream is not constrained by the tag coding procedures. Instead, the order in which code-blocks first contribute to the evolving pack-stream is explicitly signalled through inclusion coding, as described above, and this same order drives the embedding of information concerning the number of bit-planes for each code-block.

Coding passes The number of new coding passes, $\Delta z_i^l = z_i^l - z_i^{l-1} > 0$, is represented using the variable length code defined in Table 8.4.

Code bytes The number of new code bytes, ΔL_i^l, is signalled as an unsigned integer with $\beta_i + \lfloor \log_2 \Delta z_i^l \rfloor$ bits, where β_i is a state variable unique to code-block $\mathcal{B}_i$. The idea behind this representation is that we expect the number of new code bytes to be roughly proportional to the number of new coding passes, Δz_i^l. The state variable, β_i, is initialized to 3 prior to pack-stream construction. Before coding

Table 8.4. Variable length code for signalling Δz_i^l.

Δz_i^l	code word
1	0
2	10
3 through 5	11 00 through 11 10
6 through 36	1111 00000 through 1111 11110
	5-bits 5-bits
37 through 164	111111111 0000000 through 111111111 1111111
	9-bit prefix 7-bit suffix 9-bit prefix 7-bit suffix

ΔL_i^l, the value of β_i is updated as necessary using a comma code. In particular, a "0" bit is emitted if the current value of β_i is sufficient to represent the value of ΔL_i^l. Otherwise, one or more "1" bits are emitted prior to the "0" bit, where each "1" signals an increment of β_i by 1. This strategy is motivated by the fact that the number of code bytes generated in each coding pass is most often an increasing function of the number of bit-planes which have been coded.

8.5 EXPERIMENTAL COMPARISON
8.5.1 JPEG2000 VERSUS SPIHT

In this section we provide a comparison of objective compression performance between the SPIHT algorithm, introduced in Chapter 7, and the JPEG2000 algorithm. In both cases, we employ the CDF 9/7 wavelet transform kernel of Example 6.4, which is one of the two transforms defined by JPEG2000. The dyadic wavelet decomposition has $D = 5$ levels and hence 6 resolution levels, with the structure illustrated in Figure 8.3. Code-blocks of size 64×64 are used throughout these experiments.

Whereas SPIHT produces a single embedded bit-stream, a JPEG2000 pack-stream may be endowed with varying degrees of distortion scalability by controlling the number of quality layers. Accordingly, we report JPEG2000 results for each of the three different application scenarios suggested in Section 8.4.1. In the "single layer" case, a separate pack-stream must be generated for each of the tested bit-rates. In the "targeted" case, a single pack-stream is constructed containing 6 quality layers, which are optimized for each of the tested bit-rates and the additional rate of 0.0625 bps (bits per sample). In the "generic" case, a pack-stream with approximately 50 layers is constructed, where the rates of the various layers are randomly distributed over a logarithmic scale from 0.05 bps to 2.0 bps; that is, they are not targeted to the

Bike (2560x2048)   Cafe (2560x2048)   Woman (2560x2048)

Figure 8.21. High resolution ISO/IEC test images, representing natural photographic content. The original images form part of the Standard Color Image Data (SCID) described in ISO 12640 [9].

specific bit-rates being tested. In these last two cases, the packets are organized in a quality progressive sequence and each of the test bit-rates is obtained simply by truncating the pack-stream to the appropriate length.

Results are reported for 14 test images, averaged within each of five image categories. The first category consists of the high resolution ISO/IEC test images depicted in Figure 8.21. The second category consists of lower resolution ISO/IEC test images depicted in Figure 8.22. These first two categories represent natural image content. The third category contains the 512×512 test images "Lenna" and "Barbara," which have been particularly popular for image compression research. It is worth noting that colour sensor mis-registration is responsible for abnormal attenuation of the high frequency details in the "Lenna" image[13]. The fourth category consists of non-natural ISO/IEC test images, as shown in Figure 8.24. An assortment of other test images appears in the last group, depicted in Figure 8.25. All images are monochrome with 8 bits per sample.

Table 8.5 compares the mean-squared reconstruction error (MSE) associated with the various algorithms and pack-stream organizations, in terms of the Peak Signal to Noise Ratio (PSNR) measure. PSNR is inversely related to MSE through equation (1.2). Observe firstly that the

---

[13]This image was scanned from the November 1972 edition of Playboy magazine, at the University of Southern California. The monochrome test image corresponds to the luminance (Y) component in a YCbCr representation of the original RGB data.

Goldhill (576x720)   Hotel (576x720)   Tools (1200x1524)

Figure 8.22. Lower resolution ISO/IEC test images, representing natural photographic content. All three images were submitted by the UK National Body: "Goldhill" is a photograph of the village of Shaftesbury in Wiltshire, UK; "Hotel" is a scan of a holiday brochure; and "Tools" was provided by Crosfield Electronics, UK.

Lenna (512x512)   Barbara (512x512)

Figure 8.23. Popular 512×512 test images.

overhead associated with the introduction of numerous quality layers is relatively small. Even in the generic case, with about 50 quality layers (more than the number of distinct coding passes), the tag coding techniques described in Section 8.4.2 keep the overhead penalty to within about 0.15 dB in most cases.

The reader should also observe that the objective compression performance of JPEG2000 is marginally better than that of SPIHT with arithmetic coding and significantly better than the uncoded version of SPIHT. More significantly, the JPEG2000 coder is more robust to changes in image type. In particular, the performance gap between SPIHT and JPEG2000 is much larger for artificial image sources than it is for natural imagery.

Table 8.6 provides timing figures for software implementations of the SPIHT and JPEG2000 algorithms. Only the decoding process is considered (not including the inverse DWT). This is due, in part, to the

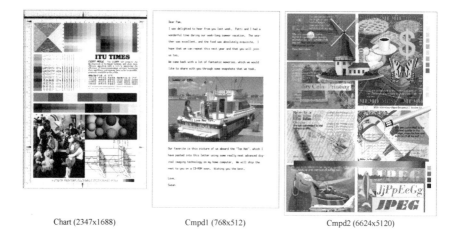

Chart (2347x1688)   Cmpd1 (768x512)   Cmpd2 (6624x5120)

Figure 8.24. Non-natural ISO/IEC test images. "Chart" is part of Standard Image Set CD-03, put out by the US National Communications System; it was originally contributed by Dennis Bodson. "Cmpd1" was produced by Majid Rabanni and Bhavan Gandhi at Eastman Kodak Company. "Cmpd2" was contributed by Phil Marchand at Xerox Corporation.

Cats (2048x3072)

Aerial2 (2048x2048)

Finger (735x755)

Figure 8.25. Miscellaneous ISO/IEC test images. "Aerial2" is provided courtesy of RECON/OPTICAL, INC., Barrington, Illinois. The "Cats" photograph was submitted by Philip J. Fennessy. "Finger" was provided by Chris Brislawn at Los Alamos National Laboratory. This last is not one of the ISO test images.

fact that the encoding time for JPEG2000 depends upon the policy used to determine the number of bit-planes which should actually be encoded for each code-block. Ideally, one would estimate this quantity

Table 8.5. Comparison of reconstructed image PSNR for SPIHT (with and without arithmetic coding) and the JPEG2000 algorithm (with various pack-stream organizations).

Category		0.125 bps	0.25 bps	0.5 bps	1.0 bps	2.0 bps
Natural	Single	24.84 dB	27.61 dB	31.35 dB	36.22 dB	42.42 dB
(large)	Targeted	-0.02 dB	-0.02 dB	-0.03 dB	-0.04 dB	-0.03 dB
	Generic	-0.07 dB	-0.09 dB	-0.14 dB	-0.16 dB	-0.11 dB
	SPIHT-AC	-0.21 dB	0.24 dB	-0.32 dB	-0.32 dB	-0.19 dB
	SPIHT-NC	-0.68 dB	-0.80 dB	-0.97 dB	-1.02 dB	-1.06 dB
Natural	Single	26.11 dB	28.64 dB	32.05 dB	36.26 dB	41.74 dB
(small)	Targeted	-0.02 dB	-0.03 dB	-0.02 dB	-0.04 dB	-0.04 dB
	Generic	-0.12 dB	-0.15 dB	-0.14 dB	-0.14 dB	-0.14 dB
	SPIHT-AC	-0.43 dB	-0.31 dB	-0.32 dB	-0.30 dB	-0.18 dB
	SPIHT-NC	-4.12 dB	-2.58 dB	-2.04 dB	-1.61 dB	-1.55 dB
Popular	Single	28.22 dB	31.27 dB	34.82 dB	38.81 dB	44.04 dB
	Targeted	-0.00 dB	-0.01 dB	-0.06 dB	-0.04 dB	-0.03 dB
	Generic	-0.17 dB	-0.11 dB	-0.16 dB	-0.17 dB	-0.12 dB
	SPIHT-AC	-0.24 dB	-0.41 dB	-0.49 dB	-0.37 dB	-0.15 dB
	SPIHT-NC	-0.63 dB	-0.80 dB	-0.91 dB	-0.83 dB	-0.81 dB
Artificial	Single	26.04 dB	29.98 dB	35.70 dB	43.32 dB	52.98 dB
	Targeted	-0.03 dB	-0.05 dB	-0.06 dB	-0.10 dB	-0.10 dB
	Generic	-0.17 dB	-0.19 dB	-0.29 dB	-0.33 dB	-0.21 dB
	SPIHT-AC[a]	-0.65 dB	-0.74 dB	-0.78 dB	-0.87 dB	-0.71 dB
	SPIHT-NC[a]	-4.01 dB	-3.41 dB	-3.54 dB	-3.30 dB	-2.93 dB
Misc.	Single	27.57 dB	30.20 dB	33.53 dB	38.16 dB	45.45 dB
	Targeted	-0.00 dB	-0.01 dB	-0.02 dB	-0.03 dB	-0.04 dB
	Generic	-0.08 dB	-0.06 dB	-0.09 dB	-0.15 dB	-0.13 dB
	SPIHT-AC	-0.16 dB	-0.23 dB	-0.28 dB	-0.36 dB	-0.31 dB
	SPIHT-NC	-0.80 dB	-0.82 dB	-0.84 dB	-1.07 dB	-1.26 dB

[a]Comparison performed using "Chart" and "Cmpd1" only; "Cmpd2" is too large for the available computer memory.

with sufficient accuracy to avoid discarding most of the encoded data during pack-stream formation. The Kakadu software supplied with this text implements a conservative heuristic for incrementally estimating the number of coding passes which are likely to be included in the final pack-stream.

SPIHT has the advantage that the encoding process may be terminated as soon as the desired bit-rate has been achieved. The price paid for this, however, is that the encoding and decoding processes require non-local access to the image transform coefficients. This, in turn, causes

Table 8.6.   Comparison of decoding CPU time (expressed as μs per pixel with a 400 MHz Pentium II processor) for JPEG2000 and SPIHT (with and without arithmetic coding). Actual CPU times are listed for the JPEG2000 algorithm, with SPIHT times expressed in terms of multiplicative factors.

Category		0.125 bps	0.25 bps	0.5 bps	1.0 bps	2.0 bps
Natural	Targeted	0.04 μs	0.06 μs	0.12 μs	0.22 μs	0.41 μs
(large)	SPIHT-AC	× 8.4	× 20.0	× 30.5	× 40.1	× 36.6
	SPIHT-NC	× 6.7	× 14.4	× 24.0	× 40.1	× 38.7
Natural	Targeted	0.03 μs	0.06 μs	0.10 μs	0.20 μs	0.39 μs
(small)	SPIHT-AC	× 3.3	× 5.1	× 8.0	× 12.5	× 14.5
	SPIHT-NC	× 0.7	× 1.7	× 3.6	× 7.4	× 14.3
Popular	Targeted	0.04 μs	0.05 μs	0.08 μs	0.19 μs	0.39 μs
	SPIHT-AC	× 1.5	× 3.1	× 3.9	× 3.7	× 5.8
	SPIHT-NC	× 0.8	× 1.2	× 1.4	× 1.7	× 4.6
Artificial	Targeted	0.04 μs	0.07 μs	0.12 μs	0.23 μs	0.43 μs
	SPIHT-AC[a]	× 5.4	× 8.7	× 13.0	× 14.4	× 19.6
	SPIHT-NC[a]	× 0.8	× 2.7	× 5.5	× 12.8	× 15.2
Misc.	Targeted	0.04 μs	0.06 μs	0.11 μs	0.20 μs	0.38 μs
	SPIHT-AC	× 6.2	× 13.8	× 21.7	× 25.6	× 30.4
	SPIHT-NC	× 4.8	× 11.1	× 18.9	× 25.1	× 32.0

[a]Comparison performed using "Chart" and "Cmpd1" only.

a dramatic reduction in throughput as the image dimensions grow, since the machine spends most of its time performing non-local memory accesses. Table 8.6 clearly demonstrates this sensitivity to image size. Interestingly, the throughput of the JPEG2000 algorithm is competitive with the uncoded version of SPIHT even when working with small images[14].

8.5.2   JPEG2000 VERSUS SBHP

In Section 8.3.4, we described a second class of embedded block coding algorithms which was carefully considered by the JPEG2000 committee as an alternative to the context adaptive coding techniques described in Section 8.3.2. These algorithms employ quad-tree coding techniques,

---

[14]Of course, these results are somewhat dependent upon the efficiency of each implementation. The JPEG2000 implementation used here is the Kakadu software supplied with this text. It has been heavily optimized for speed and memory usage. The publically available SPIHT implementation from "http://www.rpi.edu" has been used as a comparison.

Table 8.7. Comparison of reconstructed image PSNR and decoding CPU time for JPEG2000 and SBHP. Cpu times obtained using a 400MHz Pentium II processor.

Category		0.125 bps	0.25 bps	0.5 bps	1.0 bps	2.0 bps
Natural	J2K PSNR	24.83 dB	27.59 dB	31.32 dB	36.18 dB	42.39 dB
(large)	SBHP	-0.53 dB	-0.64 dB	-0.73 dB	-0.71 dB	-0.60 dB
	J2K CPU	0.04 μs	0.06 μs	0.12 μs	0.22 μs	0.41 μs
	SBHP	$\times$ 0.19	$\times$ 0.26	$\times$ 0.28	$\times$ 0.32	$\times$ 0.34
Natural	J2K PSNR	26.09 dB	28.62 dB	32.03 dB	36.23 dB	41.70 dB
(small)	SBHP	-0.81 dB	-0.66 dB	-0.72 dB	-0.68 dB	-0.71 dB
	J2K CPU	0.03 μs	0.06 μs	0.10 μs	0.20 μs	0.39 μs
	SBHP	$\times$ 0.25	$\times$ 0.25	$\times$ 0.32	$\times$ 0.31	$\times$ 0.34
Popular	J2K PSNR	28.22 dB	31.26 dB	34.76 dB	38.77 dB	44.01 dB
	SBHP	-0.96 dB	-1.05 dB	-0.92 dB	-0.73 dB	-0.62 dB
	J2K CPU	0.04 μs	0.05 μs	0.08 μs	0.19 μs	0.39 μs
	SBHP	$\times$ 0.11	$\times$ 0.27	$\times$ 0.41	$\times$ 0.34	$\times$ 0.34
Artificial	J2K PSNR	26.01 dB	29.94 dB	35.64 dB	43.22 dB	52.89 dB
	SBHP	-1.25 dB	-1.59 dB	-2.07 dB	-2.18 dB	-1.99 dB
	J2K CPU	0.04 μs	0.07 μs	0.13 μs	0.24 μs	0.44 μs
	SBHP	$\times$ 0.23	$\times$ 0.22	$\times$ 0.27	$\times$ 0.27	$\times$ 0.30
Misc.	J2K PSNR	27.56 dB	30.19 dB	33.51 dB	38.13 dB	45.41 dB
	SBHP	-0.47 dB	-0.49 dB	-0.46 dB	-0.68 dB	-0.77 dB
	J2K CPU	0.04 μs	0.06 μs	0.11 μs	0.20 μs	0.38 μs
	SBHP	$\times$ 0.17	$\times$ 0.29	$\times$ 0.29	$\times$ 0.34	$\times$ 0.35

which are adapted from the set partitioning principles in SPIHT and utilize only the simplest of variable length coding techniques.

Table 8.7 provides a comparison of objective compression and timing performance between the JPEG2000 algorithm and the SBHP algorithm[15] [94, SBHP], which was perhaps the closest contender amongst the quad-tree coding approaches. Both algorithms employ the EBCOT paradigm with 64 × 64 code-blocks and 6 targeted quality layers. As in Table 8.6, CPU times refer to the block decoder only, not the full decompression system.

While the SBHP algorithm typically loses about 0.7 dB in PSNR, its elegant simplicity is also evident. Decoding CPU times for SBHP are about 3 times smaller than those observed with the JPEG2000 algorithm at moderate to high bit-rates. Of course, a full decompressor must

---

[15] The implementation of SBHP was kindly provided by Dr. Christos Chrysafis, then with Hewlett-Packard Laboratories, Palo Alto, California.

perform other tasks besides block decoding, notably the inverse DWT. Consequently, one could not expect an SBHP-based decompressor to execute a full three times faster than a JPEG2000 decompressor. Nevertheless, at high bit-rates the block coding process does dominate overall complexity and an SBHP-based decompressor should execute substantially faster than JPEG2000. Although we only provide decoding times here, the SBHP encoder also executes faster than its JPEG2000 counterpart by a somewhat smaller factor.

As with SPIHT, the performance of SBHP is much less robust to variations in image type than JPEG2000. Average PSNR degrades by about 2 dB for artificial imagery and performance differences of up to 6 dB are reported in [94] for certain ISO/IEC test images. This lack of robustness to image type is perhaps the most significant concern which prevented the adoption of quad-tree based techniques in the JPEG2000 standard.

II
THE JPEG2000 STANDARD

Chapter 9

INTRODUCTION TO JPEG2000

9.1 HISTORICAL PERSPECTIVE

At the time of this writing, JPEG2000 is the newest international standard for still image compression. The JPEG acronym stands for "Joint Photographic Experts Group." The "Joint" here signifies that JPEG2000 is published as an ISO/IEC standard, as well as an ITU-T Recommendation. The acronyms of the previous sentence stand for International Organization for Standardization/International Electrotechnical Commission, and International Telecommunications Union-Terminal Sector,[1] respectively. The Joint Bi-level Imagery Experts Group (JBIG) has often worked closely together with the JPEG committee. In fact, the two committees typically meet in the same venues.

Throughout the development of JPEG2000, the JPEG and JBIG committees have functioned under the auspices of ISO/IEC JTC 1/SC 29/WG 1, which denotes Working Group 1 of Study Committee 29 of Joint Technical Committee 1 of ISO/IEC, hereinafter referred to as simply WG1. Before work commenced on JPEG2000, the study committee and working group (SC/WG) structure for JPEG and JBIG changed several times, including the merging of the JPEG and JBIG committees under WG1. The history of the formation for these groups and changes to their SC/WG structure is detailed in [119].

The JPEG and JBIG committees have been quite fruitful. Of course, the JPEG committee produced the extremely successful JPEG standard, which is published in four parts. Part 1 contains requirements and guide-

---

[1] The ITU-T was known formerly as the "Consultative Committee of the International Telephone and Telegraph" (CCITT).

lines for JPEG compression systems. This part contains the definition of all four JPEG modes (sequential, progressive, hierarchical, and lossless), while Part 2 deals with compliance testing. These two parts are the subject of the classic book by Pennebaker and Mitchell [119]. Parts 3 and 4 were completed after the publication of [119]. Part 3 deals with extensions (adaptive quantization, composite tiling, file format, etc.), while Part 4 deals with profiling and registration of profiles.

Also completed recently, is JPEG-LS which yields improved lossless and near-lossless compression, as compared to the lossless mode of JPEG. The JPEG standard is discussed briefly in Chapter 19 of this text, while JPEG-LS is the subject of Chapter 20.

The JBIG effort has produced two standards for the compression of bi-level imagery. These standards are known as JBIG [7] and JBIG2 [10], and are beyond the scope of this text. We do note however, that collaboration between JPEG and JBIG has been strong. This is evidenced by the fact that the QM arithmetic coder, developed for use in JBIG, was adopted as an option for entropy coding in JPEG. Similarly, the MQ arithmetic coder, developed for use in JBIG2, was adopted as the only approved entropy coder for JPEG2000. The MQ coder is discussed in Chapter 12 of this text.

This chapter provides an overview of the JPEG2000 standard, its feature set, and a brief history of the JPEG2000 standardization process. The JPEG2000 standard will be published in six parts. For the purpose of interchange it is important to have a standard with a limited number of options, so that decoders in browsers, printers, cameras, or palm-top computers can be counted on to implement all options. In this way, an encoded image will be displayable by all devices.[2] For this reason, Part 1 describes the minimal decoder and code-stream syntax required for JPEG2000, which should be used to provide maximum interchange. Additionally, Part 1 describes an optional minimal file format.

There are many applications for image compression where interchange is less important than other requirements (e.g., ability to handle a particular type of data). Thus, Part 2 consists of optional "value added" extensions, that enhance compression performance and/or enable compression of unusual data types. These extensions are not required of all implementations. It should be noted then, that images encoded with

---

[2]It is worth noting that the standard specifies only the decoder and code-stream syntax. Although informative descriptions of some encoding functions are provided in the text of the standard, there are no requirements that the encoder perform compression in any prescribed manner. This leaves room for future innovations in encoder implementations.

Part 2 technologies may not be decodable by Part 1 decoders. Part 2 also includes an enhanced file format.

Part 3 includes extensions for image sequences, and is known as "Motion JPEG2000."[3] Part 4 includes information on compliance/conformance, while Part 5 includes reference software.[4] Finally, Part 6 includes an additional file format tailored for compound documents.

At the time of this writing, only JPEG2000 Part 1 is finalized. Thus, the primary focus of this book is JPEG2000 Part 1. That being said, a high level treatment of Part 2 appears in Chapter 15. While we do not discuss Part 5, the Kakadu software implementation of Part 1 is provided on the compact disc accompanying this book.

9.1.1   THE JPEG2000 PROCESS

The JPEG2000 project was motivated by Ricoh's submission of the CREW algorithm [174, 27] to an earlier standardization effort for lossless and near-lossless compression (now known as JPEG-LS). Although LOCO-I [168] was ultimately selected as the basis for JPEG-LS, it was recognized that CREW provided a rich set of features worthy of a new standardization effort. Early in 1996, Dr. Daniel Lee of Hewlett-Packard company was named as the WG1 Convener. He oversaw the development of a proposal [1] (authored largely by Martin Boliek of Ricoh) which resulted in the approval of JPEG2000 as a new WG1 work item. Boliek was named as the JPEG2000 project editor at that time. Dr. Lee continued to serve as convenor through the development of JPEG2000.

A Call for Technical Contributions was issued in March 1997 [2], requesting compression technologies be submitted to an evaluation during the November 1997 WG1 meeting in Sydney, Australia. This call for contributions included a copy of the original proposal [1] detailing the desired feature set for JPEG2000. The most important of these desired features are listed below. We note here that each of the desired features has been achieved in JPEG2000, all within a single, tightly integrated, compression architecture and code-stream syntax. The algorithmic methods which enable these features are described in subsequent chapters.

---

[3] We note that "Motion JPEG" has been a commonly used format for the purpose of video editing (e.g., in production studios) even though never officially standardized.

[4] The JJ2000 group has produced a Java implementation of Part 1 for inclusion in Part 5. This group consists of: Swiss Federal Institute of Technology – Ecole Polytechnique Federale de Lausanne (EPFL); Canon Research Centre France (CRF); and Ericsson. The University of British Columbia (UBC) and Image Power have produced a C implementation of Part 1 for inclusion in Part 5.

In 1997, an ad hoc group on JPEG2000 requirements was established. Touradj Ebrahimi (EPFL) was appointed to chair this group. This group was tasked with the creation and maintenance of a JPEG2000 requirements document.

DESIRED FEATURES

- **Superior low bit-rate performance:** While superior performance at all bit-rates was considered desirable, improved performance at low bit-rates, with respect to JPEG, was considered to be an important requirement for JPEG2000.

  - This requirement has been met. At high bit-rates, where artifacts become just imperceptible, JPEG2000 has a compression advantage over JPEG of roughly 20% on average. At lower bit-rates, JPEG2000 has a much more significant advantage over certain modes of JPEG (See, for example, Figure 4.30).

- **Continuous-tone and bi-level compression:** Seamless compression of image components (e.g., R,G, or B), each from 1 to 16 bits deep, was desired from one unified compression architecture.

  - JPEG2000 has met this goal, providing state-of-the-art compression performance for continuous-tone gray scale and color imagery. At the time of this writing, JPEG2000 is perhaps the only standardized solution for 16+ bits per component-sample. At the other extreme, the performance of JPEG2000 on bi-level imagery (1 bit per sample) is comparable to that of the ITU-T (CCITT) G4 standard for facsimile compression [35].

- **Progressive transmission by pixel accuracy and resolution:** Progressive transmission is highly desirable when receiving imagery over slow communication links. Code-stream organizations which are progressive by "pixel accuracy" (or by "quality," or by "SNR") improve the quality of decoded imagery as more data are received. Code-stream organizations which are progressive by "resolution" increase the resolution, or size, of the decoded imagery as more data are received.

  - This goal, and more, has been achieved in JPEG2000. In addition to these two dimensions of progression, progression by spatial location, and progression by image component are also supported. These dimensions of progression can be "mixed and matched" within a single compressed code-stream.

- **Lossless and lossy compression:** Both lossless and lossy compression were desired, again from a single compression architecture. It was desired to achieve lossless compression in the natural course of progressive decoding.

  - Through the inclusion of a reversible (integer) wavelet transform, this goal has been achieved. When this transform is employed, competitive lossy and lossless compression/decompression are possible from a single compression algorithm. In fact, lossy and lossless decompression are possible from a single compressed code-stream. The lossless performance of JPEG2000 is within a few percent of state-of-the-art.

- **Random code-stream access and processing:** Spatial random access, as well as compressed domain processing were considered important features for JPEG2000.

  - JPEG2000 code-streams offer several mechanisms to support spatial random access (or region of interest access) at varying degrees of granularity. "Degree of interest" is also supported, whereby the quality of the decompressed image may be adjusted for each region of interest.

  - Compressed domain processing is also supported, including cropping, rotation, flipping, translation, and scaling.

- **Robustness to bit-errors:** It was desired that JPEG2000 should be robust to bit errors introduced by noisy communication channels (e.g., wireless).

  - This goal has been met by the inclusion of resynchronization markers, the coding of data in relatively small independent blocks, and the provision of mechanisms to detect and conceal errors within each block. The standard also supports code-stream organizations which can prove advantageous to applications requiring substantial error resilience.

- **Sequential build-up capability:** The essence of this desired feature was to allow for encoding of an image from top to bottom in a sequential fashion without the need to buffer an entire image. This is very useful for low memory implementations in scan-based systems.

  - This requirement has been met by JPEG2000. It can be implemented via tiling and/or progression by spatial location, as described in subsequent sections and chapters.

THE SYDNEY EVALUATIONS

In addition to the call for contributions, WG1 released a CD-ROM containing 40 test images to be processed and submitted for evaluation. For the evaluations, it was stipulated that compressed bit-streams and decompressed imagery be submitted for six different bit-rates (ranging from 0.0625 to 2.0 bits per sample) and for lossless encoding. Eastman Kodak computed quantitative metrics for all images and bit-rates. They also conducted a subjective evaluation involving 18 of the images at three bit-rates using evaluators from among the Sydney meeting attendees. The imagery from 24 algorithms was evaluated by ranking the perceived image quality of hard-copy prints.

Although the performance of the top third of the submitted algorithms was very good, the Wavelet/Trellis Coded Quantization (WTCQ) algorithm, submitted by SAIC and the University of Arizona (SAIC/UA), ranked first overall in both the subjective and objective evaluations. In the subjective evaluation, WTCQ ranked first at 0.25 and 0.125 bits per sample, and second at 0.0625 bits per sample. In terms of mean-squared-error, averaged over all images, WTCQ ranked first at each of the six bit-rates. Based on these results, WTCQ was selected as the reference JPEG2000 algorithm at the conclusion of the meeting. It was further decided that a series of "core experiments" would be conducted to evaluate WTCQ and other techniques in terms of the JPEG2000 desired features and in terms of algorithm complexity. Kathleen Rattell (of Booz-Allen & Hamilton) was named to chair the core experiments ad hoc group.

Results from the first round of core experiments were presented at the March 1998 WG1 meeting in Geneva, Switzerland. Based on these experiments, it was decided to create a JPEG2000 Verification Model (VM) which would lead to a reference implementation of JPEG2000. The VM would be the software in which future rounds of core experiments would be conducted. The VM would also be updated after each WG1 meeting based on the results of core experiments. Michael Marcellin (University of Arizona/SAIC) was appointed to head the VM ad hoc group, with SAIC developing and maintaining the VM software. Eric Majani (Canon-France) and Charis Christopoulos (Ericsson-Sweden) were also named as coeditors of the standard at that time. Results from round 1 core experiments were selected to modify WTCQ into the first release of the VM (VM0).

In the next few subsections, we describe briefly some of the technologies included in the various VMs. These technologies are described more thoroughly in subsequent chapters.

THE WTCQ ALGORITHM

The basic ingredients of the WTCQ algorithm are: the discrete wavelet transform, TCQ [104, 86] (using step sizes chosen via Lagrangian rate allocation), and binary arithmetic bit-plane coding of subbands. The bit-plane coding operates on TCQ indices (trellis quantized wavelet coefficients) in a way that enables successive refinement. This is accomplished by coding bit-planes in order from most to least significant. To exploit dependencies within subbands, spatial context models are used.

In general, contexts can be chosen within a subband and across subbands. The WTCQ bit-plane coder avoids the use of inter-subband contexts to maximize flexibility in scalable decoding, and to facilitate parallel implementation. The WTCQ bit-stream is scalable in both resolution and quality, although its best performance is obtained for the single rate (reached at the end of progression) at which the Lagrangian rate allocation is optimized. WTCQ also includes a "binary mode," and an adaptive classification of subband coefficients by blocks. A more complete description of WTCQ can be found in [128].

VM0 – VM2

Additions and modifications to VM0 continued over two meeting cycles, with refinements contributed by many WG1 members. VM2 supports user specified irreversible (real number) and reversible (integer) wavelet transforms, as well as user specified decomposition structures. As a simpler alternative to the Lagrangian rate allocation, a fixed quantization table ("Q-table") is included. This is analogous to the current JPEG standard [119]. When a Q-table is employed, precise rate control can still be obtained by truncating the embedded bit-stream. In addition to TCQ, scalar quantization is included in VM2.

For reversible wavelet transforms, scalar quantization with step size 1 is employed (i.e., no quantization), which allows progression to lossless in the manner of CREW or SPIHT [127]. For lossless compression of color imagery, the reversible color transform from CREW is included, as well. Rate control for integer wavelet transforms is accomplished by truncation of the embedded bit-stream, and lossless compression is available naturally from the fully decoded bit-stream.

Other features were added to the VM, often from original contributions to the Sydney meeting. Examples include, tiling, region of interest coding/decoding (University of Maryland, Mitsubishi, Ericsson), error resilience (Motorola, Texas Instruments, Sarnoff Corporation, UBC, Rockwell, University of Southern California (USC), Norwegian University of Science and Technology), approximate and exact wavelet transforms with low memory implementations (CRF, Motorola-

Australia, UBC, Hewlett-Packard, USC). For a more complete description of these technologies see [40].

Along with the additions described above, several refinements were introduced to the bit-plane coder. The major changes were improvements to the context modeling, and the de-interleaving of bit-planes into "sub-bit-planes" or "coding passes." Within each bit-plane of each subband, the bits are "de-interleaved" into three coding passes of the following types: 1) bits likely to be newly "significant," 2) "magnitude refinement" bits, and 3) bits likely to remain "insignificant" (i.e., all bits not included in the first two passes). The idea of coding passes was first presented in [113] for use with Golomb coding, and is motivated by rate-distortion concerns [92, 93]. Such concerns imply that it is desirable to have the bits with the steepest distortion-rate slopes appear first in an embedded code-stream.

As in VM0, all coding was carried out using context dependent binary arithmetic coding. The arithmetic coder employed up to and including VM2 is described in [140]. The coding passes employed in VM2 were adapted from [113] for use with arithmetic coding in [129][133]. The VM2 bit-plane coder has no inter-subband dependencies such as those used in [113] and in the zero-tree based schemes of [132][126].

VM2 also includes progressive visual weighting technology contributed by Sharp Laboratories of America [91]. This allows for each spatial frequency band to be weighted, or emphasized, differently at each truncation point within the same bit-stream. At low rates, or early truncation points, the emphasis might be on preserving the low frequency bands more accurately than the high frequency bands. At higher rates, or later truncation points, all bands might be emphasized equally. The distribution of quantization errors amongst different spatial frequency bands is then a function of the bit-rate associated with the truncated bit-stream which is actually decompressed. When reversible wavelet transforms are employed, progression to lossless is still possible, with very little change in file size as compared to the non-weighted case.

VM3 – VM5

At the November 1998 WG1 meeting in Los Angeles, David Taubman (then at Hewlett-Packard) presented EBCOT (Embedded Block Coding with Optimized Truncation) [147, 149]. EBCOT includes the idea of dividing each subband into rectangular blocks of coefficients and performing the bit-plane coding independently on these "code-blocks," rather than on entire subbands as in previous VMs. This partitioning reduces memory requirements in both hardware and software implementations, as well as providing a certain degree of spatial random access to the bit-stream. These ideas were motivated by similar concepts in [146]. EBCOT also includes an efficient syntax for forming the coding passes of multiple code-blocks into "packets," which taken together form quality "layers."

Tremendous flexibility in the formation of packets and layers is available to the implementer of an encoder. The default policy of the VM encoder is to place in each layer, the coding passes (among all coding passes not yet included in previous layers) with steepest distortion-rate slope, as estimated in the encoder. This policy aims to minimize the MSE at each point in the embedded bit-stream. Other policies are included as well.

Progressive visual weighting is achieved by modulating the distortion-rate slopes during layer formation. Another particularly interesting policy weights the distortion estimates of each block in a manner consistent with masking properties of the human visual system [149]. Thus, code-block contributions are de-emphasized in spatial regions where more distortion can be tolerated, visually. Even when this masking policy is employed, progressive transmission eventually results in lossless decompression, when reversible wavelet transforms are employed. As in the visually weighted case, the policy has little effect on the ultimate lossless file size, but can have dramatic impact on the visual quality for partial decoding at lower rates.

EBCOT was adopted for inclusion in VM3 at the Los Angeles meeting. Over the next three months, Taubman re-implemented the VM in an object-oriented manner, took a faculty position at the University of New South Wales (UNSW), and joined the VM maintenance team.

At the March 1999 WG1 meeting in Korea, the MQ coder [152] was adopted as the arithmetic coder for JPEG2000. This coder is functionally similar to the QM coder used in JBIG (and as an option in JPEG). The MQ coder has some useful bit-stream creation properties, and is used in the JBIG2 standard. Also, the MQ coder is believed to be available on a royalty and fee free basis for ISO standards. In fact, most of the companies involved in the JPEG2000 effort have generously agreed to grant royalty and fee free licenses to their technologies as required for implementation of JPEG2000.

At the July 1999 meeting in Vancouver, several changes and additions were made to the bit-plane coding employed within the EBCOT code-blocks [101]. These changes were made to reduce complexity, and to increase hardware friendliness. For the default mode (see Chapter 12), these changes have negligible effect on compression performance. Also included at the July 1999 meeting was the ability to perform geometric manipulations (e.g., rotation and/or flipping) of imagery [65].

At the same time as these changes were being made to the coding algorithms, the code-stream syntax was developed by Ricoh. This syntax is composed of markers and marker segments, compatible with those of JPEG. Features were added (Ricoh, Aerospace Corporation) as appropriate for JPEG2000.

While the code-stream syntax provides all the data necessary for decompression, applications often require additional information not present in the code-stream. As mentioned previously, one annex of JPEG2000 Part 1 contains an optional minimal file format. This file format allows for the inclusion of information such as the color space of the image samples, and intellectual property (copyright) information for the image. This optional file format is extensible, and Part 2 of the standard defines containers for many additional types of "metadata."

VM6 – VM9

Most of the inclusions to VM6 through VM9 pertain to JPEG2000 Part 2. One notable exception is the canvas coordinate system (UNSW, SAIC/UA, Ricoh, CRF) employed by all parts of JPEG2000. Another exception is the addition of two dimensions of progressivity (component and spatial location). Finally, significant changes were introduced to support error detection and concealment within code-block bit-streams (UNSW, HP).

Additions to the VM targeted to Part 2 include user selectable wavelet filters (CRF, Los Alamos National Laboratories), user selectable decomposition tree structures (SAIC/UA), and support for palettized imagery (Sharp Labs of America). Also included are decorrelating techniques for the third dimension in volumetric imagery (e.g., LANDSAT, SPOT, CT, MRI, etc.). These techniques include general linear transforms, inter-component prediction, and the wavelet transform. These technologies were contributed by Eastman Kodak, the Aerospace Corporation, the University of Arizona, and Lockheed-Martin. Also included, are generalized level shift and generalized quantizer deadzone width (MITRE Corporation).

FILE FORMAT

File format development proceded in parallel with algorithm/code-stream development. As mentioned previously, a basic file format is specified in JPEG2000 Part 1, while more advanced file formats are provided by other parts. While file format writers and readers were not implemented in the VM, several independent implementations exist. In fact, the Kakadu software accompanying this text employs the Part 1 file format. Contributions to file formats came from many organizations, including Adobe, Apple, Canon Information Systems Research Australia, Elysium, Eastman-Kodak, Net Image, Picture Elements, and Xerox.

9.2   THE JPEG2000 FEATURE SET

Previous image compression systems and/or standards have been used primarily as input-output filters within applications. That is, when an image is written or read, it is compressed or decompressed largely as a storage function. Additionally, decisions as to image quality and/or compression ratio are made at compression time. At decompression time, only the image quality, size, resolution, and spatial extent envisioned by the compressor is available to the decompressor.

For example, with JPEG baseline (sequential mode), an image is compressed using a particular quantization table. This essentially determines the quality that will be achieved at decompression time. Lower (or higher) quality decompressions are not available to the decompressor. Similarly, if the lossless mode of JPEG is employed, lossy decompression is unavailable, and high compression ratios are not generally possible. JPEG-LS provides superior lossless performance to lossless JPEG, while also supporting lossy (near lossless) compression. However, all decisions are still made at compression time, and only the image resolution and quality envisioned at compression time is available to the decompressor.

Notable exceptions to these rigid structures exist. Many modern compression systems allow for progressive and/or hierarchical decoding of compressed image data. Such systems are discussed in Chapters 7 and 8. The original JPEG standard also allows for these features. Specifically, JPEG has a "progressive mode," that allows decompression of a code-stream at any lower quality than some maximum quality determined at compress time. Additionally, the code-stream is ordered so that the "most important" bits appear earliest in the code-stream. This is particularly useful for transmission of imagery over slow communication links. As the first few bytes of data are received, a low quality rendition of the imagery can be decoded and displayed. As more bytes are received, they can be combined with previously received bytes for decoding and display of "progressively" higher quality renditions of the imagery. Progressive JPEG has recently gained acceptance in web-based applications.

Hierarchical JPEG is philosophically similar. However, rather than improving quality, additional bytes are used to successively improve the "resolution" (or size) of the decoded imagery. For example, as the first few bytes are received and decoded, a small "thumbnail" image becomes available. As more bytes are received, they are combined with previously received bytes and decoded to obtain successively larger and larger "zoomed" images. Typically, each such image has twice as many samples on each side as the previous image in the series.

As discussed above then, JPEG has four "modes" of operation: sequential, progressive, hierarchical, and lossless. Certain interactions between the modes are allowed according to the JPEG standard. For example, hierarchical and progressive modes can be mixed within the same code-stream. However, few if any implementations have exploited this ability. Also, quite different technologies are employed for the lossless and lossy modes. The lossless mode relies on predictive coding techniques (see Chapters 2 and 3), while lossy compression relies on the discrete cosine transform (Chapter 4).

A JPEG code-stream must be decoded in the fashion intended by the compressor. For example if reduced resolution is desired at the decompressor (when progressive mode was employed at the compressor), the entire image must be decompressed and then downsampled. Conversion of a code-stream from one mode to another can be difficult. Typically, such conversion must be accomplished via decompression/recompression, sometimes resulting in loss of image quality.

In what follows, we present an overview of the JPEG2000 feature set, which overcomes the limitations described above. More detailed discussions of the feature set, as well as the algorithmic details that enable these features, are discussed in subsequent chapters.

9.2.1   COMPRESS ONCE: DECOMPRESS MANY WAYS

JPEG2000 brings a new paradigm to image compression standards. The benefits of all four JPEG modes are tightly integrated in JPEG2000. The compressor decides maximum image quality, up to and including lossless. Also chosen at compression time, is maximum resolution, or size. Any image quality or size can be decompressed from the resulting code-stream, up to and including the maximums chosen at encode time.[5]

For example, suppose an image is compressed losslessly at full size. Suppose further that the resulting file is of size B_0 bytes. It is then possible to extract B_1 bytes from the file, $(B_1 < B_0)$ and decompress those B_1 bytes to obtain a lossy decompressed image. This image will be identical to the image obtained if compression were performed to B_1 bytes in the first place. Similarly, it is possible to extract B_2 bytes

---

[5] We note that it is possible to construct code-streams that substantially limit the ability of decoders to exploit the JPEG2000 feature set. While this may be necessary in certain applications, implementers are encouraged to design encoders that produce feature-rich code-streams.

from the file and decompress to obtain a reduced resolution image. The resulting image will be exactly the same as if the lower resolution version of the image were compressed to B_2 bytes in the first place.

In addition to the quality scalability and resolution scalability discussed above, JPEG2000 code-streams support spatial random access. There are several mechanisms to retrieve and decompress data from the code-stream corresponding to selected spatial regions of an image. The different mechanisms yield different granularity of access, at varying levels of difficulty. This is detailed in subsequent chapters. For now, we note that each region so accessed can be decoded at a variety of different resolutions and qualities.

Random access extends to components as well. For example, the grayscale component can be extracted from a color image. Similarly, overlay components containing text or graphics can be extracted, when present. This can be done region by region with varying qualities and resolutions.

It is important to note that in each case discussed above, it is possible to locate, extract, and decode the bytes required for the desired image product. It is *not* necessary to decode the entire code-stream and/or image. In many cases, the bytes extracted and decoded are identical to those that would be obtained if only the desired image products were compressed in the first place.

Figure 9.1 shows examples of the image products discussed above. The figure portrays an original (perhaps color) image being compressed once to form a JPEG2000 code-stream. The code-stream can be decompressed in many ways to obtain different image products. Included in the figure are two reduced resolution images. These might be appropriate for monitor display, or color printing, as hinted by the output devices portrayed there. Also included is a high resolution version of a single spatial region (the face). This is indicative of what might be displayed as the user "zooms in" on the small display. Finally, the figure shows a version of the image decompressed at full resolution. This might be a grayscale image appropriate for high resolution black and white printers.

9.2.2   COMPRESSED DOMAIN IMAGE PROCESSING/EDITING

Any of the image products discussed above can be extracted from a JPEG2000 code-stream to create a new JPEG2000 code-stream. This is not terribly surprising. Clearly, whatever can be decompressed and displayed can be recompressed and stored. However, with JPEG2000, the relevant compressed bytes can be extracted and reassembled into a compliant code-stream *without* decompressing.

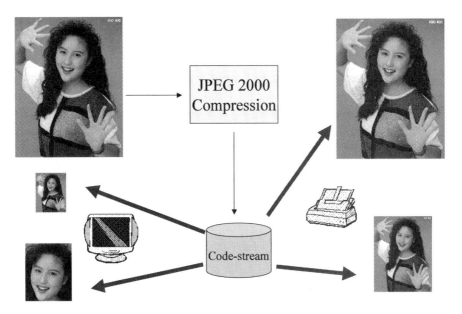

Figure 9.1. Multiple image products from a single compressed JPEG2000 code-stream.

Specifically, reduced resolution and/or reduced quality *compressed* imagery can be produced without a decompress/recompress cycle. In addition to the elegance and computational savings afforded by this capability, compression noise "build-up" is avoided. Such build-up occurs in most compression schemes when repetitive compress/decompress cycles occur.[6]

In addition to reduced quality and reduced resolution, compressed domain image cropping is possible. Cropping in the compressed domain is accomplished by accessing the compressed data associated with a given spatial region and rewriting it as a compliant code-stream. Some special processing is required around the cropped image borders, however decompression/recompression can be largely avoided.

Geometric manipulations are also supported in the (partially) compressed domain. Image rotations of 90, 180, and 270 degrees are possible. Image "mirroring" or "flipping" (top-to-bottom and/or left-to-right) can also be performed. These procedures cannot be carried out entirely in the compressed domain. Some transcoding of data is required, but the

---

[6]It is possible in some cases to avoid noise buildup in JPEG2000 even in the event that decompress/recompress cycling is performed, due to the lossless capability of JPEG2000.

cycle of inverse and forward transformation is avoided, and only small changes in distortion and bit-rate are incurred.

This idea can also be used to decode and render a flipped or rotated version of an image in approximate raster fashion. This might be useful for sending large images to low-memory scan-based printers.

9.2.3 PROGRESSION

Many types of progressive transmission are supported by JPEG2000. As mentioned previously, progressive transmission is highly desirable when receiving imagery over slow communication links. As more data is received, the rendition of the displayed imagery improves in some fashion. JPEG2000 supports progression in four dimensions: Quality, Resolution, Spatial Location, and Component.

The first dimension of progressivity in JPEG2000 is quality. As more data are received, image quality is improved. A JPEG2000 code-stream ordered for quality progression corresponds roughly to a JPEG progressive mode code-stream.

It should be noted that the image quality improves remarkably quickly with JPEG2000. An image is typically recognizable after only about 0.05 bits/sample have been received. For a 320×240 sample image, this corresponds to only 480 bytes of received data. With only 0.25 bits/sample (2,400 bytes) received, most major compression artifacts disappear. To achieve quality corresponding to no visual distortion, between 0.75 and 1.0 bits per sample are usually required. Demanding applications may require up to 2.0 bits/sample or even truly lossless decompression. We remark here again, that any quality up to and including lossless may be contained within a single compressed code-stream. Improving quality is then a simple matter of decoding more bits.

The second dimension of progressivity in JPEG2000 is resolution. In this type of progression, the first few bytes are used to represent a small "thumbnail" of the image. As more bytes are received, the resolution (or size) of the image increases by factors of 2 on each side. Eventually, the full size image is obtained. A JPEG2000 code-stream ordered for resolution progression corresponds roughly to a JPEG hierarchical mode code-stream.

The third dimension of progressivity in JPEG2000 is spatial location. With this type of progression, imagery can be received in approximately raster fashion, from top-to-bottom. This type of progression is particularly useful for memory constrained applications such as printers. It is also useful for encoding. Low memory scanners can create spatially progressive code-streams "on-the-fly," without buffering either the image or the compressed code-stream. A JPEG2000 code-stream ordered for progression by spatial location corresponds roughly to a JPEG sequential mode code-stream.

The fourth and final dimension of progressivity is the component. JPEG2000 supports images with up to 16384 components. Most images with more than 4 components are from scientific instruments (e.g., LANDSAT). More typically, images are 1 component (grayscale), 3 components (e.g., RGB, YUV, etc.), or 4 components (CMYK). Overlay components containing text or graphics are also common. Component progression controls the order in which the data corresponding to different components is decoded. With progression by component, the grayscale version of an image might first be decoded, followed by color information, followed by overlaid annotations, text, etc. This type of progression, in concert with the other progression types, can be used to effect various component interleaving strategies.

The four dimensions of progressivity are very powerful and can be "mixed-and-matched" within a single code-stream. That is, the progression type can be changed throughout the code-stream. For example, the first few bytes transmitted might contain the information for a low quality, grayscale, thumbnail image. The next few bytes might add quality, followed by color. The resolution of the thumbnail might then be increased several times so that the size is appropriate for display on a monitor. The quality could then be improved until visually lossless display is achieved. At this point, the viewer might desire to print the image. The resolution could then be increased to that appropriate for the particular printer. If the printer is black and white, the color components can be omitted from the remaining transmitted data.

The main points to be understood from this discussion are that 1) the imagery can be improved in many dimensions as more data are received, and 2) only the data required by the viewer needs to be transmitted. This can dramatically improve the latency experienced by an image browsing application. Thus, the "effective compression ratio" experienced by the client can be many times greater than the actual compression ratio as measured from the file size at the server.

We conclude this section by noting that a code-stream is transmitted to a user (or stored in a file) in one particular progression order, perhaps the one discussed in the example above. However, an existing code-stream can always be parsed and rewritten with a different progression order, without actually decompressing the image. A smart server can even construct the desired progression order on the fly in response to user requests.

9.2.4   LOW BIT-DEPTH IMAGERY

Binary valued components, (or binary valued tiles of components) can be compressed using JPEG2000. Lossless compression of such binary data can be accomplished via setting the bit depth to 1 and setting 0 levels of wavelet transform. The result of these settings is that no wavelet transform is performed, and the binary image is treated as a single bit-plane at a single resolution. This bit-plane is divided into code-blocks and subjected to context dependent arithmetic coding. Unfortunately however, scalability in quality and resolution are sacrificed. On the other hand, spatial random access is preserved.

If the wavelet transform is employed for compressing binary imagery, all the scalability and progression types discussed in the previous section are present, but with some loss in compression efficiency over the "zero-level" case. The quality of lossy decompressed binary imagery can vary widely, depending on the image content and bit-rate chosen.

Graphic imagery containing a limited number of colors can also be compressed with JPEG2000. A particularly effective approach treats the image as a single palettized component. The sample values of this component then specify image sample colors via a look-up-table (LUT). Careful construction of the LUT often results in compression performance comparable to that of JPEG-LS.

9.2.5   REGION OF INTEREST CODING

We close our discussion of the JPEG2000 feature set with a description of "region of interest coding." In previous sections we mentioned the possibility of varying quality by spatial region. Such variation can be effected at encode time, or in subsequent parsing or decode operations. This capability derives from the independence of code-blocks, and so the code-block dimensions govern the granularity of the spatial regions that can be targeted.

JPEG2000 also allows the encoder to select regions of arbitrary shape and size for preferential treatment. In this case, the ROI (Region of Interest) must be chosen at encode time and is not easily altered via parsing or decoding. For this form of ROI coding, wavelet coefficients that affect image samples within the ROI are pre-emphasized (left shifted) prior to bit-plane coding. The amount of pre-emphasis is written to the code-stream, and is used to properly realign (right shift) the ROI coefficients at decode time.

Chapter 10

SAMPLE DATA TRANSFORMATIONS

10.1 ARCHITECTURAL OVERVIEW

Part 1 of the JPEG2000 standard specifies a comparatively rigid set of transformations which are applied to the image samples to produce quantized subband sample indices; the quantization indices are then passed to the block coder. The various transformations are identified in Figure 10.1, which also indicates the major parameters which are available to control the behaviour of each type of transformation.

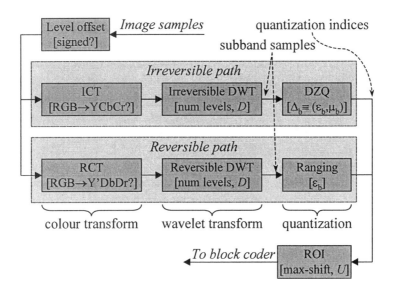

Figure 10.1. Sample data transformations.

The core operations in Figure 10.1 are those of the Discrete Wavelet Transform (DWT) and Deadzone Scalar Quantization (DZQ). As such, the compression system conforms to the common model of Figure 1.4. The additional elements identified in Figure 10.1 serve the following purposes.

Level offset: If the original B-bit image sample values are unsigned (non-negative) quantities, an offset of -2^{B-1} is added so that the samples have a signed representation in the range

$$-2^{B-1} \leq x[\mathbf{n}] < 2^{B-1} \tag{10.1}$$

If the data already conforms to this range, no adjustment is performed. When multiple image components (e.g., colour components) are being compressed together, the adjustment is made independently to each of the components. Whether or not each component's range is adjusted in this fashion is identified by a corresponding "signed" flag in the global code-stream marker, *SIZ*.

The motivation for the offset is that almost all of the subband samples produced by the DWT involve high-pass filtering and hence have a symmetric distribution about 0. Without the level offset, the LL subband would present an exception, introducing some irregularity into efficient implementations of the standard.

Evidently, the level offset is optional, since unsigned data in the range $0 \leq x[\mathbf{n}] < 2^B$ may be marked as "signed" with a bit-depth of $B+1$ bits per sample. The risk, however, is that $B+1$ may exceed the bit-depth to which a decompressor implementation is able to comply with the standard.

Colour transform: The colour transform is optional. It may be used only when three or more colour components are available and only when the first three components all have identical sizes and identical bit-depths. The assumption is that these first three image components contain the red, green and blue sample values of a colour image. The transform converts the RGB data into an "opponent" colour representation, with a luminance (or intensity) channel and two colour difference channels. This has the effect of exploiting some of the redundancy between the original colour components. In particular, colour difference components commonly account for less than 20% of the bits used to compress a colour image.

ROI (Region of Interest): JPEG2000 allows compressors to emphasize arbitrary regions of interest in the image. This is achieved by

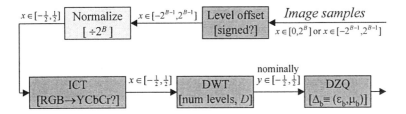

Figure 10.2. Normalized interpretation of the irreversible path, in which transformation operators are applied to sample values with a nominal range of $-\frac{1}{2}$ to $\frac{1}{2}$.

scaling up the quantization indices of more important subband samples, which effectively reduces their quantization step sizes.

10.1.1   PATHS AND NORMALIZATION

Figure 10.1 contains two transformation paths labeled "reversible" and "irreversible." The reversible path maps the B-bit sample values, $x[\mathbf{n}]$, to integer subband samples of similar precision, from which the original image may be recovered exactly. In this case, there is no quantization and the DWT must be implemented using the non-linear lifting steps described in Section 6.4.2. The reversible colour transform involves similar non-linearities. Lossless compression almost invariably requires the reversible path.

Although the reversible path is useful for lossy as well as lossless compression, the subtle non-linearities which are required for reversibility tend to damage compression performance, particularly at high bit-rates. The sample transformations associated with the irreversible path are entirely linear (prior to quantization). The irreversible path also offers the freedom to select arbitrary quantization parameters, which can be advantageous in certain applications.

An important distinction between the reversible and irreversible paths is that the sample transformations of the irreversible path are unaffected by scaling the sample values. Specifically, if we multiply the image sample values by some factor and scale the quantization step sizes by the same factor, the quantization indices delivered to the block coder will be unaffected. This is an obvious consequence of the linearity of the irreversible transformations.

The irreversible path is most easily described in terms of real-valued samples which are normalized (through division by 2^B) to the "unit range"

$$-\frac{1}{2} \leq x[\mathbf{n}] \leq \frac{1}{2}$$

The colour transform and DWT operations may be normalized so that all subband samples nominally retain this same unit range of $-\frac{1}{2}$ to $\frac{1}{2}$. Quantization step sizes, Δ_b, are specified relative to this nominal range. The concept of nominal range will be made more precise in Section 10.4.1. Figure 10.2 depicts the normalized perspective which we shall adopt for describing the elements of the irreversible path. We point out that the standard itself [12] does not describe the irreversible path from this normalized perspective; however, the description given here is equivalent and more intuitive.

It should now be apparent that the irreversible path is essentially independent of the sample bit-depth, B. For example, a 12-bit compressed image may be decompressed as an 8-bit image simply by modifying the bit-depth, B, specified in the global code-stream marker, *SIZ*. The 8 decompressed sample bits will be the 8 most significant bits of the 12 bit representation which would otherwise have been decompressed. Similarly, an 8-bit image may be decompressed as though it were a 12-bit image.

Unfortunately, this bit-depth invariance property is not shared by the reversible path[1]. Consequently, the reversible operations are best described directly as mappings from integers to integers, while the irreversible operations are best described as transformations performed on floating point numbers with a unit range of $-\frac{1}{2}$ to $\frac{1}{2}$. It is worth noting, however, that fixed point approximations to the irreversible operations may be preferred in many implementations, in which case both paths will involve closely related integer transformations.

10.2   COLOUR TRANSFORMS

As noted above, the colour transform is optional and is applied only to the first three image components, denoted $x_0[\mathbf{n}]$ through $x_2[\mathbf{n}]$. These must have identical dimensions and bit-depths. They are interpreted as the red, green and blue sample values of a colour image, respectively. Note, however, that the JPEG2000 code-stream is notionally colour blind so that the first three colour components need not necessarily represent RGB data. The colour transform may be used to exploit redundancy between any collection of three image components which satisfy the conditions mentioned above. Matters of colour interpretation are relegated to the optional file format described in Chapter 14.

---

[1] It is still possible to decompress a reversibly compressed image with a different bit-depth, but the misalignment of non-linear rounding operations will introduce additional distortion. This may be necessary if the bit-depth of the original image exceeds the numerical precision of a decompressor, as suggested in Section 18.3.3.

10.2.1 DEFINITION OF THE ICT

The irreversible colour transform is a point-wise linear operator described by weighting factors

$$\alpha_R \triangleq 0.299, \quad \alpha_G \triangleq 0.587, \quad \alpha_B \triangleq 0.114$$

and the relationships

$$x_Y[n] \triangleq \alpha_R x_R[n] + \alpha_G x_G[n] + \alpha_B x_B[n] \tag{10.2}$$

$$x_{Cb}[n] \triangleq \frac{0.5}{1 - \alpha_B}(x_B[n] - x_Y[n]) \tag{10.3}$$

$$x_{Cr}[n] \triangleq \frac{0.5}{1 - \alpha_R}(x_R[n] - x_Y[n]) \tag{10.4}$$

The above expressions are to be treated as exact. The relationship may be equivalently expressed in terms of matrix-vector multiplication, as follows

$$\begin{pmatrix} x_Y[n] \\ x_{Cb}[n] \\ x_{Cr}[n] \end{pmatrix} = \begin{pmatrix} 0.299 & 0.587 & 0.114 \\ -0.168736 & -0.331264 & 0.5 \\ 0.5 & -0.418688 & -0.081312 \end{pmatrix} \begin{pmatrix} x_R[n] \\ x_G[n] \\ x_B[n] \end{pmatrix}$$

Note, however, that the decimals used in the above matrix are only approximate.

In accordance with Figure 10.2, we identify the inputs to the transform, $x_R[n]$, $x_G[n]$ and $x_B[n]$, with the level-shifted and normalized samples, $x_0[n], x_1[n], x_2[n] \in \left[-\frac{1}{2}, \frac{1}{2}\right]$, respectively. After application of the transform, the new image components (we use the same symbols), $x_0[n]$ through $x_2[n]$, are identified with $x_Y[n]$, $x_{Cb}[n]$ and $x_{Cr}[n]$, respectively. The reader may verify that these again exhibit a unit range, with $x_k[n] \in \left[-\frac{1}{2}, \frac{1}{2}\right]$, $\forall k$.

Observe that x_Y is a weighted average of the red, green and blue colour components. As such, it may be regarded as a measure of image intensity (or luminance). The weights reflect the importance of information in the green portion of the visible spectrum to visual perception of detail. Observe also that x_{Cb} and x_{Cr} are weighted differences between the blue (respectively, red) input and the luminance. These colour differences are commonly known as "chrominance components." The human visual system is known to be substantially less sensitive to distortion in the chrominance components than the luminance component. This is especially true at higher spatial frequencies, since chrominance plays a relatively minor role in the perception of edges and fine details. These properties may be exploited through careful control of the quantization step sizes and/or code-block truncation points associated with the luminance and chrominance components. The reader is referred to Section 16.1.2 for guidance in this matter.

During decompression, the transformation must be inverted using the reciprocal relationship

$$
\begin{pmatrix} x_R [\mathbf{n}] \\ x_G [\mathbf{n}] \\ x_B [\mathbf{n}] \end{pmatrix} = \begin{pmatrix} 1 & 0 & 1.402 \\ 1 & -0.344136 & -0.714136 \\ 1 & 1.772 & 0 \end{pmatrix} \begin{pmatrix} x_Y [\mathbf{n}] \\ x_{Cb} [\mathbf{n}] \\ x_{Cr} [\mathbf{n}] \end{pmatrix}
$$

The decimals on the first and third rows of the matrix are exact. To obtain more accurate expressions for the second row of the matrix, the reader may directly invert equations (10.2), through (10.4).

Evidently, the inverse transform does not possess the range preserving property exhibited by the forward transform. Also, since the subband sample values are generally corrupted by quantization, there is no guarantee that the image components supplied to the inverse transform will strictly conform to the nominal range of $-\frac{1}{2}$ to $\frac{1}{2}$. In this way, the reconstructed sample values might not lie within the interval $\left[-\frac{1}{2}, \frac{1}{2}\right]$ suggested by Figure 10.2.

Decompressors may deal with out of range sample values by clipping them to the nearest representable value. In this case, the clipping is best applied to RGB data after inverting any colour transform, since YCbCr data which conforms to the nominal range bounds may still produce out of range RGB samples. Note, however, that the standard does not specify how a decoder should deal with out of range reconstructed sample values.

10.2.2   DEFINITION OF THE RCT

The reversible colour transform is defined by the following non-linear relations

$$
x_{Y'} [\mathbf{n}] \triangleq \left\lfloor \frac{x_R [\mathbf{n}] + 2x_G [\mathbf{n}] + x_B [\mathbf{n}]}{4} \right\rfloor
$$

$$
x_{Db} [\mathbf{n}] \triangleq x_B [\mathbf{n}] - x_G [\mathbf{n}]
$$

$$
x_{Dr} [\mathbf{n}] \triangleq x_R [\mathbf{n}] - x_G [\mathbf{n}]
$$

Here $\lfloor x \rfloor$ denotes the "floor" function, which returns the largest integer not exceeding x. Thus, all quantities are integers. In particular, the inputs to the transform, $x_R [\mathbf{n}]$, $x_G [\mathbf{n}]$ and $x_B [\mathbf{n}]$, are identified with the level shifted sample values, $x_0 [\mathbf{n}]$ through $x_2 [\mathbf{n}]$, respectively, which are B-bit signed integers in the range -2^{B-1} to $2^{B-1} - 1$. After application of the transform, the new image components (we use the same symbols), $x_0 [\mathbf{n}]$ through $x_2 [\mathbf{n}]$, are identified with $x_{Y'} [\mathbf{n}]$, $x_{Db} [\mathbf{n}]$ and

$x_{\text{Dr}}[\mathbf{n}]$, respectively. These are again integers. The luminance component, $x_0[\mathbf{n}] = x_{\text{Y}'}[\mathbf{n}]$, is another B-bit signed integer. The colour difference components, however, are $(B+1)$-bit integers in the range $1 - 2^B$ to $2^B - 1$. Implementations must be careful to accommodate these larger precision integers.

The RCT has similar features to the YCbCr transform described above. $x_{\text{Y}'}$ is approximately a weighted average of the red, green and blue colour components, with the green channel weighted more heavily than the others. Again, x_{Db} and x_{Dr} are colour difference components.

The advantage of the RCT is that it can be exactly inverted. To see this, observe that

$$\left\lfloor \frac{x_{\text{Db}}[\mathbf{n}] + x_{\text{Dr}}[\mathbf{n}]}{4} \right\rfloor = \left\lfloor \frac{x_{\text{R}}[\mathbf{n}] + x_{\text{B}}[\mathbf{n}] + 2x_{\text{G}}[\mathbf{n}]}{4} - x_{\text{G}}[\mathbf{n}] \right\rfloor$$
$$= x_{\text{Y}'}[\mathbf{n}] - x_{\text{G}}[\mathbf{n}]$$

This allows us to reconstruct $x_{\text{G}}[\mathbf{n}]$ and hence $x_{\text{B}}[\mathbf{n}]$ and $x_{\text{R}}[\mathbf{n}]$. Specifically, we have the following relationships

$$x_{\text{G}}[\mathbf{n}] = x_{\text{Y}'}[\mathbf{n}] - \left\lfloor \frac{x_{\text{Db}}[\mathbf{n}] + x_{\text{Dr}}[\mathbf{n}]}{4} \right\rfloor$$
$$x_{\text{B}}[\mathbf{n}] = x_{\text{Db}}[\mathbf{n}] + x_{\text{G}}[\mathbf{n}]$$
$$x_{\text{R}}[\mathbf{n}] = x_{\text{Dr}}[\mathbf{n}] + x_{\text{G}}[\mathbf{n}]$$

10.3 WAVELET TRANSFORM BASICS

Subband and wavelet transform fundamentals are treated extensively in Chapters 4 and 6 of this text. The purpose of this section is to provide a concise, if rather mechanistic description of the two dimensional DWT (Discrete Wavelet Transform). Here and in the following sections, we borrow results directly from the earlier chapters. The present treatment provides no insight into the design of suitable wavelet transforms; however, Part 1 of the standard offers little flexibility in this regard. The reader whose primary interest is implementing the standard may find the material presented here sufficient.

10.3.1 TWO CHANNEL BUILDING BLOCK

The two dimensional DWT is constructed from simple one dimensional building blocks which convert a finite length input sequence, $x[n]$, into two subband sequences, $y_0[n]$ and $y_1[n]$. The former is identified as the low-pass subband and its construction may be understood in terms of low-pass filtering followed by sub-sampling (discarding every second sample). Similarly, $y_1[n]$ is identified as the high-pass subband and its

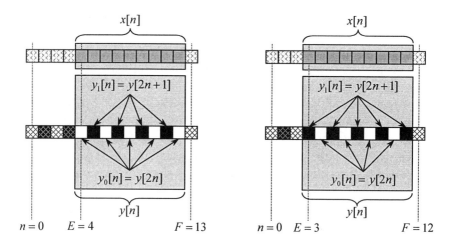

Figure 10.3. Coordinate relationships between the input sequence, $x[n]$, the interleaved subband sequence, $y[n]$, and the individual subband sequences, $y_0[n]$ and $y_1[n]$, for sequences with different regions of support.

construction may be understood in terms of high-pass filtering followed by sub-sampling. Together, the two subband sequences contain the same number of samples as $x[n]$.

It is convenient to describe the subband transform as a mapping from $x[n]$ to the single sequence, $y[n]$, which is obtained by interleaving the low- and high-pass subband samples according to

$$y[2n] = y_0[n]$$
$$y[2n+1] = y_1[n]$$

Both $x[n]$ and $y[n]$ are defined over the same interval, $E \leq n < F$, so that each contains $F - E$ samples. This means that the individual subband sequences, $y_0[n]$ and $y_1[n]$, are defined over the intervals $E^0 \leq n < F^0$ and $E^1 \leq n < F^1$, respectively, where

$$E^0 = \left\lceil \frac{E}{2} \right\rceil, \quad F^0 = \left\lceil \frac{F}{2} \right\rceil, \quad E^1 = \left\lfloor \frac{E}{2} \right\rfloor, \quad F^1 = \left\lfloor \frac{F}{2} \right\rfloor$$

More compactly, we have

$$E^b = \left\lceil \frac{E - b}{2} \right\rceil, \quad F^b = \left\lceil \frac{F - b}{2} \right\rceil, \quad \text{for } b = 0, 1 \qquad (10.5)$$

These coordinate relationships are illustrated in Figure 10.3. For odd length input sequences, the low-pass subband may have more or less samples than the high-pass subband, depending on the parity of E.

A few comments are in order concerning these general regions of support. A useful mnemonic for the notation used here is to regard E as the "entry" point and F as the "finishing" point for both $x[n]$ and $y[n]$. In the simplest case, $E = 0$ and F is the length of the input sequence. Then, either both subbands have the same length or the low-pass subband has one more sample than the high-pass subband. Most wavelet-based image compression algorithms which have been proposed in the literature have adopted this convention. This is insufficient, however, to realize a number of important features offered by the JPEG2000 standard. An important example is efficient cropping of compressed images, which is discussed further in Section 11.4. The idea of using arbitrary boundaries, E and F, to realize such capabilities was inspired by the use of this method for global motion compensation in [150].

The mapping from $x[n]$ to $y[n]$ is most easily described in terms of their symmetric extensions, $\breve{x}[n]$ and $\breve{y}[n]$. The symmetric extension of $x[n]$ is the infinite length sequence, $\breve{x}[n]$, defined by the following three relations.

$$\breve{x}[n] = x[n], \quad E \leq n < F$$
$$\breve{x}[E - n] = \breve{x}[E + n], \quad \forall n \in \mathbb{Z}$$
$$\breve{x}[F - 1 - n] = \breve{x}[F - 1 + n], \quad \forall n \in \mathbb{Z}$$

Notice that $\breve{x}[n]$ is identical to $x[n]$ over its region of support and that $\breve{x}[n]$ is symmetric about each of the end points, $n = E$ and $n = F - 1$. For sequences of length $F - E \geq 2$, the symmetric extension algorithm on Page 298 may be used to progressively construct the unique infinite length sequence, $\breve{x}[n]$, which obeys these relations. For sequences of length 1, these relations fail to define an extension; the behaviour in this case is described separately for irreversible and reversible transforms in Sections 10.4.1 and 10.4.2 below.

The relationship between $y[n]$ and its symmetric extension, $\breve{y}[n]$, is identical to that between $x[n]$ and $\breve{x}[n]$. The forward and reverse subband transforms are defined on the extended sequences by the following time-varying convolution expressions.

$$\breve{y}[n] = \sum_{i \in \mathbb{Z}} h_{n \bmod 2}^t[i]\, \breve{x}[n - i] \tag{10.6a}$$

$$\breve{x}[n] = \sum_{i \in \mathbb{Z}} \breve{y}[i]\, g_{i \bmod 2}^t[n - i] \tag{10.6b}$$

where $h_0^t[n]$ and $h_1^t[n]$ denote the low- and high-pass "translated" analysis filter impulse responses and $g_0^t[n]$ and $g_1^t[n]$ denote the low- and high-pass "translated" synthesis filter impulse responses. The term "translated" is used to distinguish these from the analysis and synthesis filters of a conventional filter bank. For simplicity, we shall henceforth refer to the translated impulse responses as the "wavelet kernels." The relationship between these wavelet kernels and the conventional filters is given by equation (6.9) although we prefer to work exclusively with the kernels when describing the standard.

The analysis kernels, $h_0^t[n]$ and $h_1^t[n]$, and the synthesis kernels, $g_0^t[n]$ and $g_1^t[n]$, are odd length sequences, symmetric about $n = 0$. That is,

$$h_b^t[n] = h_b^t[-n], \quad g_b^t[n] = g_b^t[-n], \quad \text{for } b \in \{0,1\}$$

These conditions are discussed extensively in Section 6.1.3. They also ensure that the subband transform operations described above will preserve the symmetry properties required of $\breve{x}[n]$ and $\breve{y}[n]$, as explained in Section 6.5. While Part 2 of the JPEG2000 standard is also expected to offer support for even length kernels, these entail a number of subtle complications. The interested reader is referred to Section 15.6.2 for more information on this topic.

Obviously the analysis and synthesis kernels are closely connected, since the subband synthesis operation of equation (10.6b) must invert the subband analysis operation of equation (10.6a). In fact, the analysis and synthesis kernels must be connected through equation (6.10), which we repeat here as

$$\begin{aligned} g_0^t[n] &= \alpha^{-1} (-1)^n h_1^t[n] \\ g_1^t[n] &= \alpha^{-1} (-1)^n h_0^t[n] \end{aligned} \tag{10.7}$$

The gain factor, α, is given by

$$\alpha = \frac{1}{2} \left(h_0^{\text{dc}} h_1^{\text{nyq}} + h_1^{\text{dc}} h_0^{\text{nyq}} \right) \tag{10.8}$$

where h_b^{dc} and h_b^{nyq} denote the gains of the relevant analysis filters at DC and at the Nyquist frequency. Specifically,

$$h_b^{\text{dc}} = \sum_n h_b^t[n] \text{ and } h_b^{\text{nyq}} = \sum_n (-1)^n h_b^t[n], \quad \text{for } b = 0, 1$$

With these relationships in hand, it is sufficient to supply only the DWT analysis kernels, $h_0^t[n]$ and $h_1^t[n]$, to completely define the forward and reverse mappings between $\breve{x}[n]$ and $\breve{y}[n]$.

Following the notation developed in Section 6.1.3, let the low- and high-pass analysis kernels have lengths $2L_0 + 1$ and $2L_1 + 1$, respectively. Exploiting symmetry, the analysis operation of equation (10.6a) may be

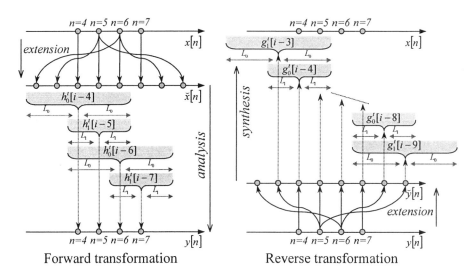

Forward transformation	Reverse transformation

Figure 10.4. Forward and reverse subband transformations, illustrated for sequences, $x[n]$, supported on $E = 4 \le n < F = 8$, and symmetric filter length parameters $L_0 = 2$ and $L_1 = 1$.

rewritten as

$$\breve{y}[n] = \sum_{i=-L_{n \bmod 2}}^{L_{n \bmod 2}} h_{n \bmod 2}^t[i]\, \breve{x}[n+i] \tag{10.9}$$

$$= \sum_{i=n-L_{n \bmod 2}}^{n+L_{n \bmod 2}} \breve{x}[i]\, h_{n \bmod 2}^t[i-n]$$

In words, $\breve{y}[n]$ is obtained by shifting (delaying) the relevant analysis kernel, h_0^t or h_1^t depending on n, so that its region of support is centered about the location n; we then take the correlation (or dot product) of this shifted version of the kernel and the extended input sequence, $\breve{x}[n]$. Of course, we need only evaluate $\breve{y}[n]$ and hence $y[n]$ over the interval $E \le n < F$. Consequently, the forward transform requires access only to extended input samples, $\breve{x}[n]$, in the range $E - L_{\max} \le n < F + L_{\max}$, where $L_{\max} = \max\{L_0, L_1\}$.[2] These forward transform operations are depicted in Figure 10.4.

---

[2] While this range of input samples is sufficient, a slightly smaller range may be possible, as seen from the example in Figure 10.4.

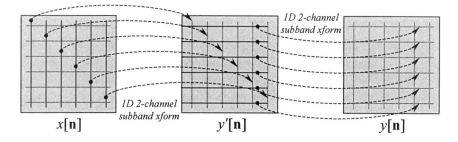

Figure 10.5. Two dimensional DWT stage implemented by separable application of one dimensional subband transforms.

The synthesis operation of equation (10.6b) may be interpreted as follows. An array holding $\breve{x}[n]$ is initialized to 0. Then, for each $i \in \mathbb{Z}$, we shift the relevant synthesis kernel, $g_0^t[n]$ or $g_1^t[n]$ depending on i, so that its region of support is centered about the location i, multiplying the shifted kernel by the value of the coefficient, $\breve{y}[i]$, and adding the resulting sequence into the evolving array, $\breve{x}[n]$. Of course, only those samples, $\breve{x}[n]$, having $E \leq n < F$ need actually be manipulated. Moreover, these $\breve{x}[n]$ can be impacted only by those subband samples, $\breve{y}[i]$, having $E - L_{\max} \leq i < F + L_{\max}$. These reverse transform operations are depicted in Figure 10.4.

10.3.2   THE 2D DWT

We now describe the construction of a D level two dimensional DWT from the one dimensional, two channel subband transforms described above. A single "DWT stage" applies the subband transform separably to the columns and then the rows of the two dimensional sequence (image component), $x[\mathbf{n}]$, yielding four subbands, $y_{0,0}[\mathbf{n}]$ through $y_{1,1}[\mathbf{n}]$. Specifically, $x[\mathbf{n}] \equiv x[n_1, n_2]$ is a finite sequence supported over the region

$$E_1 \leq n_1 < F_1, \quad E_2 \leq n_2 < F_2$$

where n_1 and n_2 denote row and column indices, respectively. Let $y[\mathbf{n}]$ be the two dimensional sequence of interleaved subband samples, defined by

$$y[2n_1 + b_1, 2n_2 + b_2] = y_{b_1, b_2}[n_1, n_2], \quad \text{for } b_1, b_2 \in \{0, 1\}$$

Then $y[\mathbf{n}]$ has the same region of support as $x[\mathbf{n}]$ and is obtained by applying the one dimensional subband transform first to each column of $x[\mathbf{n}]$ and then to each row of the result, as illustrated in Figure 10.5. Similarly, the reverse transformation from $y[\mathbf{n}]$ back to $x[\mathbf{n}]$ is obtained by applying the reverse one dimensional transform first to each row of $y[\mathbf{n}]$ and then to each column of the result.

The description provided above in terms of an interleaved subband sequence, $y[\mathbf{n}]$, is convenient because $y[\mathbf{n}]$ has exactly the same region of support as $x[\mathbf{n}]$. However, it is important to bear in mind that there are actually four distinct subbands. These may be understood as arising from the two dimensional filter bank illustrated in Figure 4.15. The subband $y_{0,0}[\mathbf{n}]$ arises from the application of the low-pass analysis kernel in both the horizontal and the vertical direction. Accordingly, we refer to $y_{0,0}$ as the LL subband. The subband $y_{0,1}[\mathbf{n}]$ involves the application of the low-pass analysis kernel in the vertical direction and the high-pass analysis kernel in the horizontal direction. We refer to this subband as the HL (horizontally high-pass) subband. The other subbands, $y_{1,0}[\mathbf{n}]$ and $y_{1,1}[\mathbf{n}]$, are identified as LH (vertically high-pass) and HH, respectively. The characteristics of these four subbands are studied more carefully in Section 4.2.5. The region of support for any of these four subbands, $y_{b_1,b_2}[\mathbf{n}]$, is given by

$$E_1^{\mathbf{b}} \leq n_1 < F_1^{\mathbf{b}} \text{ and } E_2^{\mathbf{b}} \leq n_2 < F_2^{\mathbf{b}}$$

where the bounds, $E_i^{\mathbf{b}} \equiv E_i^{[b_1,b_2]}$ and $F_i^{\mathbf{b}} \equiv F_i^{[b_1,b_2]}$, are obtained by straightforward generalization of equation (10.5) as

$$E_i^{[b_1,b_2]} = \left\lceil \frac{E_i - b_i}{2} \right\rceil, \quad F_i^{[b_1,b_2]} = \left\lceil \frac{F_i - b_i}{2} \right\rceil \qquad (10.10)$$

A D level DWT is obtained by applying D DWT stages in the manner illustrated in Figure 10.6. That is, the LL subband produced by a first DWT stage is subjected to a subsequent DWT stage and so on. The subbands produced by the dth stage of the transform are labelled $y_{0,0}^{(d)}[\mathbf{n}]$, $y_{0,1}^{(d)}[\mathbf{n}]$, $y_{1,0}^{(d)}[\mathbf{n}]$ and $y_{1,1}^{(d)}[\mathbf{n}]$, or LL_d, HL_d, LH_d and HH_d, whichever is most convenient. Note that $y_{0,0}^{(d)}[\mathbf{n}]$ (i.e., LL_d) is an intermediate subband for $d < D$, since it is further decomposed by subsequent DWT stages. These intermediate subbands are not subjected to quantization and coding.

The JPEG2000 standard supports values of D in the range $0 \leq D \leq 32$. Typical values are in the range $D = 4$ through $D = 8$ with $D = 5$ sufficient to obtain near optimal compression performance for the full resolution image. If the quantization and coding operations specified by the standard are to be performed directly on the image sample values, $x[\mathbf{n}]$, the DWT operation may be skipped by setting $D = 0$. This can be useful when compressing bi-level images or palettized colour images,

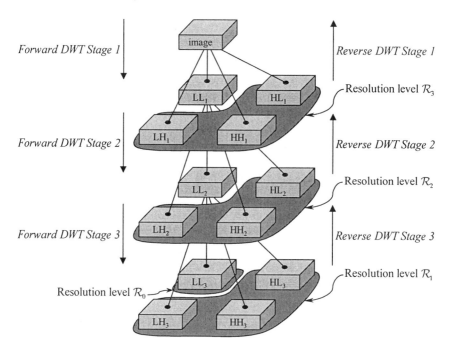

Figure 10.6.   Three level, two dimensional DWT.

as discussed in Section 16.3. Consequently, it is convenient to regard the image itself as subband LL_0; i.e., $y_{0,0}^{(0)}[\mathbf{n}] = x[\mathbf{n}]$. The region of support for any particular subband, $y_{b_1,b_2}^{(d)}[\mathbf{n}]$, may be found by iterative application of equation (10.10) as

$$E_i^{[b_1,b_2]_d} = \left\lceil \frac{E_i - 2^{d-1}b_i}{2^d} \right\rceil, \quad F_i^{[b_1,b_2]_d} = \left\lceil \frac{F_i - 2^{d-1}b_i}{2^d} \right\rceil \qquad (10.11)$$

10.3.3   RESOLUTIONS AND RESOLUTION LEVELS

In the inverse transformation, each of the D DWT stages is inverted, starting from the last DWT stage which was performed during the forward transform. After inverting r DWT stages, the decompressor recovers $LL_{D-r} = y_{0,0}^{(D-r)}[\mathbf{n}]$, which is a low-resolution version of the image. We refer to LL_{D-r} as the rth resolution of the image, with $r = 0$ corresponding to the lowest available resolution and $r = D$ corresponding to the original image resolution. The region of support and hence the

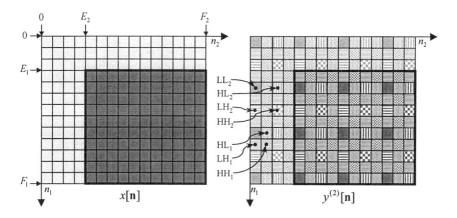

Figure 10.7.   Interleaving of subband samples from a two level DWT.

dimensions of each image resolution, $y_{0,0}^{(D-r)}[\mathbf{n}]$, may be found by substituting $b_1 = b_2 = 0$ into equation (10.11).

We refer to the collection of subbands which are required to augment the image from resolution LL_{r-1} to resolution LL_r as "resolution level" $\mathcal{R}_r$. The relationship between DWT stages, subbands and resolution levels is illustrated in Figure 10.6. Notice that there are always $D+1$ distinct resolution levels and that $\mathcal{R}_0$ always has only one subband, LL_D. When no DWT is used, the image is available only at one resolution.

10.3.4   THE INTERLEAVED PERSPECTIVE

Recall that the four subbands produced by a single stage of the DWT may be interleaved into a single sequence, call this $y^{(1)}[\mathbf{n}]$, according to

$$y^{(1)}[2n_1 + b_1, 2n_2 + b_2] = y_{b_1,b_2}^{(1)}[n_1, n_2], \quad \text{for } b_1, b_2 \in \{0, 1\}$$

where $y^{(1)}[\mathbf{n}]$ has the same region of support as $x[\mathbf{n}]$. Since the second stage of the DWT processes only $y_{0,0}^{(1)}[\mathbf{n}]$, corresponding to the even indexed elements of $y^{(1)}[\mathbf{n}]$, we may interleave the resulting subband samples into the even indexed elements of the new sequence, $y^{(2)}[\mathbf{n}]$, leaving the remaining elements of $y^{(2)}[\mathbf{n}]$ as they were in $y^{(1)}[\mathbf{n}]$. This is illustrated in Figure 10.7. Clearly, $y^{(2)}[\mathbf{n}]$ has the same region of support as $y^{(1)}[\mathbf{n}]$ and hence $x[\mathbf{n}]$.

Proceeding in this way, the dth DWT stage may be viewed as processing only those elements of $y^{(d-1)}[\mathbf{n}]$ whose indices, n_1 and n_2, are both divisible by 2^{d-1}, producing the new interleaved sequence, $y^{(d)}[\mathbf{n}]$. Each of the two dimensional interleaved sequences, $y^{(d)}[\mathbf{n}]$, has the same region of support as $x[\mathbf{n}]$; i.e., $E_1 \leq n_1 < F_1$ and $E_2 \leq n_2 < F_2$. Ultimately, all of the subbands produced by the D level DWT may be found in a single interleaved sequence, $y^{(D)}[\mathbf{n}]$, with the following index assignments

$$y^{(d)}_{b_1,b_2}[n_1, n_2] = y^{(D)}\left[2^d n_1 + 2^{d-1} b_1, 2^d n_2 + 2^{d-1} b_2\right], \quad b_1, b_2 \in \{0, 1\}$$

The perspective offered by this interleaving of the subband samples is interesting for two reasons. Firstly, it suggests that every subband sample occupies a unique position in the original image's region of support. The reader may verify that the individual subband supports given by equation (10.11) may be derived directly by restricting the coordinates, $[2^d n_1 + 2^{d-1} b_1, 2^d n_2 + 2^{d-1} b_2]$, to lie within the region $[E_1, F_1) \times [E_2, F_2)$. The fact that each subband sample conceptually occupies a unique position on the image grid forms the basis for all of the partitions defined in Chapter 11.

The interleaved subband perspective also reveals an interesting property of the DWT operator defined by JPEG2000. Specifically, let $x'[\mathbf{n}]$ be the horizontal mirror image of $x[\mathbf{n}]$ obtained by setting

$$x'[n_1, n_2] = x[n_1, -n_2]$$

The support bounds for $x'[\mathbf{n}]$ and its interleaved subband sequence, $y'^{(D)}[\mathbf{n}]$, are given by

$$E_1' = E_1, \quad F_1' = F_1, \quad E_2' = -(F_2 - 1), \quad F_2' = -(E_2 - 1)$$

Interestingly, $y'^{(D)}[\mathbf{n}]$ is itself the horizontal mirror image of $y^{(D)}[\mathbf{n}]$. To see this, consider the one dimensional building block which maps $x[n]$ to an interleaved subband sequence, $y[n]$, through symmetric extension and subband analysis. The symmetric extension operation clearly commutes with the mirror imaging operation so that

$$x'[n] = x[-n] \iff \breve{x}'[n] = \breve{x}[-n]$$

The subband analysis operation of equation (10.6a) also commutes with mirror imaging, since

$$\breve{y}'[n] = \sum_{i \in \mathbb{Z}} h^t_{n \bmod 2}[i]\, \breve{x}'[n - i]$$

$$= \sum_{i \in \mathbb{Z}} h^t_{n \bmod 2}[i]\, \breve{x}[i - n]$$

$$= \sum_{i \in \mathbb{Z}} h^t_{n \bmod 2}[-i]\, \breve{x}[(-n) - i]$$

$$= \breve{y}[-n]$$

where we have used the fact that the wavelet kernels, $h_0^t [n]$ and $h_1^t [n]$, are symmetric about $n = 0$. The result is easily extended to two dimensions and multiple DWT levels.

The above analysis obviously applies to the case of vertical mirror images as well. The fact that taking the mirror image of $x [\mathbf{n}]$ is equivalent to taking the mirror image of the interleaved subband sequence, $y^{(D)} [\mathbf{n}]$, greatly facilitates some simple geometric manipulations of JPEG2000 compressed images, as described in Section 11.4.

10.4   WAVELET TRANSFORMS
10.4.1   THE IRREVERSIBLE DWT

Recall that our convention[3] for the irreversible path is to normalize all sample values to a unit nominal range of $-\frac{1}{2}$ to $\frac{1}{2}$. To ensure that the DWT preserves the nominal range of its input samples, we normalize the DWT analysis kernels so that

$$h_0^{dc} = \sum_{n=-L_0}^{L_0} h_0^t [n] = 1, \quad h_1^{nyq} = \sum_{n=-L_1}^{L_1} (-1)^n h_1^t [n] = 1 \quad (10.12)$$

We think of h_0^{dc} as the "nominal gain" of the low-pass subband filter, while h_1^{nyq} is interpreted as the nominal gain of the high-pass filter. By selecting these nominal gains to be 1, we expect the nominal range of the subband samples to be preserved. We refer only to a "nominal" range, however, because unlikely combinations of input sample values may conspire to produce isolated excursions beyond the nominal bounds.

Although Part 2 of the JPEG2000 standard is expected to support irreversible transformations with a large class of wavelet kernels, Part 1 supports only the CDF 9/7 kernels. The low- and high-pass analysis filters have lengths 9 and 7, respectively, and are given approximately by

$$h_0^t (z) =$$
$$\begin{aligned}
&\ 0.602949018236 \\
&+\ 0.266864118443 \left(z^1 + z^{-1}\right) \\
&-\ 0.078223266529 \left(z^2 + z^{-2}\right) \\
&-\ 0.016864118443 \left(z^3 + z^{-3}\right) \\
&+\ 0.026748757411 \left(z^4 + z^{-4}\right)
\end{aligned}$$

$$h_1^t (z) =$$
$$\begin{aligned}
&\ 0.557543526229 \\
&-\ 0.295635881557 \left(z^1 + z^{-1}\right) \\
&-\ 0.028771763114 \left(z^2 + z^{-2}\right) \\
&+\ 0.045635881557 \left(z^3 + z^{-3}\right)
\end{aligned}$$

$$(10.13)$$

---

[3] The convention is useful for descriptive purposes. Floating point implementations might choose to use this convention, while fixed point implementations will work with integers which represent suitably scaled versions of the samples.

These kernels belong to the first member of the Cohen-Daubechies-Feauveau family of odd-length linear phase wavelets with maximal regularity. Example 6.4 provides a derivation of these kernels, including exact expressions which may be used if the precision of the numerical approximations above proves insufficient.

Notice that $h_0^{\text{nyq}} = h_1^{\text{dc}} = 0$. In fact, this is a necessary condition if we insist that the DWT kernels belong to an underlying biorthogonal wavelet transform. This fact is established in Theorem 6.6 and the implications of the condition for image compression are discussed in Section 6.3.3. Substituting into equation (10.8) we see that the gain factor in equation (10.7) is $\alpha = \frac{1}{2}$. The synthesis kernels, $g_0^t[n]$ and $g_1^t[n]$, are then trivially deduced from the analysis kernels. We note, however, that Part 2 of the standard supports custom wavelet kernels which need not necessarily satisfy the condition $h_0^{\text{nyq}} = h_1^{\text{dc}} = 0$. Thus, in this more general setting, α, is not guaranteed to equal $\frac{1}{2}$.

SEQUENCES OF UNIT LENGTH

Recall that the two channel subband transform described in Section 10.3.1 is defined only for sequences of length $F - E \geq 2$. For sequences of unit length (i.e., $F - E = 1$), we define the interleaved sequence of subband samples, $y[n]$, to be equal to the input sequence, $x[n]$. That is, $y[E] = x[E]$. This policy is followed regardless of whether E is even ($y[E]$ is a low-pass subband sample) or odd ($y[E]$ is a high-pass subband sample).

LIFTING IMPLEMENTATIONS

The fundamental building block of the irreversible DWT is the two channel subband transform described in Section 10.3.1. The analysis and synthesis procedures embodied by equations (10.6a) and (10.6b) may be implemented directly. In many cases, however, it may be preferable to employ the lifting procedure introduced in Section 6.4.1. Example 6.6 describes a lifting implementation of the CDF 9/7 kernels, complete with lifting step coefficients. The reader is also referred to Section 6.5.3 for a discussion of the interaction between symmetric extension and lifting.

One reason to prefer a lifting implementation is that lifting is the only vehicle for implementing reversible transformations in the JPEG2000 standard. A common lifting framework for both the reversible and irreversible paths generally simplifies the task of implementing the standard as a whole. A second reason to prefer lifting implementations is that they may be designed to require less working memory and fewer arithmetic computations than a direct implementation of equations (10.6a) and (10.6b). These implementation considerations are taken up again in Chapter 17.

10.4.2 THE REVERSIBLE DWT

The reversible DWT is implemented within exactly the same framework as the irreversible DWT, except that the analysis and synthesis operations of equations (10.6a) and (10.6b) are approximated by nonlinear operations which efficiently map integers to integers. Specifically, we employ the modified lifting procedure developed in Section 6.4.2, where linear lifting steps are replaced by their non-linear approximations, in accordance with equation (6.51), and the subband gain factors are set to 1.

Although Part 2 of the standard is expected to support reversible transformations with a large class of wavelet kernels, Part 1 defines only one reversible DWT which is derived from the spline 5/3 transform. Specifically, the analysis operation is defined by

$$\breve{y}_1[n] = \breve{y}[2n+1] = \breve{x}[2n+1] + \left\lfloor \frac{1}{2} - \frac{1}{2}\breve{x}[2n] - \frac{1}{2}\breve{x}[2n+2] \right\rfloor \quad (10.14)$$

$$\approx \breve{x}[2n+1] - \frac{1}{2}\left(\breve{x}[2n] + \breve{x}[2n+2]\right)$$

$$\breve{y}_0[n] = \breve{y}[2n] = \breve{x}[2n] + \left\lfloor \frac{1}{2} + \frac{1}{4}\breve{y}[2n-1] + \frac{1}{4}\breve{y}[2n+1] \right\rfloor$$

$$\approx \breve{x}[2n] + \frac{1}{4}\left(\breve{y}[2n-1] + \breve{y}[2n+1]\right)$$

The above expressions are obtained by direct application of the approximation embodied by equation (6.51) to the spline 5/3 lifting steps of Example 6.5. Since all quantities are integers, equation (10.14) may equivalently be expressed as

$$\breve{y}_1[n] = \breve{x}[2n+1] - \left\lfloor \frac{1}{2}\left(\breve{x}[2n] + \breve{x}[2n+2]\right) \right\rfloor$$

which more clearly reveals the interpretation of the first lifting step as "predicting" $x[2n+1]$ from its two neighbours. The synthesis operation is defined by

$$\breve{x}[2n] = \breve{y}[2n] - \left\lfloor \frac{1}{2} + \frac{1}{4}\breve{y}[2n-1] + \frac{1}{4}\breve{y}[2n+1] \right\rfloor$$

$$\breve{x}[2n+1] = \breve{y}[2n+1] - \left\lfloor \frac{1}{2} - \frac{1}{2}\breve{x}[2n] - \frac{1}{2}\breve{x}[2n+2] \right\rfloor$$

$$= \breve{y}[2n+1] + \left\lfloor \frac{1}{2}\left(\breve{x}[2n] + \breve{x}[2n+2]\right) \right\rfloor$$

The lifting steps described above are close approximations of the linear lifting steps which implement the spline 5/3 transform as in Example 6.5, except that the subband gain factors are all set to 1. The reader may verify that the corresponding linear analysis kernels are

$$h_0^t(z) = h_0(z) = -\tfrac{1}{8}z^{-2} + \tfrac{1}{4}z^{-1} + \tfrac{3}{4} + \tfrac{1}{4}z - \tfrac{1}{8}z^2$$
$$h_1^t(z) = z^{-1}h_1(z) = -\tfrac{1}{2}z^{-1} + 1 - \tfrac{1}{2}z \qquad (10.15)$$

having nominal gains of

$$h_0^{dc} = 1, \quad h_1^{nyq} = 2 \qquad (10.16)$$

As noted in Section 10.1.1, preservation of nominal range is of interest only in the irreversible path, while the important feature of the reversible path is that it should map integers to integers. Nevertheless, the nominal gains may be used to guide ranging decisions, as discussed in Section 10.5.

SEQUENCES OF UNIT LENGTH

As in the irreversible case, sequences of unit length (i.e., when $F - E = 1$) are treated differently. If E is even, so that $y[n]$ is a low-pass subband sample, we simply set $y[n] = x[n]$. If E is odd, meaning that $y[n]$ is a high-pass subband sample, we set $y[n] = 2x[n]$. Note that this procedure is consistent with the nominal gains of the low- and high-pass analysis kernels, as indicated by equation (10.16).

10.5   QUANTIZATION AND RANGING

Although this section is written to be as self-contained as possible, the reader is strongly encouraged to review the principles of Embedded Block Coding with Optimal Truncation (EBCOT) and bit-plane coding, as expounded in Sections 8.1.3, and 8.3.1.

In the irreversible path, quantization is the process by which subband samples are mapped to quantization indices for coding. Since non-trivial quantization is inherently irreversible, the reversible path contains no quantization. However, both the reversible and irreversible paths must provide the block coder with consistent interpretations for the absolute range of the integers which are to be coded; we call this "ranging." In the irreversible path, ranging is connected to the choice of quantization parameters. In the reversible path, ranging may be understood as a separate task, as indicated in Figure 10.1.

10.5.1   IRREVERSIBLE PROCESSING

Part 1 of the JPEG2000 standard requires that subband samples be subjected to deadzone scalar quantization, as described in Section 3.2.7.

The central quantization interval (i.e., the deadzone) is twice as wide as the other quantization intervals and the quantization operation is defined by the step size parameter, Δ_b, through

$$q_b\left[\mathbf{n}\right] = \text{sign}\left(y_b\left[\mathbf{n}\right]\right)\left\lfloor \frac{\left|y_b\left[\mathbf{n}\right]\right|}{\Delta_b}\right\rfloor \tag{10.17}$$

Here, $y_b\left[\mathbf{n}\right]$ denotes the samples of subband b, with a unit nominal range of $-\frac{1}{2}$ to $\frac{1}{2}$, while $q_b\left[\mathbf{n}\right]$ denotes their quantization indices. A separate step size, Δ_b, may be selected for each subband. Since the block coder is best described as operating on a sign magnitude representation of the quantization indices, it is convenient to introduce the notation $\chi_b\left[\mathbf{n}\right] \in \{-1,1\}$ for $\text{sign}\left(y_b\left[\mathbf{n}\right]\right)$ and $v_b\left[\mathbf{n}\right]$ for $\left|q_b\left[\mathbf{n}\right]\right|$. Then equation (10.17) becomes

$$\chi_q\left[\mathbf{n}\right] = \text{sign}\left(y_b\left[\mathbf{n}\right]\right), \quad v_b\left[\mathbf{n}\right] = \left\lfloor \frac{\left|y_b\left[\mathbf{n}\right]\right|}{\Delta_b}\right\rfloor \tag{10.18}$$

The step size for each subband is specified in terms of an exponent, ε_b, and a mantissa, μ_b, where

$$\Delta_b = 2^{-\varepsilon_b}\left(1 + \frac{\mu_b}{2^{11}}\right) \tag{10.19}$$

Both ε_b and μ_b are non-negative integers with the following numerical ranges,

$$0 \le \varepsilon_b < 2^5, \quad 0 \le \mu_b < 2^{11} \tag{10.20}$$

RANGING

It is important for the encoder and decoder to agree on the number of bits which will be sufficient to represent the quantization index magnitudes, $v_b\left[\mathbf{n}\right]$, in each subband, b. We denote this quantity K_b^{max} and will see exactly how it is used in Chapter 12.

To allow for the fact that subband sample values may occasionally violate our nominal range bounds, we introduce a parameter, G, and insist that all subband samples conform to the less restrictive bounds

$$-2^{G-1} < y_b\left[\mathbf{n}\right] < 2^{G-1}, \quad \forall b \tag{10.21}$$

The integer, G, is interpreted as the number of "guard bits." It is explicitly identified through code-stream markers, QCD and QCC, and may take values in the range 0 through 7. A value of 0 means that $y_b\left[\mathbf{n}\right] \in \left(-\frac{1}{2},\frac{1}{2}\right)$ so that our nominal range bounds are never violated; this is rarely sufficient. A more typical value is $G = 1$, and indeed this value is sufficient to avoid violation of equation (10.21) when the CDF 9/7 wavelet kernels are used. For a proof that G need not exceed 1, the reader is referred to Table 17.5 and the surrounding discussion in Section 17.3.3.

Combining equations (10.19) and (10.21), we see that the index magnitudes must satisfy

$$v_b[\mathbf{n}] = \left\lfloor \frac{|y_b[\mathbf{n}]|}{\Delta_b} \right\rfloor < 2^{\varepsilon_b + G - 1}$$

Accordingly, the number of magnitude bit-planes for subband b is defined to be

$$K_b^{\max} \triangleq \max\{0, \varepsilon_b + G - 1\} \qquad (10.22)$$

TYPICAL STEP SIZES

The verification model which was developed in conjunction with the JPEG2000 standard selects quantization step sizes in the following manner. Let G_b denote the squared norm of the DWT synthesis basis vectors for subband b. This quantity represents the energy (sum of squared sample values) in an image reconstructed from exactly one unit amplitude sample in subband b (i.e., setting all other samples in subband b and all samples in all other subbands to 0). G_b is thus an energy gain factor. It is identical for all samples in subband b which are located sufficiently far from the image boundaries to not be affected by the symmetric extension procedure. As explained in Chapter 5, the optimal[4] strategy is to select quantization step sizes according to

$$\Delta_b = \Delta \cdot \sqrt{\frac{1}{G_b}} \qquad (10.23)$$

where Δ is a "base step size" parameter, which may be adjusted to achieve a desired overall compressed bit-rate or level of distortion.

The reader should note that the EBCOT (Embedded Block Coding with Optimal Truncation) paradigm employed by JPEG2000 allows for truncation of the embedded bit-streams representing each code-block. This paradigm was introduced in Section 8.1.3 and forms the motivation for the block coding mechanisms which are described in Chapter 12. As a result, the effective step size associated with any particular sample, $y_i[\mathbf{j}]$, within code-block $\mathcal{B}_i$, may be written

$$\Delta_i[\mathbf{j}] = \Delta_{b_i} \cdot 2^{p_i[\mathbf{j}]} \qquad (10.24)$$

---

[4]This result depends upon a variety of assumptions concerning the rate-distortion behaviour of the quantizers in each subband and the independence of the quantization errors. Also, the optimality criterion here is the MSE (Mean Squared Error) associated with the reconstructed image.

where b_i is the index of the subband which contains code-block $\mathcal{B}_i$ and $p_i\,[\mathbf{j}]$ is the number of least significant bits which have been discarded from the quantization index magnitude, $v_i\,[\mathbf{j}]$, as a result of truncating the code-block's embedded representation. These concepts are explained in Sections 3.2.7 and 8.3.1.

Since quantization may be effectively controlled by block truncation, the selection of an appropriate step size, Δ_b, is less critical than one might imagine. The selection identified by equation (10.23) has the property that optimal code-block truncation is likely to discard a similar number of coding passes from every code-block. For distortion measures other than reconstructed image MSE, however, this is less likely to be the case.

The JPEG2000 code-stream syntax provides an abbreviated signalling method for the step size parameters, ε_b and μ_b, which is based upon equation (10.23) together with the approximation

$$G_b \approx 2^{2d_b} \tag{10.25}$$

where d_b denotes the DWT level index for subband b. Thus, if b is the subband HL_2 in Figure 10.6, then $d_b = 2$. In general, d_b is the number of two dimensional DWT stages involved in the generation of subband b. When using the abbreviated signalling method, code-stream markers QCD and QCC supply explicit step size parameters only for the lowest frequency subband, LL_D. The quantization parameters for an arbitrary subband, b, are then given by

$$\varepsilon_b = \varepsilon_{LL_D} + (d_b - D)\,, \quad \mu_b = \mu_{LL_D} \tag{10.26}$$

The approximation of equation (10.25) is quite accurate for nearly orthogonal transforms such as the CDF 9/7 transform. To see this, note firstly that an orthonormal transform (one in which $G_b = 1$ for all b) must satisfy the power complementary property of equation (6.5). Assuming a Nyquist gain of 0 for the low-pass wavelet kernel, h_0^t, this yields $h_0^{dc} = \sqrt{2}$. Similarly, an orthonormal transform must have $h_1^{nyq} = \sqrt{2}$. In our case, the wavelet kernels are normalized for unit nominal gain (see equation (10.12)) so that $h_0^{dc} = h_1^{nyq} = 1$. Thus, the one dimensional wavelet analysis kernels are smaller than those of an orthonormal transform by a factor $\sqrt{2}$ and the synthesis kernels are larger by the same factor. Equation (10.25) follows from the fact that each two dimensional DWT stage involves two one dimensional transforms and hence a scaling factor of 2 in amplitude (4 in energy) relative to the orthonormal case.

DEQUANTIZATION

Suppose subband sample y has been assigned a quantization index, $q = \chi \cdot v$. The dequantization operation consists in assigning a reconstruction value, $\hat{y}$, which lies somewhere in the corresponding quantization interval, $\mathcal{I}_q$. In the case of deadzone scalar quantization, the reconstruction procedure is expressed in equation (3.35), which we write as

$$\hat{y}_b[\mathbf{n}] = \begin{cases} 0 & \text{if } v_b[\mathbf{n}] = 0 \\ \chi_b[\mathbf{n}] \left(v_b[\mathbf{n}] + \delta_{b,v_b[\mathbf{n}]} \right) \Delta_b & \text{if } v_b[\mathbf{n}] \neq 0 \end{cases} \qquad (10.27)$$

Here, $\delta_{b,v} = \frac{1}{2}$ corresponds to mid-point reconstruction.

A JPEG2000 decompressor is at liberty to select any value for $\delta_{b,v}$ in the interval $[0, 1)$. Normally, the selected value will depend at most upon the relevant subband, b, and the quantization index magnitude, $v_b[\mathbf{n}]$. Ideally, $\delta_{b,v}$ is chosen so that $\hat{y}$ is the statistical centroid of $\mathcal{I}_q$; however, the decompressor does not usually have access to the underlying statistics for each subband. Many decompressors are expected to adopt the mid-point reconstruction rule, setting $\delta_{b,v} = \frac{1}{2}$. This is a safe choice since the compressor is most likely to base its code-block truncation policy upon this assumption (see Section 8.3.5). Experience indicates that some small improvements can be obtained by selecting a slightly smaller value (e.g., $\delta_{b,v} = \frac{3}{8}$) when v is small, especially for the higher frequency subbands.

The procedure described above is appropriate only when the original quantization indices are available at the decoder. It frequently happens that the embedded bit-streams corresponding to individual code-blocks are truncated prior to decoding. Following the notation developed above, let $p_i[\mathbf{j}]$ denote the number of least significant bits of the index magnitude, $v_i[\mathbf{j}]$, which were not decoded due to truncation of the embedded bit-stream representing code-block $\mathcal{B}_i$. Then the effective quantization step size, $\Delta_i[\mathbf{j}]$, is given by equation (10.24) and the decoder should employ the modified reconstruction rule

$$\begin{aligned} \hat{y}_i[\mathbf{j}] &= \begin{cases} 0 & \text{if } \hat{v}_i[\mathbf{j}] = 0 \\ \hat{\chi}_i[\mathbf{j}] \left(\left\lfloor \frac{\hat{v}_i[\mathbf{j}]}{2^{p_i[\mathbf{j}]}} \right\rfloor + \delta \right) \Delta_i[\mathbf{j}] & \text{if } \hat{v}_i[\mathbf{j}] \neq 0 \end{cases} \\ &= \begin{cases} 0 & \text{if } \hat{v}_i[\mathbf{j}] = 0 \\ \hat{\chi}_i[\mathbf{j}] \left(\hat{v}_i[\mathbf{j}] + 2^{p_i[\mathbf{j}]} \delta \right) \Delta_{b_i} & \text{if } \hat{v}_i[\mathbf{j}] \neq 0 \end{cases} \end{aligned} \qquad (10.28)$$

Here, $\hat{v}_i[\mathbf{j}]$ is the decoder's version of $v_i[\mathbf{j}]$, obtained by setting the $p_i[\mathbf{j}]$ undecoded LSBs to 0. Similarly, $\hat{\chi}_i[\mathbf{j}]$ is the decoder's version of $\chi_i[\mathbf{j}]$, the two being identical except possibly for samples whose decoded magnitude is $\hat{v}_i[\mathbf{j}] = 0$.

10.5.2   REVERSIBLE PROCESSING

RANGING

In the reversible path, the quantization indices supplied to the block coder are identical to the integer subband samples; i.e., $q_b[n] = y_b[n]$. A nominal range for these integer sample values may be determined from the original sample bit-depth, B, and the nominal gains of the linearized wavelet analysis kernels, as given by equation (10.16). As in the irreversible case, we add G extra guard bits to accommodate isolated excursions beyond these nominal range bounds.

With a suitable choice of G, all of the subband samples satisfy

$$-2^{B-1+X_b+G} < y_b[n] < 2^{B-1+X_b+G} \tag{10.29}$$

where X_b is an extra term to account for the fact that the nominal gains of the linearized analysis kernels are not both equal to 1; specifically,

$$X_{LL_d} = 0, \ X_{LH_d} = X_{HL_d} = 1, \text{ and } X_{HH_d} = 2, \quad \forall d$$

A sufficient number of bits to represent the sample magnitudes, $v_b[n]$, of subband b is thus

$$K_b^{\max} = B - 1 + X_b + G \tag{10.30}$$

As discussed above for the irreversible case, a typical value for G is $G = 1$ or, more conservatively, $G = 2$. Worst case bit-depth expansion in the reversible processing path is analyzed more carefully in Section 17.3.2. In particular, the entries in Table 17.4 identify lower bounds for the value of $X_b + G$. These results indicate that $G = 2$ guard bits are required to be guaranteed that equation (10.29) will not be violated.

The JPEG2000 standard defines $K_b^{\max}$ in terms of parameters G and ε_b, using exactly the same expression (i.e., equation (10.22)) as that used for the irreversible path. In this way, the block coder sees a consistent set of parameters whose interpretation does not depend upon whether the sample transformations are reversible or irreversible. In the irreversible case, ε_b is the quantization step size exponent. In the reversible case, however, ε_b is interpreted as a "ranging" parameter and it is the encoder's responsibility to set ε_b in such a way as to ensure that sufficient bits are available to represent each subband's sample magnitudes. Comparing equations (10.22) and (10.30), it is clear that a reasonable policy for the encoder is to set

$$\varepsilon_b = B + X_b \tag{10.31}$$

Note, however, that only the sum, $\varepsilon_b + G$, ultimately has any effect upon the value of $K_b^{\max}$ yielded by equation (10.22).

Before concluding our discussion of ranging, it is worth considering the impact of the reversible colour transform (RCT) on the selection of suitable ranging parameters, ε_b. As discussed in Section 10.2.2, when the first three image components are subjected to the RCT their original bit-depths must be identical; i.e., $B_0 = B_1 = B_2 = B$. After application of the RCT, the luminance component, $x_0[\mathbf{n}]$, again has bit-depth B, while the two colour difference components, $x_1[\mathbf{n}]$ and $x_2[\mathbf{n}]$, have bit-depths $B+1$. This suggests that the compressor should replace B by $B+1$ in equation (10.31) when determining suitable ranging parameters for the chrominance components. Ultimately, though, the compressor is free to select the parameters, ε_b and G, in any manner which accommodates the range of subband samples encountered in the application at hand.

DEQUANTIZATION

Although the reversible path involves no explicit quantization, the embedded bit-streams corresponding to individual code-blocks may be truncated prior to decoding and this is equivalent to quantization. As explained for the irreversible path, truncation of the embedded bit-stream representing code-block $\mathcal{B}_i$ causes the decoder to receive modified sample magnitudes, $\hat{v}_i[\mathbf{j}]$, which agree with $v_i[\mathbf{j}]$ on all but the least significant $p_i[\mathbf{j}]$ bits (these $p_i[\mathbf{j}]$ undecoded LSBs are all taken to be 0). Subband sample $y_i[\mathbf{j}]$ is thus effectively subjected to deadzone scalar quantization with a step size of $\Delta_i[\mathbf{j}] = 2^{p_i[\mathbf{j}]}$.

Following equation (10.28), we see that the decompressor should reconstruct $\hat{y}_i[\mathbf{j}]$ as

$$\hat{y}_i[\mathbf{j}] = \begin{cases} 0 & \text{if } \hat{v}_i[\mathbf{j}] = 0 \\ \hat{\chi}_i[\mathbf{j}] \left(\hat{v}_i[\mathbf{j}] + \left\lfloor 2^{p_i[\mathbf{j}]}\delta \right\rfloor \right) & \text{if } \hat{v}_i[\mathbf{j}] \neq 0 \end{cases}$$

where the decoded sign, $\hat{\chi}_i[\mathbf{j}]$, may not agree with $\chi_i[\mathbf{j}]$ at samples whose decoded magnitude is $\hat{v}_i[\mathbf{j}] = 0$. The floor function, $\lfloor \rfloor$, is introduced around the term, $2^{p_i[\mathbf{j}]}\delta$, to ensure that the reconstructed values are integers. This is important because the reversible path works only with integers. As in the irreversible case, a mid-point reconstruction policy is recommended, with $\delta = \frac{1}{2}$.

10.6   ROI ADJUSTMENTS

JPEG2000 provides mechanisms whereby the compressor may assign higher priority to certain regions of the image. This is known as "region of interest encoding," or simply ROI. The EBCOT paradigm described in Section 8.1.3 provides a crude ROI capability, since code-block contributions may be sequenced into the code-stream in a manner which effectively elevates the priority of certain spatial regions. The standard provides a second mechanism which allows the compressor to adjust the priority of an arbitrary set of subband samples, irrespective of code-block boundaries. This mechanism allows ROI capabilities to be realized over small regions with arbitrary shape. Unlike the crude method based on code-blocks, however, this fine-grain ROI capability requires modifications to the sample data transformation pipeline. As such, the prioritization cannot later be adjusted simply by re-sequencing code-block contributions.

10.6.1   PRIORITIZATION BY SCALING

The fine-grain ROI capability offered by JPEG2000 is realized by scaling the more important subband samples by 2^U for some non-negative integer, U. More precisely, the quantization indices of the more important subband samples are "up-shifted" by U before subjecting them to the block coder. Of course, the decoder must reverse this scaling operation and so it must also have a means for determining which samples have been scaled.

To see how scaling can introduce a prioritization of the subband samples, recall that the block coder produces an embedded bit-stream for each code-block, $\mathcal{B}_i$, and that the decoder generally receives a truncated version of this embedded bit-stream. Following previously established notation, let $p_i[\mathbf{j}]$ denote the number of least significant bits which are missing from the quantization index for subband sample $y_i[\mathbf{j}]$, due to truncation of the bit-stream representing code-block $\mathcal{B}_i$. The effective quantization step size for subband sample $y_i[\mathbf{j}]$ is then $2^{p_i[\mathbf{j}]}\Delta_{b_i}$ (see equation (10.24)). To allow for ROI adjustments, let $U_i[\mathbf{j}]$ denote the up-shift for sample $y_i[\mathbf{j}]$, so that the block coder processes the shifted index magnitudes,

$$\overleftarrow{v}_i[\mathbf{j}] = 2^{U_i[\mathbf{j}]} \cdot v_i[\mathbf{j}]$$

Let $\overleftarrow{p}_i[\mathbf{j}]$ denote the number of least significant bits which are missing from the shifted index magnitudes due to truncation of the bit-stream for code-block $\mathcal{B}_i$. The decompressor reconstructs $v_i[\mathbf{j}]$ from $\overleftarrow{v}_i[\mathbf{j}]$ according to

$$v_i[\mathbf{j}] = 2^{-U_i[\mathbf{j}]} \cdot \overleftarrow{v}_i[\mathbf{j}]$$

after which the number of missing magnitude LSBs is

$$p_i[\mathbf{j}] = \max\{0, \overleftarrow{p}_i[\mathbf{j}] - U_i[\mathbf{j}]\}$$

and the effective quantization step size is

$$\Delta_i[\mathbf{j}] = 2^{\max\{0, \overleftarrow{p}_i[\mathbf{j}] - U_i[\mathbf{j}]\}} \cdot \Delta_{b_i}$$

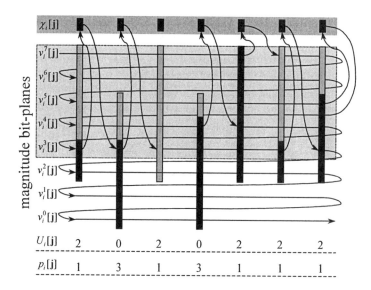

Figure 10.8. Effect of ROI scaling on the number of magnitude bit-planes available when a code-block bit-stream is truncated. Light shading identifies the magnitude bits remaining after truncation. Medium shading is used to identify leading zeros in the quantized sample magnitudes. Note that the scaling factor used here is incompatible with the max-shift method described later in Section 10.6.2.

If the code-block bit-stream is not truncated, $\overleftarrow{p}_i[\mathbf{j}] = 0$, $\forall \mathbf{j}$ and all subband samples in the block have exactly the same effective step size, Δ_{b_i}. Whenever $\overleftarrow{p}_i[\mathbf{j}] > 0$, though, the effective quantization step size is smaller for the more important samples (i.e., those with larger $U_i[\mathbf{j}]$). Figure 10.8 illustrates the impact of the up-shift, $U_i[\mathbf{j}]$, on the number of magnitude bits which are completely decoded for any given sample when the code-block's bit-stream is truncated. The simple depiction in this figure belies the fact that bit-planes are actually coded in multiple passes following a data-dependent scanning pattern. Also, the figure depicts only the case in which the block bit-stream is truncated at a bit-plane end-point so that all samples in the block have the same number of discarded LSBs, namely $\overleftarrow{p}_i[\mathbf{j}] = 3$. More generally, the $\overleftarrow{p}_i[\mathbf{j}]$ may differ by 1 from sample to sample within the block.

Although not obvious from Figure 10.8, the block coder treats all subband samples in the same way, irrespective of their up-shifts. The most significant magnitude bits of those samples which are not shifted are coded as zero, while the least significant $U_i[\mathbf{j}]$ bits of sample $y_i[\mathbf{j}]$ are also coded as zero. If the encoder and decoder were both aware of the up-shift factors, these redundant zero magnitude bits could be skipped, improving coding efficiency at the expense of increased complexity.

In JPEG2000, code-blocks may be truncated for one of a variety of reasons, all of which expose the prioritization introduced by up-shifts, $U_i[\mathbf{j}]$. Even in a single layer code-stream, the code-block bit-streams are typically truncated as a means of controlling the overall compressed bit-rate. When multiple quality layers are used to create a quality progressive code-stream, the lower quality layers generally contain heavily truncated code-block representations; in these layers, we expect the decompressor to recover non-zero quantization indices only from those samples which have large up-shift factors, $U_i[\mathbf{j}]$. As the number of received quality layers increases, we expect the decompressed image to resolve details first in those spatial regions which correspond to samples with large up-shift factors. In the case of an image which is compressed losslessly using the reversible path, the decompressor eventually receives all quality layers, containing non-truncated representations of all code-blocks.

10.6.2   THE MAX-SHIFT METHOD

While Part 2 of the JPEG2000 standard is expected to support the specification of arbitrary up-shift factors with explicit region identification, Part 1 of the standard supports only a specific instance of the ROI scaling concept, known as the "max-shift" method. Let $\mathcal{F}_b$ denote the set of more important (or "foreground") samples for subband b. In the max-shift method, each sample in $\mathcal{F}_b$ has the same up-shift, U, while the background samples have zero up-shift. Moreover, U is selected to be sufficiently large that the foreground and background samples can be distinguished based upon the decoded quantization indices alone. To guarantee this, the compressor should select

$$U \geq \max_b K_b^{\mathrm{max}} \tag{10.32}$$

where K_b^{max} is the number of bits used to represent quantization index magnitudes, $v_b[\mathbf{n}]$, for subband b. Its value is given by equation (10.22).

The single value, U, is signalled in the code-stream marker segment, RGN. The compressor implements

$$\overleftarrow{v}_b[\mathbf{n}] = 2^{U_b[\mathbf{n}]} \cdot v_b[\mathbf{n}] = \begin{cases} v_b[\mathbf{n}] & \text{if } \mathbf{n} \notin \mathcal{F}_b \\ 2^U \cdot v_b[\mathbf{n}] & \text{if } \mathbf{n} \in \mathcal{F}_b \end{cases}$$

The decompressor then implements

$$v_b[\mathbf{n}] = 2^{-U_b[\mathbf{n}]} \cdot \overleftarrow{v}_b[\mathbf{n}] = \begin{cases} \overleftarrow{v}_b[\mathbf{n}] & \text{if } \overleftarrow{v}_b[\mathbf{n}] < 2^U \\ 2^{-U} \cdot \overleftarrow{v}_b[\mathbf{n}] & \text{if } \overleftarrow{v}_b[\mathbf{n}] \geq 2^U \end{cases}$$

and

$$p_b[\mathbf{n}] = \max\{0, \overleftarrow{p}_b[\mathbf{n}] - U_b[\mathbf{n}]\}$$

$$= \begin{cases} \overleftarrow{p}_b[\mathbf{n}] & \text{if } \overleftarrow{v}_b[\mathbf{n}] < 2^U \\ \max\{0, \overleftarrow{p}_b[\mathbf{n}] - U\} & \text{if } \overleftarrow{v}_b[\mathbf{n}] \geq 2^U \end{cases}$$

In this way, the decompressor implicitly recovers the foreground membership of every subband sample with a non-zero quantization index. This works because background samples have

$$\overleftarrow{v}_b[\mathbf{n}] = v_b[\mathbf{n}] < 2^{K_b^{\max}} \leq 2^U$$

while non-zero foreground samples have $\overleftarrow{v}_b[\mathbf{n}] \geq 2^U$. There is no need to determine whether a zero-valued sample belongs to the foreground or the background.

The advantage of the max-shift method is that there is no need to send any side information to identify the locations of the foreground samples. An obvious disadvantage, however, is that U must be so large that the foreground quantization indices are entirely decoded before any information is recovered for the background quantization indices. This behaviour is explored further in Section 16.2; there we also discuss an "implicit" ROI encoding strategy, which adjusts the relative contributions of different code-blocks instead of shifting individual samples.

10.6.3   IMPACT ON CODING

When $U \neq 0$, the block coder processes the modified index magnitudes, $\overleftarrow{v}_b[\mathbf{n}]$, instead of $v_b[\mathbf{n}]$. The number of magnitude bit-planes for subband b is then

$$\overleftarrow{K}_b^{\max} = K_b^{\max} + U$$
$$= U + \max\{0, \varepsilon_b + G - 1\} \tag{10.33}$$

This number may be quite large (e.g., $\gtrsim 30$ bits) and may well exceed the bit-depth available to the decoder for representing decoded index magnitudes, $\overleftarrow{v}_b[\mathbf{n}]$. In this event, it is expected that the decoder will process as many of the most significant bits of $\overleftarrow{v}_b[\mathbf{n}]$ as possible, with any remaining bit positions implicitly set to 0. Thus, the code-block bit-streams may be effectively truncated due to limitations in the precision to which the decoder is able to represent quantization indices. In fact, this may happen even in the absence of any up-shift factors if the quantization/ranging parameter, ε_b, is too large. In the extreme case, the decoder may only be able to process the most significant U magnitude bit-planes and all of the background will be set to 0. Section 18.3.2

describes exactly what is expected of a compliant JPEG2000 decoder in regard to ROI processing.

As explained in Section 13.3.10, U may be as large as 255. However, the largest value of any practical interest is 37. To see this, note that the code-stream syntax restricts the number of guard bits to $G \leq 7$ and ε_b to the range given in equation (10.19). This means that $K_b^{\max}$ cannot exceed 37, so that equation (10.32) can always be satisfied by some value of $U \leq 37$. Compressors are strongly encouraged to adhere to this limit, which is likely to be enforced in all of the restricted code-stream profiles, as suggested by Table 18.1. The limit ensures that the number of coding passes in any given code-block can always be represented using an 8-bit number. Other quantities of interest to the decoder can also be comfortably represented using 8-bit numbers when U obeys this limit.

10.6.4 REGION MAPPING

The JPEG2000 standard imposes no restrictions on the way in which the compressor determines which subband samples to include in the foreground sets, $\mathcal{F}_b$. However, it is expected that the capability will be used to elevate the importance of specific spatial regions in the image. In this section, we briefly show how the compressor may determine the subband foreground sets, $\mathcal{F}_b$, from an image foreground set, $\mathcal{F}$. The assumption is that the image samples, $x[\mathbf{n}]$, for which $\mathbf{n} \in \mathcal{F}$, are of high priority and all subband samples which are involved in their reconstruction should belong to the respective foreground sets, $\mathcal{F}_b$.

To determine the subband samples which are involved in the reconstruction of the foreground region, $\mathcal{F}$, the compressor must conceptually trace each location in $\mathcal{F}$ back through the DWT reconstruction process employed by the decompressor. This task is considerably simplified by the structure of the DWT. As illustrated in Figure 10.6, the DWT is inverted in stages, where stage d reconstructs LL_{d-1} from subbands LL_d, HL_d, LH_d and HH_d. Thus, starting from the final stage of the reconstruction, we can first determine the foreground sets, $\mathcal{F}_{LL_1}$ through $\mathcal{F}_{HH_1}$, representing samples from subbands LL_1 through HH_1 which contribute to the reconstruction of samples in $\mathcal{F}_{LL_0} = \mathcal{F}$. Repeating the procedure for $d = 2, 3, \ldots, D$, we may progressively determine the foreground sets, $\mathcal{F}_{LL_d}$ through $\mathcal{F}_{HH_d}$, representing subband samples which contribute to the reconstruction of the samples in $\mathcal{F}_{LL_{d-1}}$.

Evidently, it is sufficient to solve the foreground mapping problem for a single DWT stage. The problem is further simplified by observing that the DWT stage is itself a separable operator, constructed from the one dimensional building blocks described in Section 10.3.1. From Figure 10.5 we see that our first task is to identify the foreground samples in $y'[\mathbf{n}]$, which contribute to the reconstruction of foreground samples in $x[\mathbf{n}]$. This reduces to a one dimensional foreground mapping problem, conducted within each image column. The second task is to deduce the foreground samples in $y[\mathbf{n}]$, which contribute to the foreground samples just identified for the intermediate sequence, $y'[\mathbf{n}]$. Again, this reduces to a one dimensional foreground mapping problem, conducted within each row.

In this way, the foreground mapping problem reduces to a one dimensional problem in which we wish to determine the foreground sets, $\mathcal{F}_0$ and $\mathcal{F}_1$, containing the indices, n, of the low- and high-pass subband samples, $y_0[n]$ and $y_1[n]$, which contribute to the reconstruction of those samples, $x[n]$, for which $n \in \mathcal{F}$. It is clear from Figure 10.4 that

$$\mathcal{F}_b = \left\{ n \in \left[E^b, F^b \right) \;\middle|\; (2n + b) + k \in \mathcal{F}, \text{ for at least one } |k| \leq L_{1-b} \right\}$$

Thus, $\mathcal{F}_b$ may be viewed as the result of a one dimensional morphological dilation of the set $\mathcal{F}$, followed by sub-sampling. Using the language of mathematical morphology, we may write

$$\mathcal{F}_b = \left\{ n \in \left[E^b, F^b \right) \;\middle|\; (2n + b) \in \mathcal{F} \oplus \mathcal{L}_b \right\}, \quad \text{for } b \in \{0, 1\}$$

where $\mathcal{L}_0$ and $\mathcal{L}_1$ are the so-called "structuring sets,"

$$\mathcal{L}_b = \{ n \mid |n| \leq L_{1-b} \}, \quad \text{for } b \in \{0, 1\}$$

Chapter 11

SAMPLE DATA PARTITIONS

In Section 10.3.4 we showed how every subband sample may be associated with a unique location in the original image. JPEG2000 defines a variety of partitions and uses this association to map the partitions from the image domain into each of its subbands. These partitions define the scope within which the sample data transformations of Chapter 10 and the block coding operations of Chapter 12 are to be performed.

When multiple image components (usually colour components) are involved, each component may have different dimensions. JPEG2000 describes these different components using a single reference grid, which we term the "canvas." Each sample in each subband of each image component has a notional location on the canvas. This allows partitions to be defined on the canvas and mapped into every subband of every image component using a consistent set of region mapping rules.

The purpose of this chapter is to describe the various partitions which are defined by JPEG2000 and to explain the region mapping rules. Together, we refer to these partitions and mapping rules as the "canvas coordinate system." The system serves to provide an organizational structure for the sample data which is transformed and coded. Although any number of organizational structures may be conceived (and many were considered during the development of the standard), the canvas coordinate system has unique properties which facilitate spatial manipulation of compressed images. We allude to these properties at appropriate points in the ensuing description, while Section 11.4 describes some of the relevant manipulations in more detail.

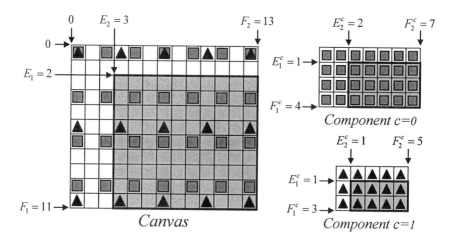

Figure 11.1. Notional placement of component samples on the canvas. Shading identifies the image region. Sub-sampling factors are $S_1^0 = 3$ and $S_2^0 = 2$ for component $c = 0$ and $S_1^1 = 5$ and $S_2^1 = 3$ for component $c = 1$.

11.1   COMPONENTS ON THE CANVAS

We use the notation $x_c[\mathbf{n}] \equiv x_c[n_1, n_2]$ for the samples of image component c, where $0 \le c < C$ and C is the total number of image components. Associated with each image component, c, are vertical and horizontal sub-sampling (or spacing) factors, S_1^c and S_2^c. These factors describe the notional location of each component sample on the canvas. Specifically, component sample $x_c[n_1, n_2]$ is associated with the canvas location $[n_1 S_1^c, n_2 S_2^c]$. This relationship is indicated in Figure 11.1.

All image components are bounded by the same region of support on the canvas. This so-called "image region" is defined by

$$E_1 \le n_1 < F_1 \text{ and } E_2 \le n_2 < F_2$$

or, equivalently,

$$\mathbf{n} \in [E_1, F_1) \times [E_2, F_2)$$

The region bounds are all **non-negative** integers, which appear in the global code-stream marker, *SIZ*. The image region may be mapped into the domain of any given image component, c, by asserting that

$$[n_1, n_2] \in [E_1^c, F_1^c) \times [E_2^c, F_2^c) \iff [S_1^c n_1, S_2^c n_2] \in [E_1, F_1) \times [E_2, F_2)$$

whence we deduce the component's region bounds as

$$E_i^c = \left\lceil \frac{E_i}{S_i^c} \right\rceil, \quad F_i^c = \left\lceil \frac{F_i}{S_i^c} \right\rceil, \quad i = 1, 2 \tag{11.1}$$

The reader is invited to verify these relationships for the particular example illustrated in Figure 11.1.

As expected, the height, $F_1^c - E_1^c$, of image component c is roughly proportional to the reciprocal of its vertical spacing factor, S_1^c. However, the exact height of each image component also depends upon the alignment of the upper and lower region bounds. Similar considerations apply in the horizontal direction. JPEG2000 supports component sub-sampling factors in the range 1 through 255. For small images, then, virtually any collection of component dimensions may be supported by appropriate selection of the sub-sampling factors and the canvas parameters. In most cases, however, the component sample spacings will be related by simple factors such as 1, 2 or 4.

As we shall see, it is possible to efficiently crop a JPEG2000 compressed image at any of its four boundaries. When this happens, any or all of the individual image components' dimensions may be affected. For example, if we crop the left hand column from the image region in Figure 11.1, E_2 becomes 4. From equation (11.1) or by inspection of the figure, $E_2^c|_{c=0}$ retains its value of 2, while $E_2^c|_{c=1}$ changes from 1 to 2. Thus, only component $c = 1$ is actually cropped. Since none of the sub-sampling factors need be 1, it can happen that cropping the image region on the canvas has no impact whatsoever on any image component.

Before closing this section, it is worth noting that the notional alignment of image component samples on the canvas need not necessarily reflect the physical sampling geometry of a multi-component image. If possible, applications are recommended to select canvas parameters which do accurately reflect the physical sampling geometry, since then spatial manipulations such as cropping are guaranteed to have a direct physical interpretation. When this is not possible, additional alignment information may be recorded in the code-stream marker segment *CRG* (Component Registration). However, applications are at liberty to ignore this informative marker segment.

11.2   TILES ON THE CANVAS

JPEG2000 allows an image to be divided into smaller rectangular regions known as "tiles," each of which is treated as a small independent image for the purpose of compression. Moreover, the parameters which control sample transformations and other aspects of the compression system may adjusted on a tile-by-tile basis. One possible application of tiles is in compressing so-called "compound documents." These are images which may contain text, graphics and photographic material in different regions. A tile which contains text or graphics might be quantized and coded directly, without any wavelet transform at all (see Section 16.3), while tiles containing photographic content are best compressed using a DWT with 5 or more levels.

Tiling is also an obvious mechanism for supporting region-of-interest access into the compressed image, since the decompressor need only have access to those tiles which overlap with its region of interest. Finally, tiling provides a simple mechanism for controlling the amount of working memory used to compress or decompress a large image.

Despite these attractions, the very fact that each tile is compressed independently introduces the possibility that the tile boundaries will be visible in the reconstructed image. This is a fundamental weakness of the JPEG compression algorithm, whose 8×8 blocks are effectively small tiles. The annoying appearance of block boundaries at moderate to high compression ratios is evidenced by Figure 4.30.

Fortunately, with JPEG2000 most applications need not resort to the use of tiling in order to bound implementation resources or support region-of-interest access. This is because these features are also supported by the EBCOT paradigm, whereby small blocks of subband samples are coded independently. We shall have more to say concerning such matters in Chapters 16 and 17.

11.2.1   THE TILE PARTITION

The tile partition is defined on the canvas through four non-negative integer parameters, T_1, T_2, Ω_1^T and Ω_2^T, which appear in the global code-stream marker, *SIZ*. They induce a partition which is anchored at $[\Omega_1^T, \Omega_2^T]$ (read "Omega" as "origin," with the super-script "T" for "tiling"), with elements of size $T_1 \times T_2$, as illustrated in Figure 11.2. We identify individual tiles by a pair of indices, $[t_1, t_2] \equiv \mathbf{t}$, where

$$0 \leq t_1 < N_1^T = \left\lceil \frac{F_1 - \Omega_1^T}{T_1} \right\rceil \quad \text{and} \quad 0 \leq t_2 < N_2^T = \left\lceil \frac{F_2 - \Omega_2^T}{T_2} \right\rceil$$

Here, N_1^T and N_2^T denote the number of tile rows and tile columns, respectively, which are required to cover the image area. The tile partition spans the region

$$\left[\Omega_1^T, \, \Omega_1^T + T_1 N_1^T \right) \times \left[\Omega_2^T, \, \Omega_2^T + T_2 N_2^T \right)$$

which covers the image region, $[E_1, F_1) \times [E_2, F_2)$, so long as

$$0 \leq \Omega_1^T \leq E_1 \quad \text{and} \quad 0 \leq \Omega_2^T \leq E_2$$

For efficiency of representation, the JPEG2000 standard requires not only that the tiles cover the image region, but also that every tile has

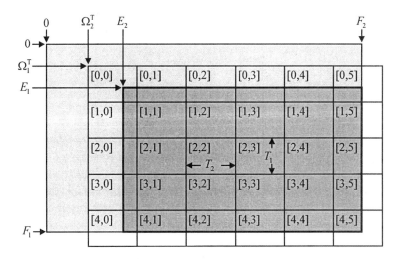

Figure 11.2.   Tile partition on the canvas. Dark shading identifies the image region.

a **non-empty** intersection with the image region on the canvas. Thus, the canvas parameters must satisfy

$$0 \le E_1 - \Omega_1^T < T_1 \text{ and } 0 \le E_2 - \Omega_2^T < T_2$$

There are thus $N_1^T \times N_2^T$ tiles which intersect with the image region. Interestingly, it can happen that some of these tiles are empty, in the sense that they contain no samples from any image component (see the discussion of Figure 11.3 in Section 11.2.2). Whether it is empty or not, each tile is assigned a unique identifier,

$$t = t_2 + t_1 N_2^T \tag{11.2}$$

which must appear within each of the tile's *SOT* (Start of Tile-Part) marker segments, as explained in Section 13.3.

In many cases, the tile anchor point, $[\Omega_1^T, \Omega_2^T]$, and the upper left hand corner of the image region, $[E_1, E_2]$, will both lie at the origin of the canvas coordinate system, $[0, 0]$. The reader may wonder why the standard supports non-zero coordinates for both $[\Omega_1^T, \Omega_2^T]$ and $[E_1, E_2]$. To understand the need for this degree of generality, suppose that a compressed image has its tile partition anchored at $\Omega_1^T = \Omega_2^T = 0$ and consider the problem of cropping the first row of tiles from the image. Since tiles are compressed independently, this should involve relatively minor modifications to the compressed code-stream.

The image region is modified by setting E_1 to T_1. To ensure that all tiles intersect with the image region, we must also add T_1 to Ω_1^T. One

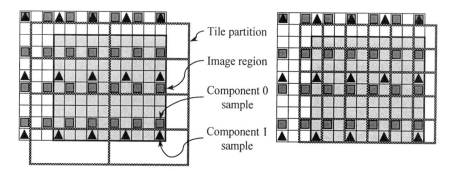

Figure 11.3. Weird tilings of a two component image. Shading identifies the image region.

might then argue that T_1 could be subtracted from all vertical canvas coordinates, Ω_1^T, E_1 and F_1, to restore the tile anchor point to the origin. However, this would alter the alignment of the image region on the canvas. Amongst other devastating effects, the alignment shift could alter the number of samples in each image component as determined by equation (11.1). Thus, operations as simple as cropping tiles away from an image cannot be supported if either $\left[\Omega_1^T, \Omega_2^T\right]$ or $[E_1, E_2]$ is forced to coincide with the canvas origin.

11.2.2   TILE-COMPONENTS AND REGIONS

The tile at location **t** occupies a region, $\left[E_1^{\mathbf{t}}, F_1^{\mathbf{t}}\right) \times \left[E_2^{\mathbf{t}}, F_2^{\mathbf{t}}\right)$, on the canvas, with region bounds given by

$$E_i^{\mathbf{t}} = \max\left\{E_i, \Omega_i^T + t_i T_i\right\}, \quad F_i^{\mathbf{t}} = \min\left\{F_i, \Omega_i^T + (t_i + 1)\, T_i\right\} \quad (11.3)$$

This region may be mapped into each image component by treating it as an image region and applying equation (11.1). The bounds for tile **t** in image component c (we call this a "tile-component") may then be expressed as

$$E_i^{\mathbf{t},c} = \left\lceil \frac{E_i^{\mathbf{t}}}{S_i^c} \right\rceil, \quad F_i^{\mathbf{t},c} = \left\lceil \frac{F_i^{\mathbf{t}}}{S_i^c} \right\rceil, \quad i = 1, 2 \qquad (11.4)$$

It is easy to see from these expressions that the tile-components constitute a partition of each component. Interestingly, though, this partition need not be regular. Figure 11.3 provides two extreme examples of this phenomenon for the two component image of Figure 11.1. In the example on the left of the figure, the nominal tile height of 3 is less than the vertical sub-sampling factor, $S_1^1 = 5$, for component 1. As a result, that component does not contribute to every tile. In the example on the right of the figure, each of the 2×2 tiles contains a sample from either both components, one component or no components at all. This is in spite of the fact that all tiles have a non-empty intersection with the image region on the canvas. The reader is reminded that even those tiles which contain no samples from any image component must each have a unique tile identifier, according to equation (11.2), and these tiles may contribute any number of tile-parts to the code-stream, all of which will be empty.

11.2.3   SUBBANDS ON THE CANVAS

The DWT and all subsequent quantization and coding steps are applied independently within each tile-component. Consider the subband, $y_{b_1,b_2}^{t,c,(d)}[\mathbf{n}]$, obtained by taking the DWT of tile $\mathbf{t}$ in component c. The indices, $b_1, b_2 \in \{0, 1\}$ identify the particular subband at DWT level d, with $[b_1, b_2] = [0, 0]$ for the LL_d subband, $[b_1, b_2] = [0, 1]$ for the HL_d (horizontally high-pass) subband, $[b_1, b_2] = [1, 0]$ for the LH_d (vertically high-pass) subband and $[b_1, b_2] = [1, 1]$ for the HH_d subband.

To find the subband's region of support we may apply equation (10.11) to the region of support for the corresponding tile-component. Using the readily verified identity,

$$\left\lceil \frac{\left\lceil \frac{i}{j} \right\rceil}{k} \right\rceil = \left\lceil \frac{\left\lceil \frac{i}{k} \right\rceil}{j} \right\rceil = \left\lceil \frac{i}{jk} \right\rceil, \quad \forall i \in \mathbb{Z}, \quad \forall j, k \in \mathbb{N} \tag{11.5}$$

we find the subband's region bounds to be

$$E_i^{t,c,[b_1,b_2]_d} = \left\lceil \frac{E_i^{t,c} - 2^{d-1}b_i}{2^d} \right\rceil = \left\lceil \frac{\left\lceil \frac{E_i^t}{S_i^c} \right\rceil - 2^{d-1}b_i}{2^d} \right\rceil$$

$$= \left\lceil \frac{E_i^t - 2^{d-1}S_i^c b_i}{2^d S_i^c} \right\rceil, \quad d \in \{1, 2, \ldots, D_{t,c}\}, \; i = 1, 2 \tag{11.6}$$

and, similarly,

$$F_i^{t,c,[b_1,b_2]_d} = \left\lceil \frac{F_i^t - 2^{d-1}S_i^c b_i}{2^d S_i^c} \right\rceil, \quad d \in \{1, 2, \ldots, D_{t,c}\}, \; i = 1, 2 \tag{11.7}$$

Note that the code-stream syntax permits a different number of DWT levels, $D_{t,c}$, to be specified for each tile-component.

Recall from Section 10.3.4 that each subband sample notionally occupies a unique location in the image from which it was generated. In particular, subband sample $y_{b_1,b_2}^{t,c,(d)}[n_1,n_2]$ is notionally co-located with image component sample $x_c\left[2^d n_1 + 2^{d-1} b_1, 2^d n_2 + 2^{d-1} b_2\right]$. But $x_c[n_1,n_2]$ occupies a notional location, $[S_1^c n_1, S_2^c n_2]$, on the canvas. Combining these two sets of coordinate correspondences, we find that subband sample $y_{b_1,b_2}^{t,c,(d)}[\mathbf{n}]$ has a notional location on the canvas at

$$\left[2^d S_1^c n_1 + 2^{d-1} S_1^c b_1,\ 2^d S_2^c n_2 + 2^{d-1} S_2^c b_2\right]$$

This association of subband samples with locations on the canvas provides an alternate and entirely equivalent mechanism for mapping regions on the canvas into regions within any given subband. To see this, observe that

$$\left[2^d S_1^c n_1 + 2^{d-1} S_1^c b_1,\ 2^d S_2^c n_2 + 2^{d-1} S_2^c b_2\right] \in \left[E_1^t, F_1^t\right) \times \left[E_2^t, F_2^t\right)$$

if and only if

$$E_i^t \le 2^d S_i^c n_i + 2^{d-1} S_i^c b_i < F_i^t, \quad i = 1, 2$$

or, equivalently,

$$\left\lceil \frac{E_i^t - 2^{d-1} S_i^c b_i}{2^d S_i^c} \right\rceil \le n_i < \left\lceil \frac{F_i^t - 2^{d-1} S_i^c b_i}{2^d S_i^c} \right\rceil, \quad i = 1, 2$$

These are exactly the same region bounds which we derived above for subband $y_{b_1,b_2}^{t,c,(d)}[\mathbf{n}]$.

11.2.4   RESOLUTIONS AND SCALING

Resolutions and resolution levels were introduced in Section 10.3.3. Applying these notions to an individual tile-component, we identify the rth resolution of tile t in component c with the intermediate subband, $\text{LL}_{D_{t,c}-r}^{t,c} \equiv y_{0,0}^{t,c,(D_{t,c}-r)}[\mathbf{n}]$, which can be reconstructed by applying only r stages of the inverse DWT (see Figure 10.6). There are $D_{t,c} + 1$ resolutions available for the tile-component. The lowest-resolution, $r = 0$, corresponds to the actual subband $\text{LL}_{D_{t,c}}^{t,c}$. The highest resolution, $r = D_{t,c}$, corresponds to the original tile-component, which may be interpreted as subband $\text{LL}_0^{t,c}$. Substituting $b_1 = b_2 = 0$ and $d = D_{t,c} - r$ into equations (11.6) and (11.7) we find the region bounds for any particular tile-component resolution as

$$E_i^{t,c,r} = \left\lceil \frac{E_i^t}{2^{(D_{t,c}-r)} S_i^c} \right\rceil, \quad F_i^{t,c,r} = \left\lceil \frac{F_i^t}{2^{(D_{t,c}-r)} S_i^c} \right\rceil, \quad i = 1, 2$$

Again, following Section 10.3.3 we use the term "resolution level" $\mathcal{R}_r^{t,c}$ to refer collectively to all subbands required to reconstruct resolution r from the lower resolution, $r - 1$, if any. Thus, $\mathcal{R}_0^{t,c}$ consists of the single subband, $\mathrm{LL}_{D_{t,c}}^{t,c} \equiv y_{0,0}^{t,c,(D_{t,c})}[\mathbf{n}]$, while each subsequent resolution level, $\mathcal{R}_r^{t,c}$, consists of the three subbands, $\mathrm{HL}_{D_{t,c}+1-r}^{t,c}$, $\mathrm{LH}_{D_{t,c}+1-r}^{t,c}$ and $\mathrm{HH}_{D_{t,c}+1-r}^{t,c}$. The subbands belonging to each resolution level are identified in Figure 10.6.

It is instructive to consider what happens if we discard the highest d resolution levels from the DWT associated with every tile-component. The remaining subbands in each tile-component may be used to reconstruct resolution $r = D_{t,c} - d$, having region of support

$$\left\lceil \frac{E_1^t}{2^d S_1^c} \right\rceil \leq n_1 < \left\lceil \frac{F_1^t}{2^d S_1^c} \right\rceil, \quad \left\lceil \frac{E_2^t}{2^d S_2^c} \right\rceil \leq n_2 < \left\lceil \frac{F_2^t}{2^d S_2^c} \right\rceil$$

These regions partition a reduced resolution version of each image component, c, whose region of support is

$$\left\lceil \frac{E_1}{2^d S_1^c} \right\rceil \leq n_1 < \left\lceil \frac{F_1}{2^d S_1^c} \right\rceil, \quad \left\lceil \frac{E_2}{2^d S_2^c} \right\rceil \leq n_2 < \left\lceil \frac{F_2}{2^d S_2^c} \right\rceil \qquad (11.8)$$

The new reduced resolution image may be properly described within the canvas coordinate system (as an image in its own right), after introducing the following modifications.

- Multiply the component sub-sampling factors, S_1^c and S_2^c, by 2^d.

- Subtract d from the number of DWT levels, $D^{t,c}$, associated with each tile-component.

- Remove any coding parameters which are specific to the discarded subbands from their respective code-stream markers (see Chapter 13).

Unfortunately, the fact that component sub-sampling factors may not exceed 255 imposes a limit on the degree to which resolution may be scaled using this method. An alternate approach may be employed so long as the tile dimensions T_1 and T_2 are both divisible by 2^d or there is only one tile. Then, exploiting the ceiling identities of equation (11.5), we may rewrite equation (11.8) as

$$\left\lceil \frac{\left\lceil \frac{E_1}{2^d} \right\rceil}{S_1^c} \right\rceil \leq n_1 < \left\lceil \frac{\left\lceil \frac{F_1}{2^d} \right\rceil}{S_1^c} \right\rceil, \quad \left\lceil \frac{\left\lceil \frac{E_2}{2^d} \right\rceil}{S_2^c} \right\rceil \leq n_2 < \left\lceil \frac{\left\lceil \frac{F_2}{2^d} \right\rceil}{S_2^c} \right\rceil$$

Multiplication of S_1^c and S_2^c by 2^d is thus equivalent to the following steps.

- Replace the image region bounds, E_1, F_1, E_2 and F_2, by the quantities, $\lceil 2^{-d}E_1 \rceil$, $\lceil 2^{-d}F_1 \rceil$, $\lceil 2^{-d}E_2 \rceil$ and $\lceil 2^{-d}F_2 \rceil$.

- Replace the tile origin coordinates, Ω_1^T and Ω_2^T, by $\lceil 2^{-d}\Omega_1^T \rceil$ and $\lceil 2^{-d}\Omega_2^T \rceil$.

- Divide the tile dimensions, T_1 and T_2, by 2^d.

Of course, a combination of these coordinate transformation methods may be employed to achieve large resolution scaling factors, when the tile dimensions are not exact powers of 2.

11.3   CODE-BLOCKS AND PRECINCTS

In JPEG2000, the subbands of each tile-component are further partitioned into code-blocks, which are then coded independently. The EBCOT paradigm described in Section 8.1.3, relies upon independent coding of relatively small blocks of subband samples.

The code-block partition is defined by three sets of parameters: a global anchor point, $[\Omega_1^C, \Omega_2^C]$ (read "Omega" as "origin," with the super-script "C" for "coding"); maximum height and width parameters, $J_1^{t,c}$ and $J_2^{t,c}$; and "precinct" dimensions, $P_1^{t,c,r}$ and $P_2^{t,c,r}$. Note that the maximum code-block dimensions, $J_1^{t,c}$ and $J_2^{t,c}$, may be different for each tile-component, while the precinct dimensions may differ for each resolution, r, of each tile-component. Both sets of dimensions are restricted to exact powers of 2 and the coding anchor point coordinates, Ω_1^C and Ω_2^C, may only take values of 0 or 1.

We shall describe the role played by precincts shortly. For the moment, it is sufficient to understand that this role requires each precinct to consist of a whole number of code-blocks. For this reason, the code-block partition is subordinated by a precinct partition and the latter must be described first.

The reader should note that the coding anchor point coordinates, Ω_1^C and Ω_2^C, are the subject of a proposed amendment to Part 1 of the standard, whose fate will not be known until after publication of this text. If the amendment is not accepted, the coding anchor points will be forced to $\Omega_1^C = \Omega_2^C = 0$ for the purposes of Part 1 of the standard and free to take on values of 1 or 0 only in Part 2 of the standard.

11.3.1   PRECINCT PARTITION

Each resolution of each tile-component is partitioned into precincts, as shown in Figure 11.4. The precinct partition differs from the tile partition in a number of important respects. Firstly, the precinct partition

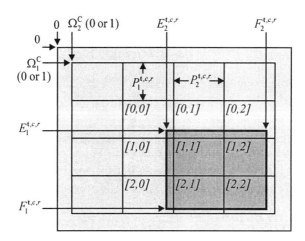

Figure 11.4.   Division of a tile-component resolution into precincts.

has no impact on sample data transformations or coding, except in the constraints which it imposes on the subordinate code-block partition, described next. Tile dimensions are arbitrary positive integers, while precinct dimensions must be exact powers of 2. Finally, whereas tile boundaries are common to all components, resolutions and subbands, this need not be true for precincts.

Each precinct is identified by a pair of indices, $[p_1, p_2] \equiv \mathbf{p}$, which range over

$$0 \le p_1 < N_1^{\mathrm{P,t},c,r} \quad \text{and} \quad 0 \le p_2 < N_2^{\mathrm{P,t},c,r} \tag{11.9}$$

Here, $N_1^{\mathrm{P,t},c,r}$ and $N_2^{\mathrm{P,t},c,r}$ identify the number of precinct rows and the number of precinct columns, respectively, which are required to cover tile $\mathbf{t}$ in component c at resolution r. These quantities may be found from

$$N_i^{\mathrm{P,t},c,r} = \begin{cases} \left\lceil \dfrac{F_i^{t,c,r} - \Omega_i^C}{P_i^{t,c,r}} \right\rceil - \left\lfloor \dfrac{E_i^{t,c,r} - \Omega_i^C}{P_i^{t,c,r}} \right\rfloor & \text{if } F_i^{t,c,r} > E_i^{t,c,r} \\ 0 & \text{if } F_i^{t,c,r} = E_i^{t,c,r} \end{cases} \tag{11.10}$$

Notice that the value of $N_i^{\mathrm{P,t},c,r}$ must be explicitly forced to zero when the relevant dimension of the tile-component resolution is zero ($F_i^{t,c,r} = E_i^{t,c,r}$).

The region associated with precinct **p** is defined by the bounds

$$E_i^{t,c,r,\mathbf{P}} = \max\left\{ E_i^{t,c,r},\ \Omega_i^C + P_i^{t,c,r}\left(p_i + \left\lfloor \frac{E_i^{t,c,r} - \Omega_i^C}{P_i^{t,c,r}} \right\rfloor \right) \right\}$$

$$F_i^{t,c,r,\mathbf{P}} = \min\left\{ F_i^{t,c,r},\ \Omega_i^C + P_i^{t,c,r}\left(p_i + 1 + \left\lfloor \frac{E_i^{t,c,r} - \Omega_i^C}{P_i^{t,c,r}} \right\rfloor \right) \right\}$$

The reader may verify that each precinct region, $\left[E_1^{t,c,r,\mathbf{P}}, F_1^{t,c,r,\mathbf{P}} \right) \times \left[E_2^{t,c,r,\mathbf{P}}, F_2^{t,c,r,\mathbf{P}} \right)$, has a non-empty intersection with the region occupied by tile t in image component c at resolution r.

11.3.2   SUBBAND PARTITIONS
INDUCED PRECINCT PARTITION OF SUBBANDS

The region occupied by any precinct of a particular tile-component resolution may be mapped into each of the subbands belonging to the corresponding resolution level. Following our now well-established region mapping conventions, we obtain

$$E_i^{t,c,r,\mathbf{p},\mathbf{b}} = \left\lceil \frac{E_i^{t,c,r,\mathbf{P}} - b_i}{2^{s_r}} \right\rceil, \qquad F_i^{t,c,r,\mathbf{p},\mathbf{b}} = \left\lceil \frac{F_i^{t,c,r,\mathbf{P}} - b_i}{2^{s_r}} \right\rceil$$

where the notation, s_r, identifies the number of two dimensional DWT stages used to recover the rth image resolution from the subbands in resolution level $\mathcal{R}_r$. By inspection of Figure 10.6, we see that

$$s_r = \begin{cases} 0 & \text{if } r = 0 \\ 1 & \text{if } r > 0 \end{cases}$$

In each subband, the precinct partition has dimensions $2^{-s_r} P_1^{t,c,r}$ and $2^{-s_r} P_2^{t,c,r}$, which are required to be integers to avoid irregularity in the partitioning of subbands. In particular, JPEG2000 insists that the precinct dimensions be exact powers of 2, no less than s_r. We express this requirement as

$$s_r \leq \log_2\left(P_i^{t,c,r} \right) \in \mathbb{Z}, \quad \text{for } i = 1, 2 \tag{11.11}$$

The reader may verify that the precinct partition of subband $\mathbf{b} \equiv [b_1, b_2]$ in resolution level $\mathcal{R}_r^{t,c}$, is anchored at the location $\left[\Omega_1^{C,\mathbf{b}}, \Omega_2^{C,\mathbf{b}} \right]$ whose coordinates also take one of the values 0 or 1 and are given by

$$\Omega_1^{C,\mathbf{b}} = \left\lceil \frac{\Omega_1^C - b_1}{2^{s_r}} \right\rceil, \qquad \Omega_2^{C,\mathbf{b}} = \left\lceil \frac{\Omega_2^C - b_2}{2^{s_r}} \right\rceil$$

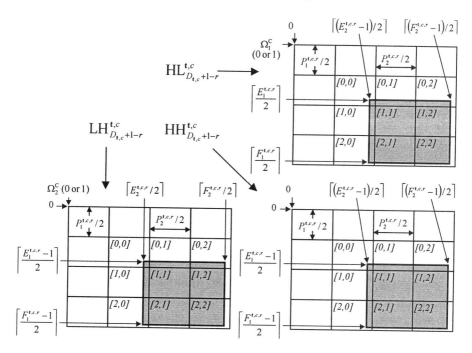

Figure 11.5. Induced precinct partitions for the three subbands in resolution level $\mathcal{R}_r^{t,c}$, for $r > 0$.

Figure 11.5 illustrates this induced precinct partition for each subband in resolution level $\mathcal{R}_r^{t,c}$, for $r > 0$.

CODE-BLOCK PARTITION OF SUBBANDS

Within each subband, precincts are further sub-divided into code-blocks. The code-block partition has the same anchor point $\left[\Omega_1^{C,b}, \Omega_2^{C,b}\right]$ as the precinct partition for the same subband. The elements of the code-block partition have dimensions $J_1^{t,c,r}$ and $J_2^{t,c,r}$, which are exact powers of 2 derived from code-stream parameters as follows

$$J_1^{t,c,r} = \min\left\{J_1^{t,c}, 2^{-s_r} P_1^{t,c,r}\right\}, \quad J_2^{t,c,r} = \min\left\{J_2^{t,c}, 2^{-s_r} P_2^{t,c,r}\right\}$$

The code-block partition is illustrated in Figure 11.6.

At one extreme, it can happen that $J_1^{t,c} \geq 2^{-s_r} P_1^{t,c,r}$ and $J_2^{t,c} \geq 2^{-s_r} P_2^{t,c,r}$, in which case each precinct contains exactly one code-block within each subband. When precincts are large, however, the dimensions of the code-block partition are determined exclusively by the parameters,

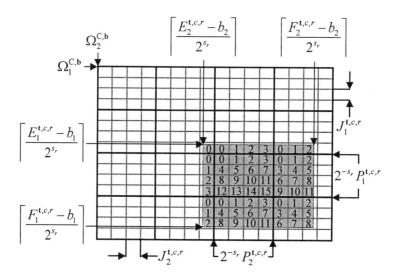

Figure 11.6. Code-block partition for subband $\mathbf{b} \equiv [b_1, b_2]$ in resolution-level $\mathcal{R}_r^{\mathbf{t},c}$.

$J_1^{\mathbf{t},c}$ and $J_2^{\mathbf{t},c}$, whose base-2 logarithm is signalled using the code-stream marker segments, *COD* and *COC*. In many cases, each resolution of each tile-component contains only one precinct. When this happens, all subbands of the relevant tile-component have code-blocks of the same size (ignoring the impact of tile boundaries). The JPEG2000 standard restricts the allowable values for $J_1^{\mathbf{t},c}$ and $J_2^{\mathbf{t},c}$ so as to ensure that no code-block may contain more than $2^{12} = 4096$ subband samples. In particular, the constraints may be expressed as

$$2^2 \le J_1^{\mathbf{t},c}, \ J_2^{\mathbf{t},c} \le 2^{10}, \quad \text{and} \quad J_1^{\mathbf{t},c} J_2^{\mathbf{t},c} \le 2^{12}$$
$$\text{where } J_1^{\mathbf{t},c} \text{ and } J_2^{\mathbf{t},c} \text{ are both powers of 2} \tag{11.12}$$

Let $N_1^{\mathbf{B},\mathbf{t},c,r,\mathbf{p},\mathbf{b}}$ and $N_2^{\mathbf{B},\mathbf{t},c,r,\mathbf{p},\mathbf{b}}$ denote the number of code-blocks in the vertical and horizontal directions, respectively, which cover precinct $\mathbf{p}$ in subband $\mathbf{b} \equiv [b_1, b_2]$ of resolution level $\mathcal{R}_r^{\mathbf{t},c}$. These quantities may be found from

$$N_i^{\mathbf{B},\mathbf{t},c,r,\mathbf{p},\mathbf{b}} = \begin{cases} \left\lceil \dfrac{F_i^{\mathbf{t},c,r,\mathbf{p},\mathbf{b}} - \Omega_i^{C,\mathbf{b}}}{J_i^{\mathbf{t},c,r}} \right\rceil - \left\lfloor \dfrac{E_i^{\mathbf{t},c,r,\mathbf{p},\mathbf{b}} - \Omega_i^{C,\mathbf{b}}}{J_i^{\mathbf{t},c,r}} \right\rfloor & \text{if } F_i^{\mathbf{t},c,r,\mathbf{p},\mathbf{b}} > E_i^{\mathbf{t},c,r,\mathbf{p},\mathbf{b}} \\ 0 & \text{if } F_i^{\mathbf{t},c,r,\mathbf{p},\mathbf{b}} = E_i^{\mathbf{t},c,r,\mathbf{p},\mathbf{b}} \end{cases}$$

For the purpose of sequencing code-block contributions into packets, the $N_1^{\mathbf{B},\mathbf{t},c,r,\mathbf{p},\mathbf{b}} \times N_2^{\mathbf{B},\mathbf{t},c,r,\mathbf{p},\mathbf{b}}$ code-blocks within precinct $\mathbf{p}$, are numbered from left to right and from top to bottom, as indicated in Figure 11.6.

When the relevant resolution level contains multiple subbands (i.e., all but $\mathcal{R}_0$), the code-block contributions from the HL subband appear first and those from the HH subband appear last.

It is important to note that one or more subbands may contribute no code-blocks whatsoever to some precincts; that is, either $N_1^{B,t,c,r,p,b} = 0$ or $N_2^{B,t,c,r,p,b} = 0$. This can happen only in precincts which lie on the boundaries of the tile-component. In fact, it can happen that a precinct is entirely empty, containing no code-blocks from any subband. This is in spite of the fact that every precinct must have a non-empty intersection with its tile-component at the relevant resolution. To see this, consider a tile-component which consists of only a single sample located at the origin of the canvas. Then $E_1^{t,c,r} = E_2^{t,c,r} = 0$ and $F_1^{t,c,r} = F_2^{t,c,r} = 1$ at all resolutions r. In this way, each resolution of the tile-component contains exactly one precinct, but there is only one code-block for the entire tile-component and this code-block belongs to the $LL_{D_{t,c}}^{t,c}$ subband in resolution level $\mathcal{R}_0^{t,c}$.

11.3.3 PRECINCTS AND PACKETS

At this point, the reader may wonder why the JPEG2000 committee found it necessary to introduce so many different partitions. The code-block partition is fundamental to the EBCOT paradigm, which in turn imparts many important features to the JPEG2000 compression algorithm (see Section 8.1.3). The tile partition provides a mechanism to divide images up into smaller independent pieces which can have a number of uses (see Section 11.2). Unlike the tile and code-block partitions, the precinct partition does not affect the transformation or coding of sample data. Instead, the precinct partition plays an important role in organizing compressed data within the code-stream.

The fundamental unit of code-stream organization is the "packet." Each precinct contributes one packet to the code-stream for every quality layer, $\mathcal{Q}_l$ (quality layers are described in Section 8.1.3). The packet for quality layer $\mathcal{Q}_l$ from precinct p of resolution level $\mathcal{R}_r^{t,c}$ contains the incremental contributions to that quality layer from all code-blocks within the scope of the precinct. The number of these code-block contributions is

$$\sum_{b \in \mathcal{R}_r^{t,c}} N_1^{B,t,c,r,p,b} N_2^{B,t,c,r,p,b} \tag{11.13}$$

Note, however, that any or all of the contributions to any particular quality layer may be empty. In the unlikely event that a precinct encompasses no code-blocks whatsoever (i.e., the above count is 0), its empty packets must nevertheless appear within the code-stream. Construction of packets is the subject of Section 12.5.

The number of quality layers in tile $\mathbf{t}$ is denoted $\Lambda_{\mathbf{t}}$. Although the number of quality layers is allowed to vary from tile to tile, compressors would do well to use the same number of quality layers in every tile where possible. This avoids ambiguity regarding the number of quality layers which should be discarded from each tile when the code-stream must be scaled down to a lower bit-rate. Since every precinct in tile $\mathbf{t}$ contributes exactly $\Lambda_{\mathbf{t}}$ packets, the total number of packets in a JPEG2000 code-stream is

$$\text{num code-stream packets} = \sum_{t_1=0}^{N_1^{\mathrm{T}}-1} \sum_{t_2=0}^{N_2^{\mathrm{T}}-1} \sum_{c=0}^{C-1} \sum_{r=0}^{D_{\mathbf{t},c}} \Lambda_{\mathbf{t}} N_1^{\mathrm{P},\mathbf{t},c,r} N_2^{\mathrm{P},\mathbf{t},c,r}$$

JPEG2000 provides a rich language for describing the sequence in which these packets actually appear within the code-stream. This is known as the packet progression sequence. A quality progressive code-stream is constructed by sequencing all packets corresponding to quality layer $\mathcal{Q}_0$ followed by all packets corresponding to layer $\mathcal{Q}_1$ and so forth. A resolution progressive code-stream is constructed by sequencing all packets from precincts of resolution $r = 0$ prior to all packets of resolution $r = 1$ and so forth. Such progressions may be realized without partitioning each tile-component resolution into any more than one precinct.

The standard also supports spatially oriented progressions, in which packets from precincts representing the upper-left portion of the image region are sequenced first, while packets representing the lower-right portion of the image region appear last. For applications requiring a spatially progressive code-stream, it is advantageous to employ relatively small precincts. Packet progressions are described thoroughly in Chapter 13.

The tag coding techniques described in Section 8.4.2, which efficiently identify code-block contributions to each quality layer, are applied independently within each precinct. This facilitates re-sequencing of packets within the code-stream. It also means that the precinct partition provides a mechanism for controlling the extent of the dependencies introduced by tag coding.

11.4   SPATIAL MANIPULATIONS

At the beginning of this chapter, we pointed out that the canvas coordinate system has properties which facilitate the efficient manipulation of compressed images. One such property is resolution scalability, which we discussed in Section 11.2.4. Indeed, an important factor in the adoption of the canvas coordinate system by the JPEG2000 committee was the desire to support resolution scalability in conjunction with arbitrary tile sizes and component sub-sampling factors. The coordinate system was also designed specifically with two other types of spatial manipulation in mind: arbitrary cropping of compressed images; and simple rotations.

11.4.1   ARBITRARY CROPPING

In Section 11.2.1 we motivated the existence of an explicit tiling anchor point, $\left[\Omega_1^T, \Omega_2^T\right]$, by considering the problem of cropping tiles away from a compressed image. This ought to be achievable in a simple manner, since tiles are compressed independently. A more interesting problem is that of cropping an arbitrary number of rows or columns from any boundary, without having to re-compress an entire row or column of tiles. Two attributes of the JPEG2000 standard make this possible: independent compression of each code-block; and arbitrary alignment of the DWT and all associated partitions with respect to a global reference point (the origin of the canvas).

From the coordinate perspective, cropping affects only the bounds, E_1, E_2, F_1 and F_2, of the image region on the canvas. It has no effect on the tiling anchor point, $\left[\Omega_1^T, \Omega_2^T\right]$, or the coding anchor point, $\left[\Omega_1^C, \Omega_2^C\right]$. Cropping also has no effect upon subband samples which are located in the interior of the image, away from any of the new boundaries. Figure 11.7 is helpful in understanding the effect of cropping on subband samples and code-blocks. The figure illustrates only the horizontal direction (e.g., the first row) of an image which is two tiles wide and is compressed using a single level $(D = 1)$ DWT. The image is cropped by 5 samples from the left and five samples from the right and the effect of this cropping is illustrated on the subband samples of the first tile.

Notice that most of the subband samples and hence most of the code-blocks are unaffected by cropping. Of the subband samples which remain after cropping, only those which are formed from samples in the symmetrically extended portion of the cropped input image must be re-calculated. In particular, if the image is cropped from the left, leaving a new horizontal boundary, E_2, the elements of the interleaved sequence of subbands which must be recalculated are those with indices, n, satisfying

$$E_2 \leq n < E_2 + L_{n \bmod 2}$$

where $2L_0 + 1$ and $2L_1 + 1$ are the lengths of the symmetric low- and high-pass wavelet analysis kernels, $h_0^t[n]$ and $h_1^t[n]$. A similar relationship applies for cropping at the opposite boundary. In the example of

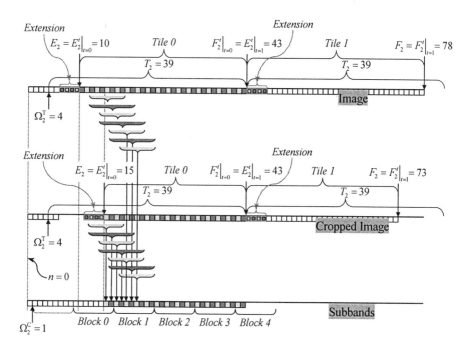

Figure 11.7.   Effect of cropping on the interleaved sequence of subbands when $D = 1$.

Figure 11.7, $2L_0 + 1 = 9$ and $2L_1 + 1 = 7$ are the lengths of the CDF 9/7 irreversible wavelet analysis kernels (see Section 10.4.1) and the leftmost 4 subband samples (2 low-pass and 2 high-pass) of the cropped image must be recomputed. The code-block width is 4 samples (maps to 8 samples in the interleaved subband sequence) and we see that the first two code-blocks of the cropped image must be re-coded. The remaining code-blocks of tile 0 may be taken from the original image as-is.

When the number of DWT levels is greater than 1, a somewhat larger number of subband samples may be affected by cropping. This is because lower frequency subband samples are formed from intermediate low-pass subbands, which are themselves affected by cropping. Nevertheless, it can be shown that within any given subband, the samples which are affected by cropping must lie within a distance δ of one of the cropped boundaries, where

$$\delta \leq \max\{L_0, L_1\}$$

That is, cropping has no impact on samples from subband $y_{b_1,b_2}^{t,c,(d)}[\mathbf{n}]$, for which

$$E_i^{t,c,[b_1,b_2]_d} + \delta \leq n_i < F_i^{t,c,[b_1 b_2]_d} - \delta, \quad i = 1, 2$$

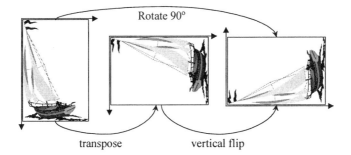

Figure 11.8.   Simple rotations from transposition and flipping.

In most cases, cropping affects only those code-blocks which are adjacent to the cropped boundary. More important than the obvious computational savings for image editing applications is the fact that JPEG2000 compressed images may be repeatedly cropped from any boundary without the accumulation of compression artifacts from multiple compress-decompress cycles. By contrast, images compressed using the existing JPEG standard may not be cropped from the top or the left by anything other than a multiple of 8 samples (the DCT block size) without complete recompression. Decompressing, shifting and re-compressing an image generally involves the accumulation of quantization errors from each compression cycle[1].

It is also worth noting that cropping may affect each image component differently. For example, cropping the last row from the image region in Figure 11.1 affects only component $c = 1$, leaving component $c = 0$ unchanged.

11.4.2   ROTATION AND FLIPPING

In this section, we consider simple geometric transformations in which the image is flipped about one of its axes, transposed or rotated by a multiple of 90°. As illustrated in Figure 11.8, rotation may be achieved by a combination of flipping and transposition. Accordingly, we restrict our attention to these two operations, which are facilitated by the canvas

---

[1] This is because image transforms are inherently not shift invariant. Cropping from the left or the top shifts the image in relation to the alignment of the non-stationary transform and so the recompressed image involves different transform samples and hence different quantization errors, largely uncorrelated with those introduced during the initial compression. Truncation (or clipping) of out-of-range sample values is another source of compression noise "build-up" from multiple compression cycles. JPEG2000 avoids these problems by allowing the image region to be arbitrarily positioned relative to the canvas on which the transform is aligned.

coordinate system and the fact that subband samples are compressed independently within each code-block.

TRANSPOSITION

From the description in Section 10.3.2, each two dimensional DWT stage is a separable combination of one dimensional row and column transformations. As such, the order in which the one dimensional operations are performed is irrelevant (at least for irreversible transforms). We may apply the one dimensional analysis operations vertically along each column and then horizontally along each row, as in Figure 10.5, or horizontally first and then vertically. Continuing this reasoning for all stages of the DWT, transposing the entire image may be seen to be equivalent to transposing each subband individually and labelling the HL_d subbands as LH_d and vice-versa.

The canvas parameters for the transposed image are obtained simply by transposing the various dimensions and anchor point coordinates. Since each individual subband is transposed and its code-block partition is also transposed, the code-blocks of the transposed image are simply transposed versions of the code-blocks of the original image. Unfortunately, to transpose each code-block, it must be fully decoded and re-coded after transposition (we call this "transcoding.") Finally, the order in which code-block contributions appear within each packet and the order of the packets themselves must be adjusted to the new geometry.

Several comments are in order regarding the transcoding of each code-block. First note that this is a local operation, requiring much less working memory than decompression and recompression of the entire image. Secondly, in some applications a non-standard decoder may be available which is able to output decoded samples from each code-block in transposed order. Such a capability might be provided in a print engine to simplify rendering in "landscape" mode.

It is important to realize that transcoding does not generally preserve the exact size (number of code bytes) or distortion associated with each truncation point in the code-block's embedded bit-stream. Thus, assuming that the application does not change the assignment of block truncation points to quality layers (see Section 8.1.3), both the size of each code-block contribution and the distortion associated with each quality layer will generally be affected by transposition. Nevertheless, these effects may only be slight. Small changes in distortion can be expected because transposition generally alters the sets of samples which are processed by each of the three bit-plane coding passes (see Section 8.3.3). If a code-block's embedded bit-stream is truncated at the

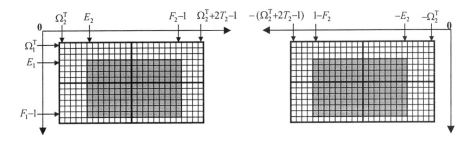

Figure 11.9.   Flipping of the image region and tile partition about the vertical axis.

end of the third coding pass of any given bit-plane then distortion for that code-block will be unaffected by transposition.

Although the irreversible DWT is insensitive to the order in which its one dimensional subband transforms are applied, this is not strictly true for the reversible DWT due to the presence of non-linear rounding operations. As a result, losslessly compressed images will remain lossless under transposition only if the DWT is completely inverted and re-applied. Thus, for reversibly compressed images, transposition and rotation are not possible without either complete recompression or the introduction of small errors (these can be removed again upon restoration of the original image geometry).

FLIPPING

Without loss of generality, we will restrict our attention to flipping the image horizontally. Suppose firstly that we flip the image on the canvas about the vertical axis, $n_2 = 0$, as shown in Figure 11.9. Specifically, each original location $\mathbf{n} \equiv [n_1, n_2]$ on the canvas is mapped to its mirror image location, $\tilde{\mathbf{n}} \equiv [n_1, -n_2]$. This means that each image component, $x_c[\mathbf{n}]$, is mapped to a flipped version, $\vec{x}_c[\mathbf{n}]$, according to

$$x_c[n_1, n_2] = \vec{x}_c[n_1, -n_2], \quad \text{for } 0 \le c < C$$

and the bounds of the flipped image region, $[E_1, F_1) \times \left[\vec{E}_2, \vec{F}_2\right)$, are given by (see the figure)

$$\vec{E}_2 = 1 - F_2, \quad \vec{F}_2 = 1 - E_2$$

Each component's region bounds map in the same way; i.e.,

$$\vec{E}_2^c = 1 - F_2^c, \quad \vec{F}_2^c = 1 - E_2^c, \quad \text{for } 0 \le c < C$$

One way to see this is by combining equation (11.1) with the fact that

$$\left\lceil \frac{1-m}{S} \right\rceil = -\left\lfloor \frac{m-1}{S} \right\rfloor = -\left(\left\lceil \frac{m}{S} \right\rceil - 1\right), \quad \forall m \in \mathbb{Z}, S \in \mathbb{N}$$

If we also arrange for the tile partition to be flipped about the vertical axis, then each tile-component in the flipped image is a mirror image of a corresponding tile-component in the original image. That is,

$$E_2^{\tilde{t},c} = 1 - F_2^{t,c}, \quad F_2^{\tilde{t},c} = 1 - E_2^{t,c}, \quad \forall t, c$$

where $\tilde{t} \equiv [t_1, -t_2]$ represents the flipped tile indices. As suggested by Figure 11.9, the new tile anchor point is given by

$$\vec{\Omega}_2^T = -\left(\Omega_2^T + N_2^T T_2 - 1\right)$$

Now recall from Section 10.3.4 that flipping a tile-component about the vertical axis, $n_2 = 0$, is equivalent to flipping its interleaved sequence of subbands about the same axis. We conclude that the subband samples of the flipped image are identical to those of the original image, except that each subband has been flipped horizontally. Finally, by adjusting the coding anchor point according to

$$\vec{\Omega}_2^C = 1 - \Omega_2^C \in \{0, 1\}$$

we ensure that the precinct and code-block partitions are all correctly flipped[2]. Then each code-block in the flipped image is simply the flipped version of a corresponding code-block in the original image. A local transcoding step is required to obtain the compressed representation of each of the flipped code-blocks. Considerations for block transcoding are identical to those discussed above for the case of image transposition. Finally, the sequence of packets within the code-stream and the order of the code-block contributions in each packet must be modified to reflect the new geometry. It is worth noting that reversible transforms do not cause any difficulties in the case of flipping, as opposed to transposition.

The operations described above fail to generate a legal set of canvas coordinates, since all coordinates are required to be non-negative integers. This difficulty may be rectified, however, by shifting the image region and tile anchor point to the right by some suitable integer, Z_2. That is,

$$\vec{E}_2 \leftarrow \vec{E}_2 + Z_2, \quad \vec{F}_2 \leftarrow \vec{F}_2 + Z_2, \quad \text{and} \quad \vec{\Omega}_2^T \leftarrow \vec{\Omega}_2^T + Z_2$$

---

[2] Note that this manipulation might not be supported by Part 1 of the JPEG2000 standard, depending on the success of a proposed amendment.

Of course, not any Z_2 will do, since arbitrary shifts of the canvas will generally alter image component dimensions, DWT alignment and code-block and precinct partitions. Any shift, Z_2, which is divisible by $2^{D_{t,c}-r}S_2^c P_2^{t,c,r}$ for each tile, t, component, c, and resolution, r, is guaranteed to be acceptable, although other shifts can sometimes be acceptable. For example, if resolution, r, of tile t, in component c, is not divided into multiple precincts, it is sufficient for the shift to be divisible by $2^{D_{t,c}-r+s_r}S_2^c J_2^{t,c}$; that is, the acceptability of the shift is determined by the code-block dimensions, rather than the precinct dimensions[3].

---

[3]In this case, one must be careful to ensure that the shift does not itself cause the resolution to straddle a precinct boundary.

Chapter 12

SAMPLE DATA CODING

In this chapter we describe the various coding operations which are defined by the JPEG2000 standard. In particular, we describe the creation of embedded bit-streams to represent each code-block, and the representation of code-block contributions to packets. The principles and many of the details of these coding operations have already been described in Chapters 2 and 8. Our primary goal in this chapter is to equip the reader who is interested in implementing the standard or working closely with an existing implementation. The reader is assumed to be familiar with the basic principles of the EBCOT paradigm, as described in Section 8.1.3, but a thorough reading of Chapter 8 is not required.

12.1   THE MQ CODER

12.1.1   MQ CODER OVERVIEW

JPEG2000 creates an embedded representation for each code-block of quantized subband samples by subjecting a sign-magnitude version of the quantization indices to a bit-plane coding procedure. The bit-plane coding procedure relies upon the availability of an underlying mechanism for efficiently mapping binary symbols to compressed data bits. The mechanism is that of arithmetic coding and the specific incarnation of arithmetic coding is known as the MQ coder.

Substantial attention has already been devoted in Section 2.3 to an exposition of the principles of practical arithmetic coding. In this section, we describe in detail the particular variant known as MQ. While the reader should find the earlier exposition enlightening, our current treatment is largely self-contained.

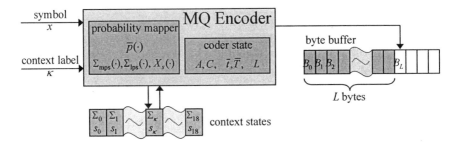

Figure 12.1.   MQ encoder.

The MQ encoder may be understood as a machine, illustrated in Figure 12.1, which maps a sequence of input symbols, $x_n \in \{0,1\}$, and associated context labels, κ_n, to a single compressed codeword. Since the compressed codeword may be only a portion of a larger compressed bit-stream, we shall refer to it as an "MQ codeword segment." The codeword segment represents anything from a single coding pass (see Section 12.2) of a code-block to all coding passes for the code-block. The codeword segment is generated incrementally as the symbol and context pairs, (x_n, κ_n), arrive from the bit-plane coder. As suggested by Figure 12.1, the incremental coding algorithm may be described in terms of three components:

1. A set of "internal" state variables, A, C, $\bar{t}$, $\bar{T}$ and L. We attach subscripts, n, to these variables whenever we wish to identify the state immediately after coding the first n symbols, $x_0, x_1, \ldots, x_{n-1}$. Here, A and C denote the interval length and lower bound registers introduced in Section 2.3, which are common to most implementations of the arithmetic coding principle. L represents the number of code bytes which have been generated so far. $\bar{T}$ is a temporary byte buffer and $\bar{t}$ is a down-counter which identifies the point $(\bar{t} = 0)$ at which partially generated code bits should be moved out of the C register and into the temporary byte buffer, $\bar{T}$.

2. A context state "file" with one pair of entries, $(\Sigma_\kappa, s_\kappa)$, for each possible context label, κ. The single bit, $s_\kappa \in \{0,1\}$, identifies the MPS (Most Probable Symbol) for the coding context labelled κ. Σ_κ is a 6-bit quantity in the range 0 through 46, which indirectly identifies the MPS probability estimate for this coding context.

   Since JPEG2000 defines only 19 contexts, the context state file might be tightly integrated with the rest of the coder in a dedicated hardware implementation (e.g., using high speed registers). Nevertheless,

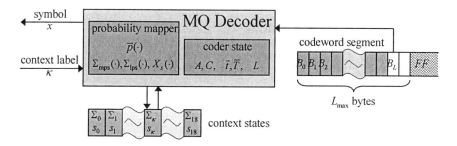

Figure 12.2. MQ decoder.

it is convenient to maintain a conceptual separation between context states and the "internal" coder state variables described above.

3. A set of probability mapping rules which are used to interpret and manipulate the context state, $(\Sigma_\kappa, s_\kappa)$, associated with the current coding context. These probability mapping rules may be understood in terms of four functions (lookup tables), which are defined by Table 2.1. The function $\bar{p}(\Sigma_\kappa)$ embodies the relationship between the state value, Σ_κ, and the LPS probability estimate for context κ. The functions, $\Sigma_{\mathrm{mps}}(\cdot)$ and $\Sigma_{\mathrm{lps}}(\cdot)$ identify the new value for Σ_κ depending on whether the coded symbol is an MPS $(x_n = s_{\kappa_n})$ or an LPS $(x_n = 1 - s_{\kappa_n})$, respectively. The function, $X_s(\cdot)$, is invoked only after coding an LPS; it indicates whether or not the MPS and LPS symbols should be exchanged, i.e., whether or not s_κ should be replaced by $1 - s_\kappa$.

The MQ decoder is illustrated in Figure 12.2. It may be understood as a machine which accepts a sequence of context labels, κ_n, from the bit-plane decoder and returns decoded symbols, $x_n \in \{0, 1\}$. Again, the decoder is implemented incrementally[1], consuming bytes from the compressed codeword segment only as necessary. As suggested by the figure, the encoder and decoder state machines possess many elements in common.

It is important to note that the codeword segment available to the decoder is not generally identical to that produced by the encoder. This is because the embedded representation of any given code-block is often truncated. When the decoder requires code bytes from beyond the

---

[1] The decoder must necessarily be implemented incrementally, since the decoded symbol, x_n, must generally be recovered before the next context label, κ_{n+1}, can be determined. The same is not true for the encoder.

end of the codeword segment, it is expected to substitute the value, FF_h (i.e., 255). The encoder determines the length, L_z, of each allowable truncation point, z, based on the assumption that the decoder will follow this policy. We discuss the length computation task separately in Section 12.3.

We say that the MQ coder is "byte oriented," because codeword segments must consist of a whole number of bytes and certain operations are performed only when a full byte of compressed data has been produced. In particular, the MQ coder adopts a "bit stuffing" approach to avoid the need for full carry resolution (see Section 2.3.1). Bit stuffing occurs whenever an FF_h byte[2] is output to the byte buffer. It inserts an extra (redundant) bit into the evolving codeword, which ensures that carry bits arising from arithmetic operations on the C register cannot propagate into bytes which have already been despatched to the byte buffer. We shall see exactly how this works in Section 12.1.2.

Since bit stuffing adds one redundant bit (one $\frac{1}{8}$ of a byte) to the evolving codeword whenever an FF_h byte is placed in the byte buffer, the cost to overall coding efficiency is one part in $8 \cdot 256 = 2^{11}$ (i.e., about 0.05%). Clearly, some combinations of code bytes cannot arise. In particular, the precise implementation of bit stuffing in the MQ coder is such that the byte following an FF_h must be in the range 00_h through $8F_h$.[3] The JPEG2000 standard takes advantage of this property by assigning values in the range $FF90_h$ through $FFFF_h$ to important delimiting code-stream markers. Examples include the SOT (Start Of Tile-Part) and SOP (Start Of Packet) markers which have values of $FF90_h$ and $FF91_h$, respectively. The provisions of the standard ensure that these markers should never occur within the compressed data itself, so they may be used to recover synchronization in the event of an error.

The MQ coder is a descendant of the multiplier-free Q coder algorithm [118]. Significant enhancements are the conditional exchange mechanism described in Section 2.3.4 and the "start up" portion of the probability estimation state machine. It is closely related to the QM coder, another descendant of the Q coder, which is specified by the JBIG standard [7] and as an option for the JPEG standard. Unlike the QM coder [119], which uses full carry resolution in conjunction with byte stuffing for error resilience, the MQ coder employs the original Q coder's bit

---

[2] We use hexadecimal notation to identify the values of 8-bit bytes and 16-bit marker codes.
[3] The reader might expect that this range should be FF00 through FF7F. A full bit of redundancy is indeed introduced by the MQ coder's bit stuffing policy. However, since carry propagation is bounded, some of the redundant codespace manifests itself in the form of unreachable sequences of code bytes prior to the FF which prompted the bit stuffing operation.

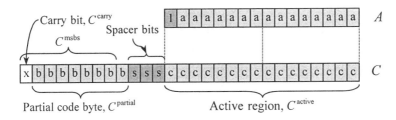

Figure 12.3.   Interpretation of MQ encoder registers.

stuffing policy, as mentioned above. It is worth noting that the JBIG2 standard [10] for bi-level image compression also adopts the MQ coder.

12.1.2   ENCODING PROCEDURES

In this section we describe the actual encoding procedures followed by the MQ coder. The algorithm is substantially similar to the conditional exchange encoder on Page 70, the most important difference being that full carry resolution is replaced by the bit stuffing policy mentioned above. The A register has a 16-bit representation and so the active portion of the C register also involves 16 bits. To avoid moving bits out of the C register whenever a renormalization shift occurs, it is convenient to work with the 28-bit representation for C, which is shown in Figure 12.3.

As explained in Section 2.3, A_n holds a normalized version of the length of the current coding interval, $[c_n, c_n + a_n) \subseteq [0, 1)$. Specifically, let b_n be the number of bits which have been shifted left out of the active region of the C register as a result of coding symbols x_0 through x_{n-1}. Then $a_n = 2^{-b_n} \left(2^{-16} A_n \right)$, with $2^{-16} A_n \in \left[\frac{1}{2}, 1 \right)$. The active region of C_n represents the 16 fraction bits of the normalized interval lower bound, $2^{b_n} c_n$. When b_n reaches the value 11, the partial code byte identified in Figure 12.3 is full and its contents are transferred to the temporary byte buffer, $\bar{T}$.

Following the arithmetic coding procedures described in Section 2.3, one would expect to initialize the MQ coder with (note that $b_0 = 0$) $A_0 = 2^{16}$ (representing $a_0 = 1$), $C_0 = 0$ and $\bar{t}_0 = 11$ (the number of left shifts before a partial code byte needs to be transferred out of C). The only problem with this initialization is that the value $A = 2^{16}$ requires a 17-bit representation for the A register; this is wasteful, since all subsequent values for A are guaranteed to lie in the range $2^{15} \le A < 2^{16}$. The MQ coder avoids this difficulty by initializing $A_0 = 2^{15}$, with an assumed value of $b_0 = -1$, so that the initial interval length is still $a_0 = 1$. It follows that the condition, $b_n = 11$, associated with the first transfer of data out of C, corresponds to 12 left shifts, so we initialize $\bar{t}_0 = 12$. We summarize the initialization procedure as follows.

MQ Encoder Initialization

$A \leftarrow 8000_\mathrm{h}$, $C \leftarrow 0$, $\bar{t} \leftarrow 12$, $\bar{T} \leftarrow 0$

$L \leftarrow -1$ (avoid transferring $\bar{T}$ to the byte buffer before it is first filled)

The main encoding procedure below is invoked to encode a single symbol x with context label κ.

MQ-Encode Procedure

Set $s = s_\kappa$ and $\bar{p} = \bar{p}\left(\Sigma_\kappa\right)$

$A \leftarrow A - \bar{p}$

If $A < \bar{p}$,

    $s \leftarrow 1 - s$ (conditional exchange of MPS and LPS)

If $x = s$,

    $C \leftarrow C + \bar{p}$ (assign MPS the upper sub-interval)

else

    $A \leftarrow \bar{p}$ (assign LPS the lower sub-interval)

If $A < 2^{15}$,

    If $x = s_\kappa$, (the symbol was a real MPS)

        $\Sigma_\kappa \leftarrow \Sigma_\mathrm{mps}\left(\Sigma_\kappa\right)$

    else (the symbol was a real LPS)

        $s_\kappa \leftarrow s_\kappa \oplus X_s\left(\Sigma_\kappa\right)$ (i.e., switch MPS/LPS if $X_s\left(\Sigma_\kappa\right) = 1$)

        $\Sigma_\kappa \leftarrow \Sigma_\mathrm{lps}\left(\Sigma_\kappa\right)$

While $A < 2^{15}$, (perform renormalization shift)

    $A \leftarrow 2A$, $C \leftarrow 2C$, $\bar{t} \leftarrow \bar{t} - 1$

    If $\bar{t} = 0$

        Transfer-Byte$\left(\bar{T}, C, L, \bar{t}\right)$

Notes:

- Although conditional exchange may swap the roles of the MPS and LPS for the purpose of sub-interval assignment, it has no impact on the operation of the probability estimation state machine, which is invoked whenever one or more renormalization shifts are required.

- More efficient implementations of the encoding algorithm (and the decoding algorithm) may be deduced by observing that conditional exchange must always be accompanied by renormalization (see equation (2.20)). Together with a reorganization of the algorithm, this fact may be used to reduce the number of tests which are performed (see Section 17.1.1).

The **Transfer-Byte** procedure transfers data out of the partial code byte portion of the C register, pushing the contents of the temporary byte buffer, $\bar{T}$, to the output byte buffer. It implements the MQ coder's bit stuffing policy.

Transfer-Byte Procedure (encoder)

If $\bar{T} =$ FF$_h$, (can't propagate any carry past $\bar{T}$; need bit stuff)
    Put-Byte$(\bar{T}, L)$
    $\bar{T} \leftarrow C^{\text{msbs}}$, $C^{\text{msbs}} \leftarrow 0$, $\bar{t} \leftarrow 7$ (transfer 7 bits plus carry)
else
    $\bar{T} \leftarrow \bar{T} + C^{\text{carry}}$ (propagate any carry bit from C into $\bar{T}$)
    $C^{\text{carry}} \leftarrow 0$ (reset the carry bit)
    Put-Byte$(\bar{T}, L)$
    If $\bar{T} =$ FF$_h$, (decoder will see this as a bit stuff; need to act accordingly)
        $\bar{T} \leftarrow C^{\text{msbs}}$, $C^{\text{msbs}} \leftarrow 0$, $\bar{t} \leftarrow 7$ (transfer 7 bits plus carry)
    else
        $\bar{T} \leftarrow C^{\text{partial}}$, $C^{\text{partial}} \leftarrow 0$, $\bar{t} \leftarrow 8$ (transfer full byte)

The **Put-Byte** procedure simply writes the contents of $\bar{T}$ to the output byte buffer, except on the first occasion that data is transferred out of the C register, when $\bar{T}$ contains no information; this event is identified by the fact that $L = -1$.

Put-Byte Procedure

If $L \geq 0$,
    $B_L \leftarrow \bar{T}$
$L \leftarrow L + 1$

BIT STUFFING AND SPACER BITS

At this point it is worth making a few additional comments concerning the bit stuffing procedure and the role of the spacer bits identified in Figure 12.3. Arithmetic operations on the C register may produce a carry bit which needs to be propagated into the code bytes which have already been transferred out of the C register. This carry bit must be added into the last transferred byte, which is stored in $\bar{T}$. To prevent the carry from propagating any further, we first check to see if $\bar{T} =$ FF$_h$. If so, we introduce an extra redundant bit into the evolving MQ codeword, transferring only the most significant 7 bits of the partial code byte out of C, which allows any carry bit to occupy the most significant bit position in the new $\bar{T}$ value.

Of course, the decoder must be able to undo the effects of the bit stuffing operation. To do so, the decoder looks for FF_h's in the codeword segment. If one is discovered, the decoder effectively applies a left shift to all remaining code bits and adds them into its working version of the interval lower bound register, C. The decoder is able to do this without risk of carry propagation, because it works with an offset version of the encoder's C register, representing the difference between the final codeword and the current interval lower bound. This approach, common to virtually all arithmetic decoder implementations, is described in Section 2.3.2. To see how the decoder manages to effect the left shift, the reader should carefully review the decoding algorithm presented in Section 12.1.3.

Since the decoder expects a bit stuff following any FF_h in the codeword segment, the encoder must also employ the bit stuffing procedure when a carry bit from C promotes an FE_h value in $\bar{T}$ to FF_h, even though further carry propagation is not possible in this case. This explains the second test for $\bar{T} = FF_h$ in the **Renormalize-Once** procedure above.

We may deduce an upper bound for the code byte which follows any FF_h by considering the maximum value for the most significant byte of C (i.e., C^{msbs}), allowing for the propagation of carry bits from future coding steps. Each transfer from the C register leaves 0's in the most significant $\bar{t} + 1$ bits. Letting S denote the number of spacer bits ($S = 3$ in Figure 12.3), the C register has a $16+S+9$ bit representation. Letting C^{trans} denote the value of C immediately after a transfer, we then have

$$C^{trans} < 2^{16+S+9-(\bar{t}+1)} = 2^{24+S-\bar{t}}$$

In the final codeword, the value, C^{final}, represented by these same bit positions is constrained by the coding interval; i.e.,

$$C^{final} \in \left[C^{trans}, C^{trans} + A^{trans} \right) \subset \left[0, 2^{24+S-\bar{t}} + 2^{16} \right)$$

Left shifting by $\bar{t}$, and taking only the most significant 8 bits of the $25+S$ bit C register leaves us with

$$C^{msbs} = \left\lfloor \frac{2^{\bar{t}} C^{final}}{2^{25+S-8}} \right\rfloor < \frac{2^{24+S} + 2^{16+\bar{t}}}{2^{17+S}} = 2^7 + 2^{\bar{t}-1-S} \le 2^7 + 2^{7-S}$$

Evidently, the spacer bits play an important role in bounding the value of the byte which follows any FF_h in an MQ codeword. In fact, this is

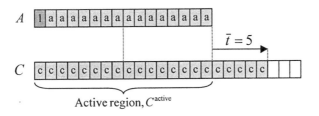

Figure 12.4.   Interpretation of MQ decoder registers.

their only function. With the specific choice $S = 3$ as in Figure 12.3, the non-inclusive upper bound is $2^7 + 2^{7-3} = 90_h$. As mentioned earlier, JPEG2000 defines important delimiting marker codes in the range $FF90_h$ through $FFFF_h$, which must not occur within any MQ codeword segment. Interestingly, the value of S has no impact on the decoding algorithm, except that the decoder is required to treat codes in the range $FF90_h$ through $FFFF_h$ as terminating markers (see below). As a result, there is nothing to prevent an encoder implementation from using more than 3 spacer bits, further restricting the range of byte values which may follow an FF_h.

TERMINATION

The encoding procedure described above is invoked repeatedly until all symbols for the relevant codeword segment have been encoded. At that point, we may flush the contents of the internal state registers, C and $\bar{T}$, to the output byte buffer. However, not all of the bits in these registers are required to construct a codeword segment which uniquely identifies the coded symbols. There are a variety of algorithms which may be employed to efficiently terminate an MQ codeword segment and so we defer discussion of termination until Section 12.3.

12.1.3   DECODING PROCEDURES

The MQ decoder is substantially similar to the conditional exchange decoding algorithm given on Page 71, the main difference being the need to undo the effects of bit stuffing in the encoder. The algorithm may be implemented using 16-bit representations for the C and A registers. However, to avoid transferring bits one-by-one from the codeword buffer to the C register, it is advantageous to work with a 24-bit representation for C, as shown in Figure 12.4. The counter, $\bar{t}$, is used to identify the number of code bits in the least significant byte of C, which have yet to be shifted into the active region. The counter is decremented each time

C is shifted left during renormalization; when $\bar{t} = 0$, a new code byte is loaded into the C register.

Prior to decoding any particular symbol, x_n, the length of the coding interval is given by $a_n = 2^{-b_n} \left(2^{-16} A_n\right)$, where b_n is the number of renormalization shifts which have been applied to the C and A registers. We are obliged to initialize $A_0 = 2^{15}$, with the interpretation that $b_0 = -1$, as we did for the encoder. Accordingly, we must initialize the C register in such a way as to ensure that its active region holds the most significant $16 + b_0 = 15$ bits of the codeword. The following initialization procedure accomplishes this goal.

MQ-Decoder Initialization

$\bar{T} \leftarrow 0,\ L \leftarrow 0,\ C \leftarrow 0$

Fill-LSBs$(C, \bar{T}, \bar{t}, L)$

$C \leftarrow C \cdot 2^{\bar{t}}$ (i.e., left shift C by $\bar{t}$ positions; we can be sure that $\bar{t} = 8$)

Fill-LSBs$(C, \bar{T}, \bar{t}, L)$

$C \leftarrow C \cdot 2^7$

$\bar{t} \leftarrow \bar{t} - 7$

$A \leftarrow 8000_\mathrm{h}$

The **Fill-LSBs** procedure usually loads the next codeword byte into the least significant byte of C, making $\bar{t} = 8$ bits available for transfer into the active portion of the register. In the event that the most recently loaded codeword byte was an $\mathrm{FF_h}$, however, the next codeword byte is left shifted before adding it into C. This fills only 7 bit positions in the least significant byte of C. Accordingly, we set $\bar{t} = 7$ so that a new codeword byte will be loaded after only 7 renormalization shifts. In this way, all subsequent codeword contributions are effectively left shifted, compensating for bit stuffing by the encoder.

Fill-LSBs Procedure

$\bar{t} \leftarrow 8$

If $(L = L_{\max})$ or $(\bar{T} = \mathrm{FF_h}$ and $B_L > 8\mathrm{F_h})$,

  $C \leftarrow C + \mathrm{FF_h}$ (codeword exhausted; fill C with 1's from now on)

else

  If $\bar{T} = \mathrm{FF_h}$,

    $\bar{t} \leftarrow 7$

  $\bar{T} \leftarrow B_L,\ L \leftarrow L + 1$

  $C \leftarrow C + \bar{T} \cdot 2^{8-\bar{t}}$

Notes:

- Once the codeword segment is exhausted, the **Fill-LSBs** procedure is expected to fill the C register with 1's indefinitely, thereby synthesizing the largest possible codeword which is consistent with the available data. Termination and length computation algorithms implemented by the encoder depend upon the fact that the decoder will behave in this way (see Section 12.3).

- The codeword segment is considered exhausted if all $L_{\max}$ bytes are read or if any marker code in the range FF90_h through FFFF_h is encountered. The most efficient compressed representations will not contain such terminating marker codes; however, the JPEG2000 standard does not forbid their appearance within a legal codeword segment. Since a standard decoder will not read past a terminating marker code, the remainder of the codeword segment may contain private or application-specific data. As an example, the compressor might insert error correction codes after an explicit terminating marker. The delimiting marker codes, FF90 (*SOT*), FF91 (*SOP*), FF92 (*EPH*), FF93 (*SOD*) and FFD9 (*EOC*), should generally be avoided, since their appearance within an MQ codeword segment may interfere with resynchronization logic in error resilient decompressors.

The main decoding procedure is shown below.

MQ Decode Procedure (returns x)

Set $s = s_\kappa$ and $\bar{p} = \bar{p}(\Sigma_\kappa)$

$A \leftarrow A - \bar{p}$

If $A < \bar{p}$,

    $s \leftarrow 1 - s$ (conditional exchange of MPS and LPS)

If $C^{\text{active}} < \bar{p}$, (compare active region of C)

    Output $x = 1 - s$

    $A \leftarrow \bar{p}$

else

    Output $x = s$

    $C^{\text{active}} \leftarrow C^{\text{active}} - \bar{p}$

If $A < 2^{15}$

    If $x = s_\kappa$, (the symbol was a real MPS)

        $\Sigma_\kappa \leftarrow \Sigma_{\text{mps}}(\Sigma_\kappa)$

    else (the symbol was a real LPS)

        $s_\kappa \leftarrow s_\kappa \oplus X_s(\Sigma_\kappa)$ (i.e., switch MPS/LPS if $X_s(\Sigma_\kappa) = 1$)

        $\Sigma_\kappa \leftarrow \Sigma_{\text{lps}}(\Sigma_\kappa)$

While $A < 2^{15}$,
       Renormalize-Once$\left(A, C, \bar{t}, \bar{T}, L\right)$

The **Renormalize-Once** procedure may be described as follows.

Renormalize-Once Procedure (decoder)

If $\bar{t} = 0$,
       Fill-LSBs$\left(C, \bar{T}, \bar{t}, L\right)$
$A \leftarrow 2A,\ C \leftarrow 2C,\ \bar{t} \leftarrow \bar{t} - 1$

12.2   EMBEDDED BLOCK CODING

The embedded block coding algorithm adopted by JPEG2000 has already been described in Chapter 8. Specifically, the bit-plane coding primitives are detailed in Section 8.3.2, while the fractional bit-plane scanning pattern is the subject of Section 8.3.3. Our goal here is to provide an algorithmic description of the coding procedures. The reader should refer to the earlier development for an explanation of the motivation behind these procedures. The JPEG2000 coder also supports a number of mode variations. These are described later in Section 12.4.

12.2.1   OVERVIEW

The block coder may be understood as a machine which processes a single code-block of quantized subband samples, having height J_1 and width J_2, producing an embedded bit-stream which consists of a whole number of bytes. In the elementary mode described here, the embedded bit-stream constitutes a single MQ codeword segment. The maximum number of samples in any block and the individual block dimensions must satisfy

$$J_1 J_2 \leq 4096, \quad J_1 \leq 1024 \text{ and } J_2 \leq 1024$$

Figure 12.5 provides a high level view of the block coding machine.

Let $y[j_1, j_2] \equiv y[\mathbf{j}]$ denote the $J_1 \times J_2$ array of subband samples which constitute the code-block. Also let the corresponding quantization indices be represented in sign-magnitude form, with the sign denoted $\chi[\mathbf{j}] \in \{-1, 1\}$ and the magnitude denoted $v[\mathbf{j}] \geq 0$. The quantization procedure which produces these sign and magnitude values is described in Section 10.5. When necessary, we attach a subscript i to these quantities so as to identify the particular code-block $\mathcal{B}_i$ under consideration. The quantized magnitudes have a $K_b^{\max}$-bit representation,

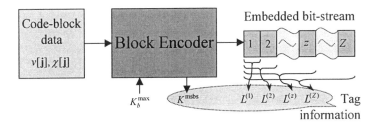

Figure 12.5.   Embedded block encoder high level perspective.

where b identifies the subband to which the code-block belongs and the value of $K_b^{\max}$ is given by equation (10.22). This value depends upon quantization and ranging parameters which may be different for each subband of each tile-component.

We point out that the notation used here agrees with that of Chapter 10 only in the absence of ROI adjustments. For consistency with the discussion of ROI adjustments in Section 10.6, the magnitude indices processed by the block coder should actually be denoted $\overleftarrow{v}\,[\mathbf{j}]$, with the number of magnitude bits denoted $\overleftarrow{K}_b^{\max}$ and given by equation (10.33). We prefer to avoid this notational clutter for the present description.

The block coder first determines the number of bits, $K \leq K_b^{\max}$, which are actually required to represent the quantized magnitudes, $v\,[\mathbf{j}]$. That is $v\,[\mathbf{j}]$ must be less than 2^K for all $\mathbf{j} \in [0, J_1) \times [0, J_2)$. Ideally, the encoder finds the smallest such K, but any $K \leq K_b^{\max}$ is acceptable. The difference,

$$K^{\mathrm{msbs}} = K_b^{\max} - K$$

represents the number of most significant magnitude bits which the decoder will take to be zero for all samples. The remaining K magnitude bits must be explicitly coded. As suggested by Figure 12.5, the value of K^{msbs} is explicitly signalled as part of the code-block's tag information (see Section 12.5).

Coding proceeds incrementally through the K magnitude bit-planes of the quantization indices, starting with the most significant bit-plane, $p = K - 1$, and working down to the least significant bit-plane, $p = 0$. More specifically, the coding proceeds via a number of passes through the entire code-block. We use the labels z and p to identify coding pass and bit-plane indices, respectively. The first coding pass, $z = 1$, represents the most significant magnitude bit-plane, $p = K - 1$, for all samples in the code-block. Thereafter, three coding passes are used to represent each successive bit-plane, so that the total number of coding passes is $Z = 3K - 2$. The binary symbols used to represent these coding

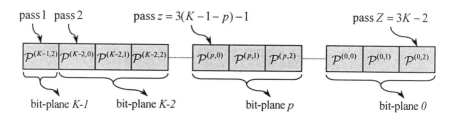

Figure 12.6.   Relationship between coding passes and bit-planes.

passes are delivered to the MQ coder, which incrementally constructs the embedded bit-stream. For each coding pass, z, the coder determines the length in bytes, $L^{(z)}$, of a prefix which is sufficient to decode all of the symbols used to represent the first z coding passes. The lengths, $L^{(z)}$, define the allowable truncation points for the code-block's embedded bit-stream. As suggested by Figure 12.5, some or all of these lengths may be signalled in packet headers as part of the code-block's tag information (see Section 12.5).

Following the notation developed in Section 8.3, let $v^{(p)}\,[\mathbf{j}]$ denote the value formed by discarding p LSBs (Least Significant Bits) from $v\,[\mathbf{j}]$, i.e.,

$$v^{(p)}\,[\mathbf{j}] = \left\lfloor \frac{v\,[\mathbf{j}]}{2^p} \right\rfloor$$

Also, let $v^p\,[\mathbf{j}]$ denote the binary digit in bit position p of $v\,[\mathbf{j}]$; this is the LSB of $v^{(p)}\,[\mathbf{j}]$. The three coding passes for bit-plane p together represent the magnitude bit, $v^p\,[\mathbf{j}]$, for every sample in the block, together with the sign, $\chi\,[\mathbf{j}]$, of any sample for which $v^p\,[\mathbf{j}]$ is the most significant non-zero magnitude bit (i.e., $v^p\,[\mathbf{j}] = 1$ and $v^{(p-1)}\,[\mathbf{j}] = 0$). Each of the $J_1 J_2$ samples is processed in exactly one of these three coding passes. The samples processed in the first coding pass of bit-plane p are identified by the set $\mathcal{P}^{(p,0)}$. Those processed in the second and third coding passes are identified by the sets $\mathcal{P}^{(p,1)}$ and $\mathcal{P}^{(p,2)}$, respectively. Figure 12.6 depicts the relationship between coding passes and bit-planes.

Decoding also proceeds incrementally, working from the first coding pass, $z = 1$ (most significant bit-plane, $p = K - 1$) through to the last available coding pass, $\hat{Z} \leq Z$. It is important to realize that the embedded bit-stream passed to the decoder may be truncated to some length, $L^{(\hat{Z})}$, which is sufficient to represent the binary symbols coded up to and including the corresponding coding pass, $\hat{Z}$. The decoder deduces both the length, $L^{(\hat{Z})}$, and the number of available coding passes, $\hat{Z}$, from information signalled in packet headers (see Section 12.5).

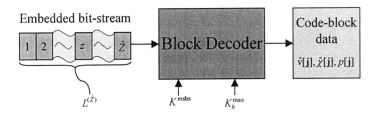

Figure 12.7.   Embedded block decoder high level perspective.

If $\hat{Z}$ corresponds to the final pass, $\mathcal{P}^{(p,2)}$, of some bit-plane, p, then all samples have $p[\mathbf{j}] = p$ missing LSBs. More generally, if coding pass $\hat{Z}$ corresponds to $\mathcal{P}^{(p,i)}$, the number of missing LSBs at sample location $\mathbf{j}$ is given by

$$p[\mathbf{j}] = \begin{cases} p & \text{if } \mathbf{j} \in \mathcal{P}^{(p,k)} \text{ for some } k \leq i \\ p+1 & \text{otherwise} \end{cases}$$

As indicated by Figure 12.7, the output of the block decoder consists of the reconstructed sign and magnitude values, $\hat{\chi}[\mathbf{j}]$ and $\hat{v}[\mathbf{j}]$, together with the number of undecoded LSBs, $p[\mathbf{j}]$, for each sample in the code-block. The decoded magnitude, $\hat{v}[\mathbf{j}]$, agrees with the encoded value, $v[\mathbf{j}]$, except possibly in its least significant $p[\mathbf{j}]$ bit positions; these undecoded bits are set to 0. The decoded sign, $\hat{\chi}[\mathbf{j}]$, agrees with $\chi[\mathbf{j}]$, except possibly when the decoded magnitude is $\hat{v}[\mathbf{j}] = 0$. These three quantities are employed by the dequantization procedure described in Section 10.5.

12.2.2   STATE INFORMATION

In addition to the internal state registers used by the MQ coder (see Section 12.1), the embedded block coder maintains four different types of state information, as shown in Figure 12.8. Although specific implementations of the encoder and decoder might employ different state variables, we find those identified in the figure to be convenient both for description and implementation.

Binary symbols are coded in one of 19 contexts whose labels, κ, are used to index the context state file. Any permutation of the labels used in the present description is acceptable, since they are used only to distinguish between contexts. Table 12.1 identifies the states with which each of the 19 contexts should be initialized. The last column in the table indicates the conditional probability estimates (for the LPS, $x = 1$) implied by these initial states. These values are taken directly from Table 2.1. It is worth noting that the last context label, $\kappa^{\text{uni}} = 18$, has

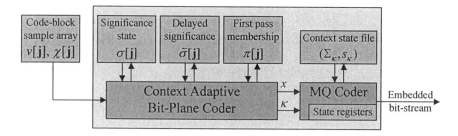

Figure 12.8.   Block diagram of the embedded block coder.

Table 12.1.   MQ context state initialization.

κ	designation	Σ_κ	s_κ	$f_X(1)$
0	κ^{sig}	4	0	≈ 0.0283
$1-8$	κ^{sig}	0	0	≈ 0.475
9	κ^{run}	3	0	≈ 0.0593
$10-14$	κ^{sign}	0	0	≈ 0.475
$15-17$	κ^{mag}	0	0	≈ 0.475
18	κ^{uni}	46	0	≈ 0.475

a special non-adaptive state. According to the state transition rules of Table 2.1, neither Σ_{18} nor s_{18} may transition to anything other than their initial values, since $\Sigma_{18} = 46$. For this reason, some implementations might choose not to allocate storage for this context.

At each sample location, $\mathbf{j}$, the embedded block coder maintains three binary state variables, $\sigma[\mathbf{j}]$, $\overleftarrow{\sigma}[\mathbf{j}]$ and $\pi[\mathbf{j}]$. The value of $\sigma[\mathbf{j}]$ is initialized to 0 and transitions to 1 when the first non-zero magnitude bit is coded for the sample. We say that the sample becomes "significant" at this point and refer to $\sigma[\mathbf{j}]$ as the sample's significance state. $\overleftarrow{\sigma}[\mathbf{j}]$ holds a delayed version of $\sigma[\mathbf{j}]$. Specifically, $\overleftarrow{\sigma}[\mathbf{j}]$ is initialized to 0 and transitions to 1 after the first magnitude refinement coding step at location $\mathbf{j}$. This occurs in the bit-plane following that in which the sample first became significant. The value of $\pi[\mathbf{j}]$ is set during the first coding pass, $\mathcal{P}^{(p,0)}$, of each bit-plane, p. If the sample is processed in that coding pass (i.e., $\mathbf{j} \in \mathcal{P}^{(p,0)}$), $\pi[\mathbf{j}]$ is set to 1. Otherwise, $\pi[\mathbf{j}]$ is set to 0. This assists in the determination of coding pass membership for subsequent coding passes.

The decoder maintains an identical set of state variables to the encoder. Moreover, the decoding procedures ensure that these state variables hold identical values to their encoder counterparts after each coding step.

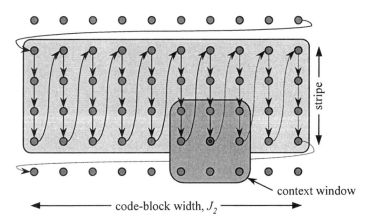

Figure 12.9. Stripe-oriented scanning pattern followed within each coding pass. This is identical to Figure 8.15, reproduced here for convenience.

12.2.3 SCAN AND NEIGHBOURHOODS

In each of its $3K - 2$ coding passes, the block coder follows a stripe-oriented scan through the code-block samples, as illustrated in Figure 12.9. Each stripe represents four rows of code-block samples, with the possible exception of the last stripe in the block. Note that stripes are always aligned with the top of the code-block, without regard for the positioning of the block within the canvas coordinate system.

Although each coding pass conceptually involves a scan through all samples in the code-block, information for any given sample location, $\mathbf{j}$, is coded only in one of the three coding passes for each bit-plane. Coding pass membership and the coding context labels, κ, are determined on the basis of state and sign information within a 3×3 neighbourhood. This neighbourhood is known as the context window and is illustrated in Figure 12.9. To facilitate the description of encoding and decoding procedures, it is convenient to define several functions on the neighbourhood quantities as follows.

Significance functions: The following functions are defined in terms of the significance state variables. For the purpose of these definitions, as well as the sign functions below, the value of $\sigma\left[\mathbf{j}\right]$ should be taken as 0 (insignificant) at any location, $\mathbf{j}$, which falls outside the block boundaries.

$$\kappa^{\mathrm{h}}\left[\mathbf{j}\right] \triangleq \sigma\left[j_1, j_2 - 1\right] + \sigma\left[j_1, j_2 + 1\right]$$
$$\kappa^{\mathrm{v}}\left[\mathbf{j}\right] \triangleq \sigma\left[j_1 - 1, j_2\right] + \sigma\left[j_1 + 1, j_2\right]$$
$$\kappa^{\mathrm{d}}\left[\mathbf{j}\right] \triangleq \sum_{k_1, k_2 \in \{-1,1\}} \sigma\left[j_1 + k_1, j_2 + k_2\right] \tag{12.1}$$

Table 8.1 defines a significance coding context label, $\kappa^{\mathrm{sig}}[\mathbf{j}]$, as a function of $\kappa^{\mathrm{h}}[\mathbf{j}]$, $\kappa^{\mathrm{v}}[\mathbf{j}]$ and $\kappa^{\mathrm{d}}[\mathbf{j}]$. Importantly, $\kappa^{\mathrm{sig}}[\mathbf{j}] = 0$ if and only if all 8 neighbouring samples are insignificant.

Sign functions: We first define horizontal and vertical sign bias functions according to

$$\chi^{\mathrm{h}}[\mathbf{j}] \triangleq \chi[j_1, j_2 - 1]\,\sigma[j_1, j_2 - 1] + \chi[j_1, j_2 + 1]\,\sigma[j_1, j_2 + 1]$$
$$\chi^{\mathrm{v}}[\mathbf{j}] \triangleq \chi[j_1 - 1, j_2]\,\sigma[j_1 - 1, j_2] + \chi[j_1 + 1, j_2]\,\sigma[j_1 + 1, j_2] \quad (12.2)$$

These are then truncated to the range -1 through 1 to form the modified quantities

$$\bar{\chi}^{\mathrm{h}}[\mathbf{j}] \triangleq \mathrm{sign}\left(\chi^{\mathrm{h}}[\mathbf{j}]\right) \min\left\{1, \left|\chi^{\mathrm{h}}[\mathbf{j}]\right|\right\}$$
$$\bar{\chi}^{\mathrm{v}}[\mathbf{j}] \triangleq \mathrm{sign}\left(\chi^{\mathrm{v}}[\mathbf{j}]\right) \min\left\{1, \left|\chi^{\mathrm{v}}[\mathbf{j}]\right|\right\}$$

Apart from the neighbourhood functions defined above, all encoding and decoding operations for the sample at location $\mathbf{j}$ depend only on the state, sign and magnitude bits for that location.

12.2.4   ENCODING PROCEDURES

We are now in a position to provide a functional description of the embedded block coder.

Embedded Block Encoder

Initialize the MQ encoder

Initialize the context states according to Table 12.1

Set $v^{\mathrm{tmp}} \leftarrow 0$

For each $\mathbf{j} \in [0, J_1) \times [0, J_2)$,

    Initialize $\sigma[\mathbf{j}] \leftarrow 0$, $\overleftarrow{\sigma}[\mathbf{j}] \leftarrow 0$ and $\pi[\mathbf{j}] \leftarrow 0$

    $v^{\mathrm{tmp}} \leftarrow v^{\mathrm{tmp}} \vee v[\mathbf{j}]$ (note that $\vee$ means "inclusive or")

Set $K \leftarrow K_b^{\mathrm{max}}$ (use $\overleftarrow{K}_b^{\mathrm{max}}$ if there are ROI adjustments)

While $K > 0$ and $2^{K-1} > v^{\mathrm{tmp}}$,

    $K \leftarrow K - 1$

For $p = K - 1, \ldots, 1, 0$,

    If $p < K - 1$,

        Perform **Encoder-Pass0** (i.e., $\mathcal{P}^{(p,0)}$)

        Perform **Encoder-Pass1** (i.e., $\mathcal{P}^{(p,1)}$)

    Perform **Encoder-Pass2** (i.e., $\mathcal{P}^{(p,2)}$)

Coding pass $\mathcal{P}^{(p,0)}$ is known as the "significance propagation pass." It processes those samples which are not currently significant ($\sigma\,[\mathbf{j}] = 0$), but have a significant neighbourhood ($\kappa^{\mathrm{sig}}\,[\mathbf{j}] \neq 0$). In each of the following pseudo-code fragments, $\kappa^{\mathrm{sig}}\,[\mathbf{j}]$ is to be understood as a function which is evaluated whenever it is required, based on the prevailing significance states, $\sigma\,[\mathbf{j}]$, together with Table 8.1 and equation (12.1). Similar considerations apply to the modified sign bias functions, $\bar{\chi}^{\mathrm{h}}\,[\mathbf{j}]$ and $\bar{\chi}^{\mathrm{v}}\,[\mathbf{j}]$.

Encoder-Pass0 Procedure (significance propagation)

For each location, $\mathbf{j}$, following the stripe-based scan of Figure 12.9,
    If $\sigma\,[\mathbf{j}] = 0$ and $\kappa^{\mathrm{sig}}\,[\mathbf{j}] > 0$,
        MQ-Encode$\left(x = v^p\,[\mathbf{j}]\,, \kappa = \kappa^{\mathrm{sig}}\,[\mathbf{j}]\right)$
        If $v^p\,[\mathbf{j}] = 1$,
            $\sigma\,[\mathbf{j}] \leftarrow 1$
            Encode-Sign$\left(\chi\,[\mathbf{j}]\,, \bar{\chi}^{\mathrm{h}}\,[\mathbf{j}]\,, \bar{\chi}^{\mathrm{v}}\,[\mathbf{j}]\right)$
        $\pi\,[\mathbf{j}] \leftarrow 1$
    else
        $\pi\,[\mathbf{j}] \leftarrow 0$

The sign coding primitive may be implemented as follows.

Encode-Sign Procedure

Determine κ^{sign} and χ^{flip} from $\bar{\chi}^{\mathrm{h}}\,[\mathbf{j}]$ and $\bar{\chi}^{\mathrm{v}}\,[\mathbf{j}]$ using Table 8.2
If $\chi\,[\mathbf{j}] \cdot \chi^{\mathrm{flip}} = 1$,
    MQ-Encode$\left(x = 0, \kappa = \kappa^{\mathrm{sign}}\right)$
else
    MQ-Encode$\left(x = 1, \kappa = \kappa^{\mathrm{sign}}\right)$

Coding pass $\mathcal{P}^{(p,1)}$ is known as the "magnitude refinement pass." It includes those samples which first became significant in a previous bit-plane; i.e., those samples which are significant ($\sigma\,[\mathbf{j}] = 1$) and were not coded in the significance propagation pass ($\pi\,[\mathbf{j}] = 0$).

Encoder-Pass1 Procedure (magnitude refinement)

For each location, $\mathbf{j}$, following the stripe-based scan of Figure 12.9,
    If $\sigma\,[\mathbf{j}] = 1$ and $\pi\,[\mathbf{j}] = 0$,
        Find κ^{mag} from $\overleftarrow{\sigma}\,[\mathbf{j}]$ and $\kappa^{\mathrm{sig}}\,[\mathbf{j}]$ using Table 8.3
        MQ-Encode$\left(x = v^p\,[\mathbf{j}]\,, \kappa = \kappa^{\mathrm{mag}}\right)$
        $\overleftarrow{\sigma}\,[\mathbf{j}] \leftarrow \sigma\,[\mathbf{j}]$

The final coding pass in each bit-plane, $\mathcal{P}^{(p,2)}$, is the "cleanup pass." It includes all samples which were passed over by $\mathcal{P}^{(p,0)}$ and $\mathcal{P}^{(p,1)}$. Since each such sample must currently be insignificant, the coding procedures used here are similar to those of the first pass, $\mathcal{P}^{(p,0)}$. However, a run mode is introduced to reduce the total number of symbols which must be coded. For a more complete discussion of the run mode, see Section 8.3.2.

Encoder-Pass2 Procedure (cleanup)

For each location, $\mathbf{j}$, following the stripe-based scan of Figure 12.9,

    If $j_1 \bmod 4 = 0$ and $j_1 \leq J_1 - 4$, (entering a full stripe column)

        $r \leftarrow -1$ (signifies not using run mode)

        If $\kappa^{\text{sig}}[j_1 + i, j_2] = 0$ for all $i \in \{0, 1, 2, 3\}$, (enter run mode)

            $r \leftarrow 0$

            While $r < 4$ and $v^p[j_1 + r, j_2] = 0$,

            $r \leftarrow r + 1$

            If $r = 4$,

                MQ-Encode$(x = 0, \kappa = \kappa^{\text{run}})$

            else (run interruption)

                MQ-Encode$(x = 1, \kappa = \kappa^{\text{run}})$

                MQ-Encode$\left(x = \left\lfloor \frac{r}{2} \right\rfloor, \kappa = \kappa^{\text{uni}}\right)$

                MQ-Encode$\left(x = r \bmod 2, \kappa = \kappa^{\text{uni}}\right)$

    If $\sigma[\mathbf{j}] = 0$ and $\pi[\mathbf{j}] = 0$,

        If $r \geq 0$,

            $r \leftarrow r - 1$ (no need to code significance)

        else

            MQ-Encode$\left(x = v^p[\mathbf{j}], \kappa = \kappa^{\text{sig}}[\mathbf{j}]\right)$

        If $v^p[\mathbf{j}] = 1$,

             $\sigma[\mathbf{j}] \leftarrow 1$

            Encode-Sign$\left(\chi[\mathbf{j}], \bar{\chi}^{\text{h}}[\mathbf{j}], \bar{\chi}^{\text{v}}[\mathbf{j}]\right)$

It is worth noting that the algorithmic steps presented above do not represent the most efficient implementation of the cleanup coding pass in software. The Kakadu implementation provided with this text, for example, avoids a number of the conditional statements suggested by the description given here.

12.2.5   DECODING PROCEDURES

A suitable set of decoding procedures may be readily derived from the encoding procedures described above. The outer processing loop differs from that of the encoder in that there is no need to compute K. This is deduced from the value of K^{msbs}, recovered from the appropriate packet header. We are also obliged to initialize the output quantities, $\hat{v}\,[\mathbf{j}]$ and $p\,[\mathbf{j}]$, to reflect the fact that nothing has yet been decoded. Finally, we must be careful to decode only those passes for which compressed data is available.

Embedded Block Decoder

Initialize the MQ decoder

Initialize the context states according to Table 12.1

For each $\mathbf{j} \in [0, J_1) \times [0, J_2)$,

    Initialize $\sigma\,[\mathbf{j}] \leftarrow 0$, $\overleftarrow{\sigma}\,[\mathbf{j}] \leftarrow 0$, $\pi\,[\mathbf{j}] \leftarrow 0$, $\hat{v}\,[\mathbf{j}] \leftarrow 0$ and $p\,[\mathbf{j}] \leftarrow K$

For $p = K - 1, \ldots, 1, 0$,

    $z = 3(K - 1 - p) - 1$

    If $p < K - 1$,

        If $z \leq \hat{Z}$,

            Perform **Decoder-Pass0** (i.e., $\mathcal{P}^{(p,0)}$)

        If $z + 1 \leq \hat{Z}$,

            Perform **Decoder-Pass1** (i.e., $\mathcal{P}^{(p,1)}$)

    If $z + 2 \leq \hat{Z}$,

        Perform **Decoder-Pass2** (i.e., $\mathcal{P}^{(p,2)}$)

The significance propagation pass may be decoded as follows.

Decoder-Pass0 Procedure (significance propagation)

For each location, $\mathbf{j}$, following the stripe-based scan of Figure 12.9,

    If $\sigma\,[\mathbf{j}] = 0$ and $\kappa^{\mathrm{sig}}\,[\mathbf{j}] > 0$,

        $p\,[\mathbf{j}] \leftarrow p$

        $\hat{v}^p\,[\mathbf{j}] \leftarrow \mathbf{MQ\text{-}Decode}\big(\kappa = \kappa^{\mathrm{sig}}\,[\mathbf{j}]\big)$

        If $\hat{v}^p\,[\mathbf{j}] = 1$,

            $\sigma\,[\mathbf{j}] \leftarrow 1$

            $\hat{\chi}\,[\mathbf{j}] \leftarrow \mathbf{Decode\text{-}Sign}\big(\bar{\chi}^{\mathrm{h}}\,[\mathbf{j}], \bar{\chi}^{\mathrm{v}}\,[\mathbf{j}]\big)$

        $\pi\,[\mathbf{j}] \leftarrow 1$

    else

        $\pi\,[\mathbf{j}] \leftarrow 0$

Sign decoding may be implemented as follows.

Decode-Sign Procedure (returns χ)

Determine κ^{sign} and χ^{flip} from $\bar{\chi}^{\mathrm{h}}\,[\mathbf{j}]$ and $\bar{\chi}^{\mathrm{v}}\,[\mathbf{j}]$ using Table 8.2

$x \leftarrow \mathbf{MQ\text{-}Decode}\big(\kappa = \kappa^{\mathrm{sign}}\big)$

If $x = 0$,

$$\chi \leftarrow \chi^{\text{flip}}$$

else

$$\chi \leftarrow -\chi^{\text{flip}}$$

For the magnitude refinement pass we have the following.

Decoder-Pass1 Procedure (magnitude refinement)

For each location, $\mathbf{j}$, following the stripe-based scan of Figure 12.9,

    If $\sigma[\mathbf{j}] = 1$ and $\pi[\mathbf{j}] = 0$,

        $p[\mathbf{j}] \leftarrow p$

        Find κ^{mag} from $\overleftarrow{\sigma}[\mathbf{j}]$ and $\kappa^{\text{sig}}[\mathbf{j}]$ using Table 8.3

        $\hat{v}^p[\mathbf{j}] \leftarrow \textbf{MQ-Decode}(\kappa = \kappa^{\text{mag}})$

        $\overleftarrow{\sigma}[\mathbf{j}] \leftarrow \sigma[\mathbf{j}]$

Finally, the cleanup pass may be decoded as follows.

Decoder-Pass2 Procedure (cleanup)

For each location, $\mathbf{j}$, following the stripe-based scan of Figure 12.9,

    If $j_1 \bmod 4 = 0$ and $j_1 \leq J_1 - 4$, (entering a full stripe column)

        $r \leftarrow -1$ (signifies not using run mode)

        If $\kappa^{\text{sig}}[j_1 + i, j_2] = 0$ for all $i \in \{0, 1, 2, 3\}$, (enter run mode)

            $x \leftarrow \textbf{MQ-Decode}(\kappa = \kappa^{\text{run}})$

            If $x = 0$,

                $r \leftarrow 4$

            else (run interruption)

                $r \leftarrow \textbf{MQ-Decode}(\kappa = \kappa^{\text{uni}})$

                $r \leftarrow 2r + \textbf{MQ-Decode}(\kappa = \kappa^{\text{uni}})$

                $\hat{v}^p[j_1 + r, j_2] \leftarrow 1$

    If $\sigma[\mathbf{j}] = 0$ and $\pi[\mathbf{j}] = 0$,

        $p[\mathbf{j}] \leftarrow p$

        If $r \geq 0$,

            $r \leftarrow r - 1$ (no need to decode significance)

        else

            $\hat{v}^p[\mathbf{j}] \leftarrow \textbf{MQ-Decode}(\kappa = \kappa^{\text{sig}}[\mathbf{j}])$

        If $\hat{v}^p[\mathbf{j}] = 1$,

            $\sigma[\mathbf{j}] \leftarrow 1$

            $\hat{\chi}[\mathbf{j}] \leftarrow \textbf{Decode-Sign}(\bar{\chi}^{\text{h}}[\mathbf{j}], \bar{\chi}^{\text{v}}[\mathbf{j}])$

12.3   MQ CODEWORD TERMINATION

As suggested by Figure 12.5, the embedded block coder is required to determine a set of truncation lengths, $L^{(z)}$, such that an $L^{(z)}$-byte prefix of the code-block's embedded bit-stream is sufficient to recover all symbols associated with coding passes 1 through z. A closely related problem is that of terminating MQ codeword segments. In particular, suppose we have a means of terminating MQ codeword segments, such that all symbols may be correctly decoded from the resulting bit-stream. We may then apply an algorithm which computes optimal truncation lengths to determine the shortest prefix of the terminated codeword segment which still allows correct decoding. For this reason, we begin in Section 12.3.1 by describing a simple and generally sub-optimal algorithm for correctly terminating MQ codeword segments. We then devote most of our effort to the problem of computing optimal truncation lengths, $L^{(z)}$.

The JPEG2000 standard places one important restriction on the termination and truncation length computation strategies which may be adopted by a compressor. Specifically, no code-block contribution to any packet may terminate with an FF_h. This restriction ensures that none of the code-stream's delimiting marker codes (these all lie in the range $FF90_h$ through $FFFF_h$) can appear as a side effect of concatenating code-block contributions to form packets. To satisfy this requirement, the length computation algorithm should make sure that code byte $B_{L^{(z)}-1}$ does not equal FF_h for any $z \in \{1, 2, \ldots, Z\}$. This does not present a practical limitation, since the MQ decoder effectively appends an FF_h to the end of the available codeword segment. Thus, in the event that $B_{L^{(z)}-1} = FF_h$, the value of $L^{(z)}$ may simply be decremented.

12.3.1   EASY TERMINATION

In Section 12.1.2, we described initialization and step by step encoding procedures for the MQ encoder. Once all symbols have been coded using these procedures, the L register holds the total number of bytes which have been output to the byte buffer (see Figure 12.1); however, the temporary byte buffer, $\bar{T}$, and the C register both hold partial codeword products, some of which must be flushed out to the byte buffer in order to construct a suitable codeword segment. In this section, we describe a simple algorithm for accomplishing this task. A key problem is that of determining the number of bits which actually need to be flushed out to the byte buffer for correct decoding. The algorithm presented here does not generally find the minimum possible code length. If desired, the algorithm described in Section 12.3.2 may be invoked later to find the smallest prefix of the terminated code-stream which is sufficient for correct decoding.

Our approach is based on the general discussion of length-indicated arithmetic codeword termination in Section 2.3.3. Recall that the codeword is essentially the binary fraction representation of the interval lower bound, c, where the final coding interval is $[c, c + a) \subseteq [0, 1)$ and c and a are related to the MQ coder's internal state variables as follows. Writing b for the number of renormalization shifts which have been applied to the coder's C and A registers, we have $2^{-16}A = 2^b a$. Also, the least significant 16 bits of C (the active region shown in Figure 12.3) are the fraction bits in the binary fraction representation of $2^b c$. As shown in Section 2.3.3, it is sufficient for the arithmetic codeword to include all of the non-fraction bits of $2^b c$, plus one bit to the right of the binary point. Since the decoder appends 1's indefinitely to the codeword bits which it receives, this is guaranteed to produce a quantity which lies within the coding interval.

From the above discussion, we conclude that it is sufficient to flush the contents of the temporary byte buffer, $\bar{T}$, and all but the least significant 15 bits of the C register out to the terminated codeword. There is no harm in including additional bits from C in the terminated codeword segment, and this is usually necessary to produce a whole number of bytes. Observing the register organization conventions of Figure 12.3, the following algorithm generates the terminated codeword segment.

Easy MQ Codeword Termination

$n^{\text{bits}} \leftarrow 27 - 15 - \bar{t}$ (the number of bits we need to flush out of C)

$C \leftarrow 2^{\bar{t}} \cdot C$ (move the next 8 available bits into the partial byte)

While $n^{\text{bits}} > 0$,

    Transfer-Byte $(\bar{T}, C, L, \bar{t})$

    $n^{\text{bits}} \leftarrow n^{\text{bits}} - \bar{t}$ (new value of $\bar{t}$ is the number of bits just transferred)

    $C \leftarrow 2^{\bar{t}} \cdot C$ (move bits into available positions for next transfer)

Transfer-Byte $(\bar{T}, C, L, \bar{t})$ (flush the byte buffer, $\bar{T}$)

Although generally sub-optimal, this algorithm does have the desirable property that the terminated codeword segment has length $L = 0$ if no symbols were coded. To see this, observe that the MQ coder is initialized with $\bar{t} = 12$ and $L = -1$. Thus $n^{\text{bits}} = 0$ and **Transfer-Byte** is invoked only once. The resulting call to **Put-Byte** produces no output since $L = -1$.

The termination algorithm described above may produce a codeword segment whose last byte, B_{L-1}, is an FF$_\text{h}$. As mentioned above, this

MQ codeword segment

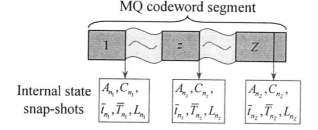

Internal state
snap-shots

Figure 12.10. Complete MQ codeword segment and internal state snap-shots from which truncation lengths, $L^{(z)}$, are determined.

is illegal and the encoder is obliged to decrement L. The previous byte cannot also be an FF_h because the MQ coder's bit stuffing policy ensures that any FF_h must be followed by a byte in the range 00_h through $8F_h$ (see Section 12.1.2). If the length computation algorithm described below is to be used to find an optimal truncation point for the terminated codeword segment, there is no need to prune terminal FF_h's at this stage.

12.3.2   TRUNCATION LENGTHS

In this section we consider the problem of finding the smallest prefix of a terminated MQ codeword segment which will allow the decoder to correctly recover all symbols from coding passes 1 through z. We denote the length (in bytes) of this prefix by $L^{(z)}$. The resulting algorithm is usually invoked for each z in the range 1 through Z, where Z is the total number of coding passes (see Figure 12.6). We refer to the $L^{(z)}$ as truncation lengths.

For convenience, we assume that a complete, terminated MQ codeword segment is generated first and the truncation lengths are generated as a post-processing step. Note, however, that it is possible to compute truncation lengths incrementally as the bit-stream is being generated; the Kakadu software supplied with this text does just that. We also assume that a "snap shot" of the MQ coder's internal state variables is taken at the end of each coding pass.

Let n_z denote the number of symbols processed by the MQ coder up to and including the end of coding pass z. The recorded MQ state variables are A_{n_z}, C_{n_z}, $\bar{t}_{n_z}$, $\bar{T}_{n_z}$ and L_{n_z}. The information available for length computation is illustrated in Figure 12.10. Note that the number of bytes in the terminated codeword segment is usually a little larger than the number of bytes, L_{n_Z}, which had been written to the byte buffer by the time the last symbol was coded.

LAZY LENGTH COMPUTATION

We begin by describing a remarkably simple and also very conservative algorithm for determining suitable truncation lengths, $L^{(z)}$. The algorithm usually produces lengths which are 1 or 2 bytes longer than they need be. Our main purpose in presenting this algorithm is to derive a simple upper bound for the set of acceptable truncation lengths. We use this upper bound in deriving an algorithm for computing optimal truncation lengths below.

The behaviour of the MQ decoder depends only upon those code bits which lie within the active region of its C register. Moreover, the tight synchronization of the encoder and decoder is such that the active regions of both coders' C registers occupy identical positions in the codeword. It follows that correct decoding of the symbols in passes 1 through z is assured provided the codeword is truncated beyond the last bit position in C_{n_z}. There are $27 - \bar{t}_{n_z}$ relevant bits in C_{n_z}, plus the 8 bits of the temporary byte buffer, $\bar{T}_{n_z}$. These must be added to the L_{n_z} bytes already in the output buffer by the end of coding pass z, before we can be certain that the truncated codeword segment will contain all information relevant to the decoding of symbols coded up to that point.

The encoder usually transfers 8 bits out of the C register on each call to the **Transfer-Byte** routine; however, it may occasionally transfer only 7 bits. Since transfers of 7 bits follow the generation of an FF_h byte and this may occur at most on every second transfer, the total number of byte transfers required to output the contents of $\bar{T}_{n_z}$ plus the $27 - \bar{t}_{n_z}$ bits of C_{n_z} is bounded by

$$F_{n_z}^{\max} = 1 + \begin{cases} 3 & \text{if } (27 - \bar{t}_{n_z}) \leq 22 \\ 4 & \text{if } (27 - \bar{t}_{n_z}) > 22 \end{cases} \tag{12.3}$$

We conclude that it is sufficient to set

$$L^{(z)} = L_{n_z} + F_{n_z}^{\max}, \quad z = 1, 2, \ldots, Z$$

bearing in mind that if $B_{L^{(z)}-1} = FF_h$ the value of $L^{(z)}$ should be decremented.

OPTIMAL LENGTH COMPUTATION

The principle behind truncation length optimization is that the bitstream synthesized by the decoder must represent a binary fraction, c, whose value lies within the coding interval, $[c_{n_z}, a_{n_z} + c_{n_z}) \subseteq [0, 1)$. The quantities a_{n_z} and c_{n_z} may be related to the MQ coder's state variables, A_{n_z} and C_{n_z}. Letting b_{n_z} denote the number of renormalization shifts of the A and C register up to the end of coding pass z, we have $2^{-16}A_{n_z} =$

$a_{n_z} 2^{b_{n_z}}$. The active region of C_{n_z} (i.e., its least significant 16 bits) holds the fraction bits of the binary fraction representing $c_{n_z} 2^{b_{n_z}}$.

It is convenient to divide the problem of truncation length optimization into two phases. Since L_{n_z} bytes have already been written to the byte buffer by the end of coding pass z, and these are unaffected by any of the coding steps performed by future coding passes, our first phase will involve the determination of the smallest integer $F \geq 0$ such that $L_{n_z} + F$ bytes of the final codeword segment are sufficient for correct decoding of the first z coding passes. We denote this minimum F by $F_{n_z}^{\min}$. Of course $F_{n_z}^{\min}$ will not exceed the value $F_{n_z}^{\max}$ given by equation (12.3). If $F_{n_z}^{\min} = 0$, it may even be possible to truncate the codeword segment further to some number of bytes smaller than L_{n_z}. We consider this in a second phase of the truncation length optimization procedure.

Subtracting the binary fraction represented by the first L_{n_z} code bytes from c_{n_z} and scaling the resulting coding interval by $2^{b_{n_z} - (19 - \bar{t}_{n_z})}$ leaves us with the "remainder" interval, $[c_r, a_r + c_r)$. Let r_F denote the quantity obtained by applying the same subtraction and scaling operations to the binary fraction represented by the final codeword segment, assuming that it is truncated after $L_{n_z} + F$ bytes. The condition required for correct decoding then becomes

$$c_r \leq r_F < c_r + a_r \tag{12.4}$$

and the interval bounds may be expressed in terms of our "snap-shot" variables as

$$c_r = 2^{-8} \bar{T}_{n_z} + 2^{-8} \cdot 2^{-(27 - \bar{t}_{n_z})} C_{n_z}$$
$$a_r = 2^{-8} \cdot 2^{-(27 - \bar{t}_{n_z})} A_{n_z}$$

The quantity r_F may be computed from the final buffered code bytes, $B_{L_{n_z}}$ through $B_{L_{n_z} + F - 1}$, by following exactly the same procedure as the MQ decoder in undoing the effects of bit stuffing. An algorithm for computing r_F is given below.

Partial Remainder Calculation

$\overline{r_F} \leftarrow 0$, $s_F \leftarrow 0$, $s \leftarrow 8$

For $i = 0$ to $F - 1$

    $s_F \leftarrow s_F + s$

    $\overline{r_F} \leftarrow \overline{r_F} + 2^{-s_F} B_{L_{n_z} + i}$

    If $B_{L_{n_z} + i} = \text{FF}_\text{h}$

        $s \leftarrow 7$

    else

$$s \leftarrow 8$$
$$r_F \leftarrow \overline{r_F} + 2^{-s_F} \times 0.1111111\cdots$$

Note that r_F is an infinite binary fraction, even though F is finite, since the decoder's policy is to append 1's indefinitely to the truncated codeword segment. Only the first s_F bits of r_F are affected by the F code bytes. These first s_F bits are identified by the finite binary fraction, $\overline{r_F}$. The remaining bits of r_F are all 1's.

It is convenient to multiply all quantities by 2^{35} so that the interval bounds in equation (12.4) become non-negative integers. Specifically, define the integer quantities C_r and A_r by

$$C_r \triangleq 2^{35} c_r = 2^{27} \overline{T}_{n_z} + 2^{\overline{t}_{n_z}} C_{n_z}$$
$$A_r \triangleq 2^{35} a_r = 2^{\overline{t}_{n_z}} A_{n_z}$$

Our objective then, is find the smallest $F \geq 0$ such that

$$C_r \leq \left(2^{35}\overline{r_F} + 2^{35-s_F} \times 0.1111111\cdots\right) < C_r + A_r$$

or, equivalently,

$$C_r \leq \left\lfloor 2^{35}\overline{r_F} + 2^{35-s_F} \times 0.1111111\cdots \right\rfloor < C_r + A_r$$

The above condition may be simplified by observing that $F_{n_z}^{\min} \leq F_{n_z}^{\max} \leq 5$. Thus, there is no need to explicitly test the condition for values of F larger than 4. Subject to this restriction, we have $s_F \leq 32$ and $R_F \triangleq 2^{35}\overline{r_F}$ is an integer. The testing condition may now be expressed entirely in terms of integer quantities as

$$C_r \leq R_F + \left(2^{35-s_F} - 1\right) < C_r + A_r$$

A complete algorithm for determining $F_{n_z}^{\min}$ is given below.

Determination of $F_{n_z}^{\min}$

$C_r \leftarrow 2^{27} \overline{T}_{n_z} + 2^{\overline{t}_{n_z}} C_{n_z}$, $A_r \leftarrow 2^{\overline{t}_{n_z}} A_{n_z}$
$R_F \leftarrow 0$, $s \leftarrow 8$, $S_F \leftarrow 35$ (note: $S_F = 35 - s_F$)
$F \leftarrow 0$
While $F < 5$ and $\left(R_F + 2^{S_F} - 1 < C_r \text{ or } R_F + 2^{S_F} - 1 \geq C_r + A_r\right)$,
    $F \leftarrow F + 1$
    If $F \leq 4$, (otherwise, no further tests will be performed)
        $S_F \leftarrow S_F - s$
        $R_F \leftarrow R_F + 2^{S_F} B_{L_{n_z}+F-1}$

If $B_{L_{n_z}+F-1} = \mathrm{FF_h}$

$$s \leftarrow 7$$

else

$$s \leftarrow 8$$

$F_{n_z}^{\min} \leftarrow F$

Since this algorithm involves 35-bit quantities, it may need to be implemented somewhat differently on platforms which support only lower precision arithmetic. Various approaches may be adopted, which we will not develop here. The Kakadu software supplied with this text demonstrates one useful implementation strategy.

As mentioned above, it may occasionally be possible to find truncated codeword segments which are actually smaller than L_{n_z}. The decoder's policy of appending 1's indefinitely to the received codeword segment effectively appends an alternating pattern of $\mathrm{FF_h}$'s and $\mathrm{7F_h}$'s beyond the truncation point. Thus, if the $L_{n_z} + F_{n_z}^{\min}$-byte prefix, already found to be sufficient for correct decoding, terminates with such an alternating pattern of $\mathrm{FF_h}$'s and $\mathrm{7F_h}$'s, these bytes may be safely removed. In fact, truncation lengths smaller than L_{n_z} are acceptable only when the discarded bytes will be re-synthesized by the decoder. Any other form of truncation must alter the binary fraction represented by these L_{n_z} code bytes. To see why this is unacceptable, we consider separately the possibility of a decrease and the possibility of an increase in this binary fraction value.

Due to the effects of bit-stuffing, truncation may actually cause the value represented by the binary fraction to decrease. For example, placing the truncation point immediately prior to an $\mathrm{FF80_h}$ will cause the decoder to synthesize $\mathrm{FF7F_h}$, $\mathrm{FF7F_h}$, ... In such cases, the binary fraction, c, synthesized by the decoder must be strictly less than the interval lower bound, c_{n_z}, so that one or more symbols will be incorrectly decoded. Otherwise (this is the more likely case), truncating to lengths less than L_{n_z}, in any manner which is inconsistent with re-synthesis of the discarded bytes at the decoder, must increase the value represented by the binary fraction, c, by at least $2^{-b_{n_z}+19-\bar{t}_{n_z}}$.[4] This increase is much larger than the size of the coding interval, $a_{n_z} < 2^{-b_{n_z}}$, so incorrect decoding is again assured. We conclude that truncation lengths smaller than L_{n_z} are achievable only when the $L_{n_z} + F_{n_z}^{\min}$-byte codeword prefix terminates with an alternating pattern of $\mathrm{FF_h}$'s and $\mathrm{7F_h}$'s. The follow-

---

[4]There are $19 - \bar{t}_{n_z}$ bit positions between the active regions of A_{n_z} and C_{n_z} and the least significant bit of $B_{L_{n_z}-1}$.

ing code fragment uses this fact to find the minimum possible truncation length, $L^{(z)}$.

Truncation Length Minimization

$L^{(z)} \leftarrow L_{n_z} + F_{n_z}^{\min}$

If $L^{(z)} \geq 1$ and $B_{L^{(z)}-1} = \text{FF}_\text{h}$,

$\qquad L^{(z)} \leftarrow L^{(z)} - 1$

While $L^{(z)} \geq 2$ and $B_{L^{(z)}-2} = \text{FF}_\text{h}$ and $B_{L^{(z)}-1} = 7\text{F}_\text{h}$,

$\qquad L^{(z)} \leftarrow L^{(z)} - 2$

Note that the first test for a terminating FF_h is required for compliance with the standard. The additional tests for trailing FF7F_h's may be skipped, with negligible impact on the expected compression efficiency.

12.4   MODE VARIATIONS

The preceding sections describe what may be termed the "default" mode for the JPEG2000 embedded block coder. This default mode processes the $Z = 3K - 2$ coding passes one by one, producing a single MQ codeword segment which may be truncated to various lengths, $L^{(z)}$, corresponding to the end of each successive coding pass, $z = 1$ through Z. The JPEG2000 standard defines several variations on this default mode. The variations are sufficiently minor that all modes may be efficiently supported within a single decoder implementation. The mode variations generally represent some sacrifice in compression efficiency (often small) in exchange for additional capabilities.

Mode variations are controlled by a collection of six binary flags which appear in the COD and COC code-stream marker segments (see Section 13.3). Each flag is a switch which turns on or off some particular attribute of the coder. Any combination of the six mode switches is permitted, although certain combinations are of particular significance. In particular, the intent is to provide support for the following three capabilities: 1) parallel encoding/decoding of the coding passes; 2) reduced complexity at high bit-rates (when K is large); and 3) enhanced error resilience.

12.4.1   INDIVIDUAL MODE SWITCHES

Table 12.4.1 identifies the six mode switches, along with their flag bits. In this section, we describe the implications of each mode switch on the behavior of the block coder. Some useful combinations are discussed later in Sections 12.4.2 and 12.4.3.

Table 12.2. Mode switches and associated flag bits for the embedded block coder.

Switch	Flag bit	Description
BYPASS	01_h	Selective MQ coder bypass
RESET	02_h	Reset context states
RESTART	04_h	Terminate and restart MQ coder
CAUSAL	08_h	Stripe-causal context formation
ERTERM	10_h	Predictable termination
SEGMARK	20_h	Segmentation marker

THE "RESET" MODE SWITCH

When the *RESET* mode is used, the 19 context states, Σ_κ and s_κ, are reset to the values identified in Table 12.1 at the beginning of each coding pass. Otherwise, the context states are initialized only once, prior to the first coding pass. Forced reset of the context states at each and every coding pass usually reduces coding efficiency somewhat, but helps to decouple the coding passes. Together with the *RESTART* and *CAUSAL* switches, this enables parallel implementation of the coding passes.

THE "RESTART" MODE SWITCH

When the *RESTART* mode is used, the MQ coder is restarted at the beginning of each coding pass. In this case, each coding pass has its own MQ codeword segment. At the end of each coding pass, the codeword segment for that pass is appropriately terminated (see Section 12.3) and the coder is re-initialized in preparation for the next coding pass. Note that MQ coder initialization does not entail resetting of the context states; this is controlled by the *RESET* switch. It is important to realize that the length of each MQ codeword segment must be explicitly signalled in the relevant packet headers. This is discussed further in Section 12.5.4.

THE "CAUSAL" MODE SWITCH

Recall that each coding pass follows a stripe-oriented scan with stripes of height 4. The *CAUSAL* mode switch introduces subtle modifications in the process used to form context labels for bit-plane coding. We say that the modified context formation process is "stripe-causal." The idea is to ensure that the samples within a given stripe may be coded without any dependence on samples from future stripes. By default, coding contexts are formed using the 3×3 context window shown in

Figure 12.9, which is affected by the coding of samples from the following row of code-block samples in previous coding passes.

The impact of the *CAUSAL* mode switch is that samples from future stripes are treated as insignificant for the purpose of forming context labels. That is, at sample locations $\mathbf{j}$, belonging to the fourth row of any stripe, the significance states $\sigma [j_1 + 1, i]$, are regarded as 0 for all i, for the purpose of evaluating the significance functions, $\kappa^v [\mathbf{j}]$ and $\kappa^d [\mathbf{j}]$, and the sign coding function, $\chi^v [\mathbf{j}]$, using equations (12.1) and (12.2).

THE "BYPASS" MODE SWITCH

The MQ coder achieves compression only when the probability estimates associated with the binary symbols being coded are highly skewed. In the most significant bit-planes of any given code-block, most symbols tend to exhibit significantly skewed probabilities. In the less significant bit-planes, however, this is less likely to be true. In particular, the symbols coded in the significance propagation and magnitude refinement passes, $\mathcal{P}^{(p,0)}$ and $\mathcal{P}^{(p,1)}$, usually exhibit nearly uniform distributions when p is small. In such cases, use of the MQ arithmetic coder is of little or no benefit. The *BYPASS* mode switch causes the MQ coder to be bypassed during these passes, for each $p < K - 4$.

The *BYPASS* mode is provided to allow reduced encoding and decoding complexity at high bit-rates, with little or no loss in compression efficiency. It should be noted, however, that loss of compression efficiency can be significant when compressing some types of artificial imagery, including text, graphics or compound documents. The *BYPASS* mode may also cause significant degradation of compression performance when used in conjunction with ROI adjustments (see Section 10.6).

In order to bypass the MQ coder in a particular pass, the current MQ codeword segment must be terminated and a raw (uncoded) segment must be introduced. This fragments the embedded bit-stream into alternate MQ and raw codeword segments, as indicated in Figure 12.11. In the event that the *RESTART* switch is also used, each and every coding pass must be terminated, creating a separate codeword segment (MQ or raw, as appropriate) for every pass. The length of every terminated codeword segment must be explicitly signalled in the relevant packet headers, as discussed in Section 12.5.4.

In raw segments the binary symbols emitted by the bit-plane coder are assembled into bytes and written directly to the segment's byte buffer. A simple bit stuffing procedure is employed to prevent the appearance of delimiting marker codes in the range $FF90_h$ (actually $FF80_h$ here) through $FFFF_h$. Immediately after assembling an FF_h byte, a redundant 0 is inserted in the most significant bit position of the next byte. The

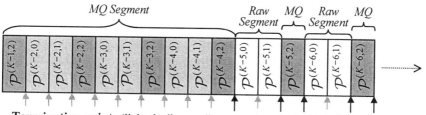

Termination points (light shading applies only when used with *RESTART* mode)

Figure 12.11. Alternate MQ and raw segments of the embedded bit-stream produced in BYPASS mode.

Emit-Raw-Symbol procedure below may be used to perform these tasks. This procedure accepts binary symbols, $x \in \{0,1\}$, and employs state variables, $\bar{T}$, $\bar{t}$ and L. The state variables have similar interpretations to their MQ coder counterparts and are initialized to $\bar{T} = 0$, $\bar{t} = 8$ and $L = 0$ at the commencement of the raw codeword segment.

Emit-Raw-Symbol

$\bar{t} \leftarrow \bar{t} - 1$

$\bar{T} \leftarrow \bar{T} + 2^{\bar{t}} x$

If $\bar{t} = 0$,

$\quad B_L \leftarrow \bar{T}, \; L \leftarrow L + 1$

$\quad \bar{t} \leftarrow 8$

$\quad$ If $\bar{T} = \text{FF}_h$,

$\quad\quad \bar{t} \leftarrow 7$

$\quad \bar{T} \leftarrow 0$

The **Emit-Raw-Symbol** procedure replaces **MQ-Encode** in the significance propagation and magnitude refinement passes described in Section 12.2.4. Context labels are ignored and so need not be formed in the coding passes affected by the *BYPASS* mode. The **Emit-Raw-Symbol** procedure also replaces **Encode-Sign** in all affected significance propagation coding passes; $\chi[\mathbf{j}] = 1$ is represented by the symbol 0, while $\chi[\mathbf{j}] = -1$ is represented by the symbol 1. The terms $\bar{\chi}^h[\mathbf{j}]$ and $\bar{\chi}^v[\mathbf{j}]$ are ignored and need not be generated.

During raw segments, the decoding procedures are modified in the obvious way, replacing **MQ-Decode** and **Decode-Sign** with the **Get-Raw-Symbol** procedure below. This procedure undoes the effects of bit stuffing. It also mimics the MQ decoder's policy of returning 1's once all available bytes are exhausted. Here, $L_{\max}$ denotes the number of available bytes for the segment and the state variables, $\bar{T}$, $\bar{t}$ and L, are all initialized to 0.

Get-Raw-Symbol (returns $x \in \{0, 1\}$)

If $\bar{t} = 0$

    $\bar{t} \leftarrow 8$

    If $L = L_{\max}$,

        $\bar{T} \leftarrow \text{FF}_h$

    else

        If $\bar{T} = \text{FF}_h$,

            $\bar{t} \leftarrow 7$

        $\bar{T} \leftarrow B_L, L \leftarrow L + 1$

$\bar{t} \leftarrow \bar{t} - 1$

$x \leftarrow \left\lfloor \frac{\bar{T}}{2^{\bar{t}}} \right\rfloor \bmod 2$

The encoder is free to terminate raw segments in any manner which will result in correct decoding, so long as no segment concludes with an FF_h. The reason for this restriction has already been explained in connection with MQ codeword termination (see Section 12.3). As for MQ coded segments, encoders may discard any terminating string of consecutive FF7F_h's from a raw segment, since these bytes will be synthesized by the decoder's **Get-Raw-Symbol** procedure. Additional bytes may also be appended, whose interpretation is not described by JPEG2000.

Raw codeword segments may represent two coding passes and the embedded bit-stream may be truncated at the end of the first of these passes. As for MQ codeword segments, the compressor is free to select truncation lengths, $L^{(z)}$, in any manner which is consistent with correct decoding of the relevant passes, subject only to the restriction that no truncation point may be preceded by an FF_h. This restriction is easily satisfied by decrementing $L^{(z)}$ whenever $B_{L^{(z)}-1} = \text{FF}_h$; the FF_h will be synthesized by the decoder's **Get-Raw-Symbol** procedure.

THE "ERTERM" MODE SWITCH

The *ERTERM* (Error Resilient Termination) mode switch is unique in that it need not have any impact on the behavior of a compliant decoder. Instead, it represents a guarantee by the encoder to implement a very specific termination procedure for MQ and raw codeword segments. While encoder's are generally free to terminate both MQ and raw segments in any manner which is consistent with correct decoding, subject to the restriction that no segment conclude with an FF_h, this freedom is surrendered if the *ERTERM* mode switch is turned on. In this case, a predictable termination policy must be employed.

If a predictable termination policy is employed by the encoder, it is possible for decoders to exploit the properties of this termination policy to detect errors which may have been introduced into either the bit-stream or the length values in packet headers. This can lead to substantial improvements in error resilience. When predictable termination is used, decoders can identify the MQ or raw codeword segment in which an error first occurs with remarkable reliability. Upon detection, the corrupt codeword segment and all subsequent segments should normally be discarded, thereby concealing the visual artifacts which would otherwise result from such corruption. Error resilient decoding is described further in Section 12.4.3. Our purpose here is to describe the predictable termination procedures for MQ and raw segments.

The predictable termination policy for MQ codeword segments is exactly the "easy termination" method described in Section 12.3.1. After invoking the algorithm given on Page 496, any trailing FF_h must be discarded. However, the segment may not be further truncated in any way. Such optional post-processing steps would violate the predictability of the termination procedure.

Predictable termination of raw segments is also quite simple. Once all symbols have been emitted using the **Emit-Raw-Symbol** procedure described above, the byte buffer, $\bar{T}$, may be partially full ($\bar{t} < 8$). In this case, $\bar{T}$ is written out as the last byte of the terminated segment after first filling the remaining bits with an alternating sequence of 0's and 1's. The most significant fill bit must be a 0. An interesting situation occurs when the last symbol emitted to the raw segment completes an FF_h byte. In this case, it is not clear whether the procedure should emit an extra byte containing the stuffing bit and termination pattern (the value of such a byte would be $2A_h$) or whether it should discard the FF_h. The JPEG2000 standard does not clearly resolve this ambiguity, so error resilient decoders should be prepared to accept both possibilities.

THE "SEGMARK" MODE SWITCH

If the *SEGMARK* switch is turned on, a string of four binary symbols must be encoded at the end of each bit-plane. Specifically, the symbol string "1010" must be delivered to the MQ coder using the context label, κ^{uni} (this is the unique non-adaptive context for coding uniformly distributed symbols). These symbols complete the third coding pass (i.e., the cleanup pass, $\mathcal{P}^{(p,2)}$) in each bit-plane, p. The decoder must be careful to consume these four symbols before proceeding to the next bit-plane[5]. An error resilient implementation of the decoder may use *SEGMARK* symbols to detect the presence of errors and take measures to conceal the effects of these errors, as discussed in Section 12.4.3.

12.4.2   MODES FOR CODER PARALLELISM

Since code-blocks are encoded and decoded independently, implementations are at liberty to process multiple blocks in parallel for enhanced throughput. We refer to this as "macroscopic parallelism," since it does not require tight synchronization between the parallel processing steps. An obvious drawback of macroscopic parallelism is that each parallel processor must maintain a separate copy of the coder state variables, including the arrays, $\sigma[\mathbf{j}]$, $\overleftarrow{\sigma}[\mathbf{j}]$ and $\pi[\mathbf{j}]$. In many cases, each parallel processor must also have access to a separate code-block sample buffer.

Opportunities for parallel implementation are enhanced by the *RESET*, *RESTART* and *CAUSAL* mode switches. If all three switches are turned on, encoder and decoder implementations may process any or all of the coding passes within a code-block in parallel. We refer to this as "microscopic parallelism," since it requires tight synchronization between the coding pass processors. In one possible implementation, each "clock[6]" advances all processors by one position in the stripe-oriented code-block scanning pattern, with the processor for coding pass z maintaining a position two stripe columns behind the processor for coding pass $z - 1$. This is illustrated in Figure 12.12.

Parallel implementation is possible because the state information used in coding pass z is unaffected by the coding steps in pass $z - 1$ which occur more than 5 samples ahead in the scanning pattern. This is a consequence of the *CAUSAL* mode switch. In the default, non-causal mode, parallel coding passes must be separated by $J_2 + 2$ stripe columns, which substantially increases implementation complexity[7]. Also essential to parallel processing of the coding passes is the fact that arithmetic coding and adaptive probability estimation proceed independently in each coding pass. These properties are introduced by the *RESTART* and *RESET* mode switches, respectively.

---

[5] The decoder may get away with ignoring the *SEGMARK* mode switch if the *RESTART* mode is effective, since then every coding pass is individually terminated.

[6] Here, we are thinking of a synchronous digital circuit, whose state transitions occur on the rising edge (or falling edge) of a master clock signal.

[7] It is actually possible for an encoder to implement all coding passes entirely independently, without specific synchronization constraints. This is because it has access to all bits of the code-block samples which determine coding contexts and coding pass membership. This possibility is not available to the decoder.

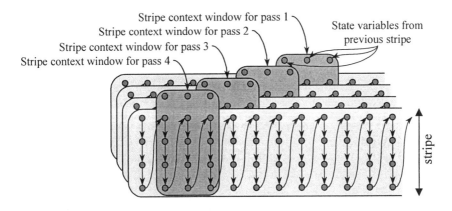

Figure 12.12. Parallel processing of coding passes with the *RESET, RESTART* and *CAUSAL* mode switches. The stripe context window is the union of the coding context windows for all samples in a single stripe column.

An additional benefit arising from the implementation of all coding passes in parallel is that there may be no need to maintain complete representations of the state arrays, $\sigma[\mathbf{j}]$, $\overleftarrow{\sigma}[\mathbf{j}]$ and $\pi[\mathbf{j}]$. However, this benefit can be realized only if sufficient resources are available to implement all Z passes simultaneously. Since the number of passes can differ substantially from block to block, many of the coding pass processors may be idle most of the time. Also, since each sample is processed in only one of the three coding passes in each bit-plane, at least two thirds of the processors can be expected to be idle in each clock period. These are consequences of the tight synchronization required for microscopic parallelism. To alleviate these concerns somewhat, it is advantageous to employ the *BYPASS* mode as well. This substantially reduces (by nearly a factor of 3) the number of parallel MQ coders which must be implemented to achieve guaranteed "sample-per-clock" throughput.

12.4.3   MODES FOR ERROR RESILIENCE

Since code-blocks are coded independently, errors may not propagate beyond the code-block whose bit-stream is corrupted. Nevertheless, a corrupted code-block bit-stream usually leads to objectionable artifacts in the decompressed image. Generally speaking, once an error occurs, the remainder of the embedded bit-stream is useless and subsequent decoding steps are likely to produce erroneous results. This is certainly true in arithmetically coded segments of the bit-stream, since a single bit error destroys synchronization between the encoder and the decoder. However, arithmetic coding is not the only source of dependencies. Even

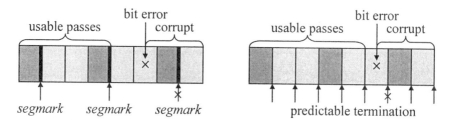

Figure 12.13.   Effect of *SEGMARK* (left) and *ERTERM* with *RESTART* (right) on error resilience.

in raw segments a single symbol error in the significance propagation pass may corrupt the $\sigma\,[\mathbf{j}]$ and $\pi\,[\mathbf{j}]$ state arrays, rendering the remainder of the bit-stream unusable.

The *SEGMARK* and *ERTERM* switches provide quite different mechanisms to enhance error resilience. Suppose firstly that the *SEGMARK* option has been used to insert the special four symbol code, "1010," at the end of each cleanup pass, $\mathcal{P}^{(p,2)}$. This scenario is illustrated on the left in Figure 12.13. A single error in bit-plane p, is likely to corrupt at least one of the four symbols at the end of pass $\mathcal{P}^{(p,2)}$. Upon detecting the corruption, an error resilient decoder should attempt to discard those coding passes which it suspects may contain errors. In the simplest case, the truncated bit-stream is then decoded over again from scratch. The result will be a lower quality rendition of the relevant subband samples, but less objectionable than the visual artifacts usually produced by decoding a corrupted bit-stream. Since the decoder has no way of knowing which of the three passes $\mathcal{P}^{(p,0)}$ through $\mathcal{P}^{(p,2)}$ contained the error, it must discard them all.

An attractive alternative is to use the *ERTERM* and *RESTART* options to create a separate predictably terminated codeword segment for each coding pass. Any error in the bit-stream is likely to leave the coder in a state which is inconsistent with the predictable termination policy. An error resilient decoder may detect this condition at the end of the coding pass in which the error occurred, using methods which we describe below. In this way, the decoder discards only those coding passes which are affected by the error. This scenario is illustrated on the right in Figure 12.13. It is worth mentioning that the *RESTART* mode need not be accompanied by *RESET*, since error detection has no dependence on the states of the probability estimation machinery.

Although the *ERTERM* and *RESTART* options provide superior error resilience to that offered by the *SEGMARK* mechanism, the overhead introduced by terminating each coding pass and explicitly signalling their lengths is larger than the cost of the four *SEGMARK* symbols per bit-plane. Both mechanisms may be used together if desired.

DETECTION OF TERMINATION INCONSISTENCIES

The following procedure may be used to detect termination inconsistencies in a predictably terminated raw codeword segment (these are generated only in *BYPASS* mode). It is invoked once all symbols have been retrieved from the segment.

Termination Consistency Check (Raw)

$x \leftarrow$ 55$_h$ (alternating string of 0's and 1's)

If $L < L_{\max}$ and $\bar{T} =$ FF$_h$ and $\bar{t} = 0$, (last byte created by bit stuffing)
$\quad \bar{T} \leftarrow B_L,\ L \leftarrow L+1,\ \bar{t} \leftarrow 8,\ x \leftarrow$ 2A$_h$

If $L \neq L_{\max}$ or $\left(\bar{T} \bmod 2^{\bar{t}} \right) \neq \left\lfloor \frac{x}{2^{8-\bar{t}}} \right\rfloor$,
$\quad$ **ERTERM-Error**

We turn our attention now to predictably terminated MQ codeword segments. Predictable termination must be performed using the "easy termination" algorithm of Section 12.3.1, which outputs sufficient bytes only to ensure that the most significant bit of the active region of C (see Figure 12.3) is included in the final codeword byte. Let $k \geq 0$ be the number of less significant bit positions which are included in this final codeword byte and let c_{msbs} be the value represented by the least significant $k+1$ bits of this byte. Thus, c_{msbs} is also the value represented by the most significant $k+1$ bits of $C_{\text{enc}}^{\text{active}}$. Figure 12.14 illustrates these relationships, along with those developed below.

Once all symbols have been correctly decoded, the bits in the active portion of the decoder's C register, $C_{\text{dec}}^{\text{active}}$, are aligned with the corresponding bits in $C_{\text{enc}}^{\text{active}}$. Now $C_{\text{dec}}^{\text{active}}$ always holds the difference between the value represented by the codeword bits and the value represented by the symbols decoded so far. Since all symbols have been decoded and the decoder's policy is to fill missing codeword bit positions with 1's, we must have

$$C_{\text{dec}}^{\text{active}} = \left(\underbrace{c_{\text{msbs}}}_{k+1 \text{ bits}} \underbrace{1111\ldots1}_{15-k \text{ bits}} \right) - C_{\text{enc}}^{\text{active}}$$

It follows that, unless an error has occurred, the most significant $k + 1$ bits of $C_{\text{dec}}^{\text{active}}$ must all be 0. These are also the most significant $k + 1$ bits of C, as illustrated in Figure 12.4. This is the principle observation underlying the decoder's error detection strategy.

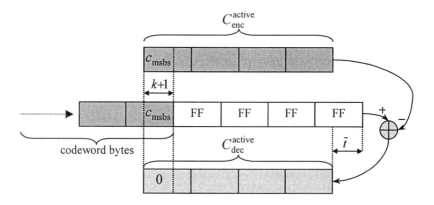

Figure 12.14.   Relationships used to derive error detection conditions for predictably terminated MQ codeword segments.

Let S denote the number of FF_h's synthesized by the **Fill-LSBs** procedure after all L_max bytes are exhausted (see Page 482). Assuming that the final byte output by the predictable termination algorithm was not an FF_h, one of the following must be true of an uncorrupted bit-stream.

- If $\bar{t} = 0$, the C register contains 16 codeword bits (including appended 1's), so exactly one FF_h must have been synthesized ($S = 1$) and the entire most significant byte of C must be 0.

- Otherwise, $1 \leq \bar{t} \leq 7$. In this case, the C register contains $16 + \bar{t}$ codeword bits (including appended 1's), so two FF_h's must have been synthesized ($S = 2$) and the $\bar{t}$ most significant bits of C must all be 0.

If the terminal byte output by the easy termination algorithm was an FF_h it must have been discarded. In this case, the scenarios described above apply with $S = 2$ and $S = 3$, respectively. The following algorithm uses these principles to test for termination inconsistencies.

Termination Consistency Check (MQ)

If $\bar{t} = 0$,
$$S \leftarrow S + 1, \; \bar{t} \leftarrow 8$$
If $S < 2$ or $S > 3$ or $\left\lfloor \frac{C}{2^{24-t}} \right\rfloor \neq 0$,
ERTERM-Error

It is worth clarifying the process by which S is determined for use in this algorithm. We may view S as a state variable which is initialized to 0 and incremented whenever the **Fill-LSBs** procedure on Page 482 synthesizes an FF_h (this happens on line 3 of the procedure). Note that a predictably terminated MQ codeword segment should never contain a termination marker code (i.e., a two byte code in the range FF90_h through FFFF_h), so that $L = L_{\max}$ is the only legitimate condition which should cause **Fill-LSBs** to synthesize an FF_h.

12.5 PACKET CONSTRUCTION
12.5.1 PACK-STREAM STRUCTURE

Following the EBCOT paradigm introduced in Section 8.1.3, each code-block's embedded bit-stream is distributed across a number of quality layers, $\mathcal{Q}_l$, where layer indices, l, run from 0 through $\Lambda - 1$. We referred to this distributed representation as a "pack-stream," since it is formed by packing individual code-block contributions together in some fashion.

The umbrella term "code-stream" refers to both the coded data and the various markers and marker segments which are used to locate and describe coding parameters and auxiliary information. Associated with each tile is a single pack-stream. In the simplest case, the pack-streams for each successive tile (there may be only one) appear one after the other in the code-stream, separated by appropriate markers. However, pack-streams may be segmented into so-called "tile-parts" and then interleaved within the final code-stream. For a thorough discussion of code-stream syntax, the reader is referred to Chapter 13. For the remainder of this discussion we restrict our attention to a single tile and hence a single pack-stream.

We write z_i^l for the number of coding passes for code-block $\mathcal{B}_i$ which may be decoded from quality layers $\mathcal{Q}_0$ through $\mathcal{Q}_l$. Code-block $\mathcal{B}_i$ contributes a total of $L_i^{(z_i^l)}$ code bytes to these layers. Its incremental contribution to any particular layer, $\mathcal{Q}_l$, thus consists of $L_i^{(z_i^l)} - L_i^{(z_i^{l-1})}$ code bytes and represents $z_i^l - z_i^{l-1}$ coding passes. Empty contributions (i.e., $z_i^l = z_i^{l-1}$) are permitted and can occur frequently in practice. Methods for calculating the truncation lengths, $L_i^{(z)}$, are discussed in Section 12.3.2, while Section 8.2 describes methods for optimizing the truncation points, z_i^l, across all code-blocks, $\mathcal{B}_i$, and quality layers, $\mathcal{Q}_l$.

The fundamental organizational unit for pack-streams is the "packet." Each packet[8] (or grouping), $\mathcal{G}_{c,r,\mathbf{p},l}$, contains the code-block contribu-

---

[8]There is no intentional connection between JPEG2000 packets and the packets used in network communications.

tions to one quality layer, $\mathcal{Q}_l$, from one image component, c, at a single DWT resolution level, $\mathcal{R}_r$, within the bounds of a single precinct, $\mathbf{p}$. Resolution level indices range from 0 to D, where D is the number of DWT levels for the relevant tile[9]. Resolution level $\mathcal{R}_0$, consists of the single subband LL_D, while each subsequent level, $\mathcal{R}_r$, consists of three subbands HL_{D-r+1}, LH_{D-r+1} and HH_{D-r+1} (see Figure 10.6). Precinct indices, $\mathbf{p} \equiv [p_1, p_2]$, range over the grid defined by equations (11.9) and (11.10). For a discussion of the role played by precincts, the reader is referred to Chapter 11.

In the simplest case, with only one image component and one precinct, each element (solid dot) in Figures 8.4a and 8.4b corresponds to a single packet. These figures illustrate quality progressive and resolution progressive packet orderings, respectively. With multiple components and multiple precincts, other interesting progressions may be supported, as discussed in Chapter 13.

12.5.2   ANATOMY OF A PACKET

Within each packet, $\mathcal{G}_{c,r,\mathbf{p},l}$, the code-block contributions are ordered in a deterministic fashion, first by subband (HL, LH then HH) within resolution level $\mathcal{R}_r$, and then in raster scan fashion within precinct $\mathbf{p}$ of each subband. There is a nominal contribution from each code-block within the packet scope; that is, each code-block within the scope of precinct $\mathbf{p}$ in resolution level $\mathcal{R}_r$ of component c. However, any or all of these nominal contributions may be empty. The total number of code-blocks within the packet scope is given by equation (11.13). Note that this number can actually be 0, in which case the packet must still be represented in the code-stream.

The very first bit of the packet, identifed as $e_{c,r,\mathbf{p},l}$, plays a special role as an "empty header" indicator. If $e_{c,r,\mathbf{p},l} = 0$, packet $\mathcal{G}_{c,r,\mathbf{p},l}$ is taken to be empty, meaning that no code-block makes any contribution to it. In this case, the entire packet consists of a single byte, whose least significant 7 bits are undefined. If $e_{c,r,\mathbf{p},l} = 1$, the size and other attributes of each code-block contribution are identified explicitly by code-block "tag" bits. It is still possible that the packet is empty, with no contributions from any code-block, but this must be explicitly identified by tag bits in the packet header. The empty header bit and the tag bits for all code-blocks within the packet scope are concatenated to form a single "packet header," which is padded to a byte boundary. The order

---

[9] The number of DWT levels may actually be different for each image component within each tile. We deliberately avoid the notational clutter associated with explicitly identifying such dependencies here.

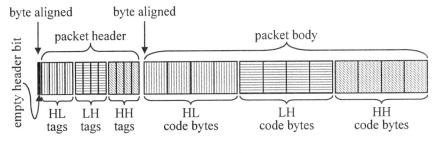

Figure 12.15. Packet structure.

of code-block tag bits in the packet header is identical to the order of the contributed code bytes which follow in the packet body. This organization is illustrated in Figure 12.15. In the example of Figure 12.15, the packet scope includes 4 code-blocks from each of the three subbands (all resolution levels except $\mathcal{R}_0$ have three subbands). In general, however, the number of code-blocks may vary somewhat from subband to subband (see Section 11.3.2).

COMMENTS ON PACKET EMPTINESS

It is worth noting that the term "empty" may be used in three different ways when refering to JPEG2000 packets. The type of "emptiness" associated with the empty header bit refers to the lack of tag bits in the packet header. That is, a packet having $e_{c,r,\mathbf{p},l} = 0$ contains no further header bits or body bytes.

A second type of "emptiness" refers to the lack of body bytes. This occurs when none of the code-blocks in the packet's scope contribute any code bytes; i.e., their nominal contributions are all empty. This may or may not be signalled by setting the empty header bit to $e_{c,r,\mathbf{p},l} = 0$. As we shall see, setting $e_{c,r,\mathbf{p},l} = 0$ can sometimes be an inefficient means of signalling this type of emptiness.

A third type of packet "emptiness" occurs when the packet's precinct contains no code-blocks whatsoever. As explained in Section 11.3.3, it can happen that a precinct has a non-empty intersection with the relevant tile-component resolution, yet contains no code-blocks. In this case, the precinct still has packets and these packets must be represented in the pack-stream. The empty header bit may be set to either $e_{c,r,\mathbf{p},l} = 1$ or $e_{c,r,\mathbf{p},l} = 0$. If it is set to 0 (empty header), the empty packet will have a one byte representation. If set to 1, the header is nominally non-empty, but the packet scope includes no code-blocks and so there will be no further header bits. Since the least significant 7 bits of the first byte are undefined, the compressor may choose to fill them with 1's. In this case, the bit stuffing procedure described below will force the inclusion of a second byte, whose most significant bit is 0. Somewhat surprisingly then, packets whose precincts encompass no code-blocks whatsoever may be represented with as many as 2 bytes.

12.5.3   PACKET HEADER

The tag bits associated with the contribution of code-block $\mathcal{B}_i$ to quality layer $\mathcal{Q}_l$ signal the following three quantities as necessary.

$$\Delta z_i^l \triangleq z_i^l - z_i^{l-1} \quad \text{(with } z_i^{-1} \triangleq 0\text{)}$$

$$\Delta L_i^l \triangleq L_i^{\left(z_i^l\right)} - L_i^{\left(z_i^{l-1}\right)}$$

$$K_i^{\text{msbs}} \triangleq \overleftarrow{K}_{b_i}^{\text{max}} - K_i$$

The number of contributed code bytes, ΔL_i^l, need not be signalled unless the number of contributed coding passes, Δz_i^l, is non-zero. The number of missing MSBs, K_i^{msbs}, is signalled only in the first layer (equivalently, the first packet) to which $\mathcal{B}_i$ makes a non-empty contribution ($\Delta z_i^l > 0$). Note that b_i is the subband to which block $\mathcal{B}_i$ belongs and $\overleftarrow{K}_b^{\text{max}}$ may be found from equation (10.33).

The techniques used to code this tag information have already been described in Section 8.4.2. Here we provide a procedural description of the packet header construction process. A bit stuffing procedure is used to pack tag bits into the packet header, so as to ensure that the packet will not contain any of the code-stream's delimiting marker codes (these all lie in the range FF90_h through FFFF_h). The bit stuffing procedure employs a temporary byte buffer, $\bar{T}$, and associated counter, $\bar{t}$, which are initialized to 0 and 8, respectively.

Emit-Tag-Bit Procedure (packs tag bit x)

$\bar{t} \leftarrow \bar{t} - 1$
$\bar{T} \leftarrow \bar{T} + 2^{\bar{t}} x$
If $\bar{t} = 0$,
    Output-Header-Byte$(\bar{T})$
    $\bar{t} \leftarrow 8$
    If $\bar{T} = \text{FF}_h$,
        $\bar{t} \leftarrow 7$
    $\bar{T} \leftarrow 0$

Once all tag bits for the header have been delivered to this procedure, the partial byte $\bar{T}$ is output as the last header byte, unless $\bar{t} = 8$. The standard does not restrict the values of the $\bar{t} < 8$ unused bits in this partial byte, except that the byte value may not be FF_h.

The header for packet $\mathcal{G}_{c,r,\mathbf{p},l}$ is constructed using the procedure given below. We use the symbol $\mathbf{n}_i$ to identify the location of code-block $\mathcal{B}_i$ within the array of code-blocks belonging to subband b_i in precinct $\mathbf{p}$. For each subband, b, the array of code-blocks has height $N_1^{B,c,r,\mathbf{p},b}$ and width $N_2^{B,c,r,\mathbf{p},b}$, as discussed in Section 11.3.2. Two tag trees (see Section 8.4.2) are defined on each such array of code-blocks.

The first tag tree, $\mathcal{W}_{c,r,\mathbf{p},b}^{\text{inclusion}}$, is used to efficiently code the index, $l_i^{\min}$, of the quality layer to which each code-block in the subband precinct first makes a non-empty contribution. Its nodes are initialized according to

$$w_{c,r,\mathbf{p},b_i}^{\text{inclusion}} [\mathbf{n}_i] = l_i^{\min} \triangleq \min \left\{ l \mid z_i^l > 0 \right\} \tag{12.5}$$

The second tag tree, $\mathcal{W}_{c,r,\mathbf{p},b}^{\text{msbs}}$, is initialized according to

$$w_{c,r,\mathbf{p},b_i}^{\text{msbs}} [\mathbf{n}_i] = K_i^{\text{msbs}} \tag{12.6}$$

Note that the tag trees are shared by all Λ packets associated with a given component, resolution level and precinct. The headers of these packets must be coded in order, starting with layer index $l = 0$ and finishing with layer index $l = \Lambda - 1$. Conceptually, the tag tree node values are initialized in accordance with equations (12.5) and (12.6) before coding the first packet header, although in practice incremental initialization is both possible and desirable (see below).

Encode-Packet-Header Procedure
Emit-Tag-Bit$(e_{c,r,\mathbf{p},l})$ (empty header bit)
If $e_{c,r,\mathbf{p},l} = 0$,
        Stop encoding
For each $\mathcal{B}_i$ in the scope of $\mathcal{G}_{c,r,\mathbf{p},l}$, in the order of Figure 12.15,
        If $z_i^{l-1} = 0$, ($\mathcal{B}_i$ has not yet contributed to any packet)
            $W^{\text{enc}} \left(l+1, \mathbf{n}_i, \mathcal{W}_{c,r,\mathbf{p},b_i}^{\text{inclusion}} \right)$ (effectively codes whether $z_i^l > 0$)
            If $z_i^l > 0$, (need to code K_i^{msbs})
                For $k = 1, 2, \ldots, K_i^{\text{msbs}} + 1$,
                    $W^{\text{enc}} \left(k, \mathbf{n}_i, \mathcal{W}_{c,r,\mathbf{p},b_i}^{\text{msbs}} \right)$
        else
            Emit-Tag-Bit$\left(x = \min \left\{ 1, \Delta z_i^l \right\} \right)$ (codes whether $\Delta z_i^l > 0$)
        If $\Delta z_i^l > 0$,
            Code Δz_i^l using VLC Table 8.4

(note that this may invoke **Emit-Tag-Bit** multiple times)
Encode-Lengths$(\mathcal{B}_i, \mathcal{G}_{c,r,\mathbf{p},l})$

We shall describe the **Encode-Lengths** procedure shortly. Before doing so, however, we present a suitable header decoding procedure.

Decode-Packet-Header Procedure

$e_{c,r,\mathbf{p},l} \leftarrow$ **Get-Tag-Bit**

If $e_{c,r,\mathbf{p},l} = 0$, (empty header)
    Stop decoding
For each $\mathcal{B}_i$ in the scope of $\mathcal{G}_{c,r,\mathbf{p},l}$, in the order of Figure 12.15,
    $\Delta z_i^l \leftarrow 0$
    If $z_i^{l-1} = 0$,
        $W^{\mathrm{dec}}\left(l+1, \mathbf{n}_i, \mathcal{W}_{c,r,\mathbf{p},b_i}^{\mathrm{inclusion}}\right)$
        If $w_{c,r,\mathbf{p},b_i}^{\mathrm{inclusion}}\left[\mathbf{n}_i\right] \leq l$,
            $\Delta z_i^l \leftarrow 1$ (decode actual value later)
            While $w_{c,r,\mathbf{p},b_i}^{\mathrm{msbs}}\left[\mathbf{n}_i\right] = \overline{w}_{c,r,\mathbf{p},b_i}^{\mathrm{msbs}}\left[\mathbf{n}_i\right]$,
                $W^{\mathrm{dec}}\left(\overline{w}_{c,r,\mathbf{p},b_i}^{\mathrm{msbs}}\left[\mathbf{n}_i\right] + 1, \mathbf{n}_i, \mathcal{W}_{c,r,\mathbf{p},b_i}^{\mathrm{msbs}}\right)$
            $K_i^{\mathrm{msbs}} \leftarrow w_{c,r,\mathbf{p},b_i}^{\mathrm{msbs}}\left[\mathbf{n}_i\right]$
    else
        $\Delta z_i^l \leftarrow$ **Get-Tag-Bit** (decode actual value later)
    If $\Delta z_i^l > 0$,
        Decode Δz_i^l using VLC Table 8.4
        (note that this may invoke **Get-Tag-Bit** multiple times)
    $z_i^l \leftarrow z_i^{l-1} + \Delta z_i^l$
    Decode-Lengths$(\mathcal{B}_i, \mathcal{G}_{c,r,\mathbf{p},l})$

INCREMENTAL TAG TREE INITIALIZATION

As mentioned above, the tag tree node values associated with any precinct are conceptually initialized prior to coding any of its packet headers. Leaf nodes are assigned in accordance with equations (12.5) and (12.6). The values at other tag tree nodes are defined following equation (8.17) to hold the minimum of their descendants' node values. In many cases, the code-block inclusion information is discovered incrementally, as quality layers are generated one-by-one using a PCRD-opt algorithm (see Section 8.2).

Fortunately, the tag tree node values can also be initialized incrementally as inclusion information becomes available. To see this, recall that the tag tree coding procedure, $W^{\text{enc}}(l+1, \mathbf{n}_i, W^{\text{inclusion}}_{c,r,\mathbf{p},b_i})$, encodes whether or not $l_i^{\min} > \bar{w}$ for each $\bar{w} \in \{1, 2, \dots, l+1\}$. Neither the code bits nor the state of the tag tree are altered in any way if we initialize the nodes with the modified values,

$$w^{\text{inclusion}}_{c,r,\mathbf{p},b_i}[\mathbf{n}_i] = \min\left\{l_i^{\min}, l+1\right\}$$

prior to coding packet headers for quality layer $\mathcal{Q}_l$. Thus, to correctly code packet headers for layer $\mathcal{Q}_l$, we need only know $l_i^{\min}$ for those code-blocks which actually contribute to the first l quality layers. This means that it is possible to update the tag tree node values incrementally prior to coding each packet, modifying only those nodes which depend on code-blocks contributing for the first time in that layer.

MORE ON EMPTY PACKETS

At this point, it is helpful to consider some subtleties of the interaction between tag tree coding and empty headers. As stated, the inclusion tag tree codes the index, $l_i^{\min}$, of the quality layer to which code-block $\mathcal{B}_i$ first makes a non-empty contribution. Suppose that the first packet (layer $l = 0$) containing this code-block has its empty header bit set to 0. Since no header bits are generated, the state of the inclusion tag tree is not updated until the next packet (layer $l = 1$). The header for this second packet contains the tag tree code bits required to indicate first that $l_i^{\min} > 0$ and then whether or not $l_i^{\min} = 1$. The first condition is redundant since emptiness of the first packet implies that no code-block can contribute until the second quality layer, i.e., $l_i^{\min} \geq 1$. Unfortunately, this redundancy is not exploited by the packet header coding rules outlined above. This was actually an oversight in the development of the standard, which was realized too late to justify correcting the algorithm[10].

The inclusion of redundant information, as described above, occurs whenever a non-empty packet follows one with an empty header. This has a number of unfortunate consequences. A tempting method for resequencing information in an existing code-stream is to add extra quality layers and use the empty header bit to selectively disperse the original packets through the larger number of quality layers. Unfortunately, introducing empty packets between original packets alters the $l_i^{\min}$ values and hence the tag tree code. This means that the original packet headers must be fully decoded and re-encoded, often requiring more bytes.

---

[10] A simple way to avoid this redundancy would have been to set the node values of the inclusion tag tree equal to the number of initial packets to which the relevant code-block makes no contribution, skipping those packets whose headers are explicitly marked as empty.

Another adverse consequence of redundant inclusion information is the possible appearance of inconsistent states. Returning to our example, suppose that the second packet contained tag tree code bits identifying the value of $l_i^{\min}$ as 0. This is meaningless, since no code-block could have contributed to the empty first packet. Nevertheless, such a condition might occur if a valid code-stream were corrupted, or if the subtle interactions described above were not fully appreciated by an implementor. The standard does not define the behaviour expected of a compliant decompressor when presented with such illegal conditions. The algorithm presented on Page 518 will implicitly convert these illegal outcomes from the inclusion tag tree decoder to the smallest legal value for $l_i^{\min}$.

When a large number of quality layers are employed it is common to find that precincts in the higher resolution levels have a significant number of initial empty packets. Interestingly, it is actually inefficient to use the empty header mechanism to identify these empty packets. To see this, observe that each initial empty packet, $\mathcal{G}_{c,r,\mathbf{p},l}$, may be represented either by setting $e_{c,r,\mathbf{p},l} = 0$ (empty header), or by setting $e_{c,r,\mathbf{p},l} = 1$ and following it with three zero-valued bits, corresponding to the root nodes of each subband's inclusion tag tree[11]. In either case, only one byte is consumed by each initial empty packet. The problem with the former approach is that the three inclusion tag tree bits from each initial empty packet are simply deferred until the first packet with a non-empty header, increasing its size accordingly.

The empty header bit provides a convenient and efficient signalling method only when one or more of the final packets for a precinct are empty. This is the simplest and most effective means of selectively reducing the number of quality layers contributed by individual precincts from an existing code-stream. Network transcoders might perform such an operation routinely.

12.5.4   LENGTH CODING

The **Encode-Lengths** procedure is responsible for coding the value of ΔL_i^l, representing the number of code bytes contributed by code-block $\mathcal{B}_i$ to quality layer $\mathcal{Q}_l$. In some cases, however, additional length information must be coded by this procedure. Let $\mathcal{Z}_i^l$ be the set of coding pass indices, z, in the range $z_i^{l-1} < z \le z_i^l$, such that z is either the last contributed coding pass ($z = z_i^l$) or the last coding pass of a

---

[11] To simplify the description, we are ignoring the lowest resolution level, $\mathcal{R}_0$, which contains only the LL_D subband.

terminated MQ or raw codeword segment in the embedded block bit-stream. Codeword segments are terminated after every coding pass when the *RESTART* mode is in force, while the *BYPASS* mode requires the termination points indicated in Figure 12.11. Let $z_{i,1}^l < z_{i,2}^l < \cdots <$ $z_{i,\|\mathcal{Z}_i^l\|}^l$ denote the elements of $\mathcal{Z}_i^l$, where $z_{i,\|\mathcal{Z}_i^l\|}^l = z_i^l$. For convenience, define $z_{i,0}^l = z_i^{l-1}$. The **Encode-Lengths** procedure encodes the length differences,

$$\Delta L_i^{\left(z_{i,j}^l\right)} \triangleq L_i^{\left(z_{i,j}^l\right)} - L_i^{\left(z_{i,j-1}^l\right)}, \quad j = 1, 2, \ldots, \left\|\mathcal{Z}_i^l\right\|$$

using an integer state variable, β_i, which is unique to each code-block. For the interpretation of this state variable, the reader is referred to the end of Section 8.4.2. Its initial value (prior to any header coding) is 3.

It is instructive to consider the length decoding procedure.

Decode-Lengths Procedure

While **Get-Tag-Bit**$() = 1$,

$\quad \beta_i \leftarrow \beta_i + 1$

For $j = 1, 2, \ldots, \left\|\mathcal{Z}_i^l\right\|$,

$\quad \beta \leftarrow \beta_i + \left\lfloor \log_2\left(z_{i,j}^l - z_{i,j-1}^l\right)\right\rfloor$ (number of bits to signal $\Delta L_i^{\left(z_{i,j}^l\right)}$)

$\quad \Delta L_i^{\left(z_{i,j}^l\right)} \leftarrow 0$

$\quad$ For $n = \beta - 1, \ldots, 1, 0$,

$\quad\quad \Delta L_i^{\left(z_{i,j}^l\right)} \leftarrow \Delta L_i^{\left(z_{i,j}^l\right)} + 2^n \cdot$ **Get-Tag-Bit**$()$

The encoding procedure has the additional task of determining a suf-ficiently large value for β_i such that $\beta = \beta_i + \left\lfloor \log_2\left(z_{i,j}^l - z_{i,j-1}^l\right)\right\rfloor$ bits is sufficient to represent the value of $\Delta L_i^{\left(z_{i,j}^l\right)}$ for each j. The standard does not prevent the encoder from selecting unnecessarily large values for β_i.

Chapter 13

CODE-STREAM SYNTAX

In this chapter, we describe the JPEG2000 code-stream syntax. Only JPEG2000 Part 1 syntax is discussed here. When applicable, Part 2 extensions are discussed in Chapter 15. This syntax provides all the information necessary for decompression of a JPEG2000 code-stream. The syntax specifies such fundamental quantities as image size, tile size, number of components, and their associated sub-sampling factors. It also specifies all parameters related to quantization and coding such as step sizes, code-block sizes, precinct sizes, as well as the transform kernel employed. All features present in the code-stream are signalled via the syntax, including the number of quality layers, the number of resolutions, progression orders, region of interest information, and whether or not error resilient encoding has been performed. Most coding parameters can be chosen on a tile-by-tile basis.

The syntax items mentioned above form only a partial list. We defer an exhaustive accounting of all syntax elements and parameters until after a high level discussion of the code-stream organization.

13.1 CODE-STREAM ORGANIZATION

In the simplest case, a JPEG2000 code-stream is structured as a main header followed by a sequence of *tile-streams*. The code-stream is terminated by a two byte marker, *EOC* (end of code-stream). This is depicted graphically in Figure 13.1.

The main header contains global information necessary for decompression of the entire code-stream. Each tile-stream consists of a tile header followed by the compressed pack-stream data for a single tile. Each tile header contains the information necessary for decompressing

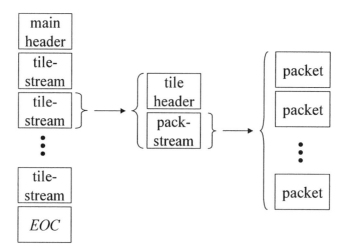

Figure 13.1.   The JPEG2000 code-stream.

the pack-stream of its associated tile. Finally, the pack-stream of a tile consists of a sequence of *packets* as defined in Chapters 8 and 11.

13.1.1   PROGRESSION

As discussed in Chapter 9, progression enables increasing quality, resolution, spatial extent, and/or color components as more bytes are decoded sequentially from the beginning of a compressed code-stream. The type of progression present in a JPEG2000 code-stream is governed by the order in which packets appear within tile-streams. As such, progression can be defined independently on a tile-by-tile basis. Accordingly, we limit our initial discussion to a single tile. We will see later that tile-streams can actually be broken at any packet boundary to form multiple *tile-parts*. Each tile-part has its own header and the tile-parts from different tiles can be interleaved within the code-stream.

As discussed in Section 12.5, a packet from a particular tile $\mathbf{t}$ is indexed by component c, resolution r, precinct $\mathbf{p}$, and quality layer l. In what follows, we denote such a packet by $\mathcal{G}_{\mathbf{t},c,r,\mathbf{p},l}$. From Chapter 11, we have $\mathbf{t} = [t_1, t_2]$ and $\mathbf{p} = [p_1, p_2]$ with

$$
\begin{aligned}
0 \le t_1 < N_1^{\mathrm{T}}; & \quad 0 \le t_2 < N_2^{\mathrm{T}}; \\
0 \le c < C; & \quad 0 \le r < D_{\mathbf{t},c} + 1; \\
0 \le p_1 < N_1^{\mathrm{P},\mathbf{t},c,r}; & \quad 0 \le p_2 < N_2^{\mathrm{P},\mathbf{t},c,r}; \text{ and} \\
0 \le l < \Lambda_{\mathbf{t}}
\end{aligned}
\tag{13.1}
$$

In these expressions, N_1^{T} and N_2^{T} are the numbers of rows and columns of tiles in the image, while C is the number of image components. Also, $D_{\mathbf{t},c} + 1$ is the number of resolutions for component c of tile $\mathbf{t}$. Equivalently, $D_{\mathbf{t},c}$ is the number of transform levels. Finally, $N_1^{\mathrm{P},\mathbf{t},c,r}$ and $N_2^{\mathrm{P},\mathbf{t},c,r}$ are the numbers of rows and columns of precincts at resolution r of component c of tile $\mathbf{t}$, as given by equation (11.10).

For the purposes of this section, it is useful to think of a packet as "one quality increment (layer) of one spatial location (precinct) of one resolution of one component of one tile." Thus, the order in which packets appear in a tile-stream governs which aspects of the tile are decoded earliest. That is, the ordering of packets within a tile-stream governs the progression properties of its associated tile.

For a given tile, the ordering of packets can be specified by a collection of nested loops. Notionally, the indices employed in these loops are c, r, $\mathbf{p}$, and l, with the hierarchy of nesting governing the progression. There are five different progression orders supported in JPEG2000. These progression orders are presented notionally here, followed below by a more rigorous definition of each:

0: Layer-Resolution-Component-Position Progression (LRCP)
> for each l
>> for each r
>>> for each c
>>>> for each $\mathbf{p}$
>>>>> include $\mathcal{G}_{\mathbf{t},c,r,\mathbf{p},l}$

This progression is primarily "progressive by quality." Since the layer index is in the outermost loop, all layer 0 packets (quality increments) appear in the code-stream for every resolution, component, and precinct before any layer 1 packets. The quality improves across the entire tile each time the layer index is incremented.[1]

1: Resolution-Layer-Component-Position Progression (RLCP)
> for each r
>> for each l
>>> for each c
>>>> for each $\mathbf{p}$
>>>>> include $\mathcal{G}_{\mathbf{t},c,r,\mathbf{p},l}$

This progression is primarily "progressive by resolution." Since the resolution index is in the outermost loop, resolution 0 is brought

---

[1]This quality increment need not be *uniform* across an entire tile, component, resolution, or even precinct. Indeed, each codeblock within each precinct may contribute a different number of coding passes to its corresponding layer.

to full quality (all layers included) in each component and precinct before any packets from resolution 1 are included.

2: Resolution-Position-Component-Layer Progression (RPCL)

for each r

for each **p**

for each c

for each l

include $\mathcal{G}_{\mathbf{t},c,r,\mathbf{p},l}$

This progression is also "progressive by resolution." In Progression 1, the progression "within a given resolution" is by layer. Here, the progression within a resolution is by position.

3: Position-Component-Resolution-Layer Progression (PCRL)

for each **p**

for each c

for each r

for each l

include $\mathcal{G}_{\mathbf{t},c,r,\mathbf{p},l}$

This progression is primarily "progressive by position," or "spatially progressive." For appropriately chosen precinct dimensions, the code-stream progresses from the top of the tile to the bottom of the tile. This progression is particularly useful for compressed data streaming in low memory scan-based systems, especially when tiling is not employed.

4: Component-Position-Resolution-Layer Progression (CPRL)

for each c

for each **p**

for each r

for each l

include $\mathcal{G}_{\mathbf{t},c,r,\mathbf{p},l}$

This progression is primarily "progressive by component," as all packets from component 0 precede all packets from component 1, etc. When one of the color transforms of Section 10.2 is employed, component 0 is the luminance component, and all packets containing grayscale information appear before any packets containing color information.

As mentioned in Chapter 9, the progression order can be changed within a tile-stream. For now, we treat only the case when a single progression order is used in a given tile. Accordingly, each index is incremented through its entire range, as given in equation (13.1).

LRCP PROGRESSION

Each tile-component of an image can have a different number of wavelet transform levels. Thus, in what follows, we define

$$D_{\mathbf{t},\max} = \max_c \{D_{\mathbf{t},c}\}$$

The LRCP progression can then be defined more precisely by the following pseudo-code:

---

for $l = 0, 1, \ldots, \Lambda_{\mathbf{t}} - 1$
    for $r = 0, 1, \ldots, D_{\mathbf{t},\max}$
        for $c = 0, 1, \ldots, C - 1$
            if $(r \leq D_{\mathbf{t},c})$
                for $p_1 = 0, 1, \ldots, N_1^{\mathrm{P},\mathbf{t},c,r} - 1$
                for $p_2 = 0, 1, \ldots, N_2^{\mathrm{P},\mathbf{t},c,r} - 1$
                    include $\mathcal{G}_{\mathbf{t},c,r,\mathbf{p},l}$

It is worth noting that the alignment of resolutions within a tile-stream is with respect to resolution 0 in each component. Specifically, for a given layer l, every resolution 0 packet of every component appears in the tile-stream before any resolution 1 packet of any component, and so on. This becomes important when the number of resolutions (number of transform levels) differs across components. In this case, at least one component, say c', will have $D_{\mathbf{t},c'} < D_{\mathbf{t},\max}$. Such components will contribute no packets to the tile-stream when $r > D_{\mathbf{t},c'}$, as indicated by the "if-statement" in the pseudo-code above.

RLCP PROGRESSION

This progression employs $D_{\mathbf{t},\max}$ as defined above to yield the following pseudo-code for creation of the tile-stream for tile $\mathbf{t}$:

---

for $r = 0, 1, \ldots, D_{\mathbf{t},\,\max}$
    for $l = 0, 1, \ldots, \Lambda_{\mathbf{t}} - 1$
        for $c = 0, 1, \ldots, C - 1$
            if $(r \leq D_{\mathbf{t},c})$
                for $p_1 = 0, 1, \ldots, N_1^{\mathrm{P},\mathbf{t},c,r} - 1$
                for $p_2 = 0, 1, \ldots, N_2^{\mathrm{P},\mathbf{t},c,r} - 1$
                    include $\mathcal{G}_{\mathbf{t},c,r,\mathbf{p},l}$

The previous comments regarding alignment of resolutions between components apply to this progression as well.

RPCL PROGRESSION

As discussed in Section 11.2.2, tile $\mathbf{t}$ occupies the region $\left[E_1^\mathbf{t}, F_1^\mathbf{t}\right) \times \left[E_2^\mathbf{t}, F_2^\mathbf{t}\right)$ on the canvas, with $E_i^\mathbf{t}$ and $F_i^\mathbf{t}$ as given in equation (11.3). Through the sub-sampling process of Section 11.2.4, this tile maps to the tile-component resolution occupying the region $\left[E_1^{\mathbf{t},c,r}, F_1^{\mathbf{t},c,r}\right) \times \left[E_2^{\mathbf{t},c,r}, F_2^{\mathbf{t},c,r}\right)$. The precinct partition of a tile-component resolution is of size $P_1^{\mathbf{t},c,r} \times P_2^{\mathbf{t},c,r}$ and is anchored at $\left[\Omega_1^{C,\mathbf{t}}, \Omega_2^{C,\mathbf{t}}\right]$.

Due to differing sub-sampling factors, differing numbers of transform levels, and differing precinct sizes by component and/or resolution, it is convenient to define the next three progressions with respect to canvas coordinates. This choice facilitates correct spatial registration of all tile-component resolutions during the packet sequencing process.

For the same reason, it is also convenient to define separate precinct counters for each tile-component resolution in the loops that define each progression. Specifically, let $\mathbf{p}^{\mathbf{t},c,r} = \left[p_1^{\mathbf{t},c,r}, p_2^{\mathbf{t},c,r}\right]$ be the two dimensional precinct counter for resolution r of component c, and define the following procedure which advances $\mathbf{p}^{\mathbf{t},c,r}$ through the precincts in raster order from left-to-right and top-to-bottom.

Advance-Counter Procedure

$p_2^{\mathbf{t},c,r} \longleftarrow p_2^{\mathbf{t},c,r} + 1$

if $p_2^{\mathbf{t},c,r} = N_2^{P,\mathbf{t},c,r}$

$\quad p_2^{\mathbf{t},c,r} \longleftarrow 0$

$\quad p_1^{\mathbf{t},c,r} \longleftarrow p_1^{\mathbf{t},c,r} + 1$

Finally, define $\mathcal{P}_i^{\mathbf{t},c,r} = S_i^c \cdot 2^{(D_{\mathbf{t},c}-r)} \cdot P_i^{\mathbf{t},c,r}$. From the discussion in Section 11.3, $\mathcal{P}_1^{\mathbf{t},c,r}$ and $\mathcal{P}_2^{\mathbf{t},c,r}$ can be seen as the precinct dimensions $P_1^{\mathbf{t},c,r}$ and $P_2^{\mathbf{t},c,r}$ projected up to the canvas coordinate system. Similarly, $\mathcal{E}_i^{\mathbf{t},c,r} = S_i^c \cdot 2^{(D_{\mathbf{t},c}-r)} \cdot E_i^{\mathbf{t},c,r}$ can be seen as the projection of the upper left corner coordinates of a tile-component resolution onto the canvas.

With these definitions, the RPCL progression is defined as:

$\mathbf{p}^{\mathbf{t},c,r} \longleftarrow \mathbf{0} \quad \forall c, r$

for $r = 0, 1, \ldots, D_{\mathbf{t}, \max}$

$\quad$ for $n_1 = E_1^\mathbf{t}, E_1^\mathbf{t} + 1, \ldots, F_1^\mathbf{t} - 1$

$\quad\quad$ for $n_2 = E_2^\mathbf{t}, E_2^\mathbf{t} + 1, \ldots, F_2^\mathbf{t} - 1$

$\quad\quad\quad$ for $c = 0, 1, \ldots, C - 1$

if $(r \leq D_{\mathbf{t},c})$

    if $\left\{ n_1 - \Omega_1^{C,\mathbf{t}} \text{ is divisible by } \mathcal{P}_1^{\mathbf{t},c,r} \right\}$ or
$$\left\{ \left(n_1 = E_1^{\mathbf{t}} \right) \text{ and } \right.$$
$$\left. \left(\mathcal{E}_1^{\mathbf{t},c,r} - \Omega_1^{C,\mathbf{t}} \text{ is not divisible by } \mathcal{P}_1^{\mathbf{t},c,r} \right) \right\}$$

        if $\left\{ n_2 - \Omega_2^{C,\mathbf{t}} \text{ is divisible by } \mathcal{P}_2^{\mathbf{t},c,r} \right\}$ or
$$\left\{ \left(n_2 = E_2^{\mathbf{t}} \right) \text{ and } \right.$$
$$\left. \left(\mathcal{E}_2^{\mathbf{t},c,r} - \Omega_2^{C,\mathbf{t}} \text{ is not divisible by } \mathcal{P}_2^{\mathbf{t},c,r} \right) \right\}$$

           if $\left(N_1^{P,\mathbf{t},c,r} \cdot N_2^{P,\mathbf{t},c,r} \neq 0 \right)$

                for $l = 0, 1, \ldots, \Lambda_{\mathbf{t}} - 1$

                    include $\mathcal{G}_{\mathbf{t},c,r,\mathbf{p},l}$

               Advance-Counter($\mathbf{p}^{\mathbf{t},c,r}$)

The specification for this progression as given above is provided for clarity and conciseness. It actually makes for a very inefficient implementation, as it scans through every canvas location within the tile, performing complicated conditional testing. Fortunately, much more efficient implementations are possible when S_1^c and S_2^c are restricted to be powers of two. Accordingly, this power of two restriction is required by JPEG2000 for this progression as well as the next progression described below.

To explore the nature of efficient implementations, we first examine the nature of the "if-statements" in the pseudocode above. To this end, we define the projection of the precinct partition from resolution r of component c onto the canvas. The anchor point for this partition is $\left[\Omega_1^{C,\mathbf{t}}, \Omega_2^{C,\mathbf{t}} \right]$, and the partition elements are of size $\mathcal{P}_1^{\mathbf{t},c,r} \times \mathcal{P}_2^{\mathbf{t},c,r}$. This projected partition is shown in Figure 13.2. The reader should verify that the usual downsampling process, when applied to this partition, yields the precinct partition as defined in Section 11.3.

We can now see that the first portion of the if-statements, which test if $n_i - \Omega_i^{C,\mathbf{t}}$ is divisible by $\mathcal{P}_i^{\mathbf{t},c,r}$, are simply identifying upper left hand corners of projected partition elements on the canvas. As mentioned previously, the reason such identification is performed at the canvas level is to ensure correct spatial registration of all tile-component resolutions during the packet sequencing process.

The second portion of the if-statements provide for special treatment of "partial precincts" which may occur along the top or left of the tile-component resolution as shown in Figure 11.4. Clearly, the test for

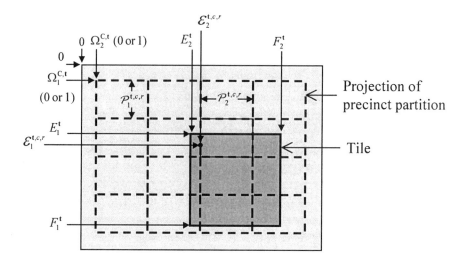

Figure 13.2.   Projection of a precinct partition onto the canvas.

$n_i = E_i^t$ addresses this issue. What may not be clear is the purpose of the last portion of the test. This is explained in the following paragraphs.

It can happen that the original tile, relative to the projected precinct partition, gives the impression that partial precincts exist along the top and/or left, when in fact, there are none. This occurs when the "misleading" samples get sub-sampled away during the process of moving from tile to tile-component resolution.

For example, let the tile have $\left[E_1^t, E_2^t\right] = [12, 15]$, $\left[\Omega_1^{C,t}, \Omega_2^{C,t}\right] = [0, 0]$, $[S_1^c, S_2^c] = [1, 1]$, and precincts of size 4×4 at resolution 0 in a one-level transform. The projection of these precincts to the canvas are then of size $\mathcal{P}_1^{t,c,r} \times \mathcal{P}_2^{t,c,r} = 8 \times 8$. The second column of these projected precincts then intersects the tile region, as shown in Figure 13.2.

This situation might lead us to believe that the second column of precincts contain samples within the tile-component resolution. On the other hand, the tile-component resolution has upper left corner at $\left[\left\lceil\frac{12}{2}\right\rceil, \left\lceil\frac{15}{2}\right\rceil\right] = [6, 8]$, while all coordinates within the second column of precincts have $4 \leq n_2 < 8$. Thus, no samples of the tile-component resolution lie within the second column of precincts. This fact can also be deduced by noting that the projected upper left hand corner of the tile-component resolution lies at $\left[\mathcal{E}_1^{t,c,r}, \mathcal{E}_2^{t,c,r}\right] = [12, 16]$. Thus the left edge of the projected tile-component resolution lies on a projected precinct boundary, as shown in Figure 13.2. In general, this condition can be detected by testing for $\mathcal{E}_i^{t,c,r} = m\mathcal{P}_i^{t,c,r} + \Omega_i^{C,t}$. The last portion of the if-statement is ruling out this condition by ensuring that $\mathcal{E}_i^{t,c,r} - \Omega_i^{C,t}$ is not divisible by $\mathcal{P}_i^{t,c,r}$.

The last item in the pseudocode for this progression that merits mention, is the test whether $N_1^{P,t,c,r} \cdot N_2^{P,t,c,r} = 0$. As discussed in Chapter 11, it is possible for a tile of non-zero size to be sub-sampled entirely away so that $E_i^{t,c,r} = F_i^{t,c,r}$. In this case, the tile-component resolution has no samples, the number of precincts is zero, and no packets should be sequenced for this tile-component-resolution.

An efficient implementation of this progression can now be devised by noting that all projected precinct partitions have $\left[\Omega_1^{C,t}, \Omega_2^{C,t}\right]$ as their common anchor point on the canvas. Further, since the precinct dimensions $P_i^{t,c,r}$ are powers of two, and the S_i^c are restricted to be powers of two for this progression, the projected precinct dimensions $\mathcal{P}_i^{t,c,r}$ must also be powers of two. It is then easy to see that $\mathcal{P}_i^{t,\min} = \min_{c,r}\left\{\mathcal{P}_i^{t,c,r}\right\}$ will be the greatest common division of all $\mathcal{P}_i^{t,c,r}$. Thus all projected precinct partition elements will "line up" on the canvas. Furthermore, with the exception of the special treatment required at the top and left tile boundaries, only canvas coordinates of the form $\left[m_1 \mathcal{P}_1^{t,\min} + \Omega_1^{C,t}, m_2 \mathcal{P}_2^{t,\min} + \Omega_2^{C,t}\right]$ need be considered for the purpose of packet sequencing. This situation is illustrated in Figure 13.3. To avoid clutter, only two tile-component resolutions are considered there.

PCRL PROGRESSION

As in the previous progression, S_1^c and S_2^c are restricted to be powers of two for each component c. Using the notation established above, the pseudocode for this progression can be written as:

$$\mathbf{p}^{t,c,r} \longleftarrow 0 \ \ \forall c, r$$

$\text{for } n_1 = E_1^t, E_1^t + 1, \ldots, F_1^t - 1$

$\text{for } n_2 = E_2^t, E_2^t + 1, \ldots, F_2^t - 1$

$\quad \text{for } c = 0, 1, \ldots, C - 1$

$\quad\quad \text{for } r = 0, 1, \ldots, D_{t,\max}$

$\quad\quad\quad \text{if } (r \leq D_{t,c})$

$\quad\quad\quad\quad \text{if } \left\{n_1 - \Omega_1^{C,t} \text{ is divisible by } \mathcal{P}_1^{t,c,r}\right\} \text{ or}$

$\quad\quad\quad\quad\quad \left\{\left(n_1 = E_1^t\right) \text{ and}\right.$

$\quad\quad\quad\quad\quad \left. \left(\mathcal{E}_1^{t,c,r} - \Omega_1^{C,t} \text{ is not divisible by } \mathcal{P}_1^{t,c,r}\right)\right\}$

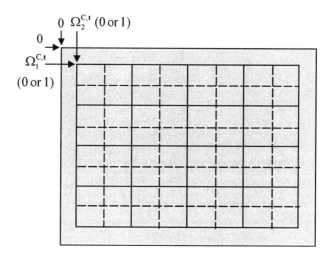

Figure 13.3.   Precinct partitions from two tile-component resolutions as projected on the canvas.   All such projected partitions are forced to "line up" by virtue of the common anchor point, and the power of two partition element sizes. The dashed lines denote partition elements of size $\mathcal{P}_i^{t,\min}$.

$$\text{if } \left\{ n_2 - \Omega_2^{C,t} \text{ is divisible by } \mathcal{P}_2^{t,c,r} \right\} \text{ or}$$
$$\left\{ \left(n_2 = E_2^t \right) \text{ and} \right.$$
$$\left. \left(\mathcal{E}_2^{t,c,r} - \Omega_2^{C,t} \text{ is not divisible by } \mathcal{P}_2^{t,c,r} \right) \right\}$$
$$\text{if } \left(N_1^{P,t,c,r} \cdot N_2^{P,t,c,r} \neq 0 \right)$$
$$\text{for } l = 0, 1, \ldots, \Lambda_t - 1$$
$$\text{include } \mathcal{G}_{t,c,r,p,l}$$
$$\textbf{Advance-Counter}(\mathbf{p}^{t,c,r})$$

The discussion from the previous progression regarding the purpose of the conditional statements and efficient implementations applies to this progression as well. It is worth noting that since the "resolution" loop lies inside the "component" loop, there is really no need for $D_{t,\max}$ and the test for $(r \leq D_{t,c})$. These could be replaced by the simpler version

$$\text{for } r = 0, 1, \ldots, D_{t,c}$$

We retain the more complicated version as it simplifies the discussion of progression order changes in Section 13.3.11.

CPRL PROGRESSION

The procedure for packet sequencing under CPRL progression is reflected in the following pseudocode. Since the "position loop" lies within the "component loop," S_1^c and S_2^c are *not* restricted to be powers of two

---

$\mathbf{p}^{t,c,r} \longleftarrow 0 \ \forall c, r$
for $c = 0, 1, \ldots, C - 1$
    for $n_1 = E_1^t, E_1^t + 1, \ldots, F_1^t - 1$
    for $n_2 = E_2^t, E_2^t + 1, \ldots, F_2^t - 1$
        for $r = 0, 1, \ldots, D_{t,\max}$
           if $(r \leq D_{t,c})$
               if $\left\{ n_1 - \Omega_1^{C,t} \text{ is divisible by } \mathcal{P}_1^{t,c,r} \right\}$ or
                   $\left\{ \left(n_1 = E_1^t \right) \text{ and} \right.$
                    $\left. \left(\mathcal{E}_1^{t,c,r} - \Omega_1^{C,t} \text{ is not divisible by } \mathcal{P}_1^{t,c,r} \right) \right\}$
               if $\left\{ n_2 - \Omega_2^{C,t} \text{ is divisible by } \mathcal{P}_2^{t,c,r} \right\}$ or
                   $\left\{ \left(n_2 = E_2^t \right) \text{ and} \right.$
                    $\left. \left(\mathcal{E}_2^{t,c,r} - \Omega_2^{C,t} \text{ is not divisible by } \mathcal{P}_2^{t,c,r} \right) \right\}$
               if $\left(N_1^{P,t,c,r} \cdot N_2^{P,t,c,r} \neq 0 \right)$
                   for $l = 0, 1, \ldots, \Lambda_t - 1$
                      include $\mathcal{G}_{t,c,r,p,l}$
                   Advance-Counter$(\mathbf{p}^{t,c,r})$

13.2   HEADERS

As discussed in Section 13.1, every JPEG2000 code-stream begins with a main header. Similarly, every tile-stream begins with a tile header. Each of these headers consists of a sequence of *markers* and *marker segments*. A marker is a two byte quantity of the form FFXX_h. That is, the first byte of every marker is hexadecimal FF, while the second byte specifies the particular marker employed. For example, the end of code-stream marker mentioned previously is $EOC = \text{FFD9}_h$. A comprehensive list of markers employed in JPEG2000 Part 1 is given in Table 13.4.

A marker segment consists of a marker followed by a parameter list. Every parameter list must contain a whole number of bytes. The first element of the parameter list is a two byte quantity that specifies the length of the parameter list, in bytes. This length includes the two bytes used for the length field itself, but not the two bytes used for the marker. All values in a marker segment are big endian. That is, for multi-byte quantities, the most significant byte appears in the code-stream first.

13.2.1   THE MAIN HEADER

The main header begins with an *SOC* (start of code-stream) marker. This marker is followed immediately by the *SIZ* (image and tile size) marker segment which specifies global information such as image size, tile size, number of components, etc. There are two more required marker segments in the main header, in addition to several optional marker segments that are allowed to appear in the main header. These required and optional marker segments can appear in any order, after the *SIZ* marker segment.

The required marker segments are *COD* (coding style default) and *QCD* (quantization default), which provide default coding and quantization parameters, respectively. For example, the *COD* marker segment contains information such as the number of transform levels, the code-block and precinct sizes, the progression order, and so forth, while the *QCD* marker segment contains quantization step size information. The optional marker segments include *COC* (coding style component) and *QCC* (quantization component), which provide the ability to override the default values specified in the *COD* and *QCD* marker segments on a component by component basis.

The remaining optional marker segments include *RGN* (region of interest), which provides information for region of interest coding, as described in Section 10.6. The *POC* (progression order change) marker segment provides the ability to change progression orders within the code-stream. The *PPM* (packed packet headers: main header) marker segment can be used to move all packet headers to the main header. The *PLM* (packet lengths: main header) and *TLM* (tile-part lengths: main header) marker segments can be used to record the lengths of compressed packets and tile-parts in the main header, respectively. These latter three markers may be useful for fast random access into the code-stream.

The last two marker segments that can appear in a main header are considered informative in that they are not needed to correctly decode image component samples. The *CRG* (component registration) marker segment can provide information about spatial registration of components, while the *COM* (comment) marker segment can contain arbitrary unstructured data.

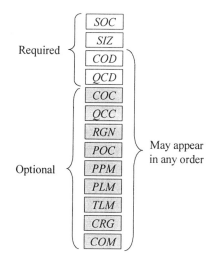

Figure 13.4. JPEG2000 main header. Shading indicates optional marker segments.

Table 13.1. Main header markers/marker segments.

Mnemonic	Value	Marker Name
SOC	$FF4F_h$	Start of code-stream
SIZ	$FF51_h$	Image and tile size
COD	$FF52_h$	Coding style default
QCD	$FF5C_h$	Quantization default
COC	$FF53_h$	Coding style component
QCC	$FF5D_h$	Quantization component
RGN	$FF5E_h$	Region of interest
POC	$FF5F_h$	Progression order change
PPM	$FF60_h$	Packed packet headers: main header
PLM	$FF57_h$	Packet lengths: main header
TLM	$FF55_h$	Tile-part lengths: main header
CRG	$FF63_h$	Component registration
COM	$FF64_h$	Comment

The organization of a JPEG2000 main header as described above, is summarized in Figure 13.4 and Table 13.1. In the figure, all but the first two markers/marker segments can appear in any order, and the shaded marker segments are all optional. Complete descriptions of all markers and marker segments appear in Section 13.3.

Figure 13.5. JPEG2000 tile header. Shaded marker segments are optional and can appear in any order.

Table 13.2.   Tile header markers/marker segments.

Mnemonic	Value	Marker Name
SOT	FF90$_h$	Start of tile
SOD	FF93$_h$	Start of data
COD	FF52$_h$	Coding style default
QCD	FF5C$_h$	Quantization default
COC	FF53$_h$	Coding style component
QCC	FF5D$_h$	Quantization component
RGN	FF5E$_h$	Region of interest
POC	FF5F$_h$	Progression order change
PPT	FF61$_h$	Packed packet headers: tile-part
PLT	FF58$_h$	Packet lengths: tile-part
COM	FF64$_h$	Comment

13.2.2   TILE HEADERS

The structure of a JPEG2000 tile header is shown in Figure 13.5 and Table 13.2. Each tile header begins with an *SOT* (start of tile) marker segment, and ends with an *SOD* (start of data) marker. There are no other required markers or marker segments in a JPEG2000 tile header.

Optional marker segments can appear between *SOT* and *SOD* marker segments in any order. As in the main header, unstructured comment data can be included in a *COM* marker segment. Other optional markers segments include *COD*, *QCD*, *COC*, *QCC*, *RGN*, and *POC*. Each of these has been discussed briefly with respect to the main header in the previous subsection. When these marker segments are used in a tile

Table 13.3.   Tile-part header markers/marker segments.

Mnemonic	Value	Marker Name
SOT	FF90$_h$	Start of tile
SOD	FF93$_h$	Start of data
POC	FF5F$_h$	Progression order change
PPT	FF61$_h$	Packed packet headers: tile-part
PLT	FF58$_h$	Packet lengths: tile-part
COM	FF64$_h$	Comment

header, they override corresponding marker segments that may appear in the main header. The scope of this override is restricted to the tile in which the marker segment appears.

There are two other optional marker segments that can appear in a tile header. Their associated markers are *PPT* (packed packet headers: tile-part) and *PLT* (packet lengths: tile-part). These marker segments are "tile specific" versions of the main header marker segments *PPM* and *PLM*. They are used for collecting packet headers together for placement within the tile header, and for recording packets lengths within the tile header, respectively. As with *PPM* and *PLM*, the *PPT* and *PLT* marker segments are useful for fast random access into the code-stream.

13.2.3   TILE-PART HEADERS

As mentioned previously, tile-streams can be broken at any packet boundary to create multiple tile-parts. We denote the number of tile-parts of a given tile t by $N_{TP,t}$. As shown in Figure 13.6, when multiple tile-parts are present, tile-part 0 has a *tile header*, while all subsequent tile-parts have *tile-part headers*. A tile-part header is constructed in exactly the same way as a tile header, except only a restricted set of marker segments are allowed. The only marker segments that may appear within a tile-part header are *POC*, *PPT*, *PLT*, and *COM*, as shown in Figure 13.7 and Table 13.3.

The following points should be stressed with regard to the remainder of this chapter. When the number of tile-parts of a particular tile t is $N_{TP,t} = 1$, "tile-stream" and "tile-part 0" are synonymous for that tile. When we refer to the tile header, we are implicitly referring to the header of tile-part 0. When we refer to a tile-part header, we are implicitly referring to the header of a tile-part other than tile-part 0. Unless explicitly stated otherwise, any statement made regarding a tile-part applies to *any* tile-part, *including* tile-part 0.

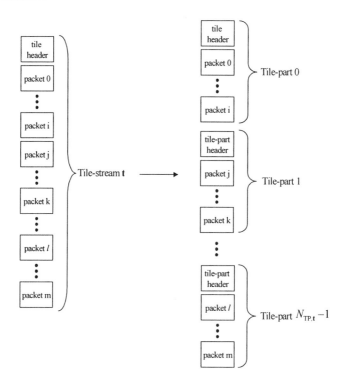

Figure 13.6.   A tile-stream broken into multiple tile-parts.

SOT
POC
PPT
PLT
COM
SOD

Figure 13.7.   JPEG2000 tile-part header. Shaded marker segments are optional and can appear in any order.

From Section 13.1.1, we know that progression is defined at the level of a tile. Tile-parts are the mechanism by which the concept of progression is extended to the level of the entire image. This is accomplished by the interleaving of tile-parts from different tiles. As an example, consider forming the tile-stream for each tile using LRCP progression. Each tile-stream might then be broken into tile-parts with each tile-part containing all packets from a single layer. Specifically, for each tile **t**, tile-part 0 might contain all layer 0 packets, tile-part 1 might contain all layer 1 packets, and so on. A quality progressive code-stream for the entire image could then be formed by including tile-part 0 for each tile, followed by tile-part 1 for each tile, and so on.

Tile-parts can be useful even when tiling is not employed; i.e., the image is comprised of a single tile. Tile-parts can be used to effect progression order changes in mid-code-stream, via inclusion of *POC* marker segments in their respective tile-part headers. Tile-parts are also useful for collecting length information, and/or packet headers into *PLT* and *PPT* marker segments, respectively. Finally, the code-stream of a very large un-tiled image may require multiple tile-parts due to certain length constraints, such as the maximum tile-part length of $2^{32} - 1$.

13.2.4   PACKET HEADERS

As detailed in Section 12.5.2, a packet consists of a packet header followed by a packet body. For the purpose of error resilient decoding, the code-stream syntax allows the optional inclusion of an *SOP* (start of packet) marker segment before some or all packets. Also, an *EPH* (end of packet header) marker can be included after each packet header.

In addition to their usefulness for error resilient decoding, *SOP* marker segments and *EPH* markers can be useful for locating packets and/or packet headers. From Section 12.1, we know that an FF_h byte can never be followed by a byte having a value greater than $8F_h$ within compressed pack-stream data. Thus, *SOP* = $FF91_h$ and *EPH* = $FF92_h$ can be located within the code-stream, when present. Care must be exercised when searching for such markers however, because values within main, tile, and tile-part headers can take on any value, *including* values between $FF90_h$ and $FFFF_h$.

13.3   MARKERS AND MARKER SEGMENTS

In this section, we describe each marker and marker segment in detail. As described previously, a marker segment consists of a marker followed by a parameter list. Only *SOC, EOC, SOD,* and *EPH* appear as isolated markers. All other markers appear only as part of marker segments. A listing of all markers, grouped by function and location within the code-stream, is given in Table 13.4. The following subsections describe each marker/marker segment in the order shown in this table.

Due to the progressive nature of JPEG2000 code-streams, it may be tempting to truncate a code-stream by simply discarding packets from the end. A robust decoder would do well to anticipate such behavior.

Table 13.4.   Code-stream markers employed by JPEG2000. Markers that appear in isolation (not as part of a marker segment) are denoted by *. Markers having values in excess of $FF8F_h$ are denoted by †. Locations where a marker can occur are denoted by M: main header, T: tile header, P: tile-part header, S: in pack-stream, E: end of code-stream.

Mnemonic	Value	Marker Name	Location	Required
Delimiting Markers				
SOC*	$FF4F_h$	Start of code-stream	M	Y
SOT†	$FF90_h$	Start of tile	T,P	Y
SOD†*	$FF93_h$	Start of data	T,P	Y
EOC†*	$FFD9_h$	End of code-stream	E	Y
Principle Markers				
SIZ	$FF51_h$	Image and tile size	M	Y
COD	$FF52_h$	Coding style default	M,T	Y
COC	$FF53_h$	Coding style component	M,T	N
QCD	$FF5C_h$	Quantization default	M,T	Y
QCC	$FF5D_h$	Quantization component	M,T	N
RGN	$FF5E_h$	Region of interest	M,T	N
POC	$FF5F_h$	Progression order change	M,T,P	N
Pointer Markers				
TLM	$FF55_h$	Tile-part lengths: main header	M	N
PLM	$FF57_h$	Packet lengths: main header	M	N
PLT	$FF58_h$	Packet lengths: tile-part	T,P	N
PPM	$FF60_h$	Packed packet headers: main header	M	N
PPT	$FF61_h$	Packed packet headers: tile-part	T,P	N
In Pack-Stream Markers				
SOP†	$FF91_h$	Start of packet	S	N
EPH†*	$FF92_h$	End of packet header	S	N
Informative Markers				
CRG	$FF63_h$	Component registration	M	N
COM	$FF64_h$	Comment	M,T,P	N

However, a valid JPEG2000 code-stream must contain correct parameter information in all marker segments. Thus, care should be taken to update all marker segments during code-stream editing operations.

13.3.1   START OF CODE-STREAM (SOC)

The *SOC* marker appears as an isolated marker, rather than as part of a marker segment. The *SOC* marker appears at the beginning of the

SOT	L_{SOT}	i_T	L_TP	i_TP	$N_{\mathrm{TP},\mathbf{t}}$

Figure 13.8.   The SOT marker segment.

main header, and comprises the first two bytes in a JPEG2000 code-stream.

13.3.2   START OF TILE (SOT)

Every tile header or tile-part header begins with an SOT marker segment. The SOT marker segment is shown graphically in Figure 13.8. This figure shows the order and size of all SOT marker segment parameters. The size of each parameter (number of bytes the parameter occupies in the code-stream) is indicated by the width of the rectangle in which it appears. The narrowest rectangles indicate one byte parameters, while the medium width rectangles indicate two byte parameters. Finally, parameters that occupy four bytes in the code-stream are indicated by the widest rectangles in the figure.

Consistent with previous discussions, the SOT marker itself comprises two bytes. Similarly, the L_{SOT} parameter is a two byte, unsigned, big endian integer that specifies the length of the SOT parameter list, in bytes. This length includes the two bytes for L_{SOT}, but not the two bytes for the SOT marker itself. From the figure, it should be clear that $L_{SOT} = 10$.

TILE INDEX

The i_T parameter is a two byte unsigned integer that specifies the *tile index*, or *tile number*, for the tile to which the SOT marker segment belongs. This tile index reflects a numbering of tiles on the canvas in raster order from left-to-right and top-to-bottom, starting from $i_\mathrm{T} = 0$. From the discussion of Section 13.1.1, it should be clear that the tile index for tile $\mathbf{t} = [t_1, t_2]$ is given by

$$i_\mathrm{T} = N_1^\mathrm{T} \cdot t_1 + t_2$$

and that

$$0 \leq i_\mathrm{T} < N_1^\mathrm{T} \cdot N_2^\mathrm{T}$$

The maximum allowable number of tiles is $N_1^\mathrm{T} \cdot N_2^\mathrm{T} = 65535$ and thus, i_T will not exceed 65534.

NUMBER OF TILE-PARTS

$N_{TP,t}$ is a one byte unsigned integer taking values from 0 to 255, inclusive. A non-zero value signals the number of tile-parts present in the code-stream for the tile **t** to which the *SOT* marker segment belongs. A value of 0 indicates that the number of tile-parts is not specified in this tile-part. $N_{TP,t}$ is allowed to be zero in any or all tile-parts of a given tile. However, all non-zero values for tile-parts of the same tile must hold exactly the same value – the total number of tile-parts for the tile. Encoders would do well to include a non-zero $N_{TP,t}$ field in at least one tile-part of each tile, to facilitate memory efficient decompression.

The number of tile-parts need not be consistent across tiles. Furthermore, it is allowable to completely omit the tile-stream of any given tile[2].

TILE-PART INDEX

The i_{TP} parameter is a one byte unsigned integer denoting the *tile-part index* or *tile-part number* for the tile-part to which this *SOT* marker segment belongs. As shown in Figure 13.6, tile-part 0 ($i_{TP} = 0$) is the only tile-part with a tile header. All subsequent tile-parts ($i_{TP} > 0$) have tile-part headers. If $N_{TP,t} > 0$, then $0 \leq i_{TP} < N_{TP,t}$. In any event, $i_{TP} \leq 254$.

When multiple tiles and tile-parts are present in the same code-stream, the tile-parts of different tiles may be interleaved. The only requirement is that tile-parts of a given tile must appear in order, possibly separated by tile-parts from other tiles. Specifically, tile-part i_{TP} from tile **t** must appear in the code-stream before tile-part $i_{TP} + 1$ from tile **t**.

TILE-PART LENGTH

The L_{TP} parameter is a four byte unsigned integer that specifies the length of the tile-part to which the *SOT* marker segment belongs. We emphasize that this length is *all-inclusive*. That is, the length includes all bytes in the tile-part from the beginning of the tile header (or tile-part header), to the last byte of the last packet in the tile part, if any.[3] Specifically, it includes the two bytes of the *SOT* marker itself. L_{TP} can range from 14 to $2^{32} - 1$, inclusive. For the last tile-part of a code-

---

[2] This fact is not entirely clear in the standard document itself.

[3] As noted above, a tile-part with no packets is perfectly legal. Such a tile-part would consist exclusively of a tile header (tile-part 0) or tile-part header (subsequent tile-parts). The length of such a tile-part is simply the *total* length of the header itself. The shortest possible tile-part then consists of only a *SOT* marker segment followed by a *SOD* marker for a total of $L_{TP} = 14$ bytes.

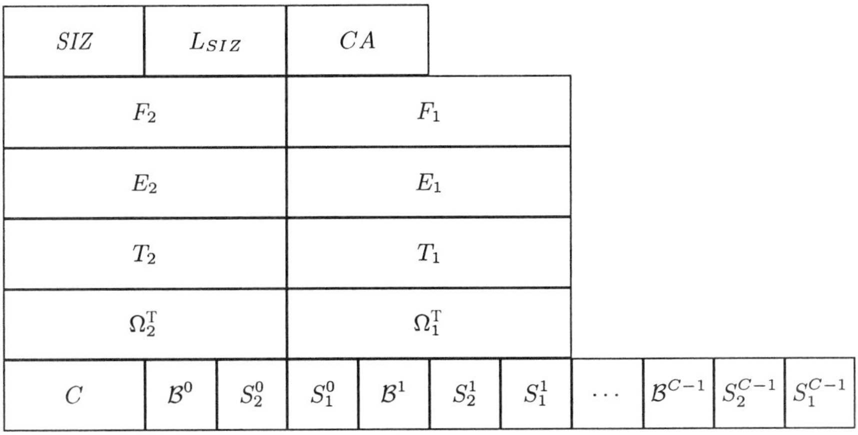

SIZ	L_{SIZ}	CA								
F_2			F_1							
E_2			E_1							
T_2			T_1							
Ω_2^{T}			Ω_1^{T}							
C	$\mathcal{B}^0$	S_2^0	S_1^0	$\mathcal{B}^1$	S_2^1	S_1^1	$\cdots$	$\mathcal{B}^{C-1}$	S_2^{C-1}	S_1^{C-1}

Figure 13.9.   The *SIZ* marker segment.

stream, $L_{\mathrm{TP}} = 0$ is also acceptable, meaning that the length of the tile-part is unspecified. This tile-part must then contain all data up to, but not including, the *EOC* marker. This exception is allowed to enable streaming compression of un-tiled imagery, in which case the compressor may begin to output the code-stream before compression is complete.

13.3.3   START OF DATA (SOD)

Every tile header or tile-part header ends with the *SOD* marker.

13.3.4   END OF CODE-STREAM (EOC)

The last two bytes in a JPEG2000 code-stream comprise the end of code-stream marker, *EOC*.

13.3.5   IMAGE AND TILE SIZE (SIZ)

The *SIZ* marker segment is required to be the first marker segment in the main header. It immediately follows the *SOC* marker. Only one *SIZ* marker segment may appear in a JPEG2000 code-stream. The *SIZ* marker segment is shown graphically in Figure 13.9. As was the case for the *SOT* marker segment, the figure shows the order and size of all parameters in a graphical fashion.

The order in which these marker segment parameters appear in the code-stream is from left-to-right, top-to-bottom in the figure. The only reason the marker segment is shown in multiple rows, is for economy of space in the text. Ideally, the marker segment would be shown as one contiguous sequence of bytes, from left-to-right.

As discussed previously, the *SIZ* marker itself occupies two bytes. Also, L_{SIZ} is a two byte unsigned integer denoting the length of the marker segment parameter list, which includes the two bytes for L_{SIZ}, but not the two bytes for the *SIZ* marker.

CAPABILITIES

In Figure 13.9, *CA* is a two byte parameter which specifies extended code-stream capabilities for JPEG2000 Part 2, and beyond. In JPEG-2000 Part 1, its value is restricted to be $CA = 0000_h$.

IMAGE SIZE AND TILING PARAMETERS

The next eight parameters are each four byte unsigned integers. As discussed in Chapter 11, $[E_1, E_2]$ and $[F_1 - 1, F_2 - 1]$ are the upper-left, and lower-right coordinates of the image region in the canvas coordinate system. Similarly, the tile size and tile partition anchor point are given by $T_1 \times T_2$ and $[\Omega_1^T, \Omega_2^T]$, respectively. Legal values for these parameters satisfy

$$0 \leq E_i, \Omega_i^T < 2^{32} - 1$$
$$1 \leq F_i, T_i < 2^{32}$$

COMPONENTS

Also from Chapter 11, are the number of components C, and the vertical and horizontal sub-sampling factors for each component S_1^c, S_2^c, $c = 0, 1, \ldots, C-1$. As indicated in Figure 13.9, C is a two byte quantity. More specifically, C is an unsigned integer satisfying $1 \leq C \leq 16384$. Similarly, the S_i^c are unsigned one byte integers, satisfying $1 \leq S_i^c \leq 255$, $i = 1, 2$.

Finally, the one byte parameters $\mathcal{B}^c$, $c = 0, 1, \ldots, C - 1$ specify the bit-depth of the image components. The most significant bit of $\mathcal{B}^c$ specifies whether component c is signed or unsigned, with 0 indicating "unsigned" and 1 indicating "signed." The remaining 7 bits are treated as an unsigned integer specifying the value of $B - 1$ as defined in Section 10.1. Specifically, unsigned components have samples satisfying $0 \leq x[\mathbf{n}] < 2^B$, while signed components have samples satisfying $-2^{B-1} \leq x[\mathbf{n}] < 2^{B-1}$. For example, $\mathcal{B}^c = 07_h$ denotes an unsigned component with samples ranging from 0 to 255, while $\mathcal{B}^c = 87_h$ denotes a signed component with samples ranging from -128 to 127. Legal values of B range from 1 to 38, inclusive; i.e., $0 \leq B - 1 \leq 37$.

COD		L_{COD}		CS	$\mathcal{O}_\mathrm{P}$	Λ_t		MC
$D_{t,c}$	E_2^{CB}	E_1^{CB}	MS	WT	E_P^0	E_P^1	$\cdots$	$E_\mathrm{P}^{D_{t,c}}$

Figure 13.10. The *COD* marker segment.

MARKER SEGMENT LENGTH

We conclude this section by noting that for a single component image ($C = 1$), the parameter list will contain $L_{SIZ} = 41$ bytes. For an image with 16384 components, $L_{SIZ} = 49190$. More generally, $L_{COD} = 38 + 3C$.

13.3.6 CODING STYLE DEFAULT (COD)

The *COD* marker segment is required in the main header. It can appear anywhere within that header after the *SIZ* marker segment. The *COD* marker segment can also appear in a tile header. No more than one *COD* marker segment can appear in any given header. The *COD* marker segment is part of a hierarchy which includes the *COC* marker segment.

As mentioned previously, the *COD* marker segment provides default coding style parameters. When it appears in the main header, the *COD* marker segment provides such parameters for all components of all tiles. These parameters can be overridden *for all tiles* of a single component using a main header *COC* marker segment. When appearing in a tile header, a *COD* marker segment overrides all main header *COD* and *COC* marker segments for all components of its respective tile. Finally a *COC* marker segment in a tile header overrides all other *COD* or *COC* marker segments for the relevant component of its respective tile. This hierarchy can be denoted by

main *COD* < main *COC* < tile *COD* < tile *COC*

where < denotes "is overridden by."

The *COD* marker segment is shown graphically in Figure 13.10. Each parameter in the *COD* parameter list is discussed below.

CODING STYLE

The *CS* parameter indicates whether *SOP* marker segments and/or *EPH* markers may be employed within the scope of the *COD* marker segment. It also specifies the coding anchor point $\left[\Omega_1^{C,t}, \Omega_2^{C,t}\right]$, and indicates whether maximal, or user specified precinct sizes are employed.

As shown by Figure 13.10, CS is a one byte parameter. We label its eight bits as $CS_7, CS_6, \ldots, CS_0$ from most significant to least significant. $CS_0 = 0$ then specifies that all precincts are of maximal size. Specifically, $P_1^{t,c,r} = P_2^{t,c,r} = 2^{15}$ for all $r = 0, 1, \ldots, D_{t,c}$. This applies for all $\mathbf{t}, c$ within the scope of the COD marker segment. Specifically the tiles and components within the scope of the marker segment are determined by the COD/COC and main/tile header hierarchy as discussed above. $CS_0 = 1$ indicates that precinct sizes are defined by the E_P^r parameters appearing later in the COD marker segment, as discussed below.

CS_1 identifies the possible usage of SOP marker segments within the scope of the COD marker segment. $CS_1 = 0$ indicates that no SOP marker segments are present. $CS_1 = 1$ indicates that SOP marker segments *may* be present. This is discussed further in Section 13.3.17.

$CS_2 = 0$ indicates that no EPH markers are present within the scope of the COD marker segment. $CS_2 = 1$ indicates that EPH markers immediately follow *every* packet header within the scope of the COD marker segment. This is discussed further in Section 13.3.18.

CS_3 and CS_4 specify the coding anchor point $\left[\Omega_1^{C,t}, \Omega_2^{C,t}\right]$. Specifically $\Omega_1^{C,t} = CS_4$ and $\Omega_2^{C,t} = CS_3$ for all tiles within the scope of the COD marker segment. At the time of this writing, JPEG2000 Part 1 restricts this anchor point to be $\left[\Omega_1^{C,t}, \Omega_2^{C,t}\right] = [0,0]$. An amendment to JPEG2000 Part 1 has been proposed to allow any anchor point

$$\left[\Omega_1^{C,t}, \Omega_2^{C,t}\right] \in \{[0,0], [0,1], [1,0], [1,1]\}$$

It is unclear whether or not this amendment will be approved.

The remaining bits in CS must be set to 0 in JPEG2000 Part 1. Specifically, $CS_5 = CS_6 = CS_7 = 0$.

PROGRESSION ORDER

The $\mathcal{O}_P$ parameter of the COD marker segment is a one byte unsigned integer. $\mathcal{O}_P$ specifies which of the five progression orders is employed within the scope of the COD marker segment. Legal values of $\mathcal{O}_P$ range from 0 to 4 inclusive. The progression order indicated by each such value is given by:

0: Layer-Resolution-Component-Position (LRCP)

1: Resolution-Layer-Component-Position (RLCP)

2: Resolution-Position-Component-Layer (RPCL)

3: Position-Component-Resolution-Layer (PCRL)

4: Component-Position-Resolution-Layer (CPRL)

In each case, the full index ranges of equation (13.1) apply.

LAYERS

The Λ_t parameter of the *COD* marker segment is a two byte unsigned integer. Λ_t can range from 1 to 65535 inclusive, and indicates the number of quality layers for all tiles within the scope of the *COD* marker segment. Every precinct within this scope must contribute exactly Λ_t packets to the code-stream

MULTI-COMPONENT TRANSFORM

The *MC* parameter is a one byte unsigned integer. Legal values are 0 and 1. $MC = 0$ indicates that no color transform is employed within the scope of the *COD* marker segment. $MC = 1$ indicates that one of the color transforms from Section 10.2 is employed on components $c = 0, 1$, and 2. As discussed in that section, when $MC = 1$, all three of these components must have identical bit-depths and sub-sampling factors.

The particular transform employed depends on the wavelet transform used. If the irreversible wavelet transform is indicated in the *WT* parameter (described below), then the irreversible color transform (ICT) is employed. Else, if the reversible wavelet transform is indicated, then the reversible color transform (RCT) is employed.

TRANSFORM LEVELS

The $D_{t,c}$ parameter specifies the number of transform levels employed for each tile-component within the scope of the *COD* marker segment. $D_{t,c}$ is an unsigned byte with legal values from 0 to 32 inclusive. A value of zero indicates that no wavelet transform is performed.

CODE-BLOCK SIZES

The E_2^{CB} and E_1^{CB} parameters are unsigned bytes, and are used to specify the nominal size $\left(J_1^{t,c} \times J_2^{t,c} \right)$ of all code-blocks within the scope of the *COD* marker segment. More specifically, these parameters are code-block size exponents. The code-block sizes themselves are computed by

$$J_1^{t,c} = 4 \cdot 2^{E_1^{\mathrm{CB}}} \qquad J_2^{t,c} = 4 \cdot 2^{E_2^{\mathrm{CB}}}$$

The constraints given in equation (11.12) indicate that legal values for E_1^{CB} and E_2^{CB} must satisfy

$$0 \le E_i^{\mathrm{CB}} \le 8 \quad \text{and} \quad 0 \le E_1^{\mathrm{CB}} + E_2^{\mathrm{CB}} \le 8$$

MODE SWITCHES

The *MS* parameter is a one byte quantity that specifies which mode variations are employed in the bit-plane coding process. This parameter applies to all code-blocks within the scope of the *COD* marker segment. Each mode switch discussed in Section 12.4 can be identified with a bit in the *MS* parameter. Denoting the bits of *MS* by $MS_7, MS_6, \ldots, MS_0$ from most to least significant, the modes are associated with the MS_i as follows:

$MS_0 = BYPASS$ mode switch

$MS_1 = RESET$ mode switch

$MS_2 = RESTART$ mode switch

$MS_3 = CAUSAL$ mode switch

$MS_4 = ERTERM$ mode switch

$MS_5 = SEGMARK$ mode switch

In each case, $MS_i = 0$ indicates that the option is *not* employed, while $MS_i = 1$ indicates that the option *is* employed. MS_6 and MS_7 are unused in JPEG2000 Part 1 and must be set to zero.

WAVELET TRANSFORM

The *WT* parameter indicates which wavelet transform is employed within the scope of the *COD* marker segment. *WT* is a one byte unsigned integer, with legal values of 0 and 1. $WT = 0$ denotes the 9/7 irreversible transform, while $WT = 1$ denotes the 5/3 reversible transform. These transforms are discussed at length in Section 10.4.

As a final note, we point out that one of these transforms must always be selected. If no transform is desired for a particular encoding function, then zero levels of transform ($D_{t,c} = 0$) can be selected.

PRECINCT SIZES

From the discussion of *Coding Style* flags above, if $CS_0 = 0$ then all precincts within the scope of the *COD* marker segment are of size $2^{15} \times 2^{15}$. In this case, no E_P^r parameters are present in the *COD* marker segment. If $CS_0 = 1$, then the $E_P^r, r = 0, 1, \ldots, D_{t,c}$ are each single byte parameters, used to specify precinct sizes. Accordingly, the *COD* parameter list length is given by

$$L_{COD} = 12 + CS_0 \cdot (D_{t,c} + 1)$$

which satisfies $12 \leq L_{COD} \leq 45$.

As discussed above, each tile-component within the scope of the *COD* marker segment has $D_{t,c}$ transform levels. Equivalently, each tile-component has $D_{t,c} + 1$ resolutions, $r = 0, 1, \ldots, D_{t,c}$. The E_P^r parameters specify the exponents of the precinct sizes for each of these resolutions. Denoting the most significant four bits of E_P^r by $E_{P,1}^r$, and the least

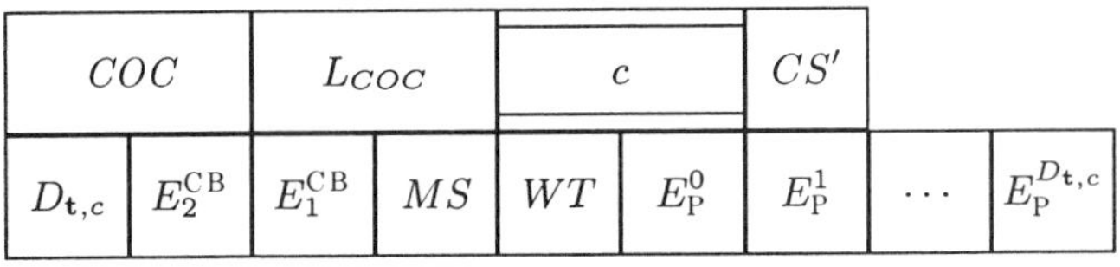

COC	L_{COC}		c	CS'			
$D_{\mathbf{t},c}$ $\;$ E_2^{CB}	E_1^{CB}	MS	WT	E_{P}^0	E_{P}^1	$\cdots$	$E_{\mathrm{P}}^{D_{\mathbf{t},c}}$

Figure 13.11. The COC marker segment.

significant four bits by $E_{\mathrm{P},2}^r$, the precincts of resolution r have heights and widths computed by

$$P_1^{\mathbf{t},c,r} = 2^{E_{\mathrm{P},1}^r} \qquad P_2^{\mathbf{t},c,r} = 2^{E_{\mathrm{P},2}^r}$$

For example, if $E_{\mathrm{P}}^r = 8A_{\mathrm{h}}$, then the precincts of resolution r are of size 256×1024 for all tile-components within the scope of the COD marker segment.

Legal values for the $E_{\mathrm{P},i}^r$ $\;i = 1,2$ range from 0 to 15, inclusive. However, $E_{\mathrm{P},i}^r = 0$ implies a precinct height or width of 1, and is only allowed for resolution $r = 0$. All other resolutions $r = 1,2,\ldots,D_{\mathbf{t},c}$ must have $E_{\mathrm{P},i}^r \geq 1$, $\;i = 1,2$.

13.3.7 CODING STYLE COMPONENT (COC)

The COC marker segment is shown in Figure 13.11, and bears a striking resemblance to the COD marker segment of Figure 13.10. In fact, the COC marker segment is identical to the COD marker segment, with two exceptions. First, certain parameters appear in the COD marker segment, but are absent from the COC marker segment. These parameters correspond to features and elements that JPEG2000 does not allow to vary on a component by component basis. Second, the COC marker segment contains a parameter, c, that specifies to which component it applies.

Recall that exactly one COD marker segment is required in the main header, and no more than one COD marker segment may appear in any tile header. On the other hand, COC marker segments are optional in every case, and multiple COC marker segments can appear in the main header and/or any given tile header, up to and including one per component. Each COC marker segment applies to only one component within its scope.

As discussed previously, a main header COC marker segment overrides the main header COD marker segment for all tiles of a specific component. A tile header COD marker segment overrides all main header COD and COC marker segments for all components within its tile. Finally, a tile header COC marker segment overrides all other COD/COC

marker segments for one specific tile-component. As before, this is denoted by

$$\text{main } COD < \text{main } COC < \text{tile } COD < \text{tile } COC$$

COMPONENT NUMBER

Each COC marker segment contains a parameter that specifies a component number to which the COC marker segment applies. If the number of components C, as specified in the SIZ marker segment, satisfies $C > 256$, then c is a two byte unsigned integer representing a component number $c \in \{0, 1, \ldots, 16383\}$. On the other hand, if $C \leq 256$, then c is a one byte unsigned integer representing a component number $c \in \{0, 1, \ldots, 255\}$. The fact that c may occupy either one or two bytes within the code-stream is denoted in Figure 13.11 by the "extra" horizontal lines above and below c in its graphical "parameter rectangle."

CODING STYLE

The CS' parameter is a stripped down version of the CS parameter as discussed for the COD marker segment. Precinct sizes are the only parameters from CS that JPEG2000 allows to vary component by component. Thus, CS' is a single byte taking values of 0 and 1 only. This is equivalent to $CS'_0 \in \{0, 1\}$. $CS'_0 = 0$ indicates maximal precincts, while $CS'_0 = 1$ indicates that precinct sizes are defined by the E^r_P parameters, appearing later in the COC marker segment.

REMAINING PARAMETERS

All other parameters in the COC marker segment, from $D_{t,c}$ to $E_P^{D_{t,c}}$ have already been described with reference to the COD marker segment. Their usage and restrictions are identical to those described in that context.

We close our discussion of the COC maker segment by noting that the length of its parameter list is

$$L_{COC} = 8 + CS'_0 \cdot (D_{t,c} + 1) + L_c$$

which satisfies $9 \leq L_{COC} \leq 43$. In this expression,

$$L_c = \begin{cases} 1 & C \leq 256 \\ 2 & C > 256 \end{cases} \tag{13.2}$$

is the number of bytes used to represent the parameter c.

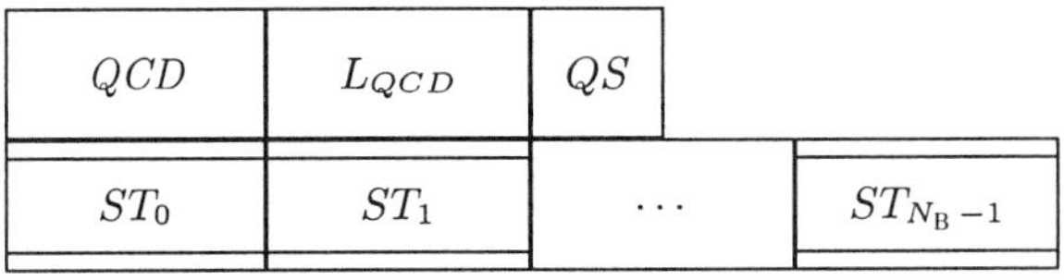

QCD	L_{QCD}	QS

ST_0	ST_1	$\cdots$	$ST_{N_B - 1}$

Figure 13.12. The *QCD* marker segment.

13.3.8 QUANTIZATION DEFAULT (QCD)

The *QCD* marker segment is required in the main header. It can appear anywhere within this header, after the *SIZ* marker segment. The *QCD* marker segment can also appear in a tile header. No more than one *QCD* marker segment can appear in any given header. The *QCD* marker segment is part of a hierarchy of marker segments including the *QCC* marker segment.

The scoping rules for *QCD/QCC* marker segments are identical to those of *COD/COC* marker segments. Specifically, the main header *QCD* marker segment applies to all tile-components of the entire image, while a tile header *QCD* marker segment applies only to tile-components within its respective tile. A main header *QCC* marker segment applies to all tile-components of a single component across the entire image, while a tile header *QCC* marker segment applies to only a single tile-component of its respective tile.

A main header *QCC* marker segment overrides the main header *QCD* marker segment for one particular component, while a tile header *QCD* marker segment overrides both types of main header marker segments for all tile-components within its tile. Finally, a tile header *QCC* marker segment overrides all three other types of marker segments for its respective tile-component. This hierarchy is denoted by:

$$\text{main } QCD < \text{main } QCC < \text{tile } QCD < \text{tile } QCC$$

The *QCD* marker segment is shown in Figure 13.12. This marker segment provides quantization parameters for all subbands within its scope. As discussed in Section 10.5, there are three parameters related to quantization in JPEG2000. First is the number of guard bits G used to protect against nominal range violations in the subband samples. Second is ε_b, which is interpreted as the exponent of the quantization step size when irreversible transforms are employed. The interpretation of ε_b is as a "ranging" parameter when the reversible wavelet transform is employed. Finally, μ_b is the quantization step size mantissa, and is only relevant for irreversible transforms.

Also discussed in Section 10.5 are two polices for signalling step sizes in the irreversible case. One policy is to explicitly signal a step size for each subband. The other policy is to signal a step size only for the lowest frequency subband, LL_D and then derive step sizes for all other subbands according to equation (10.26).

QUANTIZATION STYLE

The QS parameter occupies one byte, and indicates which style is used to signal quantization step sizes. It also indicates the number of guard bits, G. The two least significant bits QS_1QS_0, have legal values of 00, 01, and 10. Taking $Q = QS_1QS_0$ as an unsigned two bit integer, legal values are then $Q = 0, 1$, and 2. These values specify signalling of step size information as:

0: ε_b signalled for each subband (reversible transform only)

1: (ε_b, μ_b) signalled for LL_D (irreversible transform only)

2: (ε_b, μ_b) signalled for each subband (irreversible transform only)

It is worth noting that reversible or irreversible transformation can be specified on a tile-component by tile-component basis via COD/COC marker segments. Accordingly, the compressor must take care to provide QCD/QCC marker segments in an appropriate manner.

The three most significant bits of the quantization style parameter, $QS_7QS_6QS_5$, taken as an unsigned integer, specify the number of guard bits, $G \in \{0, 1, \ldots, 7\}$. The remaining bits are restricted to be zero. Specifically, $QS_4 = QS_3 = QS_2 = 0$.

STEP SIZES

The ST_n parameters are used to signal step size information. As was the case for the c parameter in the COC marker segment, the horizontal lines in the ST_n rectangles of Figure 13.12 indicate that these parameters are of variable length.

When $Q = 1$, only ST_0 is present. This parameter comprises two bytes which signal a single step size pair (ε_b, μ_b). The most significant five bits of ST_0 represent the value of ε_b. The least significant eleven bits of ST_0 represent the value of μ_b. As an example, $ST_0 = 2C13_h$ has binary representation

$$ST_0 = 0010110000010011$$

The first five bits yield $\varepsilon_b = 5$, while the remaining eleven bits yield $\mu_b = 1043$.

When $Q = 2$, there are multiple ST_n parameters. Each such parameter comprises two bytes and signals an (ε_b, μ_b) pair, in the fashion discussed in the previous paragraph. When $Q = 0$, there are also multiple ST_n parameters, however each is only one byte in length. Each such ST_n signals an ε_b. Specifically, the most significant five bits (bit 7 through bit 3) of ST_n are taken as an unsigned integer to specify an $\varepsilon_b \in \{0, 1, \ldots, 31\}$.

In the case of either $Q = 2$ or $Q = 0$, the number of ST_n parameters (denoted N_B) must be at least as large as the largest number of subbands within any tile-component in the scope of the QCD marker segment. Specifically, let $D^{\max}$ be the maximum of $D^{t,c}$ over all t, c within this scope. Then, the number of ST_n parameters must satisfy

$$N_B \geq 3 \cdot D^{\max} + 1 \qquad (13.3)$$

For each tile-component within the scope of the QCD marker segment, the ST_n parameters, $n = 0, 1, \ldots$ are applied, in order, to the subbands $LL_{D_{t,c}}$, $HL_{D_{t,c}-1}$, $LH_{D_{t,c}-1}$, $HH_{D_{t,c}-1}, \ldots$, HL_1, LH_1, HH_1. That is, ST_0 applies to $LL_{D_{t,c}}$, ST_1 applies to $HL_{D_{t,c}-1}$, and so on. The condition given in equation (13.3) guarantees that there are enough step sizes to accommodate the subbands of each tile-component. If $N_B > 3 \cdot D^{\max} + 1$, then there are actually more ST_n than required for any tile-component within scope. If this occurs, the extra parameters are to be ignored by a decoder. For this reason, it may be possible to neglect updating QCD marker segments during code-stream editing operations.

Defining

$$\mathcal{I} = \begin{cases} 0, & Q = 0 \quad \text{(reversible transform)} \\ 1, & \text{else} \quad \text{(irreversible transform)} \end{cases}$$

the QCD marker segment length and N_B are then related by

$$L_{QCD} = 3 + (1 + \mathcal{I}) \, N_B$$

As discussed above, when $Q = 1$, (derived step sizes for irreversible transforms), we have $N_B = 1$. Otherwise, N_B must satisfy equation (13.3). N_B is signalled implicitly via L_{QCD}. Legal values of L_{QCD} range from 4 to 197, inclusive.

13.3.9 QUANTIZATION COMPONENT (QCC)

The QCC marker segment is shown in Figure 13.13. The QCC marker segment is identical to the QCD marker segment, except for the insertion of the c parameter to identify the component affected by the QCC marker segment. As was the case for the COC marker segment, if the number of components satisfies $C \leq 256$, then c occupies a single byte. On the other hand, if $C > 256$, then c occupies two bytes. All discussion in the previous subsection regarding the QCD marker segment is

QCC	L_{QCC}	c	QS
ST_0	ST_1	$\cdots$	ST_{N_B-1}

Figure 13.13.   The QCC marker segment.

directly relevant to the QCC marker segment. Of course, the expression for L_{QCC} is slightly different, and is given by

$$L_{QCC} = 3 + (1 + \mathcal{I})\, N_B + L_c$$

where L_c is the number of bytes used to represent c, as given by equation (13.2).

13.3.10   REGION OF INTEREST (RGN)

The RGN marker segment is used to signal the "upshift" value U for the max-shift method of region of interest coding as described in Section 10.6. Perhaps a better mnemonic for this marker might be RGC since it follows scoping rules similar to those of the COC and QCC marker segments, as described previously. There is no "default" version of the RGN marker segment (which might have been called RGD).

Multiple RGN marker segments can appear in the main header and/or any tile header. Each RGN marker segment applies to one specific component. A main header RGN marker segment applies to all tiles of that particular component, while a tile header RGN marker segment applies only to one specific tile-component. A tile header RGN marker segment overrides any main header RGN marker segment for its particular tile and component. That is

main RGN < tile RGN

Any tile-component not falling within the scope of any RGN marker segment is assigned an upshift value of $U = 0$.

The RGN marker segment is shown in Figure 13.14. As discussed for COC and QCC marker segments, if the number of components satisfies $C \leq 256$, c is a one byte unsigned value between 0 and 255, inclusive. If $C > 256$, then c is a two byte unsigned parameter between 0 and 16383, inclusive. In either case, c specifies the component number to which this RGN marker segment applies. In JPEG2000 Part 1, RS is one byte and is restricted to be 0. Finally, U is an unsigned one byte integer specifying the region of interest upshift parameter from Section 10.6. Legal values of U range from 0 to 255 inclusive.

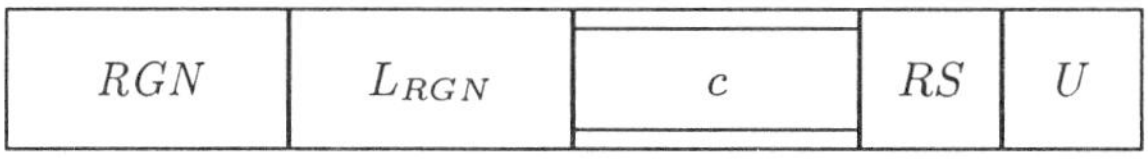

RGN	L_{RGN}	c	RS	U

Figure 13.14. The *RGN* marker segment.

The length of the *RGN* parameter list is given by

$$L_{RGN} = 4 + L_c$$

where L_c is the number of bytes used to represent c, as given by equation (13.2).

13.3.11   PROGRESSION ORDER CHANGE (POC)

The *POC* marker segment provides for changing the progression order mid-code-stream. In the absence of any *POC* marker segments, the loop counters that define packet sequencing for each tile run over their full ranges for a single progression order as described in Section 13.1.1. This single progression order is specified for each tile in a main or tile header *COD* marker segment.

If a *POC* marker segment appears in the main header, it overrides the progression specified in the main header *COD* marker segment. It *also* overrides every progression specified in any tile header *COD* marker segment. A tile header *POC* marker segment overrides the progression specified in the main header *COD* marker segment, any tile header *COD* marker segment for its respective tile, as well as any main header *POC* marker segment that may be present. This override hierarchy is denoted by

main *COD* < tile *COD* < main *POC* < tile *POC*

POC is one of the few marker segments that can occur in a tile-part header. No more than one *POC* marker segment may appear in any header, but multiple progressions can be specified in a single *POC* marker segment. If one or more tile-part header *POC* marker segments are present, there must be a *POC* marker segment in the corresponding tile header. Tile-part *POC* marker segments *do not* override the tile header *POC* marker segment. Progressions found in tile-part headers are treated as if they were actually included, in order, at the end of the list of progressions in the tile header. Not every tile-part must contain a *POC* marker segment, even when tile-part *POC* marker segments are present. But, any tile-part *POC* marker segment must appear in the code-stream before any packet it affects.

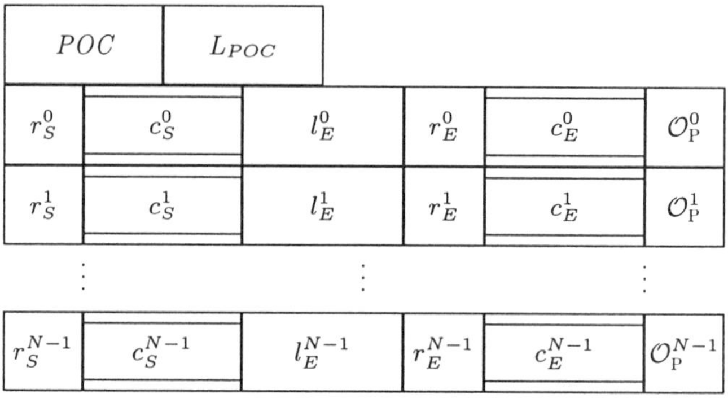

POC	L_{POC}				
r_S^0	c_S^0	l_E^0	r_E^0	c_E^0	$\mathcal{O}_P^0$
r_S^1	c_S^1	l_E^1	r_E^1	c_E^1	$\mathcal{O}_P^1$
$\vdots$		$\vdots$			$\vdots$
r_S^{N-1}	c_S^{N-1}	l_E^{N-1}	r_E^{N-1}	c_E^{N-1}	$\mathcal{O}_P^{N-1}$

Figure 13.15.   The *POC* marker segment.

The set of progressions within a given scope must be *inclusive* of all packets of that same scope. Specifically, every packet $\mathcal{G}_{t,c,r,p,l}$ for the entire image must eventually be sequenced by the collection of progressions in a main header *POC* marker segment. Similarly, every packet for a given tile must eventually be sequenced by the collection of progressions from that tile's tile header and tile-part header *POC* marker segments.

The *POC* marker segment is shown in Figure 13.15. The form of this marker segment is identical regardless of whether it appears in the main header, a tile header, or a tile-part header. As in all marker segments, the length parameter directly follows the *POC* marker itself. The length of the *POC* parameter list is given by

$$L_{POC} = 2 + (5 + 2L_c)\,N$$

where N is the number of progressions specified in the marker segment, and L_c is the number of bytes used to specify a component index, as given in equation (13.2). The number of progressions in a *POC* marker segment is signalled implicitly by the length L_{POC}.

STARTING AND ENDING INDICES

Each row, $n = 0, 1, \ldots, N-1$, in the figure specifies one progression. Each such progression consists of one progression order, and the ranges of packet indices over which it applies. Specifically, $\mathcal{O}_P^n$ is a one byte unsigned integer from 0 to 4, identifying one of the five progression orders described in Section 13.1.1.

The one byte unsigned integers r_S^n and r_E^n specify starting (inclusive) and ending (exclusive) resolutions to be used with progression order $\mathcal{O}_P^n$. Specifically, the "resolution loops" in the progression definitions of

Section 13.1.1 become

$$\text{for } r = r_S^n, r_S^n + 1, \ldots, r_E^n - 1$$

rather than the nominal version

$$\text{for } r = 0, 1, \ldots, D_{t,\max}$$

given in that section. These parameters must satisfy $0 \leq r_S^n < r_E^n \leq 33$.

Similarly, c_S^n and c_E^n specify starting and ending components for use with progression $\mathcal{O}_P^n$. The "component loops" in the progression definitions of Section 13.1.1 become

$$\text{for } c = c_S^n, c_S^n + 1, \ldots, c_E^n - 1$$

rather than the nominal version

$$\text{for } c = 0, 1, \ldots, C - 1$$

given there.

As for other marker segments, component numbers are represented using L_c bytes, where L_c is defined by equation (13.2). In the one byte case, $0 \leq c_S^n \leq 255$, while in the two byte case, $0 \leq c_S^n \leq 16383$. The treatment of c_E^n is slightly more complicated. In the one byte case, we must have either $1 \leq c_E^n \leq 255$, or $c_E^n = 0$. A value of $c_E^n = 0$ is to be interpreted as $c_E^n = 256$. The two byte case is similar, with either $1 \leq c_E^n \leq 16384$ or $c_E^n = 0$. Here, a value of $c_E^n = 0$ is interpreted as $c_E^n = 16384$. It is worth noting that in the two byte case, there are two ways to signal $c_E^n = 16384$.

The Final Draft International Standard of JPEG2000 Part 1 contains an error with respect to the treatment of c_E^n. This has been corrected in a subsequent defect report [4].

Finally, the two byte unsigned integer l_E^n is the ending layer index to be used with progression order $\mathcal{O}_P^n$. The starting index is always 0. Thus, the "layer loops" in the progression definitions of Section 13.1.1 become

$$\text{for } l = 0, 1, \ldots, l_E^n - 1$$

Legal values of l_E^n satisfy $1 \leq l_E^n \leq 65535$.

It is worth noting that there is no mechanism for specifying starting and ending precinct indices. These always run over the full range $0 \leq p_i < N_i^{P,t,c,r}$ $i = 1, 2$, as discussed in Section 13.1.1.

In what follows, we restrict our attention once more to a single tile. We emphasize that even when *POC* marker segments are present, progression orders are still implemented (i.e., packets are still sequenced) on a tile by tile basis to create tile-streams which can be divided into tile-parts. Of course, the progressions employed for a given tile are based on the *COD/POC* and main/tile hierarchy.

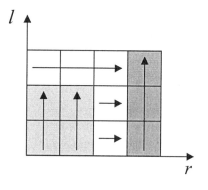

Figure 13.16.   Progression volumes for the *POC* of Figure 13.17.

PROGRESSION VOLUMES

For a specific tile **t**, we define a "progression volume" as the set of packets implied by a single progression (a row in Figure 13.15). Specifically,

$$\text{Vol}_{\mathbf{t}}^{n} = \left\{ \begin{array}{c} \mathcal{G}_{\mathbf{t},c,r,\mathbf{p},l} : c_{S}^{n} \leq c < c_{E}^{n}, r_{S}^{n} \leq r < r_{E}^{n}, \\ 0 \leq p_{i} < N_{i}^{\text{P},\mathbf{t},c,r}, 0 \leq l < l_{E}^{n} \end{array} \right\}$$

It is not required that the progression volumes $n = 0, 1, \ldots, N - 1$ be disjoint. That is, the same packet may appear in multiple progression volumes. However, even in this case, packets are not repeated in the tile-stream.

As an example, consider a tile having only one component. Assume this tile is compressed using 3 layers, 3 levels of wavelet transform, and one precinct per resolution. The resulting tile-stream will have $(D_{\mathbf{t},c} + 1)\,\Lambda_{\mathbf{t}} = 12$ packets. These packets are shown notionally as the rectangular regions in Figure 13.16. Since $c = p_1 = p_2 = 0$ for all packets of this example, the progression volume can be depicted in two dimensions, indexed by r and l, as shown in the figure.

Consider now, the *POC* marker segment of Figure 13.17. This *POC* marker segment contains three progressions. The first progression is ordered RLCP, with $0 \leq r < 2$ and $0 \leq l < 2$. The second progression is also ordered RLCP, but with $3 \leq r < 4$ and $0 \leq l < 3$. The third and final progression is ordered LRCP, with $0 \leq r < 4$ and $0 \leq l < 3$. Note that in every case $0 \leq c < 1$, consistent with the single component present in our example.

The progression volumes for the first two progressions are indicated by the shaded regions in Figure 13.16. It should be noted that these volumes are disjoint. On the other hand, the third progression vol-

POC		0017_h			
00_h	00_h	0002_h	02_h	01_h	01_h
03_h	00_h	0003_h	04_h	01_h	01_h
00_h	00_h	0003_h	04_h	01_h	00_h

Figure 13.17. Example of a POC marker segment with three different progressions.

ume encompasses all of the figure, with the first two volumes entirely contained in the third. As discussed previously, all packets must be contained in at least one progression volume. This demonstrates one method to guarantee this requirement is met.

Also as discussed previously, when the progression volumes are not disjoint, packets are not repeated. That is, packets are sequenced only in the first progression volume in which they appear. Thus, the twelve packets of this example are sequenced in the following order:

$$\mathcal{G}_{t,0,0,0,0}; \mathcal{G}_{t,0,0,0,1}; \mathcal{G}_{t,0,1,0,0}; \mathcal{G}_{t,0,1,0,1};$$
$$\mathcal{G}_{t,0,3,0,0}; \mathcal{G}_{t,0,3,0,1}; \mathcal{G}_{t,0,3,0,2}; \qquad (13.4)$$
$$\mathcal{G}_{t,0,2,0,0}; \mathcal{G}_{t,0,2,0,1}; \mathcal{G}_{t,0,0,0,2}; \mathcal{G}_{t,0,1,0,2}; \mathcal{G}_{t,0,2,0,2}$$

These progressions are indicated graphically by the arrows in Figure 13.16.

It is worth noting that the POC marker segment of our example could be divided into as many as three POC marker segments, one for each progression order. The first such marker segment must appear in the tile header. The second such marker segment could appear in any tile-part header (for this tile) that precedes the packet $\mathcal{G}_{t,0,3,0,0}$. Finally, the third marker segment must appear before $\mathcal{G}_{t,0,2,0,0}$. Conversely, POC marker segments from multiple tile-part headers can be deleted (moving their contents to the tile header POC marker segment), with no effect on the packet sequencing for that tile.

We close our discussion of POC marker segments with a few words about index bounds. It is desirable, and sometimes necessary, to allow a POC marker segment to specify index bounds outside the appropriate range for some (or all) tile-components within its scope. The necessity of this follows from the fact that there is no tile specific version of the POC marker segment, and that the number of resolutions and/or layers can differ tile-component by tile-component.

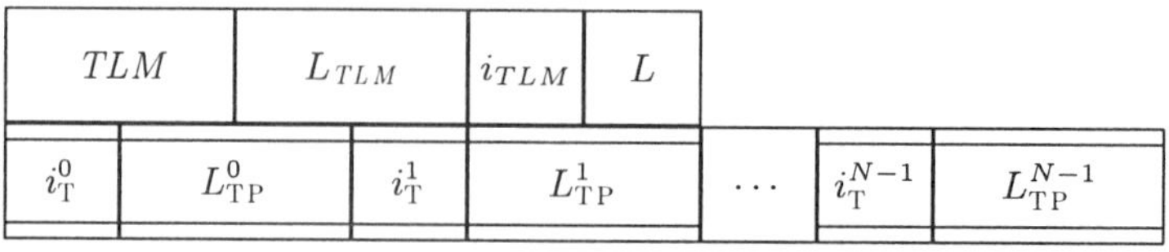

TLM	L_{TLM}	i_{TLM}	L

i_{T}^0	L_{TP}^0	i_{T}^1	L_{TP}^1	$\cdots$	i_{T}^{N-1}	L_{TP}^{N-1}

Figure 13.18.   The *TLM* marker segment.

Clearly, no packets should be sequenced for "out of bound" indices. As an example, changing the second progression of Figure 13.16 to range over $3 \le r < 9$ and $0 \le l < 5$ would not change the sequencing of packets in equation (13.4).

13.3.12   TILE-PART LENGTHS: MAIN HEADER (TLM)

TLM marker segments may be used to specify the tile-index and tile-part length of every compressed tile-part, in the order in which they appear within the code-stream. This information is useful for random access into the code-stream. For example, every tile-part of a given tile can be located and extracted from the code-stream using only information provided by *TLM* marker segments. Conceptually, this information is provided as a contiguous sequence of (tile-index, tile-part length) pairs. In practice, this sequence can be divided over multiple *TLM* marker segments. This division is always allowable, and sometimes may be necessary, since the full sequence may not fit within the maximum length of a single marker segment.[4]

TLM INDEX AND PARAMETER LENGTHS

As shown in Figure 13.18, each *TLM* marker segment has a one byte unsigned index i_{TLM} indicating its *TLM marker segment index*. These marker segment indices indicate the order in which the data from multiple *TLM* marker segments are concatenated to form the contiguous sequence of (tile-index, tile-part length) pairs. Legal values of i_{TLM} range from 0 to 255, inclusive. This i_{TLM} parameter is necessary since multiple *TLM* marker segments can appear in any order within the main header.

The i_{TLM} parameter is followed by the L parameter, which indicates the number of bytes used to represent each of the remaining parameters within the *TLM* marker segment. Labeling the bits of L as $L_7, L_6, \ldots, L_0$

---

[4]Marker segments are limited by their length parameter $L_{TLM} \le 65535$.

from most significant to least significant, L_5 and L_4 indicate the size of the i_T^n parameters, while L_6 indicates the size of the L_{TP}^n parameters. Specifically, $L_5 L_4$, taken together as an unsigned two bit integer, specify the number of bytes, L_{i_T}, used for each i_T^n parameter. Legal values for this quantity are $0 \leq L_{i_T} \leq 2$. $L_6 = 0$ indicates that each L_{TP}^n parameter is of length $L_{L_{TP}} = 2$ bytes, while $L_6 = 1$ indicates that four bytes are used for each L_{TP}^n parameter; i.e., $L_{L_{TP}} = 4$. The L parameters need not be consistent across multiple *TLM* marker segments. The remaining bits of L are all required to be zero; i.e., $L_7 = L_3 = L_2 = L_1 = L_0 = 0$.

The length of a *TLM* marker segment is then given by

$$L_{TLM} = 4 + (L_{i_T} + L_{L_{TP}}) \cdot N$$

TILE INDICES AND TILE-PART LENGTHS

For a given *TLM* marker segment, N is the number of (i_T^n, L_{TP}^n) pairs, and can be determined by a decoder from L_{TLM}. As discussed above, when multiple *TLM* marker segments are present, their contents should be concatenated in order of i_{TLM}, to get the complete sequence of (i_T^n, L_{TP}^n) pairs. The total number of such pairs is then the sum of the N over all *TLM* marker segments. In what follows, we assume that this concatenation has been done, and the total number of pairs is given by N'. Clearly, the discussion below would be unchanged if there were only a single *TLM* marker segment containing N' pairs. Note that the total number of tile-parts present in the code-stream must be N' in either case.

The (i_T^n, L_{TP}^n) pairs provide information about tile-parts in their order of appearance within the code-stream. i_T^n is the *tile index* or *tile number* of the nth such tile-part, while L_{TP}^n is its length, in bytes. This length includes all markers and marker segments within the tile-part. This tile index and length must agree with the corresponding values that appear in its *SOT* marker segment.

Legal ranges of the i_T^n and L_{TP}^n parameters are dependent upon the L parameter in the *TLM* marker segment from which they came. If L specifies that 0 bytes are used for i_T^n, then these parameters are not present. If 1 or 2 bytes are used for the i_T^n, then $0 \leq i_T^n \leq 254$ or $0 \leq i_T^n \leq 65534$, respectively. If L specifies two bytes for L_{TP}^n, then $14 \leq L_{TP}^n \leq 65535$. Similarly, for four bytes, $14 \leq L_{TP}^n \leq (2^{32} - 1)$. We recall here from Section 13.3.2, that the minimum possible length of a tile-part is 14 bytes.

We conclude our discussion of the *TLM* marker segment by noting that 0 bytes for each i_T^n is only a legal choice when there is only one

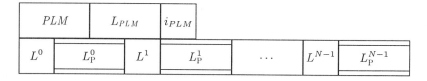

Figure 13.19.   The *PLM* marker segment.

tile-part per tile (tile-part 0). Furthermore, all tile-streams must appear *in order*, $i_\mathrm{T} = 0, 1, \ldots, N_1^\mathrm{T} \cdot N_2^\mathrm{T} - 1$, within the code-stream.

13.3.13   PACKET LENGTHS: MAIN HEADER (PLM)

The *PLM* marker segment may be used to record the lengths of all packets. These lengths must include *SOP* marker segments, if present. By providing packet length information within the main header, this marker segment can be useful for random access into the code-stream at a finer granularity than that provided by *TLM* marker segments. The price of this finer granularity is paid in higher overhead (larger compressed file size). As in the case of *TLM* marker segments, the information provided by *PLM* marker segments can be distributed over several marker segments which must be concatenated to retrieve the complete list of packet lengths. This concatenated list provides the length of every packet in the code-stream. The list is grouped by tile-part in the same order as they appear in the code-stream.

The *PLM* marker segment is shown in Figure 13.19. The i_{PLM} parameter is a one byte unsigned integer representing the index of the *PLM* marker segment relative to other *PLM* marker segments. These i_{PLM} parameters provide the order for concatenation of multiple *PLM* marker segments. Legal values are $0 \leq i_{PLM} \leq 255$, indicating that there can be up to 256 *PLM* marker segments in the main header. These *PLM* marker segments can occur in any order. In what follows, we assume that all *PLM* information has been concatenated to yield a list of (L^n, L_P^n) pairs $n = 0, 1, \ldots, N' - 1$, where N' is the sum of the N from all *PLM* marker segments.

PACKET LENGTHS

Each (L^n, L_P^n) pair provides length information for all packets within the nth tile-part, in order of appearance within the code-stream. The tile index i_T of this tile-part can be found in several ways. The easiest way is via *TLM* marker segments, if present.

OC_h	$9F_h$	62_h	$7C_h$

Figure 13.20. Example L_P^n parameter from a *PLM* marker segment. For this example, $L^n = 4$.

The L^n parameter is one unsigned byte satisfying $0 \leq L^n \leq 255$. L^n indicates the length of L_P^n, in bytes. As in previous marker segments, the variable length nature of L_P^n is indicated by the horizontal lines above and below. The L_P^n parameter itself is a sequence of bytes that provides packet lengths for all packets in the nth tile-part using a punctuated code. The MSB of each byte in L_P^n is either a comma (1), or a period (0). A period indicates that a packet length is complete, while a comma indicates that it is not. The seven LSBs of all bytes, up to and including a byte containing a period, are concatenated to form an integer representing the length of a packet.

As an example, consider the L_P^n parameter of length $L^n = 4$, as given in Figure 13.20. This L_P^n represents a tile-part with 3 packets having lengths 12, 4066, and 124 bytes, respectively. The first byte, OC_h has an MSB of 0 indicating that this is the last (in this case only) byte in a packet length. The seven LSBs then give the length of the first packet as

$$0001100 = 12$$

The next byte, $9F_h$, has an MSB of 1 indicating it is not the last byte of a packet length. However the following byte 62_h does terminate a length since its MSB is 0. The 7 least significant bits from each of these two bytes are concatenated to yield the length of the second packet as

$$\overbrace{0011111}\,\overbrace{1100010} = 4066$$

Finally, the last byte $7C_h$ has an MSB of 0, and yields the third packet length of

$$1111100 = 124$$

If a tile-part contains no packets, its $L^n = 0$ and no L_P^n is present. The length of a *PLM* marker segment must satisfy $4 \leq L_{PLM} \leq 65535$. The lower bound occurs when $N = 1$, and $L^0 = 0$.

For the purpose of breaking the full list of parameters into multiple *PLM* marker segments, the sequence may be broken after any whole number of packet lengths (possibly zero). For example, in Figure 13.20, a "seam" between concatenated *PLM* marker segments may occur after OC_h, 62_h, or $7C_h$. A seam may also occur before OC_h, but *not* after $9F_h$.

Due to this policy, the first parameter following i_{PLM} may not be an L^0 parameter when $i_{PLM} > 0$. Rather, it may be the continuation of the L_P^{N-1} parameter from the previous *PLM* marker segment, having index $i_{PLM} - 1$. In this case, the L^{N-1} from the previous *PLM* marker segment takes into account the bytes of such continuation. Figure 13.19 is not strictly applicable in this case.

LIMITATIONS

It is possible to construct a tile-part that exceeds the capabilities of the *PLM* marker segment to record its packet lengths. If one or more such tile-parts are present in the code-stream, *PLM* marker segments shall not be used. As an example of such a tile-part, consider any tile-part with more than 255 packets. Since each packet length requires at least one byte in L_P^n, the value of L^n would exceed 255 in this example. This is not possible since L^n occupies only a single byte in the code-stream.

As a final note, we comment on the interaction between *PLM*, *PLT*, *PPM*, and *PPT* marker segments. As explained in the following subsection, the *PLT* marker segment is a tile specific version of the *PLM* marker segment. *PLM* and *PLT* marker segments may appear in the same code-stream. *PPM* and *PPT* marker segments are discussed in subsequent subsections. Briefly, when either *PPM* or *PPT* marker segments are present, some or all packet headers are relocated from their respective pack-streams to main, tile, or tile-part headers. For such packets, the lengths specified in the corresponding L_P^n parameters do not include packet headers. Since the packet headers of such packets are not present in the pack-stream, only the lengths of the packet bodies are reported in the L_P^n.

13.3.14   PACKET LENGTHS: TILE-PART (PLT)

The *PLT* marker segment is a tile specific version of the *PLM* marker segment. Both *PLM* and *PLT* marker segments are optional and can both appear in the same code-stream. The packet length information for a given tile can be distributed across multiple *PLT* marker segments. These multiple *PLT* marker segments can be distributed among multiple tile-parts. Specifically the *PLT* marker segments of a tile can be distributed between the tile header and all tile-part headers of the respective tile-stream. When a *PLT* marker segment occurs in a tile-part header, it has the same effect as if it appeared in the tile header. Each *PLT* marker segment must appear in the code-stream prior to any packet for which it contains length information.

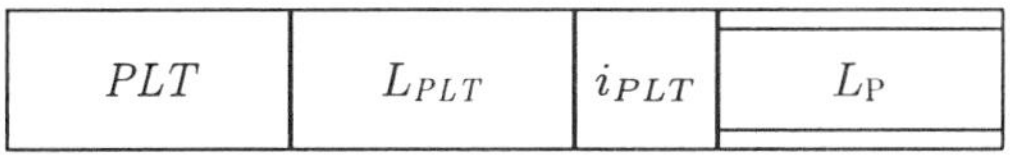

PLT	L_{PLT}	i_{PLT}	L_P

Figure 13.21. The *PLT* marker segment.

Allowing *PLT* marker segments to be distributed among tile-parts allows for a certain amount of interactive code-stream construction in client-server applications. It also allows for more efficient progressive transmission by providing packet length information on a more "as needed" basis. This is in contrast to the *PLM* marker segment where all packet length information must occur "up front," in the main header.

The *PLT* marker segment is shown in Figure 13.21. The i_{PLT} is a one byte parameter satisfying $0 \leq i_{PLT} \leq 255$ representing the index for the current *PLT* marker segment. This index specifies the order in which L_P parameters from multiple *PLT* marker segments are concatenated. The L_P parameter contains a sequence of packet lengths specified using the punctuated code described for the *PLM* marker segment. The complete concatenated list of lengths specifies the lengths of all packets in the tile-stream of the relevant tile. "Seams" in the concatenated L_P parameters may occur at the end of any "complete" packet length; i.e., after any byte having an MSB of 0.

The lengths reported in the *PLT* marker segment must be consistent with those of the *PLM* marker segment, if present. If any packet header is relocated to a *PPM* or *PPT* marker segment, its length as reported in a *PLT* marker segment must be that of its packet body. As for the *PLM* marker segment, the length of a *PLT* marker segment satisfies $4 \leq L_{PLT} \leq 65535$.

13.3.15   PACKED PACKET HEADERS: MAIN HEADER (PPM)

The *PPM* marker segment is optional and can be used to relocate packet headers from their respective pack-streams to the main header. This can be useful for certain types of random access, and for the design of error resilient communication formats. Correct decoding of a given packet header is critical to decoding not only its associated packet, but also to all subsequent packets within the same precinct. In certain scenarios, it may be beneficial to collect packet headers into the main header where they may be easier to protect. In terms of progression, the cost of this feature can be significant, since all packet headers appear in the code-stream before any compressed bit-stream data.

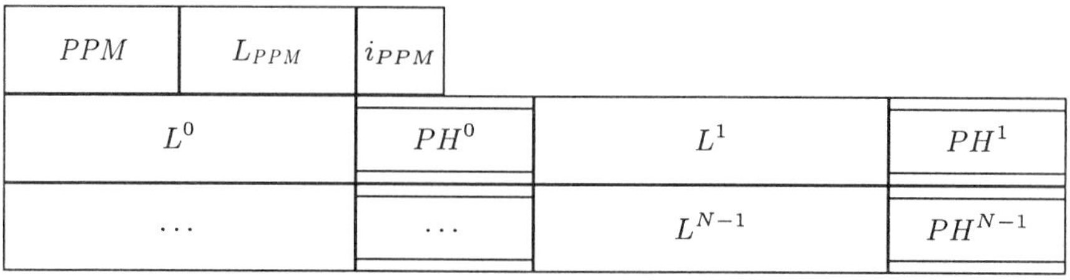

PPM	L_{PPM}	i_{PPM}		
L^0		PH^0	L^1	PH^1
$\ldots$		$\ldots$	L^{N-1}	PH^{N-1}

Figure 13.22.   The *PPM* marker segment.

If *PPM* marker segments are employed, the headers of all packets are relocated to the main header. No packet headers remain in any compressed pack-stream data, and *PPT* marker segments are not allowed. The packet headers can be distributed across multiple *PPM* marker segments. The concatenation of parameters from all such *PPM* marker segments contains the packet headers, grouped by tile-part, in order of their appearance within the code-stream.

PACKET HEADERS

The *PPM* marker segment is shown in Figure 13.22. The i_{PPM} parameter is a one byte index satisfying $0 \leq i_{PPM} \leq 255$. This index specifies the order of concatenation of multiple *PPM* marker segments. After concatenation, the complete list of (L^n, PH^n) pairs has N' elements, where N' is the sum of the N from all *PPM* marker segments. The (L^n, PH^n) pair contains packet headers for all packets in the nth tile-part, in order of appearance within the code-stream. The length of PH^n is given by the four byte parameter L^n, which satisfies $0 \leq L^n \leq 2^{32} - 1$. The PH^n parameter is a concatenation of all packet headers of tile-part n.

The number of headers present in PH^n, and their individual lengths, can be determined from *EPH* markers, if present. We note however, that the packet headers can be decoded sequentially without such prior knowledge.

The concatenation seam between multiple *PPM* marker segments can occur immediately before or after an L^n or at any packet header boundary. Thus, a *PPM* marker segment may contain any whole number (possibly zero) of complete packet headers. This allows for the continuation of PH^{N-1} data from the previous *PPM* marker segments to follow i_{PPM}, whenever $i_{PPM} > 0$. In this case, L^{N-1} from the previous *PPM* marker segment takes into account these continued bytes, and Figure 13.22 is not strictly applicable.

The length of a *PPM* marker segment satisfies $4 \leq L_{PPM} \leq 65535$. The lower bound of $L_{PPM} = 4$ can occur when $i_{PPM} > 1$ and the *PPM* marker segment contains only a single packet header, of length one byte, continued from the previous *PPM* marker segment.

The Final Draft International Standard of JPEG2000 Part 1 indicates that $7 \leq L_{PPM} \leq 65535$. At the time of this writing, this discrepancy has been noted by the JPEG committee, but has not been captured in a defect report. Encoders would do well to avoid generating *PPM* marker segments of length less than 7, while decoders would do well to expect lengths as small as 4.

As discussed in Section 13.3.18, each packet header may be followed by an *EPH* marker. When packet headers are relocated to *PPM* or *PPT* marker segments, their corresponding *EPH* markers, if any, are relocated as well. Since the presence of *EPH* can be signalled tile-by-tile in *COD*, it may not be possible for a decoder to know a priori which PH^n, if any, contain *EPH* markers. On the other hand, these markers are easily detectable since *EPH* exceeds FF8F$_h$ and cannot appear within compressed packet header data.[5]

13.3.16   PACKED PACKET HEADERS: TILE-PART (PPT)

The *PPT* marker segment is a tile specific version of the *PPM* marker segment. *PPT* marker segments are not permitted when *PPM* marker segments are employed. In the absence of *PPM* marker segments, *PPT* marker segments may be used or not used on a tile by tile basis. In any tile where *PPT* marker segments are present, all packet headers for that tile are relocated to *PPT* marker segments, and no packet headers are left within their associated pack-stream data.

The *PPT* marker segment is shown in Figure 13.23. The *PPT* marker segment index satisfies $0 \leq i_{PPT} \leq 255$, and specifies the order in which *PH* parameters from multiple *PPT* marker segments are to be concatenated. The concatenated *PH* parameters comprise the concatenation of all packet headers from the relevant tile, in order, as their corresponding packets appear in the tile-stream. The multiple *PPT* marker segments of a particular tile can be distributed among its tile and tile-part headers in arbitrary fashion. The only restriction is that each *PPT* marker segment must occur prior to any packets for which it contains headers.

Any whole number of packet headers can occur in a *PPT* marker segment. The Final Draft International Standard of JPEG2000 Part 1 specifies $4 \leq L_{PPT} \leq 65535$. As discussed in Section 13.3.2, it is possible

---

[5]In fact, compressed packet header data (as opposed to packet body data) cannot contain two consecutive bytes in the range FF80$_h$ to FFFF$_h$.

PPT	L_{PPT}	i_{PPT}	PH

Figure 13.23.   The *PPT* marker segment.

for a tile to contain no packets. In this case, $L_{PPT} = 3$ would occur. However, such a *PPT* marker segment serves no purpose and encoders should avoid generating such marker segments. On the other hand, a robust decoder would do well to handle such *PPT* marker segments appropriately.

As in the case of *PPM* marker segments, when *EPH* markers are present, they are relocated to *PPT* marker segments together with their respective packet headers. In this case, it is possible for a decoder to know a priori if *EPH* markers are present by checking both main and tile header *COD* marker segments.

13.3.17   START OF PACKET (SOP)

The *SOP* marker segment is optional and may appear immediately before each packet when indicated in a *COD* marker segment. When indicated in a main header *COD* marker segment, an *SOP* marker segment may appear before each packet in the code-stream. When indicated in a tile header *COD* marker segment, an *SOP* marker segment may appear before each packet in the relevant tile-stream.

The *SOP* marker segment may be useful for code-stream parsing and/or error resilient decoding. As discussed previously, the *SOP* marker code exceeds $\mathtt{FF8F_h}$, the maximum possible value of any two consecutive bytes of compressed pack-stream data. *SOP* markers can thus be used to detect errors arising while parsing packet headers. Upon detection of such errors, the presence of subsequent *SOP* marker segments enables resynchronization at a well-defined packet boundary, from which processing can resume.

The *SOP* marker segment is shown in Figure 13.24. i_{SOP} is a two byte counter satisfying $0 \leq i_{SOP} \leq 65535$. Each tile has its own i_{SOP} counter, even when *SOP* is signalled via the main header *COD* marker segment. The i_{SOP} counter for each tile is initialized to 0 at the beginning of the code-stream. After each packet is sequenced, the i_{SOP} counter for its tile is incremented, modulo 65536. That is, the result of incrementing 65535 is 0.

SOP marker segments are only allowed when indicated by a *COD* marker segment. Even when the relevant *COD* marker segment has its CS_1 flag set to 1, *SOP* marker segments are still optional on a packet

SOP	L_{SOP}	i_{SOP}

Figure 13.24.   The *SOP* marker segment.

by packet basis. However, if *SOP* marker segments are omitted for one or more packets, the corresponding i_{SOP} must still be incremented for each such packet.

Encoders generating error resilient code-streams would do well to not omit any *SOP* marker segments. Omission of *SOP* marker segments reduces opportunities for error detection and resynchronization. It also increases the buffering requirements of error resilient decoders. Of course, decoders are not required to make use of *SOP* marker segments in any case. The designer of a decoder may ignore *SOP* marker segments altogether, only attempt to exploit *SOP* marker segments when present for *every* packet, or anywhere in between these two extremes.

When *PPM* or *PPT* marker segments are employed, any *SOP* marker segments remain in the pack-stream with their corresponding packet bodies. When *PLM* or *PLT* marker segments are employed, the packet lengths contained therein must include the lengths of any *SOP* marker segments that may be present.

13.3.18   END OF PACKET HEADER (EPH)

End of packet header markers, *EPH*, appear as isolated markers. When indicated in a *COD* marker segment, every packet header within the scope of the *COD* marker segment must be immediately followed by an *EPH* marker. This is true for packet headers within *PPM* or *PPT* marker segments, as well as packet headers appearing in their compressed pack-streams. It is even true for packets whose precincts contain no code-block samples at all (see Section 11.3.3). *EPH* exceeds FF8F$_h$ and is "searchable" within compressed pack-stream data. Thus, *EPH* markers are useful for code-stream parsing and/or error resilient decoding.

We note that the Final Draft International Standard of JPEG2000 Part 1 listed *EPH* markers as optional, when allowed by a *COD* marker segment. This was corrected in [4] to require *EPH* markers after every packet header within the scope of the indicating *COD* marker segment.

CRG	L_{CRG}		
O_2^0	O_1^0	O_2^1	O_1^1
$\cdots$		O_2^{C-1}	O_1^{C-1}

Figure 13.25.   The CRG marker segment.

13.3.19   COMPONENT REGISTRATION (CRG)

The CRG marker segment is informative in nature. A single CRG marker segment may appear in the main header only. Component samples can always be decompressed correctly without the aid of this marker segment. However, this marker segment may be useful for rendering or display of decompressed image components.

The CRG marker segment is shown in Figure 13.25. The length of this marker segment is clearly

$$L_{CRG} = 2 + 4 \cdot C$$

where C is the number of components present in the image. O_1^c and O_2^c, $0 \le c < C$, indicate vertical and horizontal offsets that can be used for spatial registration of component c. Each offset is a two byte unsigned integer ranging between 0 and 65535, inclusive. These offsets are converted to canvas units as

$$\mathcal{O}_i^c = 2^{-16} \cdot S_i^c \cdot O_i^c \quad i = 1, 2$$

where S_1^c and S_2^c are the vertical and horizontal sub-sampling factors for component c. Clearly then, registration with accuracy finer than the grid spacing of the canvas is possible.

The offset pair $[\mathcal{O}_1^c, \mathcal{O}_2^c]$ should be applied to the coordinates of each sample of component c. As an example, consider a three component YCbCr image with 4:2:0 format, as defined in JFIF, H.261, and MPEG1. One choice for the purpose of compression/decompression is to set $E_1 = E_2 = 0$, $S_1^0 = S_2^0 = 2$ and $S_1^1 = S_2^1 = S_1^2 = S_2^2 = 4$ as shown in Figure 13.26a. In this figure, the locations of Y samples (component 0) are shown as circles, while the locations of Cb and Cr samples (components 1 and 2) are shown by squares. A CRG marker segment with $O_1^0 = O_2^0 = 0$ and $O_1^1 = O_2^1 = O_1^2 = O_2^2 = 16384$ specifies component registration (for the purpose of display) as shown in Figure 13.26b.

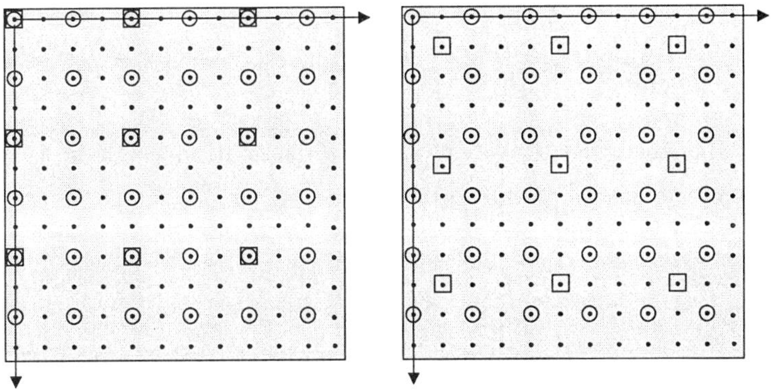

Figure 13.26. Example of the *CRG* marker segment: a) Component samples as registered on the canvas for the purposes of compression and decompression. b) Component registration as prefered for display according to the *CRG* marker segment. Shading indicates the image area of the canvas coordinate system.

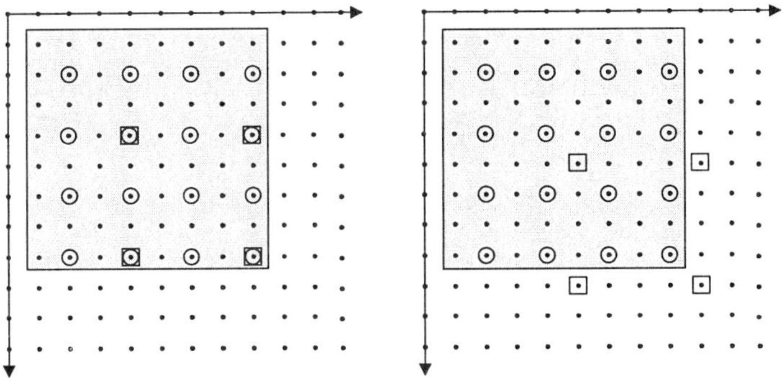

Figure 13.27. Second Example of the *CRG* marker segment: a) Component samples as registered on the canvas for the purposes of compression and decompression. b) Component registration as prefered for display according to the *CRG* marker segment. Shading indicates the image area of the canvas coordinate system.

Figure 13.27 shows another example with the same sub-sampling factors as above. In this case however, the upper left hand corner of the image area is set to $E_1 = E_2 = 1$. Additionally, the lower right hand corner is set to $F_1 = F_2 = 9$ (rather than 12, as in the previous figure). Figure 13.27a shows the location of component samples for the purpose of compression/decompression, while Figure 13.27b shows the component sample locations for the purpose of display.

COM	L_{COM}	TY	Comment Data

Figure 13.28.   The *COM* marker segment.

The undesirable behavior demonstrated by the second example can be avoid by judicious choice of canvas parameters. Specifically, let $\bar{S}_1$ be the LCM (least common multiple) of the S_1^c, $c = 0, 1, \ldots, C - 1$. Similarly, let $\bar{S}_2$ be the LCM of the S_2^c. The problem demonstrated in Figure 13.27 can then be avoided by choosing E_i and F_i to be integer multiples of $\bar{S}_i$, $i = 1, 2$, respectively.

13.3.20   COMMENT (COM)

The *COM* marker segment provides a facility for including unstructured comment information in the code-stream. The *COM* marker segment is shown in Figure 13.28. The *TY* parameter is a two byte unsigned integer. $TY = 1$ indicates that the Comment Data comprises a sequence of bytes in the form of IS 8859-15:1999 (Latin) character data [15]. $TY = 0$ indicates general binary Comment Data. No other values for *TY* are allowed in JPEG2000 Part 1. The *COM* marker segment length satisfies $5 \leq L_{COM} \leq 65535$.

Chapter 14

FILE FORMAT

As discussed in Chapter 13, all information necessary for decompressing all image component samples is specified by the JPEG2000 syntax. In this sense, a JPEG2000 code-stream is entirely self-contained. On the other hand, many applications may find certain other information useful. Such information might include color spaces, color palettes, capture resolution, display resolution, and copyright. All of these things may be included with a JPEG2000 code-stream via a JPEG2000 file format. Vendor specific information may also be included via UUIDs (Universal Unique Identifiers) and XML (eXtensible Markup Language).

As mentioned in Chapter 9, three JPEG2000 file formats are defined. These file formats may be viewed as wrappers for JPEG2000 code-streams. JPEG2000 Part 1 includes a minimal file format, while Part 2 includes compatible extensions to the Part 1 file format. Part 6 specifies a file format targeted toward document imaging. The Part 1 and Part 2 file formats are known as JP2 and JPX, respectively. The Part 6 file format is known as JPM. All three file formats are optional. Specifically, a JPEG2000 code-stream may stand alone, may be wrapped by one of the three JPEG2000 file formats, or may be contained in some other standard or proprietary file.[1]

This chapter deals exclusively with the Part 1 file format, JP2. The other file formats are beyond our scope. In file systems that employ file name extensions, '.jp2' should be used to denote JP2 files. For Macintosh files systems, JP2 files should have the type code 'jp2 '. JP2 file readers should not be case sensitive with regard to file extensions.

---

[1]Motion JPEG2000 (Part 3) is essentially a file format.

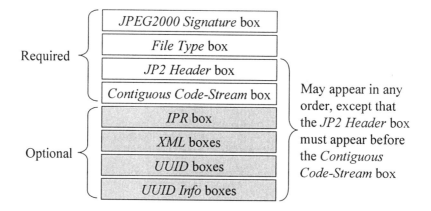

Figure 14.1.   JP2 file format structure. Shading indicates optional boxes.

14.1   FILE FORMAT ORGANIZATION

The JP2 file format is organized as a sequence of "boxes," as depicted in Figure 14.1. In this figure, each rectangle represents a box. As discussed in subsequent sections, each box has an identifier, a length, and contents. Boxes play a role in the file format similar to that of marker segments in the code-stream syntax.

Boxes appear consecutively in the file. There is no punctuation or any other type of data between boxes. As noted in the figure, there are four required boxes. The *JPEG2000 Signature* box is required as the first box in the JP2 file. The *File Type* box must follow immediately thereafter. The remaining two required boxes may appear anywhere after the *File Type* box, with the restriction that the *JP2 Header* box must appear prior to the *Contiguous Code-Stream* box. The *IPR, XML, UUID, and UUID Info* boxes are all optional and may appear in any order, anywhere after the *File Type* box. There may be multiple instances of the latter three boxes.

The *JPEG2000 Signature* box identifies the file as belonging to the JPEG2000 family of file formats. The *File Type* box identifies the file specifically as a JP2 file. The *JP2 Header* box contains information such as image size, bit-depth, resolution, and color space. The *Contiguous Code-Stream* box contains a single valid JPEG2000 code-stream. The *IPR* box contains Intellectual Property Rights information. *XML* boxes provide for the inclusion of additional structured information, while *UUID* and *UUID Info* boxes provide a mechanism for defining vendor specific extensions. Each of these boxes is discussed in more detail in subsequent sections.

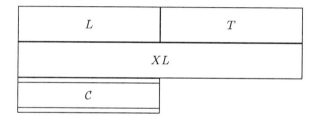

Figure 14.2.   Structure of a JP2 box.

14.1.1   THE STRUCTURE OF A BOX

Similar to the handling of marker segments in Chapter 13, we depict a box graphically as a sequence of rectangles. Each parameter in a box is contained in a field denoted by one of these rectangles. The width of a rectangle indicates the number of bytes occupied by its corresponding parameter within the file. A generic box is shown in Figure 14.2. In this figure, the *Box Type* is denoted by T. The box type serves a role in the file format similar to that of a marker in the syntax. Specifically, the box type indicates the function and format of the box. Unlike a marker however, the box type appears second within the box and occupies four bytes.

The box type is interpretable as a four character ISO 646 character string [14]. Throughout the chapter, we provide the text string surrounded in single quotes, as well as its hexadecimal equivalent. As an example, the box type for the *JP2 Header* box is 'jp2h' = $6A703268_h$. Specifically, the character codes for lower case j and p are $6A_h$ and 70_h, while the numeral 2 and lower case h are coded as 32_h and 68_h, respectively.

The length of a box is its first parameter, and is denoted by L in Figure 14.2. This length parameter is an unsigned, big endian integer occupying four bytes in the file. Legal values are $L = 0, 1$, or $8 \le L < 2^{32}$. A value of $L \ge 8$ indicates the total length of the box, including the eight bytes of L and T.

If $L = 1$, then the length of the box is given by the XL parameter, which is an 8 byte unsigned big endian integer. Legal values satisfy $16 \le XL < 2^{64}$, and include the sixteen bytes for L, T, and XL. If $L \ne 1$, the XL parameter is not present.

If $L = 0$, the length of the box was not known at the time the file was written. This is only allowable for the last box in the file. In this case, all remaining bytes, up to and including the last byte in the file, belong to this box.

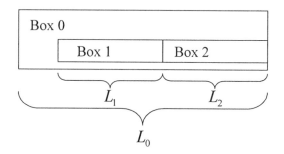

Figure 14.3. Boxes within a box. If $L_2 = 0$, then L_0 must also be 0. $L_1 = 0$ is not allowed.

The contents of the box, if any, are denoted by $\mathcal{C}$ in Figure 14.2. As in Chapter 13, the extra horizontal lines in the $\mathcal{C}$ rectangle indicate that this parameter is of variable length. The number of bytes in $\mathcal{C}$ can be determined from L and/or XL. The format of $\mathcal{C}$ varies depending on the box type. This will be described in more detail in the following sections.

Before moving on to those discussions, we note that one or more boxes may be included in the contents of another box. In this case, the "outer box" will be referred to as a *superbox*. All superboxes must follow the rules for construction of a box. In particular, the length of a superbox must convey its full length including any boxes that may be contained within. If a box has $L = 0$ then its superbox, if any, must also have $L = 0$. Furthermore, the "inner" box with $L = 0$ must appear last within the contents of its superbox. This situation is illustrated in Figure 14.3.

14.2   JP2 BOXES

This section provides detailed descriptions of each JP2 box depicted in Figure 14.1. These boxes are summarized in Table 14.1. In the table, two boxes are identified as superboxes. The boxes contained within these superboxes are described in the sections pertaining to the superboxes themselves.

14.2.1   THE JPEG2000 SIGNATURE BOX

As discussed above, the *JPEG2000 Signature* box must be the first box in a JP2 file. Only one *JPEG2000 Signature* box is allowed. This box identifies the file as belonging to the JPEG2000 family of file formats. The box type is $T = \text{'jP\ \ '} = \text{6A502020}_\text{h}$. The box length is $L = 12$, and the XL parameter is not present. The contents of the box comprise

Table 14.1. JP2 Boxes. Boxes that appear within the *JP2 Header* and *UUID Info* superboxes are introduced in subsequent sections.

Box Type	Hexadecimal Equivalent	Box Name	Required	Superbox
'jP '	$6A502020_h$	*JPEG2000 Signature*	Yes	No
'ftyp'	66747970_h	*File Type*	Yes	No
'jp2h'	$6A703268_h$	*JP2 Header*	Yes	Yes
'jp2c'	$6A703263_h$	*Contiguous Code-Stream*	Yes	No
'jp2i'	$6A703269_h$	*IPR*	No	No
'xml '	$786D6C20_h$	*XML*	No	No
'uuid'	75756964_h	*UUID*	No	No
'uinf'	$75696E66_h$	*UUID Info*	No	Yes

the four bytes $0D0A870A_h$. These bytes provide for detection of common file transmission errors. When interpreted as characters, $0D_h = <CR>$ and $0A_h = <LF>$. Thus, the common error of substituting $<CR><LF>$ for $<LF>$, or vice versa, is easily detected. Similarly, the third byte 87_h has its most significant bit set, so the common error of clearing the most significant bit during file transmission can also be detected.

14.2.2 THE FILE TYPE BOX

The *File Type* box must be the second box in a JP2 file. Only one *File Type* box shall be present in a JP2 file. Among the JPEG2000 family of file formats, the *File Type* box specifies the particular format to which a file belongs. The box type for this box is $T =$ 'ftyp' $= 66747970_h$.

The structure of the *File Type* box is shown in Figure 14.4. In the interest of space, the L, T, and XL parameters are omitted from the figure. Only the contents, C, are shown. This practice is followed throughout the remainder of this chapter.

As indicated by the figure, each parameter in the *File Type* box occupies four bytes. The Br parameter is the "Brand," and defines the specific file format employed. For JP2, $Br =$ 'jp2 ' $= 6A703220_h$. The MV parameter defines the Minor Version number for the brand. This parameter is a big endian unsigned integer. At the time of this writing, the only legal value is $MV = 0$. Conforming JP2 readers should attempt to interpret the file even when MV is not as expected.

The CL^i parameters specify a "Compatibility List" of standards to which the file conforms. In the case when $Br \neq$ 'jp2 ', but one of the CL^i parameters is given as 'jp2 ', the file is not a JP2 file, but is interpretable by JP2 readers in some fashion as intended by its creator.

Br	MV	
CL^0	$\cdots$	CL^{N-1}

Figure 14.4.   Contents of the *File Type* box.

JP2 readers are required to interpret such files. This, and similar issues are discussed further in Section 14.3.

Of course, a JP2 file must have $Br = $ 'jp2 '. Additionally, a JP2 file must have at least one CL^i parameter taking the value $CL^i = $ 'jp2 '. The number of CL^i parameters, N, can be determined from L and/or XL.

14.2.3   THE JP2 HEADER BOX

The *JP2 Header* box specifies information about the image. The box type of the *JP2 Header* box is $T = $ 'jp2h' $= $ 6A703268$_h$. As mentioned previously with respect to Figure 14.1, the *JP2 Header* box may appear anywhere after the *File Type* box. The only restriction is that it must appear before the *Contiguous Code-Stream* box. There must be one and only one *JP2 Header* box in a JP2 file.

The *JP2 Header* box is actually a superbox. It contains only other boxes. The structure of the *JP2 Header* box is shown in Figure 14.5. As in the previous section, we omit L, T, and XL, and show only the contents, C. As indicated by Figure 14.5, the *Image Header* box must appear first. All other boxes may appear in any order. Also, the *Image Header* box, the *Bits Per Component* box, and the *Color Specification* box are required. All other boxes are optional. Finally, we note that the *Resolution* box is a superbox. We discuss the boxes within the *Resolution* box in the subsection that describes the *Resolution* box itself.

The boxes contained within the *JP2 Header* box are summarized in Table 14.2, and are discussed in detail below.

THE IMAGE HEADER BOX

The *Image Header* box contains the height and width of the image, the number and depth of its components, as well as compression type, color space, and intellectual property information. The box type for the *Image Header* box is $T = $ 'ihdr' $= $ 69686472$_h$. The length of the *Image Header* box is $L = 22$ bytes and no XL parameter is present. The contents of the *Image Header* box are shown in Figure 14.6.

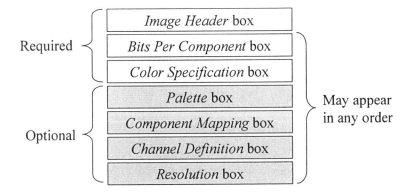

Figure 14.5.   Contents of the *JP2 Header* box.

Table 14.2. Boxes within the *JP2 Header* superbox. Boxes that appear within the *Resolution* superbox are introduced in subsequent sections.

Box Type	Hexadecimal Equivalent	Box Name	Required	Superbox
'ihdr'	69686472_h	*Image Header*	Yes	No
'bpcc'	62706363_h	*Bits Per Component*	Yes*	No
'colr'	$636F6C72_h$	*Color Specification*	Yes	No
'pclr'	$70636C72_h$	*Palette*	No	No
'cmap'	$636D6170_h$	*Component Mapping*	No	No
'cdef'	63646566_h	*Channel Definition*	No	No
'res '	72657320_h	*Resolution*	No	Yes

*If all components have the same bit depth, then the *Bits Per Component* box shall not be present.

The H and W parameters are unsigned four byte integers that specify the height and width of the image region on the canvas, as discussed in Chapter 11. Specifically, with $[E_1, E_2]$ and $[F_1 - 1, F_2 - 1]$ denoting the upper left and lower right corner of the image region, we then have $H = F_1 - E_1$ and $W = F_2 - E_2$, respectively.

The number of image components is given by C, which is a two byte unsigned integer. If all components have the same bit-depth, then the one byte parameter B specifies this common bit-depth and no *Bits Per Component* box shall be present. In this case, the interpretation of B is as described in Section 13.3.5. If the components do not all have the same bit-depth, then $B = FF_h$ and the *Bits Per Component* box must be present to specify the bit-depth component-by-component, as discussed in the next subsection.

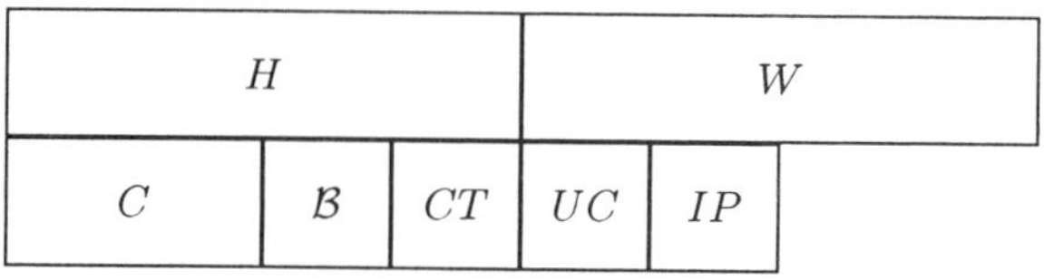

H			W	
C	$\mathcal{B}$	CT	UC	IP

Figure 14.6.   Contents of the *Image Header* box.

We note that the information provided by H, W, C, and $\mathcal{B}$ is redundant with information provided in the code-stream (see Section 13.3.5). Similarly, the bit-depths in the *Bits Per Component* box, when present, are also redundant with information provided in the code-stream. In a valid JP2 file, these redundant values must be consistent. However, a robust application may well attempt to decompress even in the case of inconsistencies. In such a situation, the values contained in the code-stream should be employed.

The CT (Compression Type) parameter is a one byte unsigned integer that specifies the algorithm used to compress the image. The only legal value in JP2 is $CT = 7$. The UC (Unknown Color space) parameter is also a one byte unsigned integer, and indicates whether the image color space is known or not. If the color space is known, and correctly specified in the *Color Specification* box (described below), then $UC = 0$. If the color space is unknown, then $UC = 1$. Other values for UC are not allowed in JP2. Even when $UC = 1$, a color space must be provided in the *Color Specification* box. The decompressed image components should be interpreted using this color space.

The IP parameter indicates whether or not Intellectual Property rights information is included in the file. This parameter is a one byte unsigned integer taking values of 0 or 1. $IP = 0$ indicates that no intellectual property rights information is present. Conversely, $IP = 1$ indicates that intellectual property rights information is present. Accordingly, an *IPR* box is present in the file if and only if $IP = 1$.

THE BITS PER COMPONENT BOX

As mentioned above, the *Bits Per Component* box specifies the bit-depth of each image component within the code-stream. The box type for the *Bits Per Component* box is $T =$ 'bpcc' $= 62706363_{\mathrm{h}}$. The contents of this box are shown in Figure 14.7. Each $\mathcal{B}^c$ parameter occupies one byte and specifies the bit-depth of component c, $c = 0, 1, \ldots, C - 1$. The interpretation of $\mathcal{B}^c$ is as described in Section 13.3.5. As mentioned previously, the values of $\mathcal{B}^c$ included in this box must be consistent

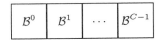

Figure 14.7.   Contents of the *Bits Per Component* box.

M	P	A	ECS	ICP

Figure 14.8.   Contents of the *Color Specification* box.

with those specified in the code-stream. If the $\mathcal{B}^c$ are identical for all $c = 0, 1, \ldots, C - 1$, this box shall not be present in the file.

THE COLOR SPECIFICATION BOX

The *Color Specification* box specifies the color space of the image. This color space applies to the "completely" decompressed image components. Specifically, the RCT or ICT (Section 10.2) are considered to be part of the compression/decompression process. Thus, the color space applies to the components obtained after the inverse RCT or ICT process, when present.

The *Color Specification* box has $T = $ 'colr' $= $ 636F6C72$_h$. The contents of this box are shown in Figure 14.8. The M parameter is a one byte unsigned integer taking values of 1 or 2, and specifies the Method by which the color space is signalled. A value of $M = 1$ indicates that the color space is signalled via the ECS parameter. In this case, the ICP parameter is not present. On the other hand, a value of $M = 2$ indicates that the color space is signalled via the ICP parameter. In this case, the ECS parameter is not present. JP2 readers must correctly interpret the color space regardless of which method is employed.

The P (precedence) and A (approximation accuracy) parameters are one byte parameters that should be set to zero in JP2. JP2 readers should ignore these parameters, whether zero or not.

Enumerated Color Spaces. The ECS (Enumerated Color Space) parameter is a four byte unsigned big endian integer used to select a color space. In JP2, only two enumerated color spaces are supported. At the time of this writing however, an amendment is in progress to add a third. The two approved color spaces are the sRGB color space, and a monochrome color space. A value of $ECS = 16$ indicates the sRGB space, while $ECS = 17$ indicates the monochrome space. The proposed third color space is a YCbCr color space. This color space is indicated by $ECS = 22$.

The sRGB color space is a non-linear, gamma corrected RGB space. Gamma correction is discussed in Section 1.1.2, while the sRGB color space itself is defined in [79]. A complete description of this color space is beyond the scope of this text. However, as mentioned in Section 1.1.2, the sRGB color space is related to a linear RGB space via

$$V_{\text{sRGB}} = \begin{cases} 12.92V_{\text{lin}}, & V_{\text{lin}} \leq 0.0031308 \\ 1.055V_{\text{lin}}^{(1/2.4)} - 0.055, & V_{\text{lin}} > 0.0031308 \end{cases} \tag{14.1}$$

where V_{lin} is an R, G, or B value in linear RGB space, and V_{sRGB} is the corresponding R, G, or B value in the sRGB color space. In equation (14.1), both V_{lin} and V_{sRGB} have a normalized range of 0.0 to 1.0.

Similarly, the monochrome color space is non-linear and related to a linear space via the sRGB non-linearity given above. Specifically equation (14.1) is applicable with V_{lin} representing a linear luminance, Y_{lin}, and V_{sRGB} representing a non-linear luminance, Y_{sRGB}.

Typically, the sRGB and monochrome color spaces are reasonable choices for legacy imagery when the exact color space is unknown. In such cases, the *UC* parameter in the *Image Header* box should be set to 1, as discussed previously.

The YCbCr color space is called sYCC and is related to the sRGB color space via the ICT, as defined in Section 10.2. When the sYCC color space is selected, the three color components in the code-stream must be unsigned. For a bit-depth of B (i.e., $\mathcal{B} = B - 1$), the Y values range from 0 to $2^B - 1$, inclusive. The Cb and Cr values range from -2^{B-1} to $2^{B-1} - 1$, but are represented with an offset of 2^{B-1} to ensure non-negativity.

The assumption in this case, is that the original image samples were acquired in the sYCC color space,[2] then compressed and decompressed via JPEG2000 without the use of the ICT or RCT. Thus, the three component sub-sampling factors (Section 11.1) need not be identical.

Restricted ICC Profiles.   When $M = 2$, the color space of the image is signalled via the *ICP* parameter of Figure 14.8. The *ICP* parameter must then contain a valid ICC Profile. ICC profiles were adopted by the International Color Consortium for the specification of color spaces. A thorough discussion of ICC profiles is beyond the scope of this text. A brief description of the profiles relevant to JP2 is included below. ICC profiles are fully defined in [78].

---

[2]This may be desirable in some cases, as the gamut of sYCC is larger that that of sRGB.

Only a restricted set of ICC profiles is supported in JP2. Specifically, only the *Monochrome Input* profiles and the *Three-Component Matrix-Based Input* profiles are supported. These profiles are defined in Clauses 6.3.1.1 and 6.3.1.2 of [78]. Briefly, the Three-Component Matrix-Based Input profiles provide necessary information to transform decompressed three-component sample data to a standardized color space. This color space is known as the Profile Connection Space (PCS). The PCS is a linear XYZ color space, known as PCS_{XYZ}. Specifically, PCS_{XYZ} represents color in terms of the tri-stimulus responses, X, Y, and Z, which should be expected from an ideal reflective print of the image, viewed under the CIE daylight illuminant "D50."

The three-component matrix-based profiles can be used with any color space which can be related to PCS_{XYZ} through point-wise non-linearities and a linear 3×3 matrix transform. Since the decompressed image samples will often belong to some type of RGB color space, we shall refer to them as R, G, and B for the purpose of this discussion. Note, however, that it is possible to describe color spaces whose channels are not properly interpreted as red, green, and blue components.

The point-wise non-linearities are known as Tone Reproduction Curves (TRC). There is one TRC for each of the three color channels and each TRC may be specified via a power law exponent or via a lookup table. Thus the *ICP* parameter of Figure 14.8 will contain a description of TRC_R, TRC_G, and TRC_B, as well as a 3×3 matrix, A. The processing required to map the decompressed image samples into PCS_{XYZ} is given by

$$R_{\text{lin}} = TRC_R(R)$$
$$G_{\text{lin}} = TRC_G(G)$$
$$B_{\text{lin}} = TRC_B(B)$$

followed by

$$\begin{bmatrix} X \\ Y \\ Z \end{bmatrix} = A \begin{bmatrix} R_{\text{lin}} \\ G_{\text{lin}} \\ B_{\text{lin}} \end{bmatrix}$$

The purpose of the PCS is to connect two profiles: an input profile, which describes the color properties of the decompressed sample values; and a display (or output) profile, which describes the color properties of the display or printing device. The connection is implemented, at least conceptually, by mapping the decompressed sample values to PCS_{XYZ} and then mapping the X, Y, and Z values to the display or rendering space.

Printer output spaces typically involve CMY or CMYK type representations, often involving complex multi-dimensional lookup tables and multiple TRC's. For monitor display, sRGB is a reasonable display profile, although one could hope to do better with a carefully calibrated monitor. A reasonable approximate transformation from PCS_{XYZ} to sRGB may be accomplished by applying

$$
\begin{bmatrix} R_{\mathrm{lin}} \\ G_{\mathrm{lin}} \\ B_{\mathrm{lin}} \end{bmatrix} = \begin{bmatrix} 3.1337 & -1.6173 & -0.4907 \\ -0.9785 & 1.9162 & 0.0334 \\ 0.0720 & -0.2290 & 1.4056 \end{bmatrix} \begin{bmatrix} X \\ Y \\ Z \end{bmatrix}
$$

followed by the point-wise non-linearity (gamma function) expressed in equation (14.1).

Conforming JP2 readers are expected to interpret any *Three-Component Matrix-Based Input* profile and supply the mapping to an appropriate rendering space through PCS_{XYZ}. For more information regarding ICC profiles, the specification [78] may be downloaded from the ICC Home Page at *http://www.color.org*.

The ICC Monochrome Input profiles are similar to those described above, except that monochrome profiles contain only a single TRC and no matrix. Application of the TRC brings the decompressed image component to the linear monochrome space, Y.

We conclude this discussion by noting that if any "private" ICC profile tags appear in the *ICP* parameter of Figure 14.8, they shall not alter the appearance of the resulting image. We also note that a decompressed R, G, or B sample may be outside the valid input range of its respective TRC. In this case, such sample values should be clipped prior to processing. Finally, we note that since JP2 restricts ICC profiles to use the XYZ connection space, the *Profile Connection Space* field in the ICC profile header must be 'XYZ ' $= 58595A20_{\mathrm{h}}$ (see [78], Clause 6.1: Header Description).

THE PALETTE BOX

The *Palette* box is used to signal a color LUT (Look Up Table). This LUT is used to convert a single decompressed component into multiple color components. These components are then interpreted according to the color space specified in the *Color Specification* box. If a *Palette* box is present, then a *Component Mapping* box must also be present. Essentially, a *Component Mapping* box provides for a renumbering of the components. This renumbering accommodates the increased number of components produced by the LUT, as compared to the number of components decompressed from the code-stream. The *Component Mapping* box is discussed further in the next section.

The box type of the *Palette* box is $T = $ 'pclr' $= 70636C72_{\mathrm{h}}$. The contents of the *Palette* box are shown in Figure 14.9. The *NE* parameter

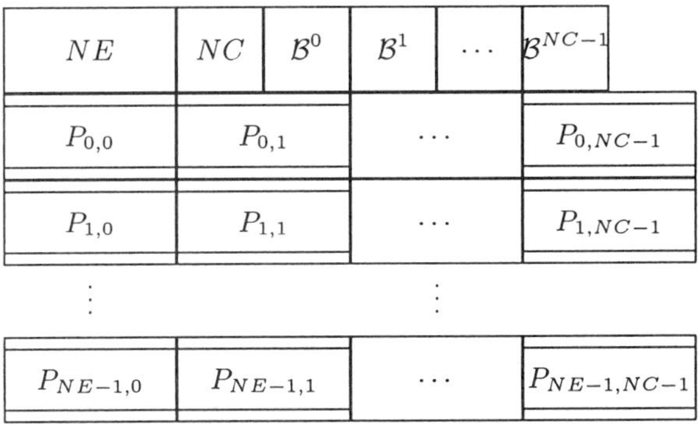

NE	NC	$\mathcal{B}^0$	$\mathcal{B}^1$	$\cdots$	$\mathcal{B}^{NC-1}$
$P_{0,0}$	$P_{0,1}$		$\cdots$		$P_{0,NC-1}$
$P_{1,0}$	$P_{1,1}$		$\cdots$		$P_{1,NC-1}$
$\vdots$			$\vdots$		
$P_{NE-1,0}$	$P_{NE-1,1}$		$\cdots$		$P_{NE-1,NC-1}$

Figure 14.9. Contents of the *Palette* box.

is a two byte unsigned integer that defines the Number of Entries in the LUT. Legal values of this parameter satisfy

$$1 \leq NE \leq 1024$$

The NC parameter is a one byte unsigned integer that signals the Number of Color components produced by application of the LUT. Equivalently, NC is the number of output values contained in each entry of the LUT. The legal range for NC is given by

$$1 \leq NC \leq 255$$

The $\mathcal{B}^j$ are single byte parameters that signal the bit-depths of the output components, $j = 0, 1, \ldots, NC - 1$. The interpretation of $\mathcal{B}^j$ is as described in Section 13.3.5.

As an example, consider the case of a 6-bit palettized representation of a 24-bit RGB (8 bits/color) image. For this case, we would have $NE = 64$, $NC = 3$, $\mathcal{B}^0 = \mathcal{B}^1 = \mathcal{B}^2 = 7$.

The ith row of parameters, $P_{i,j}$ $j = 0, 1, \ldots, NC - 1$ specifies the NC-tuple of output samples produced by the LUT in response to an input sample with value i. Each parameter of the jth column, $P_{i,j}$ $i = 0, 1, \ldots, NE - 1$ is stored as a signed integer using $\mathcal{B}^j$ bits. As discussed in Section 13.3.5, $\mathcal{B}^j = (\mathcal{B}^j \text{ AND } 7F_{\text{h}}) + 1$. If $\mathcal{B}^j$ is not a multiple of 8, the value of $P_{i,j}$ is *zero-extended* to fill the minimum number of whole bytes. For example, if $\mathcal{B}^j = 89_{\text{h}}$, then the $P_{i,j}$ are 10-bit signed quantities. Furthermore, $P_{i,j} = -1$ would be represented using two bytes as $03FF_{\text{h}}$. This is the 10-bit twos complement representation of -1, extended with 6 zeros. Robust JP2 readers would do well to ignore any such extension bits. In this way, the correct interpretation will be obtained even if a JP2 writer unwittingly uses *sign-extension* rather than zero-extension.

c^0	$\mathcal{T}^0$	j^0	c^1	$\mathcal{T}^1$	j^1
...			c^{N-1}	$\mathcal{T}^{N-1}$	j^{N-1}

Figure 14.10.   Contents of the *Component Mapping* box.

THE COMPONENT MAPPING BOX

As mentioned above, the component mapping box provides for renumbering of components to accommodate the "extra" components generated via the *Palette* box LUT. A *Component Mapping* box is present if and only if a *Palette* box is present. To avoid confusion, the word "component" will continue to refer to an image component as decompressed from the code-stream. The word "channel" will refer to a component resulting from the application of the *Palette* box LUT and *Component Mapping* box.

The *Component Mapping* box has box type $T = \text{'cmap'} = 636D6170_{\text{h}}$. The contents of this box are shown in Figure 14.10. Each set of $c^k, \mathcal{T}^k, j^k$ parameters, $k = 0, 1, \ldots, N-1$, corresponds to one channel. The number of channels N can be determined from the length of the box. The parameter c^k is a two byte unsigned integer, and indicates the component to be used in the creation of channel k. Of course, legal values for c^k satisfy $0 \le c^k \le C-1$, where C is the number of components present in the code-stream.

The $\mathcal{T}^k$ parameter is a one byte unsigned integer that specifies the Type of mapping used to obtain channel k from component c^k. $\mathcal{T}^k$ may take values of 0 and 1. A value of $\mathcal{T}^k = 0$ specifies that component k should be used directly. That is, channel k is equal to component c^k. If $\mathcal{T}^k = 1$, then channel k is obtained via the *Palette* box LUT. Specifically, channel k is obtained using column j^k of the lookup table portion of Figure 14.9. For example, if $j^k = 2$, then a sample from component c^k with value i maps to a sample in channel k with value $P_{i,2}$ (more generally, P_{i,j^k}).

If $\mathcal{T}^k = 0$, then j^k must be 0. On the other hand, if $\mathcal{T}^k = 1$, then $j^k \in \{0, 1, \ldots, NC-1\}$. In this expression, NC is the number of components produced by the *Palette* box LUT, as described in the previous section. Accordingly, each j^k parameter is a one byte unsigned integer.

It is worth noting that the JP2 specification does not require every LUT column to be referenced by the *Component Mapping* box. That is

$$\{j^0, j^1, \ldots, j^{N-1}\} \subseteq \{0, 1, \ldots, NC-1\}$$

M		
k^0	Ty^0	As^0
k^1	Ty^1	As^1
$\vdots$	$\vdots$	$\vdots$
k^{M-1}	Ty^{M-1}	As^{M-1}

Figure 14.11.   Contents of the *Channel Definition* box.

and equality need not hold. Also, there is no restriction against using the same LUT column more than once. That is, it is acceptable for $j^{k_1} = j^{k_2}$ when $k_1 \neq k_2$. Effectively then, the NC columns of the *Palette* box LUT can be used as NC independent LUTs. From this, we conclude that any "column LUT" may be applied to any unsigned image component to create an image channel. Furthermore, each component and/or column LUT can be so employed zero or more times.

THE CHANNEL DEFINITION BOX

The *Channel Definition* box provides one or more descriptions of each image channel. A *Channel Definition* box may be present, even when no *Component Mapping* box is present. In this case, components are mapped to channels in the obvious way. Specifically, channel k is set to equal to component k, $k = 0, 1, \ldots, C - 1$.

The *Channel Definition* box has box type $T = $ 'cdef' $= 63646566_{\mathrm{h}}$. The contents of this box are depicted in Figure 14.11. Each k^i, Ty^i, As^i parameter set, $i = 0, 1, \ldots, M - 1$, is comprised of three two-byte unsigned integers, and provides a description of a single channel. The number of channel descriptions, M, is a two byte unsigned integer satisfying $0 \leq M < 2^{16}$. The k^i parameter specifies the channel number to which the ith description applies.

Channel Types.   The Ty^i parameter specifies the "type" of channel k^i. Legal values of Ty^i are $0, 1, 2$, and $2^{16} - 1$. Specifically, $Ty^i = 0$ indicates that channel k^i is a color channel. The *particular* color of channel k^i is indicated via As^i as described below. A value of $Ty^i = 1$ indicates that channel k^i is an opacity channel to be applied to a color channel. The applicable color channel is indicated via the As^i parameter.

Opacity channels are required to be unsigned. The samples of an opacity channel with bit-depth B_O are interpreted in the following way. An opacity sample with value 0 indicates that its corresponding color sample is fully transparent. Conversely an opacity sample with value $2^{B_O} - 1$ indicates that its corresponding color sample is fully opaque.

When $Ty^i = 2$, channel k^i is a premultiplied opacity channel. In this case, channel k^i is an opacity channel as described above. However, the color channel to which it applies, via As^i, has been premultiplied by channel k^i. Specifically, let o be a sample from the premultiplied opacity channel, and let x be the corresponding sample from the "original" color channel. The premultiplied value, x_P, then contained in the color channel indicated by As^i, is given by

$$x_P = \left(\frac{o}{2^{B_O} - 1} \right) x$$

Finally, $Ty^i = 2^{16} - 1$ indicates that the channel type is unspecified.

Channel Associations. As mentioned above, the As^i parameter indicates the color associated with channel k^i. Legal values of this parameter are $0, 1, 2, 3$, and $2^{16} - 1$. When $Ty^i = 0$, As^i specifies the color of channel k^i itself. When $Ty^i \in \{1, 2\}$, channel k^i is an opacity channel, and As^i specifies to which color the opacity channel applies.

For RGB color spaces, As^i values of 1, 2, and 3 indicate the colors red, green, and blue respectively. In monochrome spaces, $As^i = 1$ indicates the luminance component. Similarly, in YCbCr spaces, As^i values of 1, 2, and 3 indicate Y, Cb, and Cr, respectively. Finally, $As^i = 0$ indicates that channel k^i is associated with the image as a whole, while $As^i = 2^{16} - 1$ indicates that channel k^i is not associated with any color.

As an example, consider an RGB image with R, G, and B channels numbered 0, 1, and 2, respectively. Consider also two opacity channels numbered 3 and 4. Let channel 3 apply to the red and green channels, and channel 4 apply to the blue channel. For this example, the following $[k^i, Ty^i, As^i]$ parameter sets would be present in the *Channel Definition* box: $[0, 0, 1]$, $[1, 0, 2]$, $[2, 0, 3]$, $[3, 1, 1]$, $[3, 1, 2]$, $[4, 1, 3]$.

As shown in the example above, the same channel can appear in the *Channel Definition* box more than once.[3] In such cases, a given channel must always have the same type, Ty. On the other hand, even for two different channels, the same $[Ty, As]$ pair may not appear more than once (unless both are $2^{16} - 1$). For example, adding a channel 5 to

---

[3]It should be clear then, that the number of channel descriptions may be greater than the number of channels (i.e., $M \geq N$).

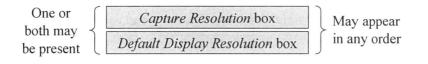

One or both may be present { [Capture Resolution box / Default Display Resolution box] } May appear in any order

Figure 14.12.   Contents of the *Resolution* box. Shading indicates optional boxes.

Table 14.3.   Boxes within the *Resolution* superbox.

Box Type	Hexadecimal Equivalent	Box Name	Required	Superbox
'resc'	72657363_h	*Capture Resolution*	No	No
'resd'	72657364_h	*Default Display Resolution*	No	No

the previous example and adding $[5, 0, 3]$ to the *Channel Definition* box would indicate both channels 2 and 5 as the blue color channel, which is not allowed.

In the previous paragraph, we noted that a given channel must only have one type, Ty. It is interesting to note however, that this does not imply that a given component may only be used for one purpose. As mentioned above, any component may be used zero or more times with any *Palette* box "column LUT" to create multiple channels. In this way, it is straightforward to define two or more channels each having exactly the same samples as one of the decompressed image components. Each of these channels may then be assigned a different type, effectively assigning multiple types (e.g., color, opacity, premultiplied opacity) to the same component. For this reason, there is currently some discussion on simply removing the restriction that a channel may have only one type. Thus, it would be prudent for a reader to expect JP2 files that do not respect the restriction of "one type per channel."

THE RESOLUTION BOX

The box type of the *Resolution* box is $T = $ 'res ' $= 72657320_h$. As shown in Figure 14.12, the *Resolution* box contains only other boxes. It may contain either one or both of the boxes summarized in Table 14.3. Specifically, it may contain a *Capture Resolution* box and/or a *Default Display Resolution* box.

The Capture Resolution Box. The *Capture Resolution* box has box type $T = $ 'resc' $= 72657363_h$, and specifies the capture resolution of the image. The contents of the *Capture Resolution* box are shown

RN_1	RD_1	RN_2	RD_2	RE_1	RE_2

Figure 14.13. Contents of the *Capture Resolution* and *Default Display Resolution* boxes.

in Figure 14.13. The RN_i and RD_i parameters are unsigned two byte integers representing "Resolution Numerators" and "Resolution Denominators," respectively. The RE_i parameters are one byte signed integers, representing "Resolution Exponents." Given these values, the vertical capture resolution (VCR) is computed as

$$VCR = \frac{RN_1}{RD_1} \times 10^{RE_1} \qquad (14.2)$$

while the horizontal capture resolution (HCR) is computed as

$$HCR = \frac{RN_2}{RD_2} \times 10^{RE_2} \qquad (14.3)$$

The units of VCR and HCR are "canvas coordinate grid points per meter." The canvas coordinate system is discussed in Chapter 11. Note that since VCR and HCR are specified independently, canvas grid points need not be square in device space. This is irrespective of any subsampling that may or may not be present (via the S_1^c and S_2^c parameters of Section 11.1). If no *Capture Resolution* box is present, then the canvas grid points are assumed to be square.

The Default Display Resolution Box. The *Default Display Resolution* box specifies a desired display resolution. The application is not required to honor the values specified in this box. The box type of this box is $T = \text{'resd'} = 72657364_h$. The structure of the *Default Display Resolution* box is identical to that of the *Capture Resolution* box, as shown in Figure 14.13. Similar to equations (14.2) and (14.3), the vertical and horizontal default display resolutions (VDR and HDR) are given by

$$VDR = \frac{RN_1}{RD_1} \times 10^{RE_1}$$

and

$$HDR = \frac{RN_2}{RD_2} \times 10^{RE_2}$$

respectively. The units of both quantities are canvas grid points per meter.

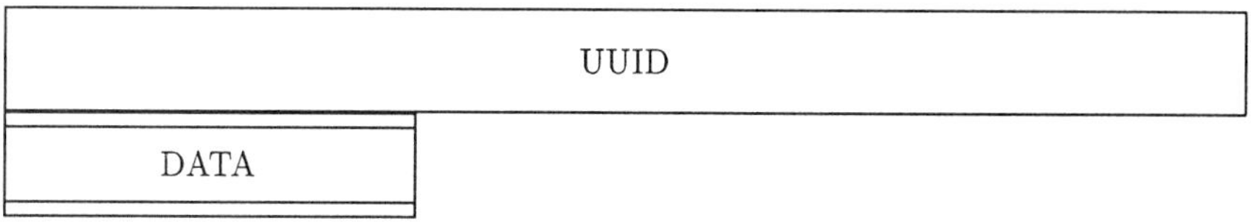

Figure 14.14. Contents of the *UUID* box. Due to its extreme width, the UUID parameter rectangle is not to scale.

We note that a JP2 writer may like to specify dimensions for the *Default Display Resolution* box even when it has no preference for display resolution. In this way, it may (indirectly) specify the image aspect ratio in an unambiguous manner.

14.2.4   THE CONTIGUOUS CODE-STREAM BOX

The box type of the *Contiguous Code-Stream* box is $T = $ 'jp2c' $=$ $6A703263_h$. The contents of this box comprise a valid JPEG2000 code-stream as described in Chapter 13.

14.2.5   THE IPR BOX

The *IPR* (Intellectual Property Rights) box has box type $T = $ 'jp2i' $= 6A703269_h$. The contents of this box are defined in JPEG2000 Part 2 and beyond. The *IPR* box may be included in a JP2 file to make the reader aware that intellectual property rights information is associated with the image content. Interpretation of this information is beyond the scope of JP2. This box is optional.

14.2.6   XML BOXES

Zero or more *XML* boxes may be included in a JP2 file. The box type of an *XML* box is $T = $ 'xml ' $= 786D6C20_h$. The contents of an *XML* box may consist of any information whatsoever, provided that it complies to the XML (eXtensible Markup Language) format. Such information shall not affect the decoding or visual appearance of the image. A discussion of XML is beyond the scope of this text. The interested reader is referred to [122] for a complete definition.

14.2.7   UUID BOXES

Zero or more *UUID* boxes may also be included in a JP2 file. The box type of the *UUID* box is $T = $ 'uuid' $= 75756964_h$. The contents of the *UUID* box are shown in Figure 14.14. The UUID parameter occupies

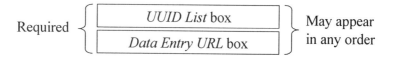

Figure 14.15.   Contents of the *UUID Info* box.

Table 14.4.   Boxes within the *UUID Info* box.

Box Type	Hexadecimal Equivalent	Box Name	Required	Superbox
'ulst'	$756C7374_h$	*UUID List*	Yes	No
'url '	$75726C20_h$	*Data Entry URL*	Yes	No

16 bytes, and is a Universal Unique Identifier (UUID) as defined in [8]. A detailed discussion of such identifiers is beyond the scope of this text. Briefly however, [8] describes a procedure for generating a unique 128-bit identifier via computer, without the need of any type of registration authority. The procedure employs the hardware address of the computer network card, a time stamp, and a pseudo-random component to ensure uniqueness.

The DATA parameter of the *UUID* box is of variable length, and may contain vendor specific information in proprietary formats. The contents of a *UUID* box shall not affect the decoding or visual appearance of the image.

14.2.8   UUID INFO BOXES

UUID Info boxes provide a mechanism to acquire additional information about the contents and format of *UUID* boxes. The *UUID Info* box has $T =$ 'uinf' $= 75696E66_h$. Zero or more *UUID Info* boxes may be present in a JP2 file. A *UUID Info* box is a superbox. It contains only other boxes. As shown in Figure 14.15, a *UUID Info* box must contain a *UUID List* box and a *Data Entry URL* box. These boxes are summarized in Table 14.4.

As detailed below, the *UUID List* box contains a list of UUIDs corresponding to *UUID* boxes for which information can be obtained. The *Data Entry URL* box contains a single URL (Uniform Resource Locator). This URL can be used to acquire information applicable to all entries of its corresponding *UUID List* box.

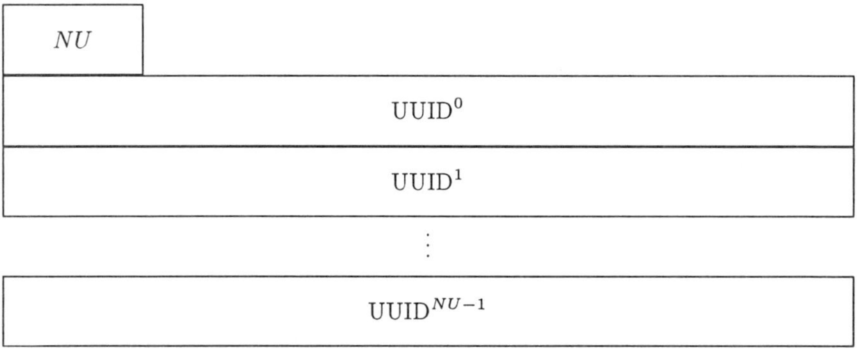

Figure 14.16. Contents of the *UUID List* box. Due to their extreme width, the UUID^i parameter rectangles are not to scale.

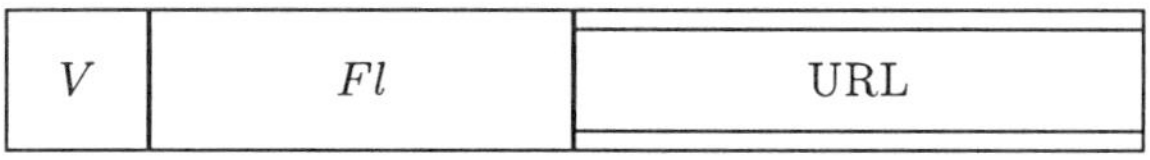

Figure 14.17. Contents of the *Data Entry URL* box.

THE UUID LIST BOX

The box type of the *UUID List* box is $T =$ 'ulst' $= 756C7374_{\mathrm{h}}$. The contents of the *UUID List* box are depicted in Figure 14.16. The *NU* parameter is a two byte unsigned integer and specifies the number of UUIDs in the box. Each UUID^i parameter, $i = 0, 1, \ldots, NU - 1$, is a 16 byte UUID. There is no required correspondence between UUIDs found in *UUID* boxes, and UUIDs found in one or more *UUID List* boxes. Specifically, some, none, or all UUIDs from *UUID* boxes may be found among the UUIDs in *UUID List* boxes. Also, multiple *UUID List* boxes may reference the same UUID, whether it is found in *UUID* boxes or not.

THE DATA ENTRY URL BOX

The *Data Entry URL* box pertains to the *UUID List* box within its *UUID Info* superbox. The box type of the *Data Entry URL* is $T =$ 'url ' $= 75726C20_{\mathrm{h}}$. The contents of this box are shown in Figure 14.17. The V (version) and Fl (flags) parameters occupy one and three bytes, respectively. In JP2, both parameters are set to zero.

The URL (Uniform Resource Locator) parameter is of variable length and is encoded as a null terminated UTF-8 character string [13]. The URL must specify a service which delivers a file (e.g., http, ftp, file, etc.). Relative URLs are allowed, in which case they are relative to the file containing the *Data Entry URL* box. The file delivered by this service provides information about certain *UUID* boxes. The specific *UUID* boxes covered are those with UUIDs included in the relevant *UUID List* box. As mentioned above, not every *UUID* box must be so covered. Also, coverage may be provided for UUIDs not present in the JP2 file.

14.3   DISCUSSION

Conforming JP2 readers must interpret all required boxes as shown in Figure 14.1. Specifically, the *JPEG2000 Signature, File Type, JP2 Header,* and *Contiguous Code-Stream* boxes must all be interpreted correctly. A conforming reader has the option to ignore the remaining boxes in Figure 14.1 (*IPR, XML, UUID, UUID Info*). As discussed in Section 14.2.3, the *JP2 Header* box is a superbox. Thus, all optional and non-optional boxes described in that section must be interpreted correctly, as well.

If a conforming JP2 reader encounters boxes not defined in JP2, these boxes should be ignored. This behavior is particularly relevant when reading a "JP2 Compatible" file. As discussed in Section 14.2.2, such a file is not a JP2 file, but contains 'jp2 ' in the compatibility list of its *File Type* box. JP2 compatible files may contain boxes not understood by a JP2 reader. Such boxes may be independent, or may be contained within superboxes.[4] By ignoring unknown boxes, these files are interpretable by a JP2 reader in some fashion as intended by the creator of the file.

Similarly, JP2 compatible files may contain multiple instances of certain boxes. In cases where JP2 files should contain only one such box, the first occurrence of the box in the file should be used. Subsequent occurrences should be ignored.

Although not mentioned previously, JP2 files are allowed to contain multiple *Contiguous Code-Stream* boxes, and multiple *Color Specification* boxes. JP2 specifies that if multiple *Color Specification* boxes are present, they must appear contiguously. That is, they must appear one after the other within the *JP2 Header* box, with no intervening boxes.

Multiple *Color Specification* boxes must all specify the same color space. However, different specification methods may be employed. For example, the sRGB color space might be indicated twice using two *Color Specification* boxes. The first box might specify sRGB using $ECS = 16$,

---

[4]The superbox itself, as well as any other "known boxes" within the superbox must still be interpreted correctly.

while the second box may employ an ICC profile giving the conversion from sRGB to XYZ.

Consistent with the discussion above, JP2 specifies that if multiple *Contiguous Code-Stream* boxes or multiple *Color Specification* boxes are present, only the first instance of each should be interpreted.

Chapter 15

PART 2 EXTENSIONS

As discussed in Chapter 9, JPEG2000 Part 1 specifies the core JPEG-2000 coding system, together with a minimal file format. Although the primary focus of this book is JPEG2000 Part 1, the purpose of the current chapter is to provide a high level overview of JPEG2000 Part 2. Part 2 specifies certain extensions to the core coding system, as well as a more extensive file format.

At the time of this writing, JPEG2000 Part 2 has not been finalized. However, most key technologies appear to have been selected. The work required to complete Part 2 lies largely in the writing and editing of the standard document itself. As such, enough is known to provide the flavor, if not the details, of JPEG2000 Part 2.

Part 2 contains extensions to allow variable level offsets and point non-linearities, both as pre/post-processing steps. It also provides for flexible deadzone sizes in scalar quantization, as well as the ability to employ trellis coded quantization. Visual masking can be employed with either scalar quantization or trellis coded quantization to obtain substantial improvements in visual quality.

Several extensions are supported with respect to the wavelet transform as applied to tile-components. Significant flexibility is available to choose the wavelet kernels employed in compression/decompression. Both odd and even length kernels (whole-sample symmetric and half-sample symmetric) are supported. The tree structure of the wavelet decomposition can also be chosen with a great deal of flexibility. Finally, the wavelet transform may be applied to overlapping "cells" and/or tiles. This latter feature allows block-based processing to be performed without the occurrence of the severe block artifacts commonly associated with such processing (see Figure 4.30).

Extended decorrelating transforms for multiple component imagery are also included in JPEG2000 Part 2. In addition to the ICT and RCT of Section 10.2, Part 2 supports general linear transforms, predictive transforms, and wavelet transforms for the decorrelation of components.

Enhanced region of interest support is also provided. In addition to the max-shift method of Section 10.6, Part 2 provides for arbitrary up-shifts, with explicit signalling of the regions to be shifted. These shifts can be applied to arbitrary rectangular and/or elliptical regions of interest.

Finally, JPEG2000 specifies an extended file format known as JPX. JPX is backward compatible with JP2 and contains many enhancements. Such enhancements include more flexibility in color space specification, opacity information, and metadata. Also included is the ability to combine multiple code-streams to obtain compositing or animation from a single JPX file. In addition to contiguous code-streams, as required by JP2 (Section 14.1), JPX allows for the inclusion of fragmented code-streams. This feature can be useful for image editing applications.

Each of the extensions outlined above are discussed briefly in the following sections. The presence of one or more of these extensions is signalled via a 1 in the most significant bit of the CA parameter in the SIZ marker segment (Section 13.3.5).

15.1   VARIABLE LEVEL OFFSET

As discussed in Section 10.1, the first processing step in JPEG2000 Part 1 is a "level offset." In this step, each unsigned component with bit-depth B is offset by subtracting 2^{B-1} from each sample so that its nominal range becomes

$$-2^{B-1} \leq x[\mathbf{n}] < 2^{B-1}$$

In Part 2, an arbitrary offset may be selected at encode time and signalled via a new marker segment, known as DCO (DC Offset). Both integer and floating-point offsets are supported and may be applied to unsigned as well as signed components. Such offsets may be selected on a tile-by-tile and/or component-by-component basis.

This variable level offset extension may serve to improve compression performance for certain images having skewed histograms. It is also helpful for transcoding images from other code-stream formats that employ variable offsets. The FBI fingerprint compression standard [155] is an example of such a format.

The presence of this particular extension is signalled via the existence of the DCO marker segment, as well as a 1 in the LSB of the CA parameter of the SIZ marker segment, (i.e., $CA_0 = 1$). The variable level offset extension may not be used in conjunction with the multiple component transform extension discussed later in Section 15.7.

15.2   NON-LINEAR POINT TRANSFORM

As discussed in Section 1.1.2, gamma correction is often useful for the display of linear color space data via non-linear display devices. Also discussed in that section, are certain perceptual advantages of gamma correction. Although most "original" images are already in gamma corrected form, some imagery is available in linear form. For such images, it may be desirable to perform forward (inverse) gamma correction as a part of the compression (decompression) process.

JPEG2000 Part 2 provides this capability via the *NLT* (Non-Linear point Transformation) marker segment. Bit 8 of the *SIZ* marker segment *CA* parameter is set (i.e., $CA_8 = 1$) to indicate the presence of this marker segment. The *NLT* marker segment can be used to specify point transforms on a tile-by-tile and/or component-by-component basis.

Two types of transforms are supported. The first type is a gamma-style non-linearity, as discussed previously. The gamma, gradient, and breakpoint parameters can all be signalled via the syntax of the *NLT* marker segment. The second type of non-linearity supported by this marker segment is an arbitrary point transform, signalled by way of a look-up table (LUT). This LUT specifies a list of input/output pairs. Each sample from a decompressed image component is then a LUT input. The resulting LUT outputs are the samples of the inverse (point) transformed image component.

15.3   VARIABLE QUANTIZATION DEADZONES

As discussed in Section 10.5, JPEG2000 Part 1 employs deadzone scalar quantization. For such a quantizer with step size Δ, the width of the deadzone, or central quantization bin, is 2Δ. As discussed in Sections 3.2.4 and 8.3.1, more general deadzone widths are possible. Specifically, a deadzone of width $2(1 - \xi)\Delta$ may be obtained via equation (3.31). Clearly then, JPEG2000 Part 1 requires $\xi = 0$ ($\tau = 0$ in equation (8.11)). On the other hand, JPEG2000 Part 2 allows any $\xi \in [-1, 1)$ to be employed. The choice of ξ can vary subband-by-subband, tile-by-tile, and component-by-component. The signalling of the chosen ξ values is accomplished via extended versions of the *QCD* and/or *QCC* marker segments. The presence of such extended marker segments is signalled via bit 1 of the *SIZ* marker segment *CA* parameter (CA_1).

Variable deadzone sizes can be used to improve the visual quality and/or mean-squared-error performance in certain types of imagery. Additionally, this feature can aid in transcoding code-streams from compression systems that employ $\xi \neq 0$ (e.g., [155]).

15.4   TRELLIS CODED QUANTIZATION

JPEG2000 Part 2 allows for the use of trellis coded quantization (TCQ) as a replacement for scalar quantization. In JPEG2000, the theoretical MSE advantage of TCQ (Section 3.5) is often seen only at high encoding rates (≥ 2 bits/sample). However, significant improvements in perceptual quality are usually present across the entire gamut of encoding rates.

The particular version of TCQ employed in JPEG2000 Part 2 is described in Section 3.5.6. Each JPEG2000 code-block is quantized independently using the same scan order as the bit-plane coder, as shown in Figure 12.9. No additional marker segments are required for the TCQ option. Indication that TCQ decoding should be employed is signalled via bit 2 of the *SIZ* marker segment CA parameter (i.e., $CA_2 = 1$).

The step sizes employed by TCQ are signalled via the QCD and/or QCC marker segments. It is worth noting that the values signalled in these marker segments are actually twice the TCQ step sizes employed by the encoder. Thus, TCQ decoders should divide these values by two before executing inverse TCQ processing. The reason for signalling twice the "correct" step size is to allow "non-TCQ" decompressors to decode TCQ code-streams in some reasonable fashion. Comparison of equations (3.37) and (3.46) reveals that inverse embedded scalar quantization using 2Δ is equivalent to approximate inverse embedded TCQ using Δ. Thus, a JPEG2000 decompressor that blindly applies inverse scalar quantization using the step sizes signalled in QCD and/or QCC will produce a reasonable decoding.

The discussion of the previous paragraph is actually only applicable if there are $p \geq 1$ missing LSBs. Clearly, if there are $p = 0$ missing LSBs, the full inverse TCQ should be employed rather than the approximate inverse of equation (3.46). If no inverse TCQ processor is available, it is actually preferable to discard the LSB prior to application of equation (3.46). This is because the state dependent sign flipping of Section 3.5.6 will make the LSBs appear to be corrupted if used in equation (3.46).

On a related note, it is worth commenting on layering in the presence of TCQ. Recall from Chapters 8 and 12 that each bit-plane is coded in three passes. Recall also that each code-block may contribute an arbitrary number of coding passes to each layer. In general, a non-connected subset of LSBs is useless for the purpose of inverse TCQ.

All LSBs of a code-block must be available before full inverse TCQ processing may occur.

For this reason, the three LSB coding passes of a given code-block should all be placed in the same packet. If rate-distortion optimization is used to drive the formation of layers (Chapter 8), the desired result may be obtained by setting the estimated distortion reduction to zero in the first two coding passes of the least significant bit-plane of each code-block.

15.5   VISUAL MASKING

The visual masking extension provides a facility for a spatially varying nonlinearity to be applied to wavelet samples prior to quantization. This nonlinearity preserves the sign of each coefficient, but modifies its magnitude according to

$$|z\,[\mathbf{j}]| = \frac{|y\,[\mathbf{j}]|^{\gamma}}{\left|1 + \frac{1}{N[\mathbf{j}]}\sum_{\mathbf{k}\in\mathcal{K}[\mathbf{j}]}|a_b \hat{y}_L\,[\mathbf{k}]|\right|^{\beta}} \qquad (15.1)$$

Using this expression, each wavelet coefficient $y\,[\mathbf{j}]$, at location $\mathbf{j}$ of a given subband b, is replaced by its processed version $z\,[\mathbf{j}]$. Equation (15.1) is valid when the wavelet transform is normalized according to equation (10.12). For other normalizations, $y\,[\mathbf{j}]$ should be normalized to satisfy equation (10.12), with $z\,[\mathbf{j}]$ then denormalized accordingly.

The $\hat{y}_L\,[\mathbf{k}]$ are locally quantized/dequantized versions of the $y\,[\mathbf{k}]$ using L bits of magnitude precision. The value of a_b is chosen to normalize out the effect of component bit-depth on the nominal range of the $\hat{y}_L\,[\mathbf{k}]$.

$\mathcal{K}\,[\mathbf{j}]$ is a neighborhood around $\mathbf{j}$ including only points to the left and above, while $N\,[\mathbf{j}]$ is the number of points included in $\mathcal{K}\,[\mathbf{j}]$. The neighborhood always contracts to obey subband boundaries. Optionally, it may also contract to obey code-block boundaries. The nature of $\mathcal{K}\,[\mathbf{j}]$ is such that the decompressor may invert the nonlinearity via

$$|y\,[\mathbf{j}]| = \left[|z\,[\mathbf{j}]|\left|1 + \frac{1}{N\,[\mathbf{j}]}\sum_{\mathbf{k}\in\mathcal{K}[\mathbf{j}]}|a_b\hat{y}_L\,[\mathbf{k}]|\right|^{\beta}\right]^{1/\gamma} \qquad (15.2)$$

in left-to-right and top-to-bottom raster order.

This parameter, as well as the nominal size of the neighborhood $\mathcal{K}\,[\mathbf{j}]$ are signalled in the code-stream via the *VMS* (Visual MaSking) marker segment. The presence of this marker segment is indicated via $CA_3 = 1$ in the *SIZ* marker segment. The *VMS* marker segment also signals whether or not the neighborhood respects code-block boundaries. Finally, γ, β, and $d_{\max}$ are signalled, where $d_{\max}$ is the maximum transform level at which visual masking is applied. The subbands processed by visual masking are then LH_d, HL_d, and HH_d $d = 1, 2, \ldots, d_{\max}$. The LL_D subband is never subjected to visual masking.

15.5.1   DISCUSSION

The nonlinearity and its inverse, given by equations (15.1) and (15.2), form an identity. Evidently, $y[\mathbf{j}]$ maps to $z[\mathbf{j}]$ then back to $y[\mathbf{j}]$. However, uniform quantization of $z[\mathbf{j}]$ is equivalent to non-uniform quantization of $y[\mathbf{j}]$. This behavior is caused by the gamma-style nonlinearity in the numerator of equation (15.1). The value of γ is constrained to lie within $(0, 1)$, with $\gamma = 0.7$ being typical.

Ignoring for the moment the denominator of equation (15.1), we see that the non-linearity in the numerator magnifies small values of $y[\mathbf{j}]$, while attenuating large values of $y[\mathbf{j}]$. Thus, after quantization larger values of $y[\mathbf{j}]$ tend to have larger errors, while smaller values of $y[\mathbf{j}]$ tend to have smaller errors. This effect can lead to perceptual quality improvements. We may think of the wavelet transform as a crude approximation to one of the cortical transforms which are commonly used to model visual masking effects (see Figure 4.28). The non-uniform quantization induced by choosing $\gamma < 1$ then closely parallels the explicit non-linear quantization of cortical subband samples which was proposed by Watson [167]. Watson's method is briefly described in Section 4.3.4.

We now turn our attention to the denominator of equation (15.1). This portion of the expression attempts to exploit visual masking more fully, by taking into account the influence of neighboring samples on the perception of quantization artifacts. In "busy" regions of the image, the denominator tends to be larger, which attentuates $z[\mathbf{j}]$, effectively increasing the quantization step size in these regions. Figure 4.29 provides a demonstration of the fact that spatial activity can suppress (mask) the visibility of quantization artifacts in the same band.

Section 16.1.4 proposes an appropriate model for the neighborhood masking phenomenon, for use in connection with masking sensitive cost functions for rate control. In the present setting, however, only causal neighbors of $y[\mathbf{j}]$ may be used to determine masking thresholds and the decoder may have relatively few decoded bits for these neighbors (at most L are used to form the $\hat{y}_L[\mathbf{k}]$ values). For this reason, a combination of the "self masking" effect represented by the numerator in equation (15.1) and the neighborhood masking effect represented by the denominator has been found to yield the greatest improvements in reconstructed image quality [176].

For a given encoding rate, use of the visual masking extension tends to increase the overall mean squared error of the decompressed imagery. At the same time, perceptual quality as gauged by human viewers, is often substantially improved.

It is worth noting that when too few bit-planes have been decoded, the accuracy of the decoded $z[\mathbf{j}]$ is poor. In turn, the accuracy of $y[\mathbf{j}]$ is poor, affecting the accuracy of $\hat{y}_L[\mathbf{k}]$, which further deteriorates the accuracy of $y[\mathbf{j}]$ after application of equation (15.2). However, when sufficient bit-planes are decoded, the $\hat{y}_L[\mathbf{k}]$ are well preserved and this problem is averted. The number of bit-planes required for this condition to be attained is governed by L.

15.6   WAVELET TRANSFORM EXTENSIONS

The Part 2 extensions to the wavelet transform are discussed in the three subsections that follow. The first subsection deals with extended decomposition structures. These structures are generalizations of the two dimensional dyadic tree structure described in Section 4.2.5. The presence of this functionality is signalled via $CA_5 = 1$ in the *SIZ* marker segment. The second subsection describes the addition of user definable wavelet kernels. This facility provides for the use of wavelet transforms other than the irreversible 9/7 and reversible 5/3 wavelet transforms discussed in Section 10.4. The presence of such transforms is indicated by $CA_6 = 1$. Finally, the third subsection defines a block-based wavelet transform. This transform employs overlapping blocks to avoid the strong artifacts produced by certain other block transforms. This transform is indicated by $CA_4 = 1$.

15.6.1   WAVELET DECOMPOSITION STRUCTURES

As discussed in Sections 4.2.5 and 10.3.2, iterative application of the separable, two dimensional wavelet transform yields a series of reduced resolution images. Specifically, for a D-level transform the original image is defined as subband LL_0, which is also known as resolution D. A one dimensional wavelet transform is applied to each column of LL_0. A one dimensional wavelet transform is then applied to each row of the result. The final result is then four subbands LL_1, LH_1, HL_1, and HH_1 as shown in Figure 15.1.

Applying the same procedure to LL_1 yields the four subbands LL_2, LH_2, HL_2, and HH_2. Iterating in this fashion eventually yields $3D + 1$ subbands LL_D, LH_d, HL_d, HH_d $d = D, D - 1, \ldots, 1$. The $D + 1$ image resolutions available for decompression from a JPEG2000 code-

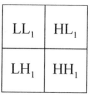

Figure 15.1. Four subbands resulting from one level of 2-D dyadic transformation.

stream are easily synthesized by inverting this process. Specifically, the resolution r image is given by LL_{D-r} $r = 0, 1, \ldots, D$. With respect to the original full resolution image, the resolution r image is reduced in size by a factor of 2^{D-r} in both width and height.

REDUCED RESOLUTION IN JPEG2000 PART 2

In JPEG2000 Part 2, this process is generalized to allow a richer class of decomposition tree structures. As in Part 1, a D level transform results in $D + 1$ resolutions. However, the transformation of rows or columns may be selectively omitted at each level. Such omissions result in unequal size reductions in the horizontal and vertical dimensions of the reduced resolution imagery. These differing size reductions impart corresponding differences in the aspect ratios of the reduced resolution images, as depicted in Figure 15.2.

As in Part 1, we begin with the original full resolution image defined as LL_0. Also as in Part 1, a compressor may choose to transform all columns followed by all rows to yield the four subbands LL_1, LH_1, HL_1, and HH_1, as depicted in Figure 15.1. On the other hand, the compressor may choose to transform just the columns. This results in only two subbands labeled by XL_1 and XH_1. Here, the X denotes the fact that no transform was performed on the rows. Clearly, given XL_1 and XH_1, inverse transformation of the columns can be performed to obtain the original image LL_0.

The subband XL_1 is a reduced resolution version of the original image. Unlike LL_1, it is reduced in resolution (size) only in the vertical dimension. Similarly, the compressor may choose to transform only the rows of LL_0 to obtain the subbands LX_1 and HX_1. Figure 15.2 shows the results of these processes. In Figure 15.2a, only column transforms are applied to yield XL_1 and XH_1. In this case, the resolution $D - 1$ image is given by subband XL_1. Similarly, Figure 15.2b shows the LX_1 and HX_1 subbands that result from row only wavelet transformation. In this case, the resolution $D - 1$ image is given by subband LX_1. As in Part 1, the transformation process is iterated on the reduced resolution

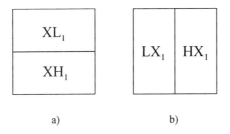

a)   b)

Figure 15.2. Subbands resulting from one level of one dimensional transformation. a) Transformation of columns only. b) Transformation of rows only.

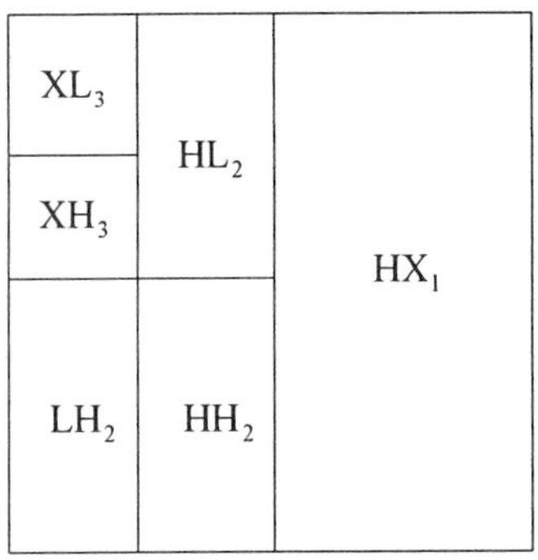

Figure 15.3. Subbands resulting from an example of a three-level non-dyadic transformation.

images to yield $D + 1$ resolutions. The choice of whether to transform rows, columns, or both can be made level-by-level.

As an example, consider a 512×512 original image. Consider further a $D = 3$ level transform, with only row transforms at the first level, both column and row transforms at second level, and only column transforms at the third level. The first level decomposes the original image LL_0 into LX_1 and HX_1, each of size 512×256. The second level decomposes LX_1 into LL_2, LH_2, HL_2, and HH_2 each of size 256×128. Finally, the third level decomposes LL_2 into XL_3 and XH_3 each of size 128×128. The complete set of subbands from this process are depicted in Figure 15.3. The $D + 1 = 4$ resolutions are XL_3, LL_2, LX_1, and LL_0 of sizes 128×128, 256×128, 512×256, and 512×512, respectively.

The number of transform levels is signalled via the *COD* and *COC* marker segments as described in Sections 13.3.6 and 13.3.7. This is unchanged from JPEG2000 Part 1. The type of transform (rows only, columns only, or both) performed at each level is signalled via a new marker segment called *DFS* (Downsampling Factor Styles). This marker segment contains a sequence of two-bit integers that specify the transform operations performed at each level. Specifically, 2 = row transforms only, 3 = column transforms only, and 1 = both.

The *DFS* marker segment can be used to select a different decomposition tree structure component-by-component, but not tile-by-tile. Specifically, all tile-components of a given component must share the same *DFS* decomposition structure. In this way, the reduced resolution tile-components are consistent across tiles.

FURTHER TRANSFORMATION OF SUBBANDS

JPEG2000 Part 2 supports further transformation of the subbands arising from the procedures discussed above. Such processing is allowed for all subbands except those required to recreate the resolution 1 image. For example, in Figure 15.3, all subbands other than XL_3 and XH_3 may be transformed further.

Four different transformation options are legal for each subband. The transformations to be applied are signalled via two-bit integers in the *ADS* marker segment (Arbitrary Decomposition Styles). As in the *DFS* marker segment, 1 = transform both rows and columns, 2 = transform only rows, and 3 = transform only columns. Additionally, the ADS marker segment uses 0 = transform neither rows nor columns. If 1, 2, or 3 is applied, this process may be repeated (one time only) on the resulting subbands. The *ADS* marker decomposition style can be changed component-by-component, as well as tile-by-tile.

As a simple example, consider the decomposition employed in the FBI fingerprint standard, as depicted in Figure 15.4. This decomposition is obtained by first performing a five level dyadic decomposition, signalled[1] via a *DFS* string 11111. One round of further transformation is then applied to the LH_1, HL_1, and HH_1 subbands. Each has both rows and columns transformed, which is signalled by three 1's in the *ADS* marker segment. At the next level, two rounds of transformation are applied to each of the LH_2 and HL_2 subbands, while neither the rows nor the columns of HH_2 are transformed. This is signalled by ten 1's followed by a single 0. Two 1's indicate the first round of row and column transformations applied to each of LH_2 and HL_2; eight 1's identify the second round of row and column transformations applied to each of the resulting subbands; and the 0 signals the lack of any further transformation for HH_2. One round of further transform is applied to LH_3, HL_3, and HH_3.

---

[1] Actually this string of 1's may be abreviated by a single 0. Since we do not discuss the syntax in detail, we give the conceptually simple description of five 1's.

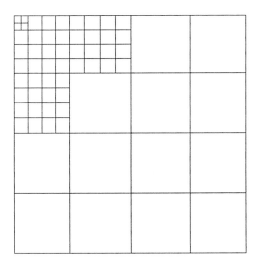

Figure 15.4. Decompostion structure employed in the FBI fingerprint compression standard.

Three 1's are included in the *ADS* marker segment to indicate that both row and column transforms are applied in each case. Although allowed, no further transformation is applied to LH_4, HL_4, and HH_4.

We close this section by noting that the simple dyadic structure allowed by JPEG2000 Part 1 is adequate for many applications. On the other hand, the more general structures of Part 2 have advantages for certain types of imagery such as fingerprints and synthetic aperture radar. For such imagery, these decompositions can provide significant improvements in compression performance. The availability of these structures also allows for easy transcoding of compressed imagery from other compression formats.

15.6.2   USER DEFINABLE WAVELET KERNELS

As discussed in Section 10.4, JPEG2000 Part 1 has two options for the wavelet transform. Specifically, either the irreversible 9/7 or the reversible 5/3 transform may be employed. In JPEG2000 Part 2, a rich family of wavelet transforms is available. Either even or odd length kernels may be selected, as well as reversible or irreversible processing. It is also intended that non-symmetric (as well as symmetric) kernels will be supported. As kernel symmetry is an important property for the compression of image components, non-symmetric kernels will be allowed only for the purpose of multi-component transformation (Section 15.7).

As discussed in Chapter 6, irreversible transforms may be implemented via convolution or lifting, while reversible transforms are only conveniently implemented via lifting. For this reason, JPEG2000 Part 2 allows the definition of alternate transforms exclusively through their lifting parameters.

Equation (6.48) gives the lifting implementation of an irreversible transform. This equation is repeated here as

$$y_{1-p(l)}^{\{l\}}[n] = y_{1-p(l)}^{\{l-1\}}[n]$$
$$y_{p(l)}^{\{l\}}[n] = y_{p(l)}^{\{l-1\}}[n] + \sum_{i} \lambda_l[i] \, y_{1-p(l)}^{\{l-1\}}[n-i] \qquad (15.3)$$

where

$$p(l) = \begin{cases} 0 & l \text{ even} \\ 1 & l \text{ odd} \end{cases}$$

$y_0^{\{l\}}[n]$ and $y_1^{\{l\}}[n]$ are the results of the lth lifting step $l = 1, 2, \ldots, L$. The initial conditions to this iteration are $y_0^{\{0\}}[n] = x[2n]$ which comprise the even indexed input samples, and $y_1^{\{0\}}[n] = x[2n+1]$ which comprise the odd indexed input samples. The final results, $y_0^{\{L\}}[n]$ and $y_1^{\{L\}}[n]$, are scaled by the gains K_0 and K_1 to yield the low and high-pass wavelet samples, respectively.

As shown in Figure 6.9, this procedure can be seen as alternately filtering the even (resp. odd) indexed samples and adding the result to the odd (resp. even) indexed samples. These operations are collectively known as high-pass (resp. low-pass) lifting steps. From the summation in equation (15.3), we see that impulse response used for filtering in the lth lifting step is given by $\lambda_l[n]$.

The reversible case is nearly identical to the irreversible case. As given by equation (6.51), the second line of equation (15.3) is modified to be

$$y_{p(l)}^{\{l\}}[n] = y_{p(l)}^{\{l-1\}}[n] + \left\lfloor \frac{1}{2} + \sum_{i} \lambda_l[i] \, y_{1-p(l)}^{\{l-1\}}[n-i] \right\rfloor \qquad (15.4)$$

We see that here, the result of the filtering steps are rounded to the nearest integer. The only other difference between the reversible and irreversible transforms is that scaling by K_0 and K_1 is omitted in the reversible case. That is $y_0^{\{L\}}[n]$ and $y_1^{\{L\}}[n]$ are directly used as the low and high-pass wavelet samples.

JPEG2000 Part 2 allows encoders to specify the parameters of equations (15.3) and (15.4). These parameters are signalled via the *ATK* marker segment (Arbitrary Transform Kernels). The number of lifting steps L is signalled and must be in the range $0 \leq L \leq 255$. The *ATK* marker segment also identifies whether the transform is reversible or irreversible, and whether it corresponds to an even symmetric, odd symmetric, or non-symmetric kernel.

For the lth lifting step, $l = 1, 2, \ldots, L$, the filter impulses response values, $\lambda_l[n]$, are signalled as either floating point values, or rational values having power of two denominators. The number of impulse response values per lifting step can range from 0 to 255, and is specified independently for each lifting step. For irreversible transforms a single gain factor, K, is signalled. The subband gain factors, K_0 and K_1, of Figure 6.9 may be deduced from K by setting $K_0 = 1/K$ and $K_1 = K/2$.[2]

Equations (15.3) and (15.4), as well as Figure 6.9 indicate that high-pass and low-pass lifting steps alternate with a high-pass step occurring first. Furthermore equation (15.4) includes a rounding factor of $1/2$ within the floor function. In JPEG2000 Part 2, more flexibility is afforded in both regards. Specifically, the lifting steps are performed in order $l = 1, 2, \ldots, L$, but each may be independently specified as being a low-pass step or a high-pass step. Also, the rounding factor employed in equation (15.4) may be signalled as a rational number with power of two denominator.

BOUNDARY HANDLING

We conclude our discussion of arbitrary kernel definitions with a word on boundary handling in JPEG2000 Part 2. As discussed in Section 10.3.1, boundary handling in JPEG2000 Part 1 is accomplished via "whole-sample" symmetric extension of the samples prior to performing the wavelet transform. In whole-sample symmetric extension, the boundary sample is not repeated at either end of the data sequence. This is illustrated in Figure 15.5a. For simplicity, only the first few extension samples are shown at each boundary.

The key feature of such an extension policy is that the symmetry survives the transformation process. The interleaved sequence of transformed samples may thus be viewed as being symmetrically extended itself. For the purpose of compression and storage, only the non-extended

---

[2] We note that the standard document indicates an assignment of $K_1 = K$ instead of $K_1 = K/2$, which is a consequence of the fact that the standard document consistently adopts a different normalization convention for the wavelet transform kernels and consequently for the quantization step sizes. The conventions used here have a number of convenient properties for description, interpretation and implementation, which we have exploited in various places.

Original
Extension   Samples   Extension

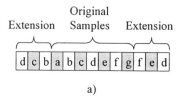

a)

Original
Extension   Samples   Extension

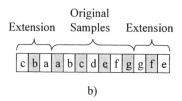

b)

Figure 15.5. Symmetric extension policies. a) Whole sample symmetry: the boundary samples are not repeated. b) Half sample symmetry; the boundary sample is repeated at each end.

transformed samples need be maintained. In this case, the number of relevant samples is the same before and after transformation.

Another property of whole-sample symmetric extension can be noted by examining the even and odd indexed subsequences. These two subsequences are differentiated by shading in Figure 15.5a. In the whole-sample symmetric case, the extended samples of each such subsequence come from the relevant subsequence itself. This property is particularly appealing in lifting implementations, where processing is performed on the odd and even indexed subsequences. In this case, a version of symmetric extension can be used within the lifting steps themselves to achieve the same effect as if the original samples were extended. For a more thorough development of these concepts, the reader is referred to Section 6.5.3.

The discussion above pertains only to odd length symmetric wavelet kernels. As mentioned in Section 6.5.2, even length kernels do not work out quite so nicely. In addition to the problems discussed there, the even and odd indexed subsequences are not extended using samples from the relevant subsequence itself. This is because the extension type that results in some type of symmetry surviving transformation is the so called half sample symmetric extension. This type of extension repeats the sample at each boundary as shown in Figure 15.5b. As before, shading indicates the even and odd subsequences. Unfortunately, the shaded subsequence is extended with samples from the unshaded subsequence, and vice versa.

To overcome the issues surrounding even length kernels, a fairly tortured procedure involving many special cases was slated for inclusion in Part 2. The special cases were based on whether the number of samples is even or odd and whether the first sample index is even or odd. In some cases, post processing was required to "move" a low-pass sample to the high-pass subband. This latter processing was employed to preserve the resolution scalability properties of the canvas system of Chapter 11.

Just prior to this writing, a new proposal was brought forward [31]. This proposal advocates always performing boundary handling within the lifting steps (with no extension of the original samples themselves). This idea is discussed in Section 6.5.3. The resulting transformation is invertible and produces the same number of samples both pre- and post-transformation. It also preserves resolution scalability in a clean manner. It is worth noting however, that for even length symmetric kernels, the wavelet samples obtained via this procedure are not identical to those that would have been obtained by symmetric extension of the original samples. For odd length symmetric kernels, the samples can be made identical under the two methods.

It appears likely that this proposal will be adopted for JPEG2000 Part 2. In addition to its benefits for even length symmetric kernels, boundary handling via lifting steps is the only reasonable approach for general (non-symmetric) kernels.

15.6.3 SINGLE SAMPLE OVERLAP TRANSFORMS

In Chapter 11, we discussed the transformation of tile-components. Each tile-component is transformed independently of all other tile-components. Such tile-based processing provides a simple method for containing implementation memory requirements and a certain degree of spatial random access. On the other hand, tile-based processing can result in objectionable block artifacts in decompressed imagery. To retain the desirable properties of tiles while combatting block artifacts, JPEG2000 Part 2 allows the use of certain overlapping block-based transforms.

OVERLAPPING TILES

One option for overlapped block-based processing allows for transformation of overlapping tiles. In this option, tiles of size $T_1 \times T_2$ are placed on a tile grid with inter-tile distance $(T_1 - 1) \times (T_2 - 1)$. This situation

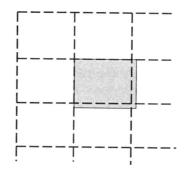

Figure 15.6. Overlapping tiles. The tile grid is shown using dashed lines. The $\mathbf{t} = [1,1]$ tile is shown via shading. It contains 1 row and column of samples from neighboring tiles.

is illustrated in Figure 15.6. Tile $\mathbf{t} = [t_1, t_2] = [1, 1]$ is shaded in the figure. Each $T_1 \times T_2$ tile is transformed and compressed as discussed in previous chapters. All overlap techniques discussed here are restricted to transforms corresponding to odd length symmetric kernels.

Upon decompression, the "interior" samples of each tile are uniquely determined, while border samples may have multiple decompressed versions. For instance, at least two versions of each sample in the first and last column and row are available (four versions of each corner sample are available). In the lossless case, all versions will be identical. In the lossy case, the various versions will differ, and JPEG2000 Part 2 specifies which version to keep based on the tile and coding partition anchor points (Chapter 11).

The height and width of the tile partition elements for this option are constrained to be powers of two. The size of the tiles themselves is then of the form $T_1 \times T_2 = \left(2^{l_1} + 1\right) \times \left(2^{l_2} + 1\right)$. The tile partition anchor point is also constrained to be of the form $\left[n_1 2^{l_1}, n_2 2^{l_2}\right]$. Finally, the sub-sampling factors of all components must be $S_i^c = 1$, $i = 1, 2$, $c = 0, 1, \ldots, C - 1$.

With these restrictions, the first and last samples of each row and column of every (overlapping) tile-component have even indices. The canvas conventions then result in the first and last sample of each row and column corresponding to a low-pass sample after transformation. It follows that every tile-component resolution also possesses this property. It has been shown empirically that this attribute substantially reduces quantization induced errors at tile boundaries. In this way, significant block artifact reduction is achieved.

The price of artifact reduction is paid in compression efficiency. The number of image samples compressed via this method is increased by a factor of $T_1 T_2 / (T_1 - 1)(T_2 - 1)$. As the tile size increases, this factor becomes insignificant. However, some memory efficiency and ease of spatial random access is then lost.

For the purpose of overlapping block transforms, JPEG2000 Part 2 actually defines an alternate boundary extension policy known as "SSO-DWT extension." While similar to regular symmetric extension, SSO-DWT extension has the important property that the boundary samples pass directly into the relevant low-pass subbands without modification. This is achieved by replacing the low-pass lifting steps which would modify these samples with point-wise scaling operations, where the scaling factors are selected to preserve the DC gain characteristics of the transform. As noted below, this modified extension policy is mandatory when the transform is used with overlapping cells. When used with overlapping tiles, the standard permits both regular symmetric extension and SSO-DWT extension at the tile boundaries.

OVERLAPPING CELLS WITHIN TILES

A modified version of the idea discussed above may be employed on small blocks within tiles. As before, only transforms corresponding to odd length symmetric kernels are allowed. To avoid confusion with other block structures in JPEG2000, these blocks are referred to as "cells." For this option, a tile-component is partitioned by a "cell" grid with elements of size $C_1 \times C_2$ where C_1 and C_2 are both powers of two. The cell grid anchor point is always the origin of the canvas. From this, we see that the upper left hand corner of each cell is of the form $[n_1 C_1, n_2 C_2] = [n_1 2^{l_1}, n_2 2^{l_2}]$.

The samples of a cell are augmented by one row and column of samples from the bottom and right to obtain an array of size $(C_1 + 1) \times (C_2 + 1)$. Such arrays overlap in the fashion of the tiles of Figure 15.6. These arrays are independently transformed exactly as though they were tiles, with the following two exceptions: 1) SSO-DWT boundary extension must be used instead of regular symmetric extension; and 2) only the wavelet samples corresponding to points within the cell are retained. The additional $C_1 + C_2 + 1$ wavelet samples are discarded.

Conceptually, after the first level of transformation, the LL subbands of all cells are placed on a cell grid with elements of size $C_1/2 \times C_2/2$. In this way a "pseudo-subband" of nominal size $T_1/2 \times T_2/2$ is formed. This process is repeated for the HL, LH, and HH subbands.

The resulting pseudo-subbands look very much like the subbands that would have been obtained via the usual wavelet transform of the full tile. In fact, the wavelet samples in the interior of the $C_1/2 \times C_2/2$ cells are identical to those that would have been obtained from the full tile-component transform. Only near the borders of the cell are there small differences between the transformed cell samples and the transformed tile samples. Even these differences are smaller than might be expected due to the same considerations discussed in the overlapped tile case. Specifically, the restrictions on the cell partition anchor point and cell sizes guarantee that all first and last rows and columns of the overlapping arrays correspond to locations of low-pass samples.

It is worth emphasizing that this procedure does not suffer from the data expansion of the tile based technique. This is due to the fact that the "extra" wavelet samples are discarded. On the other hand, cells can not then be inverse transformed independently. One row and column of wavelet samples must be "borrowed" from neighboring cells before inverse transformation to get $(C_1 + 1) \times (C_2 + 1)$ samples.

Recall that the SSO-DWT extension prevents the transform from modifying any of the samples on the overlapping cell boundaries. For this reason, wavelet samples which the inverse transform borrows from the first column of the neighboring cell to the right are identical to the samples which were discarded from the last column of the current cell. Similarly, samples borrowed from the first row of a cell's bottom neighbor are identical to those which were discarded from the last row of the current cell. After inverse transformation, the last row and column are discarded to obtain the $C_1 \times C_2$ inverse transformed cell samples. In the absence of quantization errors, this process is entirely invertible. In particular, when reversible transforms are employed, lossless compression/decompression is supported.

We discussed above the conceptual process of concatenating cell subbands to obtain tile-component pseudo-subbands. In practice however, an implementation need not accumulate entire pseudo-subbands. Assuming a dyadic decomposition tree, the LH, HL, and HH subband samples may be coded and discarded as code-blocks are filled. Similarly, the LL cells may be further transformed as soon as neighboring cells become available to contribute a "borrowed" row/column. This process can be facilitated by judicious choice of tile partition and coding partition anchor points, $[\Omega_1^T, \Omega_2^T]$ and $[\Omega_1^C, \Omega_2^C]$ (Chapter 11). In this way, code-blocks can be forced to line up advantageously with cells so that minimal transformed cell buffering is required to fill code-blocks.

15.7   MULTI-COMPONENT PROCESSING

In Chapter 10, we discussed the irreversible and reversible color transforms (ICT/RCT). Each of these transforms is applied independently at each sample location to exploit inter-component correlation in RGB color images. Such transforms, that operate across multiple image components on a point-by-point basis, are known as point transforms. JPEG 2000 Part 2 supports three types of point transforms in addition to the ICT and RCT of Part 1. These three types of point transforms are: linear block transforms, dependency transforms, and wavelet transforms.[3] When one of these three methods are employed, bit 7 of the *SIZ* marker segment CA parameter is set (i.e., $CA_7 = 1$).

As with the ICT and RCT, the goal of such transforms is to exploit inter-component correlation. Although the transforms described in this section are applicable to RGB color imagery, they are more typically intended for other types of multi-component imagery. For example, CMYK and/or LANDSAT imagery may be efficiently compressed using these methods. LANDSAT imagery typically comprises seven components, six of which are highly correlated. Certain types of hyperspectral and medical imagery may also be efficiently compressed via these transforms. Such imagery may comprise hundreds of highly correlated components.

JPEG2000 Part 2 allows the grouping of arbitrary subsets of components into "component collections." Any of the three transform types may be selected for each such collection. In some cases, multiple transforms may be performed sequentially on the same component collection. As with the ICT and RCT, the components resulting from point transformation are compressed and decompressed using the two dimensional spatial wavelet transform. The resulting spatially reconstructed components are then processed via the appropriate inverse point transform(s) to obtain the final decompressed components.

In the following subsections, we provide a brief overview of the three types of multi-component transforms. In each case, we assume that the relevant transform is applied to a component collection with C_o components, having samples at location **n**, denoted by $x_0[\mathbf{n}]$, $x_1[\mathbf{n}]$, $\ldots$, $x_{C_o-1}[\mathbf{n}]$. We use the shorthand notation

$$\mathbf{x}[\mathbf{n}] = \begin{pmatrix} x_0[\mathbf{n}] \\ x_1[\mathbf{n}] \\ \vdots \\ x_{C_o-1}[\mathbf{n}] \end{pmatrix}$$

---

[3]Strictly speaking, the non-linear transform of Section 15.2 is also a point transform. However, the non-linear transform operates on one sample of one component at a time, and is not applicable to exploitation of inter-component correlation.

to denote the C_o-vector obtained by taking one sample (at location $\mathbf{n}$) from each component. The relevant point transform is then performed on $\mathbf{x}[\mathbf{n}]$ to obtain the vector $\mathbf{y}[\mathbf{n}]$. This process is carried out independently at each sample location $\mathbf{n}$.

As described below, the size of the $\mathbf{y}[\mathbf{n}]$ vectors need not necessarily be the same as that of the $\mathbf{x}[\mathbf{n}]$ vectors. We denote the size of the $\mathbf{y}[\mathbf{n}]$ vectors by C_t. Thus, each $\mathbf{y}[\mathbf{n}]$ is of the form

$$\mathbf{y}[\mathbf{n}] = \begin{pmatrix} y_0[\mathbf{n}] \\ y_1[\mathbf{n}] \\ \vdots \\ y_{C_t-1}[\mathbf{n}] \end{pmatrix}$$

Application of the point transform at each $\mathbf{n}$ results in C_t "transformed" components. These components have samples at location $\mathbf{n}$ denoted by $y_0[\mathbf{n}]$, $y_1[\mathbf{n}]$, ..., $y_{C_t-1}[\mathbf{n}]$. These transformed components are compressed and decompressed to yield components with samples $\hat{y}_0[\mathbf{n}]$, $\hat{y}_1[\mathbf{n}]$, ..., $\hat{y}_{C_t-1}[\mathbf{n}]$. Finally, each vector

$$\hat{\mathbf{y}}[\mathbf{n}] = \begin{pmatrix} \hat{y}_0[\mathbf{n}] \\ \hat{y}_1[\mathbf{n}] \\ \vdots \\ \hat{y}_{C_t-1}[\mathbf{n}] \end{pmatrix}$$

is "inverse" transformed to get

$$\hat{\mathbf{z}}[\mathbf{n}] = \begin{pmatrix} \hat{z}_0[\mathbf{n}] \\ \hat{z}_1[\mathbf{n}] \\ \vdots \\ \hat{z}_{C_f-1}[\mathbf{n}] \end{pmatrix}$$

which comprise the samples of C_f "final" reconstructed components.

As mentioned in Chapter 9 and elsewhere, the JPEG2000 standard really specifies only the decompressor and code-stream syntax. This allows compressors room to innovate with confidence that decompressors will behave as expected. This idea is highlighted in the case of multi-component processing by the fact that C_f and C_t may differ from C_o. We will have more to say on this below.

15.7.1   LINEAR BLOCK TRANSFORMS

The linear block transform can be most easily described in terms of matrix multiplication. As an example, the ICT of Section 10.2 is a linear block transform. Specifically,

$$\mathbf{y}[\mathbf{n}] = A^t \mathbf{x}[\mathbf{n}] \tag{15.5}$$

where $\mathbf{y}[\mathbf{n}] = (x_Y[\mathbf{n}], x_{Cb}[\mathbf{n}], x_{Cr}[\mathbf{n}])^t$, $\mathbf{x}[\mathbf{n}] = (x_R[\mathbf{n}], x_G[\mathbf{n}], x_B[\mathbf{n}])^t$ and A^t is a 3×3 matrix. The operation of the inverse transform is given by

$$\mathbf{z}[\mathbf{n}] = S\hat{\mathbf{y}}[\mathbf{n}] \tag{15.6}$$

where $\hat{\mathbf{y}}[\mathbf{n}]$ is the compressed/decompressed version of $\mathbf{y}[\mathbf{n}]$ and S is also a 3×3 matrix. The particular matrices A^t and S for the ICT are given in Section 10.2, and satisfy

$$SA^t = I \tag{15.7}$$

where I is the 3×3 identity. In the absence of quantization or rounding errors, $\mathbf{z}[\mathbf{n}] = \mathbf{x}[\mathbf{n}]$. In any event, $\mathbf{z}[\mathbf{n}]$ can serve as an approximation to $\mathbf{x}[\mathbf{n}]$.

The linear block transform employed in the multi-component option of JPEG2000 Part 2 is described by equations (15.5) and (15.6). In general however, A^t is of dimension $C_t \times C_o$, while S is of dimension $C_f \times C_t$. It is not necessary that equation (15.7) be satisfied. The encoder must signal the values of the S matrix to the decoder. This is accomplished via the *MCT* marker segment (multi-component Transform). The decoder has knowledge of neither A^t nor C_o. It knows only C_t, C_f, and S, and dutifully carries out equation (15.6).

Examples indicative of the utility of such flexibility are described now. Suppose that the encoder chooses to perform the KLT (Chapter 4) on the $\mathbf{x}[\mathbf{n}]$ vectors. Suppose further that the encoder decides apriori that the last l transformed components will be discarded (Zonal Coding, Section 5.1). There is then no sense in computing and compressing these l components. This observation can be effected by simply deleting the last l rows of the $C_o \times C_o$ KLT transform matrix K^t to obtain the matrix A^t of size $C_t \times C_o$, where $C_t = C_o - l$.

The appropriate transform for application at the decoder would then be obtained by deleting the last l columns of the inverse KLT matrix K. The resulting matrix S is of size $C_f \times C_t$ where $C_f = C_o$ and $C_t = C_o - l$, as above.

As another example, the compressor might employ the full KLT with $A^t = K^t$ and $C_t = C_o$. The transform signalled for use at the decoder might consist also of the entire inverse KLT matrix K, but augmented with m additional rows at the bottom. This would yield S of size $C_f \times C_t$ where $C_t = C_o$, but $C_f = C_o + m$.

The first C_o reconstructed components would then correspond to the C_o original components at the compressor. If the bottom m rows of S

are linear combinations of the rows of K, then the last m reconstructed components correspond to linear combinations of the original components. This might be employed to force the decoder to create panchromatic components and/or R, G, and B pseudo-color components from compressed LANDSAT data when no such components were present at compression time.

Although ignored in the discussion above, the linear block transform procedure for multi-component processing allows for the addition of a constant vector to the final inverse transformed vectors. That is, equation (15.6) should be replaced by

$$\mathbf{z}\,[\mathbf{n}] = S\hat{\mathbf{y}}\,[\mathbf{n}] + \mathbf{b} \qquad\qquad (15.8)$$

Of course a corresponding subtraction may be carried out in the compressor.

The vector $\mathbf{b}$ is of dimension $C_f \times 1$ and is signalled via the *MCT* marker segment. The functionality provided by the addition of $\mathbf{b}$ is essentially the same as that provided by the variable DC offset of Section 15.1.

As a closing comment, we note that reversible versions of block transforms are also likely to be supported in JPEG2000 Part 2. Such transforms are analogous to the RCT of Section 10.2.

15.7.2   DEPENDENCY TRANSFORMS

From a decompressor point of view, the dependency transform is similar to the linear block transform. Notable differences however, are that the offset $\mathbf{b}$ is added prior to the transform, the matrix S must be lower triangular and square, and the inverse transform process is recursive.

Specifically, the inverse dependency transform process is given by

$$\mathbf{z}\,[\mathbf{n}] = S\mathbf{z}\,[\mathbf{n}] + (\hat{\mathbf{y}}\,[\mathbf{n}] + \mathbf{b}) \qquad\qquad (15.9)$$

It may seem odd that the final reconstructed component data, $\mathbf{z}\,[\mathbf{n}]$, appears on both sides of the equality. However, as mentioned above, S is restricted to be square and lower triangular. Furthermore, S must have zeros on its diagonal. As a consequence, $\mathbf{z}\,[\mathbf{n}]$ is recursively computable in the following way:

$$z_0\left[\mathbf{n}\right] = \hat{y}_0\left[\mathbf{n}\right] + b_0$$
$$z_1\left[\mathbf{n}\right] = s_{1,0}z_0\left[\mathbf{n}\right] + \hat{y}_1\left[\mathbf{n}\right] + b_1$$
$$\vdots$$
$$z_{C_f-1}\left[\mathbf{n}\right] = \sum_{j=0}^{C_f-2} s_{C_f-1,j}z_j\left[\mathbf{n}\right] + \hat{y}_{C_f-1}\left[\mathbf{n}\right] + b_{C_f-1}$$

We have already stated that S must be square. That is, $C_t = C_f$. For simplicity, we also assume that $C_o = C_f$ and that the final reconstructed components will serve as decompressed approximations of the original components. To emphasize this, we introduce the alternate notation, $\hat{x}_i\left[\mathbf{n}\right] = z_i\left[\mathbf{n}\right]$ $i = 0, 1, \ldots, C_f - 1$ and write

$$\hat{x}_i\left[\mathbf{n}\right] = \sum_{j=0}^{i-1} s_{i,j}\hat{x}_j\left[\mathbf{n}\right] + \hat{y}_i\left[\mathbf{n}\right] + b_i$$

Referring to Figure 3.7, we note that

$$\mu_i\left[\mathbf{n}\right] = \sum_{j=0}^{i-1} s_{i,j}\hat{x}_j\left[\mathbf{n}\right] + b_i$$

can be interpreted as a prediction of $x_i\left[\mathbf{n}\right]$, and that $\hat{y}_i\left[\mathbf{n}\right]$ can be interpreted as the prediction error, after compression and decompression. In this way, the dependency transform can support DPCM style coding of components.

As with the block transform discussed above, it is likely that both irreversible and reversible versions of the dependency transform will be supported.

15.7.3   WAVELET TRANSFORMS

Wavelet transforms will also be allowed as point transforms in JPEG-2000 Part 2. Little is currently settled in this regard. It is likely however, that the encoder will be allowed to signal a kernel to be used as described in Section 15.6.2. Both irreversible and reversible transforms will be supported, as well as odd length symmetric, even length symmetric, and non-symmetric kernels. Since point transforms are inherently one dimensional, all such wavelet transforms will also be one dimensional. At a minimum, D levels of dyadic decomposition will be supported in this regard. It is unclear if more general structures (Section 15.6.1) will be allowed.

15.8   REGION OF INTEREST CODING

In Section 10.6, we discussed the scaling method of ROI (Region of Interest) coding. In that discussion, the quantization index at location **j** in code-block i is denoted by y_i [**j**]. Prior to coding of the quantization indices, each y_i [**j**] is scaled by $2^{U_i[\mathbf{j}]}$. Equivalently, y_i [**j**] is "left-shifted," or "up-shifted," by U_i [**j**] bit positions. Clearly, the decoder must know the value of U_i [**j**] for each i, **j** in order to properly realign the quantization indices as part of decompression.

As discussed in Section 10.6, JPEG2000 Part 1 solves this problem by signalling a single value U in the code-stream. This value satisfies

$$U \geq \max_b K_b$$

where K_b is the number of bits used to represent quantization indices in subband b. This makes U so large that the LSB of the up-shifted indices occupy a higher bit position than the MSB of the non-up-shifted indices. Thus, the up-shifted indices are easily identified and down-shifted in the decoder.

In JPEG2000 Part 2, more flexibility is allowed in the choice of U_i [**j**]. Specifically, the location and sizes of multiple rectangular and/or elliptical regions of interest (ROI) may be signalled via an extended version of the *RGN* marker segment. A separate up-shift value is also signalled for each such ROI. Multiple regions may be specified independently within each component. The presence of this feature is indicated via $CA_9 = 1$ in the *SIZ* marker segment.

Given an ROI in a specific component, the quantization indices to be up-shifted are identified via the "region mapping" procedure discussed in Section 10.6. If a given index belongs to multiple ROIs in a given component, the largest applicable up-shift value is used.

By application of this procedure, multiple ROIs in multiple components may all be emphasized by different amounts. If eventually all coding passes are decoded, the resulting decompressed image will be as if no ROI processing were used. However, if bit-stream truncation occurs, the ROIs with larger up-shift values will be decompressed with higher fidelity than ROIs with smaller up-shift values. The "background" consists of all values not up-shifted, U_i [**j**] $= 0$, corresponding to the lowest decompressed image quality.

15.9   FILE FORMAT

As mentioned in Chapter 14, JPEG2000 Part 2 includes an extended file format. This file format is known as JPX. The JPX file format is flexible and powerful, and adds many capabilities to JP2. A full description of JPX is beyond the scope of this text. We provide only a brief overview of JPX extensions here.[4]

JPX files employ the extension ".jpx" (Macintosh file systems employ the type code 'jpx '). On the other hand, for a JPX file containing 'jp2 ' in its compatibility list, it is legal to use the ".jp2" extension. Such (mis)labeling may encourage JP2 readers to interpret such files, within the limits of their capabilities. Of course, the "true" file type may still be determined via the *Br* field of the *file type* box (Section 14.2.2). For JPX files, $Br =$ 'jpx '.

JP2 allows two methods for describing color spaces: enumeration and restricted ICC profiles. JPX adds to the list of enumerated color spaces, and includes the capability for future registration of additional enumerated color spaces. JPX also allows the use of general (non-restricted) ICC profiles.

In JP2, only a single contiguous code-stream is allowed.[5] JPX allows multiple code-streams in the same file. The inclusion of multiple code-streams allows, for example, opacity and color channel information to be stored as separate code-streams. This might ease the inclusion or modification of opacity information during file editing operations. JPX also specifies several tools for combining multiple code-streams within a JPX file. These tools are very powerful and include support for compositing as well as animation.

In JPX, code-streams may be fragmented and spread across multiple boxes. This may facilitate ease of code-stream editing. For example, single tiles may be edited without the need to rewrite the entire file. Fragmented code-streams may also be useful in the creation of scalable files. Different users may receive different subsets of code-stream fragments corresponding to different image qualities and/or resolutions.

Finally, JPX defines specific methods for the inclusion of metadata relating to creation/revision history, intellectual property rights, and descriptions of the image content.

---

[4]The description of JPX occupies more than 50% of the JPEG2000 Part 2 committee draft.
[5]More accurately, only the first contiguous code-stream is required to be read/interpreted.

III

WORKING WITH JPEG2000

Chapter 16

PERFORMANCE GUIDELINES

16.1 VISUAL OPTIMIZATIONS
16.1.1 CSF BASED OPTIMIZATIONS

In Section 4.3.4, we discussed weighted MSE (WMSE) based on the human visual system Contrast Sensitivity Function (CSF). We argued that in many cases, this WMSE is a more accurate gauge of perceptual quality than the more usual MSE. In Section 5.2, we discussed the traditional method for minimizing WMSE via rate allocation. The discussion there was based primarily on the assumption of non-scalable compression, where only "single quality" decoding is possible.

In Section 7.3, we noted that CSF weighting may also be incorporated into embedded compression schemes such as EZW or SPIHT. This is most easily accomplished by adjusting the quantization step size for each band b, in accordance with the contrast sensitivity in that band. Rather than targeting MSE by using the same step size Δ in each band (see equation (7.2)), WMSE may be targeted by multiplying each band's step size by some kind of "average" detection contrast for the band, T_b^{csf}. Specifically, we may assign step sizes Δ_b using

$$\Delta_b = \alpha^{-1} T_b^{\mathrm{csf}} \Delta = \frac{\Delta}{\sqrt{W_b^{\mathrm{csf}}}}$$

In the expression above, $b = 0, 1, \ldots, B - 1$ where B is the number of transform bands, α is an arbitrary positive constant, and W_b^{csf} is the energy weighting factor introduced in equation (4.43). It is usually convenient to select α so that $\max_b \left\{ W_b^{\mathrm{csf}} \right\} = 1$.

Although effective, this approach is limited in two fundamental ways. First, even when reversible transforms are employed, lossless coding will not generally be possible unless the values of $\sqrt{W_b^{\text{csf}}}$ are restricted to powers of 2. Second, the weighting employed is static, meaning that it cannot be later modified to meet the needs of different applications consuming the same compressed bit-stream in a scalable manner.

Both of these limitations are easy to circumvent using JPEG2000. This can be accomplished simply and effectively by replacing MSE with WMSE as the cost function which drives the formation of quality layers. Specifically, the coding pass distortion contributions, $D_i^{(z)}$, used by the PCRD-opt algorithm of Section 8.2 may be computed using the following modified form of equation (8.13).

$$D_i^{(z)} = W_{b_i}^{\text{csf}} G_{b_i} \sum_{\mathbf{j} \in \mathcal{B}_i} \left(\hat{y}_i\,[\mathbf{j}] - y_i\,[\mathbf{j}] \right)^2 \tag{16.1}$$

$$= \left(\frac{\alpha}{T_{b_i}^{\text{csf}}} \right) G_{b_i} \sum_{\mathbf{j} \in \mathcal{B}_i} \left(\hat{y}_i\,[\mathbf{j}] - y_i\,[\mathbf{j}] \right)^2$$

Here, b_i denotes the subband to which code-block $\mathcal{B}_i$ belongs, G_b is the energy gain factor associated with the synthesis waveforms in band b, and $\hat{y}_i\,[\mathbf{j}]$ is the dequantized representation of subband sample $y_i\,[\mathbf{j}]$, assuming that only the first z coding passes for the block are decoded. For simplicity, we have dropped the more explicit notation used in equation (8.13) to express the dependence of $\hat{y}_i\,[\mathbf{j}]$ on z.

By modifying only the distortion values used to construct quality layers, neither the transformed samples nor the quantization step sizes need be altered. In the remainder of this discussion, we will identify the weights in equation (16.1) as W_b instead of W_b^{csf}, since the specialization to CSF-based weighting factors is not of fundamental importance. If all bit-planes are encoded and subsequently decoded, identical images will be obtained, regardless of whether the cost function is MSE or WMSE, provided only that all W_b are non-zero.

Ignoring differences in the packet header overhead, code-streams based on MSE and WMSE differ only in the order in which compressed coding passes are included. Compared with an MSE optimized code-stream, a WMSE optimized code-stream will include coding passes from subbands with larger W_b earlier. Conversely, bands with smaller W_b values contribute their code-blocks' coding passes relatively later. Of course MSE is the special case of WMSE when all W_b are equal.

It should be clear from the discussion above that when reversible transforms are employed, lossless compression and decompression are supported regardless of the distortion measure used to create quality layers. Furthermore, weighting has no significant impact on the lossless file size.

The difference between MSE and WMSE code-streams becomes apparent, in both the reversible and irreversible cases, when decoding is halted prior to decoding all coding passes or when not all coding passes are included in the code-stream. When this occurs, relatively more coding passes will be decoded from bands with larger W_b, realizing lower distortion for these bands, in accordance with the weighting criteria.

VISUAL PROGRESSIVE WEIGHTING

The idea of visual progressive weighting (VIP) was introduced by Sharp Labs [91]. The motivation behind VIP is that early in the decoding of an embedded code-stream, visual quality may be substantially improved by imposing a fairly aggressive visual weighting strategy. However, as decoding proceeds and quality improves, less aggressive weighting may be appropriate. In fact, in the high quality regime, unweighted MSE (i.e., $W_b = 1$ $\forall b$) is typically preferred. VIP is then best thought of as a strategy that allows the weighting to change as a function of embedded encoding rate.

VIP was initially proposed in the context of single pass bit-plane coding of entire bands. A small amount of overhead was periodically included in the code-stream to signal the current set of weights. From these weights, the encoder and decoder can calculate which band is "most visually significant." This calculation is performed after each bit-plane is coded. The result is used to select the band from which the next bit-plane will be coded.

This idea was modified by Sharp Labs and SAIC/UA [144] to allow arbitrary ordering of multiple bit-plane *coding passes* within whole subbands. Again, some overhead is required to explicitly identify the ordering of coding passes. Ultimately, the process was extended to multiple coding passes of *code-blocks* by Taubman in VM3 (see Section 9.1.1). This final version enjoys the distinct advantage that it does not require any additional signalling mechanisms to identify the ordering information. This is because the quality layer mechanism in JPEG2000 already supports arbitrary ordering of the coding passes from different code-blocks.

In JPEG2000, VIP may be implemented by progressively adjusting the weights, W_b, used to define the code-block distortion contributions, $D_i^{(z)}$. The PCRD-opt algorithm of Section 8.2, depends only on the distortion-length slopes, $\lambda_i(z)$, defined by equation (8.4). The effect of W_{b_i} is simply to scale the $\lambda_i(z)$, having no impact on the convex hull set, $\mathcal{H}_i$, which defines the candidate truncation points for code-block

$\mathcal{B}_i$. For this reason, the weights need not be introduced until the point where each quality layer is actually formed, using the algorithm given on Page 344. The W_b may be changed from layer to layer following any desired strategy, without compromising the legality of the resulting code-stream.

16.1.2   WEIGHTS FOR COLOR IMAGERY

The JPEG2000 Part 1 document contains recommended values for normalized contrast sensitivity, $\sqrt{W_b^{\mathrm{csf}}}$, for use with color imagery. More precisely, the Final Draft International Standard contains a set of tables for monochrome imagery. The color tables were inadvertently omitted but will be included in the standard by means of a corrigendum. In what follows, we provide only the color tables, since the luminance values in these tables are suitable for use with monochrome imagery.

The color tables were first reported in [38] and are derived from CSF curves for Y, Cb, and Cr color components. These CSF curves were obtained experimentally according to the procedures described in [109], and are shown in Figure 16.1. Before the computation of the $\sqrt{W_b^{\mathrm{csf}}}$ as described below, each curve was normalized to have a maximum sensitivity of 1.0. As can be seen from this figure, the luminance curve has the same general shape as that in Figure 4.25.

It is common for the contrast sensitivity to vary significantly over the range of spatial frequencies occupied by a single subband. To determine a single value, $\sqrt{W_b^{\mathrm{csf}}}$, for the entire band one might simply take the average sensitivity over the relevant frequencies. A more conservative approach would be to select the peak CSF value over the frequency band. The recommended color tables described here are derived using a procedure which may be loosely understood as a combination of these two strategies. Rather than using the luminance curve directly, the sensitivity at low frequencies (below the frequency at which the curve peaks) is set to the peak value, i.e., 1.0. After making this modification, a separable CSF approximation is used to form the following simple average

$$\sqrt{W_b^{\mathrm{csf}}} = \frac{\int_{f_{b,\mathrm{vert}}^{\mathrm{low}}}^{f_{b,\mathrm{vert}}^{\mathrm{high}}} \frac{1}{T^{\mathrm{csf}}(f)} df}{\left(f_{b,\mathrm{vert}}^{\mathrm{high}} - f_{b,\mathrm{vert}}^{\mathrm{low}}\right)} \times \frac{\int_{f_{b,\mathrm{hor}}^{\mathrm{low}}}^{f_{b,\mathrm{hor}}^{\mathrm{high}}} \frac{1}{T^{\mathrm{csf}}(f)} df}{\left(f_{b,\mathrm{hor}}^{\mathrm{high}} - f_{b,\mathrm{hor}}^{\mathrm{low}}\right)}$$

Here, $f_{b,\mathrm{vert}}^{\mathrm{low}}$ and $f_{b,\mathrm{vert}}^{\mathrm{high}}$ denote the low and high vertical cut-off frequencies for band b, while $f_{b,\mathrm{hor}}^{\mathrm{low}}$ and $f_{b,\mathrm{hor}}^{\mathrm{high}}$ denote the low and high horizontal cut-off frequencies.

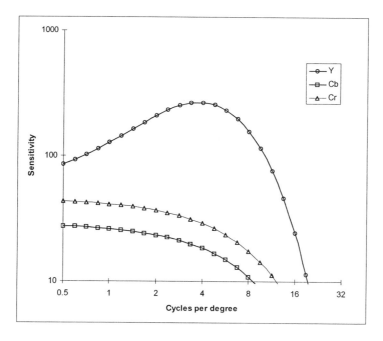

Figure 16.1.   CSF curves for YCbCr color imagery.

As discussed in Section 4.3.4, the location of the peak in a CSF curve varies according to the distance between the observer and the imagery. "Filling in the valley" to the left of this peak makes the match between assumed and actual viewing distance less critical. This modification of the CSF curve results in "extra bits" being coded at low frequencies. However, the pyramidal structure of the wavelet transform ensures that the fraction of samples receiving such extra bits is small. Generally then, the impact on compression efficiency is minor.

There will be three sets of tables given in the standard, corresponding to three assumed viewing distances. These viewing distances are 1000, 1700, and 3000 samples, respectively. For example, on a 100 sample per inch display, the 1700 tables are appropriate for a viewing distance of 17 inches. The relevant weights are included here in the form of Tables 16.1, 16.2, and 16.3. Each of these tables provide $\sqrt{W_b^{\text{csf}}}$ values for the bands of a 5 level dyadic wavelet transform. If fewer transform levels are used, the lower frequency subband weights may simply be discarded. These appear in columns to the left of each table. If more transform levels are used, the additional low frequency subbands may be assigned weights of

Table 16.1. CSF weights, $\sqrt{W_b^{\mathrm{csf}}}$, for a viewing distance of 1000 samples. The LL subbands (not shown) should be assigned a weight of 1.

		Level				
		5	4	3	2	1
	HL	1.000000	1.000000	1.000000	0.998276	0.756353
Y	LH	1.000000	1.000000	1.000000	0.998276	0.756353
	HH	1.000000	1.000000	1.000000	0.996555	0.573057
	HL	0.883196	0.793487	0.650482	0.450739	0.230503
Cb	LH	0.883196	0.793487	0.650482	0.450739	0.230503
	HH	0.833582	0.712295	0.531700	0.309177	0.113786
	HL	0.910877	0.841032	0.725657	0.552901	0.336166
Cr	LH	0.910877	0.841032	0.725657	0.552901	0.336166
	HH	0.872378	0.776180	0.625103	0.418938	0.200507

Table 16.2. CSF weights, $\sqrt{W_b^{\mathrm{csf}}}$, for a viewing distance of 1700 samples. The LL subbands (not shown) should be assigned a weight of 1.

		Level				
		5	4	3	2	1
	HL	1.000000	1.000000	1.000000	0.861593	0.307191
Y	LH	1.000000	1.000000	1.000000	0.861593	0.307191
	HH	1.000000	1.000000	1.000000	0.742342	0.108920
	HL	0.818766	0.689404	0.501652	0.280068	0.097816
Cb	LH	0.818766	0.689404	0.501652	0.280068	0.097816
	HH	0.745875	0.579220	0.362279	0.152290	0.031179
	HL	0.860885	0.757626	0.598537	0.388492	0.177435
Cr	LH	0.860885	0.757626	0.598537	0.388492	0.177435
	HH	0.803172	0.665951	0.470893	0.248566	0.077130

1. Regardless of the number of transform levels, the DC subband, LL_D, should be assigned a weight of 1 in each component.

SUB-SAMPLING IN JPEG2000

Up to this point, we have not mentioned sub-sampling of the Cb and Cr components. DCT based compression systems (e.g., JPEG, H.261, MPEG) typically employ 2:1 sub-sampling of Cb and Cr, in both the horizontal and the vertical directions. Although allowed, such sub-sampling is *not* recommended for use with JPEG2000. Superior visual performance is generally obtained without sub-sampling. This should not

Table 16.3. CSF weights, $\sqrt{W_b^{\mathrm{csf}}}$, for a viewing distance of 3000 samples. The LL subbands (not shown) should be assigned a weight of 1.

		Level				
		5	4	3	2	1
	HL	1.000000	1.000000	0.921045	0.410628	0.038487
Y	LH	1.000000	1.000000	0.921045	0.410628	0.038487
	HH	1.000000	1.000000	0.848324	0.182760	0.003075
	HL	0.717086	0.539437	0.319773	0.124021	0.023308
Cb	LH	0.717086	0.539437	0.319773	0.124021	0.023308
	HH	0.613777	0.403353	0.185609	0.044711	0.003413
	HL	0.780091	0.631632	0.428659	0.211871	0.060277
Cr	LH	0.780091	0.631632	0.428659	0.211871	0.060277
	HH	0.695128	0.509729	0.287593	0.100658	0.014977

be surprising, since the "hard sub-sampling" employed in DCT based schemes can be seen as a special case of CSF weighting, with weights W_{HL_1}, W_{LH_1} and W_{HH_1} all equal to 0. It follows that the weighting factors recommended for JPEG2000 represent a sort of "soft sub-sampling." Examining Tables 16.1, 16.2, and 16.3, we see that the weights for Cb and Cr are generally smaller than those for Y. At larger viewing distances, the trend is toward "hard sub-sampling." In fact, at a viewing distance of 3000 samples, the recommended values for $\sqrt{W_{\mathrm{HL}_1}}$, $\sqrt{W_{\mathrm{LH}_1}}$ and $\sqrt{W_{\mathrm{HH}_1}}$ are very small in *all three* components.

16.1.3 SUBJECTIVE COMPARISON OF JPEG2000 WITH JPEG

The CSF weighting tables of the previous section have been employed in visual tests conducted by Fujifilm California and Eastman Kodak [37]. These tests compared the visual quality of JPEG2000 to that of JPEG.

The JPEG2000 imagery for these tests was generated using the CDF 9/7 irreversible transform. The JPEG imagery was generated using the Independent JPEG Group implementation, cjpeg.[1] The default mode of cjpeg was employed (i.e., baseline sequential mode, see Chapter 19) and optimized Huffman tables were used. All visual comparisons were made using 24-bit color prints at 300 dpi.

The tests were conducted using six 24-bit color images with natural photographic content. The luminance components for two of these im-

---

[1] Source code for this implementation is available at http://www.ijg.org.

ages, bike and woman, are shown in Figure 8.21. Ten "reference" JPEG images were created for each of the six original images. These reference images were created by compressing and decompressing to precise rates of 0.3, 0.4, 0.5, 0.6, 0.7, 0.8, 0.9, 1.0, 1.2, and 1.4. All rates were in units of bits per color sample. For example, a rate of 0.5 bits/sample corresponds to a compression ratio of 48:1. Four JPEG2000 "test" images were also created for each original image. These test images were compressed/decompressed to rates of 0.25, 0.50, 0.75, and 1.00 bits/sample.

Visual quality testing was then carried out by six observers.[2] For a given original image, the JPEG reference prints were placed on a table in order of lowest to highest rate. Each observer was given a JPEG2000 test image, and asked to find the JPEG reference image of comparable quality. This process was repeated for each JPEG2000 rate and each original image. In this way, the rate required for JPEG to achieve the same visual quality as JPEG2000 was determined.

The average of these results over the six observers is shown in Figure 16.2. Two of the curves in this figure show results for the bike and woman images individually. The third curve shows the results averaged over all original images. Each of these three curves represents the rate required by JPEG to achieve comparable perceptual quality to that of JPEG2000. For ease of comparison, the forth curve indicates the rate required by JPEG2000 (to achieve the quality of JPEG2000). Of course this latter curve is the identity and corresponds to a straight line of slope 1.

From the figure, we see that in the case of the bike image, JPEG2000 provides a decrease in rate ranging from 14% to 47% over that of JPEG. Equivalently, JPEG requires an increase in rate between 16% and 88% to achieve equivalent visual quality to that of JPEG2000. The results for the woman image are similar, with the savings achieved by JPEG2000 ranging from 12% to 52%. Finally, we see that on average, JPEG2000 provides a reduction in required rate between 11% and 53%.

In each case, the largest improvements occur at the lower rates. This is not surprising since at low rates, the "blocking" artifacts of JPEG tend to be significantly more annoying than the "smoothing" of JPEG2000, as demonstrated in Figure 4.30. In fact, a general observation reported in [37] was that JPEG imagery tends to be "sharper" than JPEG2000 imagery at all encoding rates. Furthermore, the sharpness of JPEG increases more quickly than that of JPEG2000 as encoding rate is in-

---

[2] Subsequent testing at Fujifilm Software California using more observers yielded similar results (private communication: Troy Chinen).

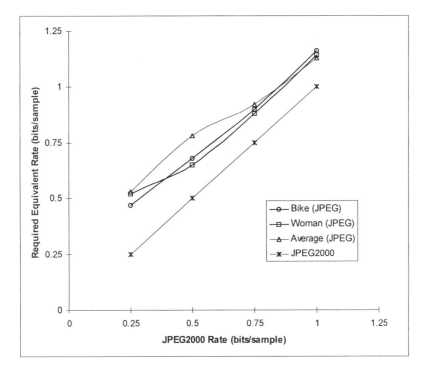

Figure 16.2.   Rate required to achieve visual quality equivalent to that of JPEG2000.

creased. On the other hand, the JPEG sharpness advantage is more than overcome by the absence of blocking artifacts in JPEG2000.

16.1.4   EXPLOITING VISUAL MASKING

As noted in Section 4.3.4, CSF data measures only the detectability of sinusoidal patterns against a uniform background. For image compression, however, we are interested in the detectability of quantization artifacts when they are superimposed on the original image. Image activity tends to mask the visibility of quantization artifacts. An example of this masking phenomenon is given in Figure 4.29. The figure provides evidence for the fact that artifacts produced by quantization errors in one subband are masked most strongly by image content from the same subband. This is known as intra-band masking and we shall henceforth consider only this effect.

MASKING SENSITIVE DISTORTION MEASURES

We saw above that CSF effects may be readily incorporated into JPEG2000 code-streams by scaling the coding pass distortion estimates, $D_i^{(z)}$, which are used to form quality layers. In particular, CSF modified distortion estimates may be found using equation (16.1). To account for the effects of visual masking, these distortion estimates may be further divided by the square of the relevant threshold elevation factor, $t_b(\mathbf{p})$, as given by equation (4.45). $t_b(\mathbf{p})$ represents the amount by which the detection threshold (amplitude) for an artifact in band b at location $\mathbf{p}$ is increased by masking, relative to the CSF detection threshold, T_b^{csf}.

For our purposes here, it is convenient to replace equation (4.45) with the almost identical formula

$$t_b(\mathbf{p}) = \sqrt{1 + (m_b(\mathbf{p}))^{2\rho}}$$

so that the modified distortion estimates become

$$D_i^{(z)} = W_{b_i}^{\text{csf}} G_{b_i} \sum_{\mathbf{j} \in \mathcal{B}_i} \left(\frac{\hat{y}_i[\mathbf{j}] - y_i[\mathbf{j}]}{t_i[\mathbf{j}]} \right)^2$$

$$= W_{b_i}^{\text{csf}} G_{b_i} \sum_{\mathbf{j} \in \mathcal{B}_i} \frac{(\hat{y}_i[\mathbf{j}] - y_i[\mathbf{j}])^2}{1 + (m_i[\mathbf{j}])^{2\rho}} \tag{16.2}$$

In these equations, $t_i[\mathbf{j}]$ denotes the threshold elevation factor at the location corresponding to subband sample $\mathbf{j}$ in code-block $\mathcal{B}_i$ and $m_i[\mathbf{j}]$ denotes the normalized masking contrast for the same location in the same band. In particular, $m_i[\mathbf{j}]$ is defined as the absolute masking contrast, $M_i[\mathbf{j}]$, divided by the CSF detection threshold, $T_{b_i}^{\text{csf}}$.

In what follows, we use the fact that the contrast (amplitude) of in-band signals is roughly preserved by the wavelet transform. As a result, $M_i[\mathbf{j}]$ may be directly measured from the amplitudes of subband samples in the neighbourhood of location $y_i[\mathbf{j}]$. This property is a consequence of the normalization convention expressed in equation (10.12), which sets the nominal gains of all wavelet analysis filters to 1. If the transform is implemented using a different convention, the differences must be reconciled through appropriate scaling of the $y_i[\mathbf{j}]$ prior to computing the masking contrast, $M_i[\mathbf{j}]$. This is particularly relevant when the reversible transform is employed, since its linearized high-pass analysis filters have a nominal gain of 2, rather than 1 (see equation (10.16)). Our discussion here is based on the unit nominal range convention outlined in Section 10.1.1, whereby the image and hence all subband samples have a unit nominal dynamic range, from $-\frac{1}{2}$ to $\frac{1}{2}$. The CSF detection thresholds, T_b^{csf}, are expressed with regard to this same normalization. To compute the masking elevation factors, $t_i [\mathbf{j}]$, for reversibly transformed images, all quantities must be appropriately rescaled.

Introduction of visual masking effects into the code-block distortion estimates used for PCRD-opt rate control was first proposed in [149]. In that work, $M_i [\mathbf{j}]$ is estimated as

$$M_i [\mathbf{j}] = \left(\frac{1}{\|\mathcal{K}_{\mathbf{j}}\|} \sum_{\mathbf{k} \in \mathcal{K}_{\mathbf{j}}} |y_i [\mathbf{k}]|^{\rho} \right)^{\frac{1}{\rho}} \tag{16.3}$$

where $\mathcal{K}_{\mathbf{j}}$ is a local neighbourhood of samples about the location $\mathbf{j}$, and $\|\mathcal{K}_{\mathbf{j}}\|$ is the number of samples in this neighbourhood. Since the exponent ρ is typically less than 1, this type of local average tends to emphasize the influence of neighbouring samples with smaller amplitudes. As such, it bears some resemblance to a geometric mean. Subjective experiments leading to the selection of this measure found the "ρ-mean" in equation (16.3) to yield more consistent visual quality than an arithmetic mean of the neighbouring sample amplitudes.

Substituting equation (16.3) into equation (16.2) and noting that $m_i [\mathbf{j}] = M_i [\mathbf{j}] / T_{b_i}^{\mathrm{csf}}$, we obtain the following distortion estimates.

$$
\begin{aligned}
D_i^{(z)} &= W_{b_i}^{\mathrm{csf}} G_{b_i} \sum_{\mathbf{j} \in \mathcal{B}_i} \frac{(\hat{y}_i [\mathbf{j}] - y_i [\mathbf{j}])^2}{1 + \left(T_{b_i}^{\mathrm{csf}} \right)^{-2\rho} \left(\frac{1}{\|\mathcal{K}_{\mathbf{j}}\|} \sum_{\mathbf{k} \in \mathcal{K}_{\mathbf{j}}} |y_i [\mathbf{k}]|^{\rho} \right)^2} \\
&= \alpha^2 \frac{G_{b_i}}{\left(T_{b_i}^{\mathrm{csf}} \right)^2} \sum_{\mathbf{j} \in \mathcal{B}_i} \frac{(\hat{y}_i [\mathbf{j}] - y_i [\mathbf{j}])^2}{1 + \left(T_{b_i}^{\mathrm{csf}} \right)^{-2\rho} \left(\frac{1}{\|\mathcal{K}_{\mathbf{j}}\|} \sum_{\mathbf{k} \in \mathcal{K}_{\mathbf{j}}} |y_i [\mathbf{k}]|^{\rho} \right)^2}
\end{aligned}
\tag{16.4}
$$

Here, we have used the fact that T_b^{csf} and W_b^{csf} are related through equation (4.43), with α an arbitrary (irrelevant) constant.

An important observation concerning equation (16.4) is that the masking sensitive distortion measure is substantially less sensitive to exact knowledge of the contrast sensitivity thresholds, T_b^{csf}, than the CSF-based distortion measure of equation (16.1). This is because the term $\left(T_{b_i}^{\mathrm{csf}} \right)^{-2\rho}$ inside the summation partially cancels the term $\left(T_{b_i}^{\mathrm{csf}} \right)^2$ outside the summation. Almost perfect cancellation is obtained at high masking levels and with exponents, ρ, close to 1. This property is most convenient, since in many applications we may have little knowledge concerning the viewing conditions and hence the contrast sensitivity thresholds. Notable among such applications are those in which the viewer is permitted to interactively "zoom" into or out of the image.

Equation (16.4) was first used in [149] with the CSF detection contrasts, T_b^{csf}, all set to a small constant, thereby avoiding any explicit dependence on viewing distance at all. Visual experiments leading to this work suggested an exponent of $\rho = \frac{1}{2}$. This is a little lower than the range of 0.6 to 1.0 suggested by the discussion of visual masking in Section 4.3.4. Some promising visual improvements were obtained in this work, particularly with photographic images containing human portraits. These observations were supported by separate visual experiments performed by Sharp Labs of America [3]. It is worth noting that in all of these experiments, the local averages of equation (16.3), are actually computed over disjoint sub-blocks of size 8×8 and the neighbourhood, $\mathcal{K}_{\mathbf{j}}$, is identified with the particular sub-block which contains location $\mathbf{j}$. This modification substantially reduces the computational complexity of the procedure.

ALTERNATIVE APPROACHES

An obvious limitation of visual optimization schemes which operate on the distortion values used by the PCRD-opt algorithm, is that it is only possible to alter the number of coding passes contributed by whole code-blocks to any given quality layer. As a result, the use of smaller code-blocks can often yield superior image quality, with 32×32 being a preferred size for masking-sensitive distortion measures. Even then, a 32×32 code-block in the LH_1, HL_1 or HH_1 subbands occupies roughly 64×64 samples in the relevant image component, while code-blocks in lower resolution subbands occupy much larger regions. For this reason, masking-sensitive distortion measures are largely ineffective when applied to images which are small, or contain many small regions with differing statistics.

These problems may be avoided by using masking attributes to directly modulate the quantization of individual samples. One such scheme is proposed by Höntsch and Karam [76], in which the masking contrast is estimated from a causal neighbourhood of previously quantized and coded subband samples. The masking contrast must be estimated at both the encoder and the decoder. Unfortunately, in a scalable setting, the encoder cannot know ahead of time how many bit-planes will actually be available to the decoder, for use in estimating the masking contrast. Also, the masking contrast may only be computed from causal neighbours.

These difficulties are partially addressed by the visual masking extension mechanisms supported by JPEG2000 Part 2, as explained in Section 15.5. This extension was conceived in light of the observations made above concerning distortion-based masking optimizations [3]. The reader should note the close resemblance between the denominators of equations (15.1) and (16.4). The reader should also note that the Part 2 extension requires the decoder to explicitly compute masking contrasts. On the other hand, distortion-based masking optimizations affect only the encoder and are fully compatible with Part 1 decoders. Nevertheless, the Part 2 masking extension can offer significant improvements in visual performance over distortion-based masking alone, particularly at higher bit-rates or with smaller images.

16.2 REGION OF INTEREST ENCODING

In some applications, an image region of particular interest can be identified at the point when the image is encoded, where this region is to be assigned higher priority in the encoding process. As an example, a radiologist might identify regions of interest within X-ray images of a patient, requesting that these regions be encoded losslessly when the image is compressed and transferred to a pathologist. More generally, it is expected that the identified regions of interest should have smaller distortion than the image as a whole. When the code-stream possesses multiple quality layers, it is expected that this property should hold for each of the reduced bit-rate streams obtained by discarding one or more of the final quality layers. In particular, when a quality progressive code-stream is transmitted incrementally to a client (e.g., the pathologist), the image quality is expected to improve most rapidly within the region of interest. Once the entire code-stream has been received, the client may be able to reconstruct a lossless representation of entire image, but lossless reconstruction of the region of interest should be possible at an earlier stage, when only part of the code-stream has been received.

JPEG2000 Part 1 provides two quite different mechanisms for assigning higher priority to regions of interest at encode time. We refer to these mechanisms collectively as ROI encoding schemes and we refer to the regions of interest themselves as the "foreground," with the remainder of the image identified as "background." The first ROI encoding mechanism is the max-shift method described in Section 10.6.2. A second ROI encoding mechanism may be constructed by modulating the cost function which drives the construction of quality layers. We refer to this second method as "implicit" ROI encoding, since there is no explicit information in the code-stream to suggest that the encoder has treated any region of the image differently to another. JPEG2000 Part 2 provides more comprehensive support for ROI encoding, as explained in Section 15.8.

16.2.1   MAX-SHIFT ROI ENCODING

JPEG2000 provides an "explicit" mechanism for assigning higher priority to arbitrary regions of an image. This explicit mechanism is known as the "max-shift" method and is described in detail in Section 10.6.2. The encoder scales up those subband samples which are involved in reconstructing the foreground region. The scale factor must be a power of 2 so that the scaling process is equivalent to left shifting the magnitude bit-planes by a single quantity U (up-shift), which is signalled using the *RGN* marker segment. All compliant decoders are expected to restore the scaled subband samples to their correct levels, which effectively reduces the quantization step size associated with those samples by 2^{-U}. As a result, the foreground region is reconstructed with higher fidelity than the image as a whole.

In JPEG2000 Part 1 the shift, U, must be sufficiently large that the decoder can distinguish between scaled and unscaled subband samples solely on the basis of their coded magnitude bits. The relevant conditions are explained in Section 10.6.2. For our purpose here, it is sufficient to appreciate that U must generally be at least as large as the number of magnitude bit-planes in any subband (see equation (10.32)). As an example, consider reversible compression of 8-bit colour imagery, using the RCT (Reversible Colour Transform). Under these conditions, the number of magnitude bit-planes in the HH_d bands of the chrominance components may need to be as large as 12 to avoid any risk of numerical overflow or underflow problems (see Table 17.4 and the surrounding discussion). Consequently, the up-shift must also satisfy $U \geq 12$. This means that the effective quantization step size associated with the foreground is at least 4096 times smaller than that for the background samples!

Although the max-shift method elegantly circumvents any need to explicitly describe the shape of the foreground region, the price payed for this convenience is the "all-or-nothing" effect demonstrated by the solid curve in Figure 16.3. In the example of Figure 16.3, the monochrome 8-bit image Café (Figure 8.21) is losslessly compressed in a quality progressive code-stream, using 30 quality layers and an up-shift of $U = 12$. A single rectangular region of interest occupies one quarter of the image area and is centred in the middle of the image. As quality layers are decompressed one by one, the quality (PSNR) of the reconstructed image is calculated separately for the foreground region and the image as a whole. As expected, the foreground quality improves rapidly, with little change in the background, until all foreground information has been received. In this example, the foreground is recovered losslessly at an overall bit-rate of 1.73 bits/sample.

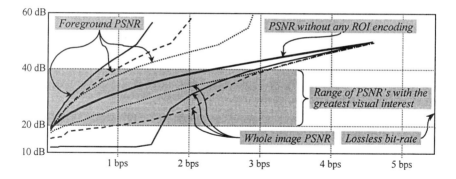

Figure 16.3. Rate-distortion characteristics for the foreground and for the whole image, with both the max-shift (solid line) and implicit (dashed and dotted lines) ROI encoding strategies.

If the background is of any interest at all, this behaviour may be too extreme. The client must wait to receive a very large number of compressed bytes before the background even becomes recognizable. Another drawback of the max-shift method is that the decoder must be capable of processing a large number of bit-planes if the background is to be fully decoded. As explained in Section 18.3.4, decompressors whose capabilities extend only to Compliance Class-0 may not be able to recover any information from the background whatsoever.

16.2.2 IMPLICIT ROI ENCODING

The EBCOT paradigm at the heart of JPEG2000 provides an alternative approach to ROI encoding. As explained in Sections 8.1.3 and 12.5, each quality layer in the code-stream comprises an arbitrary contribution from the embedded bit-stream of each code-block of each subband. One way to increase the quality associated with a region of interest is to include relatively larger contributions from the code-blocks which are involved in its reconstruction. This may be achieved by increasing the coding pass distortion estimates, $D_i^{(z)}$, which drive the PCRD-opt algorithm of Section 8.2, for every code-block $\mathcal{B}_i$ whose samples contribute to the reconstruction of the foreground. We refer to this as "implicit" ROI encoding, since code-streams constructed in this way contain no explicit indication that ROI encoding has taken place. In the discussion which follows, we use the term "foreground block" to refer to any code-block whose distortion values are scaled to reflect its role in reconstructing the region of interest.

Implicit ROI encoding has both advantages and disadvantages relative to the max-shift method described above. One key advantage is that the scaling of foreground block distortion estimates may be adjusted to match the "degree of interest." In this way, the foreground region need not be given absolute priority over the background, avoiding the "all-or-nothing" behaviour which accompanies the max-shift method. This is illustrated in Figure 16.3, where the dashed and dotted curves correspond to two different distortion scaling factors. The dotted curves are generated by scaling the MSE distortion of all foreground blocks by 64. The dashed curves correspond to an MSE scaling factor of 4096.

The effect of these MSE scaling factors is similar to shifting the magnitude bit-planes by 3 and 6 bit positions, respectively, which may be compared with the $U = 12$ bit shift required by the max-shift method. Figure 16.3 clearly reveals the effect of these different scaling factors on the ratio between foreground and background quality. In each case, the background becomes recognizable at much lower bit-rates than it does with the max-shift method. On the other hand, lossless reconstruction of the foreground is achieved at correspondingly higher bit-rates: 2.41 and 3.91 bits/sample for the dashed and dotted curves respectively.

Another advantage of implicit ROI encoding is that it has no impact on the number of bit-planes which must be processed for complete decoding of the image. In particular, even Class-0 decompressors should be capable of fully decoding an appropriately constructed, losslessly compressed 8-bit image.

One significant disadvantage of implicit ROI encoding is that adjustments may be made only on a block by block basis. This is exactly the same limitation which confounds the visual masking optimizations described in Section 16.1.4. As in that case, we find here that the preferred code-block size for implicit ROI encoding is 32×32, rather than the maximum size of 64×64 supported by the standard. The dashed and dotted curves in Figure 16.3 were both generated using 32×32 blocks, while the solid curve (max-shift method) was generated using 64×64 code-blocks. The reduced code-block size has a slight detrimental effect on the final lossless compressed bit-rate, which is 0.08 bits/sample larger than that obtained with the max-shift method.

It should be noted that the example of Figure 16.3 involves particularly simple geometry, with a single large rectangular region of interest. The superior foreground/background discrimination offered by the max-shift method may become more significant when working with smaller and/or more complex foreground regions. In any event, the mechanism of choice for ROI encoding is clearly very much dependent on the requirements of the application. The interested reader is recommended to the Kakadu software supplied with this text, which implements both of the ROI encoding strategies described here and permits the specification of arbitrary foreground regions through an auxiliary "mask" image.

16.3 BI-LEVEL IMAGERY

As mentioned in Section 9.1.1, one of the desired features for JPEG-2000 was efficient compression of both bi-level and continuous tone imagery, using a single algorithm. In practice, compression of continuous tone images has been the principle focus driving the development of JPEG2000. Nevertheless, subject to appropriate selection of the coding parameters, JPEG2000 is able to efficiently compress bi-level images, as well as other types of low bit-depth imagery. In this section, we provide some indications and suggestions in regard to this capability. We specifically consider only the case of bi-level images here. However, we note that JPEG2000 has been found to yield comparable performance to JPEG-LS and substantially better performance than GIF when losslessly compressing low bit-depth (e.g., 2 or 4 bits per sample) palettized images, after suitable re-arrangement of the palette indices [177].

Bi-level images have sample values $x[\mathbf{n}] \in \{0,1\}$. As such, they have an unsigned $B = 1$ bit representation, which must be level adjusted to conform to equation (10.1). This means that an input value of $x[\mathbf{n}] = 0$ is transformed to -1 prior to coding, while an input value of $x[\mathbf{n}] = 1$ becomes 0. This subtlety is easily overlooked when implementing the JPEG2000 standard.

Since we are interested in comparing bi-level compression efficiency with that of JBIG, it is worth noting that both algorithms are based on binary arithmetic coding. For raw coding efficiency, JPEG2000 is best used without any wavelet transform, setting the number of DWT levels to $D = 0$. In this case, the principle differences between JBIG and JPEG2000 lie in the structure and number of distinct probability modeling contexts used by the two algorithms. Figure 2.13 illustrates the two fixed context neighbourhood models defined by the JBIG standard. In each case, 10 neighbouring sample values are used to index into a table of 1024 separate adaptive probability models which drive the arithmetic coding procedure.

JPEG2000's embedded block coding algorithm codes bi-level image data using exactly $Z = 1$ coding pass[3], a "cleanup pass." All of the information is coded using the significance coding primitive, whose 8-sample context neighbourhood is illustrated in Figure 8.11. Following

---

[3]This follows from equation (8.12), noting that the number of magnitude bit-planes is $K = 1$.

the stripe-oriented scan of Figure 8.15, four of these 8 neighbours are guaranteed to be insignificant and one may deduce that the significance coding context label, $\kappa^{\text{sig}}[\mathbf{j}]$, may take on only 7 of the values described by Table 8.1. One additional coding context arises in connection with the run mode, for a total of only 9 different adaptive probability models. Unlike JBIG, the probability models are also reset at the beginning of each new code-block.

These limitations are principally responsible for the somewhat poorer performance of JPEG2000 in comparison with JBIG. The JPEG2000 block coder must also code the sign of every significant sample. These are the samples whose original value was $x[\mathbf{n}] = 0$, which were level adjusted to -1; their sign is invariably negative. The deterministic nature of the sign information is rapidly learned by the adaptive probability estimation machinery, but the learning penalty described in Section 2.3.5 may be incurred within each of the 5 sign coding contexts which can actually occur[4]. This overhead may be minimized by assigning $x[\mathbf{n}] = 1$ to the more prevalent (background) samples and $x[\mathbf{n}] = 0$ to the less prevalent (foreground) samples. The same policy tends to increase the frequency and length of insignificance runs and hence the efficiency of the significance coding primitive's run-mode.

The effect of this somewhat unnatural policy[5] is revealed by the first column of compression ratios reported in Table 16.4. These results are based on eight CCITT facsimile test charts[6]; in each case, the background is white so that higher compression is achieved by assigning black and white samples the values 0 and 1, respectively. The performance of JPEG2000 is very similar to that of the CCITT facsimile compression standard G4 [35]. The reader will notice that JBIG outperforms JPEG2000, producing files which are about 30% smaller at lower resolutions and about 37% smaller at higher resolutions, when raw compression efficiency alone is of interest. The algorithms have roughly similar levels of complexity, although it could be argued that the much smaller number of coding contexts used by JPEG2000 make it more amenable to hardware implementation[7]. Applications which require compressed do-

---

[4]The sign coding contexts which can occur are those for which $\bar{\chi}^{\text{h}}[\mathbf{j}], \bar{\chi}^{\text{v}}[\mathbf{j}] \in \{0, -1\}$ in Table 8.2.

[5]It is arguably more natural to assign $x[\mathbf{n}] = 0$ to the background samples and $x[\mathbf{n}] = 1$ to the foreground.

[6]These are documents 1 through 8 from the Standard Image Set CD-03, put out by the US National Communications System.

[7]JPEG2000 ostensibly requires the buffering of an entire row of code-blocks when used with applications which supply or consume image data in line by line fashion. However, since only one coding pass is involved, the stripe-oriented scanning pattern used by JPEG2000's block

Table 16.4. Average compression ratios for JBIG and JPEG2000 compression of eight CCITT facsimile test charts scanned at various resolutions. Image dimensions are 2339 × 1728, 4677 × 3456 and 7016 × 5184 at 200, 400 and 600 dpi, respectively. B/W means that 0 is assigned to black and 1 is assigned to white. W/B means that 0 is assigned to white and 1 to black.

Resolution	Method	Number of Resolution Levels					
		1	2	3	4	5	6
200 dpi	JBIG	21.2	20.4	19.6	19.2	19.0	18.8
	J2K B/W	14.8	9.2	7.8	7.6	7.5	7.5
	J2K W/B	12.5	8.5	7.5	7.2	7.1	7.0
400 dpi	JBIG	41.2	42.1	41.0	39.9	39.4	39.1
	J2K B/W	25.8	16.2	13.8	12.9	12.8	12.7
	J2K W/B	20.4	15.3	13.2	12.5	12.3	12.2
600 dpi	JBIG	53.9	56.5	56.2	55.0	54.2	53.9
	J2K B/W	33.9	21.4	18.4	17.3	17.0	16.9
	J2K W/B	25.8	20.3	18.0	17.1	16.7	16.6

main rotation, flipping or cropping can take advantage of the fact that JPEG2000 codes the image in independent blocks.

The last five columns in Table 16.4 report the degradation in compression efficiency suffered by JBIG and JPEG2000 as resolution scalability is introduced. Both JBIG and JPEG2000 allow the image to be coded as a multi-resolution hierarchy. JBIG's multi-resolution transform is carefully matched to the bi-level image compression task, so that compression efficiency is hardly affected by the introduction of resolution scalability. In JPEG2000 Part 1, resolution scalability may be introduced through use of the reversible 5/3 wavelet transform, which is designed primarily for efficient compression of continuous tone imagery. Evidently, the use of this transform has a significant adverse impact on the compression of bi-level images.

---

coder allows efficient implementations to work with as little as four buffered image lines, rendering its memory requirements similar to those of JBIG.

Chapter 17

IMPLEMENTATION CONSIDERATIONS

17.1 BLOCK CODING: SOFTWARE

Chapter 12 contains a complete procedural description of the embedded block coding algorithm, which may be implemented directly in software with relatively little effort. In this section, we describe non-obvious implementation techniques which may be used to substantially improve execution speed in comparison with a direct implementation of the coding procedures.

These techniques are important because block coding is the most computationally demanding task in an efficient implementation of the JPEG2000 standard. The Kakadu implementation supplied with this text relies heavily upon the techniques described here, supplementing them with a variety of other optimization strategies.

17.1.1 MQ CODER TRICKS

In Section 12.1.2, we mentioned that the MQ encoder algorithm on Page 478 could be modified so as to reduce the number of comparisons ("If" statements). This is important for software implementations, since conditional branches tend to disrupt the CPU's execution pipeline[1]. The key to reducing the number of comparisons is to observe that conditional exchange must always be accompanied by renormalization (see equation (2.20)). The MQ encoder can then be implemented as follows.

MQ-Encode Procedure

Set $\bar{p} = \bar{p}(\Sigma_\kappa)$

---

[1] The degree to which this is true depends on the CPU architecture.

$$A \leftarrow A - \bar{p}$$

If $x = s_\kappa$, (coding an MPS)

    If $A \geq 2^{15}$, (no renormalization and hence no conditional exchange)

$$C \leftarrow C + \bar{p}$$

    else

        If $A < \bar{p}$, (conditional exchange)

$$A \leftarrow \bar{p}$$

        else

$$C \leftarrow C + \bar{p}$$
$$\Sigma_\kappa \leftarrow \Sigma_{\mathrm{mps}}(\Sigma_\kappa)$$

        Do, (perform renormalization shift)

$$A \leftarrow 2A, \ C \leftarrow 2C, \ \bar{t} \leftarrow \bar{t} - 1$$
           If $\bar{t} = 0$, **Transfer-Byte**$(\bar{T}, C, L, \bar{t})$

        while $A < 2^{15}$

else (coding an LPS; renormalization is inevitable)

    If $A < \bar{p}$, (conditional exchange)

$$C \leftarrow C + \bar{p}$$

    else

$$A \leftarrow \bar{p}$$
$$s_\kappa \leftarrow s_\kappa \oplus X_s(\Sigma_\kappa), \ \Sigma_\kappa \leftarrow \Sigma_{\mathrm{lps}}(\Sigma_\kappa)$$

    Do, (perform renormalization shift)

$$A \leftarrow 2A, \ C \leftarrow 2C, \ \bar{t} \leftarrow \bar{t} - 1$$
        If $\bar{t} = 0$, **Transfer-Byte**$(\bar{T}, C, L, \bar{t})$

    while $A < 2^{15}$

As it turns out, most symbols which are coded have highly skewed probabilities so that $\bar{p}$ is small and renormalization is required relatively infrequently. As a result, only the first two "If" statements in the above algorithm are likely to be executed frequently. The same observation allows us to modify the MQ decoding procedure on Page 483 as follows.

MQ Decode Procedure (returns x)

Set $\bar{p} = \bar{p}(\Sigma_\kappa)$

$x \leftarrow s_\kappa$ (set to MPS for now, since this is most likely)

$$A \leftarrow A - \bar{p}$$

If $C^{\mathrm{active}} \geq \bar{p}$, (upper sub-interval selected)

$$C^{\mathrm{active}} \leftarrow C^{\mathrm{active}} - \bar{p}$$

    If $A < 2^{15}$, (need renormalization and perhaps conditional exchange)

        If $A < \bar{p}$, (conditional exchange, LPS decoded)

$$x \leftarrow 1 - s_\kappa$$
$$s_\kappa \leftarrow s_\kappa \oplus X_s(\Sigma_\kappa), \ \Sigma_\kappa \leftarrow \Sigma_{\mathrm{lps}}(\Sigma_\kappa)$$

      else (MPS decoded)

          $\Sigma_\kappa \leftarrow \Sigma_{\mathrm{mps}}(\Sigma_\kappa)$

          Do, **Renormalize-Once**$(A, C, \bar{t}, \bar{T}, L)$, while $A < 2^{15}$

    else (lower sub-interval selected; renormalization is inevitable)

      If $A < \bar{p}$, (conditional exchange, MPS decoded)

          $\Sigma_\kappa \leftarrow \Sigma_{\mathrm{mps}}(\Sigma_\kappa)$

      else (LPS decoded)

          $x \leftarrow 1 - s_\kappa$

          $s_\kappa \leftarrow s_\kappa \oplus X_s(\Sigma_\kappa)$, $\Sigma_\kappa \leftarrow \Sigma_{\mathrm{lps}}(\Sigma_\kappa)$

    $A \leftarrow \bar{p}$

    Do, **Renormalize-Once**$(A, C, \bar{t}, \bar{T}, L)$, while $A < 2^{15}$

As mentioned above we expect $\bar{p}$ to be small and renormalization events to be rare most of the time. For most symbols then, the test for $C^{\mathrm{active}} \geq \bar{p}$ succeeds and the test for $A < 2^{15}$ fails. We use the term CDP (Common Decoding Path) to refer to this common scenario.

The number of tests associated with the CDP may be further reduced from two to one. To see this, observe that C^{active} and A are both decremented by $\bar{p}$ within the CDP. Now define a new state variable, D, to be the minimum of $A - 2^{15}$ and C^{active}. The CDP symbols are those which leave D non-negative after subtraction of $\bar{p}$ and so we have only to test D. In fact, there is no need to explicitly decrement A, C and D all by $\bar{p}$ when a CDP symbol occurs. The algorithm below defers adjustments to A and C until the next non-CDP symbol. Notice that CDP symbols require only one table lookup, one subtraction and one comparison!

MQ Decode Procedure (one test CDP) (returns x)

Set $\bar{p} = \bar{p}(\Sigma_\kappa)$

$x \leftarrow s_\kappa$ (set to MPS for now, since this is most likely)

$D \leftarrow D - \bar{p}$

If $D < 0$, (non-CDP decoding)

    $A \leftarrow A + D$

    $C^{\mathrm{active}} \leftarrow C^{\mathrm{active}} + D$

    If $C^{\mathrm{active}} \geq 0$, (upper sub-interval selected, must have $A < 2^{15}$)

      If $A < \bar{p}$, (conditional exchange, LPS decoded)

          $x \leftarrow 1 - s_\kappa$

          $s_\kappa \leftarrow s_\kappa \oplus X_s(\Sigma_\kappa)$, $\Sigma_\kappa \leftarrow \Sigma_{\mathrm{lps}}(\Sigma_\kappa)$

      else (MPS decoded)

          $\Sigma_\kappa \leftarrow \Sigma_{\mathrm{mps}}(\Sigma_\kappa)$

      Do, **Renormalize-Once**$(A, C, \bar{t}, \bar{T}, L)$, while $A < 2^{15}$

    else (lower sub-interval selected; renormalization is inevitable)

$$C^{\text{active}} \leftarrow C^{\text{active}} + \bar{p}$$

If $A < \bar{p}$, (conditional exchange, MPS decoded)

$$\Sigma_\kappa \leftarrow \Sigma_{\text{mps}}\left(\Sigma_\kappa\right)$$

else (LPS decoded)

$$x \leftarrow 1 - s_\kappa$$
$$s_\kappa \leftarrow s_\kappa \oplus X_s\left(\Sigma_\kappa\right), \ \Sigma_\kappa \leftarrow \Sigma_{\text{lps}}\left(\Sigma_\kappa\right)$$
$$A \leftarrow \bar{p}$$

Do, **Renormalize-Once**$\left(A, C, \bar{t}, \bar{T}, L\right)$, while $A < 2^{15}$

$$D \leftarrow \min\left\{C^{\text{active}}, A - 2^{15}\right\}$$

$A \leftarrow A - D$ (we will add D back again at the next non-CDP symbol)

$$C^{\text{active}} \leftarrow C^{\text{active}} - D$$

Similar methods may be employed to reduce the number of CDP comparisons to 1 during encoding, as well as decoding. The Kakadu software supplied with this text demonstrates such techniques.

17.1.2   STATE BROADCASTING

The bit-plane coding procedures described in Section 12.2 require construction of a context label, $\kappa^{\text{sig}}[\mathbf{j}]$, at every sample location, $\mathbf{j}$, in both the significance propagation pass and the cleanup coding pass. $\kappa^{\text{sig}}[\mathbf{j}]$ is a function of the significance states associated with the location's eight neighbours. A direct implementation might require 8 memory accesses, plus a variety of logical operations and tests to construct each label, $\kappa^{\text{sig}}[\mathbf{j}]$.

Fortunately, it is possible to dramatically reduce the complexity of these operations by observing that the value of $\kappa^{\text{sig}}[\mathbf{j}]$ can only change when one of the location's eight neighbours becomes significant. At overall bit-rates of 1 bit per sample or less, most samples never become significant (this is argued at the end of Section 4.3.1). As a result, it is much more efficient to "broadcast" the effect of significance state transitions to each of the affected neighbours (8 broadcasts, once per significant sample) than to access the state of every neighbour in every individual coding step of the significance propagation and cleanup passes (8 accesses, twice per bit-plane, at every location).

An additional advantage of the state broadcasting approach is that the sign of a newly significant sample can often be broadcast to the eight neighbouring locations without any additional cost in CPU instructions. This greatly simplifies the construction of sign coding context labels, $\kappa^{\text{sign}}[\mathbf{j}]$, when they are needed.

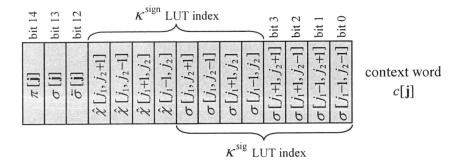

Figure 17.1. 15-bit context word for simple state broadcasting.

SIMPLE BROADCASTING

In a simple incarnation of the state broadcasting principle, each location j is assigned a separate "context word," $c[j]$, containing 15 state bits. Two of these bits represent the location's own state variables, $\sigma[j]$ and $\pi[j]$, whose interpretation is defined in Section 12.2. The remaining bits of the context word maintain replicas of the significance states for all eight neighbours, together with sign information from the horizontal and vertical neighbours. One appropriate organization for the context word is illustrated in Figure 17.1. Note that $\hat{\chi}[j]$ is a "sign state variable," which holds 1 if the sample at location j is both significant ($\sigma[j] = 1$) and negative. Otherwise, it holds 0. This information is available to both the encoder and decoder, unlike the actual sign, $\chi[j]$, which is encoded only for significant samples.

When information is coded at some particular sample location, only that sample's context word need be updated, except when the sample becomes significant. This happens at most once per sample (usually much less often) and involves setting the relevant significance bit (and possibly the sign bit) in each of the significant sample's neighbours. Now observe that $\kappa^{\text{sig}}[j]$ is a function of the least significant 8 bits (bit positions 0 through 7) of the context word, $c[j]$. Also, $\kappa^{\text{sign}}[j]$ is a function of bit positions 4 through 11. Each function may thus be implemented using a 256 byte lookup table (LUT).

ENHANCEMENTS

The number of memory accesses may be significantly reduced by keeping the state bits for multiple sample locations in a single context word. In fact, it is possible to represent the most important state information for an entire stripe column (i.e., four sample locations) within a single

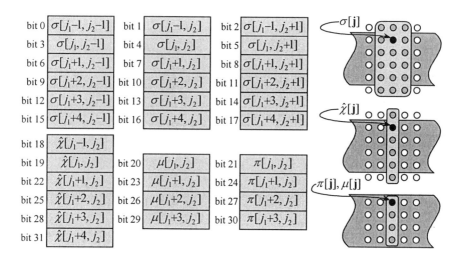

Figure 17.2. Context word, $c[\mathbf{j}]$, used to store state information for coding all four samples in a stripe column, where $\mathbf{j}$ is the location of the first sample in the column.

32-bit context word. This is accomplished by sharing common state bits, as shown in the example of Figure 17.2. The figure depicts the particular organization employed by the Kakadu software supplied with this text. Context word $c[\mathbf{j}]$ is used for all samples in the stripe column whose first sample is at location $\mathbf{j}$; that is, j_1 must be a multiple of 4. The first 18 bits of the context word hold the significance state of the stripe column's samples and their immediate neighbours. The context label, $\kappa^{\mathrm{sig}}[\mathbf{j}]$, is a function of the least significant 9 bits of $c[\mathbf{j}]$; the function is implemented using a 512 byte LUT. More generally, the context label, $\kappa^{\mathrm{sig}}[j_1 + i, j_2]$, for any $i \in \{0, 1, 2, 3\}$, may be obtained after downshifting $c[\mathbf{j}]$ by $3i$ bit positions and using the least significant 9 bits of the result to index the LUT.

Unfortunately, it is not possible to maintain the sign bits for all relevant neighbours within a single 32-bit context word. Nevertheless, the sign coding context labels, $\kappa^{\mathrm{sign}}[j_1 + 0, j_2]$ through $\kappa^{\mathrm{sign}}[j_1 + 3, j_2]$, may be constructed with relatively few CPU instructions, using $c[\mathbf{j}]$ together with $c[j_1, j_2 \pm 1]$ to index a 256 byte LUT. For details of this construction, the reader is referred to the Kakadu source code. Note that sign coding context labels must be formed only when a sample becomes significant, which happens at most once and often not at all.

The delayed significance state, $\overleftarrow{\sigma}[\mathbf{j}]$, described in Section 12.2 is conspicuously absent from Figure 17.2. There is little or no benefit in explicitly storing this state variable for software implementations. This

Table 17.1. Coding pass membership tests and state initialization.

	$\pi\,[\mathbf{j}]$	$\mu\,[\mathbf{j}]$	$\sigma\,[\mathbf{j}]$	$\hat{\chi}\,[\mathbf{j}]$	$\kappa^{\mathrm{sig}}\,[\mathbf{j}]$
Initialize ($\mathbf{j}$ in block)	0	0	0	0	0
Initialize ($\mathbf{j}$ out of bounds)	0	0	0	1	0
Membership test, $\mathbf{j} \in \mathcal{P}^{(p,0)}$	x[a]	x	0	0	> 0
Membership test, $\mathbf{j} \in \mathcal{P}^{(p,1)}$	x	1	x	x	x
Membership test, $\mathbf{j} \in \mathcal{P}^{(p,2)}$	0	x	0	0	x

[a] "x" means "don't care."

is because a software implementation generally has access to all of the previously coded (or decoded) magnitude bits, from which $\overleftarrow{\sigma}\,[\mathbf{j}]$ may be deduced directly (see Page 360). By contrast, access to the previous magnitude bits is either very costly or impossible in certain efficient hardware implementations (see Section 17.2).

The context word structure in Figure 17.2 is carefully designed to allow coding pass membership to be determined with remarkably few instructions. This is accomplished via the two state variables, $\pi\,[\mathbf{j}]$ and $\mu\,[\mathbf{j}]$. As defined in Section 12.2, $\pi\,[\mathbf{j}]$ indicates whether or not $\mathbf{j}$ has been found to belong to the significance propagation pass, $\mathcal{P}^{(p,0)}$, of the current bit-plane, p. The new state variable, $\mu\,[\mathbf{j}]$, indicates whether or not $\mathbf{j}$ belongs to the magnitude refinement pass, $\mathcal{P}^{(p,2)}$.

It is worth explaining exactly how these state variables are updated and used to determine coding pass membership. At the start of each bit-plane p (equivalently, upon completion of bit-plane $p - 1$), we set $\mu\,[\mathbf{j}]$ equal to $\sigma\,[\mathbf{j}]$. This marks all samples which became significant in a previous bit-plane as members of the magnitude refinement pass for the current bit-plane and all subsequent bit-planes. At the same time, we reset $\pi\,[\mathbf{j}]$ to 0. $\pi\,[\mathbf{j}]$ is set to 1 in the significance propagation pass, $\mathcal{P}^{(p,0)}$, if and only if location $\mathbf{j}$ is found to belong to that pass.

Conditions used to test for membership in each of the coding passes are given in Table 17.1. These tests may be accomplished with simple masking (logical "and") and testing operations on the 32-bit context word for the relevant stripe column. It is also possible to perform the tests simultaneously across all samples in the stripe column, using only one or two CPU instructions. This can be advantageous for the significance propagation and magnitude refinement passes, $\mathcal{P}^{(p,0)}$ and $\mathcal{P}^{(p,1)}$, which are often only sparsely occupied. For example, if $c\,[\mathbf{j}] = 0$, none of the four samples in the stripe column starting at location $\mathbf{j}$ belongs to $\mathcal{P}^{(p,0)}$. Similarly, if bits 20, 23, 26 and 29 of $c\,[\mathbf{j}]$ are all zero, none of the four samples in the stripe column belongs to $\mathcal{P}^{(p,1)}$.

The appearance of the sign state bit, $\hat{\chi}[\mathbf{j}]$, in Table 17.1 may seem unnecessary. In particular, if $\mathbf{j} \in \mathcal{P}^{(p,0)}$ or $\mathbf{j} \in \mathcal{P}^{(p,2)}$ then $\sigma[\mathbf{j}]$ must be 0 (insignificant) and so $\hat{\chi}[\mathbf{j}]$ must still be 0. The inclusion of the condition, $\hat{\chi}[\mathbf{j}] = 0$, in the membership tests for these coding passes provides an efficient mechanism (no cost in CPU instructions) for skipping over locations which lie outside the code-block boundaries. It can happen that the last stripe of the code-block contains less than 4 rows. Rather than explicitly testing for these few "out of bounds" samples (this would incur a significant penalty in CPU resources), the samples are simply excluded from membership in any of the three coding passes. This is accomplished by initializing out of bounds locations with the otherwise impossible state configuration identified in Table 17.1.

The use of a single context word for the entire stripe column has many advantages. The context word may be kept in a register which is loaded from memory only once per stripe column. Often, the state bits for multiple samples may be manipulated simultaneously. As explained above, an entire stripe column may be efficiently skipped when none of its samples belong to the relevant coding pass. Finally, the test for run mode in the cleanup coding pass, $\mathcal{P}^{(p,2)}$, may be implemented simply by checking whether or not $c[\mathbf{j}] = 0$. For a greater appreciation of these advantages, the reader should consult the Kakadu source code supplied with this text.

17.1.3   DEQUANTIZATION SIGNALLING

According to the description provided in Section 12.2.5, the embedded block decoder produces decoded magnitude and sign bits, $\hat{v}[\mathbf{j}]$ and $\hat{\chi}[\mathbf{j}]$, along with an indicator, $p[\mathbf{j}]$, of the number of least significant magnitude bits which were not decoded. These quantities are supplied to the dequantization procedure described by equation (10.28).

There is a convenient way to simplify the dequantization process while also avoiding the need to maintain a separate array, $p[\mathbf{j}]$. Specifically, the Kakadu software supplied with this text employs the modified sign-magnitude representation shown in Figure 17.3. The sign bit, $\hat{\chi}[\mathbf{j}]$, is the most significant bit in the word. Any word size may be used, but 16 and 32 bit integers are obvious choices. $K_b^{\max}$ is the maximum number of magnitude bits required to represent the quantization index associated with any sample in subband b. The value of $K_b^{\max}$ is given by equation (10.22). If ROI adjustments are used, $K_b^{\max}$ should be replaced by $\overleftarrow{K}_b^{\max}$, whose value is given by equation (10.33).

The quantized sample magnitudes are stored in the most significant $K_b^{\max}$ bits of the word, after the sign bit. The least significant $p[\mathbf{j}]$ of

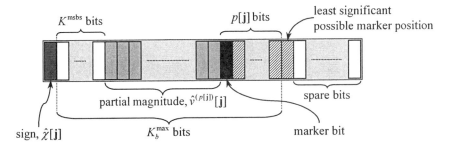

Figure 17.3. Modified sign-magnitude representation for signalling partially decoded quantization indices together with the number of undecoded LSBs, $p[j]$.

these bits are not decoded. Rather than setting these undecoded bits to 0, however, the decoder can be modified[2] to set the most significant undecoded magnitude bit to 1 whenever $\hat{v}[j] \neq 0$. This is identified as the "marker bit" in Figure 17.3. The marker bit holds 1 unless $\hat{v}[j] = 0$, in which case it holds 0. If all of the originally encoded bit-planes are decoded, $p[j] = 0$ and the marker bit appears immediately after the $K_b^{\max}$ magnitude bits. To accommodate the least significant possible marker position, a word size of at least $2 + K_b^{\max}$ bits is required ($2 + \overleftarrow{K}_b^{\max}$ bits if there are ROI adjustments). It may occasionally be necessary to discard some of the encoded bit-planes in order to satisfy this condition.

The marker bit serves to implicitly identify the value of $p[j]$ whenever $\hat{v}[j] \neq 0$ – there is no need to recover $p[j]$ if $\hat{v}[j] = 0$. If desired, the dequantizer may recover $p[j]$ by finding the least significant non-zero bit position in the word, ignoring the "spare bits" shown in the figure. This may be accomplished quite efficiently with the aid of a small lookup table. On the other hand, if mid-point reconstruction is desired ($\delta = \frac{1}{2}$), the word already holds a binary fraction representation of the quantity $\hat{v}[j] + 2^{p[j]}\delta$ required by equation (10.28). In this case, all that is required for dequantization is to multiply the value by an appropriate step size parameter.

17.2 BLOCK CODING: HARDWARE

Our purpose in this section is to suggest design strategies and to indicate some potential trade-offs between complexity and throughput. We

---

[2]The modification need have no effect on the number of CPU instructions executed by the decoder.

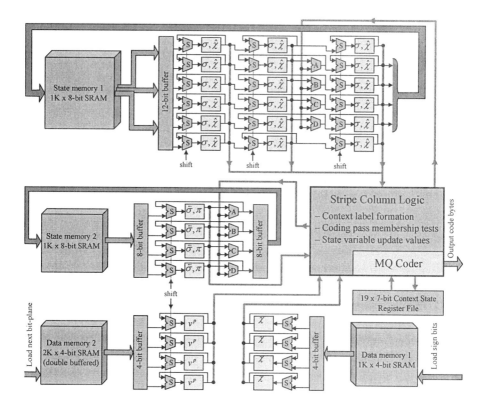

Figure 17.4. Embedded block encoder architecture. Decoder architecture is almost identical, with appropriate data flows reversed.

consider the block coder (encoder or decoder) in isolation. Its interaction with other elements in a complete system (most notably the DWT) is considered in Section 17.5.

17.2.1   EXAMPLE ARCHITECTURE

To provide a context for our discussion of hardware complexity, we must first describe a sample block coding architecture. One possible architecture is depicted in Figure 17.4. The general approach is suitable for both encoding and decoding, except that the flow of sign and magnitude bit-plane information must be reversed for decoding, along with the flow of compressed code bytes. In the discussion which follows, state variables $\sigma\left[\mathbf{j}\right]$, $\pi\left[\mathbf{j}\right]$ and $\overleftarrow{\sigma}\left[\mathbf{j}\right]$ all have exactly the same interpretation as in Section 12.2.2.

STATE AND DATA MEMORIES

For encoding, we assume that the quantized subband samples for a single code-block are all available in sign-magnitude form and that they may be accessed bit-plane by bit-plane. The sign bits and the most significant magnitude bit-plane are first loaded into internal memories (data memories 1 and 2). These memories are sized to support the maximum legal code-block area for JPEG2000 (4096 samples). The coder processes the first bit-plane, $p = K - 1$, in coding pass $\mathcal{P}^{(K-1,2)}$, while loading the next magnitude bit-plane into data memory 2. It processes this next bit-plane, $p = K - 2$, in coding passes $\mathcal{P}^{(K-2,0)}$ through $\mathcal{P}^{(K-2,2)}$, while loading magnitude bit-plane $K - 3$ into data memory 2 and so forth.

Whenever a sample becomes significant, its sign bit, $\chi[\mathbf{j}]$, is coded and written to the sign state variable, $\hat{\chi}[\mathbf{j}]$. It is not strictly necessary to maintain two distinct copies of the sign information for encoding; however, this helps to unify the encoding and decoding architectures. Using this approach, all state variables which are used to form coding contexts and determine coding pass membership are stored in private memories (state memories 1 and 2), which are not directly loaded with code-block sample data. These state memories hold exactly the same quantities during encoding and decoding.

For decoding, the process is reversed. Magnitude bit-planes are generated one by one as the embedded bit-stream is decoded. These magnitude bit-planes are written out through the double buffered bit-plane store (data memory 2) once they become available. Whenever a sample becomes significant, its sign is decoded and stored in the relevant sign state bit, $\hat{\chi}[\mathbf{j}]$. Once decoding is complete, the sign state bits are transferred from state memory 1 to data memory 1, whence they are written out. Alternatively, the sign state bits are copied to data memory 1 during each coding pass (or just the last coding pass), so that the memory is ready to be written out as soon as decoding is complete.

Recall that the decoder is responsible for informing the dequantizer of the number of undecoded bit-planes, $p[\mathbf{j}]$, for each code-block sample location, $\mathbf{j}$. This information may be deduced from the contents of the two state memories. In particular, if the last decoded pass was $\mathcal{P}^{(p,0)}$, then $p[\mathbf{j}] = p + 1 - \pi[\mathbf{j}]$, since state variable $\pi[\mathbf{j}]$ holds 1 if $\mathbf{j} \in \mathcal{P}^{(p,0)}$. If the last decoded pass was $\mathcal{P}^{(p,1)}$, then $p[\mathbf{j}] = p + 1 - (\sigma[\mathbf{j}] \mid \pi[\mathbf{j}])$. Otherwise, the last decoded pass was $\mathcal{P}^{(p,2)}$ and $p[\mathbf{j}] = p$ for all $\mathbf{j}$. There are any number of ways of signalling this information to the dequantizer. One appealing approach is to use the marker bit technique described in Section 17.1.3.

STRIPE ORIENTED PROCESSING

The sample architecture sketched in Figure 17.4 is heavily influenced by the stripe oriented scanning pattern of Figure 12.9. Data and state information are transferred to and from the relevant memories, one stripe column at a time. State memory 1 maintains those state variables which must be accessed in spatial neighbourhoods; i.e., $\sigma[j]$ and $\hat{\chi}[j]$. The spatial neighbourhood for a single stripe column is 6 rows high and 3 columns wide[3].

To minimize memory accesses, we maintain the 6×3 neighbourhood of significance and sign state bits in a kind of shift register. When the stripe column advances one position to the right, 6 new significance bits and 6 new sign bits are pushed into the shift register from state memory 1, via an intermediate 12-bit buffer. At the same time, the 4 significance bits and 4 sign bits of the current stripe column which "pop out" the end of the shift register must be saved back to state memory 1. Any or all of these state bits may have been changed.

In order to load the 12-bit buffer from which the 6×3 neighbourhood shift register is filled, three separate 8-bit words must be accessed from state memory 1. This is because each 8-bit word holds the significance and sign state bits for a single stripe column and the 6×3 neighbourhood includes parts of three stripe columns. An additional access is required to write back the state bits which may have been modified[4]. Thus, assuming that the memory can be accessed once per clock cycle, we cannot process a stripe column in less than 4 clock cycles. This is quite sufficient for the simple timing model described below. Enhancements for higher throughput are described in Section 17.2.2.

The remaining state bits, $\pi[j]$ and $\overleftarrow{\sigma}[j]$, are maintained by a second state memory and are also moved through a kind of shift register one stripe column at a time. There is no reason why the two state memories cannot be merged into a single $1K \times 16$-bit memory. This has the advantage of simplifying the control circuitry. On the other hand, the quantities in state memory 2 require less buffering and only half the memory access bandwidth of those in state memory 1. This is because no spatial neighbourhoods are involved. By merging the memories, the cost in buffering and increased memory bandwidth (and hence power consumption) may outweigh the benefits of simplified control.

---

[3] Actually, the sign state bits, $\hat{\chi}[j]$, are not required at the corners of this 6×3 window.
[4] This may be skipped in the magnitude refinement pass if desired.

SIMPLE TIMING

The architecture in Figure 17.4 represents a synchronous circuit. The lightly shaded boxes are D flip-flops (in most cases, each box stands for two flip-flops), whose inputs are transferred to their outputs on the rising (or falling) edge of every clock. The inputs are controlled by multiplexers. The multiplexers labeled "S" are used to control the shift register function, advancing the current stripe column one position to the right. Of course, at the start of each stripe, some additional clock cycles will be required to load the various flip-flops and buffers in preparation for processing the first column of that stripe. The multiplexers labeled "A," "B," "C" and "D" are used to update state variables while processing the stripe column.

In the most straightforward design, a single sample is processed in each clock cycle so that the stripe column advances once every four clocks. When processing the first sample in a stripe column, multiplexers "A" are used to update the state variables for that sample alone. Similarly, multiplexers "B" are used to update the state variables associated with the second row of each stripe and so forth.

The block labeled "stripe column logic" in Figure 17.4 determines context labels, $\kappa^{\text{sig}}[\mathbf{j}]$, $\kappa^{\text{sign}}[\mathbf{j}]$, and $\kappa^{\text{mag}}[\mathbf{j}]$, for the relevant stripe column sample, following the rules identified in Tables 8.1, 8.2 and 8.3. It also implements the logic to determine whether or not each sample belongs to the current coding pass and, in the cleanup pass, to identify and process the run mode. These logic operations are quite simple and should incur relatively little latency.

For samples which belong to the current coding pass, the relevant context labels and symbols are sent to the MQ coder, which incrementally generates and outputs code bytes for the current codeword segment. Logic for the MQ coder is also relatively simple and most symbols can be coded in a single clock cycle without any difficulty. When a highly improbable symbol is coded (of course, this can happen only rarely), the MQ coder may require a large number of renormalization shifts. If renormalization is implemented serially (rather than through a costly barrel shifting circuit), such symbols may require multiple clock cycles to code. For this and potentially other reasons, the coding logic may need to be prepared to wait for one or more clock cycles until the MQ coder has completed its task. When a sample first becomes significant, an additional symbol is sent to the MQ coder, representing the sample's sign. Again, this may be the cause of a "wait state." Fortunately, neither of these events occur frequently.

It is worth noting that the MQ encoder need not be tightly synchronized with the bit-plane coding logic. Symbol and context label pairs,

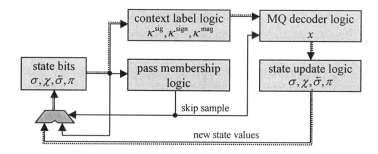

Figure 17.5. State update dependencies for the embedded block decoder, showing the critical path (dotted outline) which limits the maximum achievable clock rate.

(x_n, κ_n), may be temporarily queued prior to processing by the MQ coder (see Section 12.1.1). Such a queue may be used to absorb variations in the rate of symbol production, thereby reducing the likelihood of wait states. Information for any given sample location is coded in only one of the three bit-plane coding passes. Consequently, the MQ coder is only required to sustain an average throughput of about 1 symbol per 3 clock cycles. In this way, symbol queuing introduces the possibility of simplified MQ coder designs which might be allowed a full 2 clock cycles to process each symbol.

Timing considerations for decoding are similar to those for encoding, with one very important difference: the MQ decoder must be tightly synchronized with the bit-plane coding logic. In particular, the context label required to correctly decode the next symbol depends upon state bits which may not be valid until the completion of the previous decoding step. Figure 17.5 illustrates the relevant dependencies.

The maximum achievable clock rate is limited by signal propagation delay through the critical path (dotted line) shown in the Figure 17.5. This critical path includes the MQ decoding logic. By contrast, the corresponding critical state update path for the encoder does not include the MQ encoder at all. The extra latency for decoding is essentially the propagation delay through a 16-bit subtractor, plus the time required to address the relevant context state register. If very high clock rates are to be supported, an extra clock cycle might need to be allocated for symbol decoding.

17.2.2   THROUGHPUT ENHANCEMENTS

The simple timing model described above requires approximately four clock cycles to process each stripe column in each coding pass. The maximum average throughput for such an implementation is thus $1/Z_{\mathrm{avg}}$ im-

Table 17.2. Useful statistics for estimating the throughput of block encoding and decoding hardware. Results obtained using artificial imagery are shown in parentheses.

Bit-rate	$Z_{\max}$	Z_{avg}	R_{sym}	R_{empty}
0.125 bps	17 (17)	1.3 (1.9)	0.26 (0.35)	0.17 (0.22)
0.25 bps	22 (22)	2.5 (3.6)	0.50 (0.65)	0.33 (0.45)
0.5 bps	22 (25)	4.5 (6.2)	0.93 (1.18)	0.59 (0.82)
1.0 bps	25 (28)	7.3 (10.7)	1.66 (2.14)	0.94 (1.42)
2.0 bps	28 (28)	11.1 (16.1)	2.89 (3.61)	1.40 (2.11)
lossless	28 (28)	18.4 (18.5)	5.77 (4.38)	2.24 (2.41)

age samples per clock cycle, where Z_{avg} is the average number of coding passes per code-block. To achieve this average throughput, the block coder (or decoder) needs to be kept continuously active. We address this concern in Section 17.5.

Table 17.2 provides actual values for Z_{avg}, as a function of the overall code-stream bit-rate. The results in the table are obtained from two ensembles of test images. Entries without parentheses are obtained by averaging individual results from the three large photographic test images shown in Figure 8.21. The entries in parentheses are obtained in the same manner, but with the artificial image sources shown in Figure 8.24. In each case, a quality progressive lossless code-stream is truncated to the bit-rate of interest and the average number of coding passes remaining for each code-block is determined. While the results are oriented toward decoding they are also useful for predicting encoding performance. The table suggests that a high quality compressed image, having a bit-rate of 1 bit/sample, could be processed with a throughput of perhaps 7 to 11 clock cycles per sample, while truly lossless compression is about twice as demanding.

Table 17.2 contains a number of additional statistics which should be of interest to implementors of the standard. The second column indicates the maximum number of coding passes, $Z_{\max}$, taken over all code-blocks and all test images in the relevant ensemble (natural or artificial imagery). The worst case block coder throughput is $1/Z_{\max}$ image samples per clock cycle. This may be taken as a pessimistic estimate of the overall throughput for systems which demand that subband samples be processed at a constant rate. To avoid such pessimistic throughputs, most systems should provide at least some buffering to absorb variations in the code-block processing rate. We consider this further in Section 17.5. The remaining columns in Table 17.2 provide statistics which are relevant to the throughput enhancements suggested below.

CONCURRENT MEMBERSHIP TESTING

The simple timing model described in Section 17.2.1 devotes at least one clock cycle to each sample location in each coding pass; an extra clock cycle is usually required to process the sign bit of newly significant samples. On average then, nearly $\frac{2}{3}$ of the clock cycles are wasted in the sense that nothing is encoded or decoded. Also note that the cleanup pass contains a run mode which can code all four samples in a stripe column with a single symbol. For this reason, more than $\frac{2}{3}$ of the clock cycles are usually wasted.

If we were able to eliminate all of these wasted clock cycles then the block coder would only need to devote one clock cycle to each binary symbol actually coded (or decoded). The fourth column of Table 17.2 provides observed statistics for the average number of MQ coder symbols per image sample, R_{sym}. The conditions used to generate these results have already been explained above. Notice that R_{sym} is four or five times smaller than Z_{avg} at most bit-rates of interest. This suggests that we should be able to substantially improve the block coding throughput.

Coding pass membership tests for any given sample location involve only a few logic gates. Multiple tests can be performed concurrently to identify the next member location, at least within a single stripe column. In this way, each clock cycle can process the next unprocessed sample location which actually belongs to the current coding pass, so long as the stripe column contains such a location. With this approach, the memory architecture of Figure 17.4 must be modified slightly to allow the stripe column position to advance once per clock cycle if necessary[5]. In this case, a clock cycle need only be wasted when a stripe column is "empty," meaning that none of its sample locations belong to the current coding pass.

The fifth column in Table 17.2 provides observed statistics for the average number of empty stripe columns, R_{empty}, expressed as a fraction of the number of image samples. An enhanced implementation of the form described above should be able to achieve an average throughput of $1/(R_{\mathrm{sym}}+R_{\mathrm{empty}})$ image samples per clock cycle. Consider, for example, a high quality compressed image with a bit-rate of 1 bit/sample. The average throughput for photographic images at this bit-rate should be about 0.4 samples per clock cycle, which is approximately 3 times higher

---

[5] The worst case memory transaction bandwidth of 4 bytes per clock cycle for state memory 1 could be accommodated with extra buffering and a larger memory word size. Specifically, state memory 1 might be changed from a 1K×8 format to a 256 × 32 format. Similarly, the worst case memory bandwidth of 2 bytes per clock cycle for state memory 2 could be accommodated by changing from a 1K×8 format to a 512×16 format. Such changes typically increase the chip area occupied by the memories.

than the throughput achieved with the simple timing approach. For truly lossless compression or decompression, an average throughput of $\frac{1}{8}$ sample/clock appears to be achievable.

The impact of R_{empty} on these throughput figures may be reduced if we are prepared to further modify the implementation to process multiple stripe columns simultaneously. It is not currently clear whether the additional complexity of such modifications can be justified in practice.

CONCURRENT SAMPLE ENCODING

Recall that there is no need to tightly synchronize the bit-plane coding and MQ symbol coding activities for encoding. This fact may be exploited to realize further throughput improvements. The bit-plane coding operations themselves produce a string of symbol and context pairs, (x_n, κ_n), which may be queued for processing by the MQ encoder. The stripe column logic block in Figure 17.4 could perform all of the bit-plane coding operations for a single stripe column concurrently, generating anywhere from 0 to 10 symbols in a single clock cycle. The extreme case of 10 symbols occurs only in the cleanup pass where the run mode is used and immediately interrupted by four significant samples.

Comparing the third and fourth columns of Table 17.2, we see that each coding pass contributes approximately one symbol for every four samples (i.e., one symbol per stripe column) on average. Thus, a more realistic implementation of the concurrent processing idea might limit the number of symbols which can be generated in a single clock cycle to 2 or 3, with little penalty to the overall throughput. For the purpose of this discussion, we will assume that the MQ encoder is able to consistently process one symbol per clock cycle; a barrel shifting renormalization network would be required to guarantee this. If the symbol queue is sufficiently long, overall block coding throughput will be determined by the minimum of the symbol processing rate and the bit-plane coding rate. The first bound gives a throughput of $1/R_{\text{sym}}$ samples per clock cycle, while the second bound gives a throughput of $4/Z_{\text{avg}}$ samples per clock cycle.

From Table 17.2 we see that the symbol processing bound tends to dominate at high bit-rates, where speed is most important. Evidently, an implementation based on concurrent symbol processing should be able to operate approximately $1\frac{1}{2}$ times faster than one which relies only on concurrent membership testing. Note, however, that the symbol queue may need to be quite large in order to realize this performance. We have not specifically explored the interaction between queue length and processing throughput.

CONCURRENT MQ SYMBOL PROCESSING

The concurrent sample processing techniques described above are available only to block encoders. This is because bit-plane decoding and MQ decoding operations cannot be decoupled, as indicated by Figure 17.5. Concurrent sample decoding is possible only if multiple MQ symbols can be decoded within a single clock cycle. The possibility of concurrent MQ symbol processing is also of interest for improving encoder throughput.

Arithmetic coding is generally considered to be an inherently sequential operation. It turns out, however, that the structure of the MQ coder admits the possibility of encoding or decoding c symbols concurrently, provided the first $c-1$ of these symbols do not induce a renormalization event[6] ($A < 2^{15}$).

We consider concurrent encoding first. Let x_n through x_{n+c-1} be the c symbols to be processed concurrently. The first $c-1$ of these symbols must all be MPS's, since every LPS induces a renormalization event. Write $D_{n,c-1}$ for the cumulative impact of the first $c-1$ symbols on the A and C registers, i.e.,

$$D_{n,c-1} = \sum_{i=0}^{c-2} \bar{p}\left(\Sigma_{\kappa_{n+i}}\right)$$

The idea is to construct a logic network which finds the largest value of $c > 0$ such that $A_n - D_{n,c-1} \geq 2^{15}$ and $x_{n+i} = s_{\kappa_{n+i}}$ for $0 \leq i < c-1$ (i.e., the first $c-1$ symbols are MPS's). To encode c symbols concurrently, we subtract $D_{n,c-1}$ from A and add $D_{n,c-1}$ to C_{active} as precursors to the processing of the cth symbol, x_{n+c-1}. All of this might be done in a single clock cycle, bearing in mind that there will be a practical limit, $c_{\max}$, to the number of symbols which can be processed concurrently. In fact, the concurrent symbol coder must be able to simultaneously access c entries from the context state register file, in order to discover $\Sigma_{\kappa_{n+i}}$ and the MPS identity, $s_{\kappa_{n+i}}$, for each i. For this reason a realistic value for $c_{\max}$ might be only 2 or 3.

Table 17.3 indicates the average concurrency available for MQ encoding in JPEG2000. Specifically, for various values of the practical concurrency limit, $c_{\max}$, the table identifies the average number of symbols which may be coded concurrently, as a function of the overall code-

---

[6]Most of the more complex MQ coder operations occur only in the event of a renormalization. Most notably, state transitions for the probability estimation machinery are driven by renormalization events. The contents of the context state register file remain stable between renormalization events and this is the key property required for practical implementations of concurrent symbol processing.

Table 17.3. Average symbol encoding concurrency, subject to a constraint, c_{max}, on the maximum number of concurrently processed symbols. Results in parentheses are obtained using artificial imagery.

Bit-rate	$c_{max} = 2$	$c_{max} = 3$	$c_{max} = 4$	$c_{max} = \infty$
0.125 bps	1.63 (1.72)	2.01 (2.24)	2.26 (2.63)	3.06 (4.28)
0.25 bps	1.61 (1.71)	1.98 (2.21)	2.21 (2.56)	2.93 (3.99)
0.5 bps	1.58 (1.67)	1.92 (2.12)	2.11 (2.42)	2.67 (3.48)
1.0 bps	1.53 (1.62)	1.81 (2.00)	1.96 (2.25)	2.33 (3.05)
2.0 bps	1.46 (1.54)	1.66 (1.82)	1.76 (1.98)	1.97 (2.47)
lossless	1.34 (1.51)	1.44 (1.77)	1.48 (1.91)	1.56 (2.33)

stream bit-rate. Results are obtained under the same conditions as those used to generate Table 17.2. We are most interested in the amount of concurrency available at high bit-rates, where speed is most important, and with small values of c_{max}, where logic complexity may be acceptable. The tabulated figures suggest that concurrency factors between 1.5 and 2 might be achieved. These improvements may be used to increase the rate at which a symbol queue is cleared, allowing for smaller queues and/or higher overall throughput.

Concurrent symbol decoding follows essentially the same principles as concurrent encoding. So long as $A_n - D_{n,c-1} \geq 2^{15}$ and $C_n^{active} - D_{n,c-1} \geq 0$, the first $c-1$ symbols must all be MPS's. Provided these $c-1$ symbols do not represent sign information or cause an insignificant sample to become significant[7], the c symbols from x_n to x_{n+c-1} can all be decoded concurrently. The impact of the first $c-1$ symbols is simply to subtract $D_{n,c-1}$ from both A and C^{active}.

Unfortunately, it is difficult to take advantage of concurrent symbol processing within the tight synchronization constraints of the block decoder. If the block decoding logic is confined to work with a single stripe column, as has generally been assumed, concurrent decoding techniques cannot improve the average throughput by more than about 15%. The reason for this disappointing result is that the average number of symbols per stripe column is close to 1, leaving little opportunity for concurrency within a single stripe column. If the decoding logic is able to process stripe columns in pairs, larger gains are possible. However, the corresponding increase in logic complexity may be hard to justify. In fact, increased logic latency may limit the clock rate so as to adversely affect

---

[7]Sign decoding and significance transitions are the only events which can alter the context labels associated with neighbouring samples. We need to know the context labels of all c concurrently decoded samples ahead of time.

overall performance. Nevertheless, our experience shows that concurrent symbol decoding techniques can be advantageously deployed at least in software implementations, such as that embodied by the Kakadu tools supplied with this text.

17.2.3   OPPORTUNITIES FOR PARALLELISM

From the foregoing discussion, we may conclude that the average throughput which can be expected for an efficient block coding implementation at high bit-rates (e.g., 2 bps to lossless) is in the range $\frac{1}{4}$ to $\frac{1}{8}$ subband samples per clock cycle. To achieve higher throughputs, multiple block coders may be deployed in parallel. Following the architecture of Figure 17.4, each block coder (encoder or decoder) consumes 3.5 kB of on-chip memory, in addition to the coding and control logic. Of course, the memories in Figure 17.4 are sized for the maximum possible code-block area of 4096 samples (1024 stripe columns). While encoders are at liberty to select a smaller block size so as to reduce on-chip memory requirements, decoders must be prepared to handle the maximum code-block size. This may possibly be restricted to 32×32 for the purpose of a restricted "Profile-0," as explained in Section 18.2.

An alternative form of parallelism is supported when the *RESET*, *RESTART* and *CAUSAL* mode switches are asserted, as discussed in Section 12.4.2. In this case, it is possible to implement multiple coding passes in parallel. The principle advantage of this is that parallelism can be achieved without duplication of the internal memory resources. On the other hand, tight synchronization between parallel coding pass processors (see Section 12.4.2) can prevent exploitation of the various throughput enhancements described in Section 17.2.2. For this reason, coding pass parallelism is not likely to be worthwhile unless a large number of coding pass processors are implemented in parallel. At this stage, we have not performed a thorough comparison of the implementation costs associated with the two different forms of parallelism described here.

Coding pass parallelism is particularly interesting for applications which require hard throughput guarantees (e.g., one sample per clock cycle). This is especially important where lossless performance is also required, since an encoder or decoder not encumbered by the lossless requirement can always discard coding passes to satisfy hard throughput constraints. Unfortunately, to provide such guarantees a very large number of coding pass processors must be implemented, most of which will be idle most of the time.

17.2.4 DISTORTION ESTIMATION

One of the benefits of independent block coding is that each code-block's embedded bit-stream may be independently truncated in accordance with the importance of its contents. Each coding pass, $\mathcal{P}^{(p,k)}$, of each code-block, $\mathcal{B}_i$, reduces some measure of the reconstructed image distortion by some amount, $\Delta D_i^{(p,k)}$, in exchange for some increase in the length of the embedded bit-stream, $\Delta L_i^{(p,k)}$. These quantities may be used to drive the PCRD-opt (Post Compression Rate-Distortion optimization) algorithm developed in Section 8.2.1.

When the distortion measure is based on MSE, the quantities $\Delta D_i^{(p,k)}$ may be computed using the techniques described in Section 8.3.5. In particular, $\Delta D_i^{(p,k)}$ may be computed incrementally as the coding pass $\mathcal{P}^{(p,k)}$ proceeds, with the aid of two small lookup tables (e.g., 16 elements each) and access to several magnitude bit-planes below (less significant than) the current bit-plane, p. To accommodate these extra bit-planes, an additional 3 data memories might be added to the architecture shown in Figure 17.4, increasing the total on-chip memory from 3.5 kB to 5 kB. The extra bit-planes allow the quantization error in each sample to be estimated before and after each coding pass, which should be equally important for other distortion measures, not necessarily based on MSE.

As an alternative to explicit computation of the distortion change in each coding pass, various estimation techniques may be employed. For example, the block coder might simply count the number of samples which become significant in each of the significance propagation and cleanup coding passes. Subsequently, these counts could be multiplied by the mean distortion reduction expected for a newly significant sample. Similarly, in the magnitude refinement pass the coder would count the number of member samples and multiply by an expected per sample distortion reduction. This approach avoids the need for any additional on-chip memory.

The expected distortion changes mentioned above may be deduced from the probability distribution of the relevant subband samples, if it is available. In a sophisticated implementation, probability distributions might be adaptively estimated, either from the output of the wavelet transform or from the counts themselves. In this last approach, the counts output by the block coder for all coding passes can be used to fit a parametric probability model (e.g., generalized Gaussian). Most systems with this level of complexity include an embedded CPU which could be used to perform these estimation functions.

17.3   DWT NUMERICS

Without a doubt, the two most substantial elements in any implementation of the JPEG2000 compression standard are the embedded block coder and the DWT (Discrete Wavelet Transform). This is the first of two sections which focus specifically on implementation considerations for the DWT. The focus of this first section is on numerical representations for the subband sample values and various intermediate results. In particular, we consider the numerical precision required to implement reversible transforms and we also consider efficient fixed point approximations for irreversible transforms. Section 17.4 is concerned with implementation structures which efficiently utilize critical memory resources: memory size; and memory transaction bandwidth.

17.3.1   BIBO ANALYSIS GAIN

We begin our study of numerical implementation requirements by examining the BIBO (Bounded Input Bounded Output) gain of the DWT analysis system. The BIBO gain of a linear operator is the ratio between the maximum absolute value of the output samples and the maximum absolute value of the input samples. For the purpose of this analysis, we shall ignore the modifications introduced by symmetric extension at the boundaries of the image. We shall also ignore the small non-linearities introduced by the integer rounding steps in reversible transforms. Each subband sample may then be expressed as a linear combination of the image sample values, $x\,[\mathbf{n}]$, with

$$y_{\mathbf{b}}^{(d)}\,[\mathbf{q}] = \sum_{\mathbf{n}} x\,[\mathbf{n}]\, a_{\mathbf{b}}^{(d)}\left[\mathbf{n} - 2^d \mathbf{q}\right]$$

Here, $d = 1, 2, \ldots, D$ is the DWT level index and $\mathbf{b} \equiv [b_1, b_2]$ is the subband identifier; $[0,0]$ is the LL subband, $[0,1]$ is the HL (horizontally high-pass) subband, $[1,0]$ is the LH (vertically high-pass) subband and $[1,1]$ is the HH subband. We refer to $a_{\mathbf{b}}^{(d)}\,[\mathbf{n}] \equiv \mathbf{a}_{\mathbf{b}}^{(d)}$ as the analysis vector (or sequence) for subband $\mathbf{b}$ in level d. The BIBO analysis gain for this subband is given by

$$\beta_{\mathbf{b}}^{(d)} \triangleq \sum_{\mathbf{n}} \left| a_{\mathbf{b}}^{(d)}\,[\mathbf{n}] \right| \tag{17.1}$$

The two dimensional DWT structures supported by JPEG2000 all involve separable analysis sequences,

$$a_{b_1,b_2}^{(d)}\,[\mathbf{n}] = a_{b_1}^{(d)}\,[n_1] \cdot a_{b_2}^{(d)}\,[n_2] \tag{17.2}$$

where the one dimensional DWT sequences, $a_b^{(d)}\,[n]$, may be found from the low- and high-pass analysis filter impulse responses, $h_0\,[n]$ and $h_1\,[n]$, using the following relations:

$$a_b^{(1)}[n] = h_b[-n]$$
$$a_b^{(d)}[n] = \sum_k a_b^{(d-1)}[k]\, h_0[2k-n], \quad d > 1 \qquad b = 0,1 \qquad (17.3)$$

This construction is analogous to that of the corresponding synthesis sequences in equation (4.39). From equations (17.1) and (17.2), it follows that the two dimensional BIBO analysis gains may be expressed as the product of the relevant one dimensional BIBO gains, i.e.,

$$\beta_{\mathbf{b}}^{(d)} = \sum_{n_1} \left| a_{b_1}^{(d)}[n_1] \right| \cdot \sum_{n_2} \left| a_{b_2}^{(d)}[n_2] \right| = \beta_{b_1}^{(d)} \cdot \beta_{b_2}^{(d)} \qquad (17.4)$$

17.3.2   REVERSIBLE TRANSFORMS

Reversible transforms map integers to integers. As discussed in Section 6.4.2, such transforms are necessarily non-linear. However, the reversible transforms used by JPEG2000 are very closely related to linear wavelet transforms. The non-linear rounding in each lifting step constitutes a small perturbation of the linear subband samples.

If the original image sample values are B-bit integers in the range $-2^{B-1} \le x[\mathbf{n}] < 2^{B-1}$, the subband samples in band $\mathbf{b}$ of DWT level d are also signed integers, bounded by

$$\left| y_{\mathbf{b}}^{(d)}[\mathbf{q}] \right| \le \beta_{\mathbf{b}}^{(d)} 2^{B-1} + \varepsilon_{\mathbf{b}}^{(d)}$$

where $\beta_{\mathbf{b}}^{(d)}$ is the BIBO analysis gain for the linearized wavelet transform, and $\varepsilon_{\mathbf{b}}^{(d)}$ bounds the effect of the non-linear rounding used in each lifting step. For brevity, we shall not explicitly model this non-linear perturbation, but note that it has negligible impact on the number of bits required to represent the subband sample values, except when the original image bit-depth is very small (e.g., bi-level imagery). Thus, we may take $\left\lceil \log_2 \beta_{\mathbf{b}}^{(d)} \right\rceil$ as a worst-case bit-depth expansion for the subband samples, relative to the original image samples.

Part 1 of the JPEG2000 standard defines only one reversible DWT, whose lifting steps are given in Section 10.4.2. This transform is derived from the spline 5/3 wavelet kernels; its linearized analysis filters are given in equation (10.15). Using these in equations (17.4) and (17.3), we obtain the bit-depth expansion figures shown in Table 17.4.

Evidently, if the original image samples are B-bit integers, $(B+4)$-bit integers are sufficient to represent the reversibly transformed subband samples without risk of numerical overflow or underflow. Thus, for example, 16-bit integers are sufficient to represent reversibly transformed

Table 17.4. Bit-depth expansion figures (not rounded) for each subband in the 2D reversible DWT, based on the Spline 5/3 wavelet kernels.

DWT level, d	$\log_2 \beta_{LL}^{(d)}$	$\log_2 \beta_{LH}^{(d)} = \log_2 \beta_{HL}^{(d)}$	$\log_2 \beta_{HH}^{(d)}$
1	1.170 bits	1.585 bits	2.000 bits
2	1.401 bits	2.022 bits	2.644 bits
3	1.510 bits	2.214 bits	2.919 bits
4	1.525 bits	2.250 bits	2.976 bits
5	1.542 bits	2.267 bits	2.991 bits
15	1.557 bits	2.298 bits	3.039 bits

images with as many as 12 bits per sample. If the RCT (Reversible Colour Transform) is also used, the bit-depth of the two chrominance channels is expanded by exactly 1 bit prior to the reversible DWT. In this case, reversible transformation of 12 bit per sample colour image data would require a 17-bit representation for the HH subbands of the chrominance channels, at DWT levels beyond $d = 5$.

NOTE ON CONVERGENCE OF BIBO GAINS AS $D \to \infty$

The fact that the entries in each column of Table 17.4 appear to converge to limiting values may be understood as an inevitable consequence of the fact that the linearized transform is a true wavelet transform with compact support. As explained in Section 6.3.1, the analysis kernels of a biorthogonal wavelet transform generate the dual wavelet and scaling functions when iterated indefinitely. In fact, the analysis sequence, $a_0^{(d)}[n]$, is directly related to a bandlimited approximation, $^d\tilde{\varphi}(t)$, of the dual scaling function, $\tilde{\varphi}(t)$, having bandwidth $2^d\pi$. Specifically,

$$a_0^{(d)}[n] = 2^{-d} \cdot {}^d\tilde{\varphi}[n] = 2^{-d} \, {}^d\tilde{\varphi}(t)\Big|_{t=2^{-d}n}$$

This may easily be seen from the close connection between the recursive relations in equation (17.3) and those developed in the proof of Theorem 6.6.

Since $^d\tilde{\varphi}(t)$ converges to $\tilde{\varphi}(t)$, the regularity of wavelet scaling functions and their duals ensures that

$$\lim_{d\to\infty} \sum_n \left| a_0^{(d)}[n] \right| = \lim_{d\to\infty} 2^{-d} \sum_n \left| \tilde{\varphi}\left(2^{-d}n\right) \right| = \int_{-\infty}^{\infty} |\tilde{\varphi}(t)| \, dt$$

The integral on the right hand side of the above equation is guaranteed to exist, since all square integrable functions with compact support (non-zero over only a finite interval) are also absolutely integrable. A similar relationship exists between $\sum_n \left| a_1^{(d)} [n] \right|$ and the dual wavelet function, $\tilde{\psi}(t)$.

SYNTHESIS CONSIDERATIONS

In the foregoing discussion, we have considered only the forward DWT analysis transform. If the code-stream contains a truly lossless representation of the subband samples, the synthesis steps will exactly invert the analysis steps. On the other hand, if embedded code-block bit-streams have been truncated, the subband samples are effectively quantized and these quantization effects will propagate through the synthesis system. It is conceivable that quantization effects would be so large as to cause unexpected overflow in the numerical representation selected for the intermediate LL subbands (see Figure 10.6).

Fortunately, the LL subbands exhibit the smallest bit-depth expansion. Without quantization, Table 17.4 suggests that $(B + 2)$ bits are more than sufficient to represent all intermediate LL subband samples. Consequently, an implementation which allocates $(B + 4)$ bits for the representation of all subband samples would be highly unlikely to experience numerical overflow. Nevertheless, we note that JPEG2000 does not restrict the truncation of code-block bit-streams. We have not made a careful study of the potential for truncation-induced overflow in the implementation of reversible DWT synthesis.

17.3.3 FIXED POINT IRREVERSIBLE TRANSFORMS

In the previous section we exploited the fact that reversible transforms map integers to integers to determine the number of precision bits required to implement the transform exactly. Irreversible transforms do not have this property and cannot generally be implemented exactly. In particular, the CDF 9/7 transform has irrational coefficients. Irreversible transforms are sometimes called "floating point" transforms, because floating point arithmetic is a natural choice for their implementation. In resource critical applications, however, fixed point implementations are usually more desirable. This can be true for both hardware and software implementations. The Kakadu software supplied with this text provides a working demonstration of an efficient fixed point implementation of the irreversible DWT.

The fixed point representation (approximation) of a real-valued quantity, x, is written

$$x \approx 2^{-F} \bar{x}$$

where $\bar{x}$ is an R-bit integer and F may be interpreted as the number of fraction bits of x which are preserved by the representation. Equivalently, F is the position of the "binary point" in the fixed point representation. The value of F is fixed; i.e., it does not depend upon the value being represented. It is desirable to choose F as large as possible, while avoiding the possibility of overflow or underflow. Assuming signed quantities, this condition may be expressed as

$$-2^{R-1} \leq 2^F x < 2^{R-1}, \quad \forall x$$

Equivalently, for a given representation precision, R, we wish to minimize the number of integer bits[8], $R - F$, subject to the constraint

$$2^{R-F} > 2\,|x|, \quad \forall x$$

In Chapter 10 we adopted the convention that the irreversible processing path in JPEG2000 would be described in terms of real-valued quantities, consistently normalized to lie within a "unit nominal range" of $-\frac{1}{2}$ to $\frac{1}{2}$. To this end, the low- and high-pass DWT analysis kernels are normalized so as to have unit gain at DC and the Nyquist frequency, respectively. These normalization conventions are expressed in equation (10.12). Although the nominal range bounds are rarely exceeded, unlikely combinations of image sample values can conspire to create temporary excursions. In particular, subband samples in band **b** of DWT level d may range over the interval,

$$-\frac{1}{2}\beta_{\mathbf{b}}^{(d)} \leq y_{\mathbf{b}}^{(d)}\,[\mathbf{n}] \leq \frac{1}{2}\beta_{\mathbf{b}}^{(d)}$$

where $\beta_{\mathbf{b}}^{(d)}$ is the BIBO gain of the relevant DWT analysis sequence. It follows that the fixed point representation for this subband must satisfy

$$R - F > \log_2 \beta_{\mathbf{b}}^{(d)}$$

These absolute bounds are given in Table 17.5, for the specific case of the CDF 9/7 analysis kernels. The entries in the table are obtained by substituting the analysis filter taps from equation (10.13) into equations (17.4) and (17.3). Notice that there is little variation between the subbands, so a single fixed point representation is appropriate for all subbands. Also notice that subband sample values never exceed the nominal range bounds by as much as one bit (a factor of 2), so representations involving one integer bit[9] and $F = R - 1$ fraction bits may be

---

[8] We may interpret the most significant $R - F$ bits of $\bar{x}$ as holding the integer part of x.
[9] Recall that the integer part of the representation conveys both magnitude and sign information, so this one integer bit is actually the sign bit.

Table 17.5. Unrounded bound on the number of integer bits, $R - F$, required for fixed point representation of subband samples generated using the CDF 9/7 wavelet kernels.

DWT level, d	$\log_2 \beta_{LL}^{(d)}$	$\log_2 \beta_{LH}^{(d)} = \log_2 \beta_{HL}^{(d)}$	$\log_2 \beta_{HH}^{(d)}$
1	0.930 bits	0.841 bits	0.752 bits
2	0.829 bits	0.807 bits	0.785 bits
3	0.772 bits	0.737 bits	0.703 bits
4	0.763 bits	0.686 bits	0.609 bits
5	0.757 bits	0.679 bits	0.600 bits
15	0.755 bits	0.674 bits	0.592 bits

used. These results also demonstrate that $G = 1$ is a sufficient number of quantization guard bits to avoid violation of equation (10.21).

It is worth noting that the irreversible colour transform has a BIBO analysis gain of 1, so that its presence has no effect on the number of integer bits required by fixed point representations of the subband samples.

CONSIDERATIONS FOR LIFTING IMPLEMENTATIONS

Up to this point, we have considered only the representation of the subband samples themselves. In a complete implementation, intermediate quantities may be formed and their representation is also important. Lifting implementations are particularly interesting, because they help to unify the reversible and irreversible processing paths. The lifting framework is also helpful in realizing memory efficient implementations of the DWT, as we shall see in Section 17.4. In particular, these implementations require the outputs of some lifting steps to be stored in memory for later use. In this section, we provide some results to assist implementors in selecting appropriate representations for these intermediate quantities.

Since every lifting step is a linear operator, the outputs of each lifting step may be expressed as linear combinations of the original image sample values and the BIBO gain may be evaluated. Let $a_{\text{step-}l}^{(d)}[n]$ denote the linear operator mapping an input sequence to the output of the lth lifting step, in the dth level of a one dimensional DWT. Also, let $a_L^{(d)}[n]$ and $a_H^{(d)}[n]$ denote the linear operators associated with the low- and high-pass subbands at level d. Table 17.6 provides BIBO gains for these one dimensional operators.

We assume that each 2D DWT stage is implemented by performing first the vertical analysis operations and then the horizontal analysis

Table 17.6.   Unrounded bounds on the number of integer bits, $R - F$, required at the output of each analysis lifting step and each subband in a 1D irreversible DWT, using the CDF 9/7 wavelet kernels.

d	$\log_2 \beta_{\text{step-1}}^{(d)}$	$\log_2 \beta_{\text{step-2}}^{(d)}$	$\log_2 \beta_{\text{step-3}}^{(d)}$	$\log_2 \beta_{\text{step-4}}^{(d)}$	$\log_2 \beta_{\text{L}}^{(d)}$	$\log_2 \beta_{\text{H}}^{(d)}$
1	2.061 bits	0.528 bits	1.077 bits	0.764 bits	0.465 bits	0.376 bits
2	2.197 bits	0.879 bits	1.094 bits	0.713 bits	0.415 bits	0.392 bits
3	2.198 bits	0.814 bits	1.052 bits	0.685 bits	0.386 bits	0.351 bits
4	2.166 bits	0.800 bits	1.006 bits	0.681 bits	0.382 bits	0.305 bits
5	2.152 bits	0.787 bits	1.001 bits	0.678 bits	0.379 bits	0.300 bits
15	2.150 bits	0.784 bits	0.997 bits	0.677 bits	0.378 bits	0.296 bits

operations, as described in Section 10.3.2. This sequence of operations is mandatory for reversible transforms, so it is also the most natural choice for irreversible transforms. Synthesis is performed in the reverse order. Following this convention, the linear operator corresponding to the output of the lth vertical analysis lifting step at DWT level d may be expressed as

$$a_{\text{step-}l}^{(d)^{\text{vert}}} [n_1, n_2] = a_{\text{step-}l}^{(d)} [n_1] \cdot a_{\text{L}}^{(d-1)} [n_2]$$

Note that the lifting procedure generates vertically low- and high-pass subbands together. These vertical subbands are then each subjected to horizontal lifting steps. The linear operators associated with the output of each horizontal lifting step, when processing the vertically low- and high-pass subbands at DWT level d, are given by

$$a_{\text{step-}l}^{(d)^{\text{hor-v-low}}} [n_1, n_2] = a_{\text{L}}^{(d)} [n_1] \cdot a_{\text{step-}l}^{(d)} [n_2]$$

and

$$a_{\text{step-}l}^{(d)^{\text{hor-v-high}}} [n_1, n_2] = a_{\text{H}}^{(d)} [n_1] \cdot a_{\text{step-}l}^{(d)} [n_2]$$

Since the two dimensional linear operators are all separable, their BIBO gains are the products of the relevant one dimensional BIBO gains, as supplied in Table 17.6. Table 17.7 provides a convenient summary of the two dimensional BIBO analysis gains to the output of each different type of lifting step. Specifically, the entries in the table correspond to

$$\beta_{\text{step-}l}^{(d)^{\max}} = \max \left\{ \beta_{\text{step-}l}^{(d)^{\text{vert}}}, \beta_{\text{step-}l}^{(d)^{\text{hor-v-low}}}, \beta_{\text{step-}l}^{(d)^{\text{hor-v-high}}} \right\}$$

From the table, we may conclude that $R - F = 3$ integer bits are sufficient to avoid overflow, when using a single fixed point representation for the

Table 17.7. Unrounded bounds on the number of maximum number of integer bits, $R - F$, required at the output of each type of lifting step in a 2D irreversible DWT, using the CDF 9/7 wavelet kernels.

DWT level, d	$\log_2 \beta_{\text{step-1}}^{(d)\max}$	$\log_2 \beta_{\text{step-2}}^{(d)\max}$	$\log_2 \beta_{\text{step-3}}^{(d)\max}$	$\log_2 \beta_{\text{step-4}}^{(d)\max}$
1	2.526 bits	0.993 bits	1.542 bits	1.229 bits
2	2.662 bits	1.344 bits	1.559 bits	1.178 bits
3	2.613 bits	1.228 bits	1.467 bits	1.099 bits
4	2.552 bits	1.186 bits	1.392 bits	1.067 bits
5	2.534 bits	1.169 bits	1.383 bits	1.059 bits
15	2.527 bits	1.161 bits	1.375 bits	1.054 bits

image samples, the subband samples and all intermediate lifting step outputs[10].

SELECTING THE NUMBER OF FRACTION BITS

Fixed point representations may be understood as uniformly quantizing the subband sample values, where the relevant quantization step size is 2^{-F}. One way to measure the significance of these quantization effects is to map the quantization (approximation) errors into the image domain and measure the resulting MSE (Mean Squared Error). Other error measures may be considered, but we shall confine our attention to MSE here.

For large F, the variance of the error in any given subband sample is well approximated by $\frac{1}{12}2^{-2F}$ (see equation (3.27)). For sufficiently large F, it is also reasonable to assume that the approximation errors are uncorrelated. Following equation (4.37), the equivalent image MSE, $\sigma_{\delta X}^2$, may then be expressed as

$$\sigma_{\delta X}^2 = \frac{1}{12}2^{-2F}\sum_{d=1}^{D}\sum_{b_1=0}^{1}\sum_{b_2=0}^{1}2^{-2d}G_{\mathbf{b}_d} \tag{17.5}$$

Here, $G_{\mathbf{b}_d}$ is the squared norm of the synthesis vectors associated with subband $\mathbf{b}$ of DWT level d. We have already encountered these $G_{\mathbf{b}_d}$ factors in connection with distortion estimation for rate control (see

---

[10] Note that intermediate results involved in the implementation of each individual lifting step may require larger precision again, although such considerations are highly dependent upon the implementation structure selected. The reader should find it instructive to examine the fixed point processing paths provided by the Kakadu software supplied with this text; the software uses two different processing conventions for fixed point operations, depending on the machine architecture.

Table 17.8. Equivalent image MSE (expressed in terms of PSNR) associated with the fixed point representation of subband sample values, having a unit nominal range and F fraction bits.

DWT levels, D	$F = 10$	$F = 11$	$F = 12$	$F = 13$	$F = 14$	$F = 15$
1	64.9 dB	70.9 dB	76.9 dB	82.9 dB	89.0 dB	95.0 dB
2	61.9 dB	67.9 dB	74.0 dB	80.0 dB	86.0 dB	92.0 dB
3	60.0 dB	66.1 dB	72.1 dB	78.1 dB	84.1 dB	90.1 dB
4	58.7 dB	64.7 dB	70.7 dB	76.8 dB	82.8 dB	88.8 dB
5	57.6 dB	63.7 dB	69.7 dB	75.7 dB	81.7 dB	87.8 dB
15	52.7 dB	58.7 dB	64.7 dB	70.7 dB	76.8 dB	82.8 dB

Section 8.3.5) and their computation is the subject of equations (4.39) and (4.40). The factors 2^{-2d} in equation (17.5) arise from the fact that each subband in level d of the two dimensional DWT has 2^{-2d} times as many samples as the original image.

Note that equation (17.5) includes the effects of representing the intermediate LL_d subbands using the same F-bit fixed point representation. In most implementations, there will be additional intermediate results which are also approximated using a fixed point representation. These effects could be incorporated into the formula, together with appropriate synthesis weighting factors, yielding MSE values several times larger than those predicted by equation (17.5). In this brief treatment, however, our purpose is only to provide a rough indication of the dependence of $\sigma_{\delta X}^2$ on the number of fraction bits, F, and the number of DWT levels, D.

Table 17.8 provides computed values for $\sigma_{\delta X}^2$, using the CDF 9/7 wavelet kernels defined by Part 1 of the JPEG2000 standard. The MSE values are expressed in terms of PSNR (Peak Signal to Noise Ratio). Since we are working with normalized image samples in the range $-\frac{1}{2} \leq x < \frac{1}{2}$, the PSNR is given by

$$\text{PSNR} = 10 \log_{10} \frac{(x_{\max} - x_{\min})^2}{\sigma_{\delta X}^2} = -10 \log_{10} \sigma_{\delta X}^2$$

The dependence of $\sigma_{\delta X}^2$ on D may be understood from the fact that $G_{b_d} \approx 2^{2d}$.[11] Substituting into equation (17.5) we see that $\sigma_{\delta X}^2$ grows approximately linearly with D, meaning that the equivalent image PSNR should decrease roughly as $10 \log_{10} D$. This is readily confirmed by inspecting the entries in the table.

---

[11] See equation (10.25) and the associated discussion.

The entries in Table 17.8 may be compared with typical PSNR's experienced in lossy image compression (see Table 8.5, for example). Experience with 8-bit imagery shows that visually lossless compression usually occurs with PSNR's of about 40 dB, although this can be quite dependent on image content and viewing conditions and may not be sufficient for the most demanding applications. Accordingly, it is reasonable to suppose that implementations involving $R = 16$ or even fewer precision bits should usually be quite sufficient to render the approximation error negligible, even when the additional errors associated with fixed point representation of intermediate quantities are taken into account.

17.4 DWT STRUCTURES

In this section, we consider implementation structures for the DWT which are able to efficiently utilize memory resources. Our focus is on applications involving very large images, where memory size must be carefully managed. We are also concerned with applications demanding high throughput so that memory transaction bandwidth becomes a critical resource which must be managed. It is convenient for the moment to consider the DWT in isolation. Its interaction with the block coder in a complete system is a matter of great interest, which we defer to Section 17.5.

17.4.1 PIPELINING OF DWT STAGES

The two dimensional DWT is most easily described in terms of the iterative application of individual DWT stages. Each DWT stage is responsible for transforming a two dimensional array of image samples into four subband images, denoted LL (low-pass), HL (horizontally high-pass), LH (vertically high-pass) and HH. Each of these subbands has essentially half the height and half the width of the input image.

The complete transform is obtained by repeatedly passing the LL subband produced by each DWT stage into a further DWT stage. This DWT structure is illustrated in Figure 10.6. A D level DWT involves a total of D DWT stages and produces a total of $3D + 1$ subbands. The subbands produced by the dth DWT stage are denoted LL_d through HH_d and the input to the dth stage is LL_{d-1}. The LL_d are all intermediate subbands, except for the final LL_D subband. It is convenient to think of the original image as a 0th level intermediate subband, LL_0, since it is the input to the 1st DWT stage.

For a sequential stage by stage implementation of the DWT, sufficient memory must be available to buffer the output of each DWT stage before moving on to the next. This may be acceptable for applications

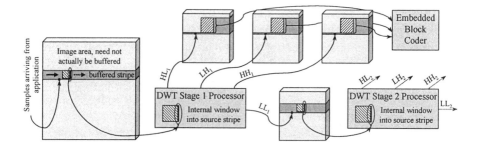

Figure 17.6. DWT pipeline. Dark shading identifies sample data buffered in external memory. Patterned shading identifies regions mapped into internal processor memory.

which already have sufficient memory to buffer the entire input image. In many cases, however, the image samples arrive incrementally from a file or scanner, with insufficient resources to buffer the image or its subbands. A dual problem exists during decompression, where the application consumes the decompressed image samples incrementally (e.g., for printing or writing to a file) without sufficient resources to buffer the image or its subbands.

Fortunately, there is no need to implement the DWT stages one at a time. Each DWT stage is essentially a type of filtering operation and may be implemented incrementally by advancing a "moving window" through the two dimensional image sample array. The DWT stages may then be pipelined, with the dth stage incrementally consuming the samples of subband LL_{d-1}, as they appear at the output of the preceding stage. A DWT pipeline is illustrated in Figure 17.6. A similar pipeline may be used to invert the DWT during decompression. In this case, the dth inverse DWT stage incrementally produces the samples of subband LL_d, which are consumed by stage $d-1$.

We use the term "pipeline" somewhat loosely here to refer to any implementation in which all of the DWT stages are active simultaneously or at least intermittently. We do not insist that each DWT stage have its own distinct processing hardware. In fact, most practical implementations will share one or two processing engines amongst the stages. The advantage of pipelining is that the intermediate subbands, LL_d, need not be fully buffered in memory. Instead, the system need only provide sufficient resources to satisfy the demands of the moving window processing in each DWT stage. We shall analyze these memory resources more carefully in Section 17.4.2.

Also important to the success of DWT pipelining is the fact that the completed subbands can be quantized and coded incrementally. Once sufficient subband samples have been produced to form code-blocks, these samples may be delivered to the embedded block coding engine. The system must provide sufficient memory to buffer subband samples until complete code-blocks can be formed.

APPLICATIONS REQUIRING PIPELINING

We have already mentioned some applications for which pipelining is important. Scanners output the image samples in a line by line fashion. Similarly, most printers consume image samples line by line. We refer to these as line-based applications. Some printers may print in vertical strips from left to right. We view these devices also as line-based applications, since the implementation techniques described in this chapter may easily be applied column-wise, rather than line by line.

Pipelining is important whenever the entire image cannot be buffered in memory. Many satellite imaging systems employ a linear sensor which scans the earth's surface as the satellite revolves about the earth. The resulting images can be enormous, with a vertical dimension (the scanning dimension) which is potentially unbounded. Pipelining of some form is unavoidable[12] in this case, since no amount of memory can suffice to buffer the entire image.

More generally, pipelining is an essential tool for applying the DWT to any data set with one unbounded dimension. Examples include the application of the one dimensional DWT to audio waveforms and application of the three dimensional DWT to video sequences. Incremental processing and pipelining were used by Taubman and Zakhor in their work on 3D wavelet video compression [150]. In that work pipelining is applied in the temporal dimension (this is the scanning direction for video). An analysis of the latency and memory requirements of the pipelined DWT may be found in [146, pp. 184-187].[13]

17.4.2   MEMORY AND BANDWIDTH

In this section, we consider the incremental processing performed within each DWT stage. We are most interested in the impact of differ-

---

[12]Recall that we are using the term "pipelining" rather loosly, referring to any system in which the DWT stages are active simultaneously or intermittently. There are many ways to realize this.

[13]The analysis in [146] is performed only for 2 tap Haar filters. Longer wavelet filters were also implemented and investigated for temporal video transformation by both Ohm and Taubman, as noted in [110] and [150].

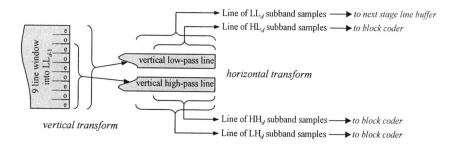

Figure 17.7.   Incremental DWT analysis processing.

ent processing strategies on two physical resources: memory and memory bandwidth. We develop the concept of an M-line DWT implementation and show how the parameter M may be used to establish a trade-off between the total amount of memory which must be provided by the system and the I/O bandwidth which this memory must support.

In a complete system, the subband samples produced by the DWT must usually be buffered in memory until complete code-blocks can be formed and delivered to the coder. Similarly, the output of a block decoder must usually be buffered prior to consumption by the inverse DWT. We regard this buffering of subband samples as a matter of system integration and defer further discussion of it until Section 17.5. In the present section, we completely ignore the costs associated with buffering subband samples for coding.

DWT ANALYSIS

Figure 17.7 illustrates the buffering required for a direct implementation of the dth DWT analysis stage, using the CDF 9/7 wavelet kernels for the purpose of example. By "direct," we mean that each one dimensional transform stage generates each of its subband samples directly from its input samples, as in the inner product formulation of equation (10.6a). The number of input samples required to form a single subband sample is equal to the region of support of the relevant wavelet kernel. By contrast, we shall later consider implementations based on lifting, which must allocate resources to buffer the results from intermediate lifting steps.

The vertical analysis processing is performed first, producing one line of vertical low-pass subband samples and one line of vertical high-pass subband samples. These lines are further subjected to horizontal analysis, generating a single line for each of the subbands LL_d through HH_d.

In this example the low-pass filter has 9 taps, so the DWT stage requires access to 9 lines from LL_{d-1} in order to produce a single line of vertical low-pass subband samples. As indicated in the figure, this 9 line buffer is also sufficient to generate a new line from the vertical high-pass subband. These relationships may be deduced directly from the description of the one dimensional DWT given in Section 10.3.1. More generally, for any set of odd length symmetric DWT kernels, $2L_{max} + 1$ lines of LL_{d-1} are sufficient to produce one new line for each of the level d subbands, LL_d through HH_d. Here $L_{max} = \max\{L_0, L_1\}$ and $2L_0 + 1$ and $2L_1 + 1$ are the lengths of the low- and high-pass DWT kernels, $h_0^t[n]$ and $h_1^t[n]$.

There is no need to actually store the vertical subband lines in memory. Instead, the vertical subband samples may be immediately subjected to horizontal subband transformation. The implementation might maintain a small (horizontal) window into each of the two vertical subband lines which are being generated. Such on-chip processing resources are considered more carefully in Section 17.4.3.

We are now in a position to determine the amount of memory required by the entire DWT processing system. We assume that the implementation involves a single processing engine, which is used to implement each DWT stage in turn. Each application of the engine at DWT stage d processes a swath of $2L_{max} + 1$ lines from LL_{d-1}, producing one new line for each of subbands LL_d through HH_d. The $2L_{max} + 1$ line buffer for subband LL_{d-1} is then advanced by two line positions. To fill the two new empty line positions, the engine must be invoked twice at DWT stage $d - 1$. This requires 4 invocations at stage $d - 2$ and so forth.

Since we do not count any memory required to buffer subband samples for coding, we need only consider the $2L_{max} + 1$ line buffers for each of the intermediate subbands, LL_d, $d = 0, 1, \ldots, D - 1$. Each line of LL_d contains $2^{-d}N_2$ samples, where N_2 is the width of the original image. For simplicity, we assume that the original image samples have an 8-bit representation and that a 16-bit representation is sufficient for all other intermediate subband samples. The analysis conducted in Section 17.3 suggests that this is a somewhat conservative assumption. The buffer size is then given by

$$S_{\mathrm{DWT}}^{(1)} = (2L_{max} + 1)\left(1 + 2\sum_{d=1}^{D-1} 2^{-d}\right) N_2 \lesssim [3(2L_{max} + 1)]N_2 \text{ bytes}$$

While memory size is important, in many cases an even more important resource is memory bandwidth. We assume that the DWT processing engine mentioned above is implemented on a silicon chip (e.g., an ASIC or an FPGA) which is connected to an external memory (e.g., a

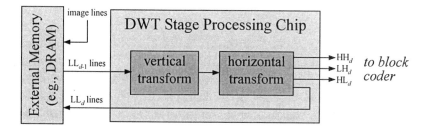

Figure 17.8.   Communication between DWT stage processing and external memory.

DRAM). We further assume that the input image is sufficiently large to render on-chip buffering of whole image lines impractical. Image and intermediate subband line buffers are stored in the external memory, which must be at least $S_{\text{DWT}}^{(1)}$ bytes in size. This configuration is illustrated in Figure 17.8.

We measure memory bandwidth in terms of the average number of byte transactions per original image sample. One byte transaction is required to write each original image sample into the portion of the buffer assigned to LL_0. Once each new pair of image lines has arrived, $2L_{\text{max}} + 1$ lines of LL_0 are read by the DWT engine. It follows that one write and $\frac{1}{2}(2L_{\text{max}} + 1)$ reads are performed for each LL_0 sample. Each of the intermediate LL_d subbands has only 2^{-2d} times as many samples as the original image, for each of which we also have one write and $\frac{1}{2}(2L_{\text{max}} + 1)$ reads. Since each of these subbands is represented with 2 bytes per sample, the total memory bandwidth is given by

$$B_{\text{DWT}}^{(1)} = \left(L_{\text{max}} + \frac{3}{2}\right) \cdot \left(1 + 2\sum_{d=1}^{D-1} 2^{-2d}\right)$$

$$\lesssim \frac{5}{3}\left(L_{\text{max}} + \frac{3}{2}\right) \text{ byte transactions/sample}$$

As it stands, this rather high memory bandwidth may be a cause of concern for many applications. To address this concern, we may modify the implementation to produce M lines of each of the LL_d through HH_d subbands at once, while reading from a moving window containing $2M + 2L_{\text{max}} - 1$ lines in LL_{d-1}. The implementation considered hitherto corresponds to the case $M = 1$, having the minimum memory size and the maximum memory bandwidth. For general M, the memory size and

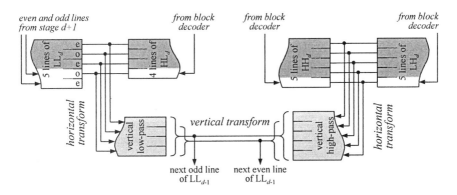

Figure 17.9. Incremental DWT synthesis processing.

bandwidth are easily shown to satisfy

$$S_{\mathrm{DWT}}^{(M)} \lesssim 3 \left(2M + 2L_{\max} - 1\right) N_2 \text{ bytes}$$

$$B_{\mathrm{DWT}}^{(M)} \lesssim \tfrac{5}{3} \left(2 + \tfrac{2L_{\max}-1}{2M}\right) \text{ byte transactions/sample}$$

(17.6)

As an example, using the CDF 9/7 wavelet kernels ($L_{\max} = 4$), the minimum memory case, $M = 1$, yields a memory bandwidth of 9.2 byte transactions per sample; this may be reduced to 4.1 byte transactions per sample by selecting $M = 8$. At the same time, the memory size grows by a factor of about $2\tfrac{1}{2}$, from $27N_2$ bytes to $69N_2$ bytes. Evidently, the parameter M allows us to trade memory size for memory bandwidth. In the limit as M becomes very large, the bandwidth approaches an absolute lower bound of $3\tfrac{1}{3}$ byte transactions per sample. Of course, the complexity of the processing engine (the chip) also grows with M. This is examined in Section 17.4.3.

DWT SYNTHESIS

Figure 17.9 illustrates the buffering required for a direct implementation of the dth DWT synthesis stage. Buffer sizes shown in the figure are selected to match the requirements of the CDF 9/7 transform. We assume that DWT synthesis stages are also implemented by a single "engine" (or chip) interacting with an external memory, which buffers subband and image lines as necessary. The operation of the engine parallels that used for DWT analysis.

When applied to DWT stage d, the synthesis engine produces two new lines of LL_{d-1} samples. To do so, it requires access to a small moving window into each of the four subbands, LL_d through HH_d. In the specific case of the CDF 9/7 transform, 4 lines from the vertically low-pass subbands, LL_d and HL_d, and 5 lines from the vertically high-pass subbands, LH_d and HH_d, are required to synthesize two new lines for LL_{d-1}. Note that a 5 line buffer must be maintained into the LL_d subband to accommodate the arrival of two lines at a time from the synthesis engine at stage $d+1$.

Horizontal subband synthesis is performed incrementally on each pair of corresponding lines from the LL_d and HL_d subbands and each pair of corresponding lines from the LH_d and HH_d subbands. The resulting vertical subband lines are supplied to the vertical synthesis engine as soon as they are generated; the vertical synthesis engine then outputs two new LL_{d-1} lines in sample by sample fashion.

As for DWT analysis, the synthesis implementation may easily be generalized to consume M new lines from each of subbands LL_d through HH_d, producing $2M$ lines of subband LL_{d-1} in each pass across the image. The total number of lines required from the two horizontal high-pass subbands HL_d and HH_d, when taken together, is $2M + 2L_{\max} - 1$.[14] Since each subband's moving window is advanced M lines after each pass, the first $2L_{\max} - 1$ lines of this $2M + 2L_{\max} - 1$ line window must overlap between passes. Similarly, the total number of lines which must be read from subbands LL_d and LH_d together is also $2M + 2L_{\max} - 1$, of which $2L_{\max} - 1$ overlap between passes. The overlapping lines are identified by darker shading in Figure 17.9. The total number of samples associated with these overlapping lines is $2^{1-d}(2L_{\max} - 1)N_2$.

Recall that we generally regard the buffering of decoded subband samples prior to DWT synthesis as a system integration cost. This cost is estimated in Section 17.5. However, the system integration cost does not include the cost of storing (memory size) and accessing (reading a second time) the overlapping lines. We ascribe these to the DWT synthesis engine. Assuming a 2 byte representation for each subband sample, the memory bandwidth associated with reading the overlapping lines at level d is

$$2\frac{(2L_{\max} - 1)2^{1-d} \cdot 2^{-d}}{M} \text{ byte transactions per image sample}$$

In addition to overlapping lines, we must provide storage for $2M$ lines of each of the intermediate subbands LL_{d-1}, since this is the number of lines which are generated by each pass of the synthesis engine at stage

---

[14]The actual distribution of these $2M + 2L_{\max} - 1$ lines between the HL_d and HH_d subbands depends upon the lengths of the low- and high-pass synthesis filters. To simplify the analysis it is convenient to consider the memory resources associated with the vertically low- and high-pass subbands together.

d. The DWT engine is responsible for reading and writing each non-overlapping LL_{d-1} sample once, including the image samples in LL_0. Based on these considerations, we obtain the following resource bounds for DWT synthesis.

$$S_{\text{IDWT}}^{(M)} \lesssim [6M + 4(2L_{\max} - 1)] N_2 \text{ bytes}$$

$$B_{\text{IDWT}}^{(M)} \lesssim \tfrac{5}{3} \left(2 + \tfrac{2L_{\max}-1}{2M}\right) + \tfrac{2L_{\max}-1}{2M} \text{ byte transactions/sample}$$
(17.7)

Both the memory size and the memory bandwidth requirements appear to be slightly larger for the inverse transform than the forward transform, although these differences become negligible as M becomes large. We hasten to point out, however, that our assumption that 2 bytes are required to buffer each subband sample is extremely conservative. This is because the samples of all coded subbands (i.e., not the intermediate LL subbands) may be efficiently stored in quantized form. The actual number of bits which must be stored in external memory for each coded subband sample is analyzed more carefully in Section 17.5.2. Table 17.11 indicates that we need only store an average of 6 bits per subband sample, assuming that lossless performance is required, while many fewer bits are required for lossy compressed images. Using this bit depth for the coded subband samples, the memory bandwidth becomes

$$B_{\text{IDWT}}^{(M)} \lesssim \frac{5}{3} \left(2 + \frac{2L_{\max} - 1}{2M}\right) - \frac{1}{4} \frac{2L_{\max} - 1}{2M} \text{ byte transactions/sample}$$

which is marginally better than that for the forward transform.

LIFTING IMPLEMENTATIONS

Our discussion of DWT implementation structures to this point has been based on the direct implementation described in Section 10.3. Reversible transforms require a subtle non-linear modification to the simple inner product expression of equation (10.9), which is best understood and implemented in the context of lifting. Lifting is described briefly in Section 10.4.2 and more extensively in Section 6.4. Lifting and reversibility have no impact on the spatial regions of support associated with the vertical and horizontal transform operators. Consequently, the buffering techniques and memory resource analyses above may be applied to both reversible and irreversible transforms. Nevertheless, it is interesting to consider what benefits might be derived by exploiting the lifting framework more directly in designing the memory structure of the transform.

For simplicity, we shall restrict our discussion here to the lifting analysis and synthesis state machines depicted in Figures 6.13 and 6.14. As shown in Section 6.4.4, these state machines are suitable for implementing any set of symmetric wavelet kernels with odd, least dissimilar lengths. The CDF 9/7 and spline 5/3 kernels described by Part 1 of the JPEG2000 standard both conform to this model. As explained in Section 6.5.3, symmetric extension at the image boundaries may also be understood in a particularly simple and elegant manner within the context of these lifting state machines.

The one dimensional operator implemented by the analysis machine of Figure 6.13 involves L_{max} lifting steps, with one state variable, Σ_l, for each step. The figure depicts $L_{max} = 4$ lifting steps, which is the number required to implement the CDF 9/7 transforms. The lifting step filters (darkly shaded boxes in the figure) add their two inputs and multiply the result by some constant, λ_l. For reversible transforms, the result must also be rounded to an integer. In some cases, such as the spline 5/3 transform, the multipliers are trivial powers of 2. The lifting state machine in Figure 6.13 does not include the subband normalization factors, K_0 and K_1, which appear in the more general lifting structure of Figure 6.9. These factors are only required for irreversible transforms, where they may be conveniently folded into the subband quantization step sizes.

Figure 6.13 suggests an alternate strategy for implementing the vertical transform, in which the L_{max} state variables for each column are themselves buffered in external memory. In this case, the total number of line buffers required for the dth stage of DWT analysis is $L_{max} + 2$. Two line buffers correspond to the next two unprocessed image lines from the LL_{d-1} subband, while the remaining L_{max} line buffers are used to store state information. This is illustrated in Figure 17.10. As with direct implementations, the vertically low- and high-pass subband lines produced by the vertical analysis machinery need not actually be buffered in memory. Instead, they are immediately subjected to horizontal analysis. This may be performed using a similar lifting state machine.

Evidently, the lifting implementation of Figure 17.10 requires fewer line buffers ($L_{max} + 2$) than the direct implementation of Figure 17.7 ($2L_{max} + 1$). Memory savings are reduced somewhat by the fact that some of the state line buffers must be maintained with higher numeric precision than the original image samples. This can also have an adverse impact on memory bandwidth, since the state buffers must be both read and written during each pass of the processing engine. Fortunately, the first state variable (or line buffer) Σ_1, is somewhat different to the others in that it is not affected by the processing. Thus, only $L_{max} - 1$ state buffers require a higher precision representation than the image samples

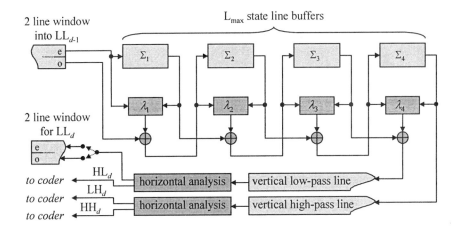

Figure 17.10. Incremental DWT analysis using the lifting state machine of Figure 6.13.

and only these buffers must be written during each pass of the DWT processing engine, while all $L_{\max}$ buffers must be read.

In keeping with the conventions adopted in Section 17.4.2, let us assume that all sample values are assigned a 2 byte representation, excepting only the original image samples, which have a 1 byte representation. We may then evaluate the overall memory size and bandwidth resources required by a lifting implementation of the M-line processing paradigm. We find that

$$
\begin{aligned}
S_{\text{DWT-lift}}^{(M)} &\lesssim [6M + 4L_{\max} - 1] \, N_2 \text{ bytes} \\
B_{\text{DWT-lift}}^{(M)} &\lesssim \tfrac{5}{3} \left(2 + \tfrac{2L_{\max}-1}{2M}\right) + \tfrac{L_{\max}-1}{M} \text{ bytes/sample}
\end{aligned}
\tag{17.8}
$$

Notice that the memory size is slightly reduced, while the memory bandwidth is slightly increased, relative to the direct implementation strategy described previously (see equation (17.6)).

Perhaps the most important benefit of the lifting approach is that DWT synthesis and analysis operations are almost identical. The close relationship between DWT analysis and synthesis may be appreciated by comparing the corresponding state machines shown in Figures 6.13 and 6.14. Lifting implementations of the analysis and synthesis systems have identical memory organizations and identical memory bandwidth requirements. By contrast, the memory organizations associated with the direct implementations shown in Figures 17.7 and 17.9 are quite different.

MULTI-STAGE IMPLEMENTATIONS

Up to this point our memory and bandwidth calculations have been based on the assumption that the wavelet transform is implemented using a single-stage processor. This processor is applied to each of the D stages using an appropriate schedule. With this approach, intermediate LL subband samples must be buffered in external memory, which has a significant impact on memory size and bandwidth. As an alternative, we may consider implementing two or more DWT stages in parallel, tightly synchronized so that the intermediate LL samples output by the first stage may be consumed by the second stage without intermediate buffering.

There is probably little to be gained by implementing more than two stages in parallel. Consequently, we shall assume that the DWT processing engine processes stages $2d - 1$ and $2d$ together, for each $d = 1, 2, \ldots, \lceil D/2 \rceil$. Again, we may describe a family of such implementations known as M-line dual-stage transforms, which "simultaneously" produce (or consume) M lines of each of the lowest frequency subbands, LL_{2d} through HH_{2d}, and $2M$ lines of each of the intermediate frequency subbands, HL_{2d-1}, LH_{2d-1} and HH_{2d-1}.

We begin by considering the dual-stage generalization of the direct analysis implementation shown in Figure 17.7. When applied to stages $2d - 1$ and $2d$, the dual-stage DWT engine reads from a moving window consisting of $4M + 2L_{\max} - 1$ lines of LL_{2d-2} samples. This allows it to produce $2M$ lines for each of subbands LL_{2d-1} through HH_{2d-1}. In order to produce the M lines required for each of the four lowest frequency subbands, the engine requires access to $2M + 2L_{\max} - 1$ lines of LL_{2d-1}. The last $2M$ of these lines are being produced by the first stage and need not be buffered. Thus, only $2L_{\max} - 1$ lines must be buffered for subbands LL_{2d-1}, while $4M + 2L_{\max} - 1$ lines must be buffered for subbands LL_{2d-2}, $d = 1, 2, \ldots$. After some algebra, we find that the total memory size is bounded by

$$S_{\text{DWT-2}}^{(M)} \lesssim \left[6\frac{2}{3}M + 3\left(2L_{\max} - 1\right) \right] N_2 \text{ bytes}$$

To determine the memory bandwidth, observe that the engine performs one pass through stages $2d - 1$ and $2d$, for every $2^{2d}M$ image lines. In the process, it reads $4M + 2L_{\max} - 1$ lines from LL_{2d-2}, writes M lines to LL_{2d}, reads $2L_{\max} - 1$ lines from LL_{2d-1} and writes $\min\{2M, 2L_{\max} - 1\}$ lines back to LL_{2d-1}. Putting this all together and being careful to account for the cost of the application writing image

Table 17.9. External memory size and bandwidth for single- and dual-stage direct implementations of the forward DWT. The inverse DWT is similar, with slightly larger values for memory size and bandwidth.

M	$S_{\text{DWT}}^{(M)}$	$B_{\text{DWT}}^{(M)}$	$S_{\text{DWT-2}}^{(M)}$	$B_{\text{DWT-2}}^{(M)}$
1	$27N_2$ bytes	9.2 bytes/sample	$28N_2$ bytes	6.65 bytes/sample
2	$33N_2$ bytes	6.3 bytes/sample	$34N_2$ bytes	4.7 bytes/sample
4	$45N_2$ bytes	4.8 bytes/sample	$48N_2$ bytes	3.7 bytes/sample
8	$69N_2$ bytes	4.1 bytes/sample	$74N_2$ bytes	3.0 bytes/sample
16	$117N_2$ bytes	3.7 bytes/sample	$128N_2$ bytes	2.6 bytes/sample

samples into the LL_0 buffer, we deduce that

$$B_{\text{DWT-2}}^{(M)} \lesssim 2\frac{4}{15} + \frac{1}{60M} \left[33\left(2L_{\max} - 1\right) + 16\min\{2M, 2L_{\max} - 1\}\right]$$

$$\lesssim 2\frac{4}{15} + \frac{1}{M}\left(1\frac{19}{30}L_{\max} - \frac{49}{60}\right) \text{ byte transactions/sample}$$

Table 17.9 compares the external memory sizes and bandwidths associated with single- and dual-stage direct implementations of the DWT analysis operations, with various values for the parameter M. The CDF 9/7 transform is assumed in these calculations so that $L_{\max} = 4$. Notice that the dual-stage transform significantly reduces memory bandwidth with very little cost in memory size. On the other hand, the internal implementation of the DWT processing engine is significantly more complex, requiring two separate DWT stage processors, the first of which must be sized to process $4M$ lines together. We emphasize the fact that our memory bandwidth figures all include the cost of initially writing the image samples into the buffer for processing.

The lifting structure of Figure 17.10 may of course also be generalized to process multiple DWT stages together. The memory and bandwidth requirements associated with a dual-stage M-line lifting implementation can be shown to satisfy

$$S_{\text{DWT-2-lift}}^{(M)} \lesssim \left[6\tfrac{2}{3}M + 4L_{\max} - 1\right] N_2 \text{ bytes}$$

$$B_{\text{DWT-2-lift}}^{(M)} \lesssim 2\tfrac{4}{15} + \tfrac{1}{M}\left(2\tfrac{2}{15}L_{\max} - \tfrac{47}{60}\right) \text{ byte transactions/sample}$$

When $L_{\max} = 4$ this represents a saving of 8 line buffers, at a cost of $\frac{2.03}{M}$ byte transactions/sample relative to the dual-stage implementation described above. Again, the most significant benefit of the lifting structure is that analysis and synthesis implementations are virtually identical.

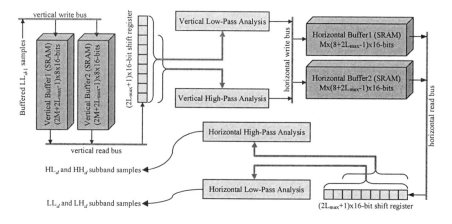

Figure 17.11.   Single-stage DWT analysis engine.

17.4.3   ON-CHIP RESOURCES

The purpose of this section is to provide some preliminary assessment of the complexity of an M-line single-stage DWT analysis engine. This is the circuit which implements the vertical and horizontal transformations shown in Figure 17.8. The synthesis engine and lifting implementations have similar levels of complexity and will not be explicitly discussed here. We confine our discussion of complexity to on-chip buffering requirements.

Figure 17.11 indicates an appropriate structure for the M-line processing engine. When applied to the dth DWT stage, the engine reads buffered samples from the intermediate LL_{d-1} subband into an internal vertical buffer. The buffered samples are then used to feed a vertical shift register, which supplies the $(2L_{\max} + 1)$ sample window actually required to produce new low- and high-pass vertical subband samples. In the example of Figure 17.11, the internal vertical buffer memory is divided into two banks, one of which is filled with LL_{d-1} samples while the other is being read into the vertical shift register.

The LL_{d-1} samples could conceptually be read directly from external memory into the vertical shift register; however, this would unduly constrain external memory access patterns for most practical applications. We have chosen to size the vertical buffer memories so that they do not dramatically affect the total on-chip memory cost, while providing substantial flexibility in regard to external memory access patterns. In this example, each of the two vertical buffer memories can hold 8 columns from the $2M + 2L_{\max} - 1$ buffered lines of subband LL_{d-1}. This per-

Table 17.10.   On-chip memory estimates for single- and dual-stage implementations of the forward DWT. The inverse DWT is similar.

M	single-stage engine	dual-stage engine
1	350 bytes	820 bytes
2	470 bytes	1.2×10^3 bytes
4	720 bytes	1.9×10^3 bytes
8	1.2×10^3 bytes	3.4×10^3 bytes
16	2.2×10^3 bytes	6.4×10^3 bytes

mits multiple contiguous external memory accesses[15]. A line-interleaved memory organization may be used to further improve memory access locality.

The remainder of the circuit is reasonably straightforward. A second double buffered memory system maintains horizontal windows of $8 + 2L_{max} - 1$ samples into each of the $2M$ vertical subband lines which are being produced by the engine. The first M lines are written into the first memory, while the second M lines are being read out of the second memory into a horizontal shift register. The horizontal shift register feeds the horizontal subband analysis blocks. Note that a horizontal buffer size of $8 + 2L_{max} - 1$ is sufficient to process 8 new columns from each of the vertical subband lines, which is exactly the number of new columns available from the vertical buffer memory.

For a complete implementation, some internal buffering should also be provided for the generated subband samples, to decouple the internal processing from external memory access patterns. Subband sample buffering is considered separately in Section 17.5.1.

Table 17.10 indicates the total amount of on-chip memory required by the DWT engine for different values of the parameter M. The last column of the table provides a rough estimate of the memory required to implement a dual-stage processing engine. The benefit of dual-stage implementations is a significant reduction in the external memory bandwidth, as discussed at the end of Section 17.4.2. We base these estimates on the fact that an M-line dual-stage processor is basically the cascade of two single-stage processors, generating $2M$ and M lines of their respective subbands in each pass across the image.

As noted previously, reversible transforms may be implemented using essentially the same structure, since reversibility does not alter the spa-

---

[15] Non-contiguous memory accesses can significantly reduce the bandwidth achieved when working with DRAM technology.

tial support of the horizontal and vertical analysis operators. On the other hand, the lifting framework suggests a number of potential modifications to the internal machinery depicted in Figure 17.11. The lifting state machine of Figure 6.13 may be used to replace the shift registers and analysis blocks shown in the figure, with obvious benefits: instead of a 9 sample shift register and two quite complex blocks of arithmetic logic, one needs only 4 state variables, 8 adders and 4 multipliers. Moreover, reversible and irreversible transforms may both be implemented with essentially the same circuit. If the external memory structure associated with the vertical transform is based on this same lifting state machine, as in Figure 17.10, the internal structure of the DWT engine must be modified accordingly. Again, the most significant advantage of lifting implementations is their ability to unify the designs of the forward and the inverse transform.

17.5   SYSTEM CONSIDERATIONS

The purpose of this section is to unify our discussions of block coding and DWT implementations in Sections 17.2 and 17.4. In particular, we must account for the resources required to interface these two sub-systems. To simplify matters, we ignore the storage and I/O requirements of the compressed data itself. When significant compression is achieved, input and output of the code-block bit-streams should not substantially impact memory bandwidth consumption. This is less likely to be the case at very high bit-rates or when truly lossless compression is expected. In many applications, the embedded code-block bit-streams might be completely buffered in memory. However, JPEG2000 also provides progression orders which support compressed data streaming (see Section 13.1.1). With these progression orders, the total amount of memory required to buffer code-block bit-streams can be small in comparison to that required to implement the rest of the system.

17.5.1   CODED DATA BUFFERING

The incremental DWT implementations described in Section 17.4 produce subband samples M lines at a time. On the other hand, the block coder consumes these subband samples in blocks, whose maximum height is described by the parameter, $J_1^{\max}$. Likewise, block decoding produces samples in blocks, which must subsequently be consumed M lines at a time for DWT synthesis. Typical values for $J_1^{\max}$ are 32 or 64, although some memory constrained compressors might select a much smaller value such as $J_1^{\max} = 8$.

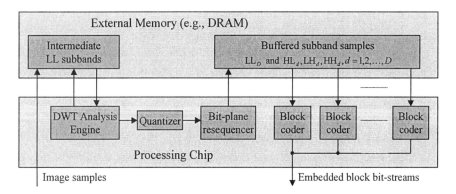

Figure 17.12. Loosely synchronized JPEG2000 compressor.

For some applications, it may be possible to arrange for M to be a multiple of $J_1^{\max}$, in which case the subband samples can be quantized and delivered directly to a collection of block coders as they emerge from the DWT processing engine. Unfortunately, this requires tight synchronization between the block coding and wavelet transform implementations. Tight synchronization may in turn prevent block coders from realizing the average throughputs suggested in Section 17.2.2. The amount of on-chip and external memory required to implement an M-line transform with $M \geq J_1^{\max}$ may also prove prohibitive.

For these reasons, we shall restrict our attention to loosely synchronized implementations, in which quantized subband samples are buffered in external memory. A pool of block coders (or decoders) processes the buffered subband samples. The number of block coders which are actually implemented may be tailored to suit the application's throughput requirements. Such an arrangement is illustrated in Figure 17.12.

In the system of Figure 17.12, subband samples generated by the DWT analysis engine are quantized immediately. The quantization indices are temporarily buffered in on-chip memory and written out in sign-magnitude form, bit-plane by bit-plane. This is important, since the block coders expect to read the quantization indices bit-plane by bit-plane, rather than sample by sample (see Section 17.2.1).

As an example, the "bit-plane resequencer" block in Figure 17.12 might collect 64 quantization indices[16] from each of the subbands being generated by the DWT analysis engine. The 64 sign bits would then be

---

[16] We could pick any number here, instead of 64. Larger values impose a larger internal buffering cost, while smaller values reduce the regularity of external memory transactions.

written out in 8 contiguous bytes, followed by each of the magnitude bit-planes in turn. Allowing for double buffering and up to 16 bits for each quantization index, such a resequencer would require 256 bytes of on-chip memory for each subband. If a single-stage DWT engine is employed, three subbands are generated together and the resequencer requires 768 bytes. The resequencer for a dual-stage DWT engine would consume 1.5 kB of on-chip memory. In some cases, even larger resequencing buffers might be used to improve external memory access patterns.

Assuming that $M < J_1^{\max}$, quantized samples for a full row of code-blocks must be accumulated in external memory for each subband before they can be coded. It follows that storage must be allocated for

$$\left\lceil \sum_{d=1}^{D} 2^{-d} \cdot 3 J_1^{\max} + 2^{-D} J_1^{\max} \right\rceil N_2 \lesssim \left\lceil 3 J_1^{\max} \right\rceil N_2 \text{ samples}$$

In practice, the external memory must be able to accommodate at least twice this number of samples so as to partially decouple the block coding and DWT analysis operations from one another. This is important if the average block coding throughputs suggested in Section 17.2.2 are to be realized.

The required subband buffer memory depends on the number of bits needed to represent each quantized subband sample. If we wish to accommodate entirely lossless compression with 8-bit/sample images, a 12-bit representation is sufficient for all subband samples produced by the reversible DWT (see Table 17.4). This leads to the following "worst case" estimates of the memory size and bandwidth required to buffer subband samples for coding (or after decoding). The memory size figure is based on double buffering, while the bandwidth figure is based on the assumption that all 12 bits of each sample must be written to and read from external memory.

$$S_{\text{coder}}^{\max} \approx 6 J_1^{\max} N_2 \cdot \tfrac{12}{8} = \left\lceil 9 J_1^{\max} \right\rceil N_2 \text{ bytes}$$

$$B_{\text{coder}}^{\max} = 2 \times \tfrac{12}{8} = 3 \text{ byte transactions/sample}$$

$$(17.9)$$

The resources required for lossy compression can be significantly smaller than these worst case figures, depending upon the quantization step sizes involved.

17.5.2   BANDWIDTH REDUCTION

Memory bandwidth may be substantially reduced by restricting external memory transfers to only those bit-planes which are actually significant. To see how this may be accomplished, recall that the bit-plane

Table 17.11. Average number of bits for each quantized subband sample which must be buffered in external memory, assuming 8×8 blocking by the bit-plane resequencer..

bit-rate:	0.25 bps	0.5 bps	1.0 bps	2.0 bps	lossless
bits/sample:	0.79	1.43	2.46	3.79	5.99
B_{coder} (bytes/sample):	0.20	0.36	0.61	0.95	1.50

resequencer buffers groups of quantized subband samples – 64 in our example. Now suppose that the resequencer determines the number of magnitude bit-planes, K, required to represent this group of 64 samples, storing the value of K to external memory, along with the sign bits (unless $K = 0$) and the K significant magnitude bit-planes. The relevant block coder later retrieves the value of K and reads only the necessary bit-planes. Schemes of this form are possible precisely because the quantized subband samples are stored bit-plane by bit-plane rather than sample by sample. The K values also allow the block coder to efficiently determine the first bit-plane at which to start coding.

The same approach may be used to conserve bandwidth during decompression, in which case a suitable value for K is readily deduced on-the-fly by the block decoder. The decoder has the opportunity to save additional bandwidth, since there is no need for it to write out any undecoded least significant bit-planes. It is common for code-block bit-streams to be truncated prior to decompression, leaving one or more undecoded least significant bit-planes. In a simple implementation, the block decoder would signal the number of undecoded bit-planes, U, in addition to the value of K, for each group of samples – 64 samples in our example. The bit-plane resequencer for the DWT synthesis engine need only retrieve those bit-planes which have been decoded. Unfortunately, when the last decoded coding pass is not a cleanup pass, $\mathcal{P}^{(p,2)}$, the decoder must also write a marker bit for each sample to indicate whether or not that sample was coded in the last bit-plane. See Section 17.1.3 for a description of the marker bit method.

To demonstrate the effectiveness of these techniques, we count the number of bit-planes which must actually be buffered in external memory for each 8×8 block of subband samples. The results reported in Table 17.11 are obtained during decompression of the three high resolution test images shown in Figure 8.21. In each case, the code-stream contains a quality progressive lossless representation of the image, which is truncated to various bit-rates of interest. Although these results apply specifically to decompression, the lossless result applies equally to compression, since there are no undecoded bit-planes and there is no need for marker bits in this case. For convenience, we report the memory bandwidth, B_{coder}, required to read and write each of the buffered bit-planes. Notice that this figure is much smaller than the worst case estimate, $B_{\text{coder}}^{\text{max}}$, given in equation (17.9).

17.5.3   PUTTING IT ALL TOGETHER

We conclude this chapter by selecting one particular system configuration and calculating its resource requirements and throughput expectations. We assume maximum code-block dimensions of $J_1^{\text{max}} = J_2^{\text{max}} = 32$, which might be the limits set for "Profile-0" code-streams (see Table 18.1). We select a dual-stage M-line implementation of the DWT with $M = 8$, and a pool of 4 block coders executing in parallel. We allow for the most complex wavelet kernels supported by Part 1 of the JPEG2000 standard, having $L_{\text{max}} = 4$. External buffering of subband samples is provided with twice the minimum amount of memory so as to substantially decouple the DWT analysis and block coding sub-systems. We also allow for double buffering of the image samples supplied by the application; this adds $4MN_2$ bytes to the size of the LL_0 buffer, since the dual-stage DWT engine processes $4M$ image lines together.

The on-chip memory requirements consist of $4 \times \frac{3.5}{4}$ kB for the block coders (32×32 code-blocks require $\frac{1}{4}$ the on-chip memory shown in Figure 17.4), 3.4 kB for the DWT engine (from Table 17.10) and 1.5 kB for bit-plane resequencing: a total of 8.4 kB. External memory must be sized to accommodate $74N_2$ bytes for the DWT (from Table 17.9), $284N_2$ bytes for subband sample buffering (from equation (17.9)), and $32N_2$ bytes for double buffering of image samples. This generous external memory allowance is equivalent to 390 original image lines. The external memory bandwidth consists of 3.0 bytes/sample for the DWT (from Table 17.9) and at most 1.5 bytes/sample for the intermediate buffering of quantized subband samples (from Table 17.11): a total of 4.5 byte transactions/sample. Finally, a lossless throughput of about one sample every two clock cycles could be expected; this is based on the realistic assumption that the 4 block coders perform concurrent membership testing, but none of the more advanced techniques discussed in Section 17.2.2. Of course, higher throughputs could be achieved by including more parallel block coders on the chip.

It is worth noting that the estimated external memory bandwidth of 4.5 byte transactions/sample includes the cost of writing every image sample to the LL_0 buffer and subsequently retrieving it. If the application is able to supply image samples on demand, in groups of $4M$ lines at a time (recall that we are using a dual-stage M-line transform), the external memory bandwidth may be reduced to about 2.5 byte transactions/sample.

Decompressors can generally achieve somewhat lower memory bandwidth and significantly higher throughput when truly lossless performance is not required. This is because code-block bit-steams are often truncated between compression and decompression. Moreover, decompressors may judiciously discard coding passes in order to satisfy throughput or memory bandwidth constraints.

Throughout this chapter, we have consistently ignored the possible existence of tiles. Although tiles may be treated as independent images, it might not always be appropriate to process tiles sequentially. For line-based applications this would require sufficient memory to buffer an entire row of tiles, which might not be possible if the tile height is too large. It is possible to incrementally decompress an entire row of tiles together. However, substantial resources may be spent "thrashing" between the tiles if they are too narrow. Ideally then, if tiles are used at all they should not be tall and narrow. If the application is column-based rather than line-based, tiles which are short and wide will be the worst offenders. Sensitive to these concerns, JPEG2000 defines profiles which restrict the allowable tile dimensions. If tiles are used (otherwise, the entire image is one tile), Profile-0 requires them to be square, with dimensions no larger than 128. Profile-1 also insists on square tiles, with a less restrictive size, if the image is to be tiled at all. Code-stream profiles are discussed further in Section 18.2.

17.6   AVAILABLE HARDWARE

At the time of this writing, a number of companies are known to be developing JPEG2000 chip sets. However, to the best of our knowledge, preliminary technical data has been released only for the ADV-JP2000 chip [19] by Analog Devices Inc. This chip implements both the DWT and the block coding algorithm. Each of these sub-systems supports both forward (compression) and reverse (decompression) processing. Other, less computationally intensive aspects of the JPEG2000 algorithm are expected to be implemented as software, in an embedded or general purpose processor which communicates with the ADV-JP2000 chip. These include layer formation, collecting code-blocks into precincts, packet header coding, packet sequencing and marker segment generation. Such a hybrid hardware/software model is eminently reasonable and likely to be adopted by other implementations.

This first chip from Analog Devices does not implement all of the features required of a compliant JPEG2000 decompressor. Amongst various less significant limitations, the chip is only capable of processing images in tiles, having a maximum size of 256 × 256. As explained in Section 18.2, all conforming decompressors must support untiled images. Nevertheless, as a compressor, the chip may be used to generate conforming JPEG2000 code-streams and, as a decompressor, it is able to reconstruct its own compressed images. This behaviour may be quite sufficient for a number of applications, including digital still and video cameras.

The ADV-JP2000 implements a single block coding/decoding engine, which supports the full 64 × 64 block size. There is thus plenty of room for future throughput enhancements, through parallel processing of multiple code-blocks. Nevertheless, the chip boasts a typical throughput of about 10 Megapixels/second for lossless compression of 4:2:2 sub-sampled colour imagery, at an internal clock rate of 150 MHz. Throughput for lossy compression/decompression can be significantly higher. These throughput figures are consistent with the predictions offered in Section 17.2.2 for block coder implementations which employ concurrent membership testing on a stripe-column basis[17]; we have been led to believe that the ADV-JP2000 is based on similar principles.

---

[17]Since 4:2:2 sub-sampled colour images have a total of 2 samples per pixel, a throughput of 10 Megapixels/second at 150 MHz means 7.5 clock cycles per sample, which is almost identical to the $R_{sym} + R_{empty} = 8$ clock cycles predicted for lossless compression of natural imagery using Table 17.2.

Chapter 18

COMPLIANCE

Perhaps the most distinctive feature of JPEG2000 is its emphasis on and support for scalability. An existing code-stream may be accessed at a reduced resolution, a reduced level of quality (higher compression), a reduced number of components and even over a reduced spatial region. Moreover, the standard supports a rich family of information progression sequences. Information may be reordered without introducing additional distortion, thereby enabling a single compressed representation to serve the needs of a diverse range of applications.

To exploit the scalability of JPEG2000, it is expected that decompressors will often not decode all of the information which was originally incorporated into the code-stream by the compressor. In fact, it is desirable to allow decompressor implementations to ignore information which is not of interest to their target application. While this flexibility is one of the strengths of JPEG2000, it also renders inappropriate some of the conventional compliance testing methodologies which have been applied to non-scalable or less scalable compression standards.

At one extreme, decompressor implementations might be allowed to decode any portion of the code-stream which is of interest to them. At the other extreme, they might be required to correctly decode the entire code-stream. The first of these extreme approaches offers content providers and consumers no guarantee concerning the quality of the resulting imagery. The other approach is also inappropriate, for two reasons: 1) it offers the implementor no guarantee concerning the resources which may be required; and 2) in many cases the code-stream may contain information which is of no interest to the application.

At the time of this writing, compliance definitions and testing procedures have not been finalized. In this chapter we summarize the philosophy and testing methodologies which are expected to underpin Part 4 of the standard; Part 4 is concerned with compliance testing. The material presented here is based on a new working document for Part 4, which was created at the March 2001 WG1 meeting in Singapore.

18.1   A SYSTEM OF GUARANTEES

Compliant implementations of the decompressor are not obliged to decode the entire code-stream with which they are presented. They are, however, required to guarantee performance up to one of a number of so-called "compliance classes." The relevant *guarantees* are connected with key resources required by an implementation. They may be interpreted as a contract by the implementation to recover, decode and transform a well-defined minimal subset of the information contained in the code-stream. This contract is described in a manner which scales with the maximum output image size to which the decompressor claims compliance. The contract may be exploited by content providers to optimize recovered image quality over a family of decompressors, according to their compliance classes. The decompressor guarantees which constitute these classes are described more carefully in Section 18.3.

Unfortunately, legal JPEG2000 code-streams may involve adverse parameter choices, which could prevent a reasonable resource constrained decompressor from recovering any useful information whatsoever. To encourage the deployment of cost effective implementations, the standard defines a number of profiles which explicitly restrict such adverse choices. These profiles may be understood as offering *guarantees* to the implementor of a compliant decompressor, concerning the "nastiness" of the code-stream which may be presented to them. The role of profiles in JPEG2000 is limited only to restricting parameter choices which might completely "break" a resource limited implementation. Decompressors are not generally obliged to process all of the information in the code-stream with which they are presented; their obligations extend only to the guarantees which they themselves are offering to content providers, as embodied by the relevant compliance class. Code-stream profiles are the subject of Section 18.2 below.

By and large, the JPEG2000 standard places few restrictions on how a compressor should use the flexibility offered by legal code-streams. Nevertheless, there are a number of *guarantees* which compressors might be expected to offer their respective applications. Certainly, a compliant compressor is obliged to create legal code-streams, as described in Chapter 13. It may also be required to create code-streams conforming to one or another of the restricted profiles described below. If lossless performance is claimed, the compressor must be capable of losslessly compressing images of a given size, bit-depth and number of image components. If error resilience is required, the compressor must also be capable of implementing the error resilient termination procedures for MQ and/or raw codeword segments, exactly as described in Section 12.4.1.

Together, this system of inter-working guarantees serves to protect decompressors from malicious code-streams, to protect content providers from uncooperative or inadequate decompressors, and to preserve the flexibility and scalability of JPEG2000 code-streams. Conformance to the various guarantees is externally testable, so that it is possible to identify the non-compliant element in a system which does not behave as expected. At the same time, content providers are free to include information in the compressed code-stream, which most compliant decompressors will not be obliged to reconstruct. By encouraging this practice, JPEG2000 code-streams should be able to serve the needs of a wide range of applications.

18.2 CODE-STREAM PROFILES

Code-stream profiles represent restricted subsets of the JPEG2000 syntax. The reader may be familiar with the notion of profiles, as used in the various MPEG video compression standards, where they play a substantial role. In JPEG2000, however, profiles play a much less significant role, since compliant decompressors are not required to recover all of the information in the code-stream. JPEG2000 profiles exist only to ensure that reasonable implementations will be able to recover at least *some* of the information in the code-stream. Hard restrictions of any form on the JPEG2000 code-stream syntax have the potential to jeopardize scalability and/or inter-operability. For this reason, profiles have been the subject of much debate and the material presented in this section may be subject to some further revision.

Although profiles may be understood as part of the compliance regime, they are unlikely to be described in Part 4 of the standard. Instead, the WG1 committee opted to include profiles as an amendment [66] to Part 1, primarily for reasons of expediency. Table 18.1 summarizes the restrictions embodied in this amendment, with a number of minor modifications which are quite likely to be adopted. Notice that only two profiles are defined: a "Profile-0;" and a "Profile-1." Code-streams conforming only to the unrestricted JPEG2000 syntax described in Chapter 13 may be identified as "Profile-2."

Decoders may determine in advance whether the code-stream conforms to one of the restricted profiles by inspecting the capabilities word, *CA*, in the *SIZ* marker segment (see Section 13.3.5). A value of 0001_h for this 16-bit quantity indicates that the code-stream con-

Table 18.1.   Restricted code-stream profiles.

Type of restriction	Profile-0	Profile-1
Code-block dimensions	*Option A: $J_1^{t,c}, J_2^{t,c} = 32$ Option B: $J_1^{t,c}, J_2^{t,c} = 64$ Option C: $J_1^{t,c}, J_2^{t,c} \leq 64$	$J_1^{t,c}, J_2^{t,c} \leq 64$
Coder mode switches	Disallowed (except $ERTERM$)	No restriction
Tile dimensions	$T_1 = T_2 \in \{64, 128\}$ or no tiling, i.e., $\Omega_i^{\mathrm{T}} + T_i \geq F_i, \quad i = 1, 2$	$T_1 = T_2 \leq 1024$ or no tiling, i.e., $\Omega_i^{\mathrm{T}} + T_i \geq F_i, \quad i = 1, 2$
Sub-sampling factors	$S_1^c, S_2^c \in \{1, 2\}$	No restriction
Canvas coordinates	$0 \leq E_i, F_i, T_i, \Omega_i^{\mathrm{T}} < 2^{31}$	$0 \leq E_i, F_i, T_i, \Omega_i^{\mathrm{T}} < 2^{31}$
LL_D subband size	$\left\lceil \frac{F_1^c}{2^{D_c}} \right\rceil - \left\lceil \frac{E_1^c}{2^{D_c}} \right\rceil \leq 120$ $\left\lceil \frac{F_2^c}{2^{D_c}} \right\rceil - \left\lceil \frac{E_2^c}{2^{D_c}} \right\rceil \leq 160$ for $c = 0, 1, 2$	$\left\lceil \frac{F_1^c}{2^{D_c}} \right\rceil - \left\lceil \frac{E_1^c}{2^{D_c}} \right\rceil \leq 120$ $\left\lceil \frac{F_2^c}{2^{D_c}} \right\rceil - \left\lceil \frac{E_2^c}{2^{D_c}} \right\rceil \leq 160$ for $c = 0, 1, 2$
PPM/PPT markers COD/COC markers QCD/QCC markers	Disallowed Main header only Main header only	No restriction No restriction No restriction
RGN up-shift parameter	$U \leq 37$	$U \leq 37$

*Three options are still being considered for code-block dimensions in Profile-0.

forms to Profile-0, while a value of 0002_h indicates that the code-stream conforms to Profile-1. The value 0000_h, means that any legitimate Part 1 code-stream may follow.

EXPLANATION OF PROFILES

Some justification for the restrictions on tile dimensions has already been given in Section 17.5.3. As noted in Table 18.1, restrictions on code-block dimensions for Profile-0 are still under consideration. Larger code-blocks yield somewhat higher coding efficiency, while offering less flexibility for rate control and spatial access to the image. Following the discussion in Sections 17.2.3 and 17.5.3, larger code-blocks have adverse implications for on- and off-chip memory requirements. On the other hand, larger code-blocks reduce the amount of memory required for managing state information for each block. This is of particular interest during compression, where the lengths and distortion contributions for each coding pass may need to be stored. Since many of these trade-offs are highly dependent on implementation strategies and application requirements, the weaker conditions associated with Option C are appealing from the perspective of maximizing inter-operability. In that case, Profile-0 and Profile-1 would be fully compatible at the code-block level.

It is worth noting that any lower bound imposed on the nominal code-block dimensions, $J_1^{t,c}$ and $J_2^{t,c}$, may be overridden by the selection of sufficiently small precinct dimensions, to which no restriction currently applies. Similarly, although Profile-0 and Profile-1 both restrict tile dimensions to be square (if the image is tiled), there is no requirement that component sub-sampling factors be identical in both directions. Consequently, when mapped into any particular image component, the elements of the tile partition might no longer be square.

The LL_D (DC subband) size restriction prescribed by both Profile-0 and Profile-1 should not be understood as limiting the size of the compressed image. The requirement can always be satisfied by employing a sufficiently large number of DWT levels, D. The notation, D_c, introduced in Table 18.1 refers to the minimum number of DWT levels in any tile of component c, i.e.,

$$D_c \triangleq \min_t D_{t,c}$$

The LL_D size restriction ensures that a decompressor will be able to reconstruct a low resolution version (thumbnail) of any or all of the first three image components, whose height is no larger than 120 samples and whose width is no larger than 160 samples. Compressors would do well to respect this constraint even when not targeting Profile-0 or Profile-1.

The reader may wonder why the two profiles identified in Table 18.1 restrict the various canvas coordinates to 31-bit unsigned integers, while the *SIZ* marker signals these quantities as 32-bit unsigned quantities. This minor restriction should have no impact on most practical uses of the standard, but may simplify the reliable representation and manipulation of canvas coordinates using 32-bit integer arithmetic. Signed integer representations of the canvas coordinates are particularly appealing for implementations which support geometric manipulation of the image, as it is being decompressed. Capabilities such as this are offered by the Kakadu software supplied with this text.

For an explanation of the restriction of the *RGN* marker segment's up-shift parameter to the range $U \leq 37$, the reader is referred to Section 10.6.3.

IMPACT ON JPEG2000 FEATURES

The profile restrictions appearing in Table 18.1 have relatively little impact on the key features offered by JPEG2000. Perhaps the most significant exception to this is the disabling in Profile-0 of all block coder mode switches: *BYPASS, RESET, RESTART, CAUSAL,* and *SEGMARK*. As explained in Section 12.4.3, error detection and concealment during block decoding relies on either the *SEGMARK* mode or, preferably, the combination of both *ERTERM* and *RESTART* mode switches. Neither of these constructions is compatible with Profile-0 as it currently stands.

Profile-1 should be sufficient to meet the needs of the vast majority of applications. To maximize inter-operability, therefore, all software implementations are strongly encouraged to support at least Profile-1. Hardware decompressors should also aim to support Profile-1 code-streams unless the cost proves prohibitive. Fortunately lossless or near-lossless transcoding between profiles is often possible. For example, a Profile-1 code-stream which uses 64×64 code-blocks may be transcoded into one which uses 32×32 code-blocks, with negligible or no additional distortion[1]. Block coder mode switches may also be added or removed without incurring distortion, except in the case of the *CAUSAL* switch. On the other hand, tile dimensions cannot generally be changed without introducing more significant levels of distortion. Inter-operability across profiles can usually be improved by avoiding tiling, since all profiles must support untiled images.

18.3   DECOMPRESSOR GUARANTEES

Decompressor guarantees are made in connection only with one or more of the code-stream profiles discussed in Section 18.2. The guarantees are expressed in relation to image dimensions, H (height) and W (width), and a number of components, C, at which the implementation claims compliance. The values, H, W and C are not dependent in any way on the code-stream which is actually being decompressed. Nor do they impose restrictions of any form on the code-streams which the implementation must be able to process in a compliant manner. They refer only to the claimed capabilities of the decompressor. The dimensions H, W and C at which compliance claims may be made and tested are identified as "compliance levels," as outlined in Section 18.3.4.

The guarantees (or obligations) expected of a compliant decompressor are described in three categories, each of which is loosely associated with one of the sub-systems which might be found in a typical implementation. The categories are as follows.

---

[1] Coding pass membership may be altered for samples on the boundaries of the 32×32 code-blocks and the effects may propagate to some interior samples, leading to a small increase in distortion when the code-block terminates with anything other than a cleanup pass.

Parsing obligations: Implementations are expected to be able to recover all of the code-block contributions from the code-stream which are relevant to their claimed resolution, $H \times W$, and number of components, C. They are released from this obligation only in the event that a so-called "parser quit" condition is encountered. This condition is developed further in Section 18.3.1 below.

Block decoding obligations: Except as described below, decompressors are expected to decode all of the compressed bits which are available for code-blocks belonging to their claimed resolution, $H \times W$, and number of components, C. Specifically, these are the code-blocks which belong to any resolution r, of any tile $\mathbf{t}$, in any image component c, such that

$$
c < C, \qquad \left\lceil \frac{F_1}{2^{D_{\mathbf{t},c}-r} S_1^c} \right\rceil - \left\lfloor \frac{E_1}{2^{D_{\mathbf{t},c}-r} S_1^c} \right\rfloor \leq H
$$

$$
\text{and} \qquad \left\lceil \frac{F_2}{2^{D_{\mathbf{t},c}-r} S_2^c} \right\rceil - \left\lfloor \frac{E_2}{2^{D_{\mathbf{t},c}-r} S_2^c} \right\rfloor \leq W
$$

(18.1)

The reader is referred to Chapter 11 for a thorough explanation of the notation used here.

Decoders are not obliged to decode compressed bits representing anything other than the most significant $K^{\min}$ magnitude bit-planes of any code-block. This is explained further in Section 18.3.2 below. Of course, there is also no need to decode compressed bits which are not recovered from the code-stream as a result of the "parser quit" condition being reached. Apart from these two exceptions, all relevant compressed bits must be correctly decoded.

Sample transformation obligations: Decompressors are obliged to implement both the reversible spline 5/3 transform and the irreversible CDF 9/7 transform to a prescribed level of accuracy. They are also obliged to implement dequantization and colour transformation to a prescribed level of accuracy. These accuracies are expressed in terms of two quantities, $P^{\min}$ (irreversible precision) and $B^{\min}$ (reversible bit-depth), as developed in Section 18.3.3 below. Note that the accuracy with which sample data transformations are performed is only partially responsible for the final reconstructed image quality, since compressed data may be lost during both parsing and decoding. This is true even for losslessly compressed images.

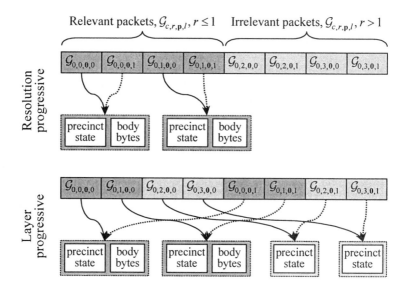

Figure 18.1.   Impact of resolution and layer progressive packet sequences on parsing resources.

18.3.1   PARSING OBLIGATIONS
BACKGROUND: PARSING RESOURCES

In order to recover compressed bits from the code-stream it is necessary to parse packet headers. The relevant packets are those belonging to tile-component resolutions which satisfy equation (18.1). Unfortunately, the resources required to locate and recover all relevant packets from the code-stream can depend upon the number and positions of any irrelevant packets encountered while searching for relevant ones. The reason for defining a "parser quit" condition is to bound the resources which implementations must commit to parsing over irrelevant packets in the code-stream. Once these bounds are exceeded compliant decompressors are at liberty to discard information from the code-stream, regardless of its relevance. Knowing these bounds, content providers may anticipate the minimum quality which will be achieved by a compliant decompressor and use this information to construct code-streams which are appropriate for the intended range of applications.

To appreciate the parsing problem, it is instructive to consider resolution and layer progressive packet sequences of a single code-stream, as illustrated in Figure 18.1. This simple example involves only one image component, $c = 0$, with two quality layers, $l = 0, 1$, four resolution levels ($D = 3$ DWT levels), and one precinct, $\mathbf{p} = 0$, per resolution. We assume the image dimensions are such that only the first two resolution levels contain relevant information, in accordance with equation (18.1). The four relevant packets are identified by darker shading in the figure, while irrelevant packets are shaded lightly.

Figure 18.1 identifies two different types of memory resources which may be associated with parsing. Precinct state memory maintains the state of tag tree nodes and various other quantities, which are specific to each precinct. Storage for this state information is required from the point at which the first packet of the precinct is encountered, until the last packet of the precinct has been parsed. It is helpful to think of the parser "opening a file" on the precinct[2] when its first packet is encountered and "closing the file" when the last packet has been parsed. The cost of maintaining this open "file" is the precinct state memory. In the absence of *SOP*, *PLM* or *PLT* marker segments, files must be opened on all precincts, whether relevant or not, up until the point at which the last relevant packet has been parsed from the code-stream. This is because the only guaranteed way to determine the length of a packet and hence find the start of the next packet is by fully parsing its header.

The second type of memory resource identified in Figure 18.1 is that required to store the compressed code bytes recovered from packet bodies. Although it may sometimes be possible to decompress and discard code bytes as they appear, it will usually be necessary to buffer most or all of the compressed data in memory before decompression of the image can commence. Of course, there is no need to preserve compressed data from the irrelevant packets.

PRECINCT STATE MEMORY

Evidently, some code-stream organizations require less parsing resources than others. As seen in Figure 18.1, resolution progressive organizations are generally preferable to layer progressive organizations, since decompressors with lower target resolutions need not allocate state memory for the higher resolution precincts, in which they have no interest. On the other hand, layer progressive organizations have incremental refinement properties which are desirable for some applications. In order to support a wide range of applications, while enabling cost effective compliant implementations, a "parser quit" condition is defined, which roughly reflects the buffering resources which might be required. Specifically, let x denote any particular location in the code-stream and let

---

[2]The analogy here is with files stored in a filing cabinet; once the file is closed, it might be discarded from the filing cabinet.

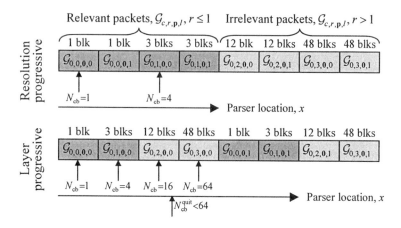

Figure 18.2. Evolution of $N_{cb}(x)$ with parser location x, showing the packets at which $N_{cb}(x)$ increases and a hypothetical "quit" point.

$N_{cb}(x)$ denote the total number of code-blocks whose precincts have been opened prior to location x. Compliant implementations are not obliged to parse beyond any point x at which $N_{cb}(x)$ exceeds a defined threshold, N_{cb}^{quit}. The obligation is the same regardless of whether the code-stream contains *SOP*, *PLM* or *PLT* marker segments, although some implementations might be able to take advantage of them to reduce memory consumption.

Figure 18.2 illustrates the evolution of $N_{cb}(x)$ as the parser advances through the code-stream, identifying those packets in which the value of $N_{cb}(x)$ increases; these are the packets which open new precincts. For the example shown in this figure, we use the same two code-stream organizations and the same set of relevant packets as in Figure 18.1. The full resolution image in this case has dimensions 256×256 and the code-block dimensions are $J_1 = J_2 = 32$, so that the subbands at DWT levels $d = 1$, 2 and 3 each consist of 16, 4 and 1 code-blocks, respectively. For convenience, Figure 18.2 explicitly identifies the number of code-blocks represented by each packet, i.e., those belonging to the packet's precinct. If $N_{cb}^{quit} < 64$, the decompressor will not be obliged to parse past the first two packets of the layer progressive code-stream, skipping the second quality layer altogether. Recall that only the first two resolution levels are considered relevant to the parser in this example.

At this point, it is helpful to consider the relationship between N_{cb} and precinct state memory resources. When a precinct p is first opened (first packet is encountered), having N_{cb}^p code-blocks, memory may be required to store the following quantities.

Inclusion tag tree state: The tag tree has one leaf node for each code-block. The number of additional nodes at higher levels in the tree is typically about $\frac{1}{3}N_{\mathrm{cb}}^p$, but is guaranteed to be strictly less than N_{cb}^p. This upper bound corresponds to the case where the precinct is much narrower than it is tall, or vice-versa. Each node must maintain a 16-bit state variable, $w^{(t)}[\mathbf{n}]$, which eventually (after everything has been decoded) represents the index of the first quality layer to which any subordinate code-block contributes. The tag tree decoding algorithm on Page 386 identifies a second 16-bit state variable, $\overline{w}^{(t)}[\mathbf{n}]$, for each node. However, since $\overline{w}^{(t)}[\mathbf{n}]$ and $w^{(t)}[\mathbf{n}]$ never differ by more than 1, the total cost per node may be reduced to as little as 17 bits. The overall cost may thus be bounded by $34N_{\mathrm{cb}}^p$ bits.

Missing MSBs tag tree state: Similar considerations apply here to those for the inclusion tag tree, with two exceptions. Firstly, the number of missing MSBs, K_b^{msbs}, for any code-block $\mathcal{B}_b$, may not exceed 37 unless ROI adjustments are involved, in which case it may not exceed 292 (74 if the restriction $U \leq 37$ appearing Table 18.1 applies)[3]. Secondly, persistent state information need not be maintained for the leaf nodes of the tree, since all decoding steps for K_b^{msbs} are executed in the first packet to which block $\mathcal{B}_b$ contributes. We conclude that the state memory may be bounded by $9N_{\mathrm{cb}}^p$ bits ($7N_{\mathrm{cb}}^p$ if the restriction $U \leq 37$ applies).

Length signalling state: As described in Section 12.5.4, the number of bytes associated with non-empty code-block contributions is signalled using a state machine, having state variable β_b, which is unique to each code-block $\mathcal{B}_b$. Assuming at most 5 bits to represent this state variable, the total cost becomes $5N_{\mathrm{cb}}^p$ bits.

Number of included coding passes: If the *BYPASS* mode switch is used without *RESTART*, the number of lengths which must be signalled in the packet header for each contributing code-block depends upon the number of coding passes which have already been contributed by that code-block in previous packets. Details of the length signalling algorithm are described in Section 12.5.4. It is sufficient here to note that length signalling depends upon the identity of the most recently contributed coding pass (significance propagation, magnitude refinement, or cleanup) and whether the bit-plane

---

[3]The limit of 37 may be deduced from equation (10.22), noting that the maximum value of G (guard bits) is 7 and ε (quantization exponent or reversible ranging parameter) is 31. When ROI adjustments are involved, the limit may be deduced from equation (10.33), noting that the maximum value of U (upshift) is 255 (37 for Profile-0 and Profile-1) code-streams.

to which that pass belongs is $p = K - 1$, $p = K - 2$, $p = K - 3$ or $p \leq K - 4$ (see Figure 12.11). In all, this information can be represented using only 4 bits, for a total cost of $4N_{\mathrm{cb}}^{p}$ bits.

Each of these quantities grows linearly with the number of code-blocks in the precinct. A highly efficient implementation may be able to expend as little as 7 bytes per code-block, although many implementations will require somewhat more memory. While this is usually small compared with the size of the compressed code bytes themselves, the precinct state memory cost can easily become burdensome when the image contains many irrelevant precincts, especially when these correspond to higher resolutions than those targeted by the implementation.

It is worth noting that $N_{\mathrm{cb}}(x)$ is a non-decreasing function of x. It is augmented whenever a new precinct is opened, but not decreased when the precinct is closed. This deliberate simplification is introduced so as not to force compliant decoders to efficiently reuse memory allocated for precincts which have since been closed. Such reuse might cause memory fragmentation and interfere with the behaviour of some implementations. The cursory memory analysis presented above is intended only to illustrate the reasoning underlying the development of the parser quit condition. Practical implementations may require much more memory and may need to store values not explicitly considered here. In particular, we have not considered the additional quantities which must be preserved for those code-blocks which are actually relevant.

Before moving on, we point out that there is actually no need to "open" a precinct until its first non-empty packet has been encountered. By "non-empty," we mean a packet which contains one or more body bytes. Empty packets include those whose first header bit is 0 (empty header) as well as those containing valid header bits which identify the body of the packet as empty. Packet emptiness is discussed thoroughly in Section 12.5.3. From that discussion, we may deduce that the first non-empty packet for any precinct with resolution $r > 0$ is the first packet, if any, whose first header byte exceeds $\mathtt{8F_h}$. For precincts with resolution $r = 0$, the first non-empty packet is that whose first header byte exceeds $\mathtt{BF_h}$. The value of $N_{\mathrm{cb}}(x)$ is incremented only when the first non-empty packet of a new precinct is encountered.

The importance of this "empty packet" condition should not be underestimated by content providers or neglected by compliant decoders. As mentioned above, if $N_{\mathrm{cb}}^{\mathrm{quit}} < 64$ in Figure 18.2, a compliant lower resolution decoder might only be able to recover the first quality layer from the layer progressive code-stream. On the other hand, if packet $\mathcal{G}_{0,3,0,0}$ were empty, the parser quit condition would not be reached until after all relevant packets had been recovered. Empty packets are commonly encountered in the higher resolution levels of a layer progressive code-stream and compressors would do well to deliberately introduce such structure, so as to exploit the limited resources offered by compliant decompressors with lower resolution capabilities.

PARSER QUIT CONDITION

In the preceding discussion, we have been concerned only with the precinct state memory. The complete parser quit condition is formulated to bound both the precinct state memory and the compressed data memory associated with relevant code-blocks. To this end, let $L_{\mathrm{body}}(x)$ denote the total number of relevant packet body bytes which have been encountered prior to location x in the code-stream. Again, a packet is deemed to be relevant if its precinct belongs to any tile-component resolution satisfying equation (18.1). A compliant implementation is permitted to "quit" once either of the following two conditions occurs.

Quit condition 1: $N_{\mathrm{cb}}(x) > N_{\mathrm{cb}}^{\mathrm{quit}}$

Quit condition 2: $L_{\mathrm{body}}(x) > L_{\mathrm{body}}^{\mathrm{quit}}$

If either of these conditions occurs, the decompressor is released from its obligation to continue parsing the code-stream. Whether it does so or not, all packet body bytes recovered from the code-stream up until such a point must be processed by the block decoder, subject to the obligations described below. It is up to the content provider (i.e., the compressor) to ensure that the more important code bytes appear early in the code-stream.

Actual values for $N_{\mathrm{cb}}^{\mathrm{quit}}$ and $L_{\mathrm{body}}^{\mathrm{quit}}$ depend upon the compliance class and level (H, W and C) at which compliance is claimed, as described in Section 18.3.4.

18.3.2   BLOCK DECODING OBLIGATIONS

The block decoder is obliged to decode all of the packet body bytes recovered while parsing the code-stream, in accordance with the obligations described above. This requirement is limited to the most significant K^{min} magnitude bit-planes of each code-block. The actual value of K^{min} depends upon the compliance class at which compliance is claimed, as identified in Table 18.2. More specifically, the block decoder must correctly decode the first

$$Z_b^{\mathrm{min}} \triangleq 3\left(K^{\mathrm{min}} - K_b^{\mathrm{msbs}}\right) - 2 \tag{18.2}$$

coding passes, if available, of any relevant code-block, $\mathcal{B}_b$. Here, K_b^{msbs} is the number of missing most significant bit-planes signalled in the appropriate packet header, as explained in Section 12.5.3. The relevant code-blocks are those belonging to any tile-component resolution which satisfies equation (18.1).

The decoder is free to decode any number of additional coding passes, but it is only obliged to decode the first $Z_b^{\min}$ passes. All decoded information must be correctly processed by subsequent dequantization and transformation operations, subject to the obligations outlined in Section 18.3.3 below.

Compliance testing procedures will involve code-streams with at most $K^{\min}$ magnitude bit-planes in any code-block. By constructing these code-streams in such a way as to avoid any limitations associated with the "parser quit" condition or sample data transformations[4], the implementation's ability to decode all $K^{\min}$ bit-planes may be probed from the outside.

IMPLICATIONS FOR CONTENT PROVIDERS

According to equation (18.2), the number of coding passes which a content provider can expect to be processed by a compliant decompressor depends upon the number of missing MSBs, K_b^{msbs}. This in turn, depends upon the number of guard bits, G, selected by the compressor. To make optimal use of the decompressor's declared capabilities, compressors should generally select the smallest value of G which is consistent with the absolute bounds of equation (10.21). For reversible processing, compressors should select the smallest value of $G + \varepsilon_b$ which is consistent with equations (10.29) and (10.31). Appropriate choices are discussed in Section 10.5.

If the "max-shift" method is used to encode user-defined regions of interest, the number of bit-planes required to represent the quantized subband samples may be increased by U, the up-shift value signalled in the RGN marker segment. This explained in Section 10.6.3. The most significant $K_b^{\max}$ magnitude bit-planes of each code-block in subband b represent foreground information (i.e., the region of interest). Of the remaining U magnitude bit-planes, only the least significant $K_b^{\max}$ can possibly contain any information concerning the background samples. Accordingly, compliant decoders can only be expected to recover information from the background region in subband b if

$$K_b^{\max} + (U - K_b^{\max}) < K^{\min}$$

---

[4] As an example, the DWT might not be used ($D = 0$), dequantization might be avoided by selecting the reversible processing path, and colour transformations might also be avoided.

Using equation (10.32) we conclude that no background information will be decoded unless

$$K^{\min} > \max_b K_b^{\max}$$

In practice, $K^{\min}$ will need to be at least 2 or 3 bits larger than the largest $K_b^{\max}$ value if a recognizable reconstruction of the background region is to be required.

In view of these considerations, content providers interested in ROI capabilities should endeavour to minimize the largest $K_b^{\max}$ value. If reversible transformation is selected, this largest $K_b^{\max}$ is largely independent of the number of DWT levels. A bound of $B + 3$ bits may be deduced from Table 17.4, where B is the image bit-depth. If the irreversible 9/7 transform is used, rate-distortion optimal quantization step sizes roughly halve with each additional DWT level. Accordingly, the maximum $K_b^{\max}$ grows linearly with the number of DWT levels, D. This suggests that content providers interested in ROI capabilities should use the reversible transform, unless the number of DWT levels is small – in that case, quantization step sizes should be selected as large as possible[5].

18.3.3 TRANSFORMATION OBLIGATIONS
REVERSIBLE PROCESSING

Compliant decompressors are obliged to implement the reversible 5/3 DWT exactly, for bit-depths of $B^{\min}$ bits/sample or less. The actual value of $B^{\min}$ depends upon the compliance class at which compliance is claimed, as identified in Table 18.2. If the reversible colour transform (RCT) has been employed, the implementation must be able to perform both the inverse DWT and the inverse RCT exactly, for bit-depths of $B^{\min}$ bits/sample or less. As shown in Section 17.3.2, internal representations involving $B^{\min} + 5$ bits ($B^{\min} + 4$ if the RCT is not used) should be sufficient to satisfy these reversible processing obligations, regardless of the number of DWT levels, D.

When presented with code-streams which require reversible processing to bit-depths in excess of $B^{\min}$ bits per sample, it is not currently clear what a compliant implementation may or may not be obliged to do. One reasonable policy would be to treat the image as though it had a lower bit-depth, discarding any block coding passes which are inconsistent with the lower bit-depth. The reconstructed image samples would then approximate a scaled (down-shifted) version of the original image samples. The approximation could be improved by working with

---

[5] We note that very large compressed images which satisfy the LL_D size requirements of Profile-0 and Profile-1 (see Table 18.1) may need a large number of DWT levels.

linearized versions of the reversible DWT and colour transformation operators whenever the bit-depth exceeds the implementation's capacity for exact reversible processing.

IRREVERSIBLE PROCESSING

As in Chapter 10, it is convenient to regard irreversible processing in terms of real-valued sample values, normalized to a nominal range of 0 to 1 (or $-\frac{1}{2}$ to $\frac{1}{2}$). The irreversible processing steps include dequantization, the inverse CDF 9/7 DWT and the irreversible colour transform, if used. For any particular set of decoded quantization indices for the subband samples, let $x_c^{\text{ref}}[\mathbf{n}]$ denote reference reconstructed sample values for image component c. These reference samples are obtained by applying the relevant irreversible processing steps using infinite precision arithmetic. Dequantization is performed using the mid-point reconstruction rule, i.e., setting $\delta = \frac{1}{2}$ in equation (10.27).

For the purpose of compliance testing, decompressors must be capable of performing dequantization using the mid-point reconstruction rule, even though other dequantization policies are allowed by the standard. Let $x_c[\mathbf{n}]$ denote the actual image sample values recovered by the implementation using this dequantization policy. Compliance is assessed with respect to the error, $\delta x_c[\mathbf{n}]$, between $x_c[\mathbf{n}]$ and $x_c^{\text{ref}}[\mathbf{n}]$, both of which are understood as real-valued quantities, having the normalized representation mentioned above, with unit nominal range. The actual means used by the implementation to represent $x_c[\mathbf{n}]$ is not specified; reasonable candidates include both floating point and fixed point representations. In any event, $x_c^{\text{ref}}[\mathbf{n}]$ is a real-valued quantity, so approximations associated with the representation of $x_c[\mathbf{n}]$ do contribute to the error, $\delta x_c[\mathbf{n}]$. As explained in Section 10.1.1, the irreversible processing path is actually independent of the bit-depth B, specified in the *SIZ* marker.

At the time of this writing, it is not clear exactly how bounds on the irreversible processing error $\delta x_c[\mathbf{n}]$, will be specified. In the remainder of this section we indicate one possible approach, based on the MSE (Mean Squared Error),

$$\sigma_{\delta X}^2 = \frac{1}{CN_1N_2} \sum_{c=0}^{C-1} \sum_{n_1=0}^{N_1-1} \sum_{n_2=0}^{N_2-1} \left(x_c[\mathbf{n}] - x_c^{\text{ref}}[\mathbf{n}] \right)^2$$

As shown in Section 17.3.3 (see Table 17.8), fixed point implementations can be expected to exhibit approximation errors whose MSE grows roughly linearly with the number of DWT levels, D. This suggests that compliance for the irreversible DWT should also be defined in terms of an MSE bound which is proportional to D.

To accommodate approximations introduced by any irreversible colour transform, as well as finite precision representation of the decompressed sample values $x_c[\mathbf{n}]$, some error must be tolerated even when $D = 0$. A simple formulation might be

$$\sigma_{\delta X}^2 \leq \sigma_{\text{fixed}}^2 + D_{\max}\sigma_{\text{var}}^2$$

where σ_{fixed}^2 allows for colour transformation and representation errors, σ_{var}^2 allows for DWT errors and $D_{\max}$ denotes the maximum number of DWT levels in any tile-component of the image. The condition may equivalently be formulated as a bound on PSNR, with

$$\text{PSNR} = -10\log_{10}\sigma_{\delta X}^2$$
$$\geq P^{\min} - 10\log_{10}\left(D_{\max} + D_{\text{off}}\right)$$

where $P^{\min}$ may be interpreted as the minimum "PSNR per DWT level" and D_{off} is a level offset term, defined by

$$D_{\text{off}} \triangleq \frac{\sigma_{\text{fixed}}^2}{\sigma_{\text{var}}^2}$$

In this way, compliance can be defined in terms of the two quantities, $P^{\min}$ and D_{off}.

Possible values for $P^{\min}$ and D_{off} are listed in Table 18.2. It is instructive to consider how these values were derived. Class-0, for example, is intended to serve the needs of typical 8-bit image compression applications. In this case, many implementations will represent the decompressed sample values using 8-bit quantities, as assumed in Section 17.4.2, even if the *SIZ* marker specifies a much larger bit-depth. The representation of real-valued quantities in the range 0 to 1 (or $-\frac{1}{2}$ to $\frac{1}{2}$) using 8 bits is analogous to uniform quantization and incurs an MSE of about $\frac{1}{12} \cdot 2^{-16}$. The value of σ_{fixed}^2 must be somewhat larger to accommodate approximation errors in the implementation of any inverse colour transform. The numbers in Table 18.2 are obtained by setting

$$\sigma_{\text{fixed}}^2 = 2 \times \frac{1}{12} \cdot 2^{-16} \quad \text{and} \quad \sigma_{\text{var}}^2 = \frac{1}{5}\sigma_{\text{fixed}}^2$$

which gives us

$$D_{\text{off}} = 5 \quad \text{and} \quad P^{\min} = 63 \text{ dB}$$

Before leaving this section, it is worth noting that it is possible to independently test the accuracy with which irreversible transforms are implemented. In particular, such tests can be conducted by supplying test code-streams with a resolution progressive organization (so that the

Table 18.2.   Compliance class definitions.

Parameter	Class-0	Class-1	Class-2
$N_{\text{cb}}^{\text{quit}}$ (code-blocks)	$\left(\frac{HW}{256} + 128\right) C$	$\left(\frac{HW}{256} + 128\right) C$	$\left(\frac{HW}{256} + 128\right) C$
$L_{\text{body}}^{\text{quit}}$ (bytes) H *(high)*	HWC	$\frac{3}{2} HWC$	$3HWC$
$L_{\text{body}}^{\text{quit}}$ (bytes) T *(typical)*	$\frac{1}{8} HWC$	$\frac{1}{4} HWC$	$\frac{1}{2} HWC$
$L_{\text{body}}^{\text{quit}}$ (bytes) S *(sequential)*	N/A	N/A	N/A
K^{min} (bit-planes)	12	16	30
B^{min} (reversible bit-depth)	8	12	16
P^{min} (PSNR per DWT level)	63 dB	87 dB	100 dB
D_{off} (irreversible level offset)	5	5	5

parser quit condition is not encountered) and with no more than K^{min} magnitude bit-planes in any code-block (so that all coding passes must be fully decoded). It is also worth noting that compliance tests based on statistical measures such as MSE are only meaningful in the context of well defined test streams.

18.3.4   COMPLIANCE CLASSES
CLASSES AND LEVELS

As mentioned previously, Part 4 of the JPEG2000 standard is still under development at the time of this writing. The material presented in this chapter is based upon a new working document adopted for this standardization effort, but is by no means set in stone. Implementors should consult the standard itself, once available, to be certain of the exact conditions which govern compliance.

Table 18.2 summarizes the parameters associated with each of three compliance classes, denoted Class-0, Class-1 and Class-2. These parameters are themselves defined in terms of dimensions, H, W and C, which should be interpreted as the maximum image dimensions and number of components which the implementation is prepared to support. To facilitate testing, compliance may be claimed only at one of the specific sets of dimensions specified in Table 18.3. These dimensions constitute what we might consider "levels" of compliance, within any given class. Of course, an implementation claiming compliance at one level, must also be compliant at all lower levels. Notice that compliance must be claimed for $C = 4$ image components, except at the lowest spatial resolution where it is acceptable to claim only monochrome processing.

Table 18.3. Levels (dimensions) at which compliance may be claimed within each class.

Level	H	W	C
Level-0	120	160	1
Level-1	120	160	4
Level-2	240	320	4
Level-3	480	640	4
Level-4	960	1280	4
$\vdots$	$\vdots$	$\vdots$	$\vdots$

INTERPRETATION OF COMPLIANCE CLASSES

The minimum compliance class, Class-0 ensures sufficient resources to allow truly lossless performance to a bit-depth of at least 8 bits per sample[6]. This does not mean that lossless performance will be achieved, even if the code-stream contains a lossless representation of the image. It might not be achieved if the code-stream contains a large amount of irrelevant information (e.g., extra image components, which are not targeted by the particular decompressor under consideration), so that the parser quit condition, $N_{\mathrm{cb}}(x) > N_{\mathrm{cb}}^{\mathrm{quit}}$, occurs before all relevant information has been recovered.

Again, lossless decompression might not be achieved, even if the code-stream contains a lossless representation of 8-bit imagery, if the compressor employed an unnecessarily large number of guard bits, G, or unnecessarily large ranging parameters, ε_b, for some subbands, or if ROI adjustments are involved. The compressor is at liberty to make such choices and their potential impact on Class-0 decompressors is well-defined.

Class-1 guarantees sufficient resources for truly lossless performance to a bit-depth of at least 12 bits, while Class-2 guarantees resources which might be reasonable to expect from a typical software implementation. In particular, Class-2 decompressors should be able to fully decompress both foreground and background regions when ROI adjustments are applied to 8-bit imagery. Again, the image quality which is actually realized depends upon numerous choices which the compressor is at liberty to make.

---

[6] As explained in Section 17.3.2, $B + 5$ bits are sufficient to represent the subband samples associated with reversible transformation of B-bit image samples, even when the RCT is employed. Since all quantities are signed integers, the number of magnitude bits need not exceed $B + 4$. This is why Class-0 defines $B^{\mathrm{min}} = 8$ and $K^{\mathrm{min}} = 12$.

INTERPRETATION OF H, T AND S MODIFIERS

To fully specify a decompressor's compliance class, one of the modi-fiers, H (high), T (typical) or S (sequential) must be given. As indicated in Table 18.2, Class-0-T guarantees sufficient resources to decompress an image compressed to 1 bit per sample (8:1 compression for 8-bit imagery). Again, this represents a commitment on the part of the de-compressor, which may or may not be exploited by the compressor which generates the code-stream. If the code-stream contains a representation of the image (or the subbands which are relevant to the decompressor in question) whose bit-rate exceeds 1 bit/sample, the decompressor is at liberty to discard the excess bits. The quality of the resulting image will depend on the organization of the code-stream. Specifically, a layer progressive organization may yield superior image quality to a resolution progressive organization if the quit condition, $L_{\text{body}}^{\text{quit}}$, is exceeded.

The H modifier guarantees more than sufficient resources to achieve lossless decompression under most circumstances. As always, it is the content provider's responsibility to use these resources in the manner which is most appropriate to its objectives.

The S modifier is reserved for implementations which are designed specifically to work with spatially progressive (sequential) code-stream organizations. Spatially progressive organizations are those which uti-lize one of the PCRL, CPRL or RPCL packet sequences described in Section 13.1.1. When processing very large images, spatially progressive code-streams can significantly reduce the resources required to buffer compressed data. However, this is only possible if the particular pro-gression sequence matches the order in which the application consumes decompressed data. Implementations which are designed to benefit from spatially progressive code-streams may be considered compliant without being obliged to commit to buffer a well-defined portion of the relevant compressed data. On the other hand, decompressors which claim com-pliance only at Class-0-S, Class-1-S or Class-2-S are not general purpose implementations of the standard.

IV

OTHER STANDARDS

Chapter 19

JPEG

19.1   OVERVIEW

JPEG is an acronym for "Joint Photographic Experts Group," the popular name for working group WG1 of the ISO/IEC Joint Technical Committee 1, Study Committee 9 (ISO/IEC JTC1/SC29/WG1 in full). The WG1 committee has produced several international standards for image compression. The first and most well known of these is what we call the JPEG compression standard [5]. More particularly, the baseline algorithm described in [5] is the most widely known system for lossy image compression. Less well known is the fact that the JPEG standard actually describes a family of lossy image compression algorithms, along with a lossless compression algorithm.

The lossless JPEG compression algorithm has not been widely accepted. Indeed, the WG1 committee has produced a completely new lossless compression standard known as JPEG-LS. The JPEG-LS standard is discussed in Chapter 20. Other standards developed within WG1 are the advanced bi-level image compression standard known as JBIG2 [10] and the JPEG2000 standard, which is the principle focus of this text.

Work on the JPEG standard began in the mid-1980's, culminating with the release of a draft international standard in 1991. During the 1990's, the JPEG standard has become established as the primary vehicle for storing and communicating compressed images. Although the standard was developed primarily to facilitate the exchange of compressed digital images, it is also commonly used for other purposes. Scanners and printers, for example, commonly employ JPEG compression as a means of saving internal memory resources. So-called "motion

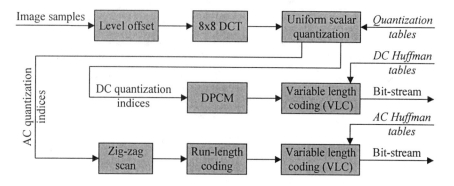

Figure 19.1.   Elements of the basic JPEG algorithm.

JPEG[1]" has long been a defacto "standard" for communicating, storing and editing digital video sequences.

BASIC ALGORITHM

Many implementations of the JPEG standard support only a basic lossy compression algorithm.. The elements of this algorithm are illustrated in Figure 19.1. Unsigned image sample values are first offset to obtain a signed representation. These samples are collected into 8×8 blocks and subjected to the DCT (Discrete Cosine Transform). The transform coefficients are individually quantized and then coded using variable length codes (also called Huffman codes). Various other techniques, such as DPCM and run-length coding are used to improve the compression efficiency.

In Section 19.2, we describe these processes in detail. For the moment, however, we note that the quantization and coding processes are controlled by parameter tables, which the compressor may choose to optimize for the image or application at hand. The tables must be explicitly recorded in the code-stream, regardless of whether any attempt is made to optimize these parameters. Interestingly, many JPEG compressors simply use the example tables described in the standard, with no attempt at customization.

For multiple component images (e.g., colour images), each component is processed in an identical fashion and the resulting bit-streams are usually interleaved. The interleaving is discussed further in Section 19.2.5.

---

[1] Motion JPEG is essentially identical to the baseline JPEG algorithm, with the exception that quantization and Huffman tables are not sent explicitly with the code-stream. Instead, motion JPEG adopts the tables which were originally given as examples in the JPEG standard.

VARIATIONS ON THE THEME

The JPEG standard supports a number of variations on the basic algorithm in Figure 19.1. Variable length coding (or Huffman coding) may be replaced by an adaptive arithmetic coder, known as the QM coder, for enhanced coding efficiency. The simple scalar quantizer may be replaced by a successive approximation quantizer, which allows image quality to be incrementally refined as the compressed data arrive. A related technique known as "spectral selection" allows the DCT coefficients to be dispersed through the code-stream. An hierarchical compression paradigm is also supported for embedding successively higher resolution versions of the image within a single file.

Part 3 of the JPEG standard [6] describes a number of extensions which can be useful to specific applications. One such extension provides support for modulating quantization step size parameters while the image is being coded. This allows the compressor to implement a single pass rate-control strategy. Another extension allows individual spatial regions to be selectively refined as the compressed data arrive.

In the brief description which follows, we make no attempt to cover all aspects of the JPEG standard. We first describe the elements of the basic algorithm more carefully. This provides a practical demonstration of a number of the fundamental techniques introduced in Chapters 2 and 4. Then, in Section 19.3 we provide a brief overview of the methods provided to construct resolution and/or distortion scalable JPEG code-streams. This is important, since one of the principle goals of the JPEG2000 standard is to offer superior scalability over that available with JPEG. For a much more comprehensive treatment of these and other topics, the reader is referred to the book by Pennebaker and Mitchell [119].

19.2   BASELINE JPEG

A "baseline JPEG" decompressor is one which supports a minimal set of features. In particular, it must be able to decompress images compressed using the basic algorithm described here. Formally, this algorithm is known as the "sequential DCT-based" mode. All lossy JPEG compression modes are based on the DCT. We will not describe the lossless algorithm here, since it is superceded by JPEG-LS (see Chapter 20).

19.2.1   SAMPLE TRANSFORMATIONS

The JPEG standard supports images with sample bit-depths of either $B = 8$ or $B = 12$. Strictly speaking, for baseline compression the bit-depth must be $B = 8$; however, it is convenient to describe the more

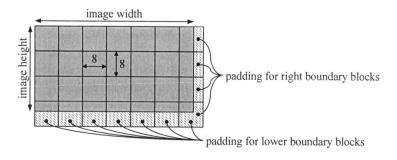

Figure 19.2.   Image partition into 8 × 8 blocks for JPEG compression.

general case here. The image samples are assumed to be unsigned quantities in the range 0 to $2^B - 1$. The "level offset" block in Figure 19.1 subtracts 2^{B-1} from every sample value so as to produce signed quantities in the range -2^{B-1} to $2^{B-1} - 1$. The purpose of this is to ensure that all of the DCT coefficients will be signed quantities with a similar dynamic range.

The image is partitioned into blocks of size 8 × 8. Each block is then independently transformed using the 8 × 8 DCT described in Section 4.1.3. If the image dimensions are not exact multiples of 8, the blocks on the lower and right hand boundaries may be only partially occupied. These boundary blocks must be padded to the full 8 × 8 block size and processed in an identical fashion to every other block. This is illustrated in Figure 19.2. The compressor is free to select the values used to pad partial boundary blocks. Of course, some choices will allow more effective compression than others. One reasonable approach is to replicate the final row or column of actual image samples into the missing locations.

Let $x_b[\mathbf{j}] \equiv x_b[j_1, j_2]$ denote the array of level shifted sample values for the bth block. Also, let $y_b[\mathbf{k}] \equiv y_b[k_1, k_2]$ denote the 8 × 8 array of DCT coefficients formed from these samples. The indices k_1 and k_2 represent vertical and horizontal frequency respectively. $y_b[\mathbf{0}]$ is the "DC" coefficient for block b. The reader may verify that

$$y_b[\mathbf{0}] = \frac{1}{8} \sum_{j_1=0}^{7} \sum_{j_2=0}^{7} x_b[\mathbf{j}]$$

so that $B + 3$ bits are required to represent the integer part of the DC coefficients. The remaining 63 coefficients in each block are called "AC" coefficients. It can be shown that $B+3$ bits are also sufficient to represent the integer part of each of these AC coefficients.

The index b identifies the sequence in which DCT blocks are processed. For the moment, we consider only one image component (i.e., a monochrome image). In this case, blocks are processed in raster order, from left to right and from top to bottom. The presence of multiple image components (e.g., colour images) can affect the scanning order, as described in Section 19.2.5. Regardless of such considerations, the first block, $b = 0$, is always located at the upper left hand corner of the image.

Each DCT coefficient is subjected to uniform scalar quantization. The quantization indices, $q_b[\mathbf{k}]$, are given by

$$q_b[\mathbf{k}] = \left\langle \frac{y_b[\mathbf{k}]}{\Delta_{\mathbf{k}}} \right\rangle, \quad 0 \leq k_1, k_2 < 8$$

where $\langle \cdot \rangle$ denotes rounding to the nearest integer. Notice that a different step size, $\Delta_{\mathbf{k}}$, may be used for each spatial frequency, $\mathbf{k}$. The quantization step sizes, $\Delta_{\mathbf{k}}$, are collected in an 8×8 array known as a "quantization table," or "Q-table." The same Q-table is used for every DCT block. When multiple image components are involved, each may have its own Q-table. The Q-table entries must be integers in the range 1 to 255. The decompressor uses a mid-point reconstruction rule to recover approximate versions of the original DCT coefficients. Specifically, the reconstructed DCT coefficients are given by

$$\hat{y}_b[\mathbf{k}] = \Delta_{\mathbf{k}} q_b[\mathbf{k}], \quad 0 \leq k_1, k_2 < 8$$

At low bit-rates, most of the AC coefficients must be quantized to 0, since every non-zero coefficient requires at least 2 bits to code (usually much more). For this reason, the compression performance at low bit-rates is significantly affected by the efficiency with which the DC coefficients are coded. JPEG uses a simple DPCM scheme to exploit some of the redundancy between the DC coefficients of adjacent blocks. The quantity which is actually coded for block b is the difference, δ_b, between $q_b[0]$ and $q_{b-1}[0]$. Specifically,

$$\delta_b = \begin{cases} q_b[0] - q_{b-1}[0] & \text{if } b > 0 \\ q_0[0] & \text{if } b = 0 \end{cases}$$

Figure 19.3 illustrates the processing of DC coefficients, $y_b[0]$, at the compressor and decompressor. According to the description given above, DPCM is a lossless tool for efficiently coding the quantization indices (see Section 2.4.2). Accordingly, the quantizer is not included inside the DPCM feedback loop.

Interestingly, the processing of DC coefficients may also be understood in terms of the lossy DPCM structure introduced in Section 3.3, where

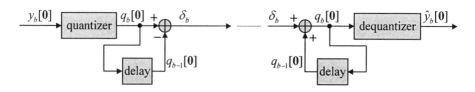

Figure 19.3.   DPCM with quantizer outside the feedback loop.

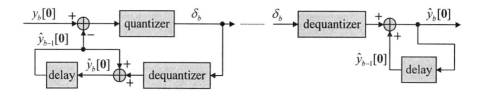

Figure 19.4.   DPCM with quantizer inside the feedback loop.

the quantizer is included inside the feedback loop. This structure is illustrated in Figure 19.4. The equivalence of these two quite different descriptions arises from the fact that the quantizer is uniform. The in-loop quantization structure of Figure 19.4 has the property that

$$\hat{y}_b[0] - \hat{y}_{b-1}[0] = \Delta_0 \delta_b$$

So long as the DPCM loop is initialized with $\hat{y}_{-1}[0] = 0$, we find that

$$\delta_b = \left\langle \frac{y_b[0] - \hat{y}_{b-1}[0]}{\Delta_0} \right\rangle = \left\langle \frac{y_b[0]}{\Delta_0} \right\rangle - \sum_{i=0}^{b-1} \delta_i$$

It follows that

$$q_b[0] = \left\langle \frac{y_b[0]}{\Delta_0} \right\rangle = \sum_{i=0}^{b} \delta_i$$

meaning that $\delta_b = q_b[0] - q_{b-1}[0]$. We conclude that the out-of-loop and in-loop configurations shown in Figures 19.3 and 19.4 are equivalent in JPEG.

19.2.2   CATEGORY CODES

As suggested by Figure 19.1, the difference symbols, δ_b, which represent the DC coefficients are to be coded using a variable length code

(VLC). Huffman's algorithm allows us to deduce a set of codewords which minimizes the overall bit-rate, as described in Section 2.2.1. Unfortunately, it is not practical to code δ_b directly with an optimal VLC, since the number of possible symbols can be very large. This is because the integer part of the DC coefficients are $B + 3$ bit quantities and the smallest quantization step size is 1. More precisely, it can be shown that the range of possible values for the magnitude, $|\delta_b|$, is

$$0 \leq |\delta_b| \leq 8 \left(2^B - 1\right) \lesssim 2^{B+3} \tag{19.1}$$

so that there are approximately 2^{B+4} possible difference symbols, δ_b.

The same problem occurs when coding quantization indices for the AC coefficients, in which case the total number of possible symbols is approximately 2^{B+3}. In both cases, JPEG addresses the problem by mapping the symbol to be coded, say s, to an ordered pair, (c, u), which we call a "category code." The "size category," c, is subjected to variable length coding, while u contains exactly c remainder bits, which are simply appended to the VLC for c without coding. The size category, c, is defined to be the smallest integer such that $|s| \leq 2^c - 1$. That is,

$$c = \lceil \log_2 (|s| + 1) \rceil$$

Evidently, when $c = 0$, s must be zero and there is no need for any remainder bits, u. In general, c represents the minimum number of bits required to hold the binary representation for $|s|$ without overflow. When $c \neq 0$, the most significant of these bits must be a 1 and so u must carry only the remaining $c - 1$ bits. When $c \neq 0$, u must also convey the sign of s. In all, then, the remainder portion of the category code, u, contains exactly c bits.

The representation adopted for u is designed to simplify certain implementations. When $s > 0$, u contains a 1 followed by the $c - 1$ least significant bits of $|s|$; equivalently, u is the c LSBs of $|s|$. When $s < 0$, however, u consists of a 0, followed by the one's complement (bit-wise negation) of the least significant $c - 1$ bits of $|s|$; equivalently, u holds the c LSBs of the one's complement of $|s|$.

Table 19.1 identifies the 2^c different symbols, s, associated with each size category, c. Category coding is efficient so long as these 2^c symbols all have similar probabilities so that little is lost by leaving them uncoded. On the other hand, the size categories can be expected to have a highly non-uniform PMF.

When category coding is applied to DC difference symbols, $B + 4$ size categories are required, in the range $0 \leq c \leq B + 3$. This is because $|\delta_b| + 1$ cannot exceed 2^{B+3} (see equation (19.1)). When applied to AC

Table 19.1. Symbols, s, associated with each size category, c. These symbols are distinguished by the c-bit uncoded remainder, u.

size category, c	symbols, s
0	0
1	$-1, +1$
2	$-3, -2, +2, +3$
3	$-7, -6, -5, -4, +4, +5, +6, +7$
$\vdots$	$\vdots$
c	$-(2^c - 1), \ldots, -2^{c-1}, +2^{c-1}, \ldots, +(2^c - 1)$
$\vdots$	$\vdots$

quantization indices, $B + 3$ size categories are sufficient. Recall that B is the image sample bit-depth, which is either 8 or 12.

19.2.3   RUN-VALUE CODING

Since variable length coding techniques cannot represent a symbol with less than 1 bit, JPEG must resort to symbol aggregation in order to achieve compressed image bit-rates below 1 bit/sample. This is done through run-length coding of the AC coefficients. The most likely outcome for the AC quantization indices is 0. As previously explained, at moderate to low bit-rates (e.g., below 1 bit/sample), most of the AC coefficients must necessarily be quantized to 0, since each non-zero coefficient will cost several bits to code.

As noted in Section 2.4.3, there are good reasons to jointly code the length of each run of zeros, together with the non-zero value which interrupts the run. We call this run-value coding. In the JPEG standard, run-value coding of the AC quantization indices is performed as follows. First, a one dimensional scan through the 63 AC coefficients of each DCT block is obtained by following the zig-zag scan shown in Figure 19.5. As noted in Section 4.1.3, this progression from low to high spatial frequencies tends to rank the DCT coefficients in order of decreasing variance. This in turn, tends to minimize the number of run-value pairs which must be coded.

Let $q[z]$ denote the sequence of AC quantization indices belonging to some DCT block, following the zig-zag scan of Figure 19.5. Thus, $q[0] = q[0, 1]$ and $q[63] = q[7, 7]$. This sequence of 63 quantization indices is replaced by a variable length sequence of run-value pairs, $(r[i], v[i])$, where the ith pair signifies a run of $r[i]$ zeros, followed by value, $v[i]$. Runs are confined to the range $0 \leq r[i] \leq 15$, so that a run of 16 zeros

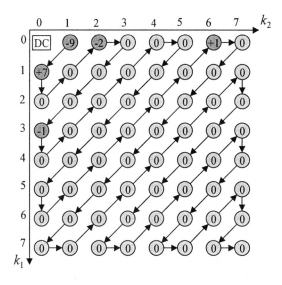

Figure 19.5. Zig-zag scan of the AC quantization indices used for run-length coding in JPEG. Numerical values are used in Example 19.1.

would be represented by the pair $(15, 0)$. Run-value pairs with $v[i] = 0$ and $r[i] < 15$ are obviously inefficient. In fact, such combinations are all illegal, with the exception of the combination $(0, 0)$. This combination is interpreted as an EOB (End of Block) marker, indicating that all remaining coefficients in the current DCT block are zero.

Each run-value pair, $(r[i], v[i])$, is converted into a triplet, $(r[i], c[i], u[i])$, where $(c[i], u[i])$ is the category code for $v[i]$. Both $r[i]$ and $c[i]$ have 4-bit representations and can be packed into an 8-bit word. This word is subjected to variable length coding, while the uncoded bits of $u[i]$ are appended to the codeword for $(r[i], c[i])$. Since there are $B + 3$ possible size categories, 16 possible run-lengths, and 14 illegal combinations $(1, 0)$ through $(14, 0)$, the total number of possible VLC codewords is $16(B + 3) - 14$. Thus, for 8-bit images there may be as many as 162 codewords, while for 12-bit images there are up to 226 distinct codewords.

Example 19.1 *The AC quantization indices shown in Figure 19.5 are represented using the following sequence of run-value pairs*

$$(0, -9), (0, +7), (2, -2), (3, -1), (15, 0), (1, +1), EOB$$

Note that the run of 17 zeros cannot be represented with a single run-value pair. The run-value pairs are converted into the following run-

category-remainder triplets

$$\underbrace{(0,4),}_{(r,c)}\underbrace{0110,}_{u}\quad \underbrace{(0,3),}_{(r,c)}\underbrace{111,}_{u}\quad \underbrace{(2,2),}_{(r,c)}\underbrace{01}_{u},$$

$$\underbrace{(3,1),}_{(r,c)}\underbrace{0}_{u},\quad \underbrace{(15,0),}_{(r,c)}\underbrace{\text{empty}}_{u},\quad \underbrace{(1,1),}_{(r,c)}\underbrace{1}_{u},\quad \underbrace{(0,0),}_{EOB}\underbrace{\text{empty}}_{u}$$

19.2.4   VARIABLE LENGTH CODING

We have already mentioned the use of variable length coding for the size category, c, associated with each DC difference symbol and for the run-category pairs, $(r[i], c[i])$, in the run-value representation of AC coefficients. In the former case, there are at most $B+4$ codewords, while in the second case there may be as many as $16(B+3)-14$ codewords. The codewords for these two cases are represented in two separate tables, known as the "DC Huffman table" and the "AC Huffman table." These tables are identified in Figure 19.1. When multiple image components are involved, each may have its own set of Huffman tables.

JPEG imposes two restrictions on the VLC codewords, which prevent Huffman's algorithm from being used directly to optimize the tables. The first restriction is that no codeword may exceed 16 bits in length. While this imposes no practical constraint on the DC Huffman table, the longest codeword in an optimal set of AC Huffman codewords could have as many as $16(B+3)-13$ bits. The impact of very long codewords on the behaviour of fast decoding algorithms has already been noted in Section 2.2.1. To satisfy the JPEG restriction, length-constrained Huffman codes must be generated using an algorithm such as that due to Voorhis [165].

The second restriction imposed by JPEG is somewhat more subtle. Let l_s denote the number of bits in the codeword for symbol s. Recall that the codeword lengths for any uniquely decodable VLC must satisfy the McMillan inequality of equation (2.11). That is, $\sum_s 2^{-l_s} \leq 1$, where s ranges over all of the codewords in the AC or DC Huffman table, as appropriate. A direct application of Huffman's optimization procedure is guaranteed to yield $\sum_s 2^{-l_s} = 1$. On the other hand, the JPEG standard insists that the inequality should be strict. That is, $\sum_s 2^{-l_s} < 1$. More specifically, the all 1's codeword is reserved for use as a prefix for extension codes. Part 3 of the JPEG standard [6] defines extension codes of this form for modifying the quantization parameters in mid-scan.

19.2.5   COMPONENTS AND SCANS

A JPEG code-stream contains one or more sequential scans through the image. Each scan may represent from 1 to 4 image components, so most colour images require only one scan. Within a scan, the bit-streams from multiple components are interleaved in a predetermined fashion. To describe this interleaving, we must first consider the effect of component sub-sampling. Moreover, to motivate component sub-sampling, we will consider colour images.

Unlike JPEG2000, the JPEG standard does not actually define any colour transform[2]. Instead, it is usually expected that colour images will have a YCbCr representation. This representation of colour is described in Section 10.2.1. The first component should contain the luminance (Y) sample values in the range 0 to $2^B - 1$. The second and third components should contain the chrominance (Cb and Cr) sample values, offset by 2^{B-1} so that they also lie in the range 0 to $2^B - 1$. After the level offset block in Figure 19.1, all components will then have a signed representation, centred about 0.

The human visual system is less sensitive to spatial details in the chrominance components than the luminance component. For this reason, the chrominance components are usually sub-sampled by a factor of 2 in each direction. JPEG supports sub-sampling in a somewhat indirect fashion by defining vertical and horizontal resolution scaling factors, F_1^i and F_2^i, for each image component, i. The actual height and width, N_1^i and N_2^i, for image component i, must satisfy the following relations

$$N_1^i = \left\lceil N_1 \frac{F_1^i}{F_1^{\max}} \right\rceil \quad \text{and} \quad N_2^i = \left\lceil N_2 \frac{F_2^i}{F_2^{\max}} \right\rceil$$

where $F_1^{\max}$ and $F_2^{\max}$ represent the maxima of the respective vertical and horizontal resolution factors, taken over all components in the image. The image dimensions, N_1 and N_2, are specified in the JPEG code-stream together with each component's resolution factors, F_1^i and F_2^i.

As an example, consider a 512×512 colour image, whose chrominance components have been sub-sampled by 2 in each direction. Then $N_1 = N_2 = 512$, $F_1^0 = F_2^0 = 2$ and $F_1^1 = F_1^2 = F_2^1 = F_2^2 = 1$. Note that F_1^i and F_2^i are not sub-sampling factors; instead, they play quite the opposite role, with larger factors for higher resolution image components.

JPEG scans are identified as either interleaved or non-interleaved. A non-interleaved scan represents only a single image component, whose DCT blocks are visited in raster order. Interleaved scans are based

---

[2]The colour representation may sometimes be specified explicitly by a containing file format.

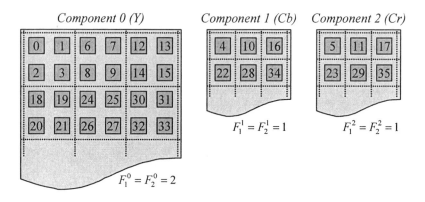

Figure 19.6. Scanning sequence for a multi-component image. Numbers indicate the position of each DCT block within the scan. The example shows a YCbCr image with chrominance components decimated by 2 in each direction. MCU's are delimited by dotted lines.

around the concept of an MCU (Minimum Coded Unit). Each MCU consists of an array of $F_1^i \times F_2^i$ DCT blocks from each component, i, to be included in the scan. Within the scan, MCU's are visited in raster order from top to bottom and left to right. Within each MCU, the $F_1^0 \times F_2^0$ blocks of component 0 are visited first, followed by the $F_1^1 \times F_2^1$ blocks of component 1 and so on. This is illustrated in Figure 19.6. The coded bits for each DCT block are interleaved in this same order. When differentially coding the DC coefficients of each DCT block, the previous DC coefficient from the same component within the scan is used as the reference.

The JPEG standard restricts the number of components in any scan to at most 4 and it also restricts the maximum number of DCT blocks in an MCU to at most 10. These restrictions serve to limit the resources required to decode a scan. They may also prevent some multi-component images from being compressed in a single scan. When this happens, multiple scans must be employed and the decompressor may need to buffer the entire decompressed image from a first scan until subsequent scans have been processed. As we shall see, multiple scans may also be used to implement various forms of scalability in JPEG.

19.3   SCALABILITY IN JPEG

Up to this point we have described what is known as JPEG's "sequential mode" of operation. This terminology comes from the fact that one or more image components are interleaved and compressed sequentially in an approximately raster scan order. Applications which supply or consume image samples in raster order need only buffer a small number of image lines, comparable to the height of an MCU. The term "JPEG baseline" refers to this sequential mode, restricted to 8-bit image samples and Huffman coding only. As mentioned previously, the standard also supports 12-bit imagery and optional arithmetic coding of the quantization indices, although the arithmetic coding option is rarely implemented.

The JPEG standard defines two additional modes, which are collectively known as "progressive JPEG." The two progressive modes each allow some property of the compressed image to be incrementally refined through multiple scans. The modes may be intermixed to obtain more interesting progressions, although this is not often done.

In addition to the progressive modes, JPEG provides a hierarchical refinement capability, whereby successively higher resolution versions of the image are compressed in a sequence of "frames." For each non-initial frame, a prediction is first formed based on the preceding frame and only the prediction residual is actually compressed. Hierarchical refinement introduces a multi-resolution capability.

19.3.1 SUCCESSIVE APPROXIMATION

Recall that the DCT coefficients are represented by quantization indices, $q_b[\mathbf{k}]$. With successive approximation, the binary representations of these indices are successively refined over two or more scans using a type of bit-plane coding. The scan essentially drops some number of least significant bits, p, from the representations of these quantization indices. This is roughly equivalent to multiplying the quantization step sizes by 2^p. Subsequent scans then each add one extra bit to the representation, roughly halving the quantization step sizes. It is worth providing some additional details of this process so as to draw appropriate connections with the bit-plane coding used by JPEG2000.

DC coefficients are treated somewhat differently from AC coefficients. For DC coefficients, the first scan codes all but the least significant p bits of $q_b[0]$. This is equivalent to using the modified quantization indices,

$$q_b^{(p)}[0] = \left\lfloor \frac{q_b[0]}{2^p} \right\rfloor = \left\lfloor \frac{\left\lfloor \frac{y_b[0]+\frac{1}{2}\Delta_0}{\Delta_0} \right\rfloor}{2^p} \right\rfloor = \left\lfloor \frac{y_b[0]+\frac{1}{2}\Delta_0}{2^p\Delta_0} \right\rfloor$$

Evidently, these modified indices represent a uniform quantization of the DC coefficients, with step size $2^p\Delta_0$, and some level adjustment of the coefficient values. The modified quantization indices are coded using exactly the same techniques (DPCM, category codes, Huffman or arithmetic coding) as in the sequential mode. There are exactly p additional scans, each of which supplies one additional LSB for all of the DC coefficients, thereby halving the effective quantization step size. These additional bits are not coded at all, leading to some loss in compression efficiency if p is large.

Successive approximation of the AC coefficients is most easily explained when the quantization indices are represented in sign-magnitude form as

$$q_b[\mathbf{k}] = \text{sign}\left(q_b[\mathbf{k}]\right) \cdot v_b[\mathbf{k}]$$

where

$$v_b[\mathbf{k}] = |q_b[\mathbf{k}]| = \left\lfloor \frac{|y_b[\mathbf{k}]|}{\Delta_{\mathbf{k}}} + \frac{1}{2} \right\rfloor$$

The first scan codes all but the least significant p bits of the magnitude, $v_b[\mathbf{k}]$, together with the sign of all coefficients for which the coded magnitude is non-zero. Equivalently, this scan codes the modified quantization indices,

$$q_b^{(p)}[\mathbf{k}] = \text{sign}\left(q_b[\mathbf{k}]\right) \cdot \left\lfloor \frac{v_b[\mathbf{k}]}{2^p} \right\rfloor$$

$$= \text{sign}\left(y_b[\mathbf{k}]\right) \cdot \left\lfloor \frac{|y_b[\mathbf{k}]|}{2^p \Delta_{\mathbf{k}}} + \frac{1}{2^p} \cdot \frac{1}{2} \right\rfloor$$

Comparing with equation (8.11), we recognize this as a deadzone quantizer with step size $2^p \Delta_{\mathbf{k}}$ and a central bin (the deadzone) which is $2 - 2^{-p}$ step sizes wide.

The second scan codes exactly one additional magnitude bit-plane from every quantization index, together with the sign of any coefficient which first becomes non-zero during this scan. After the second scan, only $p - 1$ magnitude bits are missing and the effective deadzone quantizer has a step size of $2^{p-1} \Delta_{\mathbf{k}}$ and a deadzone width of $2 - 2^{-(p-1)}$ step sizes.

As for DC coefficients, the total number of AC coefficient scans is $p + 1$. The DC and AC coefficients are required to appear in separate scans and multiple components may only be interleaved within the DC scans. This same requirement applies also to the spectral selection mode described below.

The successive approximation scans represent a family of embedded deadzone quantizers of the form indicated in Figure 8.9. JPEG2000 also uses such a family of embedded deadzone quantizers. The key differences between JPEG2000 and JPEG in this regard are as follows.

- In JPEG2000, coding is always embedded so that any number of least significant magnitude bit-planes may be discarded simply by truncating each embedded code-block bit-stream at an appropriate point. In JPEG, multiple bit-planes are usually coded in the first scan and successive approximation is often not used at all.

- In JPEG2000, the embedded quantizers all have a deadzone which is twice as wide as their step size[3]. In this way, truncating the relevant code-block's embedded bit-stream to a bit-plane boundary is equivalent to scaling the quantization step size by a power of 2. By contrast, the family of embedded quantizers induced by successive approximation in JPEG all have different relative deadzone sizes. As a result, a simple baseline decoder cannot correctly decompress even the first AC coefficient scan, despite the fact that it uses exactly the same coding techniques as the sequential mode.

- In JPEG, all DCT coefficients of all image components must be refined together, bit-plane by bit-plane. By contrast, JPEG2000 provides a much finer embedding of the information, together with an efficient mechanism for controlling the rate at which information is refined in each spatial region, frequency band and image component.

We have not yet said anything about how JPEG codes AC coefficient information in the successive approximation mode. The first scan is straightforward, using exactly the same coding techniques as the sequential mode (run-length coding, category codes, Huffman or arithmetic coding), but applying them to the modified quantization indices, $q_b^{(p)}[\mathbf{k}]$. In subsequent scans, the coding processes are significantly modified in an attempt to efficiently code the new bit-planes one at a time. For details of these processes, the reader is referred to [119]. It suffices to say here that splitting the coding of quantization indices across many scans tends to significantly reduce overall coding efficiency, except in the event that the arithmetic coding option is used.

19.3.2 SPECTRAL SELECTION

Spectral selection is the simplest of JPEG's progressive modes to understand and implement. In this case, the DCT coefficients themselves are partitioned into successive scans. The first scan codes the DC coefficient of each block. The second scan codes one or more AC coefficients of each block. Specifically, a range of AC coefficient positions, $z_1 \leq z \leq z_2$ is specified, where z represents the position of a coefficient within the zig-zag scanning order shown in Figure 19.5. Subsequent

---

[3]Part 2 of the standard is expected to provide support for different deadzone widths.

scans code additional AC coefficients until all of the AC coefficients have been processed. Typically, the scans are set up to code successively higher frequency coefficients. In this way, successive scans provide progressively higher spatial frequency information.

Spectral selection may be viewed as a method for progressively refining the image resolution, since each scan adds new spatial details. However, the type of refinement involved here does not agree well with commonly accepted notions of resolution. Rather than a blurred (i.e., low-pass filtered) rendition of the image, the initial scans yield a "blocky" structure with substantial aliasing. By contrast, JPEG2000 provides a natural family of successively lower resolution images through its use of the DWT (Discrete Wavelet Transform). The images in this family are related through the anti-aliasing and downsampling operations shown in Figure 4.16, which are the most widely accepted tools for resolution reduction.

Spectral selection has relatively little impact on the coding techniques employed, except that a set of EOB run codes is introduced to mitigate against efficiency losses. For more details, the reader is referred to [119].

19.3.3   HIERARCHICAL REFINEMENT

Whereas JPEG's progressive modes successively refine a single image through multiple scans, hierarchical refinement is achieved by compressing multiple images. JPEG calls these multiple images "frames" and we denote them f_1 through f_N. The frames usually represent successively higher resolution (i.e., larger) versions of the image, although this is not required. Each frame may be compressed using the sequential mode or any combination of the progressive modes, with its own set of Q-tables and Huffman tables. The first frame is compressed in the usual way. For each non-initial frame, f_n, the compression procedures are modified in several ways. A decompressed and suitably interpolated version of the previous frame is used to form a predictor, f_n', for frame f_n. The prediction residual, $f_n - f_n'$, is compressed instead of f_n. The level offset and DPCM operations are also skipped for non-initial frames.

Figure 19.7 illustrates the steps involved in hierarchical compression and decompression. The blocks marked "reduce" and "expand" usually implement resolution reduction and expansion by 2 in each direction. However, it is possible to preserve the horizontal and/or vertical resolution between frames. The processing performed in the "reduce" block is not specified by the standard, while that of the "expand" block is specified.

Notice that the hierarchical compressor implements a predictive feedback system, which is conceptually similar to DPCM. In particular, the

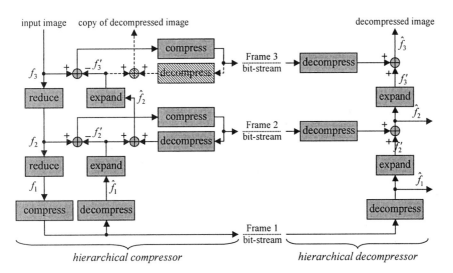

Figure 19.7. Hierarchical refinement in JPEG. Blocks shown with a patterned background need not be implemented.

compressor must determine the decompressed version, $\hat{f}_n$, of frame f_n in order to construct the same predictor, f'_{n+1}, as the decompressor. One unfortunate implication of this predictive structure is that the bit-stream representing frame f_n must be completely decompressed before any information can be reconstructed for the next (usually higher resolution) frame, f_{n+1}.

The principle motivation behind hierarchical refinement is to provide support for multi-resolution image representations. Unfortunately, this support comes at a significant cost in compression efficiency, since the total number of image samples which must be compressed is expanded by a factor of approximately $\frac{4}{3}$. Experimental results reported in [119] indicate that the overall compressed bit-rate can also expand by about 33%.

19.3.4   COMPARISON WITH JPEG2000

JPEG and JPEG2000 differ in numerous respects, including compression efficiency, complexity, scalability and "universality." Some discussion of these differences appears in Section 9.2. A subjective comparison of JPEG2000 image quality with that of JPEG appears in Section 16.1.3. In this section, we restrict our attention to those differences which are intimately connected with scalability.

Unlike JPEG2000, scalability does not come naturally to JPEG. Scalable modes are frequently not utilized, partly because they attract penalties in compression performance, complexity or both. More significantly, a resource constrained JPEG decompressor cannot generally expect to be able to recover a reduced resolution image by decompressing only a subset of the available code-stream. This is only likely to be possible if hierarchical refinement has been used. In that case, however, a more capable decompressor will not be able to decompress the full resolution image sequentially. This is unfortunate, because many applications only have sufficient memory to decompress very large images sequentially.

By contrast, the JPEG2000 standard allows and even encourages the construction of code-streams with very rich content: high resolution; high bit-rates, up to lossless; large bit-depths; many image components; etc. This is reasonable, since a resource constrained decompressor can always choose to decompress a suitable subset of the code-stream. Similarly, a bandwidth constrained communication service can always choose to deliver only a subset of the code-stream.

A key consideration in the development of the JPEG2000 standard is the separation of coding and ordering aspects of the compression system. The importance of such a separation is argued in Section 8.1.2. By contrast, the various options offered by JPEG in support of scalability have a significant impact on coding procedures and coding dependencies. If the code-stream is to support progressive refinement capabilities then it must involve multiple scans through the entire image. These scans cannot simply be collapsed back into a sequential organization for decompressors with limited memory resources. Another significant limitation is that the progressive modes cannot be used to successively improve image quality across the multiple resolutions provided by hierarchical refinement.

Finally, JPEG2000 exploits the local, fine embedding of individual code-blocks inside individual frequency bands to provide a number of interesting capabilities. Quantization can effectively be changed from block to block in order to reflect variations in the significance of different spatial regions or frequency bands and efficient rate control can be achieved in a single compression pass through the image. JPEG is able to replicate some of these capabilities in a limited fashion by resorting to the selective refinement options defined in Part 3 [6]. It is not currently clear how widely these options will be implemented.

Chapter 20

JPEG-LS

20.1 OVERVIEW

As discussed in Chapter 19, the JPEG image compression standard is actually a family of compression algorithms, of which the lossy baseline system is the most well-known. The family also includes an algorithm for lossless image compression, which is based on a collection of simple linear predictors, together with Huffman coding of the prediction residuals. By contrast with lossy JPEG, the lossless algorithm has not been widely adopted.

Lossless compression is important for a variety of applications, particularly those involving medical imagery or non-natural image content such as graphics and text. In recognition of the importance of these applications and the poor adoption of the original JPEG lossless algorithm, ISO/IEC working group JTC1/SC29/WG1 (commonly known as the "Joint Photographic Experts Group" or JPEG) began work on a new lossless image compression standard in 1995. Part 1 of the so-called JPEG-LS standard targets improved compression performance with minimal complexity. A primary objective is suitability for cost-effective hardware implementation. The algorithm draws heavily from the LOCO-I (LOw COmplexity COmpression of Images) scheme proposed by Weinberger et al. [168]. In addition to lossless compression, Part 1 also incorporates a lossy mode, termed "near lossless" compression, which is particularly suitable for applications which can tolerate a known maximum error in any given sample value.

Part 1 of the JPEG-LS standard is already complete; it is designated ISO/IEC 14495-1 [11]. Part 2 of the standard, designated ISO/IEC 14495-2, will provide mechanisms to support further improvements in compression performance with certain types of imagery, at the expense of increased complexity. It will also support enhanced near lossless compression capabilities. This second part of the standard is currently nearing completion.

The purpose of the brief exposition in this chapter is to familiarize the reader with the practice of lossless image compression. The information provided here is insufficient to implement the standard in full; for that, we would need to provide a detailed description of the marker syntax, constraints on legal parameters and so forth. Nevertheless, it should be possible to deduce the detailed structure of an implementation and hence the complexity and appropriateness of the JPEG-LS standard for applications of interest. JPEG-LS also provides a practical demonstration of a number of the coding techniques introduced in Chapter 2. Notable among these are Golomb coding (Section 2.2.2), predictive coding (Section 2.4.2), context adaptive coding (Section 2.4.1) and run-length coding (Section 2.4.3). Although not strictly necessary, the reader may find it useful to review the material in those sections.

20.1.1   CONTEXT NEIGHBOURHOOD

JPEG-LS relies upon the predictive coding techniques introduced in Section 2.4.2, together with context adaptive coding of the prediction residuals, as introduced in Section 2.4.1. Both predictive coding and context adaptive coding involve the concept of a causal neighbourhood. The neighbourhood is a collection of previously coded samples which are used to construct the prediction or coding context for the current sample value. The causal neighbourhood employed by JPEG-LS is illustrated in Figure 20.1. The image samples are coded one by one in raster order, starting from the top left sample in the image and working toward the bottom right. When coding any given sample, $x[n_1, n_2] \equiv x[\mathbf{n}]$, the neighbourhood consists of the samples,

$$x_a[\mathbf{n}] = x[n_1, n_2 - 1]$$
$$x_b[\mathbf{n}] = x[n_1 - 1, n_2]$$
$$x_c[\mathbf{n}] = x[n_1 - 1, n_2 - 1]$$
$$x_d[\mathbf{n}] = x[n_1 - 1, n_2 + 1]$$

For convenience, we shall frequently refer to the current sample and its neighbours simply as x and x_a, x_b, x_c, x_d, respectively.

20.1.2   NORMAL AND RUN MODES

Figure 20.2 identifies the major functional elements in the JPEG-LS encoder. In the normal mode, a prediction, μ_x, is formed using three

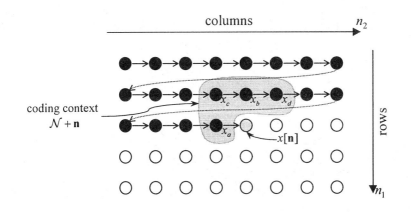

Figure 20.1.   Context neighbourhood for JPEG-LS.

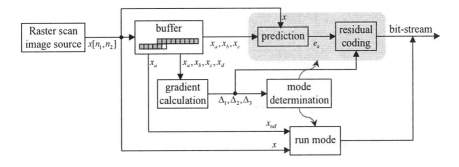

Figure 20.2.   JPEG-LS coding system.

causal neighbours, x_a, x_b and x_c. Since the decoder can form exactly the same prediction, it is sufficient to encode the residual, $e_x = x - \mu_x$, in place of the image sample, x. In practice, an additional bias term, β_x, is estimated and the residual is expressed as

$$e_x = s_x (x - \mu_x) - \beta_x \tag{20.1}$$

where $s_x \in \{1, -1\}$ is a sign reversal term. The decoder reconstructs x using

$$x = s_x(e_x + \beta_x) + \mu_x \tag{20.2}$$

The role of the bias, β_x, is to adjust the distribution of the residuals so as to maximize the efficiency with which they are subsequently coded. The role of the sign reversal term, s_x, is to halve the number of coding contexts. An adaptive Golomb code is employed to code e_x, where the Golomb parameter, $m = 2^k$, is adapted separately within each of

365 different contexts. The relevant context is identified by a context reduction function, $\lambda(\Delta_1, \Delta_2, \Delta_3)$, which depends upon the gradients,

$$
\begin{aligned}
\Delta_1 &= x_d - x_b \\
\Delta_2 &= x_b - x_c \\
\Delta_3 &= x_c - x_a
\end{aligned}
\tag{20.3}
$$

Adaptive Golomb coding and context reduction have already been introduced in Sections 2.2.2 and 2.4.1, respectively.

Good prediction and efficient residual coding are sufficient to achieve competitive compression performance with most natural image sources. Artificial imagery such as text and graphics, however, often contains extensive homogeneous regions. To capture the substantial redundancy in such images, the JPEG-LS coder provides a run mode. The coder enters this run mode whenever the neighbourhood samples are all identical; i.e., $\Delta_1 = \Delta_2 = \Delta_3 = 0$. In run mode, the value of the current sample, $x[n_1, n_2]$, is coded implicitly through a run length which identifies the number of consecutive samples, $x[n_1, n_2 + i]$, $i = 0, 1, \ldots$, which are all identical to a reference value, x_{ref}. The reference value is $x_{\mathrm{ref}} = x_a$. Through run-length coding it is possible to represent a large number of samples using relative few code bits, whereas the normal coding mode requires at least one bit to represent each sample value. Run mode also improves the execution speed of software implementations when compressing images containing substantial homogeneous regions.

20.1.3   NEAR LOSSLESS COMPRESSION

JPEG-LS supports near lossless compression, controlled through an integer-valued threshold, Θ. Specifically, Θ is the maximum permissible absolute difference between each original image sample value and its decompressed representation. When $\Theta > 0$, the algorithm is modified in a number of ways. In normal mode, the algorithm codes a quantized version of the prediction error,

$$
\hat{e}_x = \left\langle \frac{e_x}{2\Theta + 1} \right\rangle = \left\langle \frac{s_x (x - \mu_x) - \beta_x}{2\Theta + 1} \right\rangle
$$

where $\langle\rangle$ denotes rounding to the nearest integer. The decoder reconstructs

$$
\hat{x} = \mu_x + s_x \left(\beta_x + (2\Theta + 1) \hat{e}_x \right)
$$

which satisfies the maximum error constraint, since

$$
|\hat{x} - x| = |(2\Theta + 1) \hat{e}_x - e_x| < \frac{1}{2} (2\Theta + 1) \le \Theta
$$

In run mode, the run length corresponds to the number of consecutive samples, $x[n_1, n_2 + i]$, $i = 0, 1, \ldots$, which differ from a reference value, x_{ref}, by no more than Θ. The predictor, μ_x, the gradients, Δ_j and the run mode reference, x_{ref}, are all formed using the reconstructed versions of the relevant causal neighbours; i.e., $\hat{x}_a$, $\hat{x}_b$, $\hat{x}_c$ and $\hat{x}_d$ – these are the values available to the decoder. In short, with $\Theta > 0$, the JPEG-LS algorithm becomes a DPCM coding scheme, in which the prediction residual is uniformly quantized using a quantizer step size of $2\Theta + 1$. The lossless compression algorithm is obtained when $\Theta = 0$.

The lossy JPEG compression standard also involves a DPCM loop with uniform quantization of the prediction residuals. The DPCM loop in JPEG is used to compress the DC coefficients from successive DCT blocks. As discussed in Section 19.2.1, the uniform quantizer may be moved outside the DPCM loop, meaning that we may first quantize the sample values and then losslessly encode the quantization indices. This is possible in the simple DPCM loop used by JPEG, because the reconstructed DC coefficients and hence the predictor, μ_x, are always integer multiples of the step size.

By contrast, the JPEG-LS predictor incorporates a context dependent bias term, β_x, which is not generally a multiple of $2\Theta + 1$. As a result, near lossless compression in JPEG-LS is not equivalent to uniformly quantizing the original image sample values and losslessly compressing the resulting quantization indices. In normal mode, "thrashing" between the numerous coding contexts tends to randomize the bias applied to the predictor. This avoids visually disturbing contours associated with direct scalar quantization of the image sample values and reduces correlation between the reconstruction error and the image.

Henceforth, we restrict our attention to the case of pure lossless compression, with the understanding that the near lossless algorithm may be obtained by introducing the modifications described above.

20.2   NORMAL MODE CODING

20.2.1   PREDICTION

JPEG-LS incorporates a simple yet effective non-linear predictor, μ_x, which is a function of the three neighbours x_a, x_b and x_c. The predictor incorporates an "edge detecting" heuristic which selects between one of three predictors, as illustrated in Figure 20.3. If $x_c = \max\{x_a, x_b, x_c\}$ then the current sample, x, is assumed to belong to the dark side of a horizontal or vertical edge and the predictor is set to $\mu_x = \min\{x_a, x_b\}$ accordingly. Similarly, if $x_c = \min\{x_a, x_b, x_c\}$ then x is assumed to belong to the bright side of a horizontal or vertical edge and the predictor

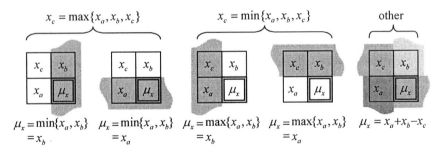

Figure 20.3.   JPEG-LS prediction.

is set to $\mu_x = \max\{x_a, x_b\}$. These edge models are most appropriate for thresholded or computer generated imagery, where samples belonging to the same side of an edge are likely to have identical values. If x_c is strictly greater than the smallest and strictly less than the largest of the three causal neighbours, the predictor is set to

$$\mu_x = x_a + x_b - x_c$$
$$= \frac{1}{2}(x_a + x_b) + \left(\frac{1}{2}(x_a + x_b) - x_c\right)$$

This predictor may be interpreted in terms of a plane passing through the 3D locations having coordinates $(n_1, n_2 - 1, x_a[\mathbf{n}])$, $(n_1 - 1, n_2, x_b[\mathbf{n}])$ and $(n_1 - 1, n_2 - 1, x_c[\mathbf{n}])$. The predictor is then the third coordinate (height) of the plane at location (n_1, n_2).

20.2.2   GOLOMB CODING OF RESIDUALS

In the absence of any bias, prediction residuals can often be approximately modeled by a symmetric, two-sided geometric distribution of the form $f_{E_x}(e_x) = \frac{1-\rho}{1+\rho}\rho^{|e_x|}$. This is illustrated in Figure 20.4. Recall that the residual which we actually code in JPEG-LS is given by $e_x = s_x(x - \mu_x) - \beta_x$. The bias compensating term, β_x, serves two roles. It adaptively compensates for any bias introduced by the non-linear predictor. It also serves to introduce a controlled bias which facilitates efficient coding of the residuals. In particular, $\beta_x \in \mathbb{Z}$ is adjusted adaptively so that the mean value of the residual satisfies

$$-1 \lesssim E[E_x] \lesssim 0$$

Figures 20.5a and 20.5b illustrate the biased distributions corresponding to $E[E_x] < -\frac{1}{2}$ and $E[E_x] > -\frac{1}{2}$, respectively.

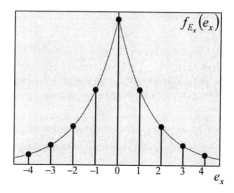

Figure 20.4. Distribution of unbiased prediction residuals.

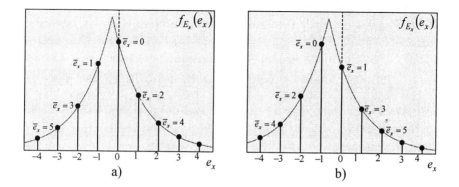

Figure 20.5. Distribution of biased prediction residuals where a) $E[E_x] > -\frac{1}{2}$ and b) $E[E_x] < -\frac{1}{2}$.

The JPEG-LS algorithm maintains four non-negative counters, A_λ, B_λ, C_λ and N_λ, for each of the 365 coding contexts, λ. We shall describe the update procedures for these counters shortly. For the moment, however, it is sufficient to accept that the following interpretations hold:

- $\frac{A_\lambda}{N_\lambda} \approx E[|E_x| \mid \lambda]$ models the mean absolute prediction residual within coding context λ.

- $\frac{B_\lambda}{N_\lambda} \approx E[E_x \mid \lambda]$ models the mean prediction residual within coding context λ.

- C_λ is the bias correction term, β_x, for any sample, x, coded in context λ.

The prediction residual, e_x, is mapped to a non-negative quantity, $\bar{e}_x$, according to

$$\bar{e}_x = \begin{cases} 2|e_x| - \text{neg}(e_x) & \text{if } \frac{B_{\lambda x}}{N_{\lambda x}} > -\frac{1}{2} \\ 2|e_x + 1| - \text{neg}(-(e_x + 1)) & \text{if } \frac{B_{\lambda x}}{N_{\lambda x}} \le -\frac{1}{2} \end{cases} \qquad (20.4)$$

where neg () is defined by

$$\text{neg}(y) = \begin{cases} 0 & \text{if } y \ge 0 \\ 1 & \text{if } y < 0 \end{cases}$$

The mapped residuals, $\bar{e}_x$, are identified in Figures 20.5a and 20.5b which correspond to the cases $\frac{B_{\lambda x}}{N_{\lambda x}} > -\frac{1}{2}$ and $\frac{B_{\lambda x}}{N_{\lambda x}} \le -\frac{1}{2}$, respectively. Notice that in both cases the positive and negative prediction residuals are interleaved in such a way that the PMF is a monotonically decreasing function of $\bar{e}_x$.

In JPEG-LS, the mapped residuals are represented using an adaptive Golomb code. The mapping described above was first used by Rice and its combination with Golomb coding of the mapped residuals is often referred to as "Golomb/Rice coding." Although $\bar{e}_x$ does not exactly follow a geometric distribution, after discarding one or more of its LSBs, we do recover an approximately geometric distribution. Thus, the use of Golomb coding is justified at least for Golomb parameters $m = 2^k$ with $k \ge 1$. In practice, the underlying distribution is not known precisely and Golomb coding is justified primarily on the basis of its simplicity and the fact that only a single coding parameter need be adapted, using simple indicators of the source statistics. In the remainder of this section, we describe the context selection, Golomb coding and parameter estimation details of the coding algorithm.

CONTEXT SELECTION

As noted earlier, residual coding is context adaptive, where the context used for the current sample is identified by a context reduction function, $\lambda(\Delta_1, \Delta_2, \Delta_3)$. Each of the local gradients, Δ_j, defined by equation (20.3) is first quantized to obtain one of 9 quantization indices, Q_j, according to

$$Q_j = \text{sign}(\Delta_j) \cdot \begin{cases} 0 & \text{if } \Delta_j = 0 \\ 1 & \text{if } 0 < |\Delta_j| < T_1 \\ 2 & \text{if } T_1 \le |\Delta_j| < T_2 \\ 3 & \text{if } T_2 \le |\Delta_j| < T_3 \\ 4 & \text{if } |\Delta_j| \ge T_3 \end{cases} \qquad (20.5)$$

Here, T_1, T_2 and T_3 are non-negative thresholds which may be signalled explicitly in the JPEG-LS code-stream[1]. Default values for 8-bit image sources are

$$T_1 = 3, \quad T_2 = 7, \quad T_3 = 21$$

Without further context merging we have a total of $9^3 = 729$ distinct contexts after quantization. This number is reduced to 365 by identifying the sign symmetric triplets, (Q_1, Q_2, Q_3) and $(-Q_1, -Q_2, -Q_3)$, with a single context model, subject to sign reversal of the prediction residual. This merging of contexts is based on the reasonable assumption that the joint statistics of the two dimensional source process, $\{X[\mathbf{n}]\}$, are likely to be identical to those of the amplitude flipped process, $\{x_{\max} - X[\mathbf{n}]\}$. The following algorithm may be used to implement the context labeling function. It generates a context label, λ_x, in the range $0 \leq \lambda_x < 365$, to be used in coding the current sample, x. Other implementations are possible, so long as a unique label is generated for each pair of sign symmetric quantized gradient triplets.

JPEG-LS Context Labeling

Compute the local gradients, Δ_j, from equations (20.3)

Compute Q_j from Δ_j for $j = 1, 2, 3$, using equation (20.5)

If $Q_1 < 0$ or $(Q_1 = 0$ and $Q_2 < 0)$ or $(Q_1 = Q_2 = 0$ and $Q_3 < 0)$

Set $Q_1 = -Q_1$, $Q_2 = -Q_2$ and $Q_3 = -Q_3$.

Set $s_x = -1$

else

Set $s_x = +1$

Set $\lambda_x = 81Q_1 + 9Q_2 + Q_3$

LENGTH-CONSTRAINED GOLOMB CODING

The mapped prediction residual, $\bar{e}_x$, is represented using a Golomb code with parameter $m_x = 2^{k_x}$. The algorithm for determining k_x is described shortly. The basic Golomb code represents $\bar{e}_x$ by appending the least significant k_x bits of $\bar{e}_x$ to a comma code which identifies the remaining most significant bits. Specifically, the code consists of $h_x =$

---

[1] Any selection of the form $1 < T_1 < T_2 < T_3$ allows each of the quantized gradients to take on all 9 possible values. In some cases, particularly with small images, it can be desirable to reduce the total number of contexts and hence the adaptive learning penalty of the coder. To this end, empty quantization bins may be created. For example, selecting $T_3 \leq T_2$ immediately reduces the total number of distinct quantization bins to 7. The convenient expression in equation (20.5) is not strictly correct for non-increasing thresholds.

$\lfloor 2^{-k_x} \bar{e}_x \rfloor$ "0"s, followed by a "1" (the comma), followed by the least significant k_x bits of $\bar{e}_x$.

The Golomb code is particularly sensitive to the choice of k_x. In particular, very long code words can be produced if the comma coded portion, h_x, is large. This value may be as large as $\bar{e}_{max}$, the maximum allowable value for $\bar{e}_x$. JPEG-LS employs a modified Golomb code which provides an escape mechanism to avoid excessively long codewords. Given a maximum codeword length, L, the regular Golomb coding procedure is employed so long as $h_x < L - \lceil \log_2 \bar{e}_{max} \rceil - 1$. Otherwise, an escape code is delivered, consisting of a string of $L - \lceil \log_2 \bar{e}_{max} \rceil - 1$ "0"s, followed by a "1". The escape code is followed by the direct binary representation of the value $\bar{e}_x - 1$, using $\lceil \log_2 \bar{e}_{max} \rceil$ bits.

In JPEG-LS, all image sample values are unsigned integers lying in the range $0 \leq x \leq x_{max}$, where the value of x_{max} is explicitly signalled in the JPEG-LS code-stream. The system ensures that the signed prediction residual, e_x, satisfies[2]

$$ -\left\lfloor \frac{x_{max} + 1}{2} \right\rfloor \leq e_x < \left\lceil \frac{x_{max} + 1}{2} \right\rceil $$

from which we may deduce that $\bar{e}_{max} = x_{max} + 1$. Also, the codeword length limit, L, is set to

$$ L = 2\left(\max\{2, \lceil \log_2 (x_{max} + 1) \rceil\} \right) + \max\{8, \lceil \log_2 (x_{max} + 1) \rceil\}) $$

PARAMETER ESTIMATION

As mentioned previously, the JPEG-LS coder maintains counters, A_λ, B_λ, C_λ and N_λ, for each of the 365 distinct coding contexts, λ. To code the current sample, the context label, λ_x, is determined following the procedure described above, after which the context-dependent bias term, β_x, is obtained as $\beta_x = C_{\lambda_x}$. This enables formation of the prediction residual, e_x. Its conversion to $\bar{e}_x$ depends upon an estimate of the mean of the prediction residual, given by the ratio $\frac{B_{\lambda_x}}{N_{\lambda_x}}$. The relevant mapping appears in equation (20.4). Finally, the Golomb coding parameter, k_x, is obtained from an estimate of the mean absolute prediction residual,

---

[2] The logic for this is omitted from the preceding description, so as to avoid unnecessary confusion. Excessively large prediction residuals are reduced immediately before the formation of $\bar{e}_x$. This is done by adding an integer multiple of $x_{max} + 1$ to the value of e_x produced by equation (20.1), so as to force it into the specified range. This does not prevent the decoder from reconstructing correct sample values. In particular, to recover recover the correct value, x, in the range 0 to x_{max}, the decoder has only to add an appropriate integer multiple of $x_{max} + 1$ to the value recovered from equation (20.2).

given by the ratio $\frac{A_{\lambda_x}}{N_{\lambda_x}}$. Specifically,

$$k_x = \left\lceil \log_2 \frac{A_{\lambda_x}}{N_{\lambda_x}} \right\rceil$$

so that

$$k_x \gtrsim \log_2 E\left[\,|E_x|\mid \lambda_x\,\right] \approx \log_2 E\left[\frac{\bar{E}_x}{2}\mid \lambda_x\right]$$

Evidently, the JPEG-LS parameter adaptation strategy is closely related to that derived in Section 2.2.2 for one-sided geometric distributions.

It is worth noting that the ratios $\frac{B_{\lambda_x}}{N_{\lambda_x}}$ and $\frac{A_{\lambda_x}}{N_{\lambda_x}}$ need never explicitly be formed. For residual mapping we are interested only in determining whether or not $2B_{\lambda_x} \leq -N_{\lambda_x}$, while for Golomb parameter estimation we have only to find the smallest k such that $2^k N_{\lambda_x} \geq A_{\lambda_x}$.

After the current sample has been coded, the counters for the relevant coding context are updated. Firstly, A_{λ_x}, B_{λ_x} and N_{λ_x} are updated according to the following procedure.

Update of Counters for Local Mean Estimation
Update $A_{\lambda_x} \leftarrow A_{\lambda_x} + |e_x|$ and $B_{\lambda_x} \leftarrow B_{\lambda_x} + e_x$
If $N_{\lambda_x} = N_{\max}$ (scale counters)
Update $A_{\lambda_x} \leftarrow \lfloor A_{\lambda_x}/2 \rfloor$, $B_{\lambda_x} \leftarrow \lfloor B_{\lambda_x}/2 \rfloor$, $N_{\lambda_x} \leftarrow \lfloor N_{\lambda_x}/2 \rfloor$
Update $N_{\lambda_x} \leftarrow N_{\lambda_x} + 1$

The renormalization point, $N_{\max}$, is a programmable parameter which is signalled through the JPEG-LS code-stream. It may be used to control the trade-off between stable parameter estimates and responsiveness to non-stationary statistics.

If $\frac{B_{\lambda_x}}{N_{\lambda_x}}$ lies in the range $-1 < \frac{B_{\lambda_x}}{N_{\lambda_x}} \leq 0$, the update procedure is complete. Otherwise, the bias value stored in C_{λ_x} is updated so as to restore the expected prediction residual to the desired interval, $(-1, 0]$. The bias value is allowed to change by at most ± 1 in each coding step, leading to the following update procedure.

Bias Update Procedure
If $\frac{B_{\lambda_x}}{N_{\lambda_x}} \leq -1$ (decrease bias by 1)
If $C_{\lambda_x} > C_{\min}$, update $C_{\lambda_x} \leftarrow C_{\lambda_x} - 1$
Update $B_{\lambda_x} \leftarrow \max\{B_{\lambda_x} + N_{\lambda_x}, 1 - N_{\lambda_x}\}$
else if $\frac{B_{\lambda_x}}{N_{\lambda_x}} > 0$ (increase bias by 1)
If $C_{\lambda_x} < C_{\max}$, update $C_{\lambda_x} \leftarrow C_{\lambda_x} + 1$

$$\text{Update } B_{\lambda_x} \leftarrow \min \left\{ B_{\lambda_x} - N_{\lambda_x}, 0 \right\}$$

The bounds, $C_{\min}$ and $C_{\max}$, are equal to -128 and 127 respectively, so that the bias parameters, C_λ, can always be stored as 8-bit quantities. The adjustment of B_{λ_x} up or down by N_{λ_x} in the above algorithm may be seen to preserve the interpretation of $\frac{B_{\lambda_x}}{N_{\lambda_x}}$ as the mean of the bias-corrected residuals. The max and min in the update relationships for B_{λ_x} ensure that $\frac{B_{\lambda_x}}{N_{\lambda_x}}$ always lies in the interval, $(-1, 0]$. This tends to reduce the impact of large transient changes in the estimated mean on the value of the value of the bias parameter.

20.3   RUN MODE CODING

20.3.1   GOLOMB CODING OF RUNS

As mentioned in Section 20.1.2, the coder enters its run mode whenever all four causal neighbours, $x_a[\mathbf{n}]$ through $x_d[\mathbf{n}]$, of the current sample, $x[\mathbf{n}]$, are identical. The assumption is that $x[\mathbf{n}]$ and possibly a large number of consecutive samples are all likely to have the same value as $x_a[\mathbf{n}]$. This value serves as the reference, x_{ref} for the duration of the run mode. Let $r[\mathbf{n}]$ denote the run length. Specifically, $r[n_1, n_2]$ is the smallest value of $r \geq 0$ such that $x[n_1, n_2 + r] \neq x_{\text{ref}}$ or $n_2 + r = N_2$, where N_2 is the width of the image. Note carefully that the run is terminated either by a sample value which is not equal to x_{ref} (interruption) or by the end of the current image row (exhaustion), whichever comes first. In the former case (interruption), both the run length and the interrupting sample value must be coded before leaving run mode. The process of interruption sample coding is discussed in Section 20.3.2. In the latter case (exhaustion), run mode terminates after the run length has been coded.

Ignoring the possibility of exhaustion, the run-lengths of an IID random process, $\{X[\mathbf{n}]\}$, possess a one-sided geometric distribution. To see this, suppose that $f_X(x_{\text{ref}}) = \rho$. Since the elements of the random process are assumed independent, we must have

$$f_R(r) = \left\{ \prod_{i=0}^{r-1} f_{X[n_1, n_2+i]}(x_{\text{ref}}) \right\} \cdot \left(1 - f_{X[n_1, n_2+r]}(x_{\text{ref}}) \right)$$
$$= (1 - \rho) \rho^r$$

In this case, the most appropriate variable length coding strategy is Golomb coding. In fact, the method was originally proposed by Golomb precisely for this purpose [71]. Although images typically contain substantial inter-sample redundancy, the IID assumption may be locally valid within the smooth regions where the run mode is important. In scanned document applications, for example, the dominant source of run interruption might be uncorrelated random noise from the scanning process. In any event, Golomb coding is attractive for its simplicity and because it lends itself to adaptive schemes. The complexity of Golomb coding is also insensitive to the size of the alphabet.

A conventional adaptive Golomb code would adjust the parameter only after coding each run, as discussed in Section 2.2.2. If the source statistics are entirely unknown a priori and subject to change frequently, superior performance may be obtained using a modified form of the Golomb code in which the parameter, m, is adapted within the run after the emission of each bit of the comma code. In JPEG-LS, the roles played by the binary digits "0" and "1" in the comma code are reversed so that "0" serves as the comma. A "1" then has the interpretation of a "hit" – the remaining run length is at least as long as m. Similarly, the terminating "0" of the comma code indicates a "miss" – the value of the Golomb parameter, m, is longer than the remaining run length. In the simple case where we constrain m to be an exact power of 2, namely $m = 2^k$, a miss is followed by the k-bit binary representation of the remaining run length.

Recall from Section 2.2.2 that an optimal value for the parameter, $m = 2^k$, is a little over half the mean run length, $E[R]$. Thus, a "typical" run should be coded with a single hit, followed by a miss and the k-bit remainder. This suggests an adaptive state machine in which equilibrium is achieved when there is one hit for every miss. Each hit increments the state index, while each miss decrements the state index, where k is a non-decreasing function of the state index. In this way, the machine adapts the value of the Golomb parameter, $m = 2^k$, after every bit of the underlying comma code. These ideas are embodied by the following run length coding algorithm. The scheme is known as a "block MELCODE"[3] [112] and was first inspired by [111].

MELCODE Run Coding Procedure
    Set $k = T[I_{\mathrm{mel}}]$ and $m = 2^k$
    While $r \geq m$
        Output a "1" (hit)
        Update remaining run length, $r \leftarrow r - m$
        Update state index, $I_{\mathrm{mel}} \leftarrow \min\{I_{\max}, I_{\mathrm{mel}} + 1\}$
        Set $k = T[I_{\mathrm{mel}}]$ and $m = 2^k$

---

[3] To the best of our knowledge, the name "MELCODE" is derived from "Mitsubishi Electric Company," where the code was first conceived.

If run interrupted (run terminates prior to end of line)
    Process the terminating miss
        Output a "0" (miss)
        Output the k-bit binary code for r
        Update state index, $I_{\text{mel}} \leftarrow \max\{0, I_{\text{mel}} - 1\}$
else (run continues for the rest of the line)
    If $r > 0$
        Output a "1" (pseudo-hit)

The MELCODE state machine in JPEG-LS contains 32 states so that $I_{\text{max}} = 31$. The state index is initialized to $I_{\text{mel}} = 0$ at the beginning of a scan, and the fixed parameter table is

$$T[I] = \begin{cases} \lfloor I/4 \rfloor & \text{if } 0 \leq I < 16 \\ \lfloor I/2 \rfloor - 4 & \text{if } 16 \leq I < 24 \\ I - 16 & \text{if } 24 \leq I < 32 \end{cases}$$

Notice the special processing for runs which are terminated by the end of the line, in which case there is no need to explicitly signal the exact run length.

20.3.2   INTERRUPTION SAMPLE CODING

When a run is interrupted by a new sample value, $x \neq x_{\text{ref}}$, the interrupting sample value must be coded. Although the interrupting sample could be coded using the methods of the regular mode, its statistics are generally quite different to those encountered in regular mode. In particular, we know a priori that $x \neq x_a$. Since there are comparatively few run interruption samples, a much smaller number of coding contexts is called for. In fact there are only two interruption coding contexts and only two causal neighbours are used, x_a and x_b, where x_b is the sample immediately above the interrupting sample, x, as in Figure 20.1. The single gradient, $\Delta = x_b - x_a$ is quantized into three regions which are then merged into two contexts using the same sign-symmetry properties as in normal mode. Thus, the sign-reversal flag is set to $s_x = \text{sign}(\Delta)$ and the context label is given by $\lambda_x = \min\{1, |\Delta|\} \in \{0, 1\}$.[4] The predictor is simply $\mu_x = x_b$ and no bias correction is employed; i.e., $\beta_x = 0$.

Coding of the prediction residual proceeds in the usual way by mapping the prediction residual, $e_x = s_x(x - \mu_x)$, into a non-negative quantity, $\bar{e}_x$, and applying the adaptive Golomb code, with parameter $m_x =$

---

[4]These context labels are not to be confused with those used in regular mode. The 365 contexts of regular mode are distinct from the 2 interruption coding contexts of run mode.

2^{k_x} derived from counters A_{λ_x} and N_{λ_x} as in normal mode. There are, however, a number of important differences from the procedure used in normal mode. These stem from the fact that no bias correction is employed and e_x must be non-zero in context $\lambda_x = 0$. Each of the two contexts maintains a scaled count, N_λ^-, of the number of negative prediction residuals which have been coded in that context. In this way, the ratio N_λ^-/N_λ is indicative of whether the prediction residual is more likely to be positive or negative. This information, together with knowledge of whether or not e_x can be zero, is used to choose between one of two schemes for mapping e_x to $\bar{e}_x$ prior to Golomb coding. For a detailed description of these variations, the reader is referred to the standard itself [11].

20.4   TYPICAL PERFORMANCE

In this section, we indicate the compression performance of the JPEG-LS system, using the reference implementation for Part 1 of the standard, with all parameters set to their recommended defaults[5]. Our goal is to indicate the degree of compression which can be expected for a variety of different image types. We examine both the lossless and near-lossless modes, with a variety of error thresholds, Θ. As described in Section 20.1.3, near-lossless mode involves embedding a uniform quantizer within the DPCM feedback loop. The thresholds, $\Theta = 0, 1, 3, 15$ and 31, for which results are reported in Table 20.1, correspond to quantizer step sizes of $\Delta = 2^n - 1$ for $n = 1$ (lossless) 2, 3, 5 and 6.

The various test images are collected into five groups. The first group consists of the high resolution ISO/IEC test images, "Bike," "Cafe" and "Woman," representing natural photographic content; the original images are depicted in Figure 8.21. The second group consists of lower resolution ISO/IEC test images, "Goldhill," "Hotel" and "Tools," also representing natural photographic content; the original images are depicted in Figure 8.22. The third group consists of the popular 512×512 test images, "Lenna" and "Barbara," which are depicted in Figure 8.23. The fourth group consists of non-natural ISO/IEC test images shown in Figure 8.24. An assortment of other test images appears in the last group; they are depicted in Figure 8.25. All images are monochrome with 8 bits per sample.

One of the most important observations from Table 20.1 is that most image sources yield lossless compression ratios of only about $1.7 : 1$. The exceptions are the non-natural image sources, "Cmpd1" and "Cmpd2,"

---

[5]A copy of this implementation may be obtained via *http://www.hpl.hp.com/loco*.

Table 20.1. JPEG2000 (lossless) and JPEG-LS compression results over a range of image types. Compressed bit-rates are expressed in bits per sample (bps).

Image	JPEG2000	$\Theta = 0$	$\Theta = 1$	$\Theta = 3$	$\Theta = 15$	$\Theta = 31$
Bike	4.54 bps ∞	4.36 bps ∞	2.85 bps 49.97 dB	1.85 bps 42.46 dB	0.73 bps 30.53 dB	0.42 bps 25.06 dB
Cafe	5.36 bps ∞	5.09 bps ∞	3.56 bps 49.95 dB	2.51 bps 42.32 dB	1.16 bps 30.01 dB	0.79 bps 24.17 dB
Woman	4.52 bps ∞	4.45 bps ∞	2.92 bps 49.91 dB	1.95 bps 42.30 dB	0.71 bps 30.73 dB	0.43 bps 24.80 dB
Goldhill	4.61 bps ∞	4.48 bps ∞	3.00 bps 49.89 dB	1.94 bps 42.23 dB	0.74 bps 30.17 dB	0.37 bps 24.66 dB
Hotel	4.60 bps ∞	4.38 bps ∞	2.87 bps 49.90 dB	1.88 bps 42.27 dB	0.73 bps 30.35 dB	0.47 bps 24.55 dB
Tools	5.47 bps ∞	5.31 bps ∞	3.74 bps 49.93 dB	2.62 bps 42.21 dB	1.10 bps 30.10 dB	0.65 bps 24.58 dB
Lenna	4.32 bps ∞	4.24 bps ∞	2.71 bps 49.89 dB	1.76 bps 42.22 dB	0.74 bps 30.15 dB	0.45 bps 24.13 dB
Barbara	4.80 bps ∞	4.86 bps ∞	3.31 bps 49.89 dB	2.28 bps 42.19 dB	1.06 bps 29.86 dB	0.72 bps 23.91 dB
Chart	3.09 bps ∞	2.84 bps ∞	1.84 bps 50.67 dB	1.23 bps 43.05 dB	0.55 bps 30.30 dB	0.37 bps 24.14 dB
Cmpd1	2.14 bps ∞	1.24 bps ∞	0.87 bps 55.77 dB	0.65 bps 40.08 dB	0.36 bps 31.43 dB	0.28 bps 29.77 dB
Cmpd2	2.56 bps ∞	1.44 bps ∞	1.11 bps 50.63 dB	0.79 bps 42.57 dB	0.38 bps 30.50 dB	0.27 bps 24.55 dB
Aerial2	5.45 bps ∞	5.29 bps ∞	3.74 bps 49.95 dB	2.61 bps 42.18 dB	1.11 bps 29.60 dB	0.58 bps 24.25 dB
Cats	2.53 bps ∞	2.57 bps ∞	1.80 bps 51.81 dB	1.25 bps 43.87 dB	0.58 bps 31.32 dB	0.36 bps 25.46 dB
Finger	4.622 bps ∞	4.623 bps ∞	3.096 bps 49.89 dB	2.163 bps 42.16 dB	0.973 bps 29.93 dB	0.669 bps 24.43 dB

which are representative of compound document imagery. For such sources, larger compression ratios are possible. The table also includes reference lossless compression results for JPEG2000, which are obtained using the Kakadu software implementation supplied with this text. The lossless JPEG2000 code-streams use the reversible 5/3 wavelet transform with 5 levels of decomposition, with a layer progressive code-stream having 7 quality layers. Evidently, JPEG-LS outperforms JPEG2000 on most images, although the performance gap is not large. Again, the exceptional images are "Cmpd1" and "Cmpd2," where JPEG-LS significantly outperforms JPEG2000. We hasten to point out that JPEG2000 offers fundamentally different features to JPEG-LS. JPEG2000 supports highly scalable code-streams, embedding efficient lossy and lossless representations within the same stream. JPEG2000 also supports resolution scalability, spatial random access, compressed domain cropping and many other features not offered by JPEG-LS. At the same time, JPEG2000 is a great deal more complex than JPEG-LS.

To achieve greater compression, the near-lossless compression mode is attractive. For this case, Table 20.1 reports both the compressed bit rate and the PSNR of the reconstructed images, for ease of comparison with similar results reported for JPEG2000 (see Table 8.5). At lower bit rates, neither the PSNR nor the visual appearance of the near-lossless compressed images is competitive with popular lossy compression alternatives. Nevertheless, the relative simplicity of the JPEG-LS algorithm renders its near-lossless compression mode attractive at high bit rates and in applications which demand a specified bound on the maximum error in any sample.

References

[1] New work item proposal: JPEG2000 image coding system. Technical Report N390, ISO/IEC JTC1/SC29/WG1, June 1996.

[2] Call for contributions for JPEG 2000 (JTC 1.29.14, 15444): Image coding system. Technical Report N505, ISO/IEC JTC1/SC29/WG1, March 1997.

[3] Report on CE V1: Exploitation of visual masking through control of individual block contributions. Technical Report N1303, ISO/IEC JTC1/SC29/WG1, June 1999.

[4] JPEG2000 Part 1 defect report. Technical Report N1980, ISO/IEC JTC1/SC29/WG1, January 2001.

[5] ISO/IEC 10918-1 and ITU-T Recommendation T.81. Information technology – digital compression and coding of continuous-tone still images: Requirements and guidelines, 1994.

[6] ISO/IEC 10918-3 and ITU-T Recommendation T.84. Information technology – digital compression and coding of continuous-tone still images: Extensions, 1996.

[7] ISO/IEC 11544 and ITU-T Recommendation T.82. JBIG bi-level image compression standard, 1993.

[8] ISO/IEC 11578. Information technology – open systems interconnection – remote procedure call, 1996.

[9] ISO 12640. Graphic technology – prepress digital data exchange – standard color image data (scid), 1995.

[10] ISO/IEC 14492 and ITU-T Recommendation T.88. JBIG2 bi-level image compression standard, 2000.

[11] ISO/IEC 14495-1 and ITU-T Recommendation T.87. Information technology – lossless and near-lossless compression of continuous-tone still images, 1999.

[12] ISO/IEC 15444-1. JPEG2000 image coding system, 2000.

[13] IETF RFC 2279. UTF-8, a transformation format of ISO 10646, January 1998.

[14] ISO/IEC 646. Information technology – ISO 7-bit coded character set for information interchange, 1991.

[15] IS 8859-15. Information technology – 8-bit single-byte coded graphic character sets - Parts 15: Latin alphabet number 9, 1999.

[16] G.P. Abousleman, M.W. Marcellin, and B.R. Hunt. Hyperspectral image compression using entropy-constrained predictive trellis coded quantization. *IEEE Trans. Image Proc.*, 6:566–573, April 1997.

[17] N.M. Abramson. *Information Theory and Coding.* McGraw-Hill, New York, 1963.

[18] N. Ahmed, T. Natarajan, and K. Rao. Discrete cosine transform. *IEEE Trans. Computers*, 23:88–93, January 1974.

[19] Analog Devices. *ADV-JP2000 JPEG2000 Co-Processor: Preliminary Technical Data*, May 2001.

[20] J. Andrew. A simple and efficient hierarchical image coder. *Proc. IEEE Int. Conf. Image Proc.*, 3:658–661, October 1997.

[21] S. Arimoto. An algorithm for calculating the capacity of an arbitrary discrete memoryless channel. *IEEE Trans. Inf. Theory*, 18:14–20, January 1972.

[22] R.B. Babat and T.E.S. Raghavan. *Nonnegative Matrices and Applications.* Cambridge University Press, 1997.

[23] R. Bellman. *Dynamic Programming.* Princeton University Press, Princeton, N.J., 1957.

[24] T. Berger. *Rate Distortion Theory.* Prentice-Hall, NJ, 1971.

[25] A. Bilgin, P.J. Sementilli, and M.W. Marcellin. Progressive image coding using trellis coded quantization. *IEEE Trans. Image Proc.*, 8:1638–1643, November 1999.

[26] R.E. Blahut. Computation of channel capacity and rate-distortion functions. *IEEE Trans. Inf. Theory*, 18:460–473, July 1972.

[27] M. Boliek, M. Gormish, E.L. Schwartz, and A.F. Keith. Decoding compression with reversible embedded wavelets (CREW) codestreams. *Journal of Electronic Imaging*, 7:402–409, July 1998.

[28] L. Bottou, P.G. Howard, and Y. Bengio. The z-coder adaptive binary coder. *Proc. IEEE Data Compression Conf. (Snowbird)*, pages 13–22, 1998.

[29] L. Breiman. The individual ergodic theorem of information theory. *Ann. Math. Stat.*, 28:809–811, 1957.

[30] L. Breiman. A correction to 'the individual ergodic theorem of information theory'. *Ann. Math. Stat.*, 31:809–810, 1960.

[31] C. Brislawn and B. Wohlberg. Boundary extensions and reversible implementation for half-sample symmetric filter banks. Technical Report N2119, ISO/IEC JTC1/SC29/WG1, March 2001.

[32] B. Buchberger. Introduction to groebner bases. In H. Schwichtenberg, editor, *Logic of Computation. Proc. Nato Advanced Study Institute*, pages 35–66. Springer-Verlag, 1997.

[33] R. Calderbank, I. Daubechies, W. Sweldens, and B. Yeo. Wavelet transforms that map integers to integers. *Applied and Computational Harmonic Analysis*, 5(3):332–369, July 1998.

[34] F.W. Campbell and J.G. Robson. Application of fourier analysis to the visibility of gratings. *Journal of Physiology (London)*, 197:551–566, 1968.

[35] CCITT. *Facsimile Coding Schemes and Coding Control Functions for Group 4 Facsimile Apparatus*, 1984. Recommendation T.6.

[36] W.H. Chen and C.H. Smith. Adaptive coding of monochrome and color images. *IEEE Trans. Commun.*, 25:1285–1292, November 1977.

[37] T. Chinen and A. Chien. Visual evaluation of JPEG2000 color image compression performance. Technical Report N1583, ISO/IEC JTC1/SC29/WG1, March 2000.

[38] T. Chinen, M. Nadenau, J. Reichel, and W. Zeng. Report on CE C03 (optimizing color image compression). Technical Report N1587, ISO/IEC JTC1/SC29/WG1, March 2000.

[39] P.A. Chou, T. Lookabaugh, and R.M. Gray. Entropy-constrained vector quantization. *IEEE Trans. Acoust. Speech and Sig. Proc.*, 37:31–42, January 1989.

[40] C. Christopoulos. JPEG2000 verification model 2.0 (technical description). Technical Report N988, ISO/IEC JTC1/SC29/WG1, October 1998.

[41] A. Cohen, I. Daubechies, and J.-C. Feauveau. Biorthogonal bases of compactly supported wavelets. *Communications on Pure and Appl. Math.*, 45(5):485–560, June 1992.

[42] J.H. Conway and N.J.A. Sloane. A lower bound on the average error of vector quantizers. *IEEE Trans. Inf. Theory*, 31:106–109, January 1985.

[43] T.M. Cover and J.A. Thomas. *Elements of Information Theory*. Wiley, New York, 1991.

[44] A. Croisier, D. Esteban, and C. Galand. Perfect channel splitting by use of interpolation/decimation/tree decomposition techniques. *Int. Conf. on Information Sciences and Systems*, pages 443–446, August 1976.

[45] A. Croisier, D. Esteban, and C. Galand. Application of quadrature mirror filters to split band voice coding systems. *Proc. Int. Conf. Acoust. Speech and Sig. Proc.*, pages 191–195, 1977.

[46] S. Daly. The visible differences predictor: An algorithm for the assessment of image fidelity. *Proc. SPIE (Human Vision, Visual Processing and Digital Display III, San Jose)*, 1666:2–15, February 1992.

[47] I. Daubechies. *Ten Lectures on Wavelets*. SIAM, Philadelphia, PA, 1992.

[48] I. Daubechies. Orthonormal bases of compactly supported wavelets. *Communications on Pure and Appl. Math.*, 41:909–996, November 1998.

[49] R.L. DeValois, D.G. Albrecht, and L.G. Thorell. Spatial frequency selectivity of cells in the macaque visual cortex. *Vision Research*, 22:545–559, 1982.

[50] R.L. DeValois, E.W. Yund, and H. Hepler. The orientation and direction selectivity of cells in macaque visual cortex. *Vision Research*, 22:531–544, 1982.

[51] Y. Du. Ein spharisch invariants Verbunddichtemodell fur Bildsignale. *Archiv fur Elektronik und Ubertragungstechnik*, 45:148–159, May 1991.

[52] P. Duhamel and M. Vetterli. Fast fourier transforms: A tutorial review and a state of the art. *Signal Proc.*, 19(4):259–299, April 1990.

[53] D.L. Duttweiler and C. Chamzas. Probability estimation in arithmetic and adaptive-huffman entropy coders. *IEEE Trans. Image Proc.*, 4(3):237–246, March 1995.

[54] H. Everett. Generalized lagrange multiplier method for solving problems of optimum allocation of resources. *Operation Res.*, 11:399–417, 1963.

[55] N. Farvardin and J.W. Modestino. Optimum quantizer performance for a class of non-Gaussian memoryless sources. *IEEE Trans. Inf. Theory*, 30:485–497, May 1984.

[56] W.A. Finamore and W.A. Pearlman. Optimal encoding of discrete-time continuous-amplitude memoryless sources with finite output alphabets. *IEEE Trans. Inf. Theory*, 26:144–155, March 1980.

[57] T.R. Fischer. Geometric source coding and vector quantization. *IEEE Trans. Inf. Theory*, 35:137–145, January 1989.

[58] T.R. Fischer. On the rate-distortion efficiency of subband coding. *IEEE Trans. Inf. Theory*, 38(2):426–428, March 1992.

[59] T.R. Fischer, M.W. Marcellin, and M. Wang. Trellis-coded vector quantization. *IEEE Trans. Inf. Theory*, 37:1551–1566, November 1991.

[60] T.R. Fischer and M. Wang. Entropy-constrained trellis-coded quantization. *IEEE Trans. Inf. Theory*, 38:415–426, March 1992.

[61] P.E. Fleischer. Sufficient conditions for achieving minimum distortion in a quantizer. *IEEE Int. Conv. Rec.*, 1:104–111, 1964.

[62] J.M. Foley and G.E. Legge. Contrast detection and near-threshold discrimination in human vision. *Vision Research*, 21:1041–1053, 1981.

[63] G.D. Forney, Jr. Convolutional codes I: Algebraic structure. *IEEE Trans. Inf. Theory*, 16:720–738, November 1970.

[64] G.D. Forney, Jr. The Viterbi algorithm. *Proc. IEEE*, 61:268–278, March 1973. (Invited Paper).

[65] Canon Research Centre France. Report on core experiment CodEff2: compressed image manipulation. Technical Report N1304, ISO/IEC JTC1/SC29/WG1, July 1999.

[66] T. (Editor) Fukuhara and N. (Co-Editor) Yasuyuki. Proposed FPDAM-1 to 15444-1. Technical Report N2091, ISO/IEC JTC1/SC29/WG1, March 2001.

[67] R.G. Gallager. *Information Theory and Reliable Communication.* Wiley, 1968.

[68] R.G. Gallager and D.V. Voorhis. Optimal source codes for geometrically distributed integer alphabets. *IEEE Trans. Inf. Theory*, 21:228–230, March 1975.

[69] A. Gersho and R.M. Gray. *Vector Quantization and Signal Compression.* Kluwer, Boston, 1992.

[70] H. Gish and J.N. Pierce. Asymptotically efficient quantizing. *IEEE Trans. Inf. Theory*, 14:676–683, September 1968.

[71] S.W. Golomb. Run-length encodings. *IEEE Trans. Inf. Theory*, 12:399–401, July 1966.

[72] I.S. Gradshteyn and I.M. Ryzhik. *Table of Integrals, Series, and Products.* Academic Press, New York, 1980.

[73] R.M. Gray. Time-invariant trellis encoding of ergodic discrete-time sources with a fidelity criterion. *IEEE Trans. Inf. Theory*, 23:71–83, January 1977.

[74] R.M. Gray. *Entropy and Information Theory*. Springer-Verlag, 1990.

[75] V.K. Heer and H.-E. Reinfelder. A comparison of reversible methods for data compression. *Proc. SPIE conference, 'Medical Imaging IV'*, 1233:354–365, 1990.

[76] I. Höntsch and L. Karam. APIC: Adaptive perceptual image coding based on subband decomposition with locally adaptive perceptual weighting. *Proc. IEEE Int. Conf. Image Proc.*, 1:37–40, 1997.

[77] D.A. Huffman. A method for the construction of minimum redundancy codes. *Proc. IRE*, 40:1098–1101, 1952.

[78] International Color Consortium (ICC). ICC profile format specification 1:1998-09, 1998.

[79] IEC TC100/61966-2.1. Colour management – default RGB colour space – sRGB, 1999.

[80] A. Islam and W.A. Pearlman. An embedded and efficient low-complexity hierarchical image coder. *Proc. SPIE Conf. Visual Comm. and Image Proc. (San Jose)*, January 1999.

[81] A.K. Jain. *Fundamentals of Digital Image Processing*. Prentice-Hall, Englewood Cliffs, N.J., 1989.

[82] N.S. Jayant and P. Noll. *Digital Coding of Waveforms*. Prentice-Hall, Englewood Cliffs, N.J., 1984.

[83] J.D. Johnston. A filter family designed for use in quadrature mirror filter banks. *Proc. Int. Conf. Acoust. Speech and Sig. Proc.*, pages 291–294, 1980.

[84] R.L. Joshi, V.J. Crump, and T.R. Fischer. Image subband coding using arithmetic coded trellis coded quantization. *IEEE Trans. Circuits Syst. Video Technol.*, 5:515–523, December 1995.

[85] R.L. Joshi, H. Jafarkhani, J.H. Kasner, T.R. Fischer, N. Farvardin, M.W. Marcellin, and R.H. Bamberger. Comparison of different methods of classification in subband coding of images. *IEEE Trans. Image Proc.*, 6:1473–1486, November 1997.

[86] J.H. Kasner, M.W. Marcellin, and B.R. Hunt. Universal trellis coded quantization. *IEEE Trans. Image Proc.*, 8:1677–1687, December 1999.

[87] R.D. Koilpillai and P.P. Vaidyanathan. Cosine-modulated FIR filter banks satisfying perfect reconstruction. *IEEE Trans. Sig. Proc.*, 40(4):770–783, April 1992.

[88] G. Langdon. Probabilistic and q-coder algorithms for binary source adaptation. *Proc. IEEE Data Compression Conf. (Snowbird)*, pages 13–22, 1991.

[89] G. Langdon and J. Rissanen. Compression of black-white images with arithmetic coding. *IEEE Trans. Commun.*, 29(6):858–867, June 1981.

[90] G.E. Legge. A power law for contrast discrimination. *Vision Research*, 21:457–467, 1981.

[91] J. Li. Visual progressive coding. Technical Report N758, ISO/IEC JTC1/SC29/WG1, March 1998.

[92] J. Li, P. Cheng, and C.-C. J. Kuo. On the improvements of embedded zerotree wavelet (EZW) coding. *Proc. SPIE: Visual Comm. and Image Proc. (Taipei)*, 2601:1490–1501, May 1995.

[93] J. Li and S. Lei. Rate-distortion optimized embedding. *Proc. Picture Coding Symposium, Berlin*, pages 201–206, September 1997.

[94] J. Liang, W.A. Pearlman, A. Islam, F. Wheeler, J. Andrew, C. Chui, J. Spring, C. Chrysafis, A. Said, and A. Drukarev. Low complexity entropy coding with set partitioning. Technical Report N1313, ISO/IEC JTC1/SC29/WG1, June 1999.

[95] J. Lim. *Two-Dimensional Signal and Image Processing*. Prentice-Hall, 1990.

[96] Y. Linde, A. Buzo, and R.M. Gray. An algorithm for vector quantizer design. *IEEE Trans. Commun.*, 28:84–95, January 1980.

[97] S.P. Lloyd. Least squares quantization in PCM. *IEEE Trans. Inf. Theory*, 28:129–137, 1982. Also, unpublished memorandum, Bell Laboratories, 1957.

[98] S. Mallat. A theory for multiresolution signal decomposition; the wavelet representation. *IEEE Trans. Patt. Anal. and Math. Intell.*, 11(7):674–693, 1989.

[99] H.S. Malvar. Modulated QMF filter banks with perfect reconstruction. *Electronic Letters*, 26:906–907, June 1990.

[100] H.S. Malvar. *Signal Processing with Lapped Transforms*. Artech House, Norwood, Massachusetts, 1992.

[101] M. Marcellin, T. Flohr, A. Bilgin, D. Taubman, E. Ordentlich, M. Weinberger, G. Seroussi, C. Chrysafis, T. Fisher, B. Banister, M. Rabbani, and R. Joshi. Reduced complexity entropy coding. Technical Report N1312, ISO/IEC JTC1/SC29/WG1, June 1999.

[102] M.W. Marcellin. On entropy-constrained trellis coded quantization. *IEEE Trans. Commun.*, pages 14–16, January 1994.

[103] M.W. Marcellin and T.R. Fischer. Generalized predictive TCQ of speech. *Communications of the ACM*, 33:11–19, January 1990.

[104] M.W. Marcellin and T.R. Fischer. Trellis coded quantization of memoryless and Gauss-Markov sources. *IEEE Trans. Commun.*, 38:82–93, January 1990.

[105] J. Max. Quantizing for minimum distortion. *IRE Trans. Inform. Theory*, 6:7–12, March 1960.

[106] R.J. McEliece. *The Theory of Information and Coding*. Addison-Wesley, 1977.

[107] B. McMillan. The basic theorems of information theory. *Ann. Math. Stat.*, 24:196–219, 1953.

[108] F. Mintzer. Filters for distortion-free two-band multirate filter banks. *IEEE Trans. Acoust. Speech and Sig. Proc.*, 33(3):626–630, June 1985.

[109] M. Nadenau and J. Reichel. Opponent color, human vision and wavelets for image compression. *Proceedings of the Seventh Color Imaging Conference*, pages 237–242, 1999.

[110] J. Ohm. Advanced packet-video coding based on layered VQ and SBC techniques. *IEEE Trans. Circ. Syst. for Video Tech.*, 3(3):208–221, June 1993.

[111] R. Ohnishi, Y. Ueno, and F. Ono. The efficient coding scheme for binary sources. *IECE of Japan*, 60-A:1114–1121, December 1977. (In Japanese).

[112] F. Ono, S. Kino, M. Yoshida, and T. Kimura. Bi-level image coding with MELCODE – comparison of block type and arithmetic type code. *Proc. Globecom'89*, pages 225–260, November 1989.

[113] E. Ordentlich, M. Weinberger, and G. Seroussi. A low-complexity modeling approach for embedded coding of wavelet coefficients. *Proc. IEEE Data Compression Conf. (Snowbird)*, pages 408–417, March 1998.

[114] P.F. Panter and W. Dite. Quantization distortion in pulse count modulation with nonuniform spacing of levels. *Proc. IRE*, pages 44–48, January 1951.

[115] A. Papoulis. *Probability, Random Variables, and Stochastic Processes*. McGraw-Hill, New York, 3rd edition, 1991.

[116] R. Pasco. *Source Coding Algorithms for Fast Data Compression*. PhD thesis, Stanford University, 1976.

[117] W. Pearlman. Performance bounds for subband coding. In John Woods, editor, *Subband Image Coding*, pages 1–41. Kluwer, 1991.

[118] W. Pennebaker, J. Mitchell, G. Langdon, and R. Arps. An overview of the basic principles of the q-coder adaptive binary arithmetic coder. *IBM J. Res. Develop.*, 32(6):717–726, November 1988.

[119] W.B. Pennebaker and J.L. Mitchell. *JPEG: Still Image Data Compression Standard*. Van Nostrand Reinhold, New York, 1992.

[120] J.P. Princen and A.B. Bradley. Analysis/synthesis filter bank design based on time domain aliasing cancellation. *IEEE Trans. Acoust. Speech and Sig. Proc.*, 34(5):1153–1161, October 1986.

[121] M. Rabbani and P.W. Jones. *Digital Image Compression Techniques*. SPIE, Bellingham, WA, 1991.

[122] WC3 Recommendation. Extensible markup language (XML) 1.0, February 1998. (Second Edition October 2000).

[123] R.C. Reininger and J.D. Gibson. Distributions of the two-dimensional DCT coefficients. *IEEE Trans. Commun.*, 31:835–839, June 1983.

[124] O. Rioul. Simple regularity criteria for subdivision schemes. *SIAM J. Math. Anal.*, 23:1544–1576, November 1992.

[125] J. Rissanen. Generalized kraft inequality and arithmetic coding. *IBM J. Res. Develop.*, 20:198–203, May 1976.

[126] A. Said and W. Pearlman. A new, fast and efficient image codec based on set partitioning in hierarchical trees. *IEEE Trans. Circ. Syst. for Video Tech.*, pages 243–250, June 1996.

[127] A. Said and W.A. Pearlman. An image multiresolution representation for lossless and lossy image compression. *IEEE Trans. Image Proc.*, 5(9):1303–1310, May 1996.

[128] P.J. Sementilli, A. Bilgin, J.H. Kasner, and M.W. Marcellin. Wavelet TCQ: submission to JPEG-2000. *Proc. SPIE, Appl. of Digital Proc. (San Diego)*, pages 2–12, July 1998.

762

[129] P.J. Sementilli and M.W. Marcellin. Scalability core experiment results. Technical Report N774, ISO/IEC JTC1/SC29/WG1, March 1998.

[130] C.E. Shannon. A mathematical theory of communication. *Bell Sys. Tech. Journal*, 27:379–423 (Part I), 623–656 (Part II), 1948. Reprinted in book form with postscript by W. Weaver, University of Illinois Press, Urbana, 1949.

[131] J.M. Shapiro. An embedded hierarchical image coder using zerotrees of wavelet coefficients. *Proc. IEEE Data Compression Conf. (Snowbird)*, pages 214–223, 1993.

[132] J.M. Shapiro. Embedded image coding using zerotrees of wavelet coefficients. *IEEE Trans. Sig. Proc.*, 41:3445–3462, December 1993.

[133] F. Sheng, A. Bilgin, P.J. Sementilli, and M.W. Marcellin. Lossy and lossless image compression using reversible integer wavelet transforms. *Proc. IEEE Int. Conf. Image Proc.*, pages 876–880, October 1998.

[134] Y. Shoham and A. Gersho. Efficient bit allocation for an arbitrary set of quantizers. *IEEE Trans. Acoust. Speech and Sig. Proc.*, 36:1445–1453, September 1988.

[135] E.P. Simoncelli and E.H. Adelson. Subband transforms. In John Woods, editor, *Subband Image Coding*, chapter Subband transforms, pages 143–192. Kluwer, 1991.

[136] M.J.T. Smith and III Barnwell, T.P. A procedure for designing exact reconstruction filter banks for tree-structured subband coders. *Proc. Int. Conf. Acoust. Speech and Sig. Proc.*, 2:27.1.1–27.1.4, 1984.

[137] M.J.T. Smith and T.P. III Barnwell. Exact reconstruction techniques for tree structured subband coders. *IEEE Trans. Acoust. Speech and Sig. Proc.*, 34:434–441, June 1986.

[138] M.J.T. Smith and S.L. Eddins. Analysis-synthesis techniques for subband image coding. *IEEE Trans. Acoust. Speech and Sig. Proc.*, 38:1446–1456, August 1990.

[139] R.A. Smith and D.J. Swift. Spatial-frequency masking and birdsall's theorem. *Journal of the Optical Society of America A*, 2:1593–1599, 1985.

[140] D. Speck. New options in radix-255 arithmetic coder. Technical Report N482R, ISO/IEC JTC1/SC29/WG1, March 1997.

[141] J. Spring, J. Andrew, and F. Chebil. Nested quadratic splitting. Technical Report N1191, ISO/IEC JTC1/SC29/WG1, March 1999.

[142] L.C. Stewart, R.M. Gray, and Y. Linde. The design of trellis waveform coders. *IEEE Trans. Commun.*, 30:702–710, February 1982.

[143] G. Strang and T. Nguyen. *Wavelets and Filter Banks*. Wellesley-Cambridge Press, Wellesley, Massachusetts, 1996.

[144] Scalability Core Experiment Sub-Group. Scalability sub-group combined results. Technical Report N846, ISO/IEC JTC1/SC29/WG1, June 1998.

[145] W. Sweldens. The lifting scheme: A custom-design construction of biorthogonal wavelets. *Applied and Computational Harmonic Analysis*, 3(2):186–200, April 996.

[146] D.S. Taubman. *Directionality and Scalability in Image and Video Compression.* PhD thesis, University of California, Berkeley, 1994.

[147] D.S. Taubman. EBCOT: Embedded block coding with optimized truncation. Technical Report N1020R, ISO/IEC JTC1/SC29/WG1, October 1998.

[148] D.S. Taubman. Embedded, independent block-based coding of subband data. Technical Report N871R, ISO/IEC JTC1/SC29/WG1, July 1998.

[149] D.S. Taubman. High performance scalable image compression with EBCOT. *IEEE Trans. Image Proc.*, 9(7):1158–1170, July 2000.

[150] D.S. Taubman and A. Zakhor. Multi-rate 3-d subband coding of video. *IEEE Trans. Image Proc.*, 3(5):572–588, September 1994.

[151] D.S. Taubman and A. Zakhor. A common framework for rate and distortion based scaling of highly scalable compressed video. *IEEE Trans. Circ. Syst. for Video Tech.*, 6(4):329–354, August 1996.

[152] I. Ueno, F. Ono, T. Yanagiya, and T. Kimura. Report on core experiment CodEff03: Arithmetic coding experiment. Technical Report N1199R, ISO/IEC JTC1/SC29/WG1, March 1999.

[153] G. Ungerboeck. Channel coding with multilevel/phase signals. *IEEE Trans. Inf. Theory*, 28:55–67, January 1982.

[154] G. Ungerboeck. Trellis-coded modulation with redundant signal sets - Part II: State of the art. *IEEE Commun. Mag.*, 25:12–21, February 1987.

[155] United States Federal Bureau of Investigation. Wavelet scalar quantizer WSQ gray-scale fingerprint image compression specifications, February 1993. Document IAFIS-IC-0110v2.

[156] M. Unser. On the approximation of the discrete karhunen-loeve transform for stationary processes. *Signal Proc.*, 5(3):229–240, May 1983.

[157] P.P. Vaidyanathan. Theory and design of m-channel maximally desimated quadrature mirror filters with arbitrary m, having the perfect reconstruction property. *IEEE Trans. Acoust. Speech and Sig. Proc.*, 35(4):476–492, April 1987.

[158] P.P. Vaidyanathan. Multirate digital filters, filter banks, polyphase networks, and applications: A tutorial. *Proc. IEEE*, 78(1):56–93, January 1990.

[159] P.P. Vaidyanathan. *Multirate Systems and Filter Banks.* Prentice-Hall, Englewood Cliffs, NJ, 1993.

[160] P.P. Vaidyanathan, T.Q. Nguyen, Z. Doğanata, and T. Saramäki. Improved technique for design of perfect reconstruction FIR QMF banks with lossless polyphase matrices. *IEEE Trans. Acoust. Speech and Sig. Proc.*, 37(7):1042–1056, July 1989.

[161] M. Vetterli. Filter banks allowing perfect reconstruction. *Signal Proc.*, 10(3):219–244, April 1986.

[162] M. Vetterli and J. Kovačević. *Wavelets and Subband Coding.* Prentice-Hall, New Jersey, 1995.

[163] M. Vetterli and D. Le Gall. Perfect reconstruction FIR filter banks: Some properties and factorizations. *IEEE Trans. Acoust. Speech and Sig. Proc.*, 37(7):1057–1071, July 1989.

[164] A.J. Viterbi and J.K. Omura. Trellis encoding of memoryless discrete-time sources with a fidelity criterion. *IEEE Trans. Inf. Theory*, 20:325–332, May 1974.

[165] D.V. Voorhis. Constructing codes with bounded codeword lengths. *IEEE Trans. Inf. Theory*, 20:288–290, 1974.

[166] D.S. Watkins. *Fundamentals of Matrix Computations*. John Wiley and Sons, New York, 1991.

[167] A.B. Watson. Efficiency of a model human image code. *Journal of the Optical Society of America A*, 4(12):2401–2417, December 1987.

[168] M.J. Weinberger, G. Seroussi, and G. Sapiro. LOCO-I: A low complexity, context-based, lossless image compression algorithm. *Proc. IEEE Data Compression Conf. (Snowbird)*, pages 140–149, April 1996.

[169] S.G. Wilson and D.W. Lytle. Trellis encoding of continuous-amplitude memoryless sources. *IEEE Trans. Inf. Theory*, 23:404–409, May 1977.

[170] W.D. Withers. A rapid entropy-coding algorithm. *Dr. Dobb's Journal*, 264:38–44, 1997.

[171] I.H. Witten, R.M. Neal, and J.G. Cleary. Arithmetic coding for data compression. *Communications of the ACM*, 30:520–540, June 1987.

[172] P. Wong. Rate distortion efficiency of subband coding with crossband prediction. *IEEE Trans. Inf. Theory*, 43(1):352–356, January 1997.

[173] X. Wu. High order context modeling and embedded conditional entropy coding of wavelet coefficients for image compression. pages 1378–1382, November 1997.

[174] A. Zandi, J.D. Allen, E.L. Schwartz, and M. Boliek. CREW: Compression with reversible embedded wavelets. *Proc. IEEE Data Compression Conf. (Snowbird)*, pages 212–221, March 1995.

[175] A. Zandi and G. Langdon. Bayesian approach to a family of fast attack priors for binary adaptive coding. *Proc. IEEE Data Compression Conf. (Snowbird)*, April 1992.

[176] Daly S. Zeng, W. and S. Lei. Point-wise extended visual masking for JPEG-2000 image compression. *Proc. IEEE Int. Conf. Image Proc.*, 1:657–660, September 2000.

[177] W. Zeng, J. Li, and S. Lei. An efficient color re-indexing scheme for palette-based compression. *Proc. IEEE Int. Conf. Image Proc.*, 3:476–479, September 2000.

Index

5. The Licensee shall have the right to use the Kakadu V2.2 source code indefinitely, subject to the TERMINATION provisions in this Agreement.

6. Unisearch may terminate this license grant, by written notice to the Licensee if the Licensee breaches any material term of this license.

7. Absent appropriate exemption certificate(s), the Licensee shall pay all taxes, duties, or customs, except for taxes based on Unisearch net income.

8. The Licensee shall not use the name, trade names or trademarks of Unisearch or any of its Affiliates in any advertising, promotional literature or any other material, whether in written, electronic, or other form, distributed to any Third Party, except in the form provided by Unisearch, and then solely for purposes of identifying Unisearch software.

9. This license is not transferable to a Third Party.

10. The Kakadu V2.2 source code is a collection of sofware tools, some of which may not be appropriate for the intended purpose. Unisearch shall have no liability for any indirect or consequential loss (whether foreseeable or otherwise and including loss of profits, loss of business, loss of opportunity, and loss of use of any computer hardware or software) resulting from the use of these tools.

   The Kakadu V2.2 source code should not be relied on as the sole basis to solve a problem whose incorrect solution could result in injury to person or property. If the software is employed in such a manner, it is at the Licensee's own risk and Unisearch explicitly disclaims all liability for such misuse to the extent allowed by law.

   Unisearch's liability for death or personal injury resulting from negligence or for any other matter in relation to which liability by law cannot be excluded or limited shall not be excluded or limited. Except as aforesaid, any other liability of Unisearch (whether in relation to breach of contract, negligence or otherwise) shall not in total exceed the amount paid to Unisearch under this agreement, for the software with respect to which the liability in question arises, as installed on any designated computer(s) or designated server(s) for which use of the software is licensed hereunder.

   Some international jurisdictions do not allow the exclusion or limitation of incidental or consequential damages, so the above exclusion or limitation may not apply to the Licensee.

11. To the extent any law, treaty, or regulation is in conflict with this Agreement, the conflicting terms of this Agreement shall be superseded only to the extent necessary by such law, treaty, or regulation. If any provision of this Agreement shall be otherwise unlawful, void, or otherwise unenforceable, that provision shall be enforced to the maximum extent permissible. In either case, the remainder of this Agreement shall not be affected.

12. This Agreement contains the entire understanding of the parties and may not be modified or amended except by written instrument, executed by authorized representatives of Unisearch and the Licensee.

ABOUT THE CD-ROM

Included with the book is a compact disc, containing documentation, binaries and all source code to the Kakadu software tools. This software provides a complete C++ implementation of JPEG2000 Part 1, demonstrating many of the principles described in the text itself. The software is frequently referenced from the text as an additional resource for understanding complex or subtle aspects of the standard. Conversely, the software makes frequent reference to this text and has been written to mesh with the terminology and notation employed therein.